Graduate Texts in Mathematics 250

T0236946

Graduate Texts in Mathematics

1 TAKEUTI/ZARING. Introduction to Axiomatic Set Theory. 2nd ed.
2 OXTOBY. Measure and Category. 2nd ed.
3 SCHAEFER. Topological Vector Spaces. 2nd ed.
4 HILTON/STAMMBACH. A Course in Homological Algebra. 2nd ed.
5 MAC LANE. Categories for the Working Mathematician. 2nd ed.
6 HUGHES/PIPER. Projective Planes.
7 J.-P. SERRE. A Course in Arithmetic.
8 TAKEUTI/ZARING. Axiomatic Set Theory.
9 HUMPHREYS. Introduction to Lie Algebras and Representation Theory.
10 COHEN. A Course in Simple Homotopy Theory.
11 CONWAY. Functions of One Complex Variable I. 2nd ed.
12 BEALS. Advanced Mathematical Analysis.
13 ANDERSON/FULLER. Rings and Categories of Modules. 2nd ed.
14 GOLUBITSKY/GUILLEMIN. Stable Mappings and Their Singularities.
15 BERBERIAN. Lectures in Functional Analysis and Operator Theory.
16 WINTER. The Structure of Fields.
17 ROSENBLATT. Random Processes. 2nd ed.
18 HALMOS. Measure Theory.
19 HALMOS. A Hilbert Space Problem Book. 2nd ed.
20 HUSEMOLLER. Fibre Bundles. 3rd ed.
21 HUMPHREYS. Linear Algebraic Groups.
22 BARNES/MACK. An Algebraic Introduction to Mathematical Logic.
23 GREUB. Linear Algebra. 4th ed.
24 HOLMES. Geometric Functional Analysis and Its Applications.
25 HEWITT/STROMBERG. Real and Abstract Analysis.
26 MANES. Algebraic Theories.
27 KELLEY. General Topology.
28 ZARISKI/SAMUEL. Commutative Algebra. Vol. I.
29 ZARISKI/SAMUEL. Commutative Algebra. Vol. II.
30 JACOBSON. Lectures in Abstract Algebra I. Basic Concepts.
31 JACOBSON. Lectures in Abstract Algebra II. Linear Algebra.
32 JACOBSON. Lectures in Abstract Algebra III. Theory of Fields and Galois Theory.
33 HIRSCH. Differential Topology.
34 SPITZER. Principles of Random Walk. 2nd ed.
35 ALEXANDER/WERMER. Several Complex Variables and Banach Algebras. 3rd ed.
36 KELLEY/NAMIOKA ET AL. Linear Topological Spaces.
37 MONK. Mathematical Logic.
38 GRAUERT/FRITZSCHE. Several Complex Variables.
39 ARVESON. An Invitation to C-Algebras.
40 KEMENY/SNELL/KNAPP. Denumerable Markov Chains. 2nd ed.
41 APOSTOL. Modular Functions and Dirichlet Series in Number Theory. 2nd ed.
42 J.-P. SERRE. Linear Representations of Finite Groups.
43 GILLMAN/JERISON. Rings of Continuous Functions.
44 KENDIG. Elementary Algebraic Geometry.
45 LOVE. Probability Theory I. 4th ed.
46 LOVE. Probability Theory II. 4th ed.
47 MOISE. Geometric Topology in Dimensions 2 and 3.
48 SACHS/WU. General Relativity for Mathematicians.
49 GRUENBERG/WEIR. Linear Geometry. 2nd ed.
50 EDWARDS. Fermats Last Theorem.
51 KLINGENBERG. A Course in Differential Geometry.
52 HARTSHORNE. Algebraic Geometry.
53 MANIN. A Course in Mathematical Logic.
54 GRAVER/WATKINS. Combinatorics with Emphasis on the Theory of Graphs.
55 BROWN/PEARCY. Introduction to Operator Theory I: Elements of Functional Analysis.
56 MASSEY. Algebraic Topology: An Introduction.
57 CROWELL/FOX. Introduction to Knot Theory.
58 KOBLITZ. p-adic Numbers, p-adic Analysis, and Zeta-Functions. 2nd ed.
59 LANG. Cyclotomic Fields.
60 ARNOLD. Mathematical Methods in Classical Mechanics. 2nd ed.
61 WHITEHEAD. Elements of Homotopy Theory.
62 KARGAPOLOV/MERIZJAKOV. Fundamentals of the Theory of Groups.
63 BOLLOBAS. Graph Theory.
64 EDWARDS. Fourier Series. Vol. I. 2nd ed.
65 WELLS. Differential Analysis on Complex Manifolds. 2nd ed.
66 WATERHOUSE. Introduction to Af. ne Group Schemes.
67 SERRE. Local Fields.
68 WEIDMANN. Linear Operators in Hilbert Spaces.
69 LANG. Cyclotomic Fields II.
70 MASSEY. Singular Homology Theory.
71 FARKAS/KRA. Riemann Surfaces. 2nd ed.
72 STILLWELL. Classical Topology and Combinatorial Group Theory. 2nd ed.
73 HUNGERFORD. Algebra.
74 DAVENPORT. Multiplicative Number Theory. 3rd ed.

(continued after index)

Loukas Grafakos

Modern Fourier Analysis

Second Edition

Springer

Loukas Grafakos
Department of Mathematics
University of Missouri
Columbia, MO 65211
USA
loukas@math.missouri.edu

ISBN: 978-1-4419-1856-7 e-ISBN: 978-0-387-09434-2
DOI 10.1007/978-0-387-09434-2

Mathematics Subject Classification (2000): 42-xx:42Axx

Printed on acid-free paper.

springer.com

Για την Ιωάννα, την Κωνσταντίνα,
και την Θεοδώρα

Preface

The great response to the publication of the book *Classical and Modern Fourier Analysis* has been very gratifying. I am delighted that Springer has offered to publish the second edition of this book in two volumes: *Classical Fourier Analysis, 2nd Edition,* and *Modern Fourier Analysis, 2nd Edition.*

These volumes are mainly addressed to graduate students who wish to study Fourier analysis. This second volume is intended to serve as a text for a second-semester course in the subject. It is designed to be a continuation of the first volume. Chapters 1–5 in the first volume contain Lebesgue spaces, Lorentz spaces and interpolation, maximal functions, Fourier transforms and distributions, an introduction to Fourier analysis on the n-torus, singular integrals of convolution type, and Littlewood–Paley theory.

Armed with the knowledge of this material, in this volume, the reader encounters more advanced topics in Fourier analysis whose development has led to important theorems. These theorems are proved in great detail and their proofs are organized to present the flow of ideas. The exercises at the end of each section enrich the material of the corresponding section and provide an opportunity to develop additional intuition and deeper comprehension. The historical notes in each chapter are intended to provide an account of past research but also to suggest directions for further investigation. The auxiliary results referred to the appendix can be located in the first volume.

A web site for the book is maintained at

http://math.missouri.edu/∼loukas/FourierAnalysis.html

I am solely responsible for any misprints, mistakes, and historical omissions in this book. Please contact me directly (loukas@math.missouri.edu) if you have corrections, comments, suggestions for improvements, or questions.

Columbia Missouri,
June 2008

Loukas Grafakos

Acknowledgements

I am very fortunate that several people have pointed out errors, misprints, and omissions in the first edition of this book. Others have clarified issues I raised concerning the material it contains. All these individuals have provided me with invaluable help that resulted in the improved exposition of the present second edition. For these reasons, I would like to express my deep appreciation and sincere gratitude to:

Marco Annoni, Pascal Auscher, Andrew Bailey, Dmitriy Bilyk, Marcin Bownik, Leonardo Colzani, Simon Cowell, Mita Das, Geoffrey Diestel, Yong Ding, Jacek Dziubanski, Wei He, Petr Honzík, Heidi Hulsizer, Philippe Jaming, Svante Janson, Ana Jiménez del Toro, John Kahl, Cornelia Kaiser, Nigel Kalton, Kim Jin Myong, Doowon Koh, Elena Koutcherik, Enrico Laeng, Sungyun Lee, Qifan Li, Chin-Cheng Lin, Liguang Liu, Stig-Olof Londen, Diego Maldonado, José María Martell, Mieczyslaw Mastylo, Parasar Mohanty, Carlo Morpurgo, Andrew Morris, Mihail Mourgoglou, Virginia Naibo, Hiro Oh, Marco Peloso, Maria Cristina Pereyra, Carlos Pérez, Humberto Rafeiro, Maria Carmen Reguera Rodríguez, Alexander Samborskiy, Andreas Seeger, Steven Senger, Sumi Seo, Christopher Shane, Shu Shen, Yoshihiro Sawano, Vladimir Stepanov, Erin Terwilleger, Rodolfo Torres, Suzanne Tourville, Ignacio Uriarte-Tuero, Kunyang Wang, Huoxiong Wu, Takashi Yamamoto, and Dachun Yang.

For their valuable suggestions, corrections, and other important assistance at different stages in the preparation of the first edition of this book, I would like to offer my deepest gratitude to the following individuals:

Georges Alexopoulos, Nakhlé Asmar, Bruno Calado, Carmen Chicone, David Cramer, Geoffrey Diestel, Jakub Duda, Brenda Frazier, Derrick Hart, Mark Hoffmann, Steven Hofmann, Helge Holden, Brian Hollenbeck, Petr Honzík, Alexander Iosevich, Tunde Jakab, Svante Janson, Ana Jiménez del Toro, Gregory Jones, Nigel Kalton, Emmanouil Katsoprinakis, Dennis Kletzing, Steven Krantz, Douglas Kurtz, George Lobell, Xiaochun Li, José María Martell, Antonios Melas, Keith Mersman, Stephen Montgomety-Smith, Andrea Nahmod, Nguyen Cong Phuc, Krzysztof Oleszkiewicz, Cristina Pereyra, Carlos Pérez, Daniel Redmond, Jorge Rivera-Noriega, Dmitriy Ryabogin, Christopher Sansing, Lynn Savino Wendel, Shih-Chi Shen, Roman Shvidkoy, Elias Stein, Atanas Stefanov, Terence Tao, Erin Terwilleger,

Christoph Thiele, Rodolfo Torres, Deanie Tourville, Nikolaos Tzirakis Don Vaught, Igor Verbitsky, Brett Wick, James Wright, Jiankai Xu, and Linqiao Zhao.

I would also like to thank all reviewers who provided me with an abundance of meaningful remarks, corrections, and suggestions for improvements. I would like to thank Springer editor Mark Spencer, Springer's digital product support personnel Frank Ganz and Frank McGuckin, and copyeditor David Kramer for their assistance during the preparation of this edition. Finally, I would like to thank Stephen Montgomery-Smith for creating the beautiful figure that appears on the cover page of this and of the previous volume.

Contents

Chapter 6
Smoothness and Function Spaces

In this chapter we study differentiability and smoothness of functions. There are several ways to interpret smoothness and numerous ways to describe it and quantify it. A fundamental fact is that smoothness can be measured and fine-tuned using the Fourier transform, and this point of view is of great importance. In fact, the investigation of the subject is based on this point. It is not surprising, therefore, that Littlewood–Paley theory plays a crucial and deep role in this study.

Certain spaces of functions are introduced to serve the purpose of measuring smoothness. The main function spaces we study are Lipschitz, Sobolev, and Hardy spaces, although the latter measure smoothness within the realm of rough distributions. Hardy spaces also serve as a substitute for L^p when $p < 1$. We also take a quick look at Besov–Lipschitz and Triebel–Lizorkin spaces, which provide an appropriate framework that unifies the scope and breadth of the subject. One of the main achievements of this chapter is the characterization of these spaces using Littlewood–Paley theory. Another major accomplishment of this chapter is the atomic characterization of these function spaces. This is obtained from the Littlewood–Paley characterization of these spaces in a single way for all of them.

Before one embarks on a study of function spaces, it is important to understand differentiability and smoothness in terms of the Fourier transform. This can be achieved using the Laplacian and the potential operators and is discussed in the first section.

6.1 Riesz Potentials, Bessel Potentials, and Fractional Integrals

Recall the Laplacian operator

$$\Delta = \partial_1^2 + \cdots + \partial_n^2 ,$$

which may act on functions or tempered distributions. The Fourier transform of a Schwartz function (or even a tempered distribution f) satisfies the following

L. Grafakos, *Modern Fourier Analysis*, DOI: 10.1007/978-0-387-09434-2_6,

identity:

$$-\widehat{\Delta(f)}(\xi) = 4\pi^2|\xi|^2\widehat{f}(\xi).$$

Motivated by this identity, we replace the exponent 2 by a complex exponent z and we define $(-\Delta)^{z/2}$ as the operator given by the multiplication with the function $(2\pi|\xi|)^z$ on the Fourier transform. More precisely, for $z \in \mathbf{C}$ and Schwartz functions f we define

$$(-\Delta)^{z/2}(f)(x) = ((2\pi|\xi|)^z\widehat{f}(\xi))^\vee(x). \tag{6.1.1}$$

Roughly speaking, the operator $(-\Delta)^{z/2}$ is acting as a derivative of order z if z is a positive integer. If z is a complex number with real part less than $-n$, then the function $|\xi|^z$ is not locally integrable on $\mathbf{R}^n$ and so (6.1.1) may not be well defined. For this reason, whenever we write (6.1.1), we assume that either $\operatorname{Re} z > -n$ or $\operatorname{Re} z \le -n$ and that $\widehat{f}$ vanishes to sufficiently high order at the origin so that the expression $|\xi|^z\widehat{f}(\xi)$ is locally integrable. Note that the family of operators $(-\Delta)^z$ satisfies the semigroup property

$$(-\Delta)^z(-\Delta)^w = (-\Delta)^{z+w}, \qquad \text{for all } z, w \in \mathbf{C},$$

when acting on spaces of suitable functions.

The operator $(-\Delta)^{z/2}$ is given by convolution with the inverse Fourier transform of $(2\pi)^z|\xi|^z$. Theorem 2.4.6 gives that this inverse Fourier transform is equal to

$$(2\pi)^z(|\xi|^z)^\vee(x) = (2\pi)^z \frac{\pi^{-\frac{z}{2}}}{\pi^{\frac{z+n}{2}}} \frac{\Gamma(\frac{n+z}{2})}{\Gamma(\frac{-z}{2})} |x|^{-z-n}. \tag{6.1.2}$$

The expression in (6.1.2) is in $L^1_{\mathrm{loc}}(\mathbf{R}^n)$ only when $-\operatorname{Re} z - n > -n$, that is when $\operatorname{Re} z < 0$. In general, (6.1.2) is a distribution. Thus only in the range $-n < \operatorname{Re} z < 0$ are both the function $|\xi|^z$ and its inverse Fourier transform locally integrable functions.

6.1.1 Riesz Potentials

When z is a negative real number, the operation $f \mapsto (-\Delta)^{z/2}(f)$ is not really "differentiating" f, but "integrating" it instead. For this reason, we introduce a slightly different notation in this case by replacing z by $-s$.

Definition 6.1.1. Let s be a complex number with $\operatorname{Re} s > 0$. The *Riesz potential* of order s is the operator

$$I_s = (-\Delta)^{-s/2}.$$

Using identity (6.1.2), we see that I_s is actually given in the form

$$I_s(f)(x) = 2^{-s}\pi^{-\frac{n}{2}} \frac{\Gamma(\frac{n-s}{2})}{\Gamma(\frac{s}{2})} \int_{\mathbf{R}^n} f(x-y)|y|^{-n+s}\,dy,$$

and the integral is convergent if f is a function in the Schwartz class.

We begin with a simple, yet interesting, remark concerning the homogeneity of the operator I_s.

Remark 6.1.2. Suppose that for s real we had an estimate

$$\big\|I_s f\big\|_{L^q(\mathbf{R}^n)} \le C(p,q,n,s)\big\|f\big\|_{L^p(\mathbf{R}^n)} \tag{6.1.3}$$

for some positive indices p,q and all $f \in L^p(\mathbf{R}^n)$. Then p and q must be related by

$$\frac{1}{p}-\frac{1}{q}=\frac{s}{n}. \tag{6.1.4}$$

This follows by applying (6.1.3) to the dilation $\delta^a(f)(x)=f(ax)$ of the function f, $a>0$, in lieu of f, for some fixed f, say $f(x)=e^{-|x|^2}$. Indeed, replacing f by $\delta^a(f)$ in (6.1.3) and carrying out some algebraic manipulations using the identity $I_s(\delta^a(f)) = a^{-s}\delta^a(I_s(f))$, we obtain

$$a^{-\frac{n}{q}-s}\big\|I_s(f)\big\|_{L^q(\mathbf{R}^n)} \le C(p,q,n,s)a^{-\frac{n}{p}}\big\|f\big\|_{L^p(\mathbf{R}^n)}. \tag{6.1.5}$$

Suppose now that $\frac{1}{p} > \frac{1}{q}+\frac{s}{n}$. Then we can write (6.1.5) as

$$\big\|I_s(f)\big\|_{L^q(\mathbf{R}^n)} \le C(p,q,n,s)a^{\frac{n}{q}-\frac{n}{p}+s}\big\|f\big\|_{L^p(\mathbf{R}^n)} \tag{6.1.6}$$

and let $a \to \infty$ to obtain that $I_s(f)=0$, a contradiction. Similarly, if $\frac{1}{p} < \frac{1}{q}+\frac{s}{n}$, we could write (6.1.5) as

$$a^{-\frac{n}{q}+\frac{n}{p}-s}\big\|I_s(f)\big\|_{L^q(\mathbf{R}^n)} \le C(p,q,n,s)\big\|f\big\|_{L^p(\mathbf{R}^n)} \tag{6.1.7}$$

and let $a \to 0$ to obtain that $\big\|f\big\|_{L^p} = \infty$, again a contradiction. It follows that (6.1.4) must necessarily hold.

We conclude that the homogeneity (or dilation structure) of an operator dictates a relationship on the indices p and q for which it (may) map L^p to L^q.

As we saw in Remark 6.1.2, if the Riesz potentials map L^p to L^q for some p,q, then we must have $q>p$. Such operators that improve the integrability of a function are called *smoothing*. The importance of the Riesz potentials lies in the fact that they are indeed smoothing operators. This is the essence of the *Hardy–Littlewood–Sobolev theorem on fractional integration*, which we now formulate and prove.

Theorem 6.1.3. *Let s be a real number with $0<s<n$ and let $1 \le p < q < \infty$ satisfy (6.1.4). Then there exist constants $C(n,s,p)<\infty$ such that for all f in $L^p(\mathbf{R}^n)$ we have*

$$\big\|I_s(f)\big\|_{L^q} \le C(n,s,p)\big\|f\big\|_{L^p}$$

when $p>1$, and also $\big\|I_s(f)\big\|_{L^{q,\infty}} \le C(n,s)\big\|f\big\|_{L^1}$ when $p=1$.

We note that the $L^p \to L^{q,\infty}$ estimate in Theorem 6.1.3 is a consequence of Theorem 1.2.13, for the kernel $|x|^{-n+s}$ of I_s lies in the space $L^{r,\infty}$ when $r = \frac{n}{n-s}$, and (1.2.15) is satisfied for this r. Applying Theorem 1.4.19, we obtain the required conclusion. Nevertheless, for the sake of the exposition, we choose to give another self-contained proof of Theorem 6.1.3.

Proof. We begin by observing that the function $I_s(f)$ is well defined whenever f is bounded and has some decay at infinity. This makes the operator I_s well defined on a dense subclass of all the L^p spaces with $p < \infty$. Second, we may assume that $f \geq 0$, since $|I_s(f)| \leq I_s(|f|)$.

Under these assumptions we write the convolution

$$\int_{\mathbf{R}^n} f(x-y)|y|^{s-n}\,dy = J_1(f)(x) + J_2(f)(x),$$

where, in the spirit of interpolation, J_1 and J_2 are defined by

$$\begin{aligned} J_1(f)(x) &= \int_{|y|<R} f(x-y)|y|^{s-n}\,dy, \\ J_2(f)(x) &= \int_{|y|\geq R} f(x-y)|y|^{s-n}\,dy, \end{aligned}$$

for some R to be determined later. Observe that J_1 is given by convolution with the function $|y|^{-n+s}\chi_{|y|<R}(y)$, which is radial, integrable, and symmetrically decreasing about the origin. It follows from Theorem 2.1.10 that

$$J_1(f)(x) \leq M(f)(x) \int_{|y|<R} |y|^{-n+s}\,dy = \frac{\omega_{n-1}}{s} R^s M(f)(x), \tag{6.1.8}$$

where M is the Hardy–Littlewood maximal function. Now Hölder's inequality gives that

$$\begin{aligned} |J_2(f)(x)| &\leq \left(\int_{|y|\geq R} (|y|^{-n+s})^{p'}\,dy \right)^{\frac{1}{p'}} \|f\|_{L^p(\mathbf{R}^n)} \\ &= \left(\frac{q\omega_{n-1}}{p'n} \right)^{\frac{1}{p'}} R^{-\frac{n}{q}} \|f\|_{L^p(\mathbf{R}^n)}, \end{aligned} \tag{6.1.9}$$

and note that this estimate is also valid when $p = 1$ (in which case $q = \frac{n}{n-s}$), provided the $L^{p'}$ norm is interpreted as the L^∞ norm and the constant $\left(\frac{q\omega_{n-1}}{p'n}\right)^{\frac{1}{p'}}$ is replaced by 1. Combining (6.1.8) and (6.1.9), we obtain that

$$I_s(f)(x) \leq C'_{n,s,p} \left(R^s M(f)(x) + R^{-\frac{n}{q}} \|f\|_{L^p} \right) \tag{6.1.10}$$

for all $R > 0$. A constant multiple of the quantity

$$R = \|f\|_{L^p}^{\frac{p}{n}} (M(f)(x))^{-\frac{p}{n}}$$

minimizes the expression on the right in (6.1.10). This choice of R yields the estimate

$$I_s(f)(x) \le C_{n,s,p} M(f)(x)^{\frac{p}{q}} \|f\|_{L^p}^{1-\frac{p}{q}}. \tag{6.1.11}$$

The required inequality for $p > 1$ follows by raising to the power q, integrating over $\mathbf{R}^n$, and using the boundedness of the Hardy–Littlewood maximal operator M on $L^p(\mathbf{R}^n)$. The case $p = 1$, $q = \frac{n}{n-s}$ also follows from (6.1.11) by the weak type $(1,1)$ property of M. Indeed,

$$\begin{aligned}
\left|\{C_{n,s,1} M(f)^{\frac{n-s}{n}} \|f\|_{L^1}^{\frac{s}{n}} > \lambda\}\right| &= \left|\left\{M(f) > \left(\frac{\lambda}{C_{n,s,1}\|f\|_{L^1}^{\frac{s}{n}}}\right)^{\frac{n}{n-s}}\right\}\right| \\
&\le 3^n \left(\frac{C_{n,s,1}\|f\|_{L^1}^{\frac{s}{n}}}{\lambda}\right)^{\frac{n}{n-s}} \|f\|_{L^1} \\
&= C(n,s)\left(\frac{\|f\|_{L^1}}{\lambda}\right)^{\frac{n}{n-s}}.
\end{aligned}$$

We now give an alternative proof of the case $p = 1$ that corresponds to $q = \frac{n}{n-s}$. Without loss of generality we may assume that $f \ge 0$ has L^1 norm 1. Once this case is proved, the general case follows by scaling. Observe that

$$\int_{\mathbf{R}^n} f(x-y)|y|^{s-n}\, dy \le \sum_{j\in\mathbf{Z}} 2^{(j-1)(s-n)} \int_{|y|\le 2^j} f(x-y)\, dy. \tag{6.1.12}$$

Let $E_\lambda = \{x : I_s(f)(x) > \lambda\}$. Then

$$\begin{aligned}
|E_\lambda| &\le \frac{1}{\lambda}\int_{E_\lambda} I_s(f)(x)\, dx \\
&= \frac{1}{\lambda}\int_{E_\lambda}\int_{\mathbf{R}^n} |y|^{s-n} f(x-y)\, dy\, dx \\
&\le \frac{1}{\lambda}\int_{E_\lambda}\sum_{j\in\mathbf{Z}} 2^{(j-1)(s-n)} \int_{|y|\le 2^j} f(x-y)\, dy\, dx \\
&= \frac{1}{\lambda}\sum_{j\in\mathbf{Z}} 2^{(j-1)(s-n)} \int_{E_\lambda}\int_{|y|\le 2^j} f(x-y)\, dy\, dx \\
&\le \frac{1}{\lambda}\sum_{j\in\mathbf{Z}} 2^{(j-1)(s-n)} \min(|E_\lambda|, v_n 2^{jn}) \\
&= \frac{1}{\lambda}\sum_{2^j > |E_\lambda|^{\frac{1}{n}}} 2^{(j-1)(s-n)}|E_\lambda| + \frac{v_n}{\lambda}\sum_{2^j \le |E_\lambda|^{\frac{1}{n}}} 2^{(j-1)(s-n)} 2^{jn} \\
&\le \frac{C}{\lambda}\left(|E_\lambda|^{\frac{s-n}{n}}|E_\lambda| + |E_\lambda|^{\frac{s}{n}}\right) \\
&= \frac{2C}{\lambda}|E_\lambda|^{\frac{s}{n}}.
\end{aligned} \tag{6.1.13}$$

It follows that $|E_\lambda|^{\frac{n-s}{n}} \le \frac{2C}{\lambda}$, which implies the weak type $(1, \frac{n}{n-s})$ estimate for I_s. Here C is a constant that depends on n and s. □

6.1.2 Bessel Potentials

While the behavior of the kernels $|x|^{-n+s}$ as $|x| \to 0$ is well suited to their smoothing properties, their decay as $|x| \to \infty$ gets worse as s increases. We can slightly adjust the Riesz potentials so that we maintain their essential behavior near zero but achieve exponential decay at infinity. The simplest way to achieve this is by replacing the "nonnegative" operator $-\Delta$ by the "strictly positive" operator $I - \Delta$. Here the terms nonnegative and strictly positive, as one may have surmised, refer to the Fourier multipliers of these operators.

Definition 6.1.4. Let s be a complex number with $0 < \operatorname{Re} s < \infty$. The *Bessel potential* of order s is the operator

$$\mathscr{J}_s = (I - \Delta)^{-s/2},$$

whose action on functions is given by

$$\mathscr{J}_s(f) = \big(\widehat{f}\,\widehat{G_s}\big)^{\vee} = f * G_s,$$

where

$$G_s(x) = \big((1 + 4\pi^2|\xi|^2)^{-s/2}\big)^{\vee}(x).$$

Let us see why this adjustment yields exponential decay for G_s at infinity.

Proposition 6.1.5. *Let $s > 0$. Then G_s is a smooth function on $\mathbf{R}^n \setminus \{0\}$ that satisfies $G_s(x) > 0$ for all $x \in \mathbf{R}^n$. Moreover, there exist positive finite constants $C(s,n), c(s,n), C_{s,n}$ such that*

$$G_s(x) \le C(s,n) e^{-\frac{|x|}{2}}, \qquad \text{when } |x| \ge 2, \tag{6.1.14}$$

and such that

$$\frac{1}{c(s,n)} \le \frac{G_s(x)}{H_s(x)} \le c(s,n), \qquad \text{when } |x| \le 2,$$

where H_s is equal to

$$H_s(x) = \begin{cases} |x|^{s-n} + 1 + O(|x|^{s-n+2}) & \text{for } 0 < s < n, \\ \log \frac{2}{|x|} + 1 + O(|x|^2) & \text{for } s = n, \\ 1 + O(|x|^{s-n}) & \text{for } s > n, \end{cases}$$

and $O(t)$ is a function with the property $|O(t)| \le C_{s,n}|t|$ for $0 \le t \le 4$.

Proof. For $A, s > 0$ we have the gamma function identity

$$A^{-\frac{s}{2}} = \frac{1}{\Gamma(\frac{s}{2})} \int_0^\infty e^{-tA} t^{\frac{s}{2}} \frac{dt}{t},$$

which we use to obtain

$$(1 + 4\pi^2|\xi|^2)^{-\frac{s}{2}} = \frac{1}{\Gamma(\frac{s}{2})} \int_0^\infty e^{-t} e^{-\pi|2\sqrt{\pi t}\,\xi|^2} t^{\frac{s}{2}} \frac{dt}{t}.$$

Note that the previous integral converges at both ends. Now take the inverse Fourier transform in ξ and use the fact that the function $e^{-\pi|\xi|^2}$ is equal to its Fourier transform (Example 2.2.9) to obtain

$$G_s(x) = \frac{(2\sqrt{\pi})^{-n}}{\Gamma(\frac{s}{2})} \int_0^\infty e^{-t} e^{-\frac{|x|^2}{4t}} t^{\frac{s-n}{2}} \frac{dt}{t}.$$

This proves that $G_s(x) > 0$ for all $x \in \mathbf{R}^n$ and that G_s is smooth on $\mathbf{R}^n \setminus \{0\}$. Now suppose $|x| \geq 2$. Then $t + \frac{|x|^2}{4t} \geq t + \frac{1}{t}$ and also $t + \frac{|x|^2}{4t} \geq |x|$. This implies that

$$-t - \frac{|x|^2}{4t} \leq -\frac{t}{2} - \frac{1}{2t} - \frac{|x|}{2},$$

from which it follows that when $|x| \geq 2$,

$$|G_s(x)| \leq \frac{(2\sqrt{\pi})^{-n}}{\Gamma(\frac{s}{2})} \left(\int_0^\infty e^{-\frac{t}{2}} e^{-\frac{1}{2t}} t^{\frac{s-n}{2}} \frac{dt}{t} \right) e^{-\frac{|x|}{2}} = C_{s,n} e^{-\frac{|x|}{2}}.$$

This proves (6.1.14).

Suppose now that $|x| \leq 2$. Write $G_s(x) = G_s^1(x) + G_s^2(x) + G_s^3(x)$, where

$$\begin{aligned}
G_s^1(x) &= \frac{(2\sqrt{\pi})^{-n}}{\Gamma(\frac{s}{2})} \int_0^{|x|^2} e^{-t'} e^{-\frac{|x|^2}{4t'}} (t')^{\frac{s-n}{2}} \frac{dt'}{t'} \\
&= |x|^{s-n} \frac{(2\sqrt{\pi})^{-n}}{\Gamma(\frac{s}{2})} \int_0^1 e^{-t|x|^2} e^{-\frac{1}{4t}} t^{\frac{s-n}{2}} \frac{dt}{t}, \\
G_s^2(x) &= \frac{(2\sqrt{\pi})^{-n}}{\Gamma(\frac{s}{2})} \int_{|x|^2}^4 e^{-t} e^{-\frac{|x|^2}{4t}} t^{\frac{s-n}{2}} \frac{dt}{t}, \\
G_s^3(x) &= \frac{(2\sqrt{\pi})^{-n}}{\Gamma(\frac{s}{2})} \int_4^\infty e^{-t} e^{-\frac{|x|^2}{4t}} t^{\frac{s-n}{2}} \frac{dt}{t}.
\end{aligned}$$

In G_s^1 we have $e^{-t|x|^2} = 1 + O(t|x|^2)$, since $t|x|^2 \leq 4$; thus we can write

$$\begin{aligned}
G_s^1(x) &= |x|^{s-n} \frac{(2\sqrt{\pi})^{-n}}{\Gamma(\frac{s}{2})} \int_0^1 e^{-\frac{1}{4t}} t^{\frac{s-n}{2}} \frac{dt}{t} + \frac{O(|x|^{s-n+2})}{\Gamma(\frac{s}{2})} \int_0^1 e^{-\frac{1}{4t}} t^{\frac{s-n}{2}} dt \\
&= c_{s,n}^1 |x|^{s-n} + O(|x|^{s-n+2}) \qquad \text{as } |x| \to 0.
\end{aligned}$$

Since $0 \le \frac{|x|^2}{4t} \le \frac{1}{4}$ and $0 \le t \le 4$ in G_s^2, we have $e^{-\frac{17}{4}} \le e^{-t-\frac{|x|^2}{4t}} \le 1$; thus as $|x| \to 0$ we obtain

$$G_s^2(x) \approx \int_{|x|^2}^4 t^{\frac{s-n}{2}} \frac{dt}{t} = \begin{cases} \frac{2}{n-s}|x|^{s-n} - \frac{2^{s-n+1}}{n-s} & \text{for } s < n, \\ 2\log\frac{2}{|x|} & \text{for } s = n, \\ \frac{1}{s-n}2^{s-n+1} - \frac{2}{s-n}|x|^{s-n} & \text{for } s > n. \end{cases}$$

Finally, we have $e^{-\frac{1}{4}} \le e^{-\frac{|x|^2}{4t}} \le 1$ in G_s^3, which yields that $G_s^3(x)$ is bounded above and below by fixed positive constants. Combining the estimates for $G_s^1(x)$, $G_s^2(x)$, and $G_s^3(x)$, we obtain the required conclusion. □

We end this section with a result analogous to that of Theorem 6.1.3 for the operator $\mathscr{J}_s$.

Corollary 6.1.6. *(a) For all $0 < s < \infty$, the operator $\mathscr{J}_s$ maps $L^r(\mathbf{R}^n)$ to itself with norm 1 for all $1 \le r \le \infty$.*
(b) Let $0 < s < n$ and $1 \le p < q < \infty$ satisfy (6.1.4). Then there exist constants $C_{p,q,n,s} < \infty$ such that for all f in $L^p(\mathbf{R}^n)$ with $p > 1$ we have

$$\big\|\mathscr{J}_s(f)\big\|_{L^q} \le C_{p,q,n,s}\big\|f\big\|_{L^p}$$

and also $\big\|\mathscr{J}_s(f)\big\|_{L^{q,\infty}} \le C_{1,q,n,s}\big\|f\big\|_{L^1}$ when $p = 1$.

Proof. (a) Since $\widehat{G_s}(0) = 1$ and $G_s > 0$, it follows that G_s has L^1 norm 1. The operator $\mathscr{J}_s$ is given by convolution with the positive function G_s, which has L^1 norm 1; thus it maps $L^r(\mathbf{R}^n)$ to itself with norm 1 for all $1 \le r \le \infty$ (see Exercise 1.2.9).
(b) In the special case $0 < s < n$ we have that the kernel G_s of $\mathscr{J}_s$ satisfies

$$G_s(x) \approx \begin{cases} |x|^{-n+s} & \text{when } |x| \le 2, \\ e^{-\frac{|x|}{2}} & \text{when } |x| \ge 2. \end{cases}$$

Then we can write

$$\begin{aligned} \mathscr{J}_s(f)(x) &\le C_{n,s}\left[\int_{|y|\le 2} |f(x-y)|\,|y|^{-n+s}\,dy + \int_{|y|\ge 2} |f(x-y)|\,e^{-\frac{|y|}{2}}\,dy\right] \\ &\le C_{n,s}\left[I_s(|f|)(x) + \int_{\mathbf{R}^n} |f(x-y)|\,e^{-\frac{|y|}{2}}\,dy\right]. \end{aligned}$$

We now use that the function $y \mapsto e^{-|y|/2}$ is in L^r for all $r < \infty$, Theorem 1.2.12 (Young's inequality), and Theorem 6.1.3 to complete the proof of the corollary. □

Exercises

6.1.1. (a) Let $0 < s,t < \infty$ be such that $s+t < n$. Show that $I_s I_t = I_{s+t}$.
(b) Prove the operator identities

$$I_s(-\Delta)^z = (-\Delta)^z I_s = I_{s-2z} = (-\Delta)^{z-\frac{s}{2}}$$

whenever $\operatorname{Re} s > 2\operatorname{Re} z$.
(c) Prove that for all $z \in \mathbf{C}$ we have

$$\langle (-\Delta)^z(f) \,|\, (-\Delta)^{-z}(g) \rangle = \langle f \,|\, g \rangle$$

whenever the Fourier transforms of f and g vanish to sufficiently high order at the origin.
(d) Given $\operatorname{Re} s > 0$, find an $\alpha \in \mathbf{C}$ such that the identity

$$\langle I_s(f) \,|\, f \rangle = \big\|(-\Delta)^\alpha(f)\big\|_{L^2}^2$$

is valid for all functions f as in part (c).

6.1.2. Use Exercise 2.2.14 to prove that for $-\infty < \alpha < n/2 < \beta < \infty$ we have

$$\big\|f\big\|_{L^\infty(\mathbf{R}^n)} \le C \big\|\Delta^{\alpha/2}(f)\big\|_{L^2(\mathbf{R}^n)}^{\frac{\beta-n/2}{\beta-\alpha}} \big\|\Delta^{\beta/2}(f)\big\|_{L^2(\mathbf{R}^n)}^{\frac{n/2-\alpha}{\beta-\alpha}},$$

where C depends only on α, n, β.

6.1.3. Show that when $0 < s < n$ we have

$$\sup_{\|f\|_{L^1(\mathbf{R}^n)}=1} \big\|I_s(f)\big\|_{L^{\frac{n}{n-s}}(\mathbf{R}^n)} = \sup_{\|f\|_{L^1(\mathbf{R}^n)}=1} \big\|\mathscr{J}_s(f)\big\|_{L^{\frac{n}{n-s}}(\mathbf{R}^n)} = \infty .$$

Thus I_s and $\mathscr{J}_s$ are not of strong type $(1, \frac{n}{n-s})$.
[*Hint:* Consider an approximate identity.]

6.1.4. Let $0 < s < n$. Consider the function $h(x) = |x|^{-s}(\log \frac{1}{|x|})^{-\frac{s}{n}(1+\delta)}$ for $|x| \le 1/e$ and zero otherwise. Prove that when $0 < \delta < \frac{n-s}{s}$ we have $h \in L^{\frac{n}{s}}(\mathbf{R}^n)$ but that $\lim_{x\to 0} I_s(h)(x) = \infty$. Conclude that I_s does not map $L^{\frac{n}{s}}(\mathbf{R}^n)$ to $L^\infty(\mathbf{R}^n)$.

6.1.5. For $1 \le p \le \infty$ and $0 < s < \infty$ define the *Bessel potential space* $\mathscr{L}_s^p(\mathbf{R}^n)$ as the space of all functions $f \in L^p(\mathbf{R}^n)$ for which there exists another function f_0 in $L^p(\mathbf{R}^n)$ such that $\mathscr{J}_s(f_0) = f$. Define a norm on these spaces by setting $\big\|f\big\|_{\mathscr{L}_s^p} = \big\|f_0\big\|_{L^p}$. Prove the following properties of these spaces:
(a) $\big\|f\big\|_{L^p} \le \big\|f\big\|_{\mathscr{L}_s^p}$; hence $\mathscr{L}_s^p(\mathbf{R}^n)$ is a subspace of $L^p(\mathbf{R}^n)$.
(b) For all $0 < t,s < \infty$ we have $G_s * G_t = G_{s+t}$ and thus

$$\mathscr{L}_s^p(\mathbf{R}^n) * \mathscr{L}_t^q(\mathbf{R}^n) \subseteq \mathscr{L}_{s+t}^r(\mathbf{R}^n),$$

where $1 \le p,q,r \le \infty$ and $\frac{1}{p}+\frac{1}{q}=\frac{1}{r}+1$.
(c) The sequence of norms $\|f\|_{\mathscr{L}_s^p}$ increases, and therefore the spaces $\mathscr{L}_s^p(\mathbf{R}^n)$ decrease as s increases.
(d) The map $\mathscr{I}_t$ is an isomorphism from the space $\mathscr{L}_s^p(\mathbf{R}^n)$ onto $\mathscr{L}_{s+t}^p(\mathbf{R}^n)$.
[Note: Note that the Bessel potential space $\mathscr{L}_s^p(\mathbf{R}^n)$ coincides with the Sobolev space $L_s^p(\mathbf{R}^n)$, introduced in Section 6.2.]

6.1.6. For $0 \le s < n$ define the *fractional maximal function*

$$M^s(f)(x) = \sup_{t>0} \frac{1}{(v_n t^n)^{\frac{n-s}{n}}} \int_{|y|\le t} |f(x-y)|\,dy,$$

where v_n is the volume of the unit ball in $\mathbf{R}^n$.
(a) Show that for some constant C we have

$$M^s(f) \le C I_s(f)$$

for all $f \ge 0$ and conclude that M^s maps L^p to L^q whenever I_s does.
(b) (*Adams [1]*) Let $s>0$, $1<p<\frac{n}{s}$, $1 \le q \le \infty$ be such that $\frac{1}{r}=\frac{1}{p}-\frac{s}{n}+\frac{sp}{nq}$. Show that there is a constant $C>0$ (depending on the previous parameters) such that for all positive functions f we have

$$\big\|I_s(f)\big\|_{L^r} \le C \big\|M^{n/p}(f)\big\|_{L^q}^{\frac{sp}{n}} \big\|f\big\|_{L^p}^{1-\frac{sp}{n}}.$$

[*Hint:* For $f \ne 0$, write $I_s(f) = I_1 + I_2$, where

$$I_1 = \int_{|x-y|\le\delta} f(y)\,|y|^{s-n}\,dy, \qquad I_2 = \int_{|x-y|>\delta} f(y)\,|y|^{s-n}\,dy.$$

Show that $I_1 \le C\delta^s M^0(f)$ and that $I_2(f) \le C\delta^{s-\frac{n}{p}} M^{n/p}(f)$. Optimize over $\delta>0$ to obtain

$$I_s(f) \le C M^{n/p}(f)^{\frac{sp}{n}} M^0(f)^{1-\frac{sp}{n}},$$

from which the required conclusion follows easily.]

6.1.7. Suppose that a function K defined on $\mathbf{R}^n$ satisfies $|K(y)| \le C(1+|y|)^{-s+n-\varepsilon}$, where $0<s<n$ and $0<C,\varepsilon<\infty$. Prove that the maximal operator

$$\sup_{t>0} t^{-n+s} \left| \int_{\mathbf{R}^n} f(x-y) K(y/t)\,dy \right|$$

maps $L^p(\mathbf{R}^n)$ to $L^q(\mathbf{R}^n)$ whenever I_s maps $L^p(\mathbf{R}^n)$ to $L^q(\mathbf{R}^n)$.
[*Hint:* Control this operator by the maximal function M^s of Exercise 6.1.6.]

6.1.8. Let $0<s<n$. Use the following steps to obtain a simpler proof of Theorem 6.1.3 based on more delicate interpolation.
(a) Prove that $\big\|I_s(\chi_E)\big\|_{L^\infty} \le |E|^{\frac{s}{n}}$ for any set E of finite measure.
(b) For any two sets E and F of finite measure show that

$$\int_F |I_s(\chi_E)(x)|\,dx \le |E|\,|F|^{\frac{s}{n}}.$$

(c) Use Exercise 1.1.12 to obtain that

$$\big\|I_s(\chi_E)\big\|_{L^{\frac{n}{n-s},\infty}} \le C_{ns}|E|.$$

(d) Use parts (a), (c), and Theorem 1.4.19 to obtain another proof of Theorem 6.1.3.
[*Hint:* Parts (a) and (b): Use that when $\lambda > 0$, the integral $\int_E |y|^{-\lambda}\,dy$ becomes largest when E is a ball centered at the origin equimeasurable to E.]

6.1.9. (*Welland [329]*) Let $0 < \alpha < n$ and suppose $0 < \varepsilon < \min(\alpha, n-\alpha)$. Show that there exists a constant depending only on α, ε, and n such that for all compactly supported bounded functions f we have

$$|I_\alpha(f)| \le C\sqrt{M^{\alpha-\varepsilon}(f)M^{\alpha+\varepsilon}(f)},$$

where $M^\beta(f)$ is the fractional maximal function of Exercise 6.1.6.
[*Hint:* Write

$$|I_\alpha(f)| \le \int_{|x-y|<s} \frac{|f(y)|\,dy}{|x-y|^{n-\alpha}} + \int_{|x-y|\ge s} \frac{|f(y)|\,dy}{|x-y|^{n-\alpha}}$$

and split each integral into a sum of integrals over annuli centered at x to obtain the estimate

$$|I_\alpha(f)| \le C\big(s^\varepsilon M^{\alpha-\varepsilon}(f) + s^{-\varepsilon}M^{\alpha+\varepsilon}(f)\big).$$

Then optimize over s.]

6.1.10. Show that the *discrete fractional integral operator*

$$\{a_j\}_{j\in\mathbf{Z}^n} \to \Big\{\sum_{k\in\mathbf{Z}^n} \frac{a_k}{(|j-k|+1)^{n-\alpha}}\Big\}_{j\in\mathbf{Z}^n}$$

maps $\ell^s(\mathbf{Z}^n)$ to $\ell^t(\mathbf{Z}^n)$ when $0 < \alpha < n$, $1 < s < t$, and

$$\frac{1}{s} - \frac{1}{t} = \frac{\alpha}{n}.$$

6.1.11. Show that the bilinear operator

$$B_\alpha(f,g)(x) = \int_{\mathbf{R}^n}\int_{\mathbf{R}^n} f(y)g(z)(|x-y|+|x-z|)^{-2n+\alpha}\,dy\,dz$$

maps $L^p(\mathbf{R}^n)\times L^q(\mathbf{R}^n)$ to $L^r(\mathbf{R}^n)$ when $1 < p, q < \infty$ and

$$\frac{1}{p} + \frac{1}{q} = \frac{\alpha}{n} + \frac{1}{r}.$$

[*Hint:* Control $B_\alpha(f,g)$ by the product of two fractional integrals.]

6.1.12. (*Grafakos and Kalton [148]/Kenig and Stein [189]*) (a) Prove that the bilinear operator

$$S(f,g)(x) = \int_{|t|\le 1} |f(x+t)g(x-t)|\,dt$$

maps $L^1(\mathbf{R}^n) \times L^1(\mathbf{R}^n)$ to $L^{\frac{1}{2}}(\mathbf{R}^n)$.
(b) For $0 < \alpha < n$ prove that the *bilinear fractional integral operator*

$$I_\alpha(f,g)(x) = \int_{\mathbf{R}^n} f(x+t)g(x-t)|t|^{-n+\alpha}\,dt$$

maps $L^1(\mathbf{R}^n) \times L^1(\mathbf{R}^n)$ to $L^{\frac{n}{2n-\alpha},\infty}(\mathbf{R}^n)$.
[*Hint:* Part (a): Write $f = \sum_{k\in\mathbf{Z}^n} f_k$, where each f_k is supported in the cube $k+[0,1]^n$ and similarly for g. Observe that the resulting double sum reduces to a single sum and use that $(\sum_j a_j)^{1/2} \le \sum_j a_j^{1/2}$ for $a_j \ge 0$. Part (b): Use part (a) and adjust the argument in (6.1.13) to a bilinear setting.]

6.2 Sobolev Spaces

In this section we study a quantitative way of measuring smoothness of functions. Sobolev spaces serve exactly this purpose. They measure the smoothness of a given function in terms of the integrability of its derivatives. We begin with the classical definition of Sobolev spaces.

Definition 6.2.1. Let k be a nonnegative integer and let $1 < p < \infty$. The *Sobolev space* $L^p_k(\mathbf{R}^n)$ is defined as the space of functions f in $L^p(\mathbf{R}^n)$ all of whose distributional derivatives $\partial^\alpha f$ are also in $L^p(\mathbf{R}^n)$ for all multi-indices α that satisfy $|\alpha| \le k$. This space is normed by the expression

$$\big\|f\big\|_{L^p_k} = \sum_{|\alpha|\le k} \big\|\partial^\alpha f\big\|_{L^p}, \tag{6.2.1}$$

where $\partial^{(0,\dots,0)} f = f$.

Sobolev spaces measure smoothness of functions. The index k indicates the "degree" of smoothness of a given function in L^p_k. As k increases the functions become smoother. Equivalently, these spaces form a decreasing sequence

$$L^p \supset L^p_1 \supset L^p_2 \supset L^p_3 \supset \cdots,$$

meaning that each $L^p_{k+1}(\mathbf{R}^n)$ is a subspace of $L^p_k(\mathbf{R}^n)$. This property, which coincides with our intuition of smoothness, is a consequence of the definition of the Sobolev norms.

We next observe that the space $L^p_k(\mathbf{R}^n)$ is complete. Indeed, if f_j is a Cauchy sequence in the norm given by (6.2.1), then $\{\partial^\alpha f_j\}_j$ are Cauchy sequences for all

$|\alpha| \le k$. By the completeness of L^p, there exist functions f_α such that $\partial^\alpha f_j \to f_\alpha$ in L^p. This implies that for all φ in the Schwartz class we have

$$(-1)^{|\alpha|}\int_{\mathbf{R}^n} f_j\,(\partial^\alpha\varphi)\,dx = \int_{\mathbf{R}^n}(\partial^\alpha f_j)\,\varphi\,dx \to \int_{\mathbf{R}^n} f_\alpha\,\varphi\,dx.$$

Since the first expression converges to

$$(-1)^{|\alpha|}\int_{\mathbf{R}^n} f_0\,(\partial^\alpha\varphi)\,dx,$$

it follows that the distributional derivative $\partial^\alpha f_0$ is f_α. This implies that $f_j \to f_0$ in $L^p_k(\mathbf{R}^n)$ and proves the completeness of this space.

Our goal in this section is to investigate relations between these spaces and the Riesz and Bessel potentials discussed in the previous section and to obtain a Littlewood–Paley characterization of them. Before we embark on this study, we note that we can extend the definition of Sobolev spaces to the case in which the index k is not necessarily an integer. In fact, we extend the definition of the spaces $L^p_k(\mathbf{R}^n)$ to the case in which the number k is real.

6.2.1 Definition and Basic Properties of General Sobolev Spaces

Definition 6.2.2. Let s be a real number and let $1 < p < \infty$. The *inhomogeneous Sobolev space* $L^p_s(\mathbf{R}^n)$ is defined as the space of all tempered distributions u in $\mathscr{S}'(\mathbf{R}^n)$ with the property that

$$((1+|\xi|^2)^{\frac{s}{2}}\widehat{u})^\vee \tag{6.2.2}$$

is an element of $L^p(\mathbf{R}^n)$. For such distributions u we define

$$\big\|u\big\|_{L^p_s} = \big\|((1+|\cdot|^2)^{\frac{s}{2}}\widehat{u})^\vee\big\|_{L^p(\mathbf{R}^n)}.$$

Note that the function $(1+|\xi|^2)^{\frac{s}{2}}$ is $\mathscr{C}^\infty$ and has at most polynomial growth at infinity. Since $\widehat{u}\in\mathscr{S}'(\mathbf{R}^n)$, the product in (6.2.2) is well defined.

Several observations are in order. First, we note that when $s=0$, $L^p_s = L^p$. It is natural to ask whether elements of L^p_s are always L^p functions. We show that this is the case when $s \ge 0$ but not when $s < 0$. We also show that the space L^p_s coincides with the space L^p_k given in Definition 6.2.1 when $s = k$ and k is an integer.

To prove that elements of L^p_s are indeed L^p functions when $s \ge 0$, we simply note that if $f_s = ((1+|\xi|^2)^{s/2}\widehat{f})^\vee$, then

$$f = \big(\widehat{f_s}(\xi)\widehat{G_s}(\xi/2\pi)\big)^\vee = f_s * (2\pi)^n G_s(2\pi(\cdot)),$$

where G_s is given in Definition 6.1.4. Thus a certain dilation of f can be expressed as the Bessel potential of itself; hence Corollary 6.1.6 yields that

$$c^{-1}\|f\|_{L^p} \le \|f_s\|_{L^p} = \|f\|_{L^p_s},$$

for some constant c.

We now prove that if $s = k$ is a nonnegative integer and $1 < p < \infty$, then the norm of the space L^p_k as given in Definition 6.2.1 is comparable to that in Definition 6.2.2. Suppose that $f \in L^p_k$ according to Definition 6.2.2. Then for all $|\alpha| \le k$ we have

$$\partial^\alpha f = c_\alpha(\widehat{f}(\xi)\xi^\alpha)^\vee = c_\alpha\left(\widehat{f}(\xi)(1+|\xi|^2)^{\frac{k}{2}}\frac{\xi^\alpha}{(1+|\xi|^2)^{\frac{k}{2}}}\right)^\vee. \tag{6.2.3}$$

Theorem 5.2.7 gives that the function

$$\frac{\xi^\alpha}{(1+|\xi|^2)^{k/2}}$$

is an L^p multiplier. Since by assumption $\big(\widehat{f}(\xi)(1+|\xi|^2)^{\frac{k}{2}}\big)^\vee$ is in $L^p(\mathbf{R}^n)$, it follows from (6.2.3) that $\partial^\alpha f$ is in L^p and also that

$$\sum_{|\alpha|\le k}\|\partial^\alpha f\|_{L^p} \le C_{p,n,k}\|((1+|\cdot|^2)^{\frac{k}{2}}\widehat{f})^\vee\|_{L^p}.$$

Conversely, suppose that $f \in L^p_k$ according to Definition 6.2.1; then

$$(1+\xi_1^2+\cdots+\xi_n^2)^{\frac{k}{2}} = \sum_{|\alpha|\le k}\frac{k!}{\alpha_1!\cdots\alpha_n!(k-|\alpha|)!}\,\xi^\alpha\,\frac{\xi^\alpha}{(1+|\xi|^2)^{\frac{k}{2}}}.$$

As we have already observed, the functions $m_\alpha(\xi) = \xi^\alpha(1+|\xi|^2)^{-\frac{k}{2}}$ are L^p multipliers whenever $|\alpha| \le k$. Since

$$((1+|\xi|^2)^{\frac{k}{2}}\widehat{f})^\vee = \sum_{|\alpha|\le k} c_{\alpha,k}\big(m_\alpha(\xi)\xi^\alpha\widehat{f}\big)^\vee = \sum_{|\alpha|\le k} c'_{\alpha,k}\big(m_\alpha(\xi)\widehat{\partial^\alpha f}\big)^\vee,$$

it follows that

$$\|(\widehat{f}(\xi)(1+|\xi|^2)^{\frac{k}{2}})^\vee\|_{L^p} \le C_{p,n,k}\sum_{|\gamma|\le k}\|(\widehat{f}(\xi)\xi^\gamma)^\vee\|_{L^p}.$$

Example 6.2.3. Every Schwartz function lies in $L^p_s(\mathbf{R}^n)$ for s real. Sobolev spaces with negative indices s can indeed contain tempered distributions that are not locally integrable functions. For example, Dirac mass at the origin δ_0 is an element of $L^p_{-s}(\mathbf{R}^n)$ for all $s > n/p'$. Indeed, when $0 < s < n$, Proposition 6.1.5 gives that G_s [i.e., the inverse Fourier transform of $(1+|\xi|^2)^{-\frac{s}{2}}$] is integrable to the power p as long as $(s-n)p > -n$ (i.e., $s > n/p'$). When $s \geq n$, G_s is integrable to any positive power.

We now continue with the *Sobolev embedding theorem.*

Theorem 6.2.4. *(a) Let $0 < s < \frac{n}{p}$ and $1 < p < \infty$. Then the Sobolev space $L_s^p(\mathbf{R}^n)$ continuously embeds in $L^q(\mathbf{R}^n)$ when*

$$\frac{1}{p} - \frac{1}{q} = \frac{s}{n}.$$

(b) Let $0 < s = \frac{n}{p}$ and $1 < p < \infty$. Then $L_s^p(\mathbf{R}^n)$ continuously embeds in $L^q(\mathbf{R}^n)$ for any $\frac{n}{s} < q < \infty$.
(c) Let $\frac{n}{p} < s < \infty$ and $1 < p < \infty$. Then every element of $L_s^p(\mathbf{R}^n)$ can be modified on a set of measure zero so that the resulting function is bounded and uniformly continuous.

Proof. (a) If $f \in L_s^p$, then $f_s(x) = ((1+|\xi|^2)^{\frac{s}{2}}\widehat{f}\,)^\vee(x)$ is in $L^p(\mathbf{R}^n)$. Thus

$$f(x) = ((1+|\xi|^2)^{-\frac{s}{2}}\widehat{f_s}\,)^\vee(x);$$

hence $f = G_s * f_s$. Since $s < n$, Proposition 6.1.5 gives that

$$|G_s(x)| \leq C_{s,n}|x|^{s-n}$$

for all $x \in \mathbf{R}^n$. This implies that $|f| = |G_s * f_s| \leq C_{s,n} I_s(|f_s|)$. Theorem 6.1.3 now yields the required conclusion

$$\big\|f\big\|_{L^q} \leq C'_{s,n}\big\|I_s(|f_s|)\big\|_{L^q} \leq C''_{s,n}\big\|f\big\|_{L_s^p}.$$

(b) Given any $\frac{n}{s} < q < \infty$ we can find $t > 1$ such that

$$1 + \frac{1}{q} = \frac{s}{n} + \frac{1}{t} = \frac{1}{p} + \frac{1}{t}.$$

Then $1 < \frac{s}{n} + \frac{1}{t}$, which implies that $(-n+s)t > -n$. Thus the function $|x|^{-n+s}\chi_{|x|\leq 2}$ is integrable to the tth power, which implies that G_s is in L^t. Since $f = G_s * f_s$, Young's inequality gives that

$$\big\|f\big\|_{L^q(\mathbf{R}^n)} \leq \big\|f_s\big\|_{L^p(\mathbf{R}^n)}\big\|G_s\big\|_{L^t(\mathbf{R}^n)} = C_{n,s}\big\|f\big\|_{L_{n/p}^p}.$$

(c) As before, $f = G_s * f_s$. If $s \geq n$, then Proposition 6.1.5 gives that the function G_s is in $L^{p'}(\mathbf{R}^n)$. Now if $n > s$, then $G_s(x)$ looks like $|x|^{-n+s}$ near zero. This function is integrable to the power p' near the origin if and only if $s > n/p$, which is what we are assuming. Thus f is given as the convolution of an L^p function and an $L^{p'}$ function, and hence it is bounded and can be identified with a uniformly continuous function (cf. Exercise 1.2.3). □

We now introduce the homogeneous Sobolev spaces $\dot{L}^p_s$. The main difference with the inhomogeneous spaces L^p_s is that elements of $\dot{L}^p_s$ may not themselves be elements of L^p. Another difference is that elements of homogeneous Sobolev spaces are not tempered distributions but equivalence classes of tempered distributions.

We would expect the homogeneous Sobolev space $\dot{L}^p_s$ to be the space of all distributions u in $\mathscr{S}'(\mathbf{R}^n)$ for which the expression

$$(|\xi|^s\widehat{u})^\vee \tag{6.2.4}$$

is an L^p function. Since the function $|\xi|^s$ is not (always) smooth at the origin, some care is needed in defining the product in (6.2.4). The idea is that when u lies in $\mathscr{S}'/\mathscr{P}$, then the value of $\widehat{u}$ at the origin is irrelevant, since we may add to $\widehat{u}$ a distribution supported at the origin and obtain another element of the equivalence class of u (Proposition 2.4.1). It is because of this irrelevance that we are allowed to multiply $\widehat{u}$ by a function that may be nonsmooth at the origin (and which has polynomial growth at infinity).

To do this, we fix a smooth function $\eta(\xi)$ on $\mathbf{R}^n$ that is equal to 1 when $|\xi|\geq 2$ and vanishes when $|\xi|\leq 1$. Then for $s\in\mathbf{R}$, $u\in\mathscr{S}'(\mathbf{R}^n)/\mathscr{P}$, and $\varphi\in\mathscr{S}(\mathbf{R}^n)$ we define

$$\left\langle |\xi|^s\widehat{u},\varphi\right\rangle = \lim_{\varepsilon\to 0}\left\langle \widehat{u},\eta(\tfrac{\xi}{\varepsilon})|\xi|^s\varphi(\xi)\right\rangle,$$

provided that the last limit exists. Note that this defines $|\xi|^s\widehat{u}$ as another element of $\mathscr{S}'/\mathscr{P}$, and this definition is independent of the function η, as follows easily from (2.3.23).

Definition 6.2.5. Let s be a real number and let $1<p<\infty$. The *homogeneous Sobolev space* $\dot{L}^p_s(\mathbf{R}^n)$ is defined as the space of all tempered distributions modulo polynomials u in $\mathscr{S}'(\mathbf{R}^n)/\mathscr{P}$ for which the expression

$$(|\xi|^s\widehat{u})^\vee$$

exists and is an $L^p(\mathbf{R}^n)$ function. For distributions u in $\dot{L}^p_s(\mathbf{R}^n)$ we define

$$\big\|u\big\|_{\dot{L}^p_s} = \big\|(|\cdot|^s\widehat{u})^\vee\big\|_{L^p(\mathbf{R}^n)}\,. \tag{6.2.5}$$

As noted earlier, to avoid working with equivalence classes of functions, we identify two distributions in $\dot{L}^p_s(\mathbf{R}^n)$ whose difference is a polynomial. In view of this identification, the quantity in (6.2.5) is a norm.

6.2.2 Littlewood–Paley Characterization of Inhomogeneous Sobolev Spaces

We now present the first main result of this section, the characterization of the inhomogeneous Sobolev spaces using Littlewood–Paley theory.

For the purposes of the next theorem we need the following setup. We fix a radial Schwartz function Ψ on $\mathbf{R}^n$ whose Fourier transform is nonnegative, supported in the annulus $1-\frac{1}{7}\le|\xi|\le 2$, equal to 1 on the smaller annulus $1\le|\xi|\le 2-\frac{2}{7}$, and satisfies $\widehat{\Psi}(\xi)+\widehat{\Psi}(\xi/2)=1$ on the annulus $1\le|\xi|\le 4-\frac{4}{7}$. This function has the property

$$\sum_{j\in\mathbf{Z}}\widehat{\Psi}(2^{-j}\xi)=1 \tag{6.2.6}$$

for all $\xi\neq 0$. We define the associated Littlewood–Paley operators Δ_j given by multiplication on the Fourier transform side by the function $\widehat{\Psi}(2^{-j}\xi)$, that is,

$$\Delta_j(f)=\Delta_j^{\Psi}(f)=\Psi_{2^{-j}}*f\,. \tag{6.2.7}$$

Notice that the support properties of the Δ_j's yield the simple identity

$$\Delta_j=\big(\Delta_{j-1}+\Delta_j+\Delta_{j+1}\big)\Delta_j$$

for all $j\in\mathbf{Z}$. We also define a Schwartz function Φ so that

$$\widehat{\Phi}(\xi)=\begin{cases}\sum_{j\le 0}\widehat{\Psi}(2^{-j}\xi) & \text{when } \xi\neq 0,\\ 1 & \text{when } \xi=0.\end{cases} \tag{6.2.8}$$

Note that $\widehat{\Phi}(\xi)$ is equal to 1 for $|\xi|\le 2-\frac{2}{7}$, vanishes when $|\xi|\ge 2$, and satisfies

$$\widehat{\Phi}(\xi)+\sum_{j=1}^{\infty}\widehat{\Psi}(2^{-j}\xi)=1 \tag{6.2.9}$$

for all ξ in $\mathbf{R}^n$. We now introduce an operator S_0 by setting

$$S_0(f)=\Phi*f\,. \tag{6.2.10}$$

Identity (6.2.9) yields the operator identity

$$S_0+\sum_{j=1}^{\infty}\Delta_j=I\,,$$

in which the series converges in $\mathscr{S}'(\mathbf{R}^n)$; see Exercise 2.3.12. (Note that $S_0(f)$ and $\Delta_j(f)$ are well defined functions when f is a tempered distribution.)

Having introduced the relevant background, we are now ready to state and prove the following result.

Theorem 6.2.6. *Let Φ, Ψ satisfy (6.2.6) and (6.2.8) and let Δ_j, S_0 be as in (6.2.7) and (6.2.10). Fix $s\in\mathbf{R}$ and all $1<p<\infty$. Then there exists a constant C_1 that depends only on n,s,p, Φ, and Ψ such that for all $f\in L^p_s$ we have*

$$\big\|S_0(f)\big\|_{L^p}+\Big\|\Big(\sum_{j=1}^{\infty}(2^{js}|\Delta_j(f)|)^2\Big)^{\frac{1}{2}}\Big\|_{L^p}\le C_1\big\|f\big\|_{L^p_s}\,. \tag{6.2.11}$$

Conversely, there exists a constant C_2 that depends on the parameters n, s, p, Φ, and Ψ such that every tempered distribution f that satisfies

$$\big\|S_0(f)\big\|_{L^p} + \Big\|\Big(\sum_{j=1}^{\infty}(2^{js}|\Delta_j(f)|)^2\Big)^{\frac{1}{2}}\Big\|_{L^p} < \infty$$

is an element of the Sobolev space L^p_s with norm

$$\big\|f\big\|_{L^p_s} \le C_2\Big(\big\|S_0(f)\big\|_{L^p} + \Big\|\Big(\sum_{j=1}^{\infty}(2^{js}|\Delta_j(f)|)^2\Big)^{\frac{1}{2}}\Big\|_{L^p}\Big). \tag{6.2.12}$$

Proof. We denote by C a generic constant that depends on the parameters n, s, p, Φ, and Ψ and that may vary in different occurrences. For a given tempered distribution f we define another tempered distribution f_s by setting

$$f_s = \big((1+|\cdot|^2)^{\frac{s}{2}}\widehat{f}\big)^{\vee},$$

so that we have $\big\|f\big\|_{L^p_s} = \big\|f_s\big\|_{L^p}$ if $f \in L^p_s$.

We first assume that the expression on the right in (6.2.12) is finite and we show that the tempered distribution f lies in the space L^p_s by controlling the L^p norm of f_s by a multiple of this expression. We begin by writing

$$f_s = \big(\widehat{\Phi}\widehat{f_s}\big)^{\vee} + \big((1-\widehat{\Phi})\widehat{f_s}\big)^{\vee},$$

and we plan to show that both quantities on the right are in L^p. Pick a smooth function with compact support η_0 that is equal to 1 on the support of $\widehat{\Phi}$. It is a simple fact that for all $s \in \mathbf{R}$ the function $(1+|\xi|^2)^{\frac{s}{2}}\eta_0(\xi)$ is in $\mathscr{M}_p(\mathbf{R}^n)$ (i.e., it is an L^p Fourier multiplier). Since

$$\big(\widehat{\Phi}\widehat{f_s}\big)^{\vee}(x) = \big\{\big((1+|\xi|^2)^{\frac{s}{2}}\eta_0(\xi)\big)\,\widehat{S_0(f)}(\xi)\big\}^{\vee}(x), \tag{6.2.13}$$

we have the estimate

$$\big\|\big(\widehat{\Phi}\widehat{f_s}\big)^{\vee}\big\|_{L^p} \le C\|S_0(f)\|_{L^p}. \tag{6.2.14}$$

We now introduce a smooth function η_∞ that vanishes in a neighborhood of the origin and is equal to 1 on the support of $1-\widehat{\Phi}$. Using Theorem 5.2.7, we can easily see that the function

$$\frac{(1+|\xi|^2)^{\frac{s}{2}}}{|\xi|^s}\eta_\infty(\xi)$$

is in $\mathscr{M}_p(\mathbf{R}^n)$ (with constant depending on n, p, η_∞, and s). Since

$$\big((1+|\xi|^2)^{\frac{s}{2}}(1-\widehat{\Phi}(\xi))\widehat{f}\big)^{\vee}(x) = \Big(\frac{(1+|\xi|^2)^{\frac{s}{2}}\eta_\infty(\xi)}{|\xi|^s}\,|\xi|^s(1-\widehat{\Phi}(\xi))\widehat{f}\Big)^{\vee}(x),$$

we obtain the estimate

$$\left\|\left((1-\widehat{\Phi})\widehat{f_s}\right)^{\vee}\right\|_{L^p} \le C\left\|f_\infty\right\|_{L^p}, \tag{6.2.15}$$

where f_∞ is another tempered distribution defined via

$$f_\infty = \left(|\xi|^s(1-\widehat{\Phi}(\xi))\widehat{f}\right)^{\vee}.$$

We are going to show that the quantity $\|f_\infty\|_{L^p}$ is finite using Littlewood–Paley theory. To achieve this, we introduce a smooth bump $\widehat{\zeta}$ supported in the annulus $\frac{1}{2} \le |\xi| \le 4$ and equal to 1 on the support of $\widehat{\Psi}$. Then we define $\widehat{\theta}(\xi) = |\xi|^s\widehat{\zeta}(\xi)$ and we introduce Littlewood–Paley operators

$$\Delta_j^\theta(g) = g * \theta_{2^{-j}},$$

where $\theta_{2^{-j}}(t) = 2^{jn}\theta(2^j t)$. Recalling that

$$1-\widehat{\Phi}(\xi) = \sum_{k\ge 1}\widehat{\Psi}(2^{-k}\xi),$$

we obtain that

$$\widehat{f_\infty} = \sum_{j=1}^{\infty}|\xi|^s\widehat{\Psi}(2^{-j}\xi)\widehat{\zeta}(2^{-j}\xi)\widehat{f} = \sum_{j=1}^{\infty}2^{js}\widehat{\Psi}(2^{-j}\xi)\widehat{\theta}(2^{-j}\xi)\widehat{f}$$

and hence

$$f_\infty = \sum_{j=1}^{\infty}\Delta_j^\theta(2^{js}\Delta_j(f)).$$

Using estimate (5.1.20), we obtain

$$\left\|f_\infty\right\|_{L^p} \le C\left\|\Big(\sum_{j=1}^{\infty}|2^{js}\Delta_j(f)|^2\Big)^{\frac{1}{2}}\right\|_{L^p} < \infty. \tag{6.2.16}$$

Combining (6.2.14), (6.2.15), and (6.2.16), we deduce the estimate in (6.2.12). (Incidentally, this argument shows that f_∞ is a function.)

To obtain the converse inequality (6.2.11) we essentially have to reverse our steps. Here we assume that $f \in L^p_s$ and we show the validity of (6.2.11). First, we have the estimate

$$\left\|S_0(f)\right\|_{L^p} \le C\left\|f_s\right\|_{L^p} = C\left\|f\right\|_{L^p_s}, \tag{6.2.17}$$

since we can obtain the Fourier transform of $S_0(f) = \Phi * f$ by multiplying $\widehat{f_s}$ by the L^p Fourier multiplier $(1+|\xi|^2)^{-\frac{s}{2}}\widehat{\Phi}(\xi)$. Second, setting $\widehat{\sigma}(\xi) = |\xi|^{-s}\widehat{\Psi}(\xi)$ and letting Δ_j^σ be the Littlewood–Paley operator associated with the bump $\widehat{\sigma}(2^{-j}\xi)$, we have

$$2^{js}\widehat{\Psi}(2^{-j}\xi)\widehat{f} = \widehat{\sigma}(2^{-j}\xi)|\xi|^s\widehat{f} = \widehat{\sigma}(2^{-j}\xi)|\xi|^s(1-\widehat{\Phi}(\xi))\widehat{f},$$

when $j \ge 2$ [since $\widehat{\Phi}$ vanishes on the support of $\widehat{\sigma}(2^{-j}\xi)$ when $j \ge 2$]. This yields the operator identity

$$2^{js}\Delta_j(f) = \Delta_j^\sigma(f_\infty). \tag{6.2.18}$$

Using identity (6.2.18) we obtain

$$\Big\|\Big(\sum_{j=2}^\infty |2^{js}\Delta_j(f)|^2\Big)^{\frac12}\Big\|_{L^p} = \Big\|\Big(\sum_{j=2}^\infty |\Delta_j^\sigma(f_\infty)|^2\Big)^{\frac12}\Big\|_{L^p} \le C\big\|f_\infty\big\|_{L^p}, \tag{6.2.19}$$

where the last inequality follows by Theorem 5.1.2. Notice that

$$f_\infty = \big(|\xi|^s(1-\widehat{\Phi}(\xi))\widehat{f}\,\big)^\vee = \Big(\frac{|\xi|^s(1-\widehat{\Phi}(\xi))}{(1+|\xi|^2)^{\frac{s}{2}}}\,\widehat{f_s}\Big)^\vee,$$

and since the function $|\xi|^s(1-\widehat{\Phi}(\xi))(1+|\xi|^2)^{-\frac{s}{2}}$ is in $\mathscr{M}_p(\mathbf{R}^n)$ by Theorem 5.2.7, it follows that

$$\big\|f_\infty\big\|_{L^p} \le C\big\|f_s\big\|_{L^p} = C\big\|f\big\|_{L^p_s},$$

which combined with (6.2.19) yields

$$\Big\|\Big(\sum_{j=2}^\infty |2^{js}\Delta_j(f)|^2\Big)^{\frac12}\Big\|_{L^p} \le C\big\|f\big\|_{L^p_s}. \tag{6.2.20}$$

Finally, we have

$$2^s\Delta_1(f) = 2^s\big(\widehat{\Psi}(\tfrac12\xi)(1+|\xi|^2)^{-\frac{s}{2}}(1+|\xi|^2)^{\frac{s}{2}}\widehat{f}\,\big)^\vee = 2^s\big(\widehat{\Psi}(\tfrac12\xi)(1+|\xi|^2)^{-\frac{s}{2}}\widehat{f_s}\,\big)^\vee,$$

and since the function $\widehat{\Psi}(\frac12\xi)(1+|\xi|^2)^{-\frac{s}{2}}$ is smooth with compact support and thus in $\mathscr{M}_p$, it follows that

$$\big\|2^s\Delta_1(f)\big\|_{L^p} \le C\big\|f_s\big\|_{L^p} = C\big\|f\big\|_{L^p_s}. \tag{6.2.21}$$

Combining estimates (6.2.17), (6.2.20), and (6.2.21), we conclude the proof of (6.2.11). □

6.2.3 Littlewood–Paley Characterization of Homogeneous Sobolev Spaces

We now state and prove the homogeneous version of the previous theorem.

Theorem 6.2.7. *Let Ψ satisfy (6.2.6) and let Δ_j be the Littlewood–Paley operator associated with Ψ. Let $s \in \mathbf{R}$ and $1 < p < \infty$. Then there exists a constant C_1 that depends only on n, s, p, and Ψ such that for all $f \in \dot{L}^p_s(\mathbf{R}^n)$ we have*

$$\Big\|\Big(\sum_{j\in\mathbf{Z}} (2^{js}|\Delta_j(f)|)^2\Big)^{\frac12}\Big\|_{L^p} \le C_1\big\|f\big\|_{\dot{L}^p_s}. \tag{6.2.22}$$

Conversely, there exists a constant C_2 that depends on the parameters n, s, p, and Ψ such that every element f of $\mathscr{S}'(\mathbf{R}^n)/\mathscr{P}$ that satisfies

$$\Big\| \Big(\sum_{j\in\mathbf{Z}} (2^{js}|\Delta_j(f)|)^2 \Big)^{\frac{1}{2}} \Big\|_{L^p} < \infty$$

lies in the homogeneous Sobolev space $\dot{L}^p_s$ and we have

$$\big\| f \big\|_{\dot{L}^p_s} \le C_2 \Big\| \Big(\sum_{j\in\mathbf{Z}} (2^{js}|\Delta_j(f)|)^2 \Big)^{\frac{1}{2}} \Big\|_{L^p}. \tag{6.2.23}$$

Proof. The proof of the theorem is similar but a bit simpler than that of Theorem 6.2.6. To obtain (6.2.22) we start with $f \in \dot{L}^p_s$ and we note that

$$2^{js}\Delta_j(f) = 2^{js}\big(|\xi|^s|\xi|^{-s}\widehat{\Psi}(2^{-j}\xi)\widehat{f}\,\big)^\vee = \big(\widehat{\sigma}(2^{-j}\xi)\widehat{f_s}\,\big)^\vee = \Delta_j^\sigma(f_s),$$

where $\widehat{\sigma}(\xi) = \widehat{\Psi}(\xi)|\xi|^{-s}$ and Δ_j^σ is the Littlewood–Paley operator given on the Fourier transform side by multiplication with the function $\widehat{\sigma}(2^{-j}\xi)$. We have

$$\Big\| \Big(\sum_{j\in\mathbf{Z}} |2^{js}\Delta_j(f)|^2 \Big)^{\frac{1}{2}} \Big\|_{L^p} = \Big\| \Big(\sum_{j\in\mathbf{Z}} |\Delta_j^\sigma(f_s)|^2 \Big)^{\frac{1}{2}} \Big\|_{L^p} \le C\big\|f_s\big\|_{L^p} = C\big\|f\big\|_{\dot{L}^p_s},$$

where the last inequality follows from Theorem 5.1.2. This proves (6.2.22).

Next we show that if the expression on the right in (6.2.23) is finite, then the distribution f in $\mathscr{S}'(\mathbf{R}^n)/\mathscr{P}$ must lie the in the homogeneous Sobolev space $\dot{L}^p_s$ with norm controlled by a multiple of this expression.

Define Littlewood–Paley operators Δ_j^η given by convolution with $\eta_{2^{-j}}$, where $\widehat{\eta}$ is a smooth bump supported in the annulus $\frac{4}{5} \le |\xi| \le 2$ that satisfies

$$\sum_{k\in\mathbf{Z}} \widehat{\eta}(2^{-k}\xi) = 1, \qquad \xi \neq 0, \tag{6.2.24}$$

or, in operator form,

$$\sum_{k\in\mathbf{Z}} \Delta_k^\eta = I,$$

where the convergence is in the sense of $\mathscr{S}'/\mathscr{P}$ in view of Exercise 2.3.12. We introduce another family of Littlewood–Paley operators Δ_j^θ given by convolution with $\theta_{2^{-j}}$, where $\widehat{\theta}(\xi) = \widehat{\eta}(\xi)|\xi|^s$. Given $f \in \mathscr{S}'(\mathbf{R}^n)/\mathscr{P}$, we set $f_s = \big(|\xi|^s\widehat{f}\,\big)^\vee$, which is also an element of $\mathscr{S}'(\mathbf{R}^n)/\mathscr{P}$. In view of (6.2.24) we can use the reverse estimate (5.1.8) in Theorem 5.1.2 to obtain for some polynomial Q,

$$\|f\|_{\dot{L}^p_s} = \|f_s - Q\|_{L^p} \le C\Big\| \Big(\sum_{j\in\mathbf{Z}} |\Delta_j^\eta(f_s)|^2 \Big)^{\frac{1}{2}} \Big\|_{L^p} = C\Big\| \Big(\sum_{j\in\mathbf{Z}} |2^{js}\Delta_j^\theta(f)|^2 \Big)^{\frac{1}{2}} \Big\|_{L^p}.$$

Recalling the definition of Δ_j (see the discussion before the statement of Theorem 6.2.6), we notice that the function

$$\widehat{\Psi}(\tfrac{1}{2}\xi)+\widehat{\Psi}(\xi)+\widehat{\Psi}(2\xi)$$

is equal to 1 on the support of $\widehat{\theta}$ (which is the same as the support of η). It follows that

$$\Delta_j^\theta = (\Delta_{j-1}+\Delta_j+\Delta_{j+1})\Delta_j^\theta .$$

We therefore have the estimate

$$\bigg\|\Big(\sum_{j\in\mathbf{Z}}|2^{js}\Delta_j^\theta(f)|^2\Big)^{\frac{1}{2}}\bigg\|_{L^p} \le \sum_{r=-1}^{1}\bigg\|\Big(\sum_{j\in\mathbf{Z}}|\Delta_j^\theta\Delta_{j+r}(2^{js}f)|^2\Big)^{\frac{1}{2}}\bigg\|_{L^p},$$

and applying Proposition 5.1.4, we can control the last expression (and thus $\|f\|_{\dot{L}^p_s}$) by a constant multiple of

$$\bigg\|\Big(\sum_{j\in\mathbf{Z}}|\Delta_j(2^{js}f)|^2\Big)^{\frac{1}{2}}\bigg\|_{L^p}.$$

This proves that the homogeneous Sobolev norm of f is controlled by a multiple of the expression in (6.2.23). In particular, the distribution f lies in the homogeneous Sobolev space $\dot{L}^p_s$. This ends the proof of the converse direction and completes the proof of the theorem. □

Exercises

6.2.1. Show that the spaces $\dot{L}^p_s$ and L^p_s are complete and that the latter are decreasing as s increases.

6.2.2. (a) Let $1<p<\infty$ and $s\in\mathbf{Z}^+$. Suppose that $f\in L^p_s(\mathbf{R}^n)$ and that φ is in $\mathscr{S}(\mathbf{R}^n)$. Prove that φf is also an element of $L^p_s(\mathbf{R}^n)$.
(b) Let v be a function whose Fourier transform is a bounded compactly supported function. Prove that if f is in $L^2_s(\mathbf{R}^n)$, then so is vf.

6.2.3. Let $s>0$ and α a fixed multi-index. Find the set of p in $(1,\infty)$ such that the distribution $\partial^\alpha\delta_0$ belongs to L^p_{-s}.

6.2.4. Let I be the identity operator, I_1 the Riesz potential of order 1, and R_j the usual Riesz transform. Prove that

$$I=\sum_{j=1}^{n} I_1 R_j \partial_j ,$$

and use this identity to obtain Theorem 6.2.4 when $s=1$.
[*Hint:* Take the Fourier transform.]

6.2.5. Let f be in L^p_s for some $1 < p < \infty$. Prove that $\partial^\alpha f$ is in $L^p_{s-|\alpha|}$.

6.2.6. Prove that for all $\mathscr{C}^1$ functions f that are supported in a ball B we have

$$|f(x)| \le \frac{1}{\omega_{n-1}} \int_B |\nabla f(y)||x-y|^{-n+1}\,dy,$$

where $\omega_{n-1} = |\mathbf{S}^{n-1}|$. For such functions obtain the local Sobolev inequality

$$\big\|f\big\|_{L^q(B)} \le C_{q,r,n}\big\|\nabla f\big\|_{L^p(B)},$$

where $1 < p < q < \infty$ and $1/p = 1/q + 1/n$.
[*Hint:* Start from $f(x) = \int_0^\infty \nabla f(x-t\theta)\cdot\theta\,dt$ and integrate over $\theta \in \mathbf{S}^{n-1}$.]

6.2.7. Show that there is a constant C such that for all $\mathscr{C}^1$ functions f that are supported in a ball B we have

$$\frac{1}{|B'|}\int_{B'} |f(x)-f(z)|\,dz \le C\int_B |\nabla f(y)||x-y|^{-n+1}\,dy$$

for all B' balls contained in B and all $x \in B'$.
[*Hint:* Start with $f(z)-f(x) = \int_0^1 \nabla f(x+t(z-x))\cdot(z-x)\,dt$.]

6.2.8. Let $1 < p < \infty$ and $s > 0$. Show that

$$f \in L^p_s \iff f \in L^p \quad \text{and} \quad f \in \dot{L}^p_s\,.$$

Conclude that $\dot{L}^p_s \cap L^p = L^p_s$ and obtain an estimate for the corresponding norms.
[*Hint:* If f is in $\dot{L}^p_s \cap L^p$ use Theorem 5.2.7 to obtain that $\big\|f\big\|_{L^p_s}$ is controlled by a multiple of the L^p norm of $(\widehat{f}(\xi)(1+|\xi|^s))^\vee$. Use the same theorem to show that $\big\|f\big\|_{\dot{L}^p_s} \le C\big\|f\big\|_{L^p_s}$.]

6.2.9. (*Gagliardo [139]/Nirenberg [249]*) Prove that all Schwartz functions on $\mathbf{R}^n$ satisfy the estimate

$$\big\|f\big\|_{L^q} \le \prod_{j=1}^n \big\|\partial_j f\big\|_{L^1}^{1/n},$$

where $1/q + 1/n = 1$.
[*Hint:* Use induction beginning with the case $n = 1$. Assuming that the inequality is valid for $n-1$, set $I_j(x_1) = \int_{\mathbf{R}^{n-1}} |\partial_j f(x_1,x')|\,dx'$ for $j = 2,\dots,n$, where $x = (x_1,x') \in \mathbf{R}\times\mathbf{R}^{n-1}$ and $I_1(x') = \int_{\mathbf{R}^1} |\partial_1 f(x_1,x')|\,dx_1$. Apply the induction hypothesis to obtain

$$\big\|f(x_1,\cdot)\big\|_{L^{q'}} \le \prod_{j=2}^n I_j(x_1)^{1/(n-1)}$$

and use that $|f|^q \le I_1(x')^{1/(n-1)}|f|$ and Hölder's inequality to calculate $\big\|f\big\|_{L^q}$.]

6.2.10. Let $f \in L^2_1(\mathbf{R}^n)$. Prove that there is a constant $c_n > 0$ such that

$$\int_{\mathbf{R}^n}\int_{\mathbf{R}^n} \frac{|f(x+t)+f(x-t)-2f(x)|^2}{|t|^{n+2}}\,dx\,dt = c_n \int_{\mathbf{R}^n} \sum_{j=1}^n |\partial_j f(x)|^2\,dx.$$

6.2.11. (*Christ [61]*) Let $0 \le \beta < \infty$ and let

$$C_0 = \int_{\mathbf{R}^n} |\widehat{g}(\xi)|^2 (1+|\xi|)^n \big(\log(2+|\xi|)\big)^{-\beta}\,d\xi\,.$$

(a) Prove that there is a constant $C(n,\beta,C_0)$ such that for every $q > 2$ we have

$$\big\|g\big\|_{L^q(\mathbf{R}^n)} \le C(n,\beta,C_0)\,q^{\frac{\beta+1}{2}}\,.$$

(b) Conclude that for any compact subset K of $\mathbf{R}^n$ we have

$$\int_K e^{|g(x)|^\gamma}\,dx < \infty$$

whenever $\gamma < \frac{2}{\beta+1}$.
[*Hint:* Part (a): For $q > 2$ control $\big\|g\big\|_{L^q(\mathbf{R}^n)}$ by $\big\|\widehat{g}\big\|_{L^{q'}(\mathbf{R}^n)}$ and apply Hölder's inequality with exponents $\frac{2}{q'}$ and $\frac{2(q-1)}{q-2}$. Part (b): Expand the exponential in a Taylor series.]

6.2.12. Suppose that $m \in L^2_s(\mathbf{R}^n)$ for some $s > \frac{n}{2}$ and let $\lambda > 0$. Define the operator T_λ by setting $\widehat{T_\lambda(f)}(\xi) = m(\lambda\xi)\widehat{f}(\xi)$. Show that there exists a constant $C = C(n,s)$ such that for all f and $u \ge 0$ and $\lambda > 0$ we have

$$\int_{\mathbf{R}^n} |T_\lambda(f)(x)|^2\,u(x)\,dx \le C\int_{\mathbf{R}^n} |f(x)|^2\,M(u)(x)\,dx\,.$$

6.3 Lipschitz Spaces

The classical definition says that a function f on $\mathbf{R}^n$ is Lipschitz (or Hölder) continuous of order $\gamma > 0$ if there is constant $C < \infty$ such that for all $x, y \in \mathbf{R}^n$ we have

$$|f(x+y) - f(x)| \le C|y|^\gamma. \tag{6.3.1}$$

It turns out that only constant functions satisfy (6.3.1) when $\gamma > 1$, and the corresponding definition needs to be suitably adjusted in this case. This is discussed in this section. The key point is that any function f that satisfies (6.3.1) possesses a certain amount of smoothness "measured" by the quantity γ. The Lipschitz norm of a function is introduced to serve this purpose, that is, to precisely quantify and exactly measure this smoothness. In this section we formalize these concepts and we explore connections they have with the orthogonality considerations of the previous chapter. The main achievement of this section is a characterization of Lipschitz spaces using Littlewood–Paley theory.

6.3.1 Introduction to Lipschitz Spaces

Definition 6.3.1. Let $0<\gamma<1$. A function f on $\mathbf{R}^n$ is said to be *Lipschitz of order* γ if it is bounded and satisfies (6.3.1) for some $C<\infty$. In this case we let

$$\big\|f\big\|_{\Lambda_\gamma(\mathbf{R}^n)} = \big\|f\big\|_{L^\infty} + \sup_{x\in\mathbf{R}^n}\ \sup_{h\in\mathbf{R}^n\setminus\{0\}} \frac{|f(x+h)-f(x)|}{|h|^\gamma}$$

and we set

$$\Lambda_\gamma(\mathbf{R}^n) = \{f:\mathbf{R}^n\to\mathbf{C}\text{ continuous}:\ \big\|f\big\|_{\Lambda_\gamma(\mathbf{R}^n)}<\infty\}.$$

Note that functions in $\Lambda_\gamma(\mathbf{R}^n)$ are automatically continuous when $\gamma<1$, so we did not need to make this part of the definition. We call $\Lambda_\gamma(\mathbf{R}^n)$ the *inhomogeneous Lipschitz space* of order γ. For reasons of uniformity we also set

$$\Lambda_0(\mathbf{R}^n) = L^\infty(\mathbf{R}^n)\cap C(\mathbf{R}^n),$$

where $C(\mathbf{R}^n)$ is the space of all continuous functions on $\mathbf{R}^n$. See Exercise 6.3.2.

Example 6.3.2. The function $h(x)=\cos(x\cdot a)$ for some fixed $a\in\mathbf{R}^n$ is in Λ_γ for all $\gamma<1$. Simply notice that $|h(x)-h(y)|\le\min(2,|a|\,|x-y|)$.

We now extend this definition to indices $\gamma\ge 1$.

Definition 6.3.3. For $h\in\mathbf{R}^n$ define the *difference operator* D_h by setting

$$D_h(f)(x) = f(x+h)-f(x)$$

for a continuous function $f:\ \mathbf{R}^n\to\mathbf{C}$. We may check that

$$\begin{aligned}
D_h^2(f)(x) &= D_h(D_hf)(x) = f(x+2h)-2f(x+h)+f(x),\\
D_h^3(f)(x) &= D_h(D_h^2f)(x) = f(x+3h)-3f(x+2h)+3f(x+h)-f(x),
\end{aligned}$$

and in general, that $D_h^{k+1}(f)=D_h^k(D_h(f))$ is given by

$$D_h^{k+1}(f)(x) = \sum_{s=0}^{k+1}(-1)^{k+1-s}\binom{k+1}{s}f(x+sh) \tag{6.3.2}$$

for a nonnegative integer k. See Exercise 6.3.3. For $\gamma>0$ define

$$\big\|f\big\|_{\Lambda_\gamma} = \big\|f\big\|_{L^\infty} + \sup_{x\in\mathbf{R}^n}\ \sup_{h\in\mathbf{R}^n\setminus\{0\}} \frac{|D_h^{[\gamma]+1}(f)(x)|}{|h|^\gamma},$$

where $[\gamma]$ denotes the integer part of γ, and set

$$\Lambda_\gamma = \{f:\mathbf{R}^n\to\mathbf{C} \text{ continuous}: \ \big\|f\big\|_{\Lambda_\gamma}<\infty\}.$$

We call $\Lambda_\gamma(\mathbf{R}^n)$ the inhomogeneous *Lipschitz space* of order $\gamma\in\mathbf{R}^+$.

For a tempered distribution u we also define another distribution $D_h^k(u)$ via the identity

$$\langle D_h^k(u),\varphi\rangle = \langle u, D_{-h}^k(\varphi)\rangle$$

for all φ in the Schwartz class.

We now define the homogeneous Lipschitz spaces. We adhere to the usual convention of using a dot on a space to indicate its homogeneous nature.

Definition 6.3.4. For $\gamma>0$ we define

$$\big\|f\big\|_{\dot\Lambda_\gamma} = \sup_{x\in\mathbf{R}^n}\ \sup_{h\in\mathbf{R}^n\setminus\{0\}} \frac{|D_h^{[\gamma]+1}(f)(x)|}{|h|^\gamma}$$

and we also let $\dot\Lambda_\gamma$ be the space of all continuous functions f on $\mathbf{R}^n$ that satisfy $\big\|f\big\|_{\dot\Lambda_\gamma}<\infty$. We call $\dot\Lambda_\gamma$ the *homogeneous Lipschitz space* of order γ. We note that elements of $\dot\Lambda_\gamma$ have at most polynomial growth at infinity and thus they are elements of $\mathscr{S}'(\mathbf{R}^n)$.

A few observations are in order here. Constant functions f satisfy $D_h(f)(x)=0$ for all $h,x\in\mathbf{R}^n$, and therefore the homogeneous quantity $\big\|\cdot\big\|_{\dot\Lambda_\gamma}$ is insensitive to constants. Similarly the expressions $D_h^{k+1}(f)$ and $\big\|f\big\|_{\dot\Lambda_\gamma}$ do not recognize polynomials of degree up to k. Moreover, polynomials are the only continuous functions with this property; see Exercise 6.3.1. This means that the quantity $\big\|f\big\|_{\dot\Lambda_\gamma}$ is not a norm but only a seminorm. To make it a norm, we need to consider functions modulo polynomials, as we did in the case of homogeneous Sobolev spaces. For this reason we think of $\dot\Lambda_\gamma$ as a subspace of $\mathscr{S}'(\mathbf{R}^n)/\mathscr{P}$.

We make use of the following proposition concerning properties of the difference operators D_h^k.

Proposition 6.3.5. *Let f be a $\mathscr{C}^m$ function on $\mathbf{R}^n$ for some $m\in\mathbf{Z}^+$. Then for all $h=(h_1,\dots,h_n)$ and $x\in\mathbf{R}^n$ the following identity holds:*

$$D_h(f)(x) = \int_0^1 \sum_{j=1}^n h_j\,(\partial_j f)(x+sh)\,ds. \tag{6.3.3}$$

More generally, we have that

$$D_h^m(f)(x) = \int_{[0,1]^m} \sum_{\substack{1 \le j_\ell \le n \\ 1 \le \ell \le m}} h_{j_1} \cdots h_{j_m} (\partial_{j_1} \cdots \partial_{j_m} f)(x + (s_1 + \cdots + s_m)h) \, ds_1 \cdots ds_m. \quad (6.3.4)$$

Proof. Identity (6.3.3) is a consequence of the fundamental theorem of calculus applied to the function $t \mapsto f((1-t)x + t(x+h))$ on $[0,1]$, while identity (6.3.4) follows by induction. □

6.3.2 Littlewood–Paley Characterization of Homogeneous Lipschitz Spaces

We now characterize the homogeneous Lipschitz spaces using the Littlewood–Paley operators Δ_j. As in the previous section, we fix a radial Schwartz function Ψ whose Fourier transform is nonnegative, supported in the annulus $1 - \frac{1}{7} \le |\xi| \le 2$, is equal to one on the annulus $1 \le |\xi| \le 2 - \frac{2}{7}$, and that satisfies

$$\sum_{j \in \mathbf{Z}} \widehat{\Psi}(2^{-j}\xi) = 1 \quad (6.3.5)$$

for all $\xi \neq 0$. The Littlewood–Paley operators $\Delta_j = \Delta_j^{\Psi}$ associated with Ψ are given by multiplication on the Fourier transform side by the smooth bump $\widehat{\Psi}(2^{-j}\xi)$.

Theorem 6.3.6. *Let Δ_j be as above and $\gamma > 0$. Then there is a constant $C = C(n,\gamma)$ such that for every f in $\dot{\Lambda}_\gamma$ we have the estimate*

$$\sup_{j \in \mathbf{Z}} 2^{j\gamma} \big\| \Delta_j(f) \big\|_{L^\infty} \le C \big\| f \big\|_{\dot{\Lambda}_\gamma}. \quad (6.3.6)$$

Conversely, every element f of $\mathscr{S}'(\mathbf{R}^n)/\mathscr{P}$ that satisfies

$$\sup_{j \in \mathbf{Z}} 2^{j\gamma} \big\| \Delta_j(f) \big\|_{L^\infty} < \infty \quad (6.3.7)$$

is an element of $\dot{\Lambda}_\gamma$ with norm

$$\big\| f \big\|_{\dot{\Lambda}_\gamma} \le C' \sup_{j \in \mathbf{Z}} 2^{j\gamma} \big\| \Delta_j(f) \big\|_{L^\infty} \quad (6.3.8)$$

for some constant $C' = C'(n,\gamma)$.

Note that condition (6.3.7) remains invariant if a polynomial is added to the function f; this is consistent with the analogous property of the mapping $f \mapsto \big\| f \big\|_{\dot{\Lambda}_\gamma}$.

Proof. We begin with the proof of (6.3.8). Let $k = [\gamma]$ be the integer part of γ. Let us pick a Schwartz function η on $\mathbf{R}^n$ whose Fourier transform is nonnegative, supported in the annulus $\frac{4}{5} \le |\xi| \le 2$, and that satisfies

$$\sum_{j\in\mathbf{Z}} \widehat{\eta}(2^{-j}\xi)^2 = 1 \tag{6.3.9}$$

for all $\xi \neq 0$. Associated with η, we define the Littlewood–Paley operators Δ_j^η given by multiplication on the Fourier transform side by the smooth bump $\widehat{\eta}(2^{-j}\xi)$. With Ψ as in (6.2.6) we set

$$\widehat{\Theta}(\xi) = \widehat{\Psi}(\tfrac{1}{2}\xi) + \widehat{\Psi}(\xi) + \widehat{\Psi}(2\xi),$$

and we denote by $\Delta_j^\Theta = \Delta_{j-1} + \Delta_j + \Delta_{j+1}$ the Littlewood–Paley operator given by multiplication on the Fourier transform side by the smooth bump $\widehat{\Theta}(2^{-j}\xi)$.

The fact that the previous function is equal to 1 on the support of $\widehat{\eta}$ together with the functional identity (6.3.9) yields the operator identity

$$I = \sum_{j\in\mathbf{Z}} (\Delta_j^\eta)^2 = \sum_{j\in\mathbf{Z}} \Delta_j^\Theta \Delta_j^\eta \Delta_j^\eta,$$

with convergence in the sense of the space $\mathscr{S}'(\mathbf{R}^n)/\mathscr{P}$. Since convolution is a linear operation, we have $D_h^{k+1}(F*G) = F*D_h^{k+1}(G)$, from which we deduce

$$\begin{aligned} D_h^{k+1}(f) &= \sum_{j\in\mathbf{Z}} \Delta_j^\Theta(f) * D_h^{k+1}(\eta_{2^{-j}}) * \eta_{2^{-j}} \\ &= \sum_{j\in\mathbf{Z}} D_h^{k+1}(\Delta_j^\Theta(f)) * (\eta*\eta)_{2^{-j}} \end{aligned} \tag{6.3.10}$$

for all tempered distributions f. The convergence of the series in (6.3.10) is in the sense of $\mathscr{S}'/\mathscr{P}$ in view of Exercise 5.2.2. The convergence of the series in (6.3.10) in the L^∞ norm is a consequence of condition (6.3.7) and is contained in the following argument.

Using (6.3.2), we easily obtain the estimate

$$\big\|D_h^{k+1}(\Delta_j^\Theta(f)) * (\eta*\eta)_{2^{-j}}\big\|_{L^\infty} \le 2^{k+1}\big\|\eta*\eta\big\|_{L^1}\big\|\Delta_j^\Theta(f)\big\|_{L^\infty}. \tag{6.3.11}$$

We first integrate over $(s_1,\dots,s_{k+1}) \in [0,1]^{k+1}$ the identity

$$\begin{aligned} &\sum_{r_1=1}^{n}\cdots\sum_{r_{k+1}=1}^{n} h_{r_1}\cdots h_{r_{k+1}}(\partial_{r_1}\cdots\partial_{r_{k+1}}\eta_{2^{-j}})(x+(s_1+\cdots+s_{k+1})h) \\ &\quad = 2^{j(k+1)}\sum_{r_1=1}^{n}\cdots\sum_{r_{k+1}=1}^{n} h_{r_1}\cdots h_{r_{k+1}}(\partial_{r_1}\dots\partial_{r_{k+1}}\eta)_{2^{-j}}(x+(s_1+\cdots+s_{k+1})h). \end{aligned}$$

We then use (6.3.4) with $m=k+1$, and we integrate over $x\in\mathbf{R}^n$ to obtain

$$\big\|D_h^{k+1}(\eta_{2^{-j}})\big\|_{L^1} \le 2^{j(k+1)}|h|^{k+1}\sum_{r_1=1}^{n}\cdots\sum_{r_{k+1}=1}^{n}\big\|\partial_{r_1}\cdots\partial_{r_{k+1}}\eta\big\|_{L^1}.$$

We deduce the validity of the estimate

$$\begin{aligned}
&\big\|\Delta_j^\Theta(f)*D_h^{k+1}(\eta_{2^{-j}})*\eta_{2^{-j}}\big\|_{L^\infty}\\
&\qquad\le \big\|\Delta_j^\Theta(f)\big\|_{L^\infty}\big\|D_h^{k+1}(\eta_{2^{-j}})*\eta_{2^{-j}}\big\|_{L^1}\\
&\qquad\le \big\|\Delta_j^\Theta(f)\big\|_{L^\infty}|2^jh|^{k+1}c_k\sum_{|\alpha|\le k+1}\big\|\partial^\alpha\eta\big\|_{L^1}\big\|\eta\big\|_{L^1}.
\end{aligned} \tag{6.3.12}$$

Combining (6.3.11) and (6.3.12), we obtain

$$\begin{aligned}
&\big\|\Delta_j^\Theta(f)*D_h^{k+1}(\eta_{2^{-j}})*\eta_{2^{-j}}\big\|_{L^\infty}\\
&\qquad\le C_{\eta,n,k}\big\|\Delta_j^\Theta(f)\big\|_{L^\infty}\min\big(1,|2^jh|^{k+1}\big).
\end{aligned} \tag{6.3.13}$$

We insert estimate (6.3.13) in (6.3.10) to deduce

$$\frac{\big\|D_h^{k+1}(f)\big\|_{L^\infty}}{|h|^\gamma}\le\frac{C'}{|h|^\gamma}\sum_{j\in\mathbf{Z}}2^{j\gamma}\big\|\Delta_j^\Theta(f)\big\|_{L^\infty}\min\big(2^{-j\gamma},2^{j(k+1-\gamma)}|h|^{k+1}\big),$$

from which it follows that

$$\begin{aligned}
\big\|f\big\|_{\dot\Lambda_\gamma}&\le\sup_{h\in\mathbf{R}^n\setminus\{0\}}\frac{C'}{|h|^\gamma}\sum_{j\in\mathbf{Z}}2^{j\gamma}\big\|\Delta_j^\Theta(f)\big\|_{L^\infty}\min\big(2^{-j\gamma},2^{j(k+1-\gamma)}|h|^{k+1}\big)\\
&\le C'\sup_{j\in\mathbf{Z}}2^{j\gamma}\big\|\Delta_j^\Theta(f)\big\|_{L^\infty}\sup_{h\neq0}\sum_{j\in\mathbf{Z}}\min\big(|h|^{-\gamma}2^{-j\gamma},2^{j(k+1-\gamma)}|h|^{k+1-\gamma}\big)\\
&\le C'\sup_{j\in\mathbf{Z}}2^{j\gamma}\big\|\Delta_j^\Theta(f)\big\|_{L^\infty},
\end{aligned}$$

since the last numerical series converges ($\gamma<k+1=[\gamma]+1$). This proves (6.3.8) with the difference that instead of Δ_j we have Δ_j^Θ on the right. The passage to Δ_j is a trivial matter, since $\Delta_j^\Theta=\Delta_{j-1}+\Delta_j+\Delta_{j+1}$.

Having established (6.3.8), we now turn to the proof of (6.3.6). We first consider the case $0<\gamma<1$, which is very simple. Since each Δ_j is given by convolution with a function with mean value zero, we may write

$$\begin{aligned}
\Delta_j(f)(x)&=\int_{\mathbf{R}^n}f(x-y)\Psi_{2^{-j}}(y)\,dy\\
&=\int_{\mathbf{R}^n}(f(x-y)-f(x))\Psi_{2^{-j}}(y)\,dy\\
&=2^{-j\gamma}\int_{\mathbf{R}^n}\frac{D_{-y}(f)(x)}{|y|^\gamma}|2^jy|^\gamma2^{jn}\Psi(2^jy)\,dy,
\end{aligned}$$

and the previous expression is easily seen to be controlled by a constant multiple of $2^{-j\gamma}\big\|f\big\|_{\dot\Lambda_\gamma}$. This proves (6.3.6) when $0<\gamma<1$. In the case $\gamma\ge1$ we have to work a bit harder.

As before, set $k=[\gamma]$. Notice that for Schwartz functions g we have the identity

$$D_h^{k+1}(g) = \big(\widehat{g}(\xi)\,(e^{2\pi i\xi\cdot h}-1)^{k+1}\big)^{\vee}.$$

To express $\Delta_j(g)$ in terms of $D_h^{k+1}(g)$, we need to introduce the function

$$\xi \mapsto \widehat{\Psi}(2^{-j}\xi)\,(e^{2\pi i\xi\cdot h}-1)^{-(k+1)}.$$

But as the support of $\widehat{\Psi}(2^{-j}\xi)$ may intersect the set of all ξ for which $\xi\cdot h$ is an integer, the previous function is not well defined. To deal with this problem, we pick a finite family of unit vectors $\{u_r\}_r$ so that the annulus $\frac{1}{2}\le|\xi|\le 2$ is covered by the union of sets

$$U_r = \big\{\xi \in \mathbf{R}^n : \tfrac{1}{2}\le|\xi|\le 2,\quad \tfrac{1}{4}\le|\xi\cdot u_r|\le 2\big\}.$$

Then we write $\widehat{\Psi}$ as a finite sum of smooth functions $\widehat{\Psi^{(r)}}$, where each $\widehat{\Psi^{(r)}}$ is supported in U_r. Setting

$$h_r = \frac{1}{8}2^{-j}u_r,$$

we note that

$$\begin{aligned}\Psi^{(r)}_{2^{-j}} * f &= \big(\widehat{\Psi^{(r)}}(2^{-j}\xi)\,(e^{2\pi i\xi\cdot h_r}-1)^{-(k+1)}(e^{2\pi i\xi\cdot h_r}-1)^{k+1}\widehat{f}(\xi)\big)^{\vee}\\ &= \big(\widehat{\Psi^{(r)}}(2^{-j}\xi)\,(e^{2\pi i 2^{-j}\xi\cdot\frac{1}{8}u_r}-1)^{-(k+1)}\widehat{D^{k+1}_{h_r}(f)}(\xi)\big)^{\vee}\end{aligned} \tag{6.3.14}$$

and observe that the exponential is never equal to 1, since

$$2^{-j}\xi \in U_r \implies \tfrac{1}{32}\le|2^{-j}\xi\cdot\tfrac{1}{8}u_r|\le\tfrac{1}{4}.$$

Since the function $\widehat{\zeta^{(r)}} = \widehat{\Psi^{(r)}}(\xi)\,(e^{2\pi i\xi\cdot\frac{1}{8}u_r}-1)^{-(k+1)}$ is well defined and smooth with compact support, it follows that

$$\Psi^{(r)}_{2^{-j}} * f = (\zeta^{(r)})_{2^{-j}} * D^{k+1}_{2^{-j}\frac{1}{8}u_r}(f),$$

which implies that

$$\begin{aligned}\big\|\Psi^{(r)}_{2^{-j}} * f\big\|_{L^\infty} &\le \big\|(\zeta^{(r)})_{2^{-j}}\big\|_{L^1}\big\|D^{k+1}_{2^{-j}\frac{1}{8}u_r}(f)\big\|_{L^\infty}\\ &\le \big\|\zeta^{(r)}\big\|_{L^1}\big\|f\big\|_{\dot{\Lambda}_\gamma}2^{-j\gamma}.\end{aligned}$$

Summing over the finite number of r, we obtain the estimate

$$\big\|\Delta_j(f)\big\|_{L^\infty} \le C\big\|f\big\|_{\dot{\Lambda}_\gamma}2^{-j\gamma},$$

which concludes the proof of the theorem. □

6.3.3 Littlewood–Paley Characterization of Inhomogeneous Lipschitz Spaces

We have seen that quantities involving the Littlewood–Paley operators Δ_j characterize homogeneous Lipschitz spaces. We now address the same question for inhomogeneous spaces.

As in the Littlewood–Paley characterization of inhomogeneous Sobolev spaces, we need to treat the contribution of the frequencies near zero separately. We recall the Schwartz function Φ introduced in Section 6.2.2:

$$\widehat{\Phi}(\xi) = \begin{cases} \sum_{j\le 0} \widehat{\Psi}(2^{-j}\xi) & \text{when } \xi \neq 0, \\ 1 & \text{when } \xi = 0. \end{cases} \tag{6.3.15}$$

Note that $\widehat{\Phi}(\xi)$ is equal to 1 for $|\xi| \le 2 - \frac{2}{7}$ and vanishes when $|\xi| \ge 2$. We also recall the operator $S_0(f) = \Phi * f$. One should not be surprised to find out that a result analogous to that in Theorem 6.2.6 is valid for Lipschitz spaces as well.

Theorem 6.3.7. *Let Ψ and Δ_j be as in the Theorem 6.3.6, Φ as in (6.3.15), and $\gamma > 0$. Then there is a constant $C = C(n,\gamma)$ such that for every f in Λ_γ we have the estimate*

$$\big\|S_0(f)\big\|_{L^\infty} + \sup_{j\ge 1} 2^{j\gamma}\big\|\Delta_j(f)\big\|_{L^\infty} \le C\big\|f\big\|_{\Lambda_\gamma}. \tag{6.3.16}$$

Conversely, every tempered distribution f that satisfies

$$\big\|S_0(f)\big\|_{L^\infty} + \sup_{j\ge 1} 2^{j\gamma}\big\|\Delta_j(f)\big\|_{L^\infty} < \infty \tag{6.3.17}$$

can be identified with an element of Λ_γ. Moreover, there is a constant $C' = C'(n,\gamma)$ such that for all f that satisfy (6.3.17) we have

$$\big\|f\big\|_{\Lambda_\gamma} \le C'\Big(\big\|S_0(f)\big\|_{L^\infty} + \sup_{j\ge 1} 2^{j\gamma}\big\|\Delta_j(f)\big\|_{L^\infty}\Big). \tag{6.3.18}$$

Proof. The proof of (6.3.16) is immediate, since we trivially have

$$\big\|S_0(f)\big\|_{L^\infty} = \big\|f * \Phi\big\|_{L^\infty} \le \big\|\Phi\big\|_{L^1}\big\|f\big\|_{L^\infty} \le C\big\|f\big\|_{\Lambda_\gamma}$$

and also

$$\sup_{j\ge 1} 2^{j\gamma}\big\|\Delta_j(f)\big\|_{L^\infty} \le C\big\|f\big\|_{\dot{\Lambda}_\gamma} \le C\big\|f\big\|_{\Lambda_\gamma}$$

by the previous theorem.

Therefore, the main part of the argument is contained in the proof of the converse estimate (6.3.18). Here we introduce Schwartz functions ζ, η so that

$$\widehat{\zeta}(\xi)^2 + \sum_{j=1}^{\infty} \widehat{\eta}(2^{-j}\xi)^2 = 1$$

and such that $\widehat{\eta}$ is supported in the annulus $\frac{4}{5} \le |\xi| \le 2$ and $\widehat{\zeta}$ is supported in the ball $|\xi| \le 1$. We associate Littlewood–Paley operators Δ_j^η given by convolution with the functions $\eta_{2^{-j}}$ and we also let $\Delta_j^\Theta = \Delta_{j-1} + \Delta_j + \Delta_{j+1}$. Note that $\widehat{\Phi}$ is equal to one on the support of $\widehat{\zeta}$. Moreover, $\Delta_j^\Theta \Delta_j^\eta = \Delta_j^\eta$; hence for tempered distributions f we have the identity

$$f = \zeta * \zeta * \Phi * f + \sum_{j=1}^{\infty} \eta_{2^{-j}} * \eta_{2^{-j}} * \Delta_j^\Theta(f), \tag{6.3.19}$$

where the series converges in $\mathscr{S}'(\mathbf{R}^n)$. With $k = [\gamma]$ we write

$$\frac{D_h^{k+1}(f)}{|h|^\gamma} = \zeta * \frac{D_h^{k+1}(\zeta)}{|h|^\gamma} * \Phi * f + \sum_{j=1}^{\infty} \eta_{2^{-j}} * \frac{D_h^{k+1}(\eta_{2^{-j}})}{|h|^\gamma} * \Delta_j^\Theta(f), \tag{6.3.20}$$

and we use Proposition 6.3.5 to estimate the L^∞ norm of the term $\zeta * \frac{D_h^{k+1}(\zeta)}{|h|^\gamma} * \Phi * f$ in the previous sum as follows:

$$\begin{aligned} \Big\| \zeta * \tfrac{D_h^{k+1}(\zeta)}{|h|^\gamma} * \Phi * f \Big\|_{L^\infty} &\le \Big\| \tfrac{D_h^{k+1}(\zeta)}{|h|^\gamma} \Big\|_{L^\infty} \big\| \zeta * \Phi * f \big\|_{L^1} \\ &\le C \min\big(\tfrac{1}{|h|^\gamma}, \tfrac{|h|^{k+1}}{|h|^\gamma} \big) \big\| \Phi * f \big\|_{L^\infty} \\ &\le C \big\| \Phi * f \big\|_{L^\infty} . \end{aligned} \tag{6.3.21}$$

The corresponding L^∞ estimates for $\Delta_j^\Theta(f) * \eta_{2^{-j}} * D_h^{k+1}(\eta_{2^{-j}})$ have already been obtained in (6.3.13). Indeed, we obtained

$$\big\| D_h^{k+1}(\eta_{2^{-j}}) * \eta_{2^{-j}} * \Delta_j^\Theta(f) \big\|_{L^\infty} \le C_{\eta,n,k} \big\| \Delta_j^\Theta(f) \big\|_{L^\infty} \min\big(1, |2^j h|^{k+1}\big),$$

from which it follows that

$$\begin{aligned} &\Big\| \sum_{j=1}^{\infty} \eta_{2^{-j}} * \frac{D_h^{k+1}(\eta_{2^{-j}})}{|h|^\gamma} * \Delta_j^\Theta(f) \Big\|_{L^\infty} \\ &\quad \le C \Big(\sup_{j\ge 1} 2^{j\gamma} \big\| \Delta_j^\Theta(f) \big\|_{L^\infty} \Big) \sum_{j=1}^{\infty} 2^{-j\gamma} |h|^{-\gamma} \min\big(1, |2^j h|^{k+1}\big) \\ &\quad \le C \Big(\sup_{j\ge 1} 2^{j\gamma} \big\| \Delta_j(f) \big\|_{L^\infty} \Big) \sum_{j=1}^{\infty} \min\big(|2^j h|^{-\gamma}, |2^j h|^{k+1-\gamma} \big) \\ &\quad \le C \sup_{j\ge 1} 2^{j\gamma} \big\| \Delta_j(f) \big\|_{L^\infty} , \end{aligned} \tag{6.3.22}$$

where the last series is easily seen to converge uniformly in $h \in \mathbf{R}^n$, since $k+1 = [\gamma]+1 > \gamma$. We now combine identity (6.3.20) with estimates (6.3.21) and (6.3.22) to obtain that the expression on the right in (6.3.19) has a bounded L^∞ norm. This implies that f can be identified with a bounded function that satisfies (6.3.18). □

Next, we obtain consequences of the Littlewood–Paley characterization of Lipschitz spaces. In the following corollary we identify Λ_0 with L^∞.

Corollary 6.3.8. *For $0 \le \gamma \le \delta < \infty$ there is a constant $C_{n,\gamma,\delta} < \infty$ such that for all $f \in \Lambda_\delta(\mathbf{R}^n)$ we have*

$$\big\|f\big\|_{\Lambda_\gamma} \le C_{n,\gamma,\delta}\big\|f\big\|_{\Lambda_\delta}.$$

In other words, the space $\Lambda_\delta(\mathbf{R}^n)$ can be identified with a subspace of $\Lambda_\gamma(\mathbf{R}^n)$.

Proof. If $0 < \gamma \le \delta$ and $j \ge 0$, then we must have $2^{j\gamma} \le 2^{j\delta}$ and thus

$$\sup_{j\ge 1} 2^{j\gamma}\big\|\Delta_j(f)\big\|_{L^\infty} \le \sup_{j\ge 1} 2^{j\delta}\big\|\Delta_j(f)\big\|_{L^\infty}.$$

Adding $\big\|S_0(f)\big\|_{L^\infty}$ and using Theorem 6.3.7, we obtain the required conclusion. The case $\gamma = 0$ is trivial. □

Remark 6.3.9. We proved estimates (6.3.18) and (6.3.8) using the Littlewood–Paley operators Δ_j constructed by a fixed choice of the function Ψ; Φ also depended on Ψ. It should be noted that the specific choice of the functions Ψ and Φ was unimportant in those estimates. In particular, if we know (6.3.18) and (6.3.8) for some choice of Littlewood–Paley operators $\widetilde{\Delta}_j$ and some Schwartz function $\widetilde{\Phi}$ whose Fourier transform is supported in a neighborhood of the origin, then (6.3.18) and (6.3.8) would also hold for our fixed choice of Δ_j and Φ. This situation is illustrated in the next corollary.

Corollary 6.3.10. *Let $\gamma > 0$ and let α be a multi-index with $|\alpha| < \gamma$. If $f \in \Lambda_\gamma$, then the distributional derivative $\partial^\alpha f$ (of f) lies in $\Lambda_{\gamma-|\alpha|}$. Likewise, if $f \in \dot{\Lambda}_\gamma$, then $\partial^\alpha f \in \dot{\Lambda}_{\gamma-|\alpha|}$. Precisely, we have the norm estimates*

$$\big\|\partial^\alpha f\big\|_{\Lambda_{\gamma-|\alpha|}} \le C_{\gamma,\alpha}\big\|f\big\|_{\Lambda_\gamma}, \tag{6.3.23}$$

$$\big\|\partial^\alpha f\big\|_{\dot{\Lambda}_{\gamma-|\alpha|}} \le C_{\gamma,\alpha}\big\|f\big\|_{\dot{\Lambda}_\gamma}. \tag{6.3.24}$$

In particular, elements of Λ_γ and $\dot{\Lambda}_\gamma$ are in $\mathscr{C}^\alpha$ for all $|\alpha| < \gamma$.

Proof. Let α be a multi-index with $|\alpha| < \gamma$. We denote by $\Delta_j^{\partial^\alpha\Psi}$ the Littlewood–Paley operator associated with the bump $(\partial^\alpha\Psi)_{2^{-j}}$. It is straightforward to check that the identity

$$\Delta_j(\partial^\alpha f) = 2^{j|\alpha|}\Delta_j^{\partial^\alpha\Psi}(f)$$

is valid for any tempered distribution f. Using the support properties of Ψ, we obtain

$$2^{j(\gamma-|\alpha|)}\Delta_j(\partial^\alpha f) = 2^{j\gamma}\Delta_j^{\partial^\alpha\Psi}(\Delta_{j-1}+\Delta_j+\Delta_{j+1})(f), \tag{6.3.25}$$

and from this it easily follows that

$$\sup_{j\in\mathbf{Z}} 2^{j(\gamma-|\alpha|)}\big\|\Delta_j(\partial^\alpha f)\big\|_{L^\infty} \le (2^\gamma+2)\,\big\|\partial^\alpha\Psi\big\|_{L^1}\sup_{j\in\mathbf{Z}} 2^{j\gamma}\big\|\Delta_j(f)\big\|_{L^\infty}$$

and also that

$$\sup_{j\ge 1} 2^{j(\gamma-|\alpha|)}\big\|\Delta_j(\partial^\alpha f)\big\|_{L^\infty} \le (2^\gamma+2)\,\big\|\partial^\alpha\Psi\big\|_{L^1} \sup_{j\ge 1} 2^{j\gamma}\big\|\Delta_j(f)\big\|_{L^\infty}. \tag{6.3.26}$$

Using Theorem 6.3.6, we deduce that if $f\in\dot\Lambda_\gamma$, then $\partial^\alpha f\in\dot\Lambda_{\gamma-|\alpha|}$, and we also obtain (6.3.24). To derive the inhomogeneous version, we note that

$$S_0(\partial^\alpha f) = \Phi*(\partial^\alpha f) = (\partial^\alpha\Phi*f) = \big(\partial^\alpha\Phi*(\Phi+\Psi_{2^{-1}})*f\big),$$

since the function $\widehat{\Phi}+\widehat{\Psi_{2^{-1}}}$ is equal to 1 on the support of $\widehat{\partial^\alpha\Phi}$. Taking L^∞ norms, we obtain

$$\begin{aligned}\big\|S_0(\partial^\alpha f)\big\|_{L^\infty} &\le \big\|\partial^\alpha\Phi\big\|_{L^1}\big(\big\|\Phi*f\big\|_{L^\infty}+\big\|\Psi_{2^{-1}}*f\big\|_{L^\infty}\big)\\ &\le \big\|\partial^\alpha\Phi\big\|_{L^1}\Big(\big\|S_0(f)\big\|_{L^\infty}+\sup_{j\ge 1}\big\|\Delta_j(f)\big\|_{L^\infty}\Big),\end{aligned}$$

which, combined with (6.3.26), yields $\big\|\partial^\alpha f\big\|_{\Lambda_{\gamma-|\alpha|}} \le C_{\gamma,\alpha}\big\|f\big\|_{\Lambda_\gamma}$. □

Exercises

6.3.1. Fix $k\in\mathbf{Z}^+$. Show that

$$D_h^k(f)(x)=0$$

for all x,h in $\mathbf{R}^n$ if and only if f is a polynomial of degree at most $k-1$.
[*Hint:* One direction may be proved by direct verification. For the converse direction, show that $\widehat{f}$ is supported at the origin and use Proposition 2.4.1.]

6.3.2. (a) Extend Definition 6.3.1 to the case $\gamma=0$ and show that for all continuous functions f we have

$$\big\|f\big\|_{L^\infty}\le\big\|f\big\|_{\Lambda_0}\le 3\big\|f\big\|_{L^\infty};$$

hence the space $\Lambda_0(\mathbf{R}^n)$ can be identified with $L^\infty(\mathbf{R}^n)\cap C(\mathbf{R}^n)$.
(b) Given a measurable function f on $\mathbf{R}^n$ we define

$$\big\|f\big\|_{\dot L^\infty}=\inf\big\{\big\|f+c\big\|_{L^\infty}:\ c\in\mathbf{C}\big\}.$$

Let $\dot L^\infty(\mathbf{R}^n)$ be the space of equivalent classes of bounded functions whose difference is a constant, equipped with this norm. Show that for all continuous functions f on $\mathbf{R}^n$ we have

$$\big\|f\big\|_{\dot L^\infty}\le\sup_{x,h\in\mathbf{R}^n}|f(x+h)-f(x)|\le 2\big\|f\big\|_{\dot L^\infty}.$$

In other words, $\dot\Lambda_0(\mathbf{R}^n)$ can be identified with $\dot L^\infty(\mathbf{R}^n)\cap C(\mathbf{R}^n)$.

6.3.3. (a) For a continuous function f prove the identity

$$D_h^{k+1}(f)(x) = \sum_{s=0}^{k+1} (-1)^{k+1-s} \binom{k+1}{s} f(x+sh)$$

for all $x,h \in \mathbf{R}^n$ and $k \in \mathbf{Z}^+ \cup \{0\}$.
(b) Prove that $D_h^k D_h^l = D_h^{k+l}$ for all $k,l \in \mathbf{Z}^+ \cup \{0\}$.

6.3.4. For $x \in \mathbf{R}$ let

$$f(x) = \sum_{k=1}^{\infty} 2^{-k} e^{2\pi i 2^k x}.$$

(a) Prove that $f \in \Lambda_\gamma(\mathbf{R})$ for all $0 < \gamma < 1$.
(b) Prove that there is an $A < \infty$ such that

$$\sup_{x,t \neq 0} |f(x+t) + f(x-t) - 2f(x)|\, |t|^{-1} \le A;$$

thus $f \in \Lambda_1(\mathbf{R})$.
(c) Show, however, that for all $x \in [0,1]$ we have

$$\sup_{0<|t|<1} |f(x+t) - f(x)|\, |t|^{-1} = \infty;$$

thus f is nowhere differentiable.
[*Hint:* Part (c): Use that $f(x)$ is 1-periodic and thus

$$\int_0^1 |f(x+t) - f(x)|^2 \, dx = \sum_{k=1}^{\infty} 2^{-2k} |e^{2\pi i 2^k t} - 1|^2.$$

Observe that when $2^k |t| \le \frac{1}{2}$ we have $|e^{2\pi i 2^k t} - 1| \ge 2^{k+2}|t|$.]

6.3.5. For $0 < a,b < \infty$ and $x \in \mathbf{R}$ let

$$g_{ab}(x) = \sum_{k=1}^{\infty} 2^{-ak} e^{2\pi i 2^{bk} x}.$$

Show that g_{ab} lies in $\Lambda_{\frac{a}{b}}(\mathbf{R})$.
[*Hint:* Use the estimate $|D_h^L(e^{2\pi i 2^{bk} x})| \le C \min\left(1, (2^{bk}|h|)^L\right)$ with $L = [a/b] + 1$ and split the sum into two parts.]

6.3.6. Let $\gamma > 0$ and let $k = [\gamma]$.
(a) Use Exercise 6.3.3(b) to prove that if $|D_h^k(f)(x)| \le C|h|^\gamma$ for all $x,h \in \mathbf{R}^n$, then $|D_h^{k+l}(f)(x)| \le C 2^l |h|^\gamma$ for all $l \ge 1$.
(b) Conversely, assuming that for some $l \ge 1$ we have

$$\sup_{x,h \in \mathbf{R}^n} \frac{|D_h^{k+l}(f)(x)|}{|h|^\gamma} < \infty,$$

show that $f \in \dot{\Lambda}_\gamma$.
[*Hint:* Part (b): Use (6.3.14) but replace $k+1$ by $k+l$.]

6.3.7. Let Ψ and Δ_j be as in Theorem 6.3.7. Define a continuous operator Q_t by setting

$$Q_t(f) = f * \Psi_t, \qquad \Psi_t(x) = t^{-n}\Psi(t^{-1}x).$$

Show that all tempered distributions f satisfy

$$\sup_{t>0} t^{-\gamma} \big\| Q_t(f) \big\|_{L^\infty} \approx \sup_{j \in \mathbf{Z}} 2^{j\gamma} \big\| \Delta_j(f) \big\|_{L^\infty}$$

with the interpretation that if either term is finite, then it controls the other term by a constant multiple of itself.
[*Hint:* Observe that $Q_t = Q_t(\Delta_{j-2} + \Delta_{j-1} + \Delta_j + \Delta_{j+1})$ when $2^{-j} \le t \le 2^{1-j}$.]

6.3.8. (a) Let $0 \le \gamma < 1$ and suppose that $\partial_j f \in \dot{\Lambda}_\gamma$ for all $1 \le j \le n$. Show that for some constant C we have

$$\big\| f \big\|_{\dot{\Lambda}_{\gamma+1}} \le C \sum_{j=1}^{n} \big\| \partial_j f \big\|_{\dot{\Lambda}_\gamma}$$

and conclude that $f \in \dot{\Lambda}_{\gamma+1}$.
(b) Let $\gamma \ge 0$. If we have $\partial^\alpha f \in \dot{\Lambda}_\gamma$ for all multi-indices α with $|\alpha| = r$, then there is an estimate

$$\big\| f \big\|_{\dot{\Lambda}_{\gamma+r}} \le C_\gamma \sum_{|\alpha|=r} \big\| \partial^\alpha f \big\|_{\dot{\Lambda}_\gamma},$$

and thus $f \in \dot{\Lambda}_{\gamma+r}$.
(c) Use Corollary 6.3.10 to obtain that the estimates in both (a) and (b) can be reversed.
[*Hint:* Part (a): Write

$$D_h^2(f)(x) = \int_0^1 \sum_{j=1}^{n} \big[\partial_j f(x + th + 2h) - \partial_j f(x + th + h) \big] h_j \, dt.$$

Part (b): Use induction.]

6.3.9. Introduce a difference operator

$$\mathscr{D}^\beta(f)(x) = \left[\int_{\mathbf{R}^n} \frac{|D_y^{[\beta]+1}(f)(x)|^2}{|y|^{n+2\beta}} \, dy \right]^{\frac{1}{2}},$$

where $\beta > 0$. Show that for some constant $c_0(n,\beta)$ we have

$$\big\| \mathscr{D}^\beta(f) \big\|_{L^2(\mathbf{R}^n)}^2 = c_0(n,\beta) \int_{\mathbf{R}^n} |\widehat{f}(\xi)|^2 |\xi|^{2\beta} \, d\xi$$

for all functions $f \in \dot{L}_\beta^2(\mathbf{R}^n)$.

6.4 Hardy Spaces

Having been able to characterize L^p spaces, Sobolev spaces, and Lipschitz spaces using Littlewood–Paley theory, it should not come as a surprise that the theory can be used to characterize other spaces as well. This is the case with the Hardy spaces $H^p(\mathbf{R}^n)$, which form a family of spaces with some remarkable properties in which the integrability index p can go all the way down to zero.

There exists an abundance of equivalent characterizations for Hardy spaces, of which only a few representative ones are discussed in this section. A reader interested in going through the material quickly may define the Hardy space H^p as the space of all tempered distributions f modulo polynomials for which

$$\big\|f\big\|_{H^p} = \bigg\|\Big(\sum_{j\in\mathbf{Z}} |\Delta_j(f)|^2\Big)^{\frac{1}{2}}\bigg\|_{L^p} < \infty \tag{6.4.1}$$

whenever $0 < p \le 1$. An atomic decomposition for Hardy spaces can be obtained from this definition (see Section 6.6), and once this is in hand, the analysis of these spaces is significantly simplified. For historical reasons, however, we choose to define Hardy spaces using a more classical approach, and as a result, we have to go through a considerable amount of work to obtain the characterization alluded to in (6.4.1).

6.4.1 *Definition of Hardy Spaces*

To give the definition of Hardy spaces on $\mathbf{R}^n$, we need some background. We say that a tempered distribution v is *bounded* if $\varphi * v \in L^\infty(\mathbf{R}^n)$ whenever φ is in $\mathscr{S}(\mathbf{R}^n)$. We observe that if v is a bounded tempered distribution and $h \in L^1(\mathbf{R}^n)$, then the convolution $h * v$ can be defined as a distribution via the convergent integral

$$\big\langle h * v, \varphi\big\rangle = \big\langle \widetilde{\varphi} * v, \widetilde{h}\big\rangle = \int_{\mathbf{R}^n} (\widetilde{\varphi} * v)(x)\widetilde{h}(x)\,dx,$$

where φ is a Schwartz function, and as usual, we set $\widetilde{\varphi}(x) = \varphi(-x)$.

Let us recall the Poisson kernel P introduced in (2.1.13):

$$P(x) = \frac{\Gamma(\frac{n+1}{2})}{\pi^{\frac{n+1}{2}}} \frac{1}{(1+|x|^2)^{\frac{n+1}{2}}}. \tag{6.4.2}$$

For $t > 0$, let $P_t(x) = t^{-n}P(t^{-1}x)$. If v is a bounded tempered distribution, then $P_t * v$ is a well defined distribution, since P_t is in L^1. We claim that $P_t * v$ can be identified with a well defined bounded function. To see this, write $1 = \widehat{\varphi}(\xi) + \eta(\xi)$, where $\widehat{\varphi}$ has compact support and η is a smooth function that vanishes in a neighborhood of the origin. Then the function ψ defined by $\widehat{\psi}(\xi) = e^{-2\pi|\xi|}\eta(\xi)$ is in the Schwartz class, and one has that

$$\widehat{P_t}(\xi) = e^{-2\pi t|\xi|} = e^{-2\pi t|\xi|}\widehat{\varphi}(t\xi) + \widehat{\psi}(t\xi)$$

is a sum of a compactly supported function and a Schwartz function. Then

$$P_t * v = P_t * (\varphi_t * v) + \psi_t * v,$$

but $\varphi_t * v$ and $\psi_t * v$ are bounded functions, since φ_t and ψ_t are in the Schwartz class. The last identity proves that $P_t * v$ is a bounded function.

Before we define Hardy spaces we introduce some notation.

Definition 6.4.1. Let $a, b > 0$. Let Φ be a Schwartz function and let f be a tempered distribution on $\mathbf{R}^n$. We define the *smooth maximal function of f with respect to Φ* as

$$M(f;\Phi)(x) = \sup_{t>0} |(\Phi_t * f)(x)|.$$

We define the *nontangential maximal function (with aperture a) of f with respect to Φ* as

$$M_a^*(f;\Phi)(x) = \sup_{t>0} \sup_{\substack{y\in\mathbf{R}^n \\ |y-x|\le at}} |(\Phi_t * f)(y)|.$$

We also define the *auxiliary maximal function*

$$M_b^{**}(f;\Phi)(x) = \sup_{t>0} \sup_{y\in\mathbf{R}^n} \frac{|(\Phi_t * f)(x-y)|}{(1+t^{-1}|y|)^b},$$

and we observe that

$$M(f;\Phi) \le M_a^*(f;\Phi) \le (1+a)^b M_b^{**}(f;\Phi) \tag{6.4.3}$$

for all $a, b > 0$. We note that if Φ is merely integrable, for example, if Φ is the Poisson kernel, the maximal functions $M(f;\Phi)$, $M_a^*(f;\Phi)$, and $M_b^{**}(f;\Phi)$ are well defined only for bounded tempered distributions f on $\mathbf{R}^n$.

For a fixed positive integer N and a Schwartz function φ we define the quantity

$$\mathfrak{N}_N(\varphi) = \int_{\mathbf{R}^n} (1+|x|)^N \sum_{|\alpha|\le N+1} |\partial^\alpha \varphi(x)|\, dx. \tag{6.4.4}$$

We now define

$$\mathscr{F}_N = \Big\{\varphi \in \mathscr{S}(\mathbf{R}^n) : \mathfrak{N}_N(\varphi) \le 1\Big\}, \tag{6.4.5}$$

and we also define the *grand maximal function of f (with respect to N)* as

$$\mathscr{M}_N(f)(x) = \sup_{\varphi\in\mathscr{F}_N} M_1^*(f;\varphi)(x).$$

Having introduced a variety of smooth maximal operators useful in the development of the theory, we proceed with the definition of Hardy spaces.

Definition 6.4.2. Let f be a bounded tempered distribution on $\mathbf{R}^n$ and let $0 < p < \infty$. We say that f lies in the *Hardy space* $H^p(\mathbf{R}^n)$ if the *Poisson maximal function*

$$M(f;P)(x) = \sup_{t>0} |(P_t * f)(x)| \tag{6.4.6}$$

is in $L^p(\mathbf{R}^n)$. If this is the case, we set

$$\|f\|_{H^p} = \|M(f;P)\|_{L^p}.$$

At this point we don't know whether these spaces coincide with any other known spaces for some values of p. In the next theorem we show that this is the case when $1 < p < \infty$.

Theorem 6.4.3. *(a) Let* $1 < p < \infty$. *Then every bounded tempered distribution* f *in* H^p *is an element of* L^p. *Moreover, there is a constant* $C_{n,p}$ *such that for all such* f *we have*

$$\|f\|_{L^p} \le \|f\|_{H^p} \le C_{n,p}\|f\|_{L^p},$$

and therefore $H^p(\mathbf{R}^n)$ *coincides with* $L^p(\mathbf{R}^n)$.
(b) When $p = 1$, *every element of* H^1 *is an integrable function. In other words,* $H^1(\mathbf{R}^n) \subseteq L^1(\mathbf{R}^n)$ *and for all* $f \in H^1$ *we have*

$$\|f\|_{L^1} \le \|f\|_{H^1}. \tag{6.4.7}$$

Proof. (a) Let $f \in H^p(\mathbf{R}^n)$. The set $\{P_t * f : t > 0\}$ lies in a multiple of the unit ball of L^p. By the Banach–Alaoglu–Bourbaki theorem there exists a sequence $t_j \to 0$ such that $P_{t_j} * f$ converges to some L^p function f_0 in the weak* topology of L^p. On the other hand, we see that $P_t * \varphi \to \varphi$ in $\mathscr{S}(\mathbf{R}^n)$ as $t \to 0$ for all φ in $\mathscr{S}(\mathbf{R}^n)$. Thus

$$P_t * f \to f \qquad \text{in } \mathscr{S}'(\mathbf{R}^n), \tag{6.4.8}$$

and it follows that the distribution f coincides with the L^p function f_0. Since the family $\{P_t\}_{t>0}$ is an approximate identity, Theorem 1.2.19 gives that

$$\|P_t * f - f\|_{L^p} \to 0 \qquad \text{as } t \to 0,$$

from which it follows that

$$\|f\|_{L^p} \le \Big\|\sup_{t>0} |P_t * f|\Big\|_{L^p} = \|f\|_{H^p}. \tag{6.4.9}$$

The converse inequality is a consequence of the fact that

$$\sup_{t>0} |P_t * f| \le M(f),$$

where M is the Hardy–Littlewood maximal operator. (See Corollary 2.1.12.)

(b) The case $p=1$ requires only a small modification of the case $p>1$. Embedding L^1 into the space of finite Borel measures $\mathscr{M}$ whose unit ball is weak* compact, we can extract a sequence $t_j \to 0$ such that $P_{t_j} * f$ converges to some measure μ in the topology of measures. In view of (6.4.8), it follows that the distribution f can be identified with the measure μ.

It remains to show that μ is absolutely continuous with respect to Lebesgue measure, which would imply that it coincides with some L^1 function. Let $|\mu|$ be the total variation of μ. We show that μ is absolutely continuous by showing that for all subsets E of $\mathbf{R}^n$ we have $|E|=0 \implies |\mu|(E)=0$. Given an $\varepsilon>0$, there exists a $\delta>0$ such that for any measurable subset F of $\mathbf{R}^n$ we have

$$|F|<\delta \implies \int_F \sup_{t>0} |P_t * f|\,dx < \varepsilon\,.$$

Given E with $|E|=0$, we can find an open set U such that $E \subseteq U$ and $|U|<\delta$. Then for any g continuous function supported in U we have

$$\begin{aligned}\left|\int_{\mathbf{R}^n} g\,d\mu\right| &= \lim_{j\to\infty}\left|\int_{\mathbf{R}^n} g(x)\,(P_{t_j} * f)(x)\,dx\right| \\ &\le \big\|g\big\|_{L^\infty}\int_U \sup_{t>0}|P_t * f|\,dx \\ &< \varepsilon\big\|g\big\|_{L^\infty}\,.\end{aligned}$$

But we have

$$|\mu(U)| = \sup\left\{\left|\int_{\mathbf{R}^n} g\,d\mu\right| : g \text{ continuous supported in } U \text{ with } \big\|g\big\|_{L^\infty}\le 1\right\},$$

which implies that $|\mu(U)|<\varepsilon$. Since ε was arbitrary, it follows that $|\mu|(E)=0$; hence μ is absolutely continuous with respect to Lebesgue measure. Finally, (6.4.7) is a consequence of (6.4.9), which is also valid for $p=1$. □

We may wonder whether H^1 coincides with L^1. We show in Theorem 6.7.4 that elements of H^1 have integral zero; thus H^1 is a proper subspace of L^1.

We now proceed to obtain some characterizations of these spaces.

6.4.2 Quasinorm Equivalence of Several Maximal Functions

It is a fact that all the maximal functions of the preceding subsection have comparable L^p quasinorms for all $0<p<\infty$. This is the essence of the following theorem.

Theorem 6.4.4. *Let $0<p<\infty$. Then the following statements are valid:*
(a) There exists a Schwartz function Φ with $\int_{\mathbf{R}^n}\Phi(x)\,dx \neq 0$ and a constant C_1 (which does not depend on any parameter) such that

$$\big\|M(f;\Phi)\big\|_{L^p} \le C_1\big\|f\big\|_{H^p} \tag{6.4.10}$$

for all bounded $f \in \mathscr{S}'(\mathbf{R}^n)$.
(b) For every $a > 0$ *and* Φ *in* $\mathscr{S}(\mathbf{R}^n)$ *there exists a constant* $C_2(n,p,a,\Phi)$ *such that*

$$\big\|M_a^*(f;\Phi)\big\|_{L^p} \le C_2(n,p,a,\Phi)\big\|M(f;\Phi)\big\|_{L^p} \tag{6.4.11}$$

for all $f \in \mathscr{S}'(\mathbf{R}^n)$.
(c) For every $a > 0$, $b > n/p$, *and* Φ *in* $\mathscr{S}(\mathbf{R}^n)$ *there exists a constant* $C_3(n,p,a,b,\Phi)$ *such that*

$$\big\|M_b^{**}(f;\Phi)\big\|_{L^p} \le C_3(n,p,a,b,\Phi)\big\|M_a^*(f;\Phi)\big\|_{L^p} \tag{6.4.12}$$

for all $f \in \mathscr{S}'(\mathbf{R}^n)$.
(d) For every $b > 0$ *and* Φ *in* $\mathscr{S}(\mathbf{R}^n)$ *with* $\int_{\mathbf{R}^n}\Phi(x)\,dx \neq 0$ *there exists a constant* $C_4(b,\Phi)$ *such that if* $N = [b]+1$ *we have*

$$\big\|\mathscr{M}_N(f)\big\|_{L^p} \le C_4(b,\Phi)\big\|M_b^{**}(f;\Phi)\big\|_{L^p} \tag{6.4.13}$$

for all $f \in \mathscr{S}'(\mathbf{R}^n)$.
(e) For every positive integer N *there exists a constant* $C_5(n,N)$ *such that every tempered distribution* f *with* $\big\|\mathscr{M}_N(f)\big\|_{L^p} < \infty$ *is a bounded distribution and satisfies*

$$\big\|f\big\|_{H^p} \le C_5(n,N)\big\|\mathscr{M}_N(f)\big\|_{L^p}, \tag{6.4.14}$$

that is, it lies in the Hardy space H^p.

We conclude that for $f \in H^p(\mathbf{R}^n)$, the inequality in (6.4.14) can be reversed, and therefore for *any* Schwartz function Φ with $\int_{\mathbf{R}^n}\Phi(x)\,dx \neq 0$, we have

$$\big\|M_a^*(f;\Phi)\big\|_{L^p} \le C(a,n,p,\Phi)\big\|f\big\|_{H^p}.$$

Consequently, there exists $N \in \mathbf{Z}^+$ large enough such that for $f \in \mathscr{S}'(\mathbf{R}^n)$ we have

$$\big\|\mathscr{M}_N(f)\big\|_{L^p} \approx \big\|M_b^{**}(f;\Phi)\big\|_{L^p} \approx \big\|M_a^*(f;\Phi)\big\|_{L^p} \approx \big\|M(f;\Phi)\big\|_{L^p} \approx \big\|f\big\|_{H^p}$$

for all Schwartz functions Φ with $\int_{\mathbf{R}^n}\Phi(x)\,dx \neq 0$ and constants that depend only on Φ,a,b,n,p. This furnishes a variety of characterizations for Hardy spaces.

The proof of this theorem is based on the following lemma.

Lemma 6.4.5. *Let* $m \in \mathbf{Z}^+$ *and let* Φ *in* $\mathscr{S}(\mathbf{R}^n)$ *satisfy* $\int_{\mathbf{R}^n}\Phi(x)\,dx = 1$. *Then there exists a constant* $C_0(\Phi,m)$ *such that for any* Ψ *in* $\mathscr{S}(\mathbf{R}^n)$, *there exist Schwartz functions* $\Theta^{(s)}$, $0 \le s \le 1$, *with the properties*

$$\Psi(x) = \int_0^1 (\Theta^{(s)} * \Phi_s)(x)\,ds \tag{6.4.15}$$

and

$$\int_{\mathbf{R}^n}(1+|x|)^m|\Theta^{(s)}(x)|\,dx \le C_0(\Phi,m)\,s^m\,\mathfrak{N}_m(\Psi). \tag{6.4.16}$$

Proof. We start with a smooth function ζ supported in $[0,1]$ that satisfies

$$\begin{aligned} 0 \le \zeta(s) \le & \frac{2s^m}{m!} && \text{for all } 0 \le s \le 1, \\ \zeta(s) = & \frac{s^m}{m!} && \text{for all } 0 \le s \le \frac{1}{2}, \\ \frac{d^r \zeta}{dt^r}(1) = & \; 0 && \text{for all } 0 \le r \le m+1. \end{aligned}$$

We define

$$\Theta^{(s)} = \Xi^{(s)} - \frac{d^{m+1}\zeta(s)}{ds^{m+1}} \overbrace{\Phi_s * \cdots * \Phi_s}^{m+1 \text{ terms}} * \Psi, \tag{6.4.17}$$

where

$$\Xi^{(s)} = (-1)^{m+1}\zeta(s) \frac{\partial^{m+1}}{\partial s^{m+1}} \Big(\overbrace{\Phi_s * \cdots * \Phi_s}^{m+2 \text{ terms}} \Big) * \Psi,$$

and we claim that (6.4.15) holds for this choice of $\Theta^{(s)}$. To verify this assertion, we apply $m+1$ integration by parts to write

$$\begin{aligned} \int_0^1 \Theta^{(s)} * \Phi_s \, ds = \int_0^1 \Xi^{(s)} * \Phi_s \, ds + \frac{d^m \zeta(s)}{ds^m}(0) \lim_{s \to 0+} \big(\overbrace{\Phi * \cdots * \Phi}^{m+2 \text{ terms}} \big)_s * \Psi \\ - (-1)^{m+1} \int_0^1 \zeta(s) \frac{\partial^{m+1}}{\partial s^{m+1}} \Big(\overbrace{\Phi_s * \cdots * \Phi_s}^{m+2 \text{ terms}} \Big) * \Psi \, ds, \end{aligned}$$

noting that all the boundary terms vanish except for the one in the first integration by parts at $s = 0$. The first and the third terms in the previous expression on the right add up to zero, while the second term is equal to Ψ, since Φ has integral one, which implies that the family $\{(\Phi * \cdots * \Phi)_s\}_{s>0}$ is an approximate identity as $s \to 0$. Therefore, (6.4.15) holds.

We now prove estimate (6.4.16). Let Ω be the $(m+1)$-fold convolution of Φ. For the second term on the right in (6.4.17), we note that the $(m+1)$st derivative of $\zeta(s)$ vanishes on $[0, \frac{1}{2}]$, so that we may write

$$\begin{aligned} &\int_{\mathbf{R}^n} (1+|x|)^m \left| \frac{d^{m+1}\zeta(s)}{ds^{m+1}} \right| |\Omega_s * \Psi(x)| \, dx \\ &\le C_m \chi_{[\frac{1}{2},1]}(s) \int_{\mathbf{R}^n} (1+|x|)^m \left[\int_{\mathbf{R}^n} \tfrac{1}{s^n} |\Omega(\tfrac{x-y}{s})| \, |\Psi(y)| \, dy \right] dx \\ &\le C_m \chi_{[\frac{1}{2},1]}(s) \int_{\mathbf{R}^n} \int_{\mathbf{R}^n} (1+|y+sx|)^m |\Omega(x)| \, |\Psi(y)| \, dy \, dx \\ &\le C_m \chi_{[\frac{1}{2},1]}(s) \int_{\mathbf{R}^n} \int_{\mathbf{R}^n} (1+|sx|)^m |\Omega(x)| \, (1+|y|)^m |\Psi(y)| \, dy \, dx \\ &\le C_m \chi_{[\frac{1}{2},1]}(s) \Big(\int_{\mathbf{R}^n} (1+|x|)^m |\Omega(x)| \, dx \Big) \Big(\int_{\mathbf{R}^n} (1+|y|)^m |\Psi(y)| \, dy \Big) \\ &\le C_0'(\Phi, m) \, s^m \, \mathfrak{N}_m(\Psi), \end{aligned}$$

since $\chi_{[\frac{1}{2},1]}(s) \le 2^m s^m$. To obtain a similar estimate for the first term on the right in (6.4.17), we argue as follows:

$$\begin{aligned}
&\int_{\mathbf{R}^n}(1+|x|)^m|\zeta(s)|\left|\frac{d^{m+1}(\Omega_s*\Psi)}{ds^{m+1}}(x)\right|dx \\
&= \int_{\mathbf{R}^n}(1+|x|)^m\,|\zeta(s)|\left|\frac{d^{m+1}}{ds^{m+1}}\int_{\mathbf{R}^n}\frac{1}{s^n}\Omega\Big(\frac{x-y}{s}\Big)\Psi(y)\,dy\right|dx \\
&= \int_{\mathbf{R}^n}(1+|x|)^m\,|\zeta(s)|\left|\int_{\mathbf{R}^n}\Omega(y)\frac{d^{m+1}\Psi(x-sy)}{ds^{m+1}}\,dy\right|dx \\
&\le C'_m\int_{\mathbf{R}^n}(1+|x|)^m\,|\zeta(s)|\int_{\mathbf{R}^n}|\Omega(y)|\Big[\sum_{|\alpha|\le m+1}|\partial^\alpha\Psi(x-sy)|\,|y|^{|\alpha|}\Big]\,dy\,dx \\
&\le C'_m|\zeta(s)|\int_{\mathbf{R}^n}\int_{\mathbf{R}^n}(1+|x+sy|)^m|\Omega(y)|\sum_{|\alpha|\le m+1}|\partial^\alpha\Psi(x)|\,(1+|y|)^{m+1}\,dy\,dx \\
&\le C'_m|\zeta(s)|\int_{\mathbf{R}^n}(1+|y|)^{m+1}\,|\Omega(y)|\,(1+|y|)^m\,dy\int_{\mathbf{R}^n}(1+|x|)^m\sum_{|\alpha|\le m+1}|\partial^\alpha\Psi(x)|\,dx \\
&\le C''_0(\Phi,m)\,s^m\,\mathfrak{N}_m(\Psi)\,.
\end{aligned}$$

We now set $C_0(\Phi,m) = C'_0(\Phi,m) + C''_0(\Phi,m)$ to conclude the proof of (6.4.16). □

Next, we discuss the proof of Theorem 6.4.4.

Proof. (a) We pick a continuous and integrable function $\psi(s)$ on the interval $[1,\infty)$ that decays faster than the reciprocal of any polynomial (i.e., $|\psi(s)| \le C_N s^{-N}$ for all $N>0$) such that

$$\int_1^\infty s^k\,\psi(s)\,ds = \begin{cases} 1 & \text{if } k=0, \\ 0 & \text{if } k=1,2,3,\dots. \end{cases} \tag{6.4.18}$$

Such a function exists; in fact, we may take

$$\psi(s) = \frac{e}{\pi}\frac{1}{s}\,\mathrm{Im}\left(e^{(\frac{\sqrt{2}}{2}-i\frac{\sqrt{2}}{2})(s-1)^{\frac{1}{4}}}\right). \tag{6.4.19}$$

See Exercise 6.4.4. We now define a function

$$\Phi(x) = \int_1^\infty \psi(s)P_s(x)\,ds\,, \tag{6.4.20}$$

where P_s is the Poisson kernel. The Fourier transform Φ is

$$\widehat{\Phi}(\xi) = \int_1^\infty \psi(s)\widehat{P_s}(\xi)\,ds = \int_1^\infty \psi(s)e^{-2\pi s|\xi|}\,ds$$

(cf. Exercise 2.2.11), which is easily seen to be rapidly decreasing as $|\xi|\to\infty$. The same is true for all the derivatives of $\widehat{\Phi}$. The function $\widehat{\Phi}$ is clearly smooth on $\mathbf{R}^n\setminus\{0\}$. Moreover,

$$\partial_j \widehat{\Phi}(\xi) = \sum_{k=0}^{L-1} (-2\pi)^{k+1} \frac{|\xi|^k}{k!} \frac{\xi_j}{|\xi|} \int_1^\infty s^{k+1} \psi(s)\, ds + O(|\xi|^L) = O(|\xi|^L)$$

as $|\xi| \to 0$, which implies that the distributional derivative $\partial_j \widehat{\Phi}$ is continuous at the origin. Since

$$\partial_\xi^\alpha (e^{-2\pi s|\xi|}) = s^{|\alpha|} p_\alpha(\xi) |\xi|^{-m_\alpha} e^{-2\pi s|\xi|}$$

for some $m_\alpha \in \mathbf{Z}^+$ and some polynomial p_α, choosing L sufficiently large gives that every derivative of $\widehat{\Phi}$ is also continuous at the origin. We conclude that the function $\widehat{\Phi}$ is in the Schwartz class, and thus so is Φ. It also follows from (6.4.18) and (6.4.20) that

$$\int_{\mathbf{R}^n} \Phi(x)\, dx = 1 \neq 0 .$$

Finally, we have the estimate

$$\begin{aligned} M(f;\Phi)(x) &= \sup_{t>0} |(\Phi_t * f)(x)| \\ &= \sup_{t>0} \left| \int_1^\infty \psi(s)(f * P_{ts})(x)\, ds \right| \\ &\le \int_1^\infty |\psi(s)|\, ds M(f;P)(x), \end{aligned}$$

and the required conclusion follows with $C_1 = \int_1^\infty |\psi(s)|\, ds$. Note that we actually obtained the stronger pointwise estimate

$$M(f;\Phi) \le C_1 M(f;P)$$

rather than (6.4.10).

(b) The control of the nontagential maximal function $M_a^*(\cdot\,;\Phi)$ in terms of the vertical maximal function $M(\cdot\,;\Phi)$ is the hardest and most technical part of the proof. For matters of exposition, we present the proof only in the case that $a = 1$ and we note that the case of general $a > 0$ presents only notational differences. We derive (6.4.11) as a consequence of the estimate

$$\left(\int_{\mathbf{R}^n} M_1^*(f;\Phi)^{\varepsilon,N}(x)^p dx \right)^{\frac{1}{p}} \le C_2(n,p,N,\Phi) \big\| M(f;\Phi) \big\|_{L^p}, \tag{6.4.21}$$

where N is a large enough integer depending on f, $0 < \varepsilon < 1$, and

$$M_1^*(f;\Phi)^{\varepsilon,N}(x) = \sup_{0<t<\frac{1}{\varepsilon}} \sup_{|y-x|\le t} |(\Phi_t * f)(y)| \left(\frac{t}{t+\varepsilon} \right)^N \frac{1}{(1+\varepsilon|y|)^N} .$$

Let us a fix an element f in $\mathscr{S}'(\mathbf{R}^n)$ such that $M(f;\Phi) \in L^p$. We first show that $M_1^*(f;\Phi)^{\varepsilon,N}$ lies in $L^p(\mathbf{R}^n) \cap L^\infty(\mathbf{R}^n)$. Indeed, using (2.3.22) (with $\alpha = 0$), we obtain the following estimate for some constants C_f, m, and l (depending on f):

$$\begin{aligned}
|(\Phi_t * f)(y)| &\le C_f \sum_{|\beta|\le l} \sup_{z\in\mathbf{R}^n} (|y|^m + |z|^m)|(\partial^\beta \widetilde{\Phi_t})(z)| \\
&\le C_f (1+|y|^m) \sum_{|\beta|\le l} \sup_{z\in\mathbf{R}^n} (1+|z|^m)|(\partial^\beta \Phi_t)(-z)| \\
&\le C_f \frac{(1+|y|^m)}{\min(t^n,t^{n+l})} \sum_{|\beta|\le l} \sup_{z\in\mathbf{R}^n} (1+|z|^m)|(\partial^\beta \Phi)(-z/t)| \\
&\le C_f \frac{(1+|y|)^m}{\min(t^n,t^{n+l})}(1+t^m) \sum_{|\beta|\le l} \sup_{z\in\mathbf{R}^n} (1+|z/t|^m)|(\partial^\beta \Phi)(-z/t)| \\
&\le C(f,\Phi)(1+\varepsilon|y|)^m \varepsilon^{-m}(1+t^m)(t^{-n}+t^{-n-l}).
\end{aligned}$$

Multiplying by $(\frac{t}{t+\varepsilon})^N(1+\varepsilon|y|)^{-N}$ for some $0<t<\frac{1}{\varepsilon}$ and $|y-x|<t$ yields

$$\big|(\Phi_t * f)(y)\big|\Big(\frac{t}{t+\varepsilon}\Big)^N \frac{1}{(1+\varepsilon|y|)^N} \le C(f,\Phi)\frac{\varepsilon^{-m}(1+\varepsilon^{-m})(\varepsilon^{n-N}+\varepsilon^{n+l-N})}{(1+\varepsilon|y|)^{N-m}},$$

and using that $1+\varepsilon|y|\ge\frac{1}{2}(1+\varepsilon|x|)$, we obtain for some $C(f,\Phi,\varepsilon,n,l,m,N)<\infty$,

$$M_1^*(f;\Phi)^{\varepsilon,N}(x) \le \frac{C(f,\Phi,\varepsilon,n,l,m,N)}{(1+\varepsilon|x|)^{N-m}}.$$

Taking $N>(m+n)/p$, we deduce that $M_1^*(f;\Phi)^{\varepsilon,N}$ lies in $L^p(\mathbf{R}^n)\cap L^\infty(\mathbf{R}^n)$.

We now introduce a parameter $L>0$ and functions

$$U(f;\Phi)^{\varepsilon,N}(x) = \sup_{0<t<\frac{1}{\varepsilon}} \sup_{|y-x|<t} t\big|\nabla(\Phi_t * f)(y)\big|\Big(\frac{t}{t+\varepsilon}\Big)^N \frac{1}{(1+\varepsilon|y|)^N}$$

and

$$V(f;\Phi)^{\varepsilon,N,L}(x) = \sup_{0<t<\frac{1}{\varepsilon}} \sup_{y\in\mathbf{R}^n} \big|(\Phi_t * f)(y)\big|\Big(\frac{t}{t+\varepsilon}\Big)^N \frac{1}{(1+\varepsilon|y|)^N}\Big(\frac{t}{t+|x-y|}\Big)^L.$$

We fix an integer $L>n/p$. We need the norm estimate

$$\big\|V(f;\Phi)^{\varepsilon,N,L}\big\|_{L^p} \le C_{n,p}\big\|M_1^*(f;\Phi)^{\varepsilon,N}\big\|_{L^p} \tag{6.4.22}$$

and the pointwise estimate

$$U(f;\Phi)^{\varepsilon,N} \le A(\Phi,N,n,p)\,V(f;\Phi)^{\varepsilon,N,L}, \tag{6.4.23}$$

where

$$A(\Phi,N,n,p) = 2^L C_0(\partial_j\Phi;N+L)\,\mathfrak{N}_{N+L}(\partial_j\Phi).$$

To prove (6.4.22) we observe that when $z\in B(y,t)\subseteq B(x,|x-y|+t)$ we have

$$\big|(\Phi_t * f)(y)\big|\Big(\frac{t}{t+\varepsilon}\Big)^N \frac{1}{(1+\varepsilon|y|)^N} \le M_1^*(f;\Phi)^{\varepsilon,N}(z)\,,$$

from which it follows that for any $0 < q < \infty$ and $y \in \mathbf{R}^n$,

$$\begin{aligned}
&\big|(\Phi_t * f)(y)\big|\Big(\frac{t}{t+\varepsilon}\Big)^N \frac{1}{(1+\varepsilon|y|)^N}\\
&\quad\le \bigg(\frac{1}{|B(y,t)|}\int_{B(y,t)} M_1^*(f;\Phi)^{\varepsilon,N}(z)^q\,dz\bigg)^{\frac{1}{q}}\\
&\quad\le \Big(\frac{|x-y|+t}{t}\Big)^{\frac{n}{q}}\bigg(\frac{1}{|B(x,|x-y|+t)|}\int_{B(x,|x-y|+t)} M_1^*(f;\Phi)^{\varepsilon,N}(z)^q\,dz\bigg)^{\frac{1}{q}}\\
&\quad\le \Big(\frac{|x-y|+t}{t}\Big)^{L} M\Big(\big[M_1^*(f;\Phi)^{\varepsilon,N}\big]^q\Big)^{\frac{1}{q}}(x)\,,
\end{aligned}$$

where we used that $L > n/p$. We now take $0 < q < p$ and we use the boundedness of the Hardy–Littlewood maximal operator M on $L^{p/q}$ to obtain (6.4.22).

In proving (6.4.23), we may assume that Φ has integral 1; otherwise we can multiply Φ by a suitable constant to arrange for this to happen. We note that

$$t\big|\nabla(\Phi_t * f)\big| = \big|(\nabla\Phi)_t * f\big| \le \sqrt{n}\sum_{j=1}^n \big|(\partial_j\Phi)_t * f\big|\,,$$

and it suffices to work with each partial derivative $\partial_j\Phi$ of Φ. Using Lemma 6.4.5 we write

$$\partial_j\Phi = \int_0^1 \Theta^{(s)} * \Phi_s\,ds$$

for suitable Schwartz functions $\Theta^{(s)}$. Fix $x \in \mathbf{R}^n$, $t > 0$, and y with $|y-x| < t < 1/\varepsilon$. Then we have

$$\begin{aligned}
&\big|\big((\partial_j\Phi)_t * f\big)(y)\big|\Big(\frac{t}{t+\varepsilon}\Big)^N \frac{1}{(1+\varepsilon|y|)^N}\\
&\quad= \Big(\frac{t}{t+\varepsilon}\Big)^N \frac{1}{(1+\varepsilon|y|)^N}\bigg|\int_0^1 \big((\Theta^{(s)})_t * \Phi_{st} * f\big)(y)\,ds\bigg| \qquad (6.4.24)\\
&\quad\le \Big(\frac{t}{t+\varepsilon}\Big)^N \int_0^1\int_{\mathbf{R}^n} t^{-n}\big|\Theta^{(s)}(t^{-1}z)\big|\,\frac{\big|\big(\Phi_{st} * f\big)(y-z)\big|}{(1+\varepsilon|y|)^N}\,dz\,ds\,.
\end{aligned}$$

Inserting the factor 1 written as

$$\Big(\frac{ts}{ts+|x-(y-z)|}\Big)^L\Big(\frac{ts}{ts+\varepsilon}\Big)^N\Big(\frac{ts+|x-(y-z)|}{ts}\Big)^L\Big(\frac{ts+\varepsilon}{ts}\Big)^N$$

in the preceding z-integral and using that

$$\frac{1}{(1+\varepsilon|y|)^N} \le \frac{(1+\varepsilon|z|)^N}{(1+\varepsilon|y-z|)^N}$$

and the fact that $|x-y|<t<1/\varepsilon$, we obtain the estimate

$$\begin{aligned}
&\left(\frac{t}{t+\varepsilon}\right)^N \int_0^1 \int_{\mathbf{R}^n} t^{-n}\left|\Theta^{(s)}(t^{-1}z)\right| \frac{\left|(\Phi_{st}*f)(y-z)\right|}{(1+\varepsilon|y|)^N}\,dz\,ds \\
&\le V(f;\Phi)^{\varepsilon,N,L}(x)\int_0^1\int_{\mathbf{R}^n}(1+\varepsilon|z|)^N\left(\frac{ts+|x-(y-z)|}{ts}\right)^L t^{-n}\left|\Theta^{(s)}(t^{-1}z)\right|dz\,\frac{ds}{s^N} \\
&\le V(f;\Phi)^{\varepsilon,N,L}(x)\int_0^1\int_{\mathbf{R}^n} s^{-L-N}(1+\varepsilon t|z|)^N(s+1+|z|)^L\left|\Theta^{(s)}(z)\right|dz\,ds \\
&\le 2^L C_0(\partial_j\Phi;N+L)\,\mathfrak{N}_{N+L}(\partial_j\Phi)\,V(f;\Phi)^{\varepsilon,N,L}(x)
\end{aligned}$$

in view of conclusion (6.4.16) of Lemma 6.4.5. Combining this estimate with (6.4.24), we deduce (6.4.23). Having established both (6.4.22) and (6.4.23), we conclude that

$$\left\|U(f;\Phi)^{\varepsilon,N}\right\|_{L^p} \le C_{n,p}A(\Phi,N,n,p)\left\|M_1^*(f;\Phi)^{\varepsilon,N}\right\|_{L^p}. \tag{6.4.25}$$

We now set

$$E_\varepsilon = \left\{x\in\mathbf{R}^n:\ U(f;\Phi)^{\varepsilon,N}(x)\le KM_1^*(f;\Phi)^{\varepsilon,N}(x)\right\}$$

for some constant K to be determined shortly. With $A=A(\Phi,N,n,p)$ we have

$$\begin{aligned}
\int_{(E_\varepsilon)^c}\left[M_1^*(f;\Phi)^{\varepsilon,N}(x)\right]^p dx &\le \frac{1}{K^p}\int_{(E_\varepsilon)^c}\left[U(f;\Phi)^{\varepsilon,N}(x)\right]^p dx \\
&\le \frac{1}{K^p}\int_{\mathbf{R}^n}\left[U(f;\Phi)^{\varepsilon,N}(x)\right]^p dx \\
&\le \frac{C_{n,p}^p A^p}{K^p}\int_{\mathbf{R}^n}\left[M_1^*(f;\Phi)^{\varepsilon,N}(x)\right]^p dx \\
&\le \frac{1}{2}\int_{\mathbf{R}^n}\left[M_1^*(f;\Phi)^{\varepsilon,N}(x)\right]^p dx,
\end{aligned} \tag{6.4.26}$$

provided we choose K such that $K^p = 2C_{n,p}^p A^p$. Obviously $K=K(\Phi,N,n,p)$, i.e., it depends on all these variables, in particular on N, which depends on f.

It remains to estimate the contribution of the integral of $\left[M_1^*(f;\Phi)^{\varepsilon,N}(x)\right]^p$ over the set E_ε. We claim that the following pointwise estimate is valid:

$$M_1^*(f;\Phi)^{\varepsilon,N}(x)\le C_{n,N,K}\,M\big(M(f;\Phi)^r\big)^{\frac{1}{q}}(x) \tag{6.4.27}$$

for any $x\in E_\varepsilon$ and $0<q<\infty$. Note that $C_{n,N,K}$ depends on K. To prove (6.4.27) we fix $x\in E_\varepsilon$ and we also fix y such that $|y-x|<t$.

By the definition of $M_1^*(f;\Phi)^{\varepsilon,N}(x)$ there exists a point $(y_0,t)\in\mathbf{R}_+^{n+1}$ such that $|x-y_0|<t<\frac{1}{\varepsilon}$ and

$$\big|(\Phi_t * f)(y_0)\big|\Big(\frac{t}{t+\varepsilon}\Big)^N \frac{1}{(1+\varepsilon|y_0|)^N} \ge \frac{1}{2} M_1^*(f;\Phi)^{\varepsilon,N}(x). \tag{6.4.28}$$

Also by the definitions of E_ε and $U(f;\Phi)^{\varepsilon,N}$, for any $x \in E_\varepsilon$ we have

$$t\big|\nabla(\Phi_t * f)(\xi)\big|\Big(\frac{t}{t+\varepsilon}\Big)^N \frac{1}{(1+\varepsilon|\xi|)^N} \le K M_1^*(f;\Phi)^{\varepsilon,N}(x) \tag{6.4.29}$$

for all ξ satisfying $|\xi - x| < t < \frac{1}{\varepsilon}$. It follows from (6.4.28) and (6.4.29) that

$$t\big|\nabla(\Phi_t * f)(\xi)\big| \le 2K\big|(\Phi_t * f)(y_0)\big|\Big(\frac{1+\varepsilon|\xi|}{1+\varepsilon|y_0|}\Big)^N \tag{6.4.30}$$

for all ξ satisfying $|\xi - x| < t < \frac{1}{\varepsilon}$. We let z be such that $|z-x| < t$. Applying the mean value theorem and using (6.4.30), we obtain, for some ξ between y_0 and z,

$$\begin{aligned}
\big|(\Phi_t * f)(z) - (\Phi_t * f)(y_0)\big| &= \big|\nabla(\Phi_t * f)(\xi)\big|\,|z-y_0| \\
&\le \frac{2K}{t}\big|(\Phi_t * f)(\xi)\big|\Big(\frac{1+\varepsilon|\xi|}{1+\varepsilon|y_0|}\Big)^N |z-y_0| \\
&\le \frac{2^{N+1}K}{t}\big|(\Phi_t * f)(y_0)\big|\,|z-y_0| \\
&\le \frac{1}{2}\big|(\Phi_t * f)(y_0)\big|,
\end{aligned}$$

provided z also satisfies $|z-y_0| < 2^{-N-2}K^{-1}t$ in addition to $|z-x| < t$. Therefore, for z satisfying $|z-y_0| < 2^{-N-2}K^{-1}t$ and $|z-x| < t$ we have

$$\big|(\Phi_t * f)(z)\big| \ge \frac{1}{2}\big|(\Phi_t * f)(y_0)\big| \ge \frac{1}{4} M_1^*(f;\Phi)^{\varepsilon,N}(x),$$

where the last inequality uses (6.4.28). Thus we have

$$\begin{aligned}
M\big(M(f;\Phi)^q\big)(x) &\ge \frac{1}{|B(x,t)|}\int_{B(x,t)} \big[M(f;\Phi)(w)\big]^q\,dw \\
&\ge \frac{1}{|B(x,t)|}\int_{B(x,t)\cap B(y_0,2^{-N-2}K^{-1}t)} \big[M(f;\Phi)(w)\big]^q\,dw \\
&\ge \frac{1}{|B(x,t)|}\int_{B(x,t)\cap B(y_0,2^{-N-2}K^{-1}t)} \frac{1}{4^q}\big[M_1^*(f;\Phi)^{\varepsilon,N}(x)\big]^q\,dw \\
&\ge \frac{|B(x,t)\cap B(y_0,2^{-N-2}K^{-1}t)|}{|B(x,t)|}\frac{1}{4^q}\big[M_1^*(f;\Phi)^{\varepsilon,N}(x)\big]^r \\
&\ge C_{n,N,K}4^{-q}\big[M_1^*(f;\Phi)^{\varepsilon,N}(x)\big]^q,
\end{aligned}$$

where we used the simple geometric fact that if $|x-y_0| \le t$ and $\delta > 0$, then

$$\frac{|B(x,t)\cap B(y_0,\delta t)|}{|B(x,t)|}\geq c_{n,\delta}>0,$$

the minimum of this constant being obtained when $|x-y_0|=t$. See Figure 6.1.

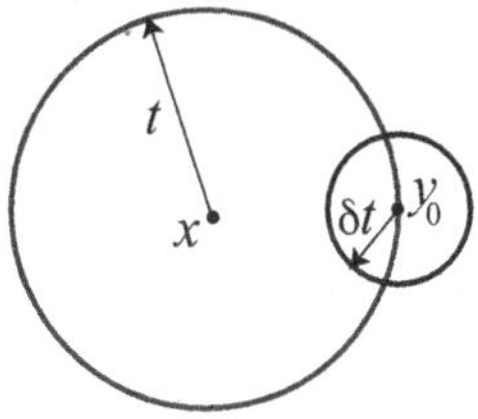

Fig. 6.1 The ball $B(y_0,\delta t)$ captures at least a fixed proportion of the ball $B(x,t)$.

This proves (6.4.27). Taking $r<p$ and applying the boundedness of the Hardy–Littlewood maximal operator yields

$$\int_{E_\varepsilon}\big[M_1^*(f;\Phi)^{\varepsilon,N}(x)\big]^p\,dx\leq C'_{\Phi,N,K,n,p}\int_{\mathbf{R}^n}M(f;\Phi)(x)^p\,dx. \tag{6.4.31}$$

Combining this estimate with (6.4.26), we obtain

$$\int_{\mathbf{R}^n}\big[M_1^*(f;\Phi)^{\varepsilon,N}\big]^p\,dx\leq C^p_{\Phi,N,K,n,p}\int_{\mathbf{R}^n}M(f;\Phi)^p\,dx+\frac{1}{2}\int_{\mathbf{R}^n}\big[M_1^*(f;\Phi)^{\varepsilon,N}\big]^p\,dx,$$

and using the fact (obtained earlier) $\big\|M_1^*(f;\Phi)^{\varepsilon,N}\big\|_{L^p}<\infty$, we obtain the required conclusion (6.4.11). This proves the inequality

$$\big\|M_1^*(f;\Phi)^{\varepsilon,N}\big\|_{L^p}\leq 2^{1/p}C_{\Phi,N,K,n,p}\big\|M(f;\Phi)\big\|_{L^p}. \tag{6.4.32}$$

The previous constant depends on f but is independent of ε. Notice that

$$M_1^*(f;\Phi)^{\varepsilon,N}(x)\geq\frac{2^{-N}}{(1+\varepsilon|x|)^N}\sup_{0<t<1/\varepsilon}\Big(\frac{t}{t+\varepsilon}\Big)^N\sup_{|y-x|<t}\big|(\Phi_t*f)(y)\big|$$

and that the preceding expression on the right increases to

$$2^{-N}M_1^*(f;\Phi)(x)$$

as $\varepsilon\downarrow 0$. Since the constant in (6.4.32) does not depend on ε, an application of the Lebesgue monotone convergence theorem yields

$$\big\|M_1^*(f;\Phi)\big\|_{L^p}\leq 2^{N+\frac{1}{p}}C_{\Phi,N,K,n,p}\big\|M(f;\Phi)\big\|_{L^p}. \tag{6.4.33}$$

The problem with this estimate is that the finite constant $2^N C_{\Phi,N,K,n,p}$ depends on N and thus on f. However, we have managed to show that under the assumption

$\|M(f;\Phi)\|_{L^p} < \infty$, one must necessarily have $\|M_1^*(f;\Phi)\|_{L^p} < \infty$. This is a significant observation that allows us now to repeat the preceding argument from the point where the functions $U(f;\phi)^{\varepsilon,N}$ and $V(f;\phi)^{\varepsilon,N,L}$ are introduced, setting $\varepsilon = N = 0$. Since the resulting constant no longer depends on the tempered distribution f, the required conclusion follows.

(c) As usual, $B(x,R)$ denotes a ball centered at x with radius R. It follows from the definition of $M_a^*(f;\Phi)$ that

$$|(\Phi_t * f)(y)| \le M_a^*(f;\Phi)(z) \qquad \text{if } z \in B(y,at).$$

But the ball $B(y,at)$ is contained in the ball $B(x,|x-y|+at)$; hence it follows that

$$\begin{aligned}
|(\Phi_t * f)(y)|^{\frac{n}{b}} &\le \frac{1}{|B(y,at)|}\int_{B(y,at)} M_a^*(f;\Phi)(z)^{\frac{n}{b}}\,dz \\
&\le \frac{1}{|B(y,at)|}\int_{B(x,|x-y|+at)} M_a^*(f;\Phi)(z)^{\frac{n}{b}}\,dz \\
&\le \left(\frac{|x-y|+at}{at}\right)^n M\big(M_a^*(f;\Phi)^{\frac{n}{b}}\big)(x) \\
&\le \max(1,a^{-n})\left(\frac{|x-y|}{t}+1\right)^n M\big(M_a^*(f;\Phi)^{\frac{n}{b}}\big)(x),
\end{aligned}$$

from which we conclude that for all $x \in \mathbf{R}^n$ we have

$$M_b^{**}(f;\Phi)(x) \le \max(1,a^{-n})\Big\{M\big(M_a^*(f;\Phi)^{\frac{n}{b}}\big)(x)\Big\}^{\frac{b}{n}}.$$

Raising to the power p and using the fact that $p > n/b$ and the boundedness of the Hardy–Littlewood maximal operator M on $L^{pb/n}$, we obtain the required conclusion (6.4.12).

(d) In proving (d) we may replace b by the integer $b_0 = [b]+1$. Let Φ be a Schwartz function with nonvanishing integral. Multiplying Φ by a constant, we can assume that Φ has integral equal to 1. Applying Lemma 6.4.5 with $m = b_0$, we write any function φ in $\mathscr{F}_N$ as

$$\varphi(y) = \int_0^1 (\Theta^{(s)} * \Phi_s)(y)\,ds$$

for some choice of Schwartz functions $\Theta^{(s)}$. Then we have

$$\varphi_t(y) = \int_0^1 ((\Theta^{(s)})_t * \Phi_{ts})(y)\,ds$$

for all $t > 0$. Fix $x \in \mathbf{R}^n$. Then for y in $B(x,t)$ we have

$$
\begin{aligned}
|(\varphi_t * f)(y)| &\le \int_0^1 \int_{\mathbf{R}^n} |(\Theta^{(s)})_t(z)|\,|(\Phi_{ts} * f)(y-z)|\,dz\,ds \\
&\le \int_0^1 \int_{\mathbf{R}^n} |(\Theta^{(s)})_t(z)|\, M_{b_0}^{**}(f;\Phi)(x)\left(\frac{|x-(y-z)|}{st}+1\right)^{b_0} dz\,ds \\
&\le \int_0^1 s^{-b_0} \int_{\mathbf{R}^n} |(\Theta^{(s)})_t(z)|\, M_{b_0}^{**}(f;\Phi)(x)\left(\frac{|x-y|}{t}+\frac{|z|}{t}+1\right)^{b_0} dz\,ds \\
&\le 2^{b_0} M_{b_0}^{**}(f;\Phi)(x) \int_0^1 s^{-b_0} \int_{\mathbf{R}^n} |\Theta^{(s)}(w)|\,\big(|w|+1\big)^{b_0}\,dw\,ds \\
&\le 2^{b_0} M_{b_0}^{**}(f;\Phi)(x) \int_0^1 s^{-b_0} C_0(\Phi,b_0)\, s^{b_0}\, \mathfrak{N}_{b_0}(\varphi)\,ds,
\end{aligned}
$$

where we applied conclusion (6.4.16) of Lemma 6.4.5. Setting $N = b_0 = [b]+1$, we obtain for y in $B(x,t)$ and $\varphi \in \mathscr{F}_N$,

$$
|(\varphi_t * f)(y)| \le 2^{b_0} C_0(\Phi,b_0)\, M_{b_0}^{**}(f;\Phi)(x).
$$

Taking the supremum over all y in $B(x,t)$, over all $t>0$, and over all φ in $\mathscr{F}_N$, we obtain the pointwise estimate

$$
\mathscr{M}_N(f)(x) \le 2^{b_0} C_0(\Phi,b_0)\, M_{b_0}^{**}(f;\Phi)(x), \qquad x \in \mathbf{R}^n,
$$

where $N = b_0 + 1$. This clearly yields (6.4.13) if we set $C_4 = 2^{b_0} C_0(\Phi,b_0)$.

(e) We fix an $f \in \mathscr{S}'(\mathbf{R}^n)$ that satisfies $\big\|\mathscr{M}_N(f)\big\|_{L^p} < \infty$ for some fixed positive integer N. To show that f is a bounded distribution, we fix a Schwartz function φ and we observe that for some positive constant $c = c_\varphi$, we have that $c\,\varphi$ is an element of $\mathscr{F}_N$ and thus $M_1^*(f;c\,\varphi) \le \mathscr{M}_N(f)$. Then

$$
\begin{aligned}
c^p\,|(\varphi * f)(x)|^p &\le \inf_{|y-x|\le 1}\ \sup_{|z-y|\le 1} |(c\,\varphi * f)(z)|^p \\
&\le \inf_{|y-x|\le 1} M_1^*(f;c\,\varphi)(y)^p \\
&\le \frac{1}{v_n}\int_{|y-x|\le 1} M_1^*(f;c\,\varphi)(y)^p\,dy \\
&\le \frac{1}{v_n}\int_{\mathbf{R}^n} M_1^*(f;c\,\varphi)(y)^p\,dy \\
&\le \frac{1}{v_n}\int_{\mathbf{R}^n} \mathscr{M}_N(f)(y)^p\,dy < \infty,
\end{aligned}
$$

which implies that $\varphi * f$ is a bounded function. We conclude that f is a bounded distribution. We now proceed to show that f is an element of H^p. We fix a smooth function with compact support θ such that

$$
\theta(x) = \begin{cases} 1 & \text{if} \quad |x| < 1, \\ 0 & \text{if} \quad |x| > 2. \end{cases}
$$

We observe that the identity

$$\begin{aligned}
P(x) &= P(x)\theta(x)+\sum_{k=1}^{\infty}\big(\theta(2^{-k}x)P(x)-\theta(2^{-(k-1)}x)P(x)\big)\\
&= P(x)\theta(x)+\frac{\Gamma(\frac{n+1}{2})}{\pi^{\frac{n+1}{2}}}\sum_{k=1}^{\infty}2^{-k}\left(\frac{\theta(\cdot)-\theta(2(\cdot))}{(2^{-2k}+|\cdot|^2)^{\frac{n+1}{2}}}\right)_{2^k}(x)
\end{aligned}$$

is valid for all $x\in\mathbf{R}^n$. Setting

$$\Phi^{(k)}(x)=\big(\theta(x)-\theta(2x)\big)\frac{1}{(2^{-2k}+|x|^2)^{\frac{n+1}{2}}},$$

we note that for some fixed constant $c_0=c_0(n,N)$, the functions $c_0\,\theta P$ and $c_0\Phi^{(k)}$ lie in $\mathscr{F}_N$ uniformly in $k=1,2,3,\dots$. Combining this observation with the identity for $P(x)$ obtained earlier, we conclude that

$$\begin{aligned}
\sup_{t>0}|P_t*f| &\le \sup_{t>0}|(\theta P)_t*f|+\frac{1}{c_0}\frac{\Gamma(\frac{n+1}{2})}{\pi^{\frac{n+1}{2}}}\sup_{t>0}\sum_{k=1}^{\infty}2^{-k}\big|(c_0\Phi^{(k)})_{2^kt}*f\big|\\
&\le C_5(n,N)\mathscr{M}_N(f),
\end{aligned}$$

which proves the required conclusion (6.4.14).

We observe that the last estimate also yields the stronger estimate

$$M_1^*(f;P)(x)=\sup_{t>0}\sup_{\substack{y\in\mathbf{R}^n\\|y-x|\le at}}|(P_t*f)(y)|\le C_5(n,N)\mathscr{M}_N(f)(x). \tag{6.4.34}$$

It follows that the quasinorm $\big\|M_1^*(f;P)\big\|_{L^p(\mathbf{R}^n)}$ is also equivalent to $\big\|f\big\|_{H^p}$. This fact is very useful. □

Remark 6.4.6. To simplify the understanding of the equivalences just proved, a first-time reader may wish to define the H^p quasinorm of a distribution f as

$$\big\|f\big\|_{H^p}=\big\|M_1^*(f;P)\big\|_{L^p}$$

and then study only the implications (a) $\Longrightarrow$ (c), (c) $\Longrightarrow$ (d), (d) $\Longrightarrow$ (e), and (e) $\Longrightarrow$ (a) in the proof of Theorem 6.4.4. In this way one avoids passing through the statement in part (b). For many applications, the identification of $\big\|f\big\|_{H^p}$ with $\big\|M_1^*(f;\Phi)\big\|_{L^p}$ for some Schwartz function Φ (with nonvanishing integral) suffices.

We also remark that the proof of Theorem 6.4.4 yields

$$\big\|f\big\|_{H^p(\mathbf{R}^n)}\approx\big\|\mathscr{M}_N(f)\big\|_{L^p(\mathbf{R}^n)},$$

where $N=[\frac{n}{p}]+1$.

6.4.3 Consequences of the Characterizations of Hardy Spaces

In this subsection we look at a few consequences of Theorem 6.4.4. In many applications we need to be working with dense subspaces of H^p. It turns out that both $H^p \cap L^2$ and $H^p \cap L^1$ are dense in H^p.

Proposition 6.4.7. *Let $0 < p \le 1$ and let r satisfy $p \le r \le \infty$. Then $L^r \cap H^p$ is dense in H^p. Hence, $H^p \cap L^2$ and $H^p \cap L^1$ are dense in H^p.*

Proof. Let f be a distribution in $H^p(\mathbf{R}^n)$. Recall the Poisson kernel $P(x)$ and set $N = [\frac{n}{p}] + 1$. For any fixed $x \in \mathbf{R}^n$ and $t > 0$ we have

$$|(P_t * f)(x)| \le M_1^*(f;P)(y) \le C\mathscr{M}_N(f)(y) \tag{6.4.35}$$

for any $|y - x| \le t$. Indeed, the first estimate in (6.4.35) follows from the definition of $M_1^*(f;P)$, and the second estimate by (6.4.34). Raising (6.4.35) to the power p and averaging over the ball $B(x,t)$, we obtain

$$|(P_t * f)(x)|^p \le \frac{C^p}{v_n t^n} \int_{B(x,t)} \mathscr{M}_N(f)(y)^p \, dy \le \frac{C_1^p}{t^n} \|f\|_{H^p}^p .$$

It follows that the function $P_t * f$ is in $L^\infty(\mathbf{R}^n)$ with norm at most a constant multiple of $t^{-n/p}\|f\|_{H^p}$. Moreover, this function is also in $L^p(\mathbf{R}^n)$, since it is controlled by $M(f;P)$. Therefore, the functions $P_t * f$ lie in $L^r(\mathbf{R}^n)$ for all $r \le p \le \infty$. It remains to show that $P_t * f$ also lie in H^p and that $P_t * f \to f$ in H^p as $t \to 0$.

To see that $P_t * f$ lies in H^p, we use the semigroup formula $P_t * P_s = P_{t+s}$ for the Poisson kernel, which is a consequence of the fact that $\widehat{P_t}(\xi) = e^{-2\pi t|\xi|}$ by applying the Fourier transform. Therefore, for any $t > 0$ we have

$$\sup_{s>0} |P_s * P_t * f| = \sup_{s>0} |P_{s+t} * f| \le \sup_{s>0} |P_s * f| ,$$

which implies that

$$\big\|P_t * f\big\|_{H^p} \le \big\|f\big\|_{H^p}$$

for all $t > 0$. We now need to show that $P_t * f \to f$ in H^p as $t \to 0$. This will be a consequence of the Lebesgue dominated convergence theorem once we know that

$$\sup_{s>0} |(P_s * P_t * f - P_s * f)(x)| \to 0 \qquad \text{as} \quad t \to 0 \tag{6.4.36}$$

pointwise for all $x \in \mathbf{R}^n$ and also

$$\sup_{s>0} |P_s * P_t * f - P_s * f| \le 2 \sup_{s>0} |P_s * f| \in L^p(\mathbf{R}^n) . \tag{6.4.37}$$

Statement (6.4.37) is a trivial consequence of the Poisson semigroup formula. As far as (6.4.36) is concerned, we note that for all $x \in \mathbf{R}^n$ the function

$$s \mapsto |(P_s * P_t * f)(x) - (P_s * f)(x)| = |(P_{s+t} * f)(x) - (P_s * f)(x)|$$

is bounded by a constant multiple of $s^{-n/p}$ and therefore tends to zero as $s \to \infty$. Given any $\varepsilon > 0$, there exists an $M > 0$ such that for all $t > 0$ we have

$$\sup_{s>M} |(P_s * P_t * f - P_s * f)(x)| < \frac{\varepsilon}{2}. \tag{6.4.38}$$

Moreover, the function $t \mapsto \sup_{0 \le s \le M} |(P_s * P_t * f - P_s * f)(x)|$ is continuous in t. Therefore, there exists a $t_0 > 0$ such that for $t < t_0$ we have

$$\sup_{0 \le s \le M} |(P_s * P_t * f - P_s * f)(x)| < \frac{\varepsilon}{2}. \tag{6.4.39}$$

Combining (6.4.38) and (6.4.39) proves (6.4.36). □

Next we observe the following consequence of Theorem 6.4.4.

Corollary 6.4.8. *For any two Schwartz functions Φ and Θ with nonvanishing integral we have*

$$\big\| \sup_{t>0} |\Theta_t * f| \big\|_{L^p} \approx \big\| \sup_{t>0} |\Phi_t * f| \big\|_{L^p} \approx \big\| f \big\|_{H^p}$$

for all $f \in \mathscr{S}'(\mathbf{R}^n)$, with constants depending only on n, p, Φ, and Θ.

Proof. See the discussion after Theorem 6.4.4. □

Next we define a *norm* on Schwartz functions relevant in the theory of Hardy spaces:

$$\mathfrak{N}_N(\varphi; x_0, R) = \int_{\mathbf{R}^n} \left(1 + \left|\frac{x - x_0}{R}\right|\right)^N \sum_{|\alpha| \le N+1} R^{|\alpha|} |\partial^\alpha \varphi(x)| \, dx.$$

Note that $\mathfrak{N}_N(\varphi; 0, 1) = \mathfrak{N}_N(\varphi)$.

Corollary 6.4.9. *(a) For any $0 < p \le 1$, any $f \in H^p(\mathbf{R}^n)$, and any $\varphi \in \mathscr{S}(\mathbf{R}^n)$ we have*

$$\big|\langle f, \varphi \rangle\big| \le \mathfrak{N}_N(\varphi) \inf_{|z| \le 1} \mathscr{M}_N(f)(z), \tag{6.4.40}$$

where $N = \left[\frac{n}{p}\right] + 1$. More generally, for any $x_0 \in \mathbf{R}^n$ and $R > 0$ we have

$$\big|\langle f, \varphi \rangle\big| \le \mathfrak{N}_N(\varphi; x_0, R) \inf_{|z - x_0| \le R} \mathscr{M}_N(f)(z). \tag{6.4.41}$$

(b) Let $0 < p \le 1$ and $p \le r \le \infty$. For any $f \in H^p$ we have the estimate

$$\big\| \varphi * f \big\|_{L^r} \le C(p,n) \mathfrak{N}_N(\varphi) \big\| f \big\|_{H^p},$$

where $N = [n/p] + 1$.

Proof. (a) Set $\psi(x) = \varphi(-Rx + x_0)$. It follows directly from Definition 6.4.1 that for any fixed z with $|z - x_0| \le R$ we have

$$\begin{aligned}
\big|\langle f,\varphi\rangle\big| &= R^n|(f*\psi_R)(x_0)| \\
&\le \sup_{y:\,|y-z|\le R} R^n|(f*\psi_R)(y)| \\
&\le R^n\bigg[\int_{\mathbf{R}^n}(1+|w|)^N\sum_{|\alpha|\le N+1}|\partial^\alpha\psi(w)|\,dw\bigg]\mathscr{M}_N(f)(z),
\end{aligned}$$

from which the second assertion in the corollary follows easily by the change of variables $x=-Rw+x_0$. Taking the infimum over all z with $|z-x_0|\le R$ yields the required conclusion.

(b) For any fixed $x\in\mathbf{R}^n$ and $t>0$ we have

$$|(\varphi*f)(x)|\le\mathfrak{N}_N(\varphi)M_1^*\Big(f;\frac{\varphi}{\mathfrak{N}_N(\varphi)}\Big)(y)\le\mathfrak{N}_N(\varphi)\mathscr{M}_N(f)(y) \tag{6.4.42}$$

for all y satisfying $|y-x|\le 1$. Hence

$$|(\varphi*f)(x)|^p\le\frac{\mathfrak{N}_N(\varphi)^p}{|B(x,1)|}\int_{B(x,1)}\mathscr{M}_N(f)^p(y)\,dy\le\mathfrak{N}_N(\varphi)^pC_{p,n}^p\big\|f\big\|_{H^p}^p.$$

This implies that $\big\|\varphi*f\big\|_{L^\infty}\le C_{p,n}\mathfrak{N}_N(\varphi)\big\|f\big\|_{H^p}$. Choosing $y=x$ in (6.4.42) and then taking L^p quasinorms yields a similar estimate for $\big\|\varphi*f\big\|_{L^p}$. By interpolation we deduce $\big\|\varphi*f\big\|_{L^r}\le\mathfrak{N}_N(\varphi)\big\|f\big\|_{H^p}$. □

Proposition 6.4.10. *Let $0<p\le 1$. Then the following statements are valid:*
(a) Convergence in H^p implies convergence in $\mathscr{S}'$.
(b) H^p is a complete quasinormed metrizable space.

Proof. Part (a) says that if a sequence f_j tends to f in $H^p(\mathbf{R}^n)$, then $f_j\to f$ in $\mathscr{S}'(\mathbf{R}^n)$. But this easily follows from the estimate

$$\big|\langle f,\varphi\rangle\big|\le C_\varphi\inf_{|z|\le 1}\mathscr{M}_N(f)(z)\le\frac{C_\varphi}{v_n}\int_{\mathbf{R}^n}\mathscr{M}_N(f)^p\,dz\le C_\varphi C_{n,p}\big\|f\big\|_{H^p}^p,$$

which is a direct consequence of (6.4.40) for all φ in $\mathscr{S}(\mathbf{R}^n)$. As before, here $N=[\frac{n}{p}]+1$.

To obtain the statement in (b), we first observe that the map $(f,g)\mapsto\big\|f-g\big\|_{H^p}^p$ is a metric on H^p that generates the same topology as the quasinorm $f\mapsto\big\|f\big\|_{H^p}$. To show that H^p is a complete space, it suffices to show that for any sequence of functions f_j that satisfies

$$\sum_j\int_{\mathbf{R}^n}\mathscr{M}_N(f_j)^p\,dx<\infty,$$

the series $\sum_j f_j$ converges in $H^p(\mathbf{R}^n)$. The partial sums of this series are Cauchy in $H^p(\mathbf{R}^n)$ and therefore are Cauchy in $\mathscr{S}'(\mathbf{R}^n)$ by part (a). It follows that the sequence $\sum_{-k}^k f_j$ converges to some tempered distribution f in $\mathscr{S}'(\mathbf{R}^n)$. Sublinearity gives

$$\int_{\mathbf{R}^n} \mathscr{M}_N(f)^p\,dx = \int_{\mathbf{R}^n} \mathscr{M}_N\Big(\sum_j f_j\Big)^p dx \le \sum_j \int_{\mathbf{R}^n} \mathscr{M}_N(f_j)^p\,dx < \infty,$$

which implies that $f \in H^p$. Finally,

$$\int_{\mathbf{R}^n} \mathscr{M}_N\Big(f - \sum_{j=-k}^{k} f_j\Big)^p dx \le \sum_{|j|\ge k+1} \int_{\mathbf{R}^n} \mathscr{M}_N(f_j)^p\,dx \to 0$$

as $k \to \infty$; thus the series converges in H^p. □

6.4.4 Vector-Valued H^p and Its Characterizations

We now obtain a vector-valued analogue of Theorem 6.4.4 crucial in the characterization of Hardy spaces using Littlewood–Paley theory. To state this analogue we need to extend the definitions of the maximal operators to sequences of distributions. Let $a, b > 0$ and let Φ be a Schwartz function on $\mathbf{R}^n$. In accordance with Definition 6.4.1, we give the following sequence of definitions.

Definition 6.4.11. For a sequence $\vec{f} = \{f_j\}_{j\in\mathbf{Z}}$ of tempered distributions on $\mathbf{R}^n$ we define the *smooth maximal function of $\vec{f}$ with respect to Φ* as

$$M(\vec{f};\Phi)(x) = \sup_{t>0} \big\|\{(\Phi_t * f_j)(x)\}_j\big\|_{\ell^2}.$$

We define the *nontangential maximal function (with aperture a) of f with respect to Φ* as

$$M_a^*(\vec{f};\Phi)(x) = \sup_{t>0} \sup_{\substack{y\in\mathbf{R}^n \\ |y-x|\le at}} \big\|\{(\Phi_t * f_j)(y)\}_j\big\|_{\ell^2}.$$

We also define the *auxiliary maximal function*

$$M_b^{**}(\vec{f};\Phi)(x) = \sup_{t>0} \sup_{y\in\mathbf{R}^n} \frac{\big\|\{(\Phi_t * f_j)(x-y)\}_j\big\|_{\ell^2}}{(1+t^{-1}|y|)^b}.$$

We note that if the function Φ is not assumed to be Schwartz but merely integrable, for example, if Φ is the Poisson kernel, the maximal functions $M(\vec{f};\Phi)$, $M_a^*(\vec{f};\Phi)$, and $M_b^{**}(\vec{f};\Phi)$ are well defined for sequences $\vec{f} = \{f_j\}_j$ whose terms are bounded tempered distributions on $\mathbf{R}^n$.

For a fixed positive integer N we define the *grand maximal function of $\vec{f}$ (with respect to N)* as

$$\mathscr{M}_N(\vec{f}) = \sup_{\varphi\in\mathscr{F}_N} M_1^*(\vec{f};\varphi), \tag{6.4.43}$$

where

$$\mathscr{F}_N = \Big\{\varphi \in \mathscr{S}(\mathbf{R}^n) : \mathfrak{N}_N(\varphi) \le 1\Big\}$$

is as defined in (6.4.5).

We note that as in the scalar case, we have the sequence of simple inequalities

$$M(\vec{f};\Phi) \le M_a^*(\vec{f};\Phi) \le (1+a)^b M_b^{**}(\vec{f};\Phi). \tag{6.4.44}$$

We now define the vector-valued Hardy space $H^p(\mathbf{R}^n,\ell^2)$.

Definition 6.4.12. Let $\vec{f} = \{f_j\}_j$ be a sequence of bounded tempered distributions on $\mathbf{R}^n$ and let $0 < p < \infty$. We say that $\vec{f}$ lies in the vector-valued Hardy space $H^p(\mathbf{R}^n,\ell^2)$ if the *Poisson maximal function*

$$M(\vec{f};P)(x) = \sup_{t>0} \big\|\{(P_t * f_j)(x)\}_j\big\|_{\ell^2}$$

lies in $L^p(\mathbf{R}^n)$. If this is the case, we set

$$\big\|\vec{f}\,\big\|_{H^p(\mathbf{R}^n,\ell^2)} = \big\|M(\vec{f};P)\big\|_{L^p(\mathbf{R}^n)} = \bigg\|\sup_{\varepsilon>0}\Big(\sum_j |f_j * P_\varepsilon|^2\Big)^{\frac{1}{2}}\bigg\|_{L^p(\mathbf{R}^n)}.$$

The next theorem provides a vector-valued analogue of Theorem 6.4.4.

Theorem 6.4.13. *Let $0 < p < \infty$. Then the following statements are valid:*
(a) There exists a Schwartz function Φ with $\int_{\mathbf{R}^n}\Phi(x)\,dx \neq 0$ and a constant C_1 (which does not depend on any parameters) such that

$$\big\|M(\vec{f};\Phi)\big\|_{L^p(\mathbf{R}^n,\ell^2)} \le C_1 \big\|\vec{f}\,\big\|_{H^p(\mathbf{R}^n,\ell^2)} \tag{6.4.45}$$

for every sequence $\vec{f} = \{f_j\}_j$ of tempered distributions.
(b) For every $a > 0$ and Φ in $\mathscr{S}(\mathbf{R}^n)$ there exists a constant $C_2(n,p,a,\Phi)$ such that

$$\big\|M_a^*(\vec{f};\Phi)\big\|_{L^p(\mathbf{R}^n,\ell^2)} \le C_2(n,p,a,\Phi)\big\|M(\vec{f};\Phi)\big\|_{L^p(\mathbf{R}^n,\ell^2)} \tag{6.4.46}$$

for every sequence $\vec{f} = \{f_j\}_j$ of tempered distributions.
(c) For every $a > 0$, $b > n/p$, and Φ in $\mathscr{S}(\mathbf{R}^n)$ there exists a constant $C_3(n,p,a,b,\Phi)$ such that

$$\big\|M_b^{**}(\vec{f};\Phi)\big\|_{L^p(\mathbf{R}^n,\ell^2)} \le C_3(n,p,a,b,\Phi)\big\|M_a^*(\vec{f};\Phi)\big\|_{L^p(\mathbf{R}^n,\ell^2)} \tag{6.4.47}$$

for every sequence $\vec{f} = \{f_j\}_j$ of tempered distributions.
(d) For every $b > 0$ and Φ in $\mathscr{S}(\mathbf{R}^n)$ with $\int_{\mathbf{R}^n}\Phi(x)\,dx \neq 0$ there exists a constant $C_4(b,\Phi)$ such that if $N = [\frac{n}{p}]+1$ we have

$$\big\|\mathscr{M}_N(\vec{f})\big\|_{L^p(\mathbf{R}^n,\ell^2)} \le C_4(b,\Phi)\big\|M_b^{**}(\vec{f};\Phi)\big\|_{L^p(\mathbf{R}^n,\ell^2)} \tag{6.4.48}$$

for every sequence $\vec{f} = \{f_j\}_j$ of tempered distributions.
(e) For every positive integer N there exists a constant $C_5(n,N)$ such that every

sequence $\vec{f} = \{f_j\}_j$ *of tempered distributions that satisfies* $\big\|\mathcal{M}_N(\vec{f})\big\|_{L^p(\mathbf{R}^n,\ell^2)} < \infty$ *consists of bounded distributions and satisfies*

$$\big\|\vec{f}\,\big\|_{H^p(\mathbf{R}^n,\ell^2)} \le C_5(n,N)\big\|\mathcal{M}_N(\vec{f})\big\|_{L^p(\mathbf{R}^n,\ell^2)}, \tag{6.4.49}$$

that is, it lies in the Hardy space $H^p(\mathbf{R}^n,\ell^2)$.

Proof. The proof of this theorem is obtained via a step-by-step repetition of the proof of Theorem 6.4.4 in which the scalar absolute values are replaced by ℓ^2 norms. This is small notational change in our point of view but yields a significant improvement of the scalar version of the theorem. Moreover, this perspective provides an example of the power of Hilbert space techniques. The verification of the details of this step-by-step repetition of the proof of Theorem 6.4.4 are left to the reader. □

We end this subsection by observing the validity of the following vector-valued analogue of (6.4.41):

$$\Big(\sum_j |\langle f_j,\varphi\rangle|^2\Big)^{\frac{1}{2}} \le \mathfrak{N}_N(\varphi;x_0,R)\inf_{|z-x_0|\le R}\mathcal{M}_N(\vec{f})(z). \tag{6.4.50}$$

The proof of (6.4.50) is identical to the corresponding estimate for scalar-valued functions. Set $\psi(x) = \varphi(-Rx+x_0)$. It follows directly from Definition 6.4.11 that for any fixed z with $|z-x_0| \le R$ we have

$$\begin{aligned}\Big(\sum_j |\langle f_j,\varphi\rangle|^2\Big)^{\frac{1}{2}} &= R^n\big\|\{(f_j*\psi_R)(x_0)\}_j\big\|_{\ell^2}\\ &\le \sup_{y:\,|y-z|\le R} R^n\big\|\{(f_j*\psi_R)(y)\}_j\big\|_{\ell^2}\\ &\le R^n\mathfrak{N}_N(\psi)\mathcal{M}_N(\vec{f})(z),\end{aligned}$$

which, combined with the observation

$$R^n\mathfrak{N}_N(\psi) = \mathfrak{N}_N(\varphi;x_0,R),$$

yields the required conclusion by taking the infimum over all z with $|z-x_0| \le R$.

6.4.5 Singular Integrals on Hardy Spaces

To obtain the Littlewood–Paley characterization of Hardy spaces, we need a multiplier theorem for vector-valued Hardy spaces.

Suppose that $K_j(x)$ is a family of functions defined on $\mathbf{R}^n\setminus\{0\}$ that satisfies the following: There exist constants $A,B<\infty$ and an integer N such that for all multi-indices α with $|\alpha| \le N$ we have

$$\Big|\sum_{j\in\mathbf{Z}}\partial^\alpha K_j(x)\Big|\le A\,|x|^{-n-|\alpha|}<\infty \tag{6.4.51}$$

and also

$$\sup_{\xi\in\mathbf{R}^n}\Big|\sum_{j\in\mathbf{Z}}\widehat{K_j}(\xi)\Big|\le B<\infty. \tag{6.4.52}$$

Theorem 6.4.14. *Suppose that a sequence of kernels $\{K_j\}_j$ satisfies (6.4.51) and (6.4.52) with $N=[\frac{n}{p}]+1$, for some $0<p\le 1$. Then there exists a constant $C_{n,p}$ that depends only on the dimension n and on p such that for all sequences of tempered distributions $\{f_j\}_j$ we have the estimate*

$$\Big\|\sum_j K_j*f_j\Big\|_{H^p(\mathbf{R}^n)}\le C_{n,p}(A+B)\big\|\{f_j\}_j\big\|_{H^p(\mathbf{R}^n,\ell^2)}\,.$$

Proof. We fix a smooth positive function Φ supported in the unit ball $B(0,1)$ with $\int_{\mathbf{R}^n}\Phi(x)\,dx=1$ and we consider the sequence of smooth maximal functions

$$M\Big(\sum_j K_j*f_j;\Phi\Big)=\sup_{\varepsilon>0}\Big|\Phi_\varepsilon*\sum_j K_j*f_j\Big|\,,$$

which will be shown to be an element of $L^p(\mathbf{R}^n,\ell^2)$. We work with a fixed sequence of integrable functions $\vec{f}=\{f_j\}_j$, since such functions are dense in $L^p(\mathbf{R}^n,\ell^2)$ in view of Proposition 6.4.7.

We now fix a $\lambda>0$ and we set $N=[\frac{n}{p}]+1$. We also fix $\gamma>0$ to be chosen later and we define the set

$$\Omega_\lambda=\{x\in\mathbf{R}^n:\ \mathscr{M}_N(\vec{f})(x)>\gamma\lambda\}.$$

The set Ω_λ is open, and we may use the Whitney decomposition (Appendix J) to write it is a union of cubes Q_k such that

(a) $\bigcup_k Q_k=\Omega_\lambda$ and the Q_k's have disjoint interiors;

(b) $\sqrt{n}\,\ell(Q_k)\le\operatorname{dist}(Q_k,(\Omega_\lambda)^c)\le 4\sqrt{n}\,\ell(Q_k)$.

We denote by $c(Q_k)$ the center of the cube Q_k. For each k we set

$$d_k=\operatorname{dist}(Q_k,(\Omega_\lambda)^c)+2\sqrt{n}\,\ell(Q_k)\approx\ell(Q_k)\,,$$

so that

$$B(c(Q_k),d_k)\cap(\Omega_\lambda)^c\neq\emptyset\,.$$

We now introduce a partition of unity $\{\varphi_k\}_k$ adapted to the sequence of cubes $\{Q_k\}_k$ such that

(c) $\chi_{\Omega_\lambda}=\sum_k\varphi_k$ and each φ_k satisfies $0\le\varphi_k\le 1$;

(d) each φ_k is supported in $\frac{6}{5}Q_k$ and satisfies $\int_{\mathbf{R}^n}\varphi_k\,dx\approx d_k^n$;

(e) $\|\partial^\alpha \varphi_k\|_{L^\infty} \le C_\alpha d_k^{-|\alpha|}$ for all multi-indices α and some constants C_α.

We decompose each f_j as

$$f_j = g_j + \sum_k b_{j,k},$$

where g_j is the *good function* of the decomposition given by

$$g_j = f_j \chi_{\mathbf{R}^n \setminus \Omega_\lambda} + \sum_k \frac{\int_{\mathbf{R}^n} f_j \varphi_k \, dx}{\int_{\mathbf{R}^n} \varphi_k \, dx} \varphi_k$$

and $b_j = \sum_k b_{j,k}$ is the *bad function* of the decomposition given by

$$b_{j,k} = \left(f_j - \frac{\int_{\mathbf{R}^n} f_j \varphi_k \, dx}{\int_{\mathbf{R}^n} \varphi_k \, dx} \right) \varphi_k .$$

We note that each $b_{j,k}$ has integral zero. We define $\vec{g} = \{g_j\}_j$ and $\vec{b} = \{b_j\}_j$. At this point we appeal to (6.4.50) and to properties (d) and (e) to obtain

$$\left(\sum_j \left| \frac{\int_{\mathbf{R}^n} f_j \varphi_k \, dx}{\int_{\mathbf{R}^n} \varphi_k \, dx} \right|^2 \right)^{\frac{1}{2}} \le \frac{\mathfrak{N}_N(\varphi_k; c(Q_k), d_k)}{\int_{\mathbf{R}^n} \varphi_k \, dx} \inf_{|z - c(Q_k)| \le d_k} \mathscr{M}_N(\vec{f})(z) . \tag{6.4.53}$$

But since

$$\frac{\mathfrak{N}_N(\varphi_k; c(Q_k), d_k)}{\int_{\mathbf{R}^n} \varphi_k \, dx} \le \left[\int_{Q_k} \left(1 + \frac{|x - c(Q_k)|}{d_k} \right)^N \sum_{|\alpha| \le N+1} \frac{d_k^{|\alpha|} C_\alpha d_k^{-|\alpha|}}{\int_{\mathbf{R}^n} \varphi_k \, dx} dx \right] \le C_{N,n} ,$$

it follows that (6.4.53) is at most a constant multiple of λ, since the ball $B(c(Q_k), d_k)$ meets the complement of Ω_λ. We conclude that

$$\|\vec{g}\|_{L^\infty(\Omega_\lambda, \ell^2)} \le C_{N,n} \gamma \lambda . \tag{6.4.54}$$

We now turn to estimating $M(\sum_j K_j * b_{j,k}; \Phi)$. For fixed k and $\varepsilon > 0$ we have

$$\begin{aligned}
&\Big(\Phi_\varepsilon * \sum_j K_j * b_{j,k}\Big)(x) \\
&= \int_{\mathbf{R}^n} \Phi_\varepsilon * \sum_j K_j(x-y) \left[f_j(y)\varphi_k(y) - \frac{\int_{\mathbf{R}^n} f_j \varphi_k \, dx}{\int_{\mathbf{R}^n} \varphi_k \, dx} \varphi_k(y) \right] dy \\
&= \int_{\mathbf{R}^n} \sum_j \left\{ (\Phi_\varepsilon * K_j)(x-z) - \int_{\mathbf{R}^n} (\Phi_\varepsilon * K_j)(x-y) \frac{\varphi_k(y)}{\int_{\mathbf{R}^n} \varphi_k \, dx} dy \right\} \varphi_k(z) f_j(z) \, dz \\
&= \int_{\mathbf{R}^n} \sum_j R_{j,k}(x,z) \varphi_k(z) f_j(z) \, dz ,
\end{aligned}$$

where we set $R_{j,k}(x,z)$ for the expression inside the curly brackets. Using (6.4.41), we obtain

$$\left|\int_{\mathbf{R}^n}\sum_j R_{j,k}(x,z)\varphi_k(z)f_j(z)\,dz\right|$$
$$\leq \sum_j \mathfrak{N}_N(R_{j,k}(x,\cdot)\varphi_k;c(Q_k),d_k)\inf_{|z-c(Q_k)|\leq d_k}\mathscr{M}_N(f_j)(z) \tag{6.4.55}$$
$$\leq \sum_j \mathfrak{N}_N(R_{j,k}(x,\cdot)\varphi_k;c(Q_k),d_k)\inf_{|z-c(Q_k)|\leq d_k}\mathscr{M}_N(\vec{f})(z).$$

Since $\varphi_k(z)$ is supported in $\frac{6}{5}Q_k$, the term $(1+\frac{|z-c(Q_k)|}{d_k})^N$ contributes only a constant factor in the integral defining $\mathfrak{N}_N(R_{j,k}(x,\cdot)\varphi_k;c(Q_k),d_k)$, and we obtain

$$\mathfrak{N}_N(R_{j,k}(x,\cdot)\varphi_k;c(Q_k),d_k)$$
$$\leq C_{N,n}\int_{\frac{6}{5}Q_k}\sum_{|\alpha|\leq N+1} d_k^{|\alpha|+n}\left|\frac{\partial^\alpha}{\partial z^\alpha}\big(R_{j,k}(x,z)\varphi_k(z)\big)\right|dz. \tag{6.4.56}$$

For notational convenience we set $K_j^\varepsilon = \Phi_\varepsilon * K_j$. We observe that the family $\{K_j^\varepsilon\}_j$ satisfies (6.4.51) and (6.4.52) with constants A' and B' that are only multiples of A and B, respectively, uniformly in ε. We now obtain a pointwise estimate for $\mathfrak{N}_N(R_{j,k}(x,\cdot)\varphi_k;c(Q_k),d_k)$ when $x\in\mathbf{R}^n\setminus\Omega_\lambda$. We have

$$R_{j,k}(x,z)\varphi_k(z) = \int_{\mathbf{R}^n}\varphi_k(z)\Big\{K_j^\varepsilon(x-z)-K_j^\varepsilon(x-y)\Big\}\frac{\varphi_k(y)\,dy}{\int_{\mathbf{R}^n}\varphi_k\,dx},$$

from which it follows that

$$\left|\frac{\partial^\alpha}{\partial z^\alpha}R_{j,k}(x,z)\varphi_k(z)\right| \leq \int_{\mathbf{R}^n}\left|\frac{\partial^\alpha}{\partial z^\alpha}\Big\{\varphi_k(z)\Big[K_j^\varepsilon(x-z)-K_j^\varepsilon(x-y)\Big]\Big\}\right|\frac{\varphi_k(y)\,dy}{\int_{\mathbf{R}^n}\varphi_k\,dx}.$$

Using hypothesis (6.4.51), we can now easily obtain the estimate

$$\sum_j\left|\frac{\partial^\alpha}{\partial z^\alpha}\Big\{\varphi_k(z)\Big\{K_{i,j}^\varepsilon(x-z)-K_{i,j}^\varepsilon(x-y)\Big\}\Big\}\right| \leq C_{N,n}A\frac{d_k d_k^{-|\alpha|}}{|x-c(Q_k)|^{n+1}}$$

for all $|\alpha|\leq N$ and for $x\in\mathbf{R}^n\setminus\Omega_\lambda$, since for such x we have $|x-c(Q_k)|\geq c_n d_k$. It follows that

$$d_k^{|\alpha|+n}\sum_j\left|\frac{\partial^\alpha}{\partial z^\alpha}\{R_{j,k}(x,z)\varphi_k(z)\}\right| \leq C_{N,n}A\,d_k^n\left(\frac{d_k}{|x-c(Q_k)|^{n+1}}\right).$$

Inserting this estimate in the summation of (6.4.56) over all j yields

$$\sum_j \mathfrak{N}_N(R_{j,k}(x,\cdot)\varphi_k;c(Q_k),d_k) \leq C_{N,n}A\left(\frac{d_k^{n+1}}{|x-c(Q_k)|^{n+1}}\right). \tag{6.4.57}$$

Combining (6.4.57) with (6.4.55) gives for $x\in\mathbf{R}^n\setminus\Omega_\lambda$,

$$\sum_j \left| \int_{\mathbf{R}^n} R_{i,j,k}(x,z)\varphi_k(z) f_j(z)\,dz \right| \le \frac{C_{N,n} A\, d_k^{n+1}}{|x-c(Q_k)|^{n+1}} \inf_{|z-c(Q_k)|\le d_k} \mathscr{M}_N(\vec{f})(z).$$

This provides the estimate

$$\sup_{\varepsilon>0} \Big| \sum_j (K_j^\varepsilon * b_{j,k})(x) \Big| \le \frac{C_{N,n} A\, d_k^{n+1}}{|x-c(Q_k)|^{n+1}} \gamma\lambda$$

for all $x \in \mathbf{R}^n \setminus \Omega_\lambda$, since the ball $B(c(Q_k), d_k)$ intersects $(\Omega_\lambda)^c$. Summing over k results in

$$M\Big(\sum_j K_j * b_j; \Phi\Big)(x) \le \sum_k \frac{C_{N,n} A\, \gamma\lambda\, d_k^{n+1}}{|x-c(Q_k)|^{n+1}} \le \sum_k \frac{C_{N,n} A\, \gamma\lambda\, d_k^{n+1}}{(d_k + |x-c(Q_k)|)^{n+1}}$$

for all $x \in (\Omega_\lambda)^c$. The last sum is known as the *Marcinkiewicz function*. It is a simple fact that

$$\int_{\mathbf{R}^n} \sum_k \frac{d_k^{n+1}}{(d_k + |x-c(Q_k)|)^{n+1}}\, dx \le C_n \sum_k |Q_k| = C_n\, |\Omega_\lambda|;$$

see Exercise 4.6.6. We have therefore shown that

$$\int_{\mathbf{R}^n} M(\vec{K} * \vec{b}; \Phi)(x)\, dx \le C_{N,n} A \gamma\lambda\, |\Omega_\lambda|, \tag{6.4.58}$$

where we used the notation $\vec{K} * \vec{b} = \sum_j K_j * b_j$.

We now combine the information we have acquired so far. First we have

$$\big|\{M(\vec{K} * \vec{f}; \Phi) > \lambda\}\big| \le \big|\{M(\vec{K} * \vec{g}; \Phi) > \tfrac{\lambda}{2}\}\big| + \big|\{M(\vec{K} * \vec{b}; \Phi) > \tfrac{\lambda}{2}\}\big|.$$

For the good function $\vec{g}$ we have the estimate

$$\begin{aligned}
\big|\{M(\vec{K} * \vec{g}; \Phi) > \tfrac{\lambda}{2}\}\big| &\le \frac{4}{\lambda^2} \int_{\mathbf{R}^n} M(\vec{K} * \vec{g}; \Phi)(x)^2\, dx \\
&\le \frac{4}{\lambda^2} \sum_j \int_{\mathbf{R}^n} M(K_j * g_j)(x)^2\, dx \\
&\le \frac{C_n B^2}{\lambda^2} \int_{\mathbf{R}^n} \sum_j |g_j(x)|^2\, dx \\
&\le \frac{C_n B^2}{\lambda^2} \int_{\Omega_\lambda} \sum_j |g_j(x)|^2\, dx + \frac{C_n B^2}{\lambda^2} \int_{(\Omega_\lambda)^c} \sum_j |f_j(x)|^2\, dx \\
&\le B^2 C_{N,n} \gamma^2\, |\Omega_\lambda| + \frac{C_n B^2}{\lambda^2} \int_{(\Omega_\lambda)^c} \mathscr{M}_N(\vec{f})(x)^2\, dx,
\end{aligned}$$

where we used Corollary 2.1.12, the L^2 boundedness of the Hardy–Littlewood maximal operator, hypothesis (6.4.52), the fact that $f_j = g_j$ on $(\Omega_\lambda)^c$, estimate (6.4.54), and the fact that $\|\vec{f}\|_{\ell^2} \le \mathscr{M}_N(\vec{f})$ in the sequence of estimates.

On the other hand, estimate (6.4.58) and Chebyshev's inequality gives

$$\big|\{M(\vec{K}*\vec{b};\Phi) > \tfrac{\lambda}{2}\}\big| \le C_{N,n} A\,\gamma |\Omega_\lambda|,$$

which, combined with the previously obtained estimate for $\vec{g}$, gives

$$\big|\{M(\vec{K}*\vec{f};\Phi) > \lambda\}\big| \le C_{N,n}(A\gamma + B^2\gamma^2)\,|\Omega_\lambda| + \frac{C_n B^2}{\lambda^2}\int_{(\Omega_\lambda)^c} \mathscr{M}_N(\vec{f})(x)^2\,dx.$$

Multiplying this estimate by $p\lambda^{p-1}$, recalling that $\Omega_\lambda = \{\mathscr{M}_N(\vec{f}) > \gamma\lambda\}$, and integrating in λ from 0 to ∞, we can easily obtain

$$\big\|M(\vec{K}*\vec{f};\Phi)\big\|^p_{L^p(\mathbf{R}^n,\ell^2)} \le C_{N,n}(A\gamma^{1-p} + B^2\gamma^{2-p})\big\|\mathscr{M}_N(\vec{f})\big\|^p_{L^p(\mathbf{R}^n,\ell^2)}. \tag{6.4.59}$$

Choosing $\gamma = (A+B)^{-1}$ and recalling that $N = [\frac{n}{p}]+1$ gives the required conclusion for some constant $C_{n,p}$ that depends only on n and p.

Finally, use density to extend this estimate to all $\vec{f}$ in $H^p(\mathbf{R}^n,\ell^2)$. □

6.4.6 The Littlewood–Paley Characterization of Hardy Spaces

We discuss an important characterization of Hardy spaces in terms of Littlewood–Paley square functions. The vector-valued Hardy spaces and the action of singular integrals on them are crucial tools in obtaining this characterization.

We first set up the notation. We fix a radial Schwartz function Ψ on $\mathbf{R}^n$ whose Fourier transform is nonnegative, supported in the annulus $\frac{1}{2}+\frac{1}{10} \le |\xi| \le 2-\frac{1}{10}$, and satisfies

$$\sum_{j\in\mathbf{Z}} \widehat{\Psi}(2^{-j}\xi) = 1 \tag{6.4.60}$$

for all $\xi \neq 0$. Associated with this bump, we define the Littlewood–Paley operators Δ_j given by multiplication on the Fourier transform side by the function $\widehat{\Psi}(2^{-j}\xi)$, that is,

$$\Delta_j(f) = \Delta_j^\Psi(f) = \Psi_{2^{-j}} * f. \tag{6.4.61}$$

We have the following.

Theorem 6.4.15. *Let Ψ be a radial Schwartz function on $\mathbf{R}^n$ whose Fourier transform is nonnegative, supported in $\frac{1}{2}+\frac{1}{10} \le |\xi| \le 2-\frac{1}{10}$, and satisfies* (6.4.60)*. Let Δ_j be the Littlewood–Paley operators associated with Ψ and let $0 < p \le 1$. Then there exists a constant $C = C_{n,p,\Psi}$ such that for all $f \in H^p(\mathbf{R}^n)$ we have*

$$\Big\| \Big(\sum_{j \in \mathbf{Z}} |\Delta_j(f)|^2 \Big)^{\frac{1}{2}} \Big\|_{L^p} \le C \big\| f \big\|_{H^p} . \tag{6.4.62}$$

Conversely, suppose that a tempered distribution f satisfies

$$\Big\| \Big(\sum_{j \in \mathbf{Z}} |\Delta_j(f)|^2 \Big)^{\frac{1}{2}} \Big\|_{L^p} < \infty . \tag{6.4.63}$$

Then there exists a unique polynomial $Q(x)$ such that $f - Q$ lies in the Hardy space H^p and satisfies the estimate

$$\frac{1}{C} \big\| f - Q \big\|_{H^p} \le \Big\| \Big(\sum_{j \in \mathbf{Z}} |\Delta_j(f)|^2 \Big)^{\frac{1}{2}} \Big\|_{L^p} . \tag{6.4.64}$$

Proof. We fix $\Phi \in \mathscr{S}(\mathbf{R}^n)$ with integral equal to 1 and we take $f \in H^p \cap L^1$ and M in $\mathbf{Z}^+$. Let r_j be the Rademacher functions, introduced in Appendix C.1, reindexed so that their index set is the set of all integers (not the set of nonnegative integers). We begin with the estimate

$$\Big| \sum_{j=-M}^{M} r_j(\omega) \Delta_j(f) \Big| \le \sup_{\varepsilon > 0} \Big| \Phi_\varepsilon * \sum_{j=-M}^{M} r_j(\omega) \Delta_j(f) \Big| ,$$

which holds since $\{\Phi_\varepsilon\}_{\varepsilon>0}$ is an approximate identity. We raise this inequality to the power p, we integrate over $x \in \mathbf{R}^n$ and $\omega \in [0,1]$, and we use the maximal function characterization of H^p [Theorem 6.4.4 (a)] to obtain

$$\int_0^1 \int_{\mathbf{R}^n} \Big| \sum_{j=-M}^{M} r_j(\omega) \Delta_j(f)(x) \Big|^p dx\, d\omega \le C_{p,n}^p \int_0^1 \Big\| \sum_{j=-M}^{M} r_j(\omega) \Delta_j(f) \Big\|_{H^p}^p d\omega .$$

The lower inequality for the Rademacher functions in Appendix C.2 gives

$$\int_{\mathbf{R}^n} \Big(\sum_{j=-M}^{M} |\Delta_j(f)(x)|^2 \Big)^{\frac{p}{2}} dx \le C_p^p C_{p,n}^p \int_0^1 \Big\| \sum_{j=-M}^{M} r_j(\omega) \Delta_j(f) \Big\|_{H^p}^p d\omega ,$$

where the second estimate is a consequence of Theorem 6.4.14 (we need only the scalar version here), since the kernel

$$\sum_{k=-M}^{M} r_k(\omega) \Psi_{2^{-k}}(x)$$

satisfies (6.4.51) and (6.4.52) with constants A and B depending only on n and Ψ (and, in particular, independent of M). We have now proved that

$$\Big\| \Big(\sum_{j=-M}^{M} |\Delta_j(f)|^2 \Big)^{\frac{1}{2}} \Big\|_{L^p} \le C_{n,p,\Psi} \big\| f \big\|_{H^p} ,$$

from which (6.4.62) follows directly by letting $M \to \infty$. We have now established (6.4.62) for $f \in H^p \cap L^1$. Using density, we can extend this estimate to all $f \in H^p$.

To obtain the converse estimate, for $r \in \{0,1,2\}$ we consider the sets

$$3\mathbf{Z}+r=\{3k+r:\ k\in\mathbf{Z}\},$$

and we observe that for $j,k \in 3\mathbf{Z}+r$ the Fourier transforms of $\Delta_j(f)$ and $\Delta_k(f)$ are disjoint if $j \neq k$. We fix a Schwartz function η whose Fourier transform is compactly supported away from the origin so that for all $j,k \in 3\mathbf{Z}$ we have

$$\Delta_j^\eta \Delta_k = \begin{cases} \Delta_j & \text{when } j=k, \\ 0 & \text{when } j\neq k, \end{cases} \tag{6.4.65}$$

where Δ_j^η is the Littlewood–Paley operator associated with the bump η, that is, $\Delta_j^\eta(f) = f * \eta_{2^{-j}}$. It follows from Theorem 6.4.14 that the map

$$\{f_j\}_{j\in\mathbf{Z}} \to \sum_{j\in 3\mathbf{Z}} \Delta_j^\eta(f_j)$$

maps $H^p(\mathbf{R}^n,\ell^2)$ to $H^p(\mathbf{R}^n)$. Indeed, we can see easily that

$$\Big|\sum_{j\in 3\mathbf{Z}} \widehat{\eta}(2^{-j}\xi)\Big| \le B$$

and

$$\sum_{j\in 3\mathbf{Z}} \big|\partial^\alpha(2^{jn}\eta(2^j x))\big| \le A_\alpha |x|^{-n-|\alpha|}$$

for all multi-indices α and for constants depending only on B and A_α. Applying this estimate with $f_j = \Delta_j(f)$ and using (6.4.65) yields the estimate

$$\Big\|\sum_{j\in 3\mathbf{Z}} \Delta_j(f)\Big\|_{H^p} \le C_{n,p,\Psi}\Big\|\Big(\sum_{j\in 3\mathbf{Z}} |\Delta_j(f)|^2\Big)^{\frac{1}{2}}\Big\|_{L^p}$$

for all distributions f that satisfy (6.4.63). Applying the same idea with $3\mathbf{Z}+1$ and $3\mathbf{Z}+2$ replacing $3\mathbf{Z}$ and summing the corresponding estimates gives

$$\Big\|\sum_{j\in \mathbf{Z}} \Delta_j(f)\Big\|_{H^p} \le 3^{\frac{1}{p}} C_{n,p,\Psi}\Big\|\Big(\sum_{j\in \mathbf{Z}} |\Delta_j(f)|^2\Big)^{\frac{1}{2}}\Big\|_{L^p}.$$

But note that $f - \sum_j \Delta_j(f)$ is equal to a polynomial $Q(x)$, since its Fourier transform is supported at the origin. It follows that $f - Q$ lies in H^p and satisfies (6.4.64). □

We show in the next section that the square function characterization of H^p is independent of the choice of the underlying function Ψ.

Exercises

6.4.1. Prove that if v is a bounded tempered distribution and h_1, h_2 are in $\mathscr{S}(\mathbf{R}^n)$, then

$$(h_1 * h_2) * v = h_1 * (h_2 * v).$$

6.4.2. (a) Show that the H^1 norm remains invariant under the L^1 dilation $f_t(x) = t^{-n} f(t^{-1}x)$.
(b) Show that the H^p norm remains invariant under the L^p dilation $t^{n-n/p} f_t(x)$ interpreted in the sense of distributions.

6.4.3. (a) Let $1 < q \le \infty$ and let g in $L^q(\mathbf{R}^n)$ be a compactly supported function with integral zero. Show that g lies in the Hardy space $H^1(\mathbf{R}^n)$.
(b) Prove the same conclusion when L^q is replaced by $L\log^+ L$.
[*Hint:* Part (a): Pick a $\mathscr{C}_0^\infty$ function Φ supported in the unit ball with nonvanishing integral and suppose that the support of g is contained in the ball $B(0,R)$. For $|x| \le 2R$ we have that $M(f;\Phi)(x) \le C_\Phi M(g)(x)$, and since $M(g)$ lies in L^q, it also lies in $L^1(B(0,2R))$. For $|x| > 2R$, write $(\Phi_t * g)(x) = \int_{\mathbf{R}^n} \big(\Phi_t(x-y) - \Phi_t(x)\big) g(y)\,dy$ and use the mean value theorem to estimate this expression by $t^{-n-1} \big\|\nabla\Phi\big\|_{L^\infty} \big\|g\big\|_{L^1} \le |x|^{-n-1} C_\Phi \big\|g\big\|_{L^q}$, since $t \ge |x-y| \ge |x| - |y| \ge |x|/2$ whenever $|x| \ge 2R$ and $|y| \le R$. Thus $M(f;\Phi)$ lies in $L^1(\mathbf{R}^n)$. Part (b): Use Exercise 2.1.4(a) to deduce that $M(g)$ is integrable over $B(0,2R)$.]

6.4.4. Show that the function $\psi(s)$ defined in (6.4.19) is continuous and integrable over $[1,\infty)$, decays faster than the reciprocal of any polynomial, and satisfies (6.4.18), that is,

$$\int_1^\infty s^k\, \psi(s)\, ds = \begin{cases} 1 & \text{if } k = 0, \\ 0 & \text{if } k = 1,2,3,\dots. \end{cases}$$

[*Hint:* Apply Cauchy's theorem over a suitable contour.]

6.4.5. Let $0 < a < \infty$ be fixed. Show that a bounded tempered distribution f lies in H^p if and only if the nontangential Poisson maximal function

$$M_a^*(f;P)(x) = \sup_{t>0} \sup_{\substack{y \in \mathbf{R}^n \\ |y-x| \le at}} |(P_t * f)(y)|$$

lies in L^p, and in this case we have $\big\|f\big\|_{H^p} \approx \big\|M_a^*(f;P)\big\|_{L^p}$.
[*Hint:* Observe that $M(f;P)$ can be replaced with $M_a^*(f;P)$ in the proof of parts (a) and (e) of Theorem 6.4.4).]

6.4.6. Show that for every integrable function g with mean value zero and support inside a ball B, we have $M(g;\Phi) \in L^p((3B)^c)$ for $p > n/(n+1)$. Here Φ is in $\mathscr{S}$.

6.4.7. Show that the space of all Schwartz functions whose Fourier transform is supported away from a neighborhood of the origin is dense in H^p.
[*Hint:* Use the square function characterization of H^p.]

6.4.8. (a) Suppose that $f \in H^p(\mathbf{R}^n)$ for some $0 < p \le 1$ and Φ in $\mathscr{S}(\mathbf{R}^n)$. Then show that for all $t > 0$ the function $\Phi_t * f$ belongs to $L^r(\mathbf{R}^n)$ for all $p \le r \le \infty$. Find an estimate for the L^r norm of $\Phi_t * f$ in terms of $\|f\|_{H^p}$ and $t > 0$.
(b) Let $0 < p \le 1$. Show that there exists a constant $C_{n,p}$ such that for all f in $H^p(\mathbf{R}^n) \cap L^1(\mathbf{R}^n)$ we have

$$|\widehat{f}(\xi)| \le C_{n,p} |\xi|^{\frac{n}{p}-n} \|f\|_{H^p}.$$

[*Hint:* Obtain that

$$\|\Phi_t * f\|_{L^1} \le C t^{-n/p+n} \|f\|_{H^p},$$

using an idea from the proof of Proposition 6.4.7.]

6.4.9. Show that $H^p(\mathbf{R}^n, \ell^2) = L^p(\mathbf{R}^n, \ell^2)$ whenever $1 < p < \infty$ and that $H^1(\mathbf{R}^n, \ell^2)$ is contained in $L^1(\mathbf{R}^n, \ell^2)$.

6.4.10. For a sequence of tempered distributions $\vec{f} = \{f_j\}_j$, define the following variant of the grand maximal function:

$$\widetilde{\mathcal{M}}_N(\vec{f})(x) = \sup_{\{\varphi_j\}_j \in \mathscr{F}_N} \sup_{\varepsilon > 0} \sup_{\substack{y \in \mathbf{R}^n \\ |y-x|<\varepsilon}} \Big(\sum_j |((\varphi_j)_\varepsilon * f_j)(y)|^2 \Big)^{\frac{1}{2}},$$

where $N \ge [\frac{n}{p}] + 1$ and

$$\mathscr{F}_N = \Big\{ \{\varphi_j\}_j \in \mathscr{S}(\mathbf{R}^n) : \sum_j \mathfrak{N}_N(\varphi_j) \le 1 \Big\}.$$

Show that for all sequences of tempered distributions $\vec{f} = \{f_j\}_j$ we have

$$\|\widetilde{\mathcal{M}}_N(\vec{f})\|_{L^p(\mathbf{R}^n, \ell^2)} \approx \|\mathcal{M}_N(\vec{f})\|_{L^p(\mathbf{R}^n, \ell^2)}$$

with constants depending only on n and p.
[*Hint:* Fix Φ in $\mathscr{S}(\mathbf{R}^n)$ with integral 1. Using Lemma 6.4.5, write

$$(\varphi_j)_t(y) = \int_0^1 ((\Theta_j^{(s)})_t * \Phi_{ts})(y)\, ds$$

and apply a vector-valued extension of the proof of part (d) of Theorem 6.4.4 to obtain the pointwise estimate

$$\widetilde{\mathcal{M}}_N(\vec{f}) \le C_{n,p} M_m^{**}(\vec{f}; \Phi),$$

where $m > n/p$.]

6.5 Besov–Lipschitz and Triebel–Lizorkin Spaces

The main achievement of the previous sections was the remarkable characterization of Sobolev, Lipschitz, and Hardy spaces using the Littlewood–Paley operators Δ_j. These characterizations motivate the introduction of classes of spaces defined in terms of expressions involving the operators Δ_j. These scales furnish a general framework within which one can launch a study of function spaces from a unified perspective.

We have encountered two expressions involving the operators Δ_j in the characterizations of the function spaces obtained in the previous sections. Some spaces were characterized by an L^p norm of the Littlewood–Paley square function

$$\Big(\sum_j |2^{j\alpha}\Delta_j(f)|^2\Big)^{\frac{1}{2}},$$

and other spaces were characterized by an ℓ^q norm of the sequence of quantities $\big\|2^{j\alpha}\Delta_j(f)\big\|_{L^p}$. Examples of spaces in the first case are the homogeneous Sobolev spaces, Hardy spaces, and, naturally, L^p spaces. We have studied only one example of spaces in the second category, the Lipschitz spaces, in which case $p = q = \infty$. These examples motivate the introduction of two fundamental scales of function spaces, called the Triebel–Lizorkin and Besov–Lipschitz spaces, respectively.

6.5.1 Introduction of Function Spaces

Before we give the pertinent definitions, we recall the setup that we developed in Section 6.2 and used in Section 6.3. Throughout this section we fix a radial Schwartz function Ψ on $\mathbf{R}^n$ whose Fourier transform is nonnegative, is supported in the annulus $1-\frac{1}{7} \le |\xi| \le 2$, is equal to one on the smaller annulus $1 \le |\xi| \le 2-\frac{2}{7}$, and satisfies

$$\sum_{j\in\mathbf{Z}} \widehat{\Psi}(2^{-j}\xi) = 1, \qquad \xi \neq 0. \tag{6.5.1}$$

Associated with this bump, we define the Littlewood–Paley operators $\Delta_j = \Delta_j^{\Psi}$ given by multiplication on the Fourier transform side by the function $\widehat{\Psi}(2^{-j}\xi)$. We also define a Schwartz function Φ such that

$$\widehat{\Phi}(\xi) = \begin{cases} \sum_{j\le 0} \widehat{\Psi}(2^{-j}\xi) & \text{when } \xi \neq 0, \\ 1 & \text{when } \xi = 0. \end{cases} \tag{6.5.2}$$

Note that $\widehat{\Phi}(\xi)$ is equal to 1 for $|\xi| \le 2-\frac{2}{7}$ and vanishes when $|\xi| \ge 2$. It follows from these definitions that

$$S_0 + \sum_{j=1}^{\infty} \Delta_j = I, \tag{6.5.3}$$

where $S_0 = S_0^{\Psi}$ is the operator given by convolution with the bump Φ and the convergence of the series in (6.5.3) is in $\mathscr{S}'(\mathbf{R}^n)$. Moreover, we also have the identity

$$\sum_{j\in\mathbf{Z}} \Delta_j = I, \tag{6.5.4}$$

where the convergence of the series in (6.5.4) is in the sense of $\mathscr{S}'(\mathbf{R}^n)/\mathscr{P}$.

Definition 6.5.1. Let $\alpha \in \mathbf{R}$ and $0 < p,q \le \infty$. For $f \in \mathscr{S}'(\mathbf{R}^n)$ we set

$$\big\|f\big\|_{B_p^{\alpha,q}} = \big\|S_0(f)\big\|_{L^p} + \Big(\sum_{j=1}^{\infty} \big(2^{j\alpha}\big\|\Delta_j(f)\big\|_{L^p}\big)^q\Big)^{\frac{1}{q}}$$

with the obvious modification when $p,q = \infty$. When $p,q < \infty$ we also set

$$\big\|f\big\|_{F_p^{\alpha,q}} = \big\|S_0(f)\big\|_{L^p} + \Big\|\Big(\sum_{j=1}^{\infty} \big(2^{j\alpha}|\Delta_j(f)|\big)^q\Big)^{\frac{1}{q}}\Big\|_{L^p}.$$

The space of all tempered distributions f for which the quantity $\big\|f\big\|_{B_p^{\alpha,q}}$ is finite is called the (inhomogeneous) *Besov–Lipschitz* space with indices α, p, q and is denoted by $B_p^{\alpha,q}$. The space of all tempered distributions f for which the quantity $\big\|f\big\|_{F_p^{\alpha,q}}$ is finite is called the (inhomogeneous) *Triebel–Lizorkin* space with indices α, p, q and is denoted by $F_p^{\alpha,q}$.

We now define the corresponding homogeneous versions of these spaces. For an element f of $\mathscr{S}'(\mathbf{R}^n)/\mathscr{P}$ we let

$$\big\|f\big\|_{\dot B_p^{\alpha,q}} = \Big(\sum_{j\in\mathbf{Z}} \big(2^{j\alpha}\big\|\Delta_j(f)\big\|_{L^p}\big)^q\Big)^{\frac{1}{q}}$$

and

$$\big\|f\big\|_{\dot F_p^{\alpha,q}} = \Big\|\Big(\sum_{j\in\mathbf{Z}} \big(2^{j\alpha}|\Delta_j(f)|\big)^q\Big)^{\frac{1}{q}}\Big\|_{L^p}.$$

The space of all f in $\mathscr{S}'(\mathbf{R}^n)/\mathscr{P}$ for which the quantity $\big\|f\big\|_{\dot B_p^{\alpha,q}}$ is finite is called the (homogeneous) *Besov–Lipschitz* space with indices α, p, q and is denoted by $\dot B_p^{\alpha,q}$. The space of f in $\mathscr{S}'(\mathbf{R}^n)/\mathscr{P}$ such that $\big\|f\big\|_{\dot F_p^{\alpha,q}} < \infty$ is called the (homogeneous) *Triebel–Lizorkin* space with indices α, p, q and is denoted by $\dot F_p^{\alpha,q}$.

We now make several observations related to these definitions. First we note that the expressions $\big\|\cdot\big\|_{F_p^{\alpha,q}}$, $\big\|\cdot\big\|_{\dot F_p^{\alpha,q}}$, $\big\|\cdot\big\|_{B_p^{\alpha,q}}$, and $\big\|\cdot\big\|_{\dot B_p^{\alpha,q}}$ are built in terms of L^p quasinorms of ℓ^q quasinorms of $2^{j\alpha}\Delta_j$ or ℓ^q quasinorms of L^p quasinorms of the same expressions. As a result, we can see that these quantities satisfy the triangle inequality with a constant (which may be taken to be 1 when $1 \le p,q < \infty$). To determine whether these quantities are indeed quasinorms, we need to check whether the following property holds:

$$\|f\|_X = 0 \implies f = 0, \tag{6.5.5}$$

where X is one of the $\dot{F}_p^{\alpha,q}$, $F_p^{\alpha,q}$, $\dot{B}_p^{\alpha,q}$, and $B_p^{\alpha,q}$. Since these are spaces of distributions, the identity $f = 0$ in (6.5.5) should be interpreted in the sense of distributions. If $\|f\|_X = 0$ for some inhomogeneous space X, then $S_0(f) = 0$ and $\Delta_j(f) = 0$ for all $j \geq 1$. Using (6.5.3), we conclude that $f = 0$; thus the quantities $\|\cdot\|_{F_p^{\alpha,q}}$ and $\|\cdot\|_{B_p^{\alpha,q}}$ are indeed quasinorms. Let us investigate what happens when $\|f\|_X = 0$ for some homogeneous space X. In this case we must have $\Delta_j(f) = 0$, and using (6.5.4) we conclude that $\widehat{f}$ must be supported at the origin. Proposition 2.4.1 yields that f must be a polynomial and thus f must be zero (since distributions whose difference is a polynomial are identified in homogeneous spaces).

Remark 6.5.2. We interpret the previous definition in certain cases. According to what we have seen so far, we have

$$\begin{aligned}
\dot{F}_p^{0,2} &\approx F_p^{0,2} \approx L^p, && 1 < p < \infty,\\
\dot{F}_p^{0,2} &\approx H^p, && 0 < p \leq 1,\\
F_p^{s,2} &\approx L_s^p, && 1 < p < \infty,\\
\dot{F}_p^{s,2} &\approx \dot{L}_s^p, && 1 < p < \infty,\\
B_\infty^{\gamma,\infty} &\approx \Lambda_\gamma, && \gamma > 0,\\
\dot{B}_\infty^{\gamma,\infty} &\approx \dot{\Lambda}_\gamma, && \gamma > 0,
\end{aligned}$$

where $\approx$ indicates that the corresponding norms are equivalent.

Although in this text we restrict attention to the case $p < \infty$, it is noteworthy mentioning that when $p = \infty$, $\dot{F}_\infty^{0,q}$ can be defined as the space of all $f \in \mathscr{S}'/\mathscr{P}$ that satisfy

$$\|f\|_{\dot{F}_\infty^{\alpha,q}} = \sup_{Q \text{ dyadic cube}} \int_Q \frac{1}{|Q|}\bigg(\sum_{j=-\log_2 \ell(Q)}^{\infty} (2^{j\alpha}|\Delta_j(f)|)^q\bigg)^{\frac{1}{q}} < \infty.$$

In the particular case $q = 2$ and $\alpha = 0$, the space obtained in this way is called *BMO* and coincides with the space introduced and studied in Chapter 7; this space serves as a substitute for L^∞ and plays a fundamental role in analysis. It should now be clear that several important spaces in analysis can be thought of as elements of the scale of Triebel–Lizorkin spaces.

It would have been more natural to denote Besov–Lipschitz and Triebel–Lizorkin spaces by $B_{\alpha,q}^p$ and $F_{\alpha,q}^p$ to maintain the upper and lower placements of the corresponding indices analogous to those in the previously defined Lebesgue, Sobolev, Lipschitz, and Hardy spaces. However, the notation in Definition 6.5.1 is more or less prevalent in the field of function spaces, and we adhere to it.

6.5.2 Equivalence of Definitions

It is not clear from the definitions whether the finiteness of the quasinorms defining the spaces $B_p^{\alpha,q}$, $F_p^{\alpha,q}$,$\dot{B}_p^{\alpha,q}$, and $\dot{F}_p^{\alpha,q}$ depends on the choice of the function Ψ (recall that Φ is determined by Ψ). We show that if Ω is another function that satisfies (6.5.1) and Θ is defined in terms of Ω in the same way that Φ is defined in terms of Ψ, [i.e., via (6.5.2)], then the norms defined in Definition 6.5.1 with respect to the pairs (Φ,Ψ) and (Θ,Ω) are comparable. To prove this we need the following lemma.

Lemma 6.5.3. *Let $0 < c_0 < \infty$ and $0 < r < \infty$. Then there exist constants C_1 and C_2 (which depend only on n, c_0, and r) such that for all $t > 0$ and for all $\mathscr{C}^1$ functions u on $\mathbf{R}^n$ whose Fourier transform is supported in the ball $|\xi| \le c_0 t$ and that satisfy $|u(z)| \le B(1+|z|)^{\frac{n}{r}}$ for some $B > 0$ we have the estimate*

$$\sup_{z\in\mathbf{R}^n} \frac{1}{t}\frac{|\nabla u(x-z)|}{(1+t|z|)^{\frac{n}{r}}} \le C_1 \sup_{z\in\mathbf{R}^n} \frac{|u(x-z)|}{(1+t|z|)^{\frac{n}{r}}} \le C_2 M(|u|^r)(x)^{\frac{1}{r}}, \tag{6.5.6}$$

where M denotes the Hardy–Littlewood maximal operator. (The constants C_1 and C_2 are independent of B.)

Proof. Select a Schwartz function ψ whose Fourier transform is supported in the ball $|\xi| \le 2c_0$ and is equal to 1 on the smaller ball $|\xi| \le c_0$. Then $\widehat{\psi}(\frac{\xi}{t})$ is equal to 1 on the support of $\widehat{u}$ and we can write

$$u(x-z) = \int_{\mathbf{R}^n} t^n \psi(t(x-z-y))u(y)\,dy.$$

Taking partial derivatives and using that ψ is a Schwartz function, we obtain

$$|\nabla u(x-z)| \le C_N \int_{\mathbf{R}^n} t^{n+1}(1+t|x-z-y|)^{-N}|u(y)|\,dy,$$

where N is arbitrarily large. Using that for all $x,y,z \in \mathbf{R}^n$ we have

$$1 \le (1+t|x-z-y|)^{\frac{n}{r}} \frac{(1+t|z|)^{\frac{n}{r}}}{(1+t|x-y|)^{\frac{n}{r}}},$$

we obtain

$$\frac{1}{t}\frac{|\nabla u(x-z)|}{(1+t|z|)^{\frac{n}{r}}} \le C_N \int_{\mathbf{R}^n} t^n(1+t|x-z-y|)^{\frac{n}{r}-N}\frac{|u(y)|}{(1+t|x-y|)^{\frac{n}{r}}}\,dy,$$

from which the first estimate in (6.5.6) follows easily.

Let $|y| \le \delta$ for some $\delta > 0$ to be chosen later. We now use the mean value theorem to write

$$u(x-z) = (\nabla u)(x-z-\xi_y)\cdot y + u(x-z-y)$$

for some ξ_y satisfying $|\xi_y| \le |y| \le \delta$. This implies that

$$|u(x-z)| \le \sup_{|w|\le|z|+\delta} |(\nabla u)(x-w)|\,\delta + |u(x-z-y)|.$$

Raising to the power r, averaging over the ball $|y| \le \delta$, and then raising to the power $\frac{1}{r}$ yields

$$|u(x-z)| \le c_r \left[\sup_{|w|\le|z|+\delta} |(\nabla u)(x-w)|\,\delta + \left(\frac{1}{v_n \delta^n} \int_{|y|\le\delta} |u(x-z-y)|^r dy \right)^{\frac{1}{r}} \right]$$

with $c_r = \max(2^{1/r}, 2^r)$. Here v_n is the volume of the unit ball in $\mathbf{R}^n$. Then

$$\frac{|u(x-z)|}{(1+t|z|)^{\frac{n}{r}}} \le c_r \left[\sup_{|w|\le|z|+\delta} \frac{|(\nabla u)(x-w)|}{(1+t|z|)^{\frac{n}{r}}}\,\delta\, \frac{\left(\frac{1}{v_n \delta^n} \int_{|y|\le\delta+|z|} |u(x-y)|^r dy \right)^{\frac{1}{r}}}{(1+t|z|)^{\frac{n}{r}}} \right].$$

We now set $\delta = \varepsilon/t$ for some $\varepsilon \le 1$. Then we have

$$|w| \le |z| + \frac{\varepsilon}{t} \implies \frac{1}{1+t|z|} \le \frac{2}{1+t|w|},$$

and we can use this to obtain the estimate

$$\frac{|u(x-z)|}{(1+t|z|)^{\frac{n}{r}}} \le c_{r,n} \left[\sup_{w\in\mathbf{R}^n} \frac{1}{t} \frac{|(\nabla u)(x-w)|}{(1+t|w|)^{\frac{n}{r}}}\,\varepsilon\, \frac{\left(\frac{t^n}{v_n \varepsilon^n} \int_{|y|\le\frac{1}{t}+|z|} |u(x-y)|^r dy \right)^{\frac{1}{r}}}{(1+t|z|)^{\frac{n}{r}}} \right]$$

with $c_{r,n} = \max(2^{1/r}, 2^r)2^{n/r}$. It follows that

$$\sup_{z\in\mathbf{R}^n} \frac{|u(x-z)|}{(1+t|z|)^{\frac{n}{r}}} \le c_{r,n} \left[\sup_{w\in\mathbf{R}^n} \frac{1}{t} \frac{|(\nabla u)(x-w)|}{(1+t|w|)^{\frac{n}{r}}}\,\varepsilon + \varepsilon^{-\frac{n}{r}} M(|u|^r)(x)^{\frac{1}{r}} \right].$$

Taking $\varepsilon = \frac{1}{2}(c_{r,n}C_1)^{-1}$, where C_1 is the constant in (6.5.6), we obtain the second estimate in (6.5.6) with $C_2 = 2\varepsilon^{-n/r}$. At this step we used the hypothesis that

$$\sup_{z\in\mathbf{R}^n} \frac{|u(x-z)|}{(1+t|z|)^{\frac{n}{r}}} \le \sup_{z\in\mathbf{R}^n} \frac{B(1+|x|+|z|)^{\frac{n}{r}}}{(1+t|z|)^{\frac{n}{r}}} < \infty.$$

This concludes the proof of the lemma. □

Remark 6.5.4. The reader is reminded that $\widehat{u}$ in Lemma 6.5.3 may not be a function; for example, this is the case when u is a polynomial (say of degree $[n/r]$). If $\widehat{u}$ were an integrable function, then u would be a bounded function, and condition $|u(x)| \le B(1+|x|)^{\frac{n}{r}}$ would not be needed.

We now return to a point alluded to earlier, that changing Ψ by another bump Ω that satisfies similar properties yields equivalent norms for the function spaces given in Definition 6.5.1. Suppose that Ω is another bump whose Fourier transform is supported in the annulus $1-\frac{1}{7} \le |\xi| \le 2$ and that satisfies (6.5.1). The support properties of Ψ and Ω imply the identity

$$\Delta_j^{\Omega} = \Delta_j^{\Omega}(\Delta_{j-1}^{\Psi} + \Delta_j^{\Psi} + \Delta_{j+1}^{\Psi}). \tag{6.5.7}$$

Let $0 < p < \infty$ and pick $r < p$ and $N > \frac{n}{r} + n$. Then we have

$$\begin{aligned} \big|\Delta_j^{\Omega}\Delta_j^{\Psi}(f)(x)\big| &\le C_{N,\Omega} \int_{\mathbf{R}^n} \frac{\big|\Delta_j^{\Psi}(f)(x-z)\big|}{(1+2^j|z|)^{\frac{n}{r}}} \frac{2^{jn}dz}{(1+2^j|z|)^{N-\frac{n}{r}}} \\ &\le C_{N,\Omega} \sup_{z\in\mathbf{R}^n} \frac{\big|\Delta_j^{\Psi}(f)(x-z)\big|}{(1+2^j|z|)^{\frac{n}{r}}} \int_{\mathbf{R}^n} \frac{2^{jn}dz}{(1+2^j|z|)^{N-\frac{n}{r}}} \\ &\le C_{N,r,\Omega}\big(M(|\Delta_j^{\Psi}(f)|^r)(x)\big)^{\frac{1}{r}} \end{aligned} \tag{6.5.8}$$

where we applied Lemma 6.5.3. The same estimate is also valid for $\Delta_j^{\Omega}\Delta_{j\pm 1}^{\Psi}(f)$ and thus for $\Delta_j^{\Omega}(f)$, in view of identity (6.5.7). Armed with this observation and recalling that $r < p$, the boundedness of the Hardy–Littlewood maximal operator on $L^{p/r}$ yields that the homogeneous Besov–Lipschitz norm defined in terms of the bump Ω is controlled by a constant multiple of the corresponding Besov–Lipschitz norm defined in terms of Ψ. A similar argument applies for the inhomogeneous Besov–Lipschitz norms. The equivalence constants depend on Ψ, Ω, n, p, q, and α.

The corresponding equivalence of norms for Triebel–Lizorkin spaces is more difficult to obtain, and it is a consequence of the characterization of these spaces proved later.

Definition 6.5.5. For $b > 0$ and $j \in \mathbf{R}$ we introduce the notation

$$M_{b,j}^{**}(f;\Psi)(x) = \sup_{y\in\mathbf{R}^n} \frac{|(\Psi_{2^{-j}} * f)(x-y)|}{(1+2^j|y|)^b},$$

so that we have

$$M_b^{**}(f;\Psi) = \sup_{t>0} M_{b,t}^{**}(f;\Psi),$$

in accordance with the notation in the previous section. The function $M_b^{**}(f;\Psi)$ is called the *Peetre maximal function of f (with respect to Ψ).*

We clearly have

$$|\Delta_j^{\Psi}(f)| \le M_{b,j}^{**}(f;\Psi),$$

but the next result shows that a certain converse is also valid.

Theorem 6.5.6. *Let $b > n(\min(p,q))^{-1}$ and $0 < p, q < \infty$. Let Ψ and Ω be Schwartz functions whose Fourier transforms are supported in the annulus $\frac{1}{2} \le |\xi| \le 2$ and satisfy (6.5.1). Then we have*

$$\Big\|\Big(\sum_{j\in\mathbf{Z}}|2^{j\alpha}M^{**}_{b,j}(f;\Omega)|^q\Big)^{\frac{1}{q}}\Big\|_{L^p}\le C\Big\|\Big(\sum_{j\in\mathbf{Z}}|2^{j\alpha}\Delta_j^{\Psi}(f)|^q\Big)^{\frac{1}{q}}\Big\|_{L^p} \tag{6.5.9}$$

for all $f\in\mathscr{S}'(\mathbf{R}^n)$, *where* $C=C_{\alpha,p,q,n,b,\Psi,\Omega}$.

Proof. We start with a Schwartz function Θ whose Fourier transform is nonnegative, supported in the annulus $1-\frac{2}{7}\le|\xi|\le 2$, and satisfies

$$\sum_{j\in\mathbf{Z}}\widehat{\Theta}(2^{-j}\xi)^2=1,\qquad \xi\in\mathbf{R}^n\setminus\{0\}. \tag{6.5.10}$$

Using (6.5.10), we have

$$\Omega_{2^{-k}}*f=\sum_{j\in\mathbf{Z}}(\Omega_{2^{-k}}*\Theta_{2^{-j}})*(\Theta_{2^{-j}}*f).$$

It follows that

$$\begin{aligned}
&2^{k\alpha}\frac{|(\Omega_{2^{-k}}*f)(x-z)|}{(1+2^k|z|)^b}\\
&\le\sum_{j\in\mathbf{Z}}2^{k\alpha}\int_{\mathbf{R}^n}|(\Omega_{2^{-k}}*\Theta_{2^{-j}})(y)|\,\frac{|(\Theta_{2^{-j}}*f)(x-z-y)|}{(1+2^k|z|)^b}\,dy\\
&=\sum_{j\in\mathbf{Z}}2^{k\alpha}\int_{\mathbf{R}^n}2^{kn}|(\Omega*\Theta_{2^{-(j-k)}})(2^ky)|\,\frac{(1+2^j|y+z|)^b}{(1+2^k|z|)^b}\,\frac{|(\Theta_{2^{-j}}*f)(x-z-y)|}{(1+2^j|y+z|)^b}\,dy\\
&\le\sum_{j\in\mathbf{Z}}2^{k\alpha}\int_{\mathbf{R}^n}|(\Omega*\Theta_{2^{-(j-k)}})(y)|\,\frac{(1+2^j|2^{-k}y+z|)^b}{(1+2^k|z|)^b}\,\frac{|(\Theta_{2^{-j}}*f)(x-z-y)|}{(1+2^j|y+z|)^b}\,dy\\
&\le\sum_{j\in\mathbf{Z}}2^{(k-j)\alpha}\int_{\mathbf{R}^n}|(\Omega*\Theta_{2^{-(j-k)}})(y)|\,\frac{(1+2^{j-k}|y|+2^j|z|)^b}{(1+2^k|z|)^b}\,dy\,2^{j\alpha}M^{**}_{b,j}(f;\Theta)(x)\\
&\le\sum_{j\in\mathbf{Z}}2^{(k-j)\alpha}\int_{\mathbf{R}^n}|(\Omega*\Theta_{2^{-(j-k)}})(y)|\,(1+2^{j-k})^b(1+2^{j-k}|y|)^b dy\,2^{j\alpha}M^{**}_{b,j}(f;\Theta)(x).
\end{aligned}$$

We conclude that

$$2^{k\alpha}M^{**}_{b,k}(f;\Omega)(x)\le\sum_{j\in\mathbf{Z}}V_{k-j}\,2^{j\alpha}M^{**}_{b,j}(f;\Theta)(x), \tag{6.5.11}$$

where

$$V_j=2^{-j\alpha}(1+2^j)^b\int_{\mathbf{R}^n}|(\Omega*\Theta_{2^{-j}})(y)|\,(1+2^j|y|)^b\,dy.$$

We now use the facts that both Ω and Θ have vanishing moments of all orders and the result in Appendix K.2 to obtain

$$|(\Omega*\Theta_{2^{-j}})(y)|\le C_{L,N,n,\Theta,\Omega}\frac{2^{-|j|L}}{(1+2^{\min(0,j)}|y|)^N}$$

for all $L,N>0$. We deduce the estimate

$$|V_j| \le C_{L,M,n,\Theta,\Omega} 2^{-|j|M}$$

for all M sufficiently large, which, in turn, yields the estimate

$$\sum_{j\in\mathbf{Z}} |V_j|^{\min(1,q)} < \infty .$$

We deduce from (6.5.11) that for all $x \in \mathbf{R}^n$ we have

$$\big\|\{2^{k\alpha} M^{**}_{b,k}(f;\Omega)(x)\}_k\big\|_{\ell^q} \le C_{\alpha,p,q,n,\Psi,\Omega} \big\|\{2^{k\alpha} M^{**}_{b,k}(f;\Theta)(x)\}_k\big\|_{\ell^q} .$$

We now appeal to Lemma 6.5.3, which gives

$$2^{k\alpha} M^{**}_{b,k}(f;\Theta) \le C 2^{k\alpha} M(|\Delta_k^\Theta(f)|^r)^{\frac{1}{r}} = CM(|2^{k\alpha}\Delta_k^\Theta(f)|^r)^{\frac{1}{r}}$$

with $b=n/r$. We choose $r<\min(p,q)$. We use the $L^{p/r}(\mathbf{R}^n,\ell^{q/r})$ to $L^{p/r}(\mathbf{R}^n,\ell^{q/r})$ boundedness of the Hardy–Littlewood maximal operator, Theorem 4.6.6, to complete the proof of (6.5.9) with the exception that the function Ψ on the right-hand side of (6.5.9) is replaced by Θ. The passage to Ψ is a simple matter (at least when $p \ge 1$), since

$$\Delta_j^\Psi = \Delta_j^\Psi \left(\Delta_{j-1}^\Theta + \Delta_j^\Theta + \Delta_{j+1}^\Theta\right).$$

For general $0<p<\infty$ the conclusion follows with the use of (6.5.8). □

We obtain as a corollary that a different choice of bumps gives equivalent Triebel–Lizorkin norms.

Corollary 6.5.7. *Let Ψ, Ω be Schwartz functions whose Fourier transforms are supported in the annulus $1-\frac{1}{7} \le |\xi| \le 2$ and satisfy (6.5.1). Let Φ be as in (6.5.2) and let*

$$\widehat{\Theta}(\xi) = \begin{cases} \sum_{j\le 0} \widehat{\Omega}(2^{-j}\xi) & \text{when } \xi \neq 0, \\ 1 & \text{when } \xi = 0. \end{cases}$$

Then the Triebel–Lizorkin quasinorms defined with respect to the pairs (Ψ,Φ) and (Ω,Θ) are equivalent.

Proof. We note that the quantity on the left in (6.5.9) is greater than or equal to

$$\Big\|\Big(\sum_{j\in\mathbf{Z}} |2^{j\alpha}\Delta_j^\Omega(f)|^q\Big)^{\frac{1}{q}}\Big\|_{L^p}$$

for all $f \in \mathscr{S}'(\mathbf{R}^n)$. This shows that the homogeneous Triebel–Lizorkin norm defined using Ω is bounded by a constant multiple of that defined using Ψ. This proves the equivalence of norms in the homogeneous case.

In the case of the inhomogeneous spaces, we let S_0^Ψ and S_0^Ω be the operators given by convolution with the bumps Φ and Θ, respectively (recall that these are defined in terms of Ψ and Ω). Then for $f \in \mathscr{S}'(\mathbf{R}^n)$ we have

$$\Theta * f = \Theta * (\Phi * f) + \Theta * (\Psi_{2^{-1}} * f), \tag{6.5.12}$$

since the Fourier transform of the function $\Phi + \Psi_{2^{-1}}$ is equal to 1 on the support of $\widehat{\Theta}$. Applying Lemma 6.5.3 (with $t = 1$), we obtain that

$$|\Theta * (\Phi * f)| \le C_r M(|\Phi * f|^r)^{\frac{1}{r}}$$

and also

$$|\Theta * (\Psi_{2^{-1}} * f)| \le C_r M(|\Psi_{2^{-1}} * f|^r)^{\frac{1}{r}}$$

for any $0 < r < \infty$. Picking $r < p$, we obtain that

$$\big\|\Theta * (\Phi * f)\big\|_{L^p} \le C \big\|S_0^{\Psi}(f)\big\|_{L^p}$$

and also

$$\big\|\Theta * (\Psi_{2^{-1}} * f)\big\|_{L^p} \le C \big\|\Delta_1^{\Psi}(f)\big\|_{L^p}.$$

Combining the last two estimates with (6.5.12), we obtain that $\big\|S_0^{\Omega}(f)\big\|_{L^p}$ is controlled by a multiple of the Triebel–Lizorkin norm of f defined using Ψ. This gives the equivalence of norms in the inhomogeneous case. □

Several other properties of these spaces are discussed in the exercises that follow.

Exercises

6.5.1. Let $0 < q_0 \le q_1 < \infty$, $0 < p < \infty$, $\varepsilon > 0$, and $\alpha \in \mathbf{R}$. Prove the embeddings

$$B_p^{\alpha,q_0} \subseteq B_p^{\alpha,q_1},$$
$$F_p^{\alpha,q_0} \subseteq F_p^{\alpha,q_1},$$
$$B_p^{\alpha+\varepsilon,q_0} \subseteq B_p^{\alpha,q_1},$$
$$F_p^{\alpha+\varepsilon,q_0} \subseteq F_p^{\alpha,q_1},$$

where p and q_1 are allowed to be infinite in the case of Besov spaces.

6.5.2. Let $0 < q < \infty$, $0 < p < \infty$, and $\alpha \in \mathbf{R}$. Show that

$$B_p^{\alpha,\min(p,q)} \subseteq F_p^{\alpha,q} \subseteq B_p^{\alpha,\max(p,q)}.$$

[*Hint:* Consider the cases $p \ge q$ and $p < q$ and use the triangle inequality in the spaces $L^{p/q}$ and $\ell^{q/p}$, respectively.]

6.5.3. (a) Let $0 < p,q \le \infty$ and $\alpha \in \mathbf{R}$. Show that $\mathscr{S}(\mathbf{R}^n)$ is continuously embedded in $B_p^{\alpha,q}(\mathbf{R}^n)$ and that the latter is continuously embedded in $\mathscr{S}'(\mathbf{R}^n)$.
(b) Obtain the same conclusion for $F_p^{\alpha,q}(\mathbf{R}^n)$ when $p,q < \infty$.

6.5.4. $0 < p,q < \infty$ and $\alpha \in \mathbf{R}$. Show that the Schwartz functions are dense in all the spaces $B_p^{\alpha,q}(\mathbf{R}^n)$ and $F_p^{\alpha,q}(\mathbf{R}^n)$.
[*Hint:* Every Cauchy sequence $\{f_k\}_k$ in $B_p^{\alpha,q}$ is also Cauchy in $\mathscr{S}'(\mathbf{R}^n)$ and hence converges to some f in $\mathscr{S}'(\mathbf{R}^n)$. Then $\Delta_j(f_k) \to \Delta_j(f)$ in $\mathscr{S}'(\mathbf{R}^n)$. But $\Delta_j(f_k)$ is also Cauchy in L^p and therefore converges to $\Delta_j(f)$ in L^p. Argue similarly for $F_p^{\alpha,q}(\mathbf{R}^n)$.]

6.5.5. Let $\alpha \in \mathbf{R}$, let $0 < p,q < \infty$, and let $N = [\frac{n}{2} + \frac{n}{\min(p,q)}] + 1$. Assume that m is a $\mathscr{C}^N$ function on $\mathbf{R}^n \setminus \{0\}$ that satisfies

$$|\partial^\gamma m(\xi)| \le C_\gamma |\xi|^{-|\gamma|}$$

for all $|\gamma| \le N$. Show that there exists a constant C such that for all $f \in \mathscr{S}'(\mathbf{R}^n)$ we have

$$\|(m\widehat{f})^\vee\|_{B_p^{\alpha,q}} \le C\|f\|_{B_p^{\alpha,q}}.$$

[*Hint:* Pick $r < \min(p,q)$ such that $N > \frac{n}{2} + \frac{n}{r}$. Write $m = \sum_j m_j$, where $\widehat{m_j}(\xi) = \widehat{\Theta}(2^{-j}\xi)m(\xi)$ and $\widehat{\Theta}(2^{-j}\xi)$ is supported in an annulus $2^j \le |\xi| \le 2^{j+1}$. Obtain the estimate

$$\begin{aligned}\sup_{z\in\mathbf{R}^n} \frac{|(m_j\widehat{\Delta_j(f)})^\vee(x-z)|}{(1+2^j|z|)^{\frac{n}{r}}} &\le C \sup_{z\in\mathbf{R}^n} \frac{|\Delta_j(f)(x-z)|}{(1+2^j|z|)^{\frac{n}{r}}} \int_{\mathbf{R}^n} |m_j^\vee(y)|(1+2^j|y|)^{\frac{n}{r}}\,dy \\ &\le C'\left(\int_{\mathbf{R}^n} |m_j(2^j(\cdot))^\vee(y)|^2(1+|y|)^{2N}\,dy\right)^{\frac{1}{2}}.\end{aligned}$$

Then use the hypothesis on m and apply Lemma 6.5.3.]

6.5.6. (*Peetre [258]*) Let m be as in Exercise 6.5.5. Show that there exists a constant C such that for all $f \in \mathscr{S}'(\mathbf{R}^n)$ we have

$$\|(m\widehat{f})^\vee\|_{F_p^{\alpha,q}} \le C\|f\|_{F_p^{\alpha,q}}.$$

[*Hint:* Use the hint of Exercise 6.5.5 and Theorem 4.6.6.]

6.5.7. (a) Suppose that $B_{p_0}^{\alpha_0,q_0} = B_{p_1}^{\alpha_1,q_1}$ with equivalent norms. Prove that $\alpha_0 = \alpha_1$ and $p_0 = p_1$. Prove the same result for the scale of F spaces.
(b) Suppose that $B_{p_0}^{\alpha_0,q_0} = B_{p_1}^{\alpha_1,q_1}$ with equivalent norms. Prove that $q_0 = q_1$. Argue similarly with the scale of F spaces.
[*Hint:* Part (a): Test the corresponding norms on the function $\Psi(2^j x)$, where Ψ is chosen so that its Fourier transform is supported in $\frac{1}{2} \le |\xi| \le 2$. Part (b): Try a function f of the form $\widehat{f}(\xi) = \sum_{j=1}^N a_j \widehat{\varphi}(\xi_1 - 2^j, \xi_2, \dots, \xi_n)$, where φ is a Schwartz function whose Fourier transform is supported in a small neighborhood of the origin.]

6.6 Atomic Decomposition

In this section we focus attention on the homogeneous Triebel–Lizorkin spaces $\dot{F}_p^{\alpha,q}$, which include the Hardy spaces discussed in Section 6.4. Most results discussed in this section are also valid for the inhomogeneous Triebel–Lizorkin spaces and for the Besov–Lipschitz via a similar or simpler analysis. We refer the interested reader to the relevant literature on the subject at the end of this chapter.

6.6.1 *The Space of Sequences* $\dot{f}_p^{\alpha,q}$

To provide further intuition in the understanding of the homogeneous Triebel–Lizorkin spaces we introduce a related space consisting of sequences of scalars. This space is denoted by $\dot{f}_p^{\alpha,q}$ and is related to $\dot{F}_p^{\alpha,q}$ in a way similar to that in which $\ell^2(\mathbf{Z})$ is related to $L^2([0,1])$.

Definition 6.6.1. Let $0 < q \leq \infty$ and $\alpha \in \mathbf{R}$. Let $\mathscr{D}$ be the set of all dyadic cubes in $\mathbf{R}^n$. We consider the set of all sequences $\{s_Q\}_{Q\in\mathscr{D}}$ such that the function

$$g^{\alpha,q}(\{s_Q\}_Q) = \Big(\sum_{Q\in\mathscr{D}} (|Q|^{-\frac{\alpha}{n}-\frac{1}{2}}\,|s_Q|\,\chi_Q)^q\Big)^{\frac{1}{q}} \tag{6.6.1}$$

is in $L^p(\mathbf{R}^n)$. For such sequences $s = \{s_Q\}_Q$ we set

$$\big\|s\big\|_{\dot{f}_p^{\alpha,q}} = \big\|g^{\alpha,q}(s)\big\|_{L^p(\mathbf{R}^n)}.$$

6.6.2 *The Smooth Atomic Decomposition of* $\dot{F}_p^{\alpha,q}$

Next, we discuss the smooth atomic decomposition of these spaces. We begin with the definition of smooth atoms on $\mathbf{R}^n$.

Definition 6.6.2. Let Q be a dyadic cube and let L be a nonnegative integer. A $\mathscr{C}^\infty$ function a_Q on $\mathbf{R}^n$ is called a *smooth L-atom for Q* if it satisfies

(a) a_Q is supported in $3Q$ (the cube concentric with Q having three times its side length);

(b) $\displaystyle\int_{\mathbf{R}^n} x^\gamma a_Q(x)\,dx = 0$ for all multi-indices $|\gamma| \leq L$;

(c) $|\partial^\gamma a_Q| \leq |Q|^{-\frac{|\gamma|}{n}-\frac{1}{2}}$ for all multi-indices γ satisfying $|\gamma| \leq L+n+1$.

The value of the constant $L+n+1$ in (c) may vary in the literature. Any sufficiently large constant depending on L will serve the purposes of the definition.

We now prove a theorem stating that elements of $\dot{F}_p^{\alpha,q}$ can be decomposed as sums of smooth atoms.

Theorem 6.6.3. *Let $0 < p,q < \infty$, $\alpha \in \mathbf{R}$, and let L be a nonnegative integer satisfying $L \geq [n\max(1,\frac{1}{p},\frac{1}{q}) - n - \alpha]$. Then there is a constant $C_{n,p,q,\alpha}$ such that for every sequence of smooth L-atoms $\{a_Q\}_{Q\in\mathscr{D}}$ and every sequence of complex scalars $\{s_Q\}_{Q\in\mathscr{D}}$ we have*

$$\Big\|\sum_{Q\in\mathscr{D}} s_Q a_Q\Big\|_{\dot{F}_p^{\alpha,q}} \leq C_{n,p,q,\alpha}\big\|\{s_Q\}_Q\big\|_{\dot{f}_p^{\alpha,q}}. \tag{6.6.2}$$

Conversely, there is a constant $C'_{n,p,q,\alpha}$ such that given any distribution f in $\dot{F}_p^{\alpha,q}$ and any $L\geq 0$, there exist a sequence of smooth L-atoms $\{a_Q\}_{Q\in\mathscr{D}}$ and a sequence of complex scalars $\{s_Q\}_{Q\in\mathscr{D}}$ such that

$$f = \sum_{Q\in\mathscr{D}} s_Q a_Q,$$

where the sum converges in $\mathscr{S}'/\mathscr{P}$ and moreover,

$$\big\|\{s_Q\}_Q\big\|_{\dot{f}_p^{\alpha,q}} \leq C'_{n,p,q,\alpha}\big\|f\big\|_{\dot{F}_p^{\alpha,q}}. \tag{6.6.3}$$

Proof. We begin with the first claim of the theorem. We let Δ_j^Ψ be the Littlewood–Paley operator associated with a Schwartz function Ψ whose Fourier transform is compactly supported away from the origin in $\mathbf{R}^n$. Let a_Q be a smooth L-atom supported in a cube $3Q$ with center c_Q and let the side length be $\ell(Q) = 2^{-\mu}$. It follows trivially from Definition 6.6.2 that a_Q satisfies

$$|\partial_y^\gamma a_Q(y)| \leq C_{N,n} 2^{-\frac{\mu n}{2}} \frac{2^{\mu|\gamma|+\mu n}}{(1+2^\mu|y-c_Q|)^N} \tag{6.6.4}$$

for all $N > 0$ and for all multi-indices γ satisfying $|\gamma| \leq L+n+1$. Moreover, the function $y \mapsto \Psi_{2^{-j}}(y-x)$ satisfies

$$|\partial_y^\delta \Psi_{2^{-j}}(y-x)| \leq C_{N,n,\delta} \frac{2^{j|\delta|+jn}}{(1+2^j|y-x|)^N} \tag{6.6.5}$$

for all $N > 0$ and for all multi-indices δ. Using first the facts that a_Q has vanishing moments of all orders up to and including $L = (L+1)-1$ and that the function $y \mapsto \Psi_{2^{-j}}(y-x)$ satisfies (6.6.5) for all multi-indices δ with $|\delta| = L$, secondly the facts that the function $y \mapsto \Psi_{2^{-j}}(y-x)$ has vanishing moments of all orders up to and including $L+n = (L+n+1)-1$ and that a_Q satisfies (6.6.4) for all multi-indices γ satisfying $|\gamma| = L+n+1$, and the result in Appendix K.2, we deduce the following estimate for all $N > 0$:

$$\big|\Delta_j^\Psi(a_Q)(x)\big| \leq C_{N,n,L'}\, 2^{-\frac{\mu n}{2}} \frac{2^{\min(j,\mu)n-|\mu-j|L'}}{(1+2^{\min(j,\mu)}|x-c_Q|)^N}, \tag{6.6.6}$$

where

$$L' = \begin{cases} L+1 & \text{when } j < \mu, \\ L+n & \text{when } \mu \le j. \end{cases}$$

Now fix $0 < b < \min(1,p,q)$ so that

$$L+1 > \tfrac{n}{b} - n - \alpha. \tag{6.6.7}$$

This can be achieved by taking b close enough to $\min(1,p,q)$, since our assumption $L \ge \left[n\max\left(1,\frac{1}{p},\frac{1}{q}\right) - n - \alpha\right]$ implies $L+1 > n\max\left(1,\frac{1}{p},\frac{1}{q}\right) - n - \alpha$.

Using Exercise 6.6.6, we obtain

$$\sum_{\substack{Q\in\mathscr{D}\\ \ell(Q)=2^{-\mu}}} \frac{|s_Q|}{(1+2^{\min(j,\mu)}|x-c_Q|)^N} \le c\, 2^{\max(\mu-j,0)\frac{n}{b}} \Big\{ M\Big(\sum_{\substack{Q\in\mathscr{D}\\ \ell(Q)=2^{-\mu}}} |s_Q|^b \chi_Q \Big)(x) \Big\}^{\frac{1}{b}}$$

whenever $N > n/b$, where M is the Hardy–Littlewood maximal operator. It follows from the preceding estimate and (6.6.6) that

$$\begin{aligned} 2^{j\alpha} \sum_{\mu\in\mathbf{Z}} \sum_{\substack{Q\in\mathscr{D}\\ \ell(Q)=2^{-\mu}}} |s_Q|\,\big|\Delta_j^{\Psi}(a_Q)(x)\big| \le C \sum_{\mu\in\mathbf{Z}} 2^{\min(j,\mu)n} 2^{-|j-\mu|L'} 2^{-\mu n} 2^{(j-\mu)\alpha} \\ \times 2^{\max(\mu-j,0)\frac{n}{b}} \Big\{ M\Big(\sum_{\substack{Q\in\mathscr{D}\\ \ell(Q)=2^{-\mu}}} \big(|s_Q|\,|Q|^{-\frac{1}{2}-\frac{\alpha}{n}}\big)^b \chi_Q \Big)(x) \Big\}^{\frac{1}{b}}. \end{aligned}$$

Raise the preceding inequality to the power q and sum over $j\in\mathbf{Z}$; then raise to the power $1/q$ and take $\|\cdot\|_{L^p}$ norms in x. We obtain

$$\|f\|_{\dot F_p^{\alpha,q}} \le \Bigg\| \Big\{ \sum_{j\in\mathbf{Z}} \Big[\sum_{\mu\in\mathbf{Z}} d(j-\mu) \Big\{ M\Big(\sum_{\substack{Q\in\mathscr{D}\\ \ell(Q)=2^{-\mu}}} \big(|s_Q|\,|Q|^{-\frac{1}{2}-\frac{\alpha}{n}}\big)^b \chi_Q \Big) \Big\}^{\frac{1}{b}} \Big]^q \Big\}^{\frac{1}{q}} \Bigg\|_{L^p},$$

where $f = \sum_{Q\in\mathscr{D}} s_Q a_Q$ and

$$d(j-\mu) = C\, 2^{\min(j-\mu,0)(n-\frac{n}{b}) + (j-\mu)\alpha - |j-\mu|L'}.$$

We now estimate the expression inside the last L^p norm by

$$\Big\{ \sum_{j\in\mathbf{Z}} d(j)^{\min(1,q)} \Big\}^{\frac{1}{\min(1,q)}} \Big\{ \sum_{\mu\in\mathbf{Z}} \Big\{ M\Big(\sum_{\substack{Q\in\mathscr{D}\\ \ell(Q)=2^{-\mu}}} \big(|s_Q|\,|Q|^{-\frac{1}{2}-\frac{\alpha}{n}}\big)^b \chi_Q \Big) \Big\}^{\frac{q}{b}} \Big\}^{\frac{1}{q}},$$

and we note that the first term is a constant in view of (6.6.7). We conclude that

$$
\begin{aligned}
\Big\|\sum_{Q\in\mathscr{D}} s_Q a_Q\Big\|_{\dot F_p^{\alpha,q}} &\le C\bigg\|\bigg\{\sum_{\mu\in\mathbf{Z}}\Big\{M\Big(\sum_{\substack{Q\in\mathscr{D}\\ \ell(Q)=2^{-\mu}}}\big(|s_Q|\,|Q|^{-\frac12-\frac{\alpha}{n}}\big)^b\chi_Q\Big)\Big\}^{\frac{q}{b}}\bigg\}^{\frac1q}\bigg\|_{L^p}\\
&= C\bigg\|\bigg\{\sum_{\mu\in\mathbf{Z}}\Big\{M\Big(\sum_{\substack{Q\in\mathscr{D}\\ \ell(Q)=2^{-\mu}}}\big(|s_Q|\,|Q|^{-\frac12-\frac{\alpha}{n}}\big)^b\chi_Q\Big)\Big\}^{\frac{q}{b}}\bigg\}^{\frac{b}{q}}\bigg\|_{L^{\frac{p}{b}}}^{\frac1b}\\
&\le C'\bigg\|\bigg\{\sum_{\mu\in\mathbf{Z}}\Big\{\sum_{\substack{Q\in\mathscr{D}\\ \ell(Q)=2^{-\mu}}}\big(|s_Q|\,|Q|^{-\frac12-\frac{\alpha}{n}}\big)^b\chi_Q\Big\}^{\frac{q}{b}}\bigg\}^{\frac{b}{q}}\bigg\|_{L^{\frac{p}{b}}}^{\frac1b}\\
&= C'\bigg\|\bigg\{\sum_{\mu\in\mathbf{Z}}\sum_{\substack{Q\in\mathscr{D}\\ \ell(Q)=2^{-\mu}}}\big(|s_Q|\,|Q|^{-\frac12-\frac{\alpha}{n}}\big)^q\chi_Q\bigg\}^{\frac1q}\bigg\|_{L^p}\\
&= C'\big\|\{s_Q\}_Q\big\|_{\dot f_p^{\alpha,q}},
\end{aligned}
$$

where in the last inequality we used Theorem 4.6.6, which is valid under the assumption $1<\frac{p}{b},\frac{q}{b}<\infty$. This proves (6.6.2).

We now turn to the converse statement of the theorem. It is not difficult to see that there exist Schwartz functions Ψ (unrelated to the previous one) and Θ such that $\widehat{\Psi}$ is supported in the annulus $\frac12\le|\xi|\le 2$, $\widehat{\Psi}$ is at least $c>0$ in the smaller annulus $\frac35\le|\xi|\le\frac53$, and Θ is supported in the ball $|x|\le 1$ and satisfies $\int_{\mathbf{R}^n}x^\gamma\Theta(x)\,dx=0$ for all $|\gamma|\le L$, such that the identity

$$
\sum_{j\in\mathbf{Z}}\widehat{\Psi}(2^{-j}\xi)\widehat{\Theta}(2^{-j}\xi)=1 \tag{6.6.8}
$$

holds for all $\xi\in\mathbf{R}^n\setminus\{0\}$. (See Exercise 6.6.1.)

Using identity (6.6.8), we can write

$$
f=\sum_{j\in\mathbf{Z}}\Psi_{2^{-j}}*\Theta_{2^{-j}}*f.
$$

Setting $\mathscr{D}_j=\{Q\in\mathscr{D}:\ \ell(Q)=2^{-j}\}$, we now have

$$
f=\sum_{j\in\mathbf{Z}}\sum_{Q\in\mathscr{D}_j}\int_Q\Theta_{2^{-j}}(x-y)(\Psi_{2^{-j}}*f)(y)\,dy=\sum_{j\in\mathbf{Z}}\sum_{Q\in\mathscr{D}_j}s_Qa_Q,
$$

where we also set

$$
s_Q=|Q|^{\frac12}\sup_{y\in Q}|(\Psi_{2^{-j}}*f)(y)|\sup_{|\gamma|\le L}\big\|\partial^\gamma\Theta\big\|_{L^1}
$$

for Q in $\mathscr{D}_j$ and

$$a_Q(x) = \frac{1}{s_Q}\int_Q \Theta_{2^{-j}}(x-y)(\Psi_{2^{-j}} * f)(y)\,dy\,.$$

It is straightforward to verify that a_Q is supported in $3Q$ and that it has vanishing moments up to and including order L. Moreover, we have

$$|\partial^\gamma a_Q| \le \frac{1}{s_Q}\big\|\partial^\gamma \Theta\big\|_{L^1} 2^{j(n+|\gamma|)} \sup_Q |\Psi_{2^{-j}} * f| \le |Q|^{-\frac{1}{2}-\frac{|\gamma|}{n}}\,,$$

which makes the function a_Q a smooth L-atom. Now note that

$$\begin{aligned}
\sum_{\ell(Q)=2^{-j}} &\big(|Q|^{-\frac{\alpha}{n}-\frac{1}{2}} s_Q \chi_Q(x)\big)^q \\
&= C \sum_{\ell(Q)=2^{-j}} \big(2^{j\alpha} \sup_{y\in Q} |(\Psi_{2^{-j}} * f)(y)| \chi_Q(x)\big)^q \\
&\le C \sup_{|z|\le\sqrt{n}2^{-j}} \big(2^{j\alpha}(1+2^j|z|)^{-b}|(\Psi_{2^{-j}} * f)(x-z)|\big)^q (1+2^j|z|)^{bq} \\
&\le C\big(2^{j\alpha} M^{**}_{b,j}(f,\Psi)(x)\big)^q,
\end{aligned}$$

where we used the fact that in the first inequality there is only one nonzero term in the sum because of the appearance of the characteristic function. Summing over all $j \in \mathbf{Z}^n$, raising to the power $1/q$, and taking L^p norms yields the estimate

$$\big\|\{s_Q\}_Q\big\|_{\dot f_p^{\alpha,q}} \le C\bigg\|\Big(\sum_{j\in\mathbf{Z}} |2^{j\alpha} M^{**}_{b,j}(f;\Psi)|^q\Big)^{\frac{1}{q}}\bigg\|_{L^p} \le C\big\|f\big\|_{\dot F_p^{\alpha,q}}\,,$$

where the last inequality follows from Theorem 6.5.6. This proves (6.6.3). □

6.6.3 The Nonsmooth Atomic Decomposition of $\dot F_p^{\alpha,q}$

We now discuss the main theorem of this section, the nonsmooth atomic decomposition of the homogeneous Triebel–Lizorkin spaces $\dot F_p^{\alpha,q}$, which in particular includes that of the Hardy spaces H^p. We begin this task with a definition.

Definition 6.6.4. Let $0 < p \le 1$ and $1 \le q \le \infty$. A sequence of complex numbers $r = \{r_Q\}_{Q\in\mathscr{D}}$ is called an ∞-atom for $\dot f_p^{\alpha,q}$ if there exists a dyadic cube Q_0 such that

(a) $r_Q = 0$ if $Q \not\subseteq Q_0$;
(b) $\big\|g^{\alpha,q}(r)\big\|_{L^\infty} \le |Q_0|^{-\frac{1}{p}}$.

We observe that every ∞-atom $r = \{r_Q\}$ for $\dot f_p^{\alpha,q}$ satisfies $\big\|r\big\|_{\dot f_p^{\alpha,q}} \le 1$. Indeed,

$$\big\|r\big\|_{\dot f_p^{\alpha,q}}^p = \int_{Q_0} |g^{\alpha,q}(r)|^p\,dx \le |Q_0|^{-1}|Q_0| = 1\,.$$

The following theorem concerns the atomic decomposition of the spaces $\dot{f}_p^{\alpha,q}$.

Theorem 6.6.5. *Suppose $\alpha \in \mathbf{R}$, $0 < q < \infty$, $0 < p < \infty$, and $s = \{s_Q\}_Q$ is in $\dot{f}_p^{\alpha,q}$. Then there exist $C_{n,p,q} > 0$, a sequence of scalars λ_j, and a sequence of ∞-atoms $r_j = \{r_{j,Q}\}_Q$ for $\dot{f}_p^{\alpha,q}$ such that*

$$s = \{s_Q\}_Q = \sum_{j=1}^{\infty} \lambda_j \{r_{j,Q}\}_Q = \sum_{j=1}^{\infty} \lambda_j r_j$$

and such that

$$\Big(\sum_{j=1}^{\infty} |\lambda_j|^p \Big)^{\frac{1}{p}} \le C_{n,p,q} \big\| s \big\|_{\dot{f}_p^{\alpha,q}} . \tag{6.6.9}$$

Proof. We fix α, p, q, and a sequence $s = \{s_Q\}_Q$ as in the statement of the theorem. For a dyadic cube R in $\mathscr{D}$ we define the function

$$g_R^{\alpha,q}(s)(x) = \Big(\sum_{\substack{Q \in \mathscr{D} \\ R \subseteq Q}} \big(|Q|^{\frac{\alpha}{n}-\frac{1}{2}} |s_Q| \chi_Q(x)\big)^q \Big)^{\frac{1}{q}}$$

and we observe that this function is constant on R. We also note that for dyadic cubes R_1 and R_2 with $R_1 \subseteq R_2$ we have

$$g_{R_2}^{\alpha,q}(s) \le g_{R_1}^{\alpha,q}(s) .$$

Finally, we observe that

$$\begin{aligned} \lim_{\substack{\ell(R) \to \infty \\ x \in R}} g_R^{\alpha,q}(s)(x) &= 0 \\ \lim_{\substack{\ell(R) \to 0 \\ x \in R}} g_R^{\alpha,q}(s)(x) &= g^{\alpha,q}(s)(x) , \end{aligned}$$

where $g^{\alpha,q}(s)$ is the function defined in (6.6.1).

For $k \in \mathbf{Z}$ we set

$$\mathscr{A}_k = \Big\{ R \in \mathscr{D} : \, g_R^{\alpha,q}(s)(x) > 2^k \quad \text{for all } x \in R \Big\} .$$

We note that $\mathscr{A}_{k+1} \subseteq \mathscr{A}_k$ for all k in $\mathbf{Z}$ and that

$$\{ x \in \mathbf{R}^n : \, g^{\alpha,q}(s)(x) > 2^k \} = \bigcup_{R \in \mathscr{A}_k} R . \tag{6.6.10}$$

Moreover, we have for all $k \in \mathbf{Z}$,

$$\Big(\sum_{Q \in \mathscr{D} \setminus \mathscr{A}_k} \big(|Q|^{-\frac{\alpha}{n}-\frac{1}{2}} |s_Q| \chi_Q(x)\big)^q \Big)^{\frac{1}{q}} \le 2^k , \qquad \text{for all } x \in \mathbf{R}^n . \tag{6.6.11}$$

To prove (6.6.11) we assume that $g^{\alpha,q}(s)(x) > 2^k$; otherwise, the conclusion is trivial. Then there exists a maximal dyadic cube $R_{\max}$ in $\mathscr{A}_k$ such that $x \in R_{\max}$. Letting R_0 be the unique dyadic cube that contains $R_{\max}$ and has twice its side length, we have that the left-hand side of (6.6.11) is equal to $g^{\alpha,q}_{R_0}(s)(x)$, which is at most 2^k, since R_0 is not contained in $\mathscr{A}_k$.

Since $g^{\alpha,q}(s) \in L^p(\mathbf{R}^n)$, by our assumption, and $g^{\alpha,q}(s) > 2^k$ for all $x \in Q$ if $Q \in \mathscr{A}_k$, the cubes in $\mathscr{A}_k$ must have size bounded above by some constant. We set

$$\mathscr{B}_k = \big\{ Q \in \mathscr{D} : \quad Q \text{ is a maximal dyadic cube in } \mathscr{A}_k \setminus \mathscr{A}_{k+1} \big\} .$$

For J in $\mathscr{B}_k$ we define a sequence $t(k,J) = \{t(k,J)_Q\}_{Q\in\mathscr{D}}$ by setting

$$t(k,J)_Q = \begin{cases} s_Q & \text{if } Q \subseteq J \text{ and } Q \in \mathscr{A}_k \setminus \mathscr{A}_{k+1}, \\ 0 & \text{otherwise.} \end{cases}$$

We can see that if

$$Q \notin \bigcup_{k\in\mathbf{Z}} \mathscr{A}_k , \qquad \text{then} \qquad s_Q = 0,$$

and the identity

$$s = \sum_{k\in\mathbf{Z}} \sum_{J\in\mathscr{B}_k} t(k,J) \tag{6.6.12}$$

is valid. For all $x \in \mathbf{R}^n$ we have

$$\begin{aligned} \big|g^{\alpha,q}(t(k,J))(x)\big| &= \Big(\sum_{\substack{Q\subseteq J \\ Q\in\mathscr{A}_k\setminus\mathscr{A}_{k+1}}} \big(|Q|^{-\frac{\alpha}{n}-\frac{1}{2}} |s_Q|\, \chi_Q(x)\big)^q \Big)^{\frac{1}{q}} \\ &\le \Big(\sum_{\substack{Q\subseteq J \\ Q\in\mathscr{D}\setminus\mathscr{A}_{k+1}}} \big(|Q|^{-\frac{\alpha}{n}-\frac{1}{2}} |s_Q|\, \chi_Q(x)\big)^q \Big)^{\frac{1}{q}} \\ &\le 2^{k+1} , \end{aligned} \tag{6.6.13}$$

where we used (6.6.11) in the last estimate. We define atoms $r(k,J) = \{r(k,J)_Q\}_{Q\in\mathscr{D}}$ by setting

$$r(k,J)_Q = 2^{-k-1} |J|^{-\frac{1}{p}} t(k,J)_Q , \tag{6.6.14}$$

and we also define scalars

$$\lambda_{k,J} = 2^{k+1} |J|^{\frac{1}{p}} .$$

To see that each $r(k,J)$ is an ∞-atom for $\dot{f}^{\alpha,q}_p$, we observe that $r(k,J)_Q = 0$ if $Q \nsubseteq J$ and that

$$\big|g^{\alpha,q}(t(k,J))(x)\big| \le |J|^{-\frac{1}{p}} , \qquad \text{for all } x \in \mathbf{R}^n,$$

in view of (6.6.13). Also using (6.6.12) and (6.6.14), we obtain that

$$s = \sum_{k\in\mathbf{Z}} \sum_{J\in\mathscr{B}_k} \lambda_{k,J}\, r(k,J) ,$$

which says that s can be written as a linear combination of atoms. Finally, we estimate the sum of the pth power of the coefficients $\lambda_{k,J}$. We have

$$
\begin{aligned}
\sum_{k\in\mathbf{Z}}\sum_{J\in\mathscr{B}_k}|\lambda_{k,J}|^p &= \sum_{k\in\mathbf{Z}}2^{(k+1)p}\sum_{J\in\mathscr{B}_k}|J| \\
&\le 2^p\sum_{k\in\mathbf{Z}}2^{kp}\Big|\bigcup_{Q\in\mathscr{A}_k}Q\Big| \\
&= 2^p\sum_{k\in\mathbf{Z}}2^{k(p-1)}2^k|\{x\in\mathbf{R}^n:\ g^{\alpha,q}(s)(x)>2^k\}| \\
&\le 2^p\sum_{k\in\mathbf{Z}}\int_{2^k}^{2^{k+1}}2^{k(p-1)}|\{x\in\mathbf{R}^n:\ g^{\alpha,q}(s)(x)>\tfrac{\lambda}{2}\}|\,d\lambda \\
&\le 2^p\sum_{k\in\mathbf{Z}}\int_{2^k}^{2^{k+1}}\lambda^{p-1}|\{x\in\mathbf{R}^n:\ g^{\alpha,q}(s)(x)>\tfrac{\lambda}{2}\}|\,d\lambda \\
&= \frac{2^{2p}}{p}\big\|g^{\alpha,q}(s)\big\|_{L^p}^p \\
&= \frac{2^{2p}}{p}\big\|s\big\|_{\dot{f}_p^{\alpha,q}}^p .
\end{aligned}
$$

Taking the pth root yields (6.6.9). The proof of the theorem is now complete. □

We now deduce a corollary regarding a new characterization of the space $\dot{f}_p^{\alpha,q}$.

Corollary 6.6.6. *Suppose $\alpha\in\mathbf{R}$, $0<p\le 1$, and $p\le q\le\infty$. Then we have*

$$
\big\|s\big\|_{\dot{f}_p^{\alpha,q}}\approx\inf\bigg\{\Big(\sum_{j=1}^{\infty}|\lambda_j|^p\Big)^{\frac{1}{p}}:\ s=\sum_{j=1}^{\infty}\lambda_j r_j,\quad r_j \textit{ is an } \infty\textit{-atom for } \dot{f}_p^{\alpha,q}\bigg\}.
$$

Proof. One direction in the previous estimate is a direct consequence of (6.6.9). The other direction uses the observation made after Definition 6.6.4 that every ∞-atom r for $\dot{f}_p^{\alpha,q}$ satisfies $\big\|r\big\|_{\dot{f}_p^{\alpha,q}}\le 1$ and that for $p\le 1$ and $p\le q$ the quantity $s\to\big\|s\big\|_{\dot{f}_p^{\alpha,q}}^p$ is subadditive; see Exercise 6.6.2. Then each $s=\sum_{j=1}^{\infty}\lambda_j r_j$ (with r_j ∞-atoms for $\dot{f}_p^{\alpha,q}$ and $\sum_{j=1}^{\infty}|\lambda_j|^p<\infty$) must be an element of $\dot{f}_p^{\alpha,q}$, since

$$
\Big\|\sum_{j=1}^{\infty}\lambda_j r_j\Big\|_{\dot{f}_p^{\alpha,q}}^p\le\sum_{j=1}^{\infty}|\lambda_j|^p\,\big\|r_j\big\|_{\dot{f}_p^{\alpha,q}}^p\le\sum_{j=1}^{\infty}|\lambda_j|^p<\infty .
$$

This concludes the proof of the corollary. □

The theorem we just proved allows us to obtain an atomic decomposition for the space $\dot{F}_p^{\alpha,q}$ as well. Indeed, we have the following result:

Corollary 6.6.7. *Let $\alpha\in\mathbf{R}$, $0<p\le 1$, $L\ge[\frac{n}{p}-n-\alpha]$ and let q satisfy $p\le q<\infty$. Then we have the following representation:*

$$\big\|f\big\|_{\dot{F}_p^{\alpha,q}} \approx \inf\Big\{\Big(\sum_{j=1}^{\infty}|\lambda_j|^p\Big)^{\frac{1}{p}} : f=\sum_{j=1}^{\infty}\lambda_j A_j, \quad A_j=\sum_{Q\in\mathscr{D}} r_Q a_Q, \quad a_Q \textit{ are}$$
$$\textit{smooth L-atoms for } \dot{F}_p^{\alpha,q} \textit{ and } \{r_Q\}_Q \textit{ is an } \infty\textit{-atom for } \dot{f}_p^{\alpha,q}\Big\}.$$

Proof. Let $f=\sum_{j=1}^{\infty}\lambda_j A_j$ as described previously. Using Exercise 6.6.2, we have

$$\big\|f\big\|_{\dot{F}_p^{\alpha,q}}^p \le \sum_{j=1}^{\infty}|\lambda_j|^p\big\|A_j\big\|_{\dot{F}_p^{\alpha,q}}^p \le c_{n,p}\sum_{j=1}^{\infty}|\lambda_j|^p\big\|r\big\|_{\dot{f}_p^{\alpha,q}}^p,$$

where in the last estimate we used Theorem 6.6.3. Using the fact that every ∞-atom $r=\{r_Q\}$ for $\dot{f}_p^{\alpha,q}$ satisfies $\big\|r\big\|_{\dot{f}_p^{\alpha,q}}\le 1$, we conclude that every element f in $\mathscr{S}'(\mathbf{R}^n)$ that has the form $\sum_{j=1}^{\infty}\lambda_j A_j$ lies in the homogeneous Triebel–Lizorkin space $\dot{F}_p^{\alpha,q}$ [and has norm controlled by a constant multiple of $\big(\sum_{j=1}^{\infty}|\lambda_j|^p\big)^{\frac{1}{p}}$].

Conversely, Theorem 6.6.3 gives that every element of $\dot{F}_p^{\alpha,q}$ has a smooth atomic decomposition. Then we can write

$$f=\sum_{Q\in\mathscr{D}} s_Q a_Q,$$

where each a_Q is a smooth L-atom for the cube Q. Using Theorem 6.6.5 we can now write $s=\{s_Q\}_Q$ as a sum of ∞-atoms for $\dot{f}_p^{\alpha,q}$, that is,

$$s=\sum_{j=1}^{\infty}\lambda_j r_j,$$

where

$$\Big(\sum_{j=1}^{\infty}|\lambda_j|^p\Big)^{\frac{1}{p}} \le c\big\|s\big\|_{\dot{f}_p^{\alpha,q}} \le c\big\|f\big\|_{\dot{F}_p^{\alpha,q}},$$

where the last step uses Theorem 6.6.3 again. It is simple to see that

$$f=\sum_{Q\in\mathscr{D}}\sum_{j=1}^{\infty}\lambda_j r_{j,Q}a_Q=\sum_{j=1}^{\infty}\lambda_j\Big(\sum_{Q\in\mathscr{D}} r_{j,Q}a_Q\Big),$$

and we set the expression inside the parentheses equal to A_j. □

6.6.4 Atomic Decomposition of Hardy Spaces

We now pass to one of the main theorems of this chapter, the atomic decomposition of $H^p(\mathbf{R}^n)$ for $0<p\le 1$. We begin by defining atoms for H^p.

Definition 6.6.8. Let $1<q\le\infty$. A function A is called *an L^q-atom for $H^p(\mathbf{R}^n)$* if there exists a cube Q such that

(a) A is supported in Q;

(b) $\|A\|_{L^q} \le |Q|^{\frac{1}{q}-\frac{1}{p}}$;

(c) $\int x^\gamma A(x)\,dx = 0$ for all multi-indices γ with $|\gamma| \le [\frac{n}{p} - n]$.

Notice that any L^r-atom for H^p is also an L^q-atom for H^p whenever $0 < p \le 1$ and $1 < q < r \le \infty$. It is also simple to verify that an L^q-atom A for H^p is in fact in H^p. We prove this result in the next theorem for $q = 2$, and we refer the reader to Exercise 6.6.4 for the case of a general q.

Theorem 6.6.9. *Let $0 < p \le 1$. There is a constant $C_{n,p} < \infty$ such that every L^2-atom A for $H^p(\mathbf{R}^n)$ satisfies*

$$\|A\|_{H^p} \le C_{n,p}.$$

Proof. We could prove this theorem either by showing that the smooth maximal function $M(A;\Phi)$ is in L^p or by showing that the square function $\big(\sum_j |\Delta_j(A)|^2\big)^{1/2}$ is in L^p. The operators Δ_j here are as in Theorem 5.1.2. Both proofs are similar; we present the second, and we refer to Exercise 6.6.3 for the first.

Let $A(x)$ be an atom that we assume is supported in a cube Q centered at the origin [otherwise apply the argument to the atom $A(x - c_Q)$, where c_Q is the center of Q]. We control the L^p quasinorm of $\big(\sum_j |\Delta_j(A)|^2\big)^{1/2}$ by estimating it over the cube Q^* and over $(Q^*)^c$, where $Q^* = 2\sqrt{n}\,Q$. We have

$$\bigg(\int_{Q^*} \Big(\sum_j |\Delta_j(A)|^2\Big)^{\frac{p}{2}} dx\bigg)^{\frac{1}{p}} \le \bigg(\int_{Q^*} \sum_j |\Delta_j(A)|^2\,dx\bigg)^{\frac{1}{2}} |Q^*|^{\frac{1}{p(2/p)'}}.$$

Using that the square function $f \mapsto \big(\sum_j |\Delta_j(f)|^2\big)^{\frac{1}{2}}$ is L^2 bounded, we obtain

$$\begin{aligned}\bigg(\int_{Q^*} \Big(\sum_j |\Delta_j(A)|^2\Big)^{\frac{p}{2}} dx\bigg)^{\frac{1}{p}} &\le C_n \|A\|_{L^2} |Q^*|^{\frac{1}{p(2/p)'}} \\ &\le C_n (2\sqrt{n})^{\frac{n}{p}-\frac{n}{2}} |Q|^{\frac{1}{2}-\frac{1}{p}} |Q|^{\frac{1}{p}-\frac{1}{2}} \\ &= C_n'.\end{aligned} \tag{6.6.15}$$

To estimate the contribution of the square function outside Q^*, we use the cancellation of the atoms. Let $k = [\frac{n}{p} - n] + 1$. We have

$$\begin{aligned}\Delta_j(A)(x) &= \int_Q A(y)\Psi_{2^{-j}}(x-y)\,dy \\ &= 2^{jn} \int_Q A(y) \bigg[\Psi(2^j x - 2^j y) - \sum_{|\beta| \le k-1} (\partial^\beta \Psi)(2^j x) \frac{(-2^j y)^\beta}{\beta!}\bigg] dy \\ &= 2^{jn} \int_Q A(y) \bigg[\sum_{|\beta| = k} (\partial^\beta \Psi)(2^j x - 2^j \theta y) \frac{(-2^j y)^\beta}{\beta!}\bigg] dy,\end{aligned}$$

where $0 \le \theta \le 1$. Taking absolute values, using the fact that $\partial^\beta \Psi$ are Schwartz functions, and that $|x-\theta y| \ge |x|-|y| \ge \frac{1}{2}|x|$ whenever $y \in Q$ and $x \notin Q^*$, we obtain the estimate

$$\begin{aligned}
|\Delta_j(A)(x)| &\le 2^{jn} \int_Q |A(y)| \sum_{|\beta|=k} \frac{C_N}{(1+2^j\frac{1}{2}|x|)^N} \frac{|2^j y|^k}{\beta!} dy \\
&\le \frac{C_{N,p,n} 2^{j(k+n)}}{(1+2^j|x|)^N} \left(\int_Q |A(y)|^2\, dy\right)^{\frac{1}{2}} \left(\int_Q |y|^{2k}\, dy\right)^{\frac{1}{2}} \\
&\le \frac{C'_{N,p,n} 2^{j(k+n)}}{(1+2^j|x|)^N} |Q|^{\frac{1}{2}-\frac{1}{p}} |Q|^{\frac{k}{n}+\frac{1}{2}} \\
&= \frac{C_{N,p,n} 2^{j(k+n)}}{(1+2^j|x|)^N} |Q|^{1+\frac{k}{n}-\frac{1}{p}}
\end{aligned}$$

for $x \in (Q^*)^c$. For such x we now have

$$\left(\sum_{j\in\mathbf{Z}} |\Delta_j(A)(x)|^2\right)^{\frac{1}{2}} \le C_{N,p,n} |Q|^{1+\frac{k}{n}-\frac{1}{p}} \left(\sum_{j\in\mathbf{Z}} \frac{2^{2j(k+n)}}{(1+2^j|x|)^{2N}}\right)^{\frac{1}{2}}. \tag{6.6.16}$$

It is a simple fact that the series in (6.6.16) converges. Indeed, considering the cases $2^j \le 1/|x|$ and $2^j > 1/|x|$ we see that both terms in the second series in (6.6.16) contribute at most a fixed multiple of $|x|^{-2k-2n}$. It remains to estimate the L^p quasinorm of the square root of the second series in (6.6.16) raised over $(Q^*)^c$. This is bounded by a constant multiple of

$$\left(\int_{(Q^*)^c} \frac{1}{|x|^{p(k+n)}}\, dx\right)^{\frac{1}{p}} \le C_{n,p} \left(\int_{c|Q|^{\frac{1}{n}}}^{\infty} r^{-p(k+n)+n-1}\, dr\right)^{\frac{1}{p}},$$

for some constant c, and the latter is easily seen to be bounded above by a constant multiple of $|Q|^{-1-\frac{k}{n}+\frac{1}{p}}$. Here we use the fact that $p(k+n) > n$ or, equivalently, $k > \frac{n}{p} - n$, which is certainly true, since k was chosen to be $[\frac{n}{p} - n] + 1$. Combining this estimate with that in (6.6.15), we conclude the proof of the theorem. □

We now know that L^q-atoms for H^p are indeed elements of H^p. The main result of this section is to obtain the converse (i.e., every element of H^p can be decomposed as a sum of L^2-atoms for H^p).

Applying the same idea as in Corollary 6.6.7 to H^p, we obtain the following result.

Theorem 6.6.10. *Let $0 < p \le 1$. Given a distribution $f \in H^p(\mathbf{R}^n)$, there exists a sequence of L^2-atoms for H^p, $\{A_j\}_{j=1}^\infty$, and a sequence of scalars $\{\lambda_j\}_{j=1}^\infty$ such that*

$$\sum_{j=1}^N \lambda_j A_j \to f \qquad \text{in } H^p.$$

Moreover, we have

$$\|f\|_{H^p} \approx \inf\Big\{ \big(\sum_{j=1}^{\infty} |\lambda_j|^p \big)^{\frac{1}{p}} : f = \lim_{N\to\infty} \sum_{j=1}^{N} \lambda_j A_j, \qquad (6.6.17)$$
$$A_j \textit{ are } L^2\textit{-atoms for } H^p \textit{ and the limit is taken in } H^p \Big\}.$$

Proof. Let A_j be L^2-atoms for H^p and $\sum_{j=1}^{\infty} |\lambda_j|^p < \infty$. It follows from Theorem 6.6.9 that

$$\Big\| \sum_{j=1}^{N} \lambda_j A_j \Big\|_{H^p}^p \le C_{n,p}^p \sum_{j=1}^{N} |\lambda_j|^p .$$

Thus if the sequence $\sum_{j=1}^{N} \lambda_j A_j$ converges to f in H^p, then

$$\|f\|_{H^p} \le C_{n,p} \Big(\sum_{j=1}^{\infty} |\lambda_j|^p \Big)^{\frac{1}{p}},$$

which proves the direction $\le$ in (6.6.17). The gist of the theorem is contained in the converse statement.

Using Theorem 6.6.3 (with $L = [\frac{n}{p} - n]$), we can write every element f in $\dot{F}_p^{0,2} = H^p$ as a sum of the form $f = \sum_{Q\in\mathscr{D}} s_Q a_Q$, where each a_Q is a smooth L-atom for the cube Q and $s = \{s_Q\}_{Q\in\mathscr{D}}$ is a sequence in $\dot{f}_p^{0,2}$. We now use Theorem 6.6.5 to write the sequence $s = \{s_Q\}_Q$ as

$$s = \sum_{j=1}^{\infty} \lambda_j r_j ,$$

i.e., as a sum of ∞-atoms r_j for $\dot{f}_p^{0,2}$, such that

$$\big(\sum_{j=1}^{\infty} |\lambda_j|^p \big)^{\frac{1}{p}} \le C \|s\|_{\dot{f}_p^{0,2}} \le C \|f\|_{H^p} . \qquad (6.6.18)$$

Then we have

$$f = \sum_{Q\in\mathscr{D}} s_Q a_Q = \sum_{Q\in\mathscr{D}} \sum_{j=1}^{\infty} \lambda_j r_{j,Q} a_Q = \sum_{j=1}^{\infty} \lambda_j A_j , \qquad (6.6.19)$$

where we set

$$A_j = \sum_{Q\in\mathscr{D}} r_{j,Q} a_Q \qquad (6.6.20)$$

and the series in (6.6.19) converges in $\mathscr{S}'(\mathbf{R}^n)$. Next we show that each A_j is a fixed multiple of an L^2-atom for H^p. Let us fix an index j. By the definition of the ∞-atom for $\dot{f}_p^{0,2}$, there exists a dyadic cube Q_0^j such that $r_{j,Q} = 0$ for all dyadic cubes Q not contained in Q_0^j. Then the support of each a_Q that appears in (6.6.20) is contained in $3Q$, hence in $3Q_0^j$. This implies that the function A_j is supported in $3Q_0^j$. The same is true for the function $g^{0,2}(r_j)$ defined in (6.6.1). Using this fact, we have

$$\begin{aligned}
\|A_j\|_{L^2} &\approx \|A_j\|_{\dot F_2^{0,2}}\\
&\le c\|r_j\|_{\dot f_2^{0,2}}\\
&= c\|g^{0,2}(r_j)\|_{L^2}\\
&\le c\|g^{0,2}(r_j)\|_{L^\infty}|3Q_0^j|^{\frac{1}{2}}\\
&\le c|3Q_0^j|^{-\frac{1}{p}+\frac{1}{2}}.
\end{aligned}$$

Since the series (6.6.20) defining A_j converges in L^2 and A_j is supported in some cube, this series also converges in L^1. It follows that the vanishing moment conditions of A_j are inherited from those of each a_Q. We conclude that each A_j is a fixed multiple of an L^2-atom for H^p.

Finally, we need to show that the series in (6.6.19) converges in $H^p(\mathbf{R}^n)$. But

$$\Big\|\sum_{j=N}^{M}\lambda_jA_j\Big\|_{H^p}\le C_{n,p}\Big(\sum_{j=N}^{M}|\lambda_j|^p\Big)^{\frac{1}{p}}\to 0$$

as $M,N\to\infty$ in view of the convergence of the series in (6.6.18). This implies that the series $\sum_{j=1}^{\infty}\lambda_jA_j$ is Cauchy in H^p, and since it converges to f in $\mathscr{S}'(\mathbf{R}^n)$, it must converge to f in H^p. Combining this fact with (6.6.18) yields the direction $\ge$ in (6.6.17). □

Remark 6.6.11. Property (c) in Definition 6.6.8 can be replaced by

$$\int x^\gamma A(x)\,dx=0 \qquad \text{for all multi-indices } \gamma \text{ with } |\gamma|\le L,$$

for any $L\ge[\frac{n}{p}-n]$, and the atomic decomposition of H^p holds unchanged. In fact, in the proof of Theorem 6.6.10 we may take $L\ge[\frac{n}{p}-n]$ instead of $L=[\frac{n}{p}-n]$ and then apply Theorem 6.6.3 for this L. Observe that Theorem 6.6.3 was valid for all $L\ge[\frac{n}{p}-n]$.

This observation can be very useful in certain applications.

Exercises

6.6.1. (a) Prove that there exists a Schwartz function Θ supported in the unit ball $|x|\le 1$ such that $\int_{\mathbf{R}^n}x^\gamma\Theta(x)\,dx=0$ for all multi-indices γ with $|\gamma|\le N$ and such that $|\widehat{\Theta}|\ge\frac{1}{2}$ on the annulus $\frac{1}{2}\le|\xi|\le 2$.
(b) Prove there exists a Schwartz function Ψ whose Fourier transform is supported in the annulus $\frac{1}{2}\le|\xi|\le 2$ and is at least $c>0$ in the smaller annulus $\frac{3}{5}\le|\xi|\le\frac{5}{3}$ such that we have

$$\sum_{j\in\mathbf{Z}}\widehat{\Psi}(2^{-j}\xi)\widehat{\Theta}(2^{-j}\xi)=1$$

for all $\xi \in \mathbf{R}^n \setminus \{0\}$.
[*Hint:* Part (a): Let θ be a real-valued Schwartz function supported in the ball $|x| \le 1$ and such that $\widehat{\theta}(0) = 1$. Then for some $\varepsilon > 0$ we have $\widehat{\theta}(\xi) \ge \frac{1}{2}$ for all ξ satisfying $|\xi| < 2\varepsilon < 1$. Set $\Theta = (-\Delta)^N(\theta_\varepsilon)$. Part (b): Define the function $\widehat{\Psi(\xi)} = \widehat{\eta}(\xi)\big(\sum_{j\in\mathbf{Z}} \widehat{\eta}(2^{-j}\xi)\widehat{\Theta}(2^{-j}\xi)\big)^{-1}$ for a suitable η.]

6.6.2. Let $\alpha \in \mathbf{R}$, $0 < p \le 1$, $p \le q \le +\infty$.
(a) For all f,g in $\mathscr{S}'(\mathbf{R}^n)$ show that

$$\big\|f+g\big\|^p_{\dot F^{\alpha,q}_p} \le \big\|f\big\|^p_{\dot F^{\alpha,q}_p} + \big\|g\big\|^p_{\dot F^{\alpha,q}_p}.$$

(b) For all sequences $\{s_Q\}_{Q\in\mathscr{D}}$ and $\{t_Q\}_{Q\in\mathscr{D}}$ show that

$$\big\|\{s_Q\}_Q + \{t_Q\}_Q\big\|^p_{\dot f^{\alpha,q}_p} \le \big\|\{s_Q\}_Q\big\|^p_{\dot f^{\alpha,q}_p} + \big\|\{t_Q\}_Q\big\|^p_{\dot f^{\alpha,q}_p}.$$

[*Hint:* Use $|a+b|^p \le |a|^p + |b|^p$ and apply Minkowski's inequality on $L^{q/p}$ (or on $\ell^{q/p}$).]

6.6.3. Let Φ be a smooth function supported in the unit ball of $\mathbf{R}^n$. Use the same idea as in Theorem 6.6.9 to show directly (without appealing to any other theorem) that the smooth maximal function $M(\cdot,\Phi)$ of an L^2-atom for H^p lies in L^p when $p < 1$. Recall that $M(f,\Phi) = \sup_{t>0} |\Phi_t * f|$.

6.6.4. Extend Theorem 6.6.9 to the case $1 < q \le \infty$. Precisely, prove that there is a constant $C_{n,p,q}$ such that every L^q-atom A for H^p satisfies

$$\big\|A\big\|_{H^p} \le C_{n,p,q}.$$

[*Hint:* If $1 < q < 2$, use the boundedness of the square function on L^q, and for $2 \le q \le \infty$, its boundedness on L^2.]

6.6.5. Show that the space H^p_F of all finite linear combinations of L^2-atoms for H^p is dense in H^p.
[*Hint:* Use Theorem 6.6.10.]

6.6.6. Show that for all $\mu, j \in \mathbf{Z}$, all $N,b > 0$ satisfying $N > n/b$ and $b < 1$, all scalars s_Q (indexed by dyadic cubes Q with centers c_Q), and all $x \in \mathbf{R}^n$ we have

$$\sum_{\substack{Q\in\mathscr{D}\\ \ell(Q)=2^{-\mu}}} \frac{|s_Q|}{(1+2^{\min(j,\mu)}|x-c_Q|)^N}$$

$$\le c(n,N,b)\, 2^{\max(\mu-j,0)\frac{n}{b}} \Big\{ M\Big(\sum_{\substack{Q\in\mathscr{D}\\ \ell(Q)=2^{-\mu}}} |s_Q|^b \chi_Q \Big)(x) \Big\}^{\frac{1}{b}},$$

where M is the Hardy–Littlewood maximal operator and $c(n,N,b)$ is a constant.
[*Hint:* Define $\mathscr{F}_0 = \{Q \in \mathscr{D} : \ell(Q) = 2^{-\mu},\, |c_Q - x|\,2^{\min(j,\mu)} \le 1\}$ and for $k \ge 1$ define $\mathscr{F}_k = \{Q \in \mathscr{D} : \ell(Q) = 2^{-\mu},\, 2^{k-1} < |c_Q - x|\,2^{\min(j,\mu)} \le 2^k\}$. Break up the sum on the left as a sum over the families $\mathscr{F}_k$ and use that $\sum_{Q\in\mathscr{F}_k} |s_Q| \le \big(\sum_{Q\in\mathscr{F}_k} |s_Q|^b\big)^{1/b}$ and the fact that $\big|\bigcup_{Q\in\mathscr{F}_k} Q\big| \le c_n 2^{-\min(j,\mu)n+kn}$.]

6.6.7. Let A be an L^2-atom for $H^p(\mathbf{R}^n)$ for some $0 < p < 1$. Show that there is a constant C such that for all multi-indices α with $|\alpha| \le k = [\frac{n}{p} - n]$ we have

$$\sup_{\xi\in\mathbf{R}^n} |\xi|^{|\alpha|-k-1} |(\partial^\alpha \widehat{A})(\xi)| \le C \|A\|_{L^2(\mathbf{R}^n)}^{-\frac{2p}{2-p}(\frac{k+1}{n}+\frac{1}{2})-1} .$$

[*Hint:* Subtract the Taylor polynomial of degree $k - |\alpha|$ at 0 of the function $x \mapsto e^{-2\pi i x\cdot\xi}$.]

6.6.8. Let A be an L^2-atom for $H^p(\mathbf{R}^n)$ for some $0 < p < 1$. Show that for all multi-indices α and all $1 \le r \le \infty$ there is a constant C such that

$$\big\| |\partial^\alpha \widehat{A}|^2 \big\|_{L^{r'}(\mathbf{R}^n)} \le C \|A\|_{L^2(\mathbf{R}^n)}^{-\frac{2p}{2-p}(\frac{2|\alpha|}{n}+\frac{1}{r})+2} .$$

[*Hint:* In the case $r = 1$ use the $L^1 \to L^\infty$ boundedness of the Fourier transform and in the case $r = \infty$ use Plancherel's theorem. For general r use interpolation.]

6.6.9. Let f be in $H^p(\mathbf{R}^n)$ for some $0 < p \le 1$. Then the Fourier transform of f, originally defined as a tempered distribution, is a continuous function that satisfies

$$|\widehat{f}(\xi)| \le C_{n,p} \|f\|_{\dot{H}^p(\mathbf{R}^n)} |\xi|^{\frac{n}{p}-n}$$

for some constant $C_{n,p}$ independent of f.
[*Hint:* If f is an L^2-atom for H^p, combine the estimates of Exercises 6.6.7 and 6.6.8 with $\alpha = 0$ (and $r = 1$). In general, apply Theorem 6.6.10.]

6.6.10. Let A be an L^∞-atom for $H^p(\mathbf{R}^n)$ for some $0 < p < 1$ and let $\alpha = \frac{n}{p} - n$. Show that there is a constant $C_{n,p}$ such that for all g in $\dot{\Lambda}_\alpha(\mathbf{R}^n)$ we have

$$\left| \int_{\mathbf{R}^n} A(x) g(x)\, dx \right| \le C_{n,p} \|g\|_{\dot{\Lambda}_\alpha(\mathbf{R}^n)} .$$

[*Hint:* Suppose that A is supported in a cube Q of side length $2^{-\nu}$ and center c_Q. Write the previous integrand as $\sum_j \Delta_j(A)\Delta_j(g)$ for a suitable Littlewood–Paley operator Δ_j and apply the result of Appendix K.2 to obtain the estimate

$$|\Delta_j(A)(x)| \le C_N |Q|^{-\frac{1}{p}+1} \frac{2^{\min(j,\nu)n} 2^{-|j-\nu|D}}{\big(1 + 2^{\min(j,\nu)}|x - c_Q|\big)^N} ,$$

where $D = [\alpha] + 1$ when $\nu \ge j$ and $D = 0$ when $\nu < j$. Use Theorem 6.3.6.]

6.7 Singular Integrals on Function Spaces

Our final task in this chapter is to investigate the action of singular integrals on function spaces. The emphasis of our study focuses on Hardy spaces, although with no additional effort the action of singular integrals on other function spaces can also be obtained.

6.7.1 Singular Integrals on the Hardy Space H^1

Before we discuss the main results in this topic, we review some background on singular integrals from Chapter 4.

Let $K(x)$ be a function defined away from the origin on $\mathbf{R}^n$ that satisfies the size estimate

$$\sup_{0<R<\infty} \frac{1}{R}\int_{|x|\le R} |K(x)|\,|x|\,dx \le A_1\,, \tag{6.7.1}$$

the smoothness estimate, expressed in terms of Hörmander's condition,

$$\sup_{y\in\mathbf{R}^n\setminus\{0\}} \int_{|x|\ge 2|y|} |K(x-y)-K(x)|\,dx \le A_2\,, \tag{6.7.2}$$

and the cancellation condition

$$\sup_{0<R_1<R_2<\infty} \left| \int_{R_1<|x|<R_2} K(x)\,dx \right| \le A_3\,, \tag{6.7.3}$$

for some $A_1, A_2, A_3 < \infty$. Condition (6.7.3) implies that there exists a sequence $\varepsilon_j \downarrow 0$ as $j \to \infty$ such that the following limit exists:

$$\lim_{j\to\infty} \int_{\varepsilon_j\le|x|\le 1} K(x)\,dx = L_0.$$

This gives that for a smooth and compactly supported function f on $\mathbf{R}^n$, the limit

$$\lim_{j\to\infty} \int_{|x-y|>\varepsilon_j} K(x-y)f(y)\,dy = T(f)(x) \tag{6.7.4}$$

exists and defines a linear operator T. This operator T is given by convolution with a tempered distribution W that coincides with the function K on $\mathbf{R}^n \setminus \{0\}$.

By the results of Chapter 4 we know that such a T, initially defined on $\mathscr{C}_0^\infty(\mathbf{R}^n)$, admits an extension that is L^p bounded for all $1<p<\infty$ and is also of weak type $(1,1)$. All these norms are bounded above by dimensional constant multiples of the quantity $A_1+A_2+A_3$ (cf. Theorem 4.4.1). Therefore, such a T is well defined on

$L^1(\mathbf{R}^n)$ and in particular on $H^1(\mathbf{R}^n)$, which is contained in $L^1(\mathbf{R}^n)$. We begin with the following result.

Theorem 6.7.1. *Let K satisfy (6.7.1), (6.7.2), and (6.7.3), and let T be defined as in (6.7.4). Then there is a constant C_n such that for all f in $H^1(\mathbf{R}^n)$ we have*

$$\big\|T(f)\big\|_{L^1} \le C_n(A_1+A_2+A_3)\big\|f\big\|_{H^1}. \tag{6.7.5}$$

Proof. To prove this theorem we have a powerful tool at our disposal, the atomic decomposition of $H^1(\mathbf{R}^n)$. It is therefore natural to start by checking the validity of (6.7.5) whenever f is an L^2-atom for H^1.

Since T is a convolution operator (i.e., it commutes with translations), it suffices to take the atom f supported in a cube Q centered at the origin. Let $f=a$ be such an atom, supported in Q, and let $Q^* = 2\sqrt{n}\,Q$. We write

$$\int_{\mathbf{R}^n} |T(a)(x)|\,dx = \int_{Q^*} |T(a)(x)|\,dx + \int_{(Q^*)^c} |T(a)(x)|\,dx \tag{6.7.6}$$

and we estimate each term separately. We have

$$\begin{aligned}
\int_{Q^*} |T(a)(x)|\,dx &\le |Q^*|^{\frac{1}{2}}\bigg(\int_{Q^*} |T(a)(x)|^2\,dx\bigg)^{\frac{1}{2}} \\
&\le C_n(A_1+A_2+A_3)|Q^*|^{\frac{1}{2}}\bigg(\int_Q |a(x)|^2\,dx\bigg)^{\frac{1}{2}} \\
&\le C_n(A_1+A_2+A_3)|Q^*|^{\frac{1}{2}}|Q|^{\frac{1}{2}-\frac{1}{1}} = C_n'(A_1+A_2+A_3),
\end{aligned}$$

where we used property (b) of atoms in Definition 6.6.8. Now note that if $x \notin Q^*$ and $y \in Q$, then $|x| \ge 2|y|$ and $x-y$ stays away from zero; thus $K(x-y)$ is well defined. Moreover, in this case $T(a)(x)$ can be expressed as a convergent integral of $a(y)$ against $K(x-y)$. We have

$$\begin{aligned}
\int_{(Q^*)^c} |T(a)(x)|\,dx &= \int_{(Q^*)^c} \bigg|\int_Q K(x-y)a(y)\,dy\bigg|\,dx \\
&= \int_{(Q^*)^c} \bigg|\int_Q (K(x-y)-K(x))a(y)\,dy\bigg|\,dx \\
&\le \int_Q\int_{(Q^*)^c} \big|K(x-y)-K(x)\big|\,dx\,|a(y)|\,dy \\
&\le \int_Q\int_{|x|\ge 2|y|} \big|K(x-y)-K(x)\big|\,dx\,|a(y)|\,dy \\
&\le A_2\int_Q |a(x)|\,dx \\
&\le A_2|Q|^{\frac{1}{2}}\bigg(\int_Q |a(x)|^2\,dx\bigg)^{\frac{1}{2}} \\
&\le A_2|Q|^{\frac{1}{2}}|Q|^{\frac{1}{2}-\frac{1}{1}} = A_2.
\end{aligned}$$

Combining this calculation with the previous one and inserting the final conclusions in (6.7.6) we deduce that L^2-atoms a for H^1 satisfy

$$\big\|T(a)\big\|_{L^1} \le (C_n' + 1)(A_1 + A_2 + A_3). \tag{6.7.7}$$

We now pass to general functions in H^1. In view of Theorem 6.6.10 we can write an $f \in H^1$ as

$$f = \sum_{j=1}^{\infty} \lambda_j a_j,$$

where the series converges in H^1, the a_j are L^2-atoms for H^1, and

$$\big\|f\big\|_{H^1} \approx \sum_{j=1}^{\infty} |\lambda_j|. \tag{6.7.8}$$

Since T maps L^1 to weak L^1 (Theorem 4.3.3), $T(f)$ is already a well defined $L^{1,\infty}$ function. We plan to prove that

$$T(f) = \sum_{j=1}^{\infty} \lambda_j T(a_j) \qquad \text{a.e.} \tag{6.7.9}$$

We observe that the series in (6.7.9) converges in L^1. Once (6.7.9) is established, the required conclusion (6.7.5) follows easily by taking L^1 norms in (6.7.9) and using (6.7.7) and (6.7.8).

To prove (6.7.9), we show that T is of weak type $(1,1)$. For a given $\delta > 0$ we have

$$\begin{aligned}
&\Big|\Big\{\Big|T(f) - \sum_{j=1}^{\infty} \lambda_j T(a_j)\Big| > \delta\Big\}\Big| \\
&\le \Big|\Big\{\Big|T(f) - \sum_{j=1}^{N} \lambda_j T(a_j)\Big| > \delta/2\Big\}\Big| + \Big|\Big\{\Big|\sum_{j=N+1}^{\infty} \lambda_j T(a_j)\Big| > \delta/2\Big\}\Big| \\
&\le \frac{2}{\delta}\big\|T\big\|_{L^1\to L^{1,\infty}}\Big\|f - \sum_{j=1}^{N}\lambda_j a_j\Big\|_{L^1} + \frac{2}{\delta}\Big\|\sum_{j=N+1}^{\infty}\lambda_j T(a_j)\Big\|_{L^1} \\
&\le \frac{2}{\delta}\big\|T\big\|_{L^1\to L^{1,\infty}}\Big\|f - \sum_{j=1}^{N}\lambda_j a_j\Big\|_{H^1} + \frac{2}{\delta}(C_n'+1)(A_1+A_2+A_3)\sum_{j=N+1}^{\infty}|\lambda_j|.
\end{aligned}$$

Since $\sum_{j=1}^{N}\lambda_j a_j$ converges to f in H^1 and $\sum_{j=1}^{\infty}|\lambda_j| < \infty$, both terms in the sum converge to zero as $N \to \infty$. We conclude that

$$\Big|\Big\{\Big|T(f) - \sum_{j=1}^{\infty} \lambda_j T(a_j)\Big| > \delta\Big\}\Big| = 0$$

for all $\delta > 0$, which implies (6.7.9). □

6.7.2 Singular Integrals on Besov–Lipschitz Spaces

We continue with a corollary concerning Besov–Lipschitz spaces.

Corollary 6.7.2. *Let K satisfy (6.7.1), (6.7.2), and (6.7.3), and let T be defined as in (6.7.4). Let $1 \le p \le \infty$, $0 < q \le \infty$, and $\alpha \in \mathbf{R}$. Then there is a constant $C_{n,p,q,\alpha}$ such that for all f in $\mathscr{S}(\mathbf{R}^n)$ we have*

$$\big\|T(f)\big\|_{\dot{B}_p^{\alpha,q}} \le C_n(A_1+A_2+A_3)\big\|f\big\|_{\dot{B}_p^{\alpha,q}}. \tag{6.7.10}$$

Therefore, T admits a bounded extension on all homogeneous Besov–Lipschitz spaces $\dot{B}_p^{\alpha,q}$ with $p \ge 1$, in particular, on all homogeneous Lipschitz spaces.

Proof. Let Ψ be a Schwartz function whose Fourier transform is supported in the annulus $1-\frac{1}{7} \le |\xi| \le 2$ and that satisfies

$$\sum_{j\in\mathbf{Z}} \widehat{\Psi}(2^{-j}\xi) = 1, \qquad \xi \neq 0.$$

Pick a Schwartz function ζ whose Fourier transform $\widehat{\zeta}$ is supported in the annulus $\frac{1}{4} < |\xi| < 8$ and that is equal to one on the support of $\widehat{\Psi}$. Let W be the tempered distribution that coincides with K on $\mathbf{R}^n \setminus \{0\}$ so that $T(f) = f * W$. Then we have $\zeta_{2^{-j}} * \Psi_{2^{-j}} = \Psi_{2^{-j}}$ for all j and hence

$$\begin{aligned} \big\|\Delta_j(T(f))\big\|_{L^p} &= \big\|\zeta_{2^{-j}} * \Psi_{2^{-j}} * W * f\big\|_{L^p} \\ &\le \big\|\zeta_{2^{-j}} * W\big\|_{L^1}\big\|\Delta_j(f)\big\|_{L^p}, \end{aligned} \tag{6.7.11}$$

since $1 \le p \le \infty$. It is not hard to check that the function $\zeta_{2^{-j}}$ is in H^1 with norm independent of j. Therefore, $\zeta_{2^{-j}}$ is in H^1. Using Theorem 6.7.1, we conclude that

$$\big\|T(\zeta_{2^{-j}})\big\|_{L^1} = \big\|\zeta_{2^{-j}} * W\big\|_{L^1} \le C\big\|\zeta_{2^{-j}}\big\|_{H^1} = C'.$$

Inserting this in (6.7.11), multiplying by $2^{j\alpha}$, and taking ℓ^q norms, we obtain the required conclusion. □

6.7.3 Singular Integrals on $H^p(\mathbf{R}^n)$

We are now interested in extending Theorem 6.7.1 to other H^p spaces for $p < 1$. It turns out that this is possible, provided some additional smoothness assumptions on K are imposed.

For the purposes of this subsection, we fix a function $K(x)$ on $\mathbf{R}^n \setminus \{0\}$ that satisfies $|K(x)| \le A|x|^{-n}$ for $x \neq 0$ and we assume that there is a distribution W in $\mathscr{S}'(\mathbf{R}^n)$ that coincides with K on $\mathbf{R}^n \setminus \{0\}$. We make two assumptions about the distribution W: first, that its Fourier transform $\widehat{W}$ is a bounded function, i.e., it satisfies

$$|\widehat{W}(\xi)| \le B, \qquad \xi \in \mathbf{R}^n, \tag{6.7.12}$$

for some $B < \infty$; secondly, that W is obtained from the function K as a limit of its smooth truncations. This allows us to properly define the convolution of this distribution with elements of H^p. So we fix a nonnegative smooth function η that vanishes in the unit ball of $\mathbf{R}^n$ and is equal to one outside the ball $B(0,2)$. We assume that for some sequence $\varepsilon_j \in (0,1)$ with $\varepsilon_j \downarrow 0$ the distribution W has the form

$$\langle W, \varphi \rangle = \lim_{j\to\infty} \int_{\mathbf{R}^n} K(y)\eta(y/\varepsilon_j)\varphi(y)\,dy \tag{6.7.13}$$

for all $\varphi \in \mathscr{S}(\mathbf{R}^n)$. Then we define the smoothly truncated singular integral associated with K and η by

$$T_\eta^{(\varepsilon)}(f)(x) = \int_{\mathbf{R}^n} \eta(y/\varepsilon)K(y)f(x-y)\,dy$$

for Schwartz functions f [actually the integral is absolutely convergent for every $f \in L^p$ and any $p \in [1,\infty)$]. We also define an operator T given by convolution with W by

$$T(f) = \lim_{j\to\infty} T^{(\varepsilon_j)}(f) = f * W. \tag{6.7.14}$$

This provides a representation of the operator T. If the function K satisfies condition (4.4.3), this representation is also valid pointwise almost everywhere for functions $f \in L^2$, i.e., $\lim_{j\to\infty} T^{(\varepsilon_j)}(f)(x) = T(f)(x)$ for almost all $x \in \mathbf{R}^n$. This follows from Theorem 4.4.5, Exercise 4.3.10, and Theorem 2.1.14 (since the convergence holds for Schwartz functions).

Next we define $T(f)$ for $f \in H^p$. One can write $W = W_0 + K_\infty$, where $W_0 = \Phi W$ and $K_\infty = (1-\Phi)K$, where Φ is a smooth function equal to one on the ball $B(0,1)$ and vanishing off the ball $B(0,2)$. Then for f in $H^p(\mathbf{R}^n)$, $0 < p \le 1$, we may define a tempered distribution $T(f) = W * f$ by setting

$$\langle T(f), \phi \rangle = \langle f, \phi * \widetilde{W_0} \rangle + \langle \widetilde{\phi} * f, \widetilde{K_\infty} \rangle \tag{6.7.15}$$

for ϕ in $\mathscr{S}(\mathbf{R}^n)$. The function $\phi * \widetilde{W_0}$ is in $\mathscr{S}$, so the action of f on it is well defined. Also $\widetilde{\phi} * f$ is in L^1 (see Proposition 6.4.9), while $\widetilde{K_\infty}$ is in L^∞; hence the second term on the right above represents an absolutely convergent integral. Moreover, in view of Theorem 2.3.20 and Corollary 6.4.9, both terms on the right in (6.7.15) are controlled by a finite sum of seminorms $\rho_{\alpha,\beta}(\phi)$ (cf. Definition 2.2.1). This defines $T(f)$ as a tempered distribution.

The following is an extension of Theorem 6.7.1 for $p < 1$.

Theorem 6.7.3. *Let $0 < p \le 1$ and $N = [\frac{n}{p} - n] + 1$. Let K be a $\mathscr{C}^N$ function on $\mathbf{R}^n \setminus \{0\}$ that satisfies*

$$|\partial^\beta K(x)| \le A\,|x|^{-n-|\beta|} \tag{6.7.16}$$

for all multi-indices $|\beta| \le N$ *and all* $x \neq 0$. *Let* W *be a tempered distribution that coincides with* K *on* $\mathbf{R}^n \setminus \{0\}$ *and satisfies* (6.7.12) *and* (6.7.13). *Then there is a constant* $C_{n,p}$ *such that the operator* T *defined in* (6.7.15) *satisfies, for all* $f \in H^p$,

$$\big\|T(f)\big\|_{L^p} \le C_{n,p}(A+B)\big\|f\big\|_{H^p}.$$

Proof. The proof of this theorem is based on the atomic decomposition of H^p.

We first take $f = a$ to be an L^2-atom for H^p, and without loss of generality we may assume that a is supported in a cube Q centered at the origin. We let Q^* be the cube with side length $2\sqrt{n}\,\ell(Q)$, where $\ell(Q)$ is the side length of Q. We have

$$\begin{aligned}\left(\int_{Q^*}|T(a)(x)|^p\,dx\right)^{\frac{1}{p}} &\le C|Q^*|^{\frac{1}{p}-\frac{1}{2}}\left(\int_{Q^*}|T(a)(x)|^2\,dx\right)^{\frac{1}{2}}\\ &\le C''B|Q|^{\frac{1}{p}-\frac{1}{2}}\left(\int_{Q}|a(x)|^2\,dx\right)^{\frac{1}{2}}\\ &\le C_nB|Q|^{\frac{1}{p}-\frac{1}{2}}|Q|^{\frac{1}{2}-\frac{1}{p}}\\ &= C_nB.\end{aligned}$$

For $x \notin Q^*$ and $y \in Q$, we have $|x| \ge 2|y|$, and thus $x-y$ stays away from zero and $K(x-y)$ is well defined. We have

$$T(a)(x) = \int_Q K^{(t)}(x-y)\,a(y)\,dy.$$

Recall that $N = [\frac{n}{p}-n]+1$. Using the cancellation of atoms for H^p, we deduce

$$\begin{aligned}T(a)(x) &= \int_Q a(y)K(x-y)\,dy\\ &= \int_Q a(y)\left[K(x-y) - \sum_{|\beta|\le N-1}(\partial^\beta K(x)\frac{(y)^\beta}{\beta!}\right]dy\\ &= \int_Q a(y)\left[\sum_{|\beta|=N}(\partial^\beta K(x-\theta_y y)\frac{(y)^\beta}{\beta!}\right]dy\,,\end{aligned}$$

for some $0 \le \theta_y \le 1$. Using that $|x| \ge 2|y|$ and (6.7.23), we obtain the estimate

$$|T(a)(x)| \le c_{n,N}\frac{A}{|x|^{N+n}}\int_Q |a(y)|\,|y|^{|\beta|}\,dy,$$

from which it follows that for $x \notin Q^*$ we have

$$|T(a)(x)| \le c_{n,p}\frac{A}{|x|^{N+n}}|Q|^{1+\frac{N}{n}-\frac{1}{p}}$$

via a calculation using properties of atoms (see the proof of Theorem 6.6.9). Integrating over $(Q^*)^c$, we obtain that

$$\left(\int_{(Q^*)^c}|T(a)(x)|^p dx\right)^{\frac{1}{p}} \le c_{n,p}A\,|Q|^{1+\frac{N}{n}-\frac{1}{p}}\left(\int_{(Q^*)^c}\frac{1}{|x|^{p(N+n)}}\,dx\right)^{\frac{1}{p}} \le c'_{n,p}A\,.$$

We have now shown that there exists a constant $C_{n,p}$ such that

$$\big\|T(a)\big\|_{L^p} \le C_{n,p}\,(A+B) \tag{6.7.17}$$

whenever a is an L^2-atom for H^p. We need to extend this estimate to infinite sums of atoms. To achieve this, it convenient to use operators with more regular kernels and then approximate T by such operators.

Recall the smooth function η that vanishes when $|x|\le 1$ and is equal to 1 when $|x|\ge 2$. We fix a smooth function θ with support in the unit ball having integral equal to 1. We define $\theta_\delta(x)=\delta^{-n}\theta(x/\delta)$,

$$K_{\varepsilon,\mu}(x)=K(x)\big(\eta(x/\varepsilon)-\eta(\mu x)\big)$$

and

$$K_{\delta,\varepsilon,\mu}=\theta_\delta * K_{\varepsilon,\mu}$$

for $0<10\delta<\varepsilon<(10\mu)^{-1}$. We make the following observations: first $K_{\delta,\varepsilon,\mu}$ is $\mathscr{C}^\infty$; second, it has rapid decay at infinity, and hence it is a Schwartz function; third, it satisfies (6.7:16) for all $|\beta|\le N$ with constant a multiple of A, that is, independent of δ, ε, μ. Let $T_{\delta,\varepsilon,\mu}$ be the operator given by convolution with $K_{\delta,\varepsilon,\mu}$ and let $T_\eta^{(*)}$ be the maximal smoothly truncated singular integral associated with the bump η. Then for $h\in L^2$ we have

$$\big\|T_{\delta,\varepsilon,\mu}(h)\big\|_{L^2}\le 2\big\|T_\eta^{(*)}(\theta_\delta*h)\big\|_{L^2}\le C_n\,(A+B)\big\|\theta_\delta*h\big\|_{L^2}\le C_n\,(A+B)\big\|h\big\|_{L^2};$$

hence $T_{\delta,\varepsilon,\mu}$ maps L^2 to L^2 with norm a fixed multiple of $A+B$. The proof of (6.7.17) thus yields for any L^2-atom a for H^p the estimate

$$\big\|T_{\delta,\varepsilon,\mu}(a)\big\|_{L^p}\le C'_{n,p}\,(A+B) \tag{6.7.18}$$

with a constant $C'_{n,p}$ that is independent of δ,ε,μ.

Let f be in $L^2\cap H^p$, which is a dense subspace of H^p, and suppose that $f=\sum_j\lambda_j a_j$, where a_j are L^2-atoms for H^p, the series converges in H^p, and we have

$$\sum_j|\lambda_j|^p\le C_p^p\big\|f\big\|_{H^p(\mathbf{R}^n)}^p. \tag{6.7.19}$$

We set $f_M=\sum_{j=1}^M\lambda_j a_j$. Then f_M,f are in L^2 but $f_M\to f$ in H^p; hence by Proposition 6.4.10, $f_M\to f$ in $\mathscr{S}'$. Acting on the Schwartz functions $K_{\delta,\varepsilon,\mu}(x-\cdot)$, we obtain that

$$T_{\delta,\varepsilon,\mu}(f_M)(x)\to T_{\delta,\varepsilon,\mu}(f)(x)\qquad \text{as } M\to\infty \text{ for all } x\in\mathbf{R}^n\,. \tag{6.7.20}$$

Recall the discussion in the introduction of this section defining $T = \lim_{j\to\infty} T^{(\varepsilon_j)}$ in an appropriate sense. Let $h \in L^2(\mathbf{R}^n)$. Since $h * K_{\delta,\varepsilon,\mu}$ is a continuous function, Theorem 1.2.19 (b) gives that

$$T_{\delta,\varepsilon_j,\mu}(h) \to T_\eta^{(\varepsilon_j)}(h) - T_\eta^{(1/\mu)}(h) \tag{6.7.21}$$

pointwise as $\delta \to 0$, where $T_\eta^{(\varepsilon)}$ is the smoothly truncated singular integral associated with the bump η (cf. Exercise 4.3.10). The expressions on the right in (6.7.21) are obviously pointwise bounded by $2T_\eta^{(*)}(h)$. Since $T_\eta^{(*)}$ is an L^2 bounded operator, and $T_\eta^{(\varepsilon_j)}(\psi) - T_\eta^{(1/\mu)}(\psi) \to T(\psi)$ for every $\psi \in \mathscr{S}(\mathbf{R}^n)$, it follows from Theorem 2.1.14 that $T_\eta^{(\varepsilon_j)}(h) - T_\eta^{(1/\mu)}(h) \to T(h)$ pointwise a.e. as $\varepsilon_j, \mu \to 0$. Thus $T_{\delta,\varepsilon_j,\mu}(h) \to T(h)$ pointwise a.e. as $\delta \to 0$, $\mu \to 0$, and $\varepsilon_j \to 0$ in this order. Using this fact, (6.7.20), and Fatou's lemma, we deduce for the given $f, f_M \in L^2 \cap H^p$ that

$$\big\|T(f)\big\|_{L^p}^p \le \liminf_{\delta,\mu,\varepsilon_j\to 0} \big\|T_{\delta,\varepsilon_j,\mu}(f)\big\|_{L^p}^p \le \liminf_{\delta,\mu,\varepsilon_j\to 0}\, \liminf_{M\to\infty} \big\|T_{\delta,\varepsilon_j,\mu}(f_M)\big\|_{L^p}^p .$$

The last displayed expression is at most $(C_pC'_{n,p})^p(A+B)^p\big\|f\big\|_{H^p}^p$ using the sublinearity of the pth power of the L^p norm, (6.7.18), and (6.7.19).

This proves the required assertion for $f \in H^p \cap L^2$. The case of general $f \in H^p$ follows by density and the fact that $T(f)$ is well defined for all $f \in H^p$, as observed at the beginning of this subsection. □

We discuss another version of the previous theorem in which the target space is H^p.

Theorem 6.7.4. *Under the hypotheses of Theorem 6.7.3, we have the following conclusion: there is a constant $C_{n,p}$ such that the operator T satisfies, for all $f \in H^p$,*

$$\big\|T(f)\big\|_{H^p} \le C_{n,p}(A+B)\big\|f\big\|_{H^p} .$$

Proof. The proof of this theorem provides another classical application of the atomic decomposition of H^p. However, we use the atomic decomposition only for the domain Hardy space, while it is more convenient to use the maximal (or square function) characterization of H^p for the target H^p space.

We fix a smooth function Φ supported in the unit ball $B(0,1)$ in $\mathbf{R}^n$ whose mean value is not equal to zero. For $t > 0$ we define the smooth functions

$$W^{(t)} = \Phi_t * W$$

and we observe that they satisfy

$$\sup_{t>0} \big|\widehat{W^{(t)}}(\xi)\big| \le \big\|\widehat{\Phi}\big\|_{L^\infty} B \tag{6.7.22}$$

and that

$$\sup_{t>0} |\partial^\beta W^{(t)}(x)| \le C_\Phi A\,|x|^{-n-|\beta|} \tag{6.7.23}$$

for all $|\beta| \le N$, where

$$C_\Phi = \sup_{|\gamma|\le N} \int_{\mathbf{R}^n} |\xi|^{|\gamma|} |\widehat{\Phi}(\xi)|\,d\xi\,.$$

Indeed, assertion (6.7.22) is easily verified, while assertion (6.7.23) follows from the identity

$$W^{(t)}(x) = \big((\Phi_t * W)\widehat{\ }\big)^\vee(x) = \int_{\mathbf{R}^n} e^{2\pi i x\cdot\xi}\, \widehat{W}(\xi)\,\widehat{\Phi}(t\xi)\,d\xi$$

whenever $|x| \le 2t$ and from (6.7.16) and the fact that for $|x| \ge 2t$ we have the integral representation

$$\partial^\beta W^{(t)}(x) = \int_{|y|\le t} \partial^\beta K(x-y)\,\Phi_t(y)\,dy\,.$$

We now take $f = a$ to be an L^2-atom for H^p, and without loss of generality we may assume that a is supported in a cube Q centered at the origin. We let Q^* be the cube with side length $2\sqrt{n}\,\ell(Q)$, where $\ell(Q)$ is the side length of Q. Recall the smooth maximal function $M(f;\Phi)$ from Section 6.4. Then $M(T(a);\Phi)$ is pointwise controlled by the Hardy–Littlewood maximal function of $T(a)$. Using an argument similar to that in Theorem 6.7.1, we have

$$\begin{aligned}
\Big(\int_{Q^*} |M(T(a);\Phi)(x)|^p\,dx\Big)^{\frac{1}{p}} &\le \big\|\Phi\big\|_{L^1} \Big(\int_{Q^*} |M(T(a))(x)|^p\,dx\Big)^{\frac{1}{p}} \\
&\le C|Q^*|^{\frac{1}{p}-\frac{1}{2}} \Big(\int_{Q^*} |M(T(a))(x)|^2\,dx\Big)^{\frac{1}{2}} \\
&\le C'|Q|^{\frac{1}{p}-\frac{1}{2}} \Big(\int_{\mathbf{R}^n} |T(a)(x)|^2\,dx\Big)^{\frac{1}{2}} \\
&\le C''B|Q|^{\frac{1}{p}-\frac{1}{2}} \Big(\int_{Q} |a(x)|^2\,dx\Big)^{\frac{1}{2}} \\
&\le C_n B|Q|^{\frac{1}{p}-\frac{1}{2}} |Q|^{\frac{1}{2}-\frac{1}{p}} \\
&= C_n B\,.
\end{aligned}$$

It therefore remains to estimate the contribution of $M(T(a);\Phi)$ on the complement of Q^*.

If $x \notin Q^*$ and $y \in Q$, then $|x| \ge 2|y|$ and hence $x - y \ne 0$. Thus $K(x-y)$ is well defined as an integral. We have

$$(T(a)*\Phi_t)(x) = (a * W^{(t)})(x) = \int_Q K^{(t)}(x-y)\,a(y)\,dy\,.$$

Recall that $N = [\frac{n}{p} - n] + 1$. Using the cancellation of atoms for H^p we deduce

$$\begin{aligned}(T(a) * \Phi_t)(x) &= \int_Q a(y)\Big[K^{(t)}(x-y) - \sum_{|\beta|\le N-1} (\partial^\beta K^{(t)})(x)\frac{(y)^\beta}{\beta!}\Big]dy \\ &= \int_Q a(y)\Big[\sum_{|\beta|=N} (\partial^\beta K^{(t)})(x-\theta_y y)\frac{(y)^\beta}{\beta!}\Big]dy\end{aligned}$$

for some $0 \le \theta_y \le 1$. Using that $|x| \ge 2|y|$ and (6.7.23), we obtain the estimate

$$|(T(a) * \Phi_t)(x)| \le c_{n,N}\frac{A}{|x|^{N+n}}\int_Q |a(y)|\,|y|^{|\beta|}\,dy,$$

from which it follows that for $x \notin Q^*$ we have

$$|(T(a) * \Phi_t)(x)| \le c_{n,p}\frac{A}{|x|^{N+n}}|Q|^{1+\frac{N}{n}-\frac{1}{p}}$$

via a calculation using properties of atoms (see the proof of Theorem 6.6.9). Taking the supremum over all $t > 0$ and integrating over $(Q^*)^c$, we obtain that

$$\Big(\int_{(Q^*)^c}\sup_{t>0}|(T(a) * \Phi_t)(x)|^p dx\Big)^{\frac{1}{p}} \le c_{n,p}A\,|Q|^{1+\frac{N}{n}-\frac{1}{p}}\Big(\int_{(Q^*)^c}\frac{1}{|x|^{p(N+n)}}dx\Big)^{\frac{1}{p}},$$

and the latter is easily seen to be finite and controlled by a constant multiple of A. Combining this estimate with the previously obtained estimate for the integral of $M(T(a);\Phi) = \sup_{t>0}|(T(a) * \Phi_t|$ over Q^* yields the conclusion of the theorem when $f = a$ is an atom.

We have now shown that there exists a constant $C_{n,p}$ such that

$$\big\|T(a)\big\|_{H^p} \le C_{n,p}(A+B) \tag{6.7.24}$$

whenever a is an L^2-atom for H^p. We need to extend this estimate to infinite sums of atoms.

Let f be $L^2 \cap H^p$ which is a dense subspace of H^p, and suppose that $f = \sum_j \lambda_j a_j$ for some L^2-atoms a_j for H^p, where the series converges in H^p and we have

$$\sum_j |\lambda_j|^p \le C_p^p \|f\|^p_{H^p(\mathbf{R}^n)}\,. \tag{6.7.25}$$

We let $f_M = \sum_{j=1}^M \lambda_j a_j$ and we recall the smooth truncations $T_{\delta,\varepsilon_j,\mu}$ of T. As $f_M \to f$ in H^p, Proposition 6.4.10 gives that $f_M \to f$ in $\mathscr{S}'$, and since the functions $K_{\delta,\varepsilon_j,\mu}$ are smooth with compact support, it follows that for all $\delta, \varepsilon_j, \mu$,

$$T_{\delta,\varepsilon_j,\mu}(f_M) \to T_{\delta,\varepsilon_j,\mu}(f) \qquad \text{in } \mathscr{S}' \text{ as } M \to \infty. \tag{6.7.26}$$

We show that this convergence is also valid for T. Given $\varepsilon > 0$ and a Schwartz function φ, we find $\delta_0, \varepsilon_{j_0}, \mu_0$ such that

$$\left|\langle T(f_M), \varphi\rangle - \langle T_{\delta_0,\varepsilon_{j_0},\mu_0}(f_M), \varphi\rangle\right| < \varepsilon C_p \big\|f\big\|_{H^p} \quad \text{for all } M = 1,2,\dots. \tag{6.7.27}$$

To find such $\delta_0, \varepsilon_{j_0}, \mu_0$, we write

$$\begin{aligned}
\left|\langle T(f_M), \varphi\rangle - \langle T_{\delta_0,\varepsilon_{j_0},\mu_0}(f_M), \varphi\rangle\right| &\le \left|\sum_{j=1}^{M} \lambda_j \langle (K_{\delta_0,\varepsilon_{j_0},\mu_0} - W) * a_j, \varphi\rangle\right| \\
&\le \left(\sum_{j=1}^{M} |\lambda_j|^p \left|\langle a_j, (\widetilde{K}_{\delta_0,\varepsilon_{j_0},\mu_0} - \widetilde{W}) * \varphi\rangle\right|^p\right)^{\frac{1}{p}} \\
&\le \left(\sum_{j=1}^{M} |\lambda_j|^p \big\|a_j\big\|_{L^2}^p \big\|(\widetilde{K}_{\delta_0,\varepsilon_{j_0},\mu_0} - \widetilde{W}) * \varphi\big\|_{L^2}^p\right)^{\frac{1}{p}} \\
&\le C_p \big\|f\big\|_{H^p} \big\|(K_{\delta_0,\varepsilon_{j_0},\mu_0} - W) * \widetilde{\varphi}\,\big\|_{L^2}.
\end{aligned}$$

Now pick $\delta_0, \varepsilon_{j_0}, \mu_0$ such that

$$\big\|(K_{\delta_0,\varepsilon_{j_0},\mu_0} - W) * \widetilde{\varphi}\,\big\|_{L^2} = \big\|((K_{\delta_0,\varepsilon_{j_0},\mu_0})\widehat{\ } - \widehat{W})\widehat{\widetilde{\varphi}}\,\big\|_{L^2} < \varepsilon.$$

This is possible, since this expression tends to zero when $\delta_0, \varepsilon_{j_0}, \mu_0 \to 0$ by the Lebesgue dominated convergence theorem; indeed, the functions $(K_{\delta_0,\varepsilon_{j_0},\mu_0})\widehat{\ } - \widehat{W}$ are uniformly bounded and converge pointwise to zero as $\delta_0, \varepsilon_{j_0}, \mu_0 \to 0$, while $\widehat{\widetilde{\varphi}}$ is square integrable. This proves (6.7.27).

Next we show that for this choice of $\delta_0, \varepsilon_{j_0}, \mu_0$ we also have

$$\left|\langle T_{\delta_0,\varepsilon_{j_0},\mu_0}(f), \varphi\rangle - \langle T(f), \varphi\rangle\right| < \varepsilon \big\|f\big\|_{L^2}. \tag{6.7.28}$$

This is a consequence of the Cauchy–Schwarz inequality, since

$$\left|\langle T_{\delta_0,\varepsilon_{j_0},\mu_0}(f), \varphi\rangle - \langle T(f), \varphi\rangle\right| \le \big\|((K_{\delta_0,\varepsilon_{j_0},\mu_0})\widehat{\ } - \widehat{W})\widehat{\widetilde{\varphi}}\,\big\|_{L^2} \big\|f\big\|_{L^2}.$$

Using (6.7.26) we can find an M_0 such that for $M \ge M_0$ we have

$$\left|\langle T_{\delta_0,\varepsilon_{j_0},\mu_0}(f_M), \varphi\rangle - \langle T_{\delta_0,\varepsilon_{j_0},\mu_0}(f), \varphi\rangle\right| < \varepsilon. \tag{6.7.29}$$

Combining (6.7.27), (6.7.28), and (6.7.29) for $M \ge M_0$, we obtain

$$\left|\langle T(f_M), \varphi\rangle - \langle T(f), \varphi\rangle\right| < \varepsilon\left(1 + C_p \big\|f\big\|_{H^p} + \big\|f\big\|_{L^2}\right),$$

and this implies that $T(f_M)$ converges to $T(f)$ in $\mathscr{S}'(\mathbf{R}^n)$.

Using the inequality,

$$\|T(f_M) - T(f_{M'})\|_{H^p}^p \le C_{n,p}^p (A+B)^p \sum_{M<j\le M'} |\lambda_j|^p,$$

one easily shows that the sequence $\{T(f_M)\}_M$ is Cauchy in H^p. Thus $T(f_M)$ converges in H^p to some element $G \in H^p$ as $M \to \infty$. By Proposition 6.4.10, $T(f_M)$ converges to G in $\mathscr{S}'$. But as we saw, $T(f_M)$ converges to $T(f)$ in $\mathscr{S}'$ as $M \to \infty$. Hence $T(f) = G$ and we conclude that $T(f_M)$ converges to $T(f)$ in H^p, i.e., the series $\sum_j \lambda_j T(a_j)$ converges to $T(f)$ in H^p. This allows us to estimate the H^p quasi-norm of $T(f)$ as follows:

$$\begin{aligned}\|T(f)\|_{H^p(\mathbf{R}^n)}^p &= \Big\|\sum_j \lambda_j T(a_j)\Big\|_{H^p(\mathbf{R}^n)}^p \\ &\le \sum_j |\lambda_j|^p \|T(a_j)\|_{H^p(\mathbf{R}^n)}^p \\ &\le (C'_{n,p})^p (A+B)^p \sum_j |\lambda_j|^p \\ &\le (C'_{n,p} C_p)^p (A+B)^p \|f\|_{H^p(\mathbf{R}^n)}^p .\end{aligned}$$

This concludes the proof for $f \in H^p \cap L^2$. The extension to general $f \in H^p$ follows by density and the fact that $T(f)$ is well defined for all $f \in H^p$, as observed at the beginning of this subsection. □

6.7.4 A Singular Integral Characterization of $H^1(\mathbf{R}^n)$

We showed in Section 6.7.1 that singular integrals map H^1 to L^1. In particular, the Riesz transforms have this property. In this subsection we obtain a converse to this statement. We show that if $R_j(f)$ are integrable functions for some $f \in L^1$ and all $j = 1, \dots, n$, then f must be an element of the Hardy space H^1. This provides a characterization of $H^1(\mathbf{R}^n)$ in terms of the Riesz transforms.

Theorem 6.7.5. *For $n \ge 2$, there exists a constant C_n such that for f in $L^1(\mathbf{R}^n)$ we have*

$$C_n \|f\|_{H^1} \le \|f\|_{L^1} + \sum_{k=1}^n \|R_k(f)\|_{L^1} . \tag{6.7.30}$$

When $n = 1$ the corresponding statement is

$$C_1 \|f\|_{H^1} \le \|f\|_{L^1} + \|H(f)\|_{L^1} \tag{6.7.31}$$

for all $f \in L^1(\mathbf{R})$. Naturally, these statements are interesting when the expressions on the right in (6.7.30) *and* (6.7.31) *are finite.*

Before we prove this theorem we discuss two corollaries.

Corollary 6.7.6. *An integrable function on the line lies in the Hardy space $H^1(\mathbf{R})$ if and only if its Hilbert transform is integrable. For $n \geq 2$, an integrable function on $\mathbf{R}^n$ lies in the Hardy space $H^1(\mathbf{R}^n)$ if and only its Riesz transforms are also in $L^1(\mathbf{R}^n)$.*

Proof. The corollary follows by combining Theorems 6.7.1 and 6.7.5. □

Corollary 6.7.7. *Functions in $H^1(\mathbf{R}^n)$, $n \geq 1$, have integral zero.*

Proof. Indeed, if $f \in H^1(\mathbf{R}^n)$, we must have $R_1(f) \in L^1(\mathbf{R}^n)$; thus $\widehat{R_1(f)}$ is uniformly continuous. But since

$$\widehat{R_1(f)}(\xi) = -i\widehat{f}(\xi)\frac{\xi_1}{|\xi|},$$

it follows that $\widehat{R_1(f)}$ is continuous at zero if and only if $\widehat{f}(\xi) = 0$. But this happens exactly when f has integral zero. □

We now discuss the proof of Theorem 6.7.5.

Proof. We consider the case $n \geq 2$, although the argument below also works in the case $n = 1$ with a suitable change of notation. Let P_t be the Poisson kernel. In the proof we may assume that f is real-valued, since it can be written as $f = f_1 + if_2$, where f_k are real-valued and $R_j(f_k)$ are also integrable. Given a real-valued function $f \in L^1(\mathbf{R}^n)$ such that $R_j(f)$ are integrable over $\mathbf{R}^n$ for every $j = 1, \dots, n$, we associate with it the $n+1$ functions

$$\begin{aligned} u_1(x,t) &= (P_t * R_1(f))(x), \\ \dots &= \dots, \\ u_n(x,t) &= (P_t * R_n(f))(x), \\ u_{n+1}(x,t) &= (P_t * f)(x), \end{aligned}$$

which are harmonic on the space $\mathbf{R}^{n+1}_+$ (see Example 2.1.13). It is convenient to denote the last variable t by x_{n+1}. One may check using the Fourier transform that these harmonic functions satisfy the following system:

$$\begin{aligned} \sum_{j=1}^{n+1} \frac{\partial u_j}{\partial x_j} &= 0, \\ \frac{\partial u_j}{\partial x_k} - \frac{\partial u_k}{\partial x_j} &= 0, \qquad k, j \in \{1, \dots, n+1\}, \quad k \neq j. \end{aligned} \tag{6.7.32}$$

This system of equations may also be expressed as div $F = 0$ and curl $F = \vec{0}$, where $F = (u_1, \dots, u_{n+1})$ is a vector field in $\mathbf{R}^{n+1}_+$. Note that when $n = 1$, the equations in (6.7.32) are the usual Cauchy–Riemann equations, which assert that the function $F = (u_1, u_2) = u_1 + iu_2$ is holomorphic in the upper half-space. For this reason, when

$n \geq 2$ the equations in (6.7.32) are often referred to as the *system of generalized Cauchy–Riemann equations.*

The function $|F|$ enjoys a crucial property in the study of this problem.

Lemma 6.7.8. *Let u_j be real-valued harmonic functions on $\mathbf{R}^{n+1}$ satisfying the system of equations* (6.7.32) *and let* $F = (u_1, \dots, u_{n+1})$. *Then the function*

$$|F|^q = \Big(\sum_{j=1}^{n+1} |u_j|^2\Big)^{q/2}$$

is subharmonic when $q \geq (n-1)/n$, i.e., it satisfies $\Delta(|F|^q) \geq 0$, on $\mathbf{R}^{n+1}_+$.

Lemma 6.7.9. *Let* $0 < q < p < \infty$. *Suppose that the function $|F(x,t)|^q$ defined on $\mathbf{R}^{n+1}_+$ is subharmonic and satisfies*

$$\sup_{t>0}\bigg(\int_{\mathbf{R}^n} |F(x,t)|^p \, dx\bigg)^{1/p} \leq A < \infty. \tag{6.7.33}$$

Then there is a constant $C_{n,p,q} < \infty$ such that the nontangential maximal function $|F|^*(x) = \sup_{t>0} \sup_{|y-x|<t} |F(y,t)|$, $x \in \mathbf{R}^n$, (cf. Definition 7.3.1) satisfies

$$\big\| |F|^* \big\|_{L^p(\mathbf{R}^n)} \leq C_{n,p,q} A .$$

Assuming these lemmas, whose proofs are postponed until the end of this section, we return to the proof of the theorem.

Since the Poisson kernel is an approximate identity, the function $x \mapsto u_{n+1}(x,t)$ converges to $f(x)$ in L^1 as $t \to 0$. To show that $f \in H^1(\mathbf{R}^n)$, it suffices to show that the Poisson maximal function

$$M(f;P)(x) = \sup_{t>0} |(P_t * f)(x)| = \sup_{t>0} |u_{n+1}(x,t)|$$

is integrable. But this maximal function is pointwise controlled by

$$\sup_{t>0} |F(x,t)| \leq \sup_{t>0} \Big[|(P_t * f)(x)| + \sum_{j=1}^{n} |(P_t * R_j(f))(x)| \Big],$$

and certainly it satisfies

$$\sup_{t>0} \int_{\mathbf{R}^n} |F(x,t)| \, dx \leq A_f, \tag{6.7.34}$$

where

$$A_f = \big\| f \big\|_{L^1} + \sum_{k=1}^{n} \big\| R_k(f) \big\|_{L^1} .$$

We now have

$$M(f;P)(x) \leq \sup_{t>0} |u_{n+1}(x,t)| \leq \sup_{t>0} |F(x,t)| \leq |F|^*(x), \tag{6.7.35}$$

and using Lemma 6.7.8 with $q = \frac{n-1}{n}$ and Lemma 6.7.9 with $p = 1$ we obtain that

$$\big\| |F|^* \big\|_{L^1(\mathbf{R}^n)} \le C_n A_f . \tag{6.7.36}$$

Combining (6.7.34), (6.7.35), and (6.7.36), one deduces that

$$\big\| M(f;P)(x) \big\|_{L^1(\mathbf{R}^n)} \le C_n \bigg(\big\| f \big\|_{L^1} + \sum_{k=1}^{n} \big\| R_k(f) \big\|_{L^1} \bigg),$$

from which (6.7.30) follows. This proof is also valid when $n = 1$, provided one replaces the Riesz transforms with the Hilbert transform; hence the proof of (6.7.31) is subsumed in that of (6.7.30). □

See Exercise 6.7.1 for an extension of this result to H^p for $\frac{n-1}{n} < p < 1$. We now give a proof of Lemma 6.7.8

Proof. Denoting the variable t by x_{n+1}, we have

$$\frac{\partial}{\partial x_j} |F|^q = q|F|^{q-2} \Big(F \cdot \frac{\partial F}{\partial x_j} \Big)$$

and also

$$\frac{\partial^2}{\partial x_j^2} |F|^q = q|F|^{q-2} \bigg[F \cdot \frac{\partial^2 F}{\partial x_j^2} + \frac{\partial F}{\partial x_j} \cdot \frac{\partial F}{\partial x_j} \bigg] + q(q-2)|F|^{q-4} \Big(F \cdot \frac{\partial F}{\partial x_j} \Big)^2$$

for all $j = 1, 2, \dots, n+1$. Summing over all these j's, we obtain

$$\Delta(|F|^q) = q|F|^{q-4} \bigg[|F|^2 \sum_{j=1}^{n+1} \Big| \frac{\partial F}{\partial x_j} \Big|^2 + (q-2) \sum_{j=1}^{n+1} \Big| F \cdot \frac{\partial F}{\partial x_j} \Big|^2 \bigg], \tag{6.7.37}$$

since the term containing $F \cdot \Delta(F) = \sum_{j=1}^{n+1} u_j \Delta(u_j)$ vanishes because each u_j is harmonic. The only term that could be negative in (6.7.37) is that containing the factor $q-2$ and naturally, if $q \ge 2$, the conclusion is obvious. Let us assume that $\frac{n-1}{n} \le q < 2$. Since $q \ge \frac{n-1}{n}$, we must have that $2 - q \le \frac{n+1}{n}$. Thus (6.7.37) is non-negative if

$$\sum_{j=1}^{n+1} \Big| F \cdot \frac{\partial F}{\partial x_j} \Big|^2 \le \frac{n}{n+1} |F|^2 \sum_{j=1}^{n+1} \Big| \frac{\partial F}{\partial x_j} \Big|^2 . \tag{6.7.38}$$

This is certainly valid for points (x,t) such that $F(x,t) = 0$. To prove (6.7.38) for points (x,t) with $F(x,t) \neq 0$, it suffices to show that for every vector $v \in \mathbf{R}^{n+1}$ with Euclidean norm $|v| = 1$, we have

$$\sum_{j=1}^{n+1} \Big| v \cdot \frac{\partial F}{\partial x_j} \Big|^2 \le \frac{n}{n+1} \sum_{j=1}^{n+1} \Big| \frac{\partial F}{\partial x_j} \Big|^2 . \tag{6.7.39}$$

Denoting by A the $(n+1)\times(n+1)$ matrix whose entries are $a_{j,k}=\partial u_k/\partial x_j$, we rewrite (6.7.39) as

$$|Av|^2 \le \frac{n}{n+1}\|A\|^2, \tag{6.7.40}$$

where

$$\|A\|^2 = \sum_{j=1}^{n+1}\sum_{k=1}^{n+1} |a_{j,k}|^2 .$$

By assumption, the functions u_j are real-valued and thus the numbers $a_{j,k}$ are real. In view of identities (6.7.32), the matrix A is real symmetric and has zero trace (i.e., $\sum_{j=1}^{n+1} a_{j,j}=0$). A real symmetric matrix A can be written as $A=PDP^t$, where P is an orthogonal matrix and D is a real diagonal matrix. Since orthogonal matrices preserve the Euclidean distance, estimate (6.7.40) follows from the corresponding one for a diagonal matrix D. If $A=PDP^t$, then the traces of A and D are equal; hence $\sum_{j=1}^{n+1}\lambda_j=0$, where λ_j are entries on the diagonal of D. Notice that estimate (6.7.40) with the matrix D in the place of A is equivalent to

$$\sum_{j=1}^{n+1} |\lambda_j|^2|v_j|^2 \le \frac{n}{n+1}\Big(\sum_{j=1}^{n+1}|\lambda_j|^2\Big), \tag{6.7.41}$$

where we set $v=(v_1,\dots,v_{n+1})$ and we are assuming that $|v|^2=\sum_{j=1}^{n+1}|v_j|^2=1$. Estimate (6.7.41) is certainly a consequence of

$$\sup_{1\le j\le n+1} |\lambda_j|^2 \le \frac{n}{n+1}\Big(\sum_{j=1}^{n+1}|\lambda_j|^2\Big). \tag{6.7.42}$$

But this is easy to prove. Let $|\lambda_{j_0}|=\max_{1\le j\le n+1}|\lambda_j|$. Then

$$|\lambda_{j_0}|^2 = \Big|-\sum_{j\ne j_0}\lambda_j\Big|^2 \le \Big(\sum_{j\ne j_0}|\lambda_j|\Big)^2 \le n\sum_{j\ne j_0}|\lambda_j|^2 . \tag{6.7.43}$$

Adding $n|\lambda_{j_0}|^2$ to both sides of (6.7.43), we deduce (6.7.42) and thus (6.7.38). □

We now give the proof of Lemma 6.7.9.

Proof. A consequence of the subharmonicity of $|F|^q$ is that

$$|F(x,t+\varepsilon)|^q \le (|F(\cdot,\varepsilon)|^q * P_t)(x) \tag{6.7.44}$$

for all $x\in\mathbf{R}^n$ and $t,\varepsilon>0$. To prove (6.7.44), fix $\varepsilon>0$ and consider the functions

$$U(x,t)=|F(x,t+\varepsilon)|^q, \qquad V(x,t)=(|F(\cdot,\varepsilon)|^q * P_t)(x).$$

Given $\eta>0$, we find a half-ball

$$B_{R_0}=\{(x,t)\in\mathbf{R}^{n+1}_+ :\ |x|^2+t^2<R_0^2\}$$

such that for $(x,t) \in \mathbf{R}^{n+1}_+ \setminus B_{R_0}$ we have

$$U(x,t) - V(x,t) \le \eta\,. \tag{6.7.45}$$

Suppose that this is possible. Since $U(x,0) = V(x,0)$, then (6.7.45) actually holds on the entire boundary of B_{R_0}. The function V is harmonic and U is subharmonic; thus $U - V$ is subharmonic. The maximum principle for subharmonic functions implies that (6.7.45) holds in the interior of B_{R_0}, and since it also holds on the exterior, it must be valid for all (x,t) with $x \in \mathbf{R}^n$ and $t \ge 0$. Since η was arbitrary, letting $\eta \to 0+$ implies (6.7.44).

We now prove that R_0 exists such that (6.7.45) is possible for $(x,t) \in \mathbf{R}^{n+1}_+ \setminus B_{R_0}$. Let $B((x,t),t/2)$ be the $(n+1)$-dimensional ball of radius $t/2$ centered at (x,t). The subharmonicity of $|F|^q$ is reflected in the inequality

$$|F(x,t)|^q \le \frac{1}{|B((x,t),t/2)|} \int_{B((x,t),t/2)} |F(y,s)|^q \, dy ds\,,$$

which by Hölder's inequality and the fact $p > q$ gives

$$|F(x,t)|^q \le \left(\frac{1}{|B((x,t),t/2)|} \int_{B((x,t),t/2)} |F(y,s)|^p \, dy ds \right)^{\frac{q}{p}}.$$

From this we deduce that

$$|F(x,t+\varepsilon)|^q \le \left[\frac{2^{n+1}/v_{n+1}}{(t+\varepsilon)^{n+1}} \int_{\frac{1}{2}(t+\varepsilon)}^{\frac{3}{2}(t+\varepsilon)} \int_{|y| \ge |x| - \frac{1}{2}(t+\varepsilon)} |F(y,s)|^p \, dy ds \right]^{\frac{q}{p}}. \tag{6.7.46}$$

If $t+\varepsilon \ge |x|$, using (6.7.33), we see that the expression on the right in (6.7.46) is bounded by $c'\varepsilon^{-n} A^q t^{-nq/p}$, and thus it can be made smaller than $\eta/2$ by taking $t \ge R_1$, for some $R_1 > \varepsilon$ large enough. Since $R_1 > \varepsilon$, we must have $2t \ge t + \varepsilon \ge |x|$, which implies that $t \ge |x|/2$, and thus with $R_0' = \sqrt{5} R_1$, if $|(x,t)| > R_0'$ then $t \ge R_1$. Hence, the expression in (6.7.46) can be made smaller than $\eta/2$ for $|(x,t)| > R_0'$.

If $t + \varepsilon < |x|$ we estimate the expression on the right in (6.7.46) by

$$\left(\frac{2^{n+1}}{v_{n+1}} \frac{1}{(t+\varepsilon)^{n+1}} \int_{\frac{1}{2}(t+\varepsilon)}^{\frac{3}{2}(t+\varepsilon)} \left[\int_{|y| \ge \frac{1}{2}|x|} |F(y,s)|^p \, dy \right] ds \right)^{\frac{q}{p}},$$

and we notice that the preceding expression is bounded by

$$\left(\frac{3^{n+1}}{v_{n+1}} \int_{\frac{1}{2}\varepsilon}^{\infty} \left[\int_{|y| \ge \frac{1}{2}|x|} |F(y,s)|^p \, dy \right] \frac{ds}{s^{n+1}} \right)^{\frac{q}{p}}. \tag{6.7.47}$$

Let $G_{|x|}(s)$ be the function inside the square brackets in (6.7.47). Then $G_{|x|}(s) \to 0$ as $|x| \to \infty$ for all s. The hypothesis (6.7.33) implies that $G_{|x|}$ is bounded by a constant and it is therefore integrable over the interval $\left[\frac{1}{2}\varepsilon, \infty\right)$ with respect to the measure $s^{-n-1} ds$. By the Lebesgue dominated convergence theorem we deduce that the expression in (6.7.47) converges to zero as $|x| \to \infty$ and thus it can be made smaller that $\eta/2$ for $|x| \geq R_2$, for some constant R_2. Then with $R_0'' = \sqrt{2}R_2$ we have that if $|(x,t)| \geq R_0''$ then (6.7.47) is at most $\eta/2$. Since $U - V \leq U$, we deduce the validity of (6.7.45) for $|(x,t)| > R_0 = \max(R_0', R_0'')$.

Let $r = p/q > 1$. Assumption (6.7.33) implies that the functions $x \mapsto |F(x,\varepsilon)|^q$ are in L^r uniformly in t. Since any closed ball of L^r is weak* compact, there is a sequence $\varepsilon_k \to 0$ such that $|F(x,\varepsilon_k)|^q \to h$ weakly in L^r as $k \to \infty$ to some function $h \in L^r$. Since $P_t \in L^{r'}$, this implies that

$$(|F(\cdot,\varepsilon_k)|^q * P_t)(x) \to (h * P_t)(x)$$

for all $x \in \mathbf{R}^n$. Using (6.7.44) we obtain

$$|F(x,t)|^q = \limsup_{k\to\infty} |F(x,t+\varepsilon_k)|^p \leq \limsup_{k\to\infty} \big(|F(x,\varepsilon_k)|^q * P_t\big)(x) = (h * P_t)(x),$$

which gives for all $x \in \mathbf{R}^n$,

$$|F|^*(x) \leq \Big[\sup_{t>0} \sup_{|y-x|<t} (|h| * P_t)(x)\Big]^{1/q} \leq C_n' M(h)(x)^{1/q}. \tag{6.7.48}$$

Let $g \in L^{r'}(\mathbf{R}^n)$ with $L^{r'}$ norm at most one. The weak convergence yields

$$\int_{\mathbf{R}^n} |F(x,\varepsilon_k)|^q g(x)\,dx \to \int_{\mathbf{R}^n} h(x)g(x)\,dx$$

as $k \to \infty$, and consequently we have

$$\left|\int_{\mathbf{R}^n} h(x)g(x)\,dx\right| \leq \sup_k \int_{\mathbf{R}^n} |F(x,\varepsilon_k)|^q |g(x)|\,dx \leq \|g\|_{L^{r'}} \sup_{t>0}\left(\int_{\mathbf{R}^n} |F(x,t)|^p\,dx\right)^{\frac{1}{r}}.$$

Since g is arbitrary with $L^{r'}$ norm at most one, this implies that

$$\|h\|_{L^r} \leq \sup_{t>0}\left(\int_{\mathbf{R}^n} |F(x,t)|^p\,dx\right)^{\frac{1}{r}}. \tag{6.7.49}$$

Putting things together, we have

$$\begin{aligned}
\big\||F|^*\big\|_{L^p} &\leq C_n' \big\|M(h)^{1/q}\big\|_{L^p} \\
&= C_n' \big\|M(h)\big\|_{L^r}^{1/q} \\
&= C_{n,p,q} \big\|h\big\|_{L^r}^{1/q} \\
&= C_{n,p,q} \sup_{t>0}\left(\int_{\mathbf{R}^n} |F(x,t)|^p\,dx\right)^{1/qr} \\
&\leq C_{n,p,q} A,
\end{aligned}$$

where we have used (6.7.48) and (6.7.49) in the last two displayed inequalities. □

Exercises

6.7.1. Prove the following generalization of Theorem 6.7.4. Let φ be a nonnegative Schwartz function with integral one on $\mathbf{R}^n$ and let $\frac{n-1}{n} < p < 1$. Prove that there are constants c_1, c_n, C_1, C_n such that for bounded tempered distributions f on $\mathbf{R}^n$ (cf. Section 6.4.1) we have

$$c_n \big\| f \big\|_{H^p} \le \sup_{\delta>0} \Big[\big\| \varphi_\delta * f \big\|_{L^p} + \sum_{k=1}^{n} \big\| \varphi_\delta * R_k(f) \big\|_{L^p} \Big] \le C_n \big\| f \big\|_{H^p}$$

when $n \ge 2$ and

$$c_1 \big\| f \big\|_{H^p} \le \sup_{\delta>0} \Big[\big\| \varphi_\delta * f \big\|_{L^p} + \big\| \varphi_\delta * H(f) \big\|_{L^p} \Big] \le C_1 \big\| f \big\|_{H^p}$$

when $n = 1$.

[*Hint:* One direction is a consequence of Theorem 6.7.4. For the other direction, define $F_\delta = (u_1 * \varphi_\delta, \dots, u_{n+1} * \varphi_\delta)$, where $u_j(x,t) = (P_t * R_j(f))(x)$, $j = 1, \dots, n$, and $u_{n+1}(x,t) = (P_t * f)(x)$. Each $u_j * \varphi_\delta$ is a harmonic function on $\mathbf{R}^{n+1}_+$ and continuous up to the boundary. The subharmonicity of $|F_\delta(x,t)|^p$ has as a consequence that $|F_\delta(x,t+\varepsilon)|^p \le |(F_\delta(\cdot,\varepsilon)|^p * P_t)(x)$ in view of (6.7.44). Letting $\varepsilon \to 0$ implies that $|F_\delta(x,t)|^p \le |(F_\delta(\cdot,0)|^p * P_t)(x)$, by the continuity of F_δ up to the boundary. Since $F_\delta(x,0) = (R_1(f) * \varphi_\delta, \dots, R_n(f) * \varphi_\delta, f * \varphi_\delta)$, the hypothesis that $f * \varphi_\delta, R_j(f) * \varphi_\delta$ are in L^p uniformly in $\delta > 0$ gives that $\sup_{t,\delta>0} \int_{\mathbf{R}^n} |F_\delta(x,t)|^p \, dx < \infty$. Fatou's lemma yields (6.7.33) for $F(x,t) = (u_1, \dots, u_{n+1})$. Then Lemma 6.7.9 implies the required conclusion.]

6.7.2. (a) Let h be a function on $\mathbf{R}$ such that $h(x)$ and $xh(x)$ are in $L^2(\mathbf{R})$. Show that h is integrable over $\mathbf{R}$ and satisfies

$$\big\| h \big\|_{L^1}^2 \le 8 \big\| h \big\|_{L^2} \big\| xh(x) \big\|_{L^2}.$$

(b) Suppose that g is an integrable function on $\mathbf{R}$ with vanishing integral and $g(x)$ and $xg(x)$ are in $L^2(\mathbf{R})$. Show that g lies in $H^1(\mathbf{R})$ and that for some constant C we have

$$\big\| g \big\|_{H^1}^2 \le C \big\| g \big\|_{L^2} \big\| xg(x) \big\|_{L^2}.$$

[*Hint:* Part (a): split the integral of $|h(x)|$ over the regions $|x| \le R$ and $|x| > R$ and pick a suitable R. Part (b): Show that both $H(g)$ and $H(yg(y))$ lie in L^2. But since g has vanishing integral, we have $xH(g)(x) = H(yg(y))(x)$.]

6.7.3. (a) Let H be the Hilbert transform. Prove the identity

$$H(fg - H(f)H(g)) = fH(g) + gH(f)$$

for all f,g in $\bigcup_{1\le p<\infty} L^p(\mathbf{R})$.
(b) Show that the bilinear operators

$$(f,g) \mapsto f H(g) + H(f) g,$$
$$(f,g) \mapsto f g - H(f) H(g),$$

map $L^p(\mathbf{R}) \times L^{p'}(\mathbf{R}) \to H^1(\mathbf{R})$ whenever $1 < p < \infty$.
[*Hint:* Part (a): Consider the boundary values of the product of the analytic extensions of $f + iH(f)$ and $g + iH(g)$ on the upper half-space. Part (b): Use part (a) and Theorem 6.7.5.]

6.7.4. Follow the steps given to prove the following interpolation result. Let $1 < p_1 \le \infty$ and let T be a subadditive operator that maps $H^1(\mathbf{R}^n) + L^{p_1}(\mathbf{R}^n)$ into measurable functions on $\mathbf{R}^n$. Suppose that there is $A_0 < \infty$ such that for all $f \in H^1(\mathbf{R}^n)$ we have

$$\sup_{\lambda>0} \lambda \left|\left\{x \in \mathbf{R}^n :\ |T(f)(x)| > \lambda\right\}\right| \le A_0 \left\|f\right\|_{H^1}$$

and that it also maps $L^{p_1}(\mathbf{R}^n)$ to $L^{p_1,\infty}(\mathbf{R}^n)$ with norm at most A_1. Show that for any $1 < p < p_1$, T maps $L^p(\mathbf{R}^n)$ to itself with norm at most

$$C A_0^{\frac{\frac{1}{p}-\frac{1}{p_1}}{1-\frac{1}{p_1}}} A_1^{\frac{1-\frac{1}{p}}{1-\frac{1}{p_1}}},$$

where $C = C(n,p,p_1)$.
(a) Fix $1 < q < p < p_1 < \infty$ and f and let Q_j be the family of all maximal dyadic cubes such that $\lambda^q < |Q_j|^{-1} \int_{Q_j} |f|^q\,dx$. Write $E_\lambda = \bigcup Q_j$ and note that $E_\lambda \subseteq \{M(|f|^q)^{\frac{1}{q}} > \lambda\}$ and that $|f| \le \lambda$ a.e. on $(E_\lambda)^c$. Write f as the sum of the *good function*

$$g_\lambda = f\chi_{(E_\lambda)^c} + \sum_j (\operatorname*{Avg}_{Q_j} f)\,\chi_{Q_j}$$

and the *bad function*

$$b_\lambda = \sum_j b_\lambda^j, \qquad \text{where} \qquad b_\lambda^j = (f - \operatorname*{Avg}_{Q_j} f)\,\chi_{Q_j}.$$

(b) Show that g_λ lies in $L^{p_1}(\mathbf{R}^n) \cap L^\infty(\mathbf{R}^n)$, $\left\|g_\lambda\right\|_{L^\infty} \le 2^{\frac{n}{q}}\lambda$, and that

$$\left\|g_\lambda\right\|_{L^{p_1}}^{p_1} \le \int_{|f|\le\lambda} |f(x)|^{p_1}\,dx + 2^{\frac{np_1}{q}} \lambda^{p_1} |E_\lambda| < \infty.$$

(c) Show that for $c = 2^{\frac{n}{q}+1}$, each $c^{-1}\lambda^{-1}|Q_j|^{-1} b_\lambda^j$ is an L^q-atom for H^1. Conclude that b_λ lies in $H^1(\mathbf{R}^n)$ and satisfies

$$\left\|b_\lambda\right\|_{H^1} \le c\lambda \sum_j |Q_j| \le c\lambda\,|E_\lambda| < \infty.$$

(d) Start with

$$\|T(f)\|_{L^p}^p \le p\gamma^p \int_0^\infty \lambda^{p-1} \big|\{|T(g_\lambda)| > \tfrac{1}{2}\gamma\lambda\}\big|\, d\lambda$$
$$+ p\gamma^p \int_0^\infty \lambda^{p-1} \big|\{|T(b_\lambda)| > \tfrac{1}{2}\gamma\lambda\}\big|\, d\lambda$$

and use the results in parts (b) and (c) to obtain that the preceding expression is at most $C(n,p,q,p_1)\max(A_1\gamma^{p-p_1}, \gamma^{p-1}A_0)$. Select $\gamma = A_1^{\frac{p_1}{p_1-1}} A_0^{-\frac{1}{p_1-1}}$ to obtain the required conclusion.
(e) In the case $p_1 = \infty$ we have $|T(g_\lambda)| \le A_1 2^{\frac{n}{q}}\lambda$ and pick $\gamma > 2A_1 2^{\frac{n}{q}}$ to make the integral involving g_λ vanishing.

6.7.5. Let f be an integrable function on the line whose Fourier transform is also integrable and vanishes on the negative half-line. Show that f lies in $H^1(\mathbf{R})$.

HISTORICAL NOTES

The strong type $L^p \to L^q$ estimates in Theorem 6.1.3 were obtained by Hardy and Littlewood [157] (see also [158]) when $n = 1$ and by Sobolev [285] for general n. The weak type estimate $L^1 \to L^{\frac{n}{n-s},\infty}$ first appeared in Zygmund [339]. The proof of Theorem 6.1.3 using estimate (6.1.11) is taken from Hedberg [161]. The best constants in this theorem when $p = \frac{2n}{n+s}$, $q = \frac{2n}{n-s}$, and $0 < s < n$ were precisely evaluated by Lieb [213]. A generalization of Theorem 6.1.3 for nonconvolution operators was obtained by Folland and Stein [132].

The Riesz potentials were systematically studied by Riesz [270] on $\mathbf{R}^n$ although their one-dimensional version appeared in earlier work of Weyl [330]. The Bessel potentials were introduced by Aronszajn and Smith [7] and also by Calderón [41], who was the first to observe that the potential space $\mathscr{L}_s^p$ (i.e., the Sobolev space L_s^p) coincides with the space L_k^p given in the classical Definition 6.2.1 when $s = k$ is an integer. Theorem 6.2.4 is due to Sobolev [285] when s is a positive integer. The case $p = 1$ of Sobolev's theorem (Exercise 6.2.9) was later obtained independently by Gagliardo [139] and Nirenberg [249]. We refer to the books of Adams [2], Lieb and Loss [214], and Maz'ya [229] for a more systematic study of Sobolev spaces and their use in analysis.

An early characterization of Lipschitz spaces using Littlewood–Paley type operators (built from the Poisson kernel) appears in the work of Hardy and Littlewood [160]. These and other characterizations were obtained and extensively studied in higher dimensions by Taibleson [300], [301], [302] in his extensive study. Lipschitz spaces can also be characterized via mean oscillation over cubes. This idea originated in the simultaneous but independent work of Campanato [39], [40] and Meyers [234] and led to duality theorems for these spaces. Incidentally, the predual of the space $\dot{\Lambda}_\alpha$ is the Hardy space H^p with $p = \frac{n}{n+\alpha}$, as shown by Duren, Romberg, and Shields [118] for the unit circle and by Walsh [327] for higher-dimensional spaces; see also Fefferman and Stein [130]. We refer to the book of García-Cuerva and Rubio de Francia [141] for a nice exposition of these results. An excellent expository reference on Lipschitz spaces is the article of Krantz [199].

Taibleson in his aforementioned work also studied the generalized Lipschitz spaces $\Lambda_\alpha^{p,q}$ called today Besov spaces. These spaces were named after Besov, who obtained a trace theorem and embeddings for them [24], [25]. The spaces $B_p^{\alpha,q}$, as defined in Section 6.5, were introduced by Peetre [255], although the case $p = q = 2$ was earlier considered by Hörmander [166]. The connection of Besov spaces with modern Littlewood–Paley theory was brought to the surface by Peetre [255]. The extension of the definition of Besov spaces to the case $p < 1$ is also due to Peetre [256], but there was a forerunner by Flett [131]. The spaces $F_p^{\alpha,q}$ with $1 < p,q < \infty$ were introduced by Triebel [316] and independently by Lizorkin [218]. The extension of the spaces $F_p^{\alpha,q}$ to the case $0 < p < \infty$ and $0 < q \leq \infty$ first appeared in Peetre [258], who also obtained a maximal characterization for all of these spaces. Lemma 6.5.3 originated in Peetre [258]; the version given in the text is based on a refinement of Triebel [317]. The article of Lions, Lizorkin, and Nikol'skij [216] presents an account of the treatment of the spaces $F_p^{\alpha,q}$ introduced by Triebel and Lizorkin as well as the equivalent characterizations obtained by Lions, using interpolation between Banach spaces, and by Nikol'skij, using best approximation.

The theory of Hardy spaces is vast and complicated. In classical complex analysis, the Hardy spaces H^p were spaces of analytic functions and were introduced to characterize boundary values of analytic functions on the unit disk. Precisely, the space $H^p(\mathbb{D})$ was introduced by Hardy [156] to consist of all analytic functions F on the unit disk $\mathbb{D}$ with the property that $\sup_{0<r<1}\int_0^1 |F(re^{2\pi i\theta})|^p\,d\theta < \infty$, $0 < p < \infty$. When $1 < p < \infty$, this space coincides with the space of analytic functions whose real parts are Poisson integrals of functions in $L^p(\mathbf{T}^1)$. But for $0 < p \leq 1$ this characterization fails and for several years a satisfactory characterization was missing. For a systematic treatment of these spaces we refer to the books of Duren [117] and Koosis [195].

With the illuminating work of Stein and Weiss [293] on systems of conjugate harmonic functions the road opened to higher-dimensional extensions of Hardy spaces. Burkholder, Gundy, and Silverstein [38] proved the fundamental theorem that an analytic function F lies in $H^p(\mathbf{R}^2_+)$ [i.e., $\sup_{y>0}\int_{\mathbf{R}} |F(x+iy)|^p\,dx < \infty$] if and only if the nontangential maximal function of its real part lies in $L^p(\mathbf{R})$. This result was proved using Brownian motion, but later Koosis [194] obtained another proof using complex analysis. This theorem spurred the development of the modern theory of Hardy spaces by providing the first characterization without the notion of conjugacy and indicating that Hardy spaces are intrinsically defined. The pioneering article of Fefferman and Stein [130] furnished three new characterizations of Hardy spaces: using a maximal function associated with a general approximate identity, using the grand maximal function, and using the area function of Luzin. From this point on, the role of the Poisson kernel faded into the background, when it turned out that it was not essential in the study of Hardy spaces. A previous characterization of Hardy spaces using the g-function, a radial analogue of the Luzin area function, was obtained by Calderón [42]. Two alternative characterizations of Hardy spaces were obtained by Uchiyama in terms of the generalized Littlewood–Paley g-function [319] and in terms of Fourier multipliers [320]. Necessary and sufficient conditions for systems of singular integral operators to characterize $H^1(\mathbf{R}^n)$ were also obtained by Uchiyama [318]. The characterization of H^p using Littlewood–Paley theory was observed by Peetre [257]. The case $p = 1$ was later independently obtained by Rubio de Francia, Ruiz, and Torrea [276].

The one-dimensional atomic decomposition of Hardy spaces is due to Coifman [72] and its higher-dimensional extension to Latter [206]. A simplification of some of the technical details in Latter's proof was subsequently obtained by Latter and Uchiyama [207]. Using the atomic decomposition Coifman and Weiss [86] extended the definition of Hardy spaces to more general structures. The idea of obtaining the atomic decomposition from the reproducing formula (6.6.8) goes back to Calderón [44]. Another simple proof of the L^2-atomic decomposition for H^p (starting from the nontangential Poisson maximal function) was obtained by Wilson [332]. With only a little work, one can show that L^q-atoms for H^p can be written as sums of L^∞-atoms for H^p. We refer to the book of García-Cuerva and Rubio de Francia [141] for a proof of this fact. Although finite sums of atoms are dense in H^1, an example due to Y. Meyer (contained in [233]) shows that the H^1 norm of a function may not be comparable to $\inf\sum_{j=1}^N |\lambda_j|$, where the infimum is taken over all representations of the function as finite linear combinations $\sum_{j=1}^N \lambda_j a_j$ with the a_j being L^∞-atoms for H^1. Based on this idea, Bownik [34] constructed an example of a linear functional on a dense subspace of H^1 that is uniformly bounded on L^∞-atoms for H^1 but does not extend to a bounded linear functional on the whole H^1. However, if a Banach-valued linear operator is bounded uniformly on all L^q-atoms for H^p with $1 < q < \infty$ and $0 < p \leq 1$, then it is bounded on the entire H^p as shown by Meda, Sjögren, and Vallarino [230]. This fact is also valid for quasi-Banach-valued linear operators, and when $q = 2$ it was obtained independently by Yang and Zhou [338]. A related general result says that a sublinear operator maps the Triebel–Lizorkin space $\dot{F}^s_{p,q}(\mathbf{R}^n)$ to a quasi-Banach space if and only if it is uniformly bounded on certain infinitely differentiable atoms of the space; see Liu and Yang [217]. Atomic decompositions of general function spaces were obtained in the fundamental work of Frazier and Jawerth [135], [136]. The exposition in Section 6.6 is based on the article of Frazier and Jawerth [137]. The work of these authors provides a solid manifestation that atomic decompositions are intrinsically related to Littlewood–Paley theory and not wedded to a particular space. Littlewood–Paley theory therefore provides a comprehensive and unifying perspective on function spaces.

Main references on H^p spaces and their properties are the books of Baernstein and Sawyer [12], Folland and Stein [133] in the context of homogeneous groups, Lu [219] (on which the proofs of Lemma 6.4.5 and Theorem 6.4.4 are based), Strömberg and Torchinsky [298] (on weighted Hardy spaces), and Uchiyama [321]. The articles of Calderón and Torchinsky [45], [46] develop and extend the theory of Hardy spaces to the nonisotropic setting. Hardy spaces can also be defined in terms of nonstandard convolutions, such as the "twisted convolution" on $\mathbf{R}^{2n}$. Characterizations of the space H^1 in this context have been obtained by Mauceri, Picardello, and Ricci [226]

The localized Hardy spaces h_p, $0 < p \leq 1$, were introduced by Goldberg [146] as spaces of distributions for which the maximal operator $\sup_{0<t<1} |\Phi_t * f|$ lies in $L^p(\mathbf{R}^n)$ (here Φ is a Schwartz function with nonvanishing integral). These spaces can be characterized in ways analogous to those of the homogeneous Hardy spaces H^p; in particular, they admit an atomic decomposition. It was shown by Bui [37] that the space h^p coincides with the Triebel–Lizorkin space $F^{0,2}_p(\mathbf{R}^n)$; see also Meyer [232]. For the local theory of Hardy spaces one may consult the articles of Dafni [100] and Chang, Krantz, and Stein [59].

Interpolation of operators between Hardy spaces was originally based on complex function theory; see the articles of Calderón and Zygmund [48] and Weiss [328]. The real-interpolation approach discussed in Exercise 6.7.4 can be traced in the article of Igari [174]. Interpolation between Hardy spaces was further studied and extended by Riviere and Sagher [271] and Fefferman, Riviere, and Sagher [128].

The action of singular integrals on periodic spaces was studied by Calderón and Zygmund [47]. The preservation of Lipschitz spaces under singular integral operators is due to Taibleson [299]. The case $0 < \alpha < 1$ was earlier considered by Privalov [268] for the conjugate function on the circle. Fefferman and Stein [130] were the first to show that singular integrals map Hardy spaces to themselves. The boundedness of fractional integrals on H^p was obtained by Krantz [198]. The case $p = 1$ was earlier considered by Stein and Weiss [293]. The action of multilinear singular integrals on Hardy spaces was studied by Coifman and Grafakos [75] and Grafakos and Kalton [149]. An exposition on the subject of function spaces and the action of singular integrals on them was written by Frazier, Jawerth, and Weiss [138]. For a careful study of the action of singular integrals on function spaces, we refer to the book of Torres [315]. The study of anisotropic function spaces and the action of singular integrals on them has been studied by Bownik [33]. Weighted anisotropic Hardy spaces have been studied by Bownik, Li, Yang, and Zhou [35].

Chapter 7
BMO and Carleson Measures

The space of functions of bounded mean oscillation, or *BMO*, naturally arises as the class of functions whose deviation from their means over cubes is bounded. L^∞ functions have this property, but there exist unbounded functions with bounded mean oscillation. Such functions are slowly growing, and they typically have at most logarithmic blowup. The space *BMO* shares similar properties with the space L^∞, and it often serves as a substitute for it. For instance, classical singular integrals do not map L^∞ to L^∞ but L^∞ to *BMO*. And in many instances interpolation between L^p and *BMO* works just as well between L^p and L^∞. But the role of the space *BMO* is deeper and more far-reaching than that. This space crucially arises in many situations in analysis, such as in the characterization of the L^2 boundedness of nonconvolution singular integral operators with standard kernels.

Carleson measures are among the most important tools in harmonic analysis. These measures capture essential orthogonality properties and exploit properties of extensions of functions on the upper half-space. There exists a natural and deep connection between Carleson measures and *BMO* functions; indeed, certain types of measures defined in terms of functions are Carleson if and only if the underlying functions are in *BMO*. Carleson measures are especially crucial in the study of L^2 problems, where the Fourier transform cannot be used to provide boundedness via Plancherel's theorem. The power of the Carleson measure techniques becomes apparent in Chapter 8, where they play a crucial role in the proof of several important results.

7.1 Functions of Bounded Mean Oscillation

What exactly is bounded mean oscillation and what kind of functions have this property? The mean of a (locally integrable) function over a set is another word for its average over that set. The oscillation of a function over a set is the absolute value of the difference of the function from its mean over this set. Mean oscillation is therefore the average of this oscillation over a set. A function is said to be of bounded

L. Grafakos, *Modern Fourier Analysis*, DOI: 10.1007/978-0-387-09434-2_7,

mean oscillation if its mean oscillation over all cubes is bounded. Precisely, given a locally integrable function f on $\mathbf{R}^n$ and a measurable set Q in $\mathbf{R}^n$, denote by

$$\operatorname*{Avg}_Q f = \frac{1}{|Q|}\int_Q f(x)\,dx$$

the mean (or average) of f over Q. Then the *oscillation* of f over Q is the function $|f - \operatorname{Avg}_Q f|$, and the *mean oscillation* of f over Q is

$$\frac{1}{|Q|}\int_Q \Big|f(x) - \operatorname*{Avg}_Q f\Big|\,dx.$$

7.1.1 Definition and Basic Properties of *BMO*

Definition 7.1.1. For f a complex-valued locally integrable function on $\mathbf{R}^n$, set

$$\big\|f\big\|_{BMO} = \sup_Q \frac{1}{|Q|}\int_Q \Big|f(x) - \operatorname*{Avg}_Q f\Big|\,dx,$$

where the supremum is taken over all cubes Q in $\mathbf{R}^n$. The function f is called of bounded mean oscillation if $\big\|f\big\|_{BMO} < \infty$ and $BMO(\mathbf{R}^n)$ is the set of all locally integrable functions f on $\mathbf{R}^n$ with $\big\|f\big\|_{BMO} < \infty$.

Several remarks are in order. First it is a simple fact that $BMO(\mathbf{R}^n)$ is a linear space, that is, if $f, g \in BMO(\mathbf{R}^n)$ and $\lambda \in \mathbf{C}$, then $f+g$ and λf are also in $BMO(\mathbf{R}^n)$ and

$$\begin{aligned} \big\|f+g\big\|_{BMO} &\le \big\|f\big\|_{BMO} + \big\|g\big\|_{BMO}, \\ \big\|\lambda f\big\|_{BMO} &= |\lambda|\,\big\|f\big\|_{BMO}. \end{aligned}$$

But $\big\|\;\big\|_{BMO}$ is not a norm. The problem is that if $\big\|f\big\|_{BMO} = 0$, this does not imply that $f = 0$ but that f is a constant. See Proposition 7.1.2. Moreover, every constant function c satisfies $\big\|c\big\|_{BMO} = 0$. Consequently, functions f and $f+c$ have the same BMO norms whenever c is a constant. In the sequel, we keep in mind that elements of BMO whose difference is a constant are identified. Although $\big\|\;\big\|_{BMO}$ is only a seminorm, we occasionally refer to it as a norm when there is no possibility of confusion.

We begin with a list of basic properties of BMO.

Proposition 7.1.2. *The following properties of the space* $BMO(\mathbf{R}^n)$ *are valid:*

(1) If $\big\|f\big\|_{BMO} = 0$*, then* f *is a.e. equal to a constant.*

(2) $L^\infty(\mathbf{R}^n)$ *is contained in* $BMO(\mathbf{R}^n)$ *and* $\big\|f\big\|_{BMO} \le 2\big\|f\big\|_{L^\infty}$.

(3) Suppose that there exists an $A > 0$ *such that for all cubes* Q *in* $\mathbf{R}^n$ *there exists a constant* c_Q *such that*

$$\sup_Q \frac{1}{|Q|}\int_Q |f(x)-c_Q|\,dx \le A\,. \tag{7.1.1}$$

Then $f \in BMO(\mathbf{R}^n)$ *and* $\|f\|_{BMO} \le 2A$.

(4) For all f locally integrable we have

$$\frac{1}{2}\|f\|_{BMO} \le \sup_Q \frac{1}{|Q|}\inf_{c_Q}\int_Q |f(x)-c_Q|\,dx \le \|f\|_{BMO}.$$

(5) If $f\in BMO(\mathbf{R}^n)$, $h\in\mathbf{R}^n$, *and* $\tau^h(f)$ *is given by* $\tau^h(f)(x)=f(x-h)$, *then* $\tau^h(f)$ *is also in* $BMO(\mathbf{R}^n)$ *and*

$$\|\tau^h(f)\|_{BMO} = \|f\|_{BMO}.$$

(6) If $f\in BMO(\mathbf{R}^n)$ *and* $\lambda>0$, *then the function* $\delta^\lambda(f)$ *defined by* $\delta^\lambda(f)(x) = f(\lambda x)$ *is also in* $BMO(\mathbf{R}^n)$ *and*

$$\|\delta^\lambda(f)\|_{BMO} = \|f\|_{BMO}.$$

(7) If $f\in BMO$ *then so is* $|f|$. *Similarly, if* f,g *are real-valued BMO functions, then so are* $\max(f,g)$, *and* $\min(f,g)$. *In other words, BMO is a lattice. Moreover,*

$$\begin{aligned}\||f|\|_{BMO} &\le 2\|f\|_{BMO},\\ \|\max(f,g)\|_{BMO} &\le \frac{3}{2}\Big(\|f\|_{BMO}+\|g\|_{BMO}\Big),\\ \|\min(f,g)\|_{BMO} &\le \frac{3}{2}\Big(\|f\|_{BMO}+\|g\|_{BMO}\Big).\end{aligned}$$

(8) For locally integrable functions f define

$$\|f\|_{BMO_{\text{balls}}} = \sup_B \frac{1}{|B|}\int_B \Big|f(x)-\operatorname*{Avg}_B f\Big|\,dx, \tag{7.1.2}$$

where the supremum is taken over all balls B in $\mathbf{R}^n$. *Then there are positive constants* c_n, C_n *such that*

$$c_n\|f\|_{BMO} \le \|f\|_{BMO_{\text{balls}}} \le C_n\|f\|_{BMO}.$$

Proof. To prove (1) note that f has to be a.e. equal to its average c_N over every cube $[-N,N]^n$. Since $[-N,N]^n$ is contained in $[-N-1,N+1]^n$, it follows that $c_N=c_{N+1}$ for all N. This implies the required conclusion. To prove (2) observe that

$$\operatorname*{Avg}_Q\Big|f-\operatorname*{Avg}_Q f\Big| \le 2\operatorname*{Avg}_Q|f| \le 2\|f\|_{L^\infty}.$$

For part (3) note that

$$\big|f-\operatorname*{Avg}_Q f\big| \le |f-c_Q|+\big|\operatorname*{Avg}_Q f-c_Q\big| \le |f-c_Q|+\frac{1}{|Q|}\int_Q |f(t)-c_Q|\,dt\,.$$

Averaging over Q and using (7.1.1), we obtain that $\big\|f\big\|_{BMO}\le 2A$. The lower inequality in (4) follows from (3) while the upper one is trivial. Property (5) is immediate. For (6) note that $\operatorname{Avg}_Q \delta^\lambda(f)=\operatorname{Avg}_{\lambda Q} f$ and thus

$$\frac{1}{|Q|}\int_Q \big|f(\lambda x)-\operatorname*{Avg}_Q \delta^\lambda(f)\big|\,dx=\frac{1}{|\lambda Q|}\int_{\lambda Q}\big|f(x)-\operatorname*{Avg}_{\lambda Q} f\big|\,dx.$$

Property (7) is a consequence of the easy fact that

$$\big||f|-\operatorname*{Avg}_Q |f|\big| \le \big|f-\operatorname*{Avg}_Q f\big|+\operatorname*{Avg}_Q\big|f-\operatorname*{Avg}_Q f\big|\,.$$

Also, the maximum and the minimum of two functions can be expressed in terms of the absolute value of their difference. We now turn to (8). Given any cube Q in $\mathbf{R}^n$, we let B be the smallest ball that contains it. Then $|B|/|Q|=2^{-n}v_n\sqrt{n^n}$, where v_n is the volume of the unit ball, and

$$\frac{1}{|Q|}\int_Q\big|f(x)-\operatorname*{Avg}_B f\big|\,dx\le\frac{|B|}{|Q|}\frac{1}{|B|}\int_B\big|f(x)-\operatorname*{Avg}_B f\big|\,dx\le\frac{v_n\sqrt{n^n}}{2^n}\big\|f\big\|_{BMO_{\text{balls}}}.$$

It follows from (3) that $\big\|f\big\|_{BMO}\le 2^{1-n}v_n\sqrt{n^n}\,\big\|f\big\|_{BMO_{\text{balls}}}$. To obtain the reverse conclusion, given any ball B find the smallest cube Q that contains it and argue similarly using a version of (3) for the space BMO_{balls}. □

Example 7.1.3. We indicate why $L^\infty(\mathbf{R}^n)$ is a proper subspace of $BMO(\mathbf{R}^n)$. We claim that the function $\log|x|$ is in $BMO(\mathbf{R}^n)$ but not in $L^\infty(\mathbf{R}^n)$. To prove that it is in $BMO(\mathbf{R}^n)$, for every $x_0\in\mathbf{R}^n$ and $R>0$, we must find a constant $C_{x_0,R}$ such that the average of $\big|\log|x|-C_{x_0,R}\big|$ over the ball $\{x:\ |x-x_0|\le R\}$ is uniformly bounded. Since

$$\frac{1}{v_nR^n}\int\limits_{|x-x_0|\le R}\big|\log|x|-C_{x_0,R}\big|\,dx=\frac{1}{v_n}\int\limits_{|z-R^{-1}x_0|\le 1}\big|\log|z|-C_{x_0,R}+\log R\big|\,dz,$$

we may take $C_{x_0,R}=C_{R^{-1}x_0,1}+\log R$, and things reduce to the case that $R=1$ and x_0 is arbitrary. If $R=1$ and $|x_0|\le 2$, take $C_{x_0,1}=0$ and observe that

$$\int\limits_{|x-x_0|\le 1}\big|\log|x|\big|\,dx\le\int\limits_{|x|\le 3}\big|\log|x|\big|\,dx=C.$$

When $R=1$ and $|x_0|\ge 2$, take $C_{x_0,1}=\log|x_0|$. In this case notice that

$$\frac{1}{v_n}\int\limits_{|x-x_0|\le 1}\big|\log|x|-\log|x_0|\big|\,dx=\frac{1}{v_n}\int\limits_{|x-x_0|\le 1}\left|\log\frac{|x|}{|x_0|}\right|dx\le\log 2\,,$$

since when $|x-x_0| \le 1$ and $|x_0| \ge 2$ we have that $\log \frac{|x|}{|x_0|} \le \log \frac{|x_0|+1}{|x_0|} \le \log \frac{3}{2}$ and $\log \frac{|x_0|}{|x|} \le \log \frac{|x_0|}{|x_0|-1} \le \log 2$. Thus $\log|x|$ is in BMO.

The function $\log|x|$ turns out to be a typical element of BMO, but we make this statement a bit more precise later. It is interesting to observe that an abrupt cutoff of a BMO function may not give a function in the same space.

Example 7.1.4. The function $h(x) = \chi_{x>0} \log \frac{1}{x}$ is not in $BMO(\mathbf{R})$. Indeed, the problem is at the origin. Consider the intervals $(-\varepsilon, \varepsilon)$, where $0 < \varepsilon < \frac{1}{2}$. We have that

$$\operatorname*{Avg}_{(-\varepsilon,\varepsilon)} h = \frac{1}{2\varepsilon} \int_{-\varepsilon}^{+\varepsilon} h(x)\,dx = \frac{1}{2\varepsilon} \int_0^{\varepsilon} \log \frac{1}{x}\,dx = \frac{1+\log\frac{1}{\varepsilon}}{2}.$$

But then

$$\frac{1}{2\varepsilon} \int_{-\varepsilon}^{+\varepsilon} \Big|h(x) - \operatorname*{Avg}_{(-\varepsilon,\varepsilon)} h\Big|\,dx \ge \frac{1}{2\varepsilon} \int_{-\varepsilon}^{0} \Big|\operatorname*{Avg}_{(-\varepsilon,\varepsilon)} h\Big|\,dx = \frac{1+\log\frac{1}{\varepsilon}}{4},$$

and the latter is clearly unbounded as $\varepsilon \to 0$.

Let us now look at some basic properties of BMO functions. Observe that if a cube Q_1 is contained in a cube Q_2, then

$$\begin{aligned} \Big|\operatorname*{Avg}_{Q_1} f - \operatorname*{Avg}_{Q_2} f\Big| &\le \frac{1}{|Q_1|} \int_{Q_1} \Big|f - \operatorname*{Avg}_{Q_2} f\Big|\,dx \\ &\le \frac{1}{|Q_1|} \int_{Q_2} \Big|f - \operatorname*{Avg}_{Q_2} f\Big|\,dx \\ &\le \frac{|Q_2|}{|Q_1|} \big\|f\big\|_{BMO}. \end{aligned} \tag{7.1.3}$$

The same estimate holds if the sets Q_1 and Q_2 are balls.

A version of this inequality is the first statement in the following proposition. For simplicity, we denote by $\|f\|_{BMO}$ the expression given by $\|f\|_{BMO_{\text{balls}}}$ in (7.1.2), since these quantities are comparable. For a ball B and $a > 0$, aB denotes the ball that is concentric with B and whose radius is a times the radius of B.

Proposition 7.1.5. *(i) Let f be in $BMO(\mathbf{R}^n)$. Given a ball B and a positive integer m, we have*

$$\Big|\operatorname*{Avg}_{B} f - \operatorname*{Avg}_{2^m B} f\Big| \le 2^n m \big\|f\big\|_{BMO}. \tag{7.1.4}$$

(ii) For any $\delta > 0$ there is a constant $C_{n,\delta}$ such that for any ball $B(x_0,R)$ we have

$$R^{\delta} \int_{\mathbf{R}^n} \frac{\big|f(x) - \operatorname{Avg}_{B(x_0,R)} f\big|}{(R+|x-x_0|)^{n+\delta}}\,dx \le C_{n,\delta} \big\|f\big\|_{BMO}. \tag{7.1.5}$$

An analogous estimate holds for cubes with center x_0 and side length R.
(iii) There exists a constant C_n such that for all $f \in BMO(\mathbf{R}^n)$ we have

$$\sup_{y\in\mathbf{R}^n}\sup_{t>0}\int_{\mathbf{R}^n}|f(x)-(P_t*f)(y)|P_t(x-y)\,dx\le C_n\|f\|_{BMO}. \tag{7.1.6}$$

Here P_t denotes the Poisson kernel introduced in Chapter 2.
(iv) Conversely, there is a constant C_n' such that for all $f\in L^1_{\mathrm{loc}}(\mathbf{R}^n)$ for which

$$\int_{\mathbf{R}^n}\frac{|f(x)|}{(1+|x|)^{n+1}}\,dx<\infty$$

we have

$$C_n'\|f\|_{BMO}\le\sup_{y\in\mathbf{R}^n}\sup_{t>0}\int_{\mathbf{R}^n}|f(x)-(P_t*f)(y)|P_t(x-y)\,dx. \tag{7.1.7}$$

Proof. (i) We have

$$\begin{aligned}\Big|\operatorname*{Avg}_{B}f-\operatorname*{Avg}_{2B}f\Big| &= \frac{1}{|B|}\Big|\int_B\big(f(t)-\operatorname*{Avg}_{2B}f\big)\,dt\Big|\\ &\le\frac{2^n}{|2B|}\int_{2B}\big|f(t)-\operatorname*{Avg}_{2B}f\big|\,dt\\ &\le 2^n\|f\|_{BMO}.\end{aligned}$$

Using this inequality, we derive (7.1.4) by adding and subtracting the terms

$$\operatorname*{Avg}_{2B}f,\quad\operatorname*{Avg}_{2^2B}f,\quad\dots,\quad\operatorname*{Avg}_{2^{m-1}B}f.$$

(ii) In the proof below we take $B(x_0,R)$ to be the ball $B=B(0,1)$ with radius 1 centered at the origin. Once this case is known, given a ball $B(x_0,R)$ we replace the function f by the function $f(Rx+x_0)$. When $B=B(0,1)$ we have

$$\begin{aligned}&\int_{\mathbf{R}^n}\frac{\big|f(x)-\operatorname*{Avg}_{B}f\big|}{(1+|x|)^{n+\delta}}\,dx\\ &\le\int_B\frac{\big|f(x)-\operatorname*{Avg}_{B}f\big|}{(1+|x|)^{n+\delta}}\,dx+\sum_{k=0}^{\infty}\int_{2^{k+1}B\setminus 2^kB}\frac{\big|f(x)-\operatorname*{Avg}_{2^{k+1}B}f\big|+\big|\operatorname*{Avg}_{2^{k+1}B}f-\operatorname*{Avg}_{B}f\big|}{(1+|x|)^{n+\delta}}\,dx\\ &\le\int_B\big|f(x)-\operatorname*{Avg}_{B}f\big|\,dx\\ &\qquad+\sum_{k=0}^{\infty}2^{-k(n+\delta)}\int_{2^{k+1}B}\Big(\big|f(x)-\operatorname*{Avg}_{2^{k+1}B}f\big|+\big|\operatorname*{Avg}_{2^{k+1}B}f-\operatorname*{Avg}_{B}f\big|\Big)\,dx\\ &\le v_n\|f\|_{BMO}+\sum_{k=0}^{\infty}2^{-k(n+\delta)}(1+2^n(k+1))(2^{k+1})^nv_n\|f\|_{BMO}\\ &=C_{n,\delta}'\|f\|_{BMO}.\end{aligned}$$

(iii) The proof of (7.1.6) is a reprise of the argument given in (ii). Set $B_t = B(y,t)$. We first prove a version of (7.1.6) in which the expression $(P_t * f)(y)$ is replaced by $\operatorname{Avg}_{B_t} f$. For fixed y,t we have

$$
\begin{aligned}
&\int_{\mathbf{R}^n} \frac{t\big|f(x)-\operatorname*{Avg}\limits_{B_t} f\big|}{(t^2+|x-y|^2)^{\frac{n+1}{2}}}\,dx \\
&\le \int_{B_t} \frac{t\big|f(x)-\operatorname*{Avg}\limits_{B_t} f\big|}{(t^2+|x-y|^2)^{\frac{n+1}{2}}}\,dx \\
&\qquad + \sum_{k=0}^{\infty} \int_{2^{k+1}B_t\setminus 2^k B_t} \frac{t\Big(\big|f(x)-\operatorname*{Avg}\limits_{2^{k+1}B_t} f\big| + \big|\operatorname*{Avg}\limits_{2^{k+1}B_t} f - \operatorname*{Avg}\limits_{B_t} f\big|\Big)}{(t^2+|x-y|^2)^{\frac{n+1}{2}}}\,dx \\
&\le \int_{B_t} \frac{\big|f(x)-\operatorname*{Avg}\limits_{B_t} f\big|}{t^n}\,dx \\
&\qquad + \sum_{k=0}^{\infty} \frac{2^{-k(n+1)}}{t^n} \int_{2^{k+1}B_t} \Big(\big|f(x)-\operatorname*{Avg}\limits_{2^{k+1}B_t} f\big| + \big|\operatorname*{Avg}\limits_{2^{k+1}B_t} f - \operatorname*{Avg}\limits_{B_t} f\big|\Big)\,dx \\
&\le v_n\big\|f\big\|_{BMO} + \sum_{k=0}^{\infty} 2^{-k(n+1)}\big(1+2^n(k+1)\big)(2^{k+1})^n v_n\big\|f\big\|_{BMO} \\
&= C_n\big\|f\big\|_{BMO}.
\end{aligned}
\tag{7.1.8}
$$

Using the inequality just proved, we also obtain

$$
\begin{aligned}
\int_{\mathbf{R}^n} \big|(P_t*f)(y)-\operatorname*{Avg}\limits_{B_t} f\big| P_t(x-y)\,dx &= \big|(P_t*f)(y)-\operatorname*{Avg}\limits_{B_t} f\big| \\
&\le \int_{\mathbf{R}^n} P_t(x-y)\big|f(x)-\operatorname*{Avg}\limits_{B_t} f\big|\,dx \\
&\le C_n\big\|f\big\|_{BMO},
\end{aligned}
$$

which, combined with the inequality in (7.1.8), yields (7.1.6) with constant $2C_n$.

(iv) Conversely, let A be the expression on the right in (7.1.7). For $|x-y|\le t$ we have $P_t(x-y) \ge c_n t(2t^2)^{-\frac{n+1}{2}} = c_n' t^{-n}$, which gives

$$
A \ge \int_{\mathbf{R}^n} |f(x)-(P_t*f)(y)|P_t(x-y)\,dx \ge \frac{c_n'}{t^n}\int_{|x-y|\le t} |f(x)-(P_t*f)(y)|\,dx.
$$

Proposition 7.1.2 (3) now implies that

$$
\big\|f\big\|_{BMO} \le 2A/(v_n c_n').
$$

This concludes the proof of the proposition. □

7.1.2 The John–Nirenberg Theorem

Having set down some basic facts about BMO, we now turn to a deeper property of BMO functions: their exponential integrability. We begin with a preliminary remark. As we saw in Example 7.1.3, the function $g(x) = \log|x|^{-1}$ is in $BMO(\mathbf{R}^n)$. This function is exponentially integrable over any compact subset K of $\mathbf{R}^n$ in the sense that

$$\int_K e^{c|g(x)|}\,dx < \infty$$

for any $c < n$. It turns out that this is a general property of BMO functions, and this is the content of the next theorem.

Theorem 7.1.6. *For all $f \in BMO(\mathbf{R}^n)$, for all cubes Q, and all $\alpha > 0$ we have*

$$\left|\left\{x \in Q : \left|f(x) - \operatorname*{Avg}_Q f\right| > \alpha\right\}\right| \le e\,|Q|\,e^{-A\alpha/\|f\|_{BMO}} \tag{7.1.9}$$

with $A = (2^n e)^{-1}$.

Proof. Since inequality (7.1.9) is not altered when we multiply both f and α by the same constant, it suffices to assume that $\|f\|_{BMO} = 1$. Let us now fix a closed cube Q and a constant $b > 1$ to be chosen later.

We apply the Calderón–Zygmund decomposition to the function $f - \operatorname{Avg}_Q f$ inside the cube Q. We introduce the following selection criterion for a cube R:

$$\frac{1}{|R|}\int_R \left|f(x) - \operatorname*{Avg}_Q f\right| dx > b. \tag{7.1.10}$$

Since

$$\frac{1}{|Q|}\int_Q \left|f(x) - \operatorname*{Avg}_Q f\right| dx \le \|f\|_{BMO} = 1 < b,$$

the cube Q does not satisfy the selection criterion (7.1.10). Set $Q^{(0)} = Q$ and subdivide $Q^{(0)}$ into 2^n equal closed subcubes of side length equal to half of the side length of Q. Select such a subcube R if it satisfies the selection criterion (7.1.10). Now subdivide all nonselected cubes into 2^n equal subcubes of half their side length by bisecting the sides, and select among these subcubes those that satisfy (7.1.10). Continue this process indefinitely. We obtain a countable collection of cubes $\{Q_j^{(1)}\}_j$ satisfying the following properties:

(A-1) The interior of every $Q_j^{(1)}$ is contained in $Q^{(0)}$.

(B-1) $b < |Q_j^{(1)}|^{-1}\displaystyle\int_{Q_j^{(1)}} \left|f(x) - \operatorname*{Avg}_{Q^{(0)}} f\right| dx \le 2^n b.$

(C-1) $\left|\operatorname*{Avg}_{Q_j^{(1)}} f - \operatorname*{Avg}_{Q^{(0)}} f\right| \le 2^n b.$

(D-1) $\sum_j |Q_j^{(1)}| \le \frac{1}{b} \sum_j \int_{Q_j^{(1)}} \big|f(x) - \operatorname*{Avg}_{Q^{(0)}} f\big|\,dx \le \frac{1}{b}|Q^{(0)}|.$

(E-1) $\big|f - \operatorname*{Avg}_{Q^{(0)}} f\big| \le b$ a.e. on the set $Q^{(0)} \setminus \bigcup_j Q_j^{(1)}$.

We call the cubes $Q_j^{(1)}$ of first generation. Note that the second inequality in (D-1) requires (B-1) and the fact that $Q^{(0)}$ does not satisfy (7.1.10).

We now fix a selected first-generation cube $Q_j^{(1)}$ and we introduce the following selection criterion for a cube R:

$$\frac{1}{|R|} \int_R \big|f(x) - \operatorname*{Avg}_{Q_j^{(1)}} f\big|\,dx > b. \tag{7.1.11}$$

Observe that $Q_j^{(1)}$ does not satisfy the selection criterion (7.1.11). We apply a similar Calderón–Zygmund decomposition to the function

$$f - \operatorname*{Avg}_{Q_j^{(1)}} f$$

inside the cube $Q_j^{(1)}$. Subdivide $Q_j^{(1)}$ into 2^n equal closed subcubes of side length equal to half of the side length of $Q_j^{(1)}$ by bisecting the sides, and select such a subcube R if it satisfies the selection criterion (7.1.11). Continue this process indefinitely. Also repeat this process for any other cube $Q_j^{(1)}$ of the first generation. We obtain a collection of cubes $\{Q_l^{(2)}\}_l$ of second generation each contained in some $Q_j^{(1)}$ such that versions of (A-1)–(E-1) are satisfied, with the superscript (2) replacing (1) and the superscript (1) replacing (0). We use the superscript (k) to denote the generation of the selected cubes.

For a fixed selected cube $Q_l^{(2)}$ of second generation, introduce the selection criterion

$$\frac{1}{|R|} \int_R \big|f(x) - \operatorname*{Avg}_{Q_l^{(2)}} f\big|\,dx > b$$

and repeat the previous process to obtain a collection of cubes of third generation inside $Q_l^{(2)}$. Repeat this procedure for any other cube $Q_j^{(2)}$ of the second generation. Denote by $\{Q_s^{(3)}\}_s$ the thus obtained collection of all cubes of the third generation.

We iterate this procedure indefinitely to obtain a doubly indexed family of cubes $Q_j^{(k)}$ satisfying the following properties:

(A-k) The interior of every $Q_j^{(k)}$ is contained in a unique $Q_{j'}^{(k-1)}$.

(B-k) $b < |Q_j^{(k)}|^{-1} \int_{Q_j^{(k)}} \big|f(x) - \operatorname*{Avg}_{Q_{j'}^{(k-1)}} f\big|\,dx \le 2^n b.$

(C-k) $\Big|\operatorname*{Avg}_{Q_j^{(k)}} f - \operatorname*{Avg}_{Q_{j'}^{(k-1)}} f\Big| \le 2^n b$.

(D-k) $\sum_j |Q_j^{(k)}| \le \frac{1}{b}\sum_{j'} |Q_{j'}^{(k-1)}|$.

(E-k) $\Big|f - \operatorname*{Avg}_{Q_{j'}^{(k-1)}} f\Big| \le b$ a.e. on the set $Q_{j'}^{(k-1)} \setminus \bigcup_j Q_j^{(k)}$.

We prove (A-k)–(E-k). Note that (A-k) and the lower inequality in (B-k) are satisfied by construction. The upper inequality in (B-k) is a consequence of the fact that the unique cube $Q_{j_0}^{(k)}$ with double the side length of $Q_j^{(k)}$ that contains it was not selected in the process. Now (C-k) follows from the upper inequality in (B-k). (E-k) is a consequence of the Lebesgue differentiation theorem, since for every point in $Q_{j'}^{(k-1)} \setminus \bigcup_j Q_j^{(k)}$ there is a sequence of cubes shrinking to it and the averages of

$$\Big|f - \operatorname*{Avg}_{Q_{j'}^{(k-1)}} f\Big|$$

over all these cubes is at most b. It remains to prove (D-k). We have

$$\begin{aligned}
\sum_j |Q_j^{(k)}| &< \frac{1}{b}\sum_j \int_{Q_j^{(k)}} \Big|f(x) - \operatorname*{Avg}_{Q_{j'}^{(k-1)}} f\Big|\,dx \\
&= \frac{1}{b}\sum_{j'} \sum_{j \text{ corresp. to } j'} \int_{Q_j^{(k)}} \Big|f(x) - \operatorname*{Avg}_{Q_{j'}^{(k-1)}} f\Big|\,dx \\
&\le \frac{1}{b}\sum_{j'} \int_{Q_{j'}^{(k-1)}} \Big|f(x) - \operatorname*{Avg}_{Q_{j'}^{(k-1)}} f\Big|\,dx \\
&\le \frac{1}{b}\sum_{j'} |Q_{j'}^{(k-1)}|\, \|f\|_{BMO} \\
&= \frac{1}{b}\sum_{j'} |Q_{j'}^{(k-1)}|.
\end{aligned}$$

Having established (A-k)–(E-k) we turn to some consequences. Applying (D-k) successively $k-1$ times, we obtain

$$\sum_j |Q_j^{(k)}| \le b^{-k} |Q^{(0)}|. \tag{7.1.12}$$

For any fixed j we have that $\big|\operatorname{Avg}_{Q_j^{(1)}} f - \operatorname{Avg}_{Q^{(0)}} f\big| \le 2^n b$ and $\big|f - \operatorname{Avg}_{Q_j^{(1)}} f\big| \le b$ a.e. on $Q_j^{(1)} \setminus \bigcup_l Q_l^{(2)}$. This gives

$$\Big|f - \operatorname*{Avg}_{Q^{(0)}} f\Big| \le 2^n b + b \qquad \text{a.e.} \quad \text{on} \quad Q_j^{(1)} \setminus \bigcup_l Q_l^{(2)},$$

which, combined with (E-1), yields

$$\Big|f - \operatorname*{Avg}_{Q^{(0)}} f\Big| \le 2^n 2b \qquad \text{a.e.} \quad \text{on} \quad Q^{(0)} \setminus \bigcup_l Q_l^{(2)}. \tag{7.1.13}$$

For every fixed l we also have that $\big|f - \operatorname{Avg}_{Q_l^{(2)}} f\big| \le b$ a.e. on $Q_l^{(2)} \setminus \bigcup_s Q_s^{(3)}$, which combined with $\big|\operatorname{Avg}_{Q_l^{(2)}} f - \operatorname{Avg}_{Q_{l'}^{(1)}} f\big| \le 2^n b$ and $\big|\operatorname{Avg}_{Q_{l'}^{(1)}} f - \operatorname{Avg}_{Q^{(0)}} f\big| \le 2^n b$ yields

$$\Big|f - \operatorname*{Avg}_{Q^{(0)}} f\Big| \le 2^n 3b \qquad \text{a.e.} \quad \text{on} \quad Q_l^{(2)} \setminus \bigcup_s Q_s^{(3)}.$$

In view of (7.1.13), the same estimate is valid on $Q^{(0)} \setminus \bigcup_s Q_s^{(3)}$. Continuing this reasoning, we obtain by induction that for all $k \ge 1$ we have

$$\Big|f - \operatorname*{Avg}_{Q^{(0)}} f\Big| \le 2^n kb \qquad \text{a.e.} \quad \text{on} \quad Q^{(0)} \setminus \bigcup_s Q_s^{(k)}. \tag{7.1.14}$$

This proves the almost everywhere inclusion

$$\Big\{x \in Q : \ \Big|f(x) - \operatorname*{Avg}_{Q} f\Big| > 2^n kb\Big\} \subseteq \bigcup_j Q_j^{(k)}$$

for all $k = 1, 2, 3, \ldots$. (This also holds when $k = 0$.) We now use (7.1.12) and (7.1.14) to prove (7.1.9). We fix an $\alpha > 0$. If

$$2^n kb < \alpha \le 2^n (k+1) b$$

for some $k \ge 0$, then

$$\begin{aligned}
\Big|\Big\{x \in Q : \ \Big|f - \operatorname*{Avg}_{Q} f\Big| > \alpha\Big\}\Big| &\le \Big|\Big\{x \in Q : \ \Big|f - \operatorname*{Avg}_{Q} f\Big| > 2^n kb\Big\}\Big| \\
&\le \sum_j \big|Q_j^{(k)}\big| \le \frac{1}{b^k} \big|Q^{(0)}\big| \\
&= |Q| \, e^{-k \log b} \\
&\le |Q| \, b \, e^{-\alpha \log b / (2^n b)},
\end{aligned}$$

since $-k \le 1 - \frac{\alpha}{2^n b}$. Choosing $b = e > 1$ yields (7.1.9). □

7.1.3 Consequences of Theorem 7.1.6

Having proved the important distribution inequality (7.1.9), we are now in a position to deduce from it a few corollaries.

Corollary 7.1.7. *Every BMO function is exponentially integrable over any cube. More precisely, for any $\gamma < 1/(2^n e)$, for all $f \in BMO(\mathbf{R}^n)$, and all cubes Q we have*

$$\frac{1}{|Q|}\int_Q e^{\gamma|f(x)-\mathrm{Avg}_Q f|/\|f\|_{BMO}}\,dx \le 1+\frac{2^n e^2\gamma}{1-2^n e\gamma}.$$

Proof. Using identity (1.1.7) with $\varphi(t) = e^t - 1$, we write

$$\frac{1}{|Q|}\int_Q e^h\,dx = 1+\frac{1}{|Q|}\int_Q (e^h-1)\,dx = 1+\frac{1}{|Q|}\int_0^\infty e^\alpha |\{x\in Q:\ |h(x)|>\alpha\}|\,d\alpha$$

for a measurable function h. Then we take $h = \gamma|f - \mathrm{Avg}_Q f|/\|f\|_{BMO}$ and we use inequality (7.1.9) with $\gamma < A = (2^n e)^{-1}$ to obtain

$$\frac{1}{|Q|}\int_Q e^{\gamma|f(x)-\mathrm{Avg}_Q f|/\|f\|_{BMO}}\,dx \le \int_0^\infty e^\alpha\, e\, e^{-A(\frac{\alpha}{\gamma}\|f\|_{BMO})/\|f\|_{BMO}}\,d\alpha = C_{n,\gamma},$$

where $C_{n,\gamma}$ is a unit less than the constant in the statement of the inequality. □

Another important corollary of Theorem 7.1.6 is the following.

Corollary 7.1.8. *For all $0 < p < \infty$, there exists a finite constant $B_{p,n}$ such that*

$$\sup_Q\left(\frac{1}{|Q|}\int_Q \Big|f(x)-\mathrm{Avg}_Q f\Big|^p\,dx\right)^{\frac{1}{p}} \le B_{p,n}\|f\|_{BMO(\mathbf{R}^n)}. \tag{7.1.15}$$

Proof. This result can be obtained from the one in the preceding corollary or directly in the following way:

$$\begin{aligned}\frac{1}{|Q|}\int_Q \Big|f(x)-\mathrm{Avg}_Q f\Big|^p\,dx &= \frac{p}{|Q|}\int_0^\infty \alpha^{p-1}|\{x\in Q:\ |f(x)-\mathrm{Avg}_Q f|>\alpha\}|\,d\alpha\\ &\le \frac{p}{|Q|}\,e\,|Q|\int_0^\infty \alpha^{p-1}e^{-A\alpha/\|f\|_{BMO}}\,d\alpha\\ &= p\Gamma(p)\frac{e}{A^p}\|f\|_{BMO}^p,\end{aligned}$$

where $A = (2^n e)^{-1}$. Setting $B_{p,n} = (p\Gamma(p)\frac{e}{A^p})^{\frac{1}{p}} = (p\Gamma(p))^{\frac{1}{p}}e^{\frac{1}{p}+1}2^n$, we conclude the proof of (7.1.15). □

Since the inequality in Corollary 7.1.8 can be reversed when $p > 1$ via Hölder's inequality, we obtain the following important L^p characterization of *BMO* norms.

Corollary 7.1.9. *For all* $1 < p < \infty$ *we have*

$$\sup_Q \left(\frac{1}{|Q|} \int_Q |f(x) - \operatorname{Avg}_Q f|^p \, dx \right)^{\frac{1}{p}} \approx \|f\|_{BMO}.$$

Proof. Obvious. □

Exercises

7.1.1. Prove that BMO is a complete space, that is, every BMO-Cauchy sequence converges in BMO.
[*Hint:* Use Proposition 7.1.5 (ii) to show first that such a sequence is Cauchy in L^1 of every compact set.]

7.1.2. Find an example showing that the product of two BMO functions may not be in BMO.

7.1.3. Prove that

$$\big\| |f|^\alpha \big\|_{BMO} \le 2 \|f\|_{BMO}^\alpha$$

whenever $0 < \alpha \le 1$.

7.1.4. Let f be a real-valued BMO function on $\mathbf{R}^n$. Prove that the functions

$$f_{KL}(x) = \begin{cases} K & \text{if } f(x) < K, \\ f(x) & \text{if } K \le f(x) \le L, \\ L & \text{if } f(x) > L, \end{cases}$$

satisfy $\|f_{KL}\|_{BMO} \le \frac{9}{4} \|f\|_{BMO}$.

7.1.5. Let $a > 1$, let B be a ball (or a cube) in $\mathbf{R}^n$, and let aB be a concentric ball whose radius is a times the radius of B. Show that there is a dimensional constant C_n such that for all f in BMO we have

$$\big| \operatorname{Avg}_{aB} f - \operatorname{Avg}_B f \big| \le C_n \log(a+1) \|f\|_{BMO}.$$

7.1.6. Let $a > 1$ and let f be a BMO function on $\mathbf{R}^n$. Show that there exist dimensional constants C_n, C_n' such that
(a) for all balls B_1 and B_2 in $\mathbf{R}^n$ with radius R whose centers are at distance aR we have

$$\big| \operatorname{Avg}_{B_1} f - \operatorname{Avg}_{B_2} f \big| \le C_n' \log(a+1) \|f\|_{BMO}.$$

(b) Conclude that

$$\big| \operatorname{Avg}_{(a+1)B_1} f - \operatorname{Avg}_{B_2} f \big| \le C_n \log(a+1) \|f\|_{BMO}.$$

[*Hint:* Part (a): Consider the balls $2^j B_1$ and $2^j B_2$ for $j = 0, 1, 2, \dots$ and find the smallest j such that these intersect. Use (7.1.3) and Exercise 7.1.5.]

7.1.7. Let f be locally integrable on $\mathbf{R}^n$. Suppose that there exist positive constants m and b such that for all cubes Q in $\mathbf{R}^n$ and for all $0 < p < \infty$ we have

$$\alpha \Big| \Big\{ x \in Q : \big| f(x) - \operatorname*{Avg}_Q f \big| > \alpha \Big\} \Big|^{\frac{1}{p}} \le b\, p^m\, |Q|^{\frac{1}{p}}.$$

Show that f satisfies the estimate

$$\Big| \Big\{ x \in Q : \big| f(x) - \operatorname*{Avg}_Q f \big| > \alpha \Big\} \Big| \le |Q|\, e^{-c\alpha^{1/m}}$$

with $c = (2b)^{-1/m} \log 2$.
[*Hint:* Try $p = (\alpha/2b)^{1/m}$.]

7.1.8. Prove that $\big| \log |x| \big|^p$ is not in $BMO(\mathbf{R})$ when $1 < p < \infty$.
[*Hint:* Show that if $\big| \log |x| \big|^p$ were in BMO, then estimate (7.1.9) would be violated for large α.]

7.1.9. Let $f \in BMO(\mathbf{R})$ have mean value equal to zero on the fixed interval I. Find a BMO function g on $\mathbf{R}$ such that

(1) $g = f$ on I.

(2) $g = 0$ on $\mathbf{R} \setminus \frac{4}{3} I$.

(3) $\|g\|_{BMO} \le C \|f\|_{BMO}$ for some constant C independent of f.

[*Hint:* Let I_0 be the middle third of I. Let I_1, I_2 be the middle thirds of $I \setminus I_0$. Let I_3, $I_4, \dots, I_8$ be the middle thirds of $I \setminus (I_0 \cup I_1 \cup I_2)$, etc. Also let J_k be the reflection of I_k with respect to the closest endpoint of I and set $g = \operatorname{Avg}_{I_k} f$ on J_k for $k > 1$, $g = f$ on I, and zero otherwise.]

7.2 Duality between H^1 and *BMO*

The next main result we discuss about *BMO* is a certain remarkable duality relationship with the Hardy space H^1. We show that *BMO* is the dual space of H^1. This means that every continuous linear functional on the Hardy space H^1 can be realized as integration against a fixed *BMO* function, where *integration* in this context is an abstract operation, not necessarily given by an absolutely convergent integral. Restricting our attention, however, to a dense subspace of H^1 such as the space of all finite sums of atoms, the use of the word *integration* is well justified. Indeed, for an L^2 atom for H^1 a and a *BMO* function b, the integral

$$\int |a(x)b(x)|\,dx < \infty$$

converges absolutely, since $a(x)$ is compactly supported and bounded and $b(x)$ is locally (square) integrable.

Definition 7.2.1. Denote by $H_0^1(\mathbf{R}^n)$ the set of all finite linear combinations of L^2 atoms for $H^1(\mathbf{R}^n)$. For $b \in BMO(\mathbf{R}^n)$ we define a linear functional L_b on $H_0^1(\mathbf{R}^n)$ by setting

$$L_b(g) = \int_{\mathbf{R}^n} g(x)b(x)\,dx, \qquad g \in H_0^1. \tag{7.2.1}$$

Certainly the integral in (7.2.1) converges absolutely and is well defined in this case. This definition is also valid for general functions g in $H^1(\mathbf{R}^n)$ if the BMO function b is bounded. Note that (7.2.1) remains unchanged if b is replaced by $b+c$, where c is an additive constant; this makes this integral unambiguously defined for $b \in BMO$.

To extend the definition of L_b on the entire H^1 for all functions b in BMO we need to know that

$$\big\|L_b\big\|_{H^1\to\mathbf{C}} \le C_n \big\|b\big\|_{BMO}, \qquad \text{whenever } b \text{ is bounded,} \tag{7.2.2}$$

a fact that will be proved momentarily. Assuming (7.2.2), take $b \in BMO$ and let $b_M(x) = b\chi_{|b|\le M}$ for $M = 1,2,3,\dots$. Since $\big\|b_M\big\|_{BMO} \le 3\big\|b\big\|_{BMO}$, the sequence of linear functionals $\{L_{b_M}\}_M$ lies in a multiple of the unit ball of $(H^1)^*$ and there is a subsequence L_{M_j} that converges weakly to a bounded linear functional $\widetilde{L}_b$ on H^1. This means that for all f in $H^1(\mathbf{R}^n)$ we have

$$L_{b_{M_j}}(f) \to \widetilde{L}_b(f)$$

as $j \to \infty$. Observe that for $g \in H_0^1$ we also have

$$L_{b_{M_j}}(g) \to L_b(g),$$

and since each $L_{b_{M_j}}$ satisfies (7.2.2), L_b is a bounded linear functional on H_0^1. Since H_0^1 is dense in H^1, $\widetilde{L}_b$ is the *unique* bounded extension of $\widetilde{L}_b$ on H^1.

Having set the definition of L_b, we proceed by showing the validity of (7.2.2). Let b be a bounded BMO function. Given f in H^1, find a sequence a_k of L^2 atoms for H^1 supported in cubes Q_k such that

$$f = \sum_{k=1}^{\infty} \lambda_k a_k \tag{7.2.3}$$

and

$$\sum_{k=1}^{\infty} |\lambda_k| \le 2\big\|f\big\|_{H^1}.$$

Since the series in (7.2.3) converges in H^1, it must converge in L^1, and then we have

$$
\begin{aligned}
|L_b(f)| &= \left| \int_{\mathbf{R}^n} f(x)b(x)\,dx \right| \\
&= \left| \sum_{k=1}^{\infty} \lambda_k \int_{Q_k} a_k(x)\big(b(x) - \operatorname*{Avg}_{Q_k} b\big)\,dx \right| \\
&\le \sum_{k=1}^{\infty} |\lambda_k| \, \big\|a_k\big\|_{L^2} |Q_k|^{\frac{1}{2}} \left(\frac{1}{|Q_k|} \int_{Q_k} \big|b(x) - \operatorname*{Avg}_{Q_k} b\big|^2 dx \right)^{\frac{1}{2}} \\
&\le 2 \big\|f\big\|_{H^1} B_{2,n} \big\|b\big\|_{BMO},
\end{aligned}
$$

where in the last step we used Corollary 7.1.8 and the fact that L^2 atoms for H^1 satisfy $\big\|a_k\big\|_{L^2} \le |Q_k|^{-\frac{1}{2}}$. This proves (7.2.2) for bounded functions b in *BMO*.

We have proved that every *BMO* function b gives rise to a bounded linear functional $\widetilde{L}_b$ on $H^1(\mathbf{R}^n)$ (from now on denoted by L_b) that satisfies

$$
\big\|L_b\big\|_{H^1 \to \mathbf{C}} \le C_n \big\|b\big\|_{BMO}. \tag{7.2.4}
$$

The fact that every bounded linear functional on H^1 arises in this way is the gist of the equivalence of the next theorem.

Theorem 7.2.2. *There exist finite constants C_n and C_n' such that the following statements are valid:*
(a) Given $b \in BMO(\mathbf{R}^n)$, the linear functional L_b is bounded on $H^1(\mathbf{R}^n)$ with norm at most $C_n \big\|b\big\|_{BMO}$.
(b) For every bounded linear functional L on H^1 there exists a BMO function b such that for all $f \in H_0^1$ we have $L(f) = L_b(f)$ and also

$$
\big\|b\big\|_{BMO} \le C_n' \big\|L_b\big\|_{H^1 \to \mathbf{C}}.
$$

Proof. We have already proved (a) and so it suffices to prove (b). Fix a bounded linear functional L on $H^1(\mathbf{R}^n)$ and also fix a cube Q. Consider the space $L^2(Q)$ of all square integrable functions supported in Q with norm

$$
\big\|g\big\|_{L^2(Q)} = \left(\int_Q |g(x)|^2 dx \right)^{\frac{1}{2}}.
$$

We denote by $L_0^2(Q)$ the closed subspace of $L^2(Q)$ consisting of all functions in $L^2(Q)$ with mean value zero. We show that every element in $L_0^2(Q)$ is in $H^1(\mathbf{R}^n)$ and we have the inequality

$$
\big\|g\big\|_{H^1} \le c_n |Q|^{\frac{1}{2}} \big\|g\big\|_{L^2}. \tag{7.2.5}
$$

To prove (7.2.5) we use the square function characterization of H^1. We fix a Schwartz function Ψ on $\mathbf{R}^n$ whose Fourier transform is supported in the annulus $\frac{1}{2} \le |\xi| \le 2$ and that satisfies (6.2.6) for all $\xi \neq 0$ and we let $\Delta_j(g) = \Psi_{2^{-j}} * g$. To estimate the L^1 norm of $\big(\sum_j |\Delta_j(g)|^2\big)^{1/2}$ over $\mathbf{R}^n$, consider the part of the integral over $3\sqrt{n}Q$ and the integral over $(3\sqrt{n}Q)^c$. First we use Hölder's inequality and an L^2 estimate to prove that

$$\int_{3\sqrt{n}Q}\Big(\sum_j|\Delta_j(g)(x)|^2\Big)^{\frac{1}{2}}dx\le c_n|Q|^{\frac{1}{2}}\big\|g\big\|_{L^2}.$$

Now for $x\notin 3\sqrt{n}Q$ we use the mean value property of g to obtain

$$|\Delta_j(g)(x)|\le\frac{c_n\big\|g\big\|_{L^2}2^{nj+j}|Q|^{\frac{1}{n}+\frac{1}{2}}}{(1+2^j|x-c_Q|)^{n+2}},\tag{7.2.6}$$

where c_Q is the center of Q. Estimate (7.2.6) is obtained in a way similar to that we obtained the corresponding estimate for one atom; see Theorem 6.6.9 for details. Now (7.2.6) implies that

$$\int_{(3\sqrt{n}Q)^c}\Big(\sum_j|\Delta_j(g)(x)|^2\Big)^{\frac{1}{2}}dx\le c_n|Q|^{\frac{1}{2}}\big\|g\big\|_{L^2},$$

which proves (7.2.5).

Since $L^2_0(Q)$ is a subspace of H^1, it follows from (7.2.5) that the linear functional $L:H^1\to\mathbf{C}$ is also a bounded linear functional on $L^2_0(Q)$ with norm

$$\big\|L\big\|_{L^2_0(Q)\to\mathbf{C}}\le c_n|Q|^{1/2}\big\|L\big\|_{H^1\to\mathbf{C}}.\tag{7.2.7}$$

By the Riesz representation theorem for the Hilbert space $L^2_0(Q)$, there is an element F^Q in $(L^2_0(Q))^*=L^2(Q)/\{\text{constants}\}$ such that

$$L(g)=\int_Q F^Q(x)g(x)\,dx,\tag{7.2.8}$$

for all $g\in L^2_0(Q)$, and this F^Q satisfies

$$\big\|F^Q\big\|_{L^2(Q)}\le\big\|L\big\|_{L^2_0(Q)\to\mathbf{C}}.\tag{7.2.9}$$

Thus for any cube Q in $\mathbf{R}^n$, there is square integrable function F^Q supported in Q such that (7.2.8) is satisfied. We observe that if a cube Q is contained in another cube Q', then F^Q differs from $F^{Q'}$ by a constant on Q. Indeed, for all $g\in L^2_0(Q)$ we have

$$\int_Q F^{Q'}(x)g(x)\,dx=L(g)=\int_Q F^Q(x)g(x)\,dx$$

and thus

$$\int_Q(F^{Q'}(x)-F^Q(x))g(x)\,dx=0.$$

Consequently,

$$g \to \int_Q (F^{Q'}(x) - F^Q(x))g(x)\,dx$$

is the zero functional on $L_0^2(Q)$; hence $F^{Q'} - F^Q$ must be the zero function in the space $(L_0^2(Q))^*$, i.e., $F^{Q'} - F^Q$ is a constant on Q.

Let $Q_m = [-m,m]^n$ for $m = 1,2,\dots$. We define a locally integrable function $b(x)$ on $\mathbf{R}^n$ by setting

$$b(x) = F^{Q_m}(x) - \frac{1}{|Q_1|}\int_{Q_1} F^{Q_m}(t)\,dt \tag{7.2.10}$$

whenever $x \in Q_m$. We check that this definition is unambiguous. Let $1 \le \ell < m$. Then for $x \in Q_\ell$, $b(x)$ is also defined as in (7.2.10) with ℓ in the place of m. The difference of these two functions is

$$F^{Q_m} - F^{Q_\ell} - \operatorname*{Avg}_{Q_1}(F^{Q_m} - F^{Q_\ell}) = 0,$$

since the function $F^{Q_m} - F^{Q_\ell}$ is constant in the cube Q_ℓ (which is contained in Q_m), as indicated earlier.

Next we claim that there is a locally integrable function b on $\mathbf{R}^n$ such that for any cube Q there is a constant C_Q such that

$$F^Q = b - C_Q \qquad \text{on } Q. \tag{7.2.11}$$

Indeed, given a cube Q pick the smallest m such that Q is contained in Q^m and let $C_Q = -\operatorname{Avg}_{Q_1}(F^{Q_m}) + D(Q,Q_m)$, where $D(Q,Q_m)$ is the constant value of the function $F^{Q_m} - F^Q$ on Q.

We have now found a locally integrable function b such that for all cubes Q and all $g \in L_0^2(Q)$ we have

$$\int_Q b(x)g(x)\,dx = \int_Q (F^Q(x) + C_Q)g(x)\,dx = \int_Q F^Q(x)g(x)\,dx = L(g), \tag{7.2.12}$$

as follows from (7.2.8) and (7.2.11). We conclude the proof by showing that $b \in BMO(\mathbf{R}^n)$. By (7.2.11), (7.2.9), and (7.2.7) we have

$$\begin{aligned}
\sup_Q \frac{1}{|Q|}\int_Q |b(x) - C_Q|\,dx &= \sup_Q \frac{1}{|Q|}\int_Q |F^Q(x)|\,dx \\
&\le \sup_Q |Q|^{-1}|Q|^{\frac{1}{2}}\big\|F^Q\big\|_{L^2(Q)} \\
&\le \sup_Q |Q|^{-\frac{1}{2}}\big\|L\big\|_{L_0^2(Q)\to\mathbf{C}} \\
&\le c_n\big\|L\big\|_{H^1\to\mathbf{C}} < \infty.
\end{aligned}$$

Using Proposition 7.1.2 (3), we deduce that $b \in BMO$ and $\big\|b\big\|_{BMO} \le 2c_n\big\|L\big\|_{H^1\to\mathbf{C}}$. Finally, (7.2.12) implies that

$$L(g) = \int_{\mathbf{R}^n} b(x)g(x)\,dx = L_b(g)$$

for all $g \in H^1_0(\mathbf{R}^n)$, proving that the linear functional L coincides with L_b on a dense subspace of H^1. Consequently, $L = L_b$, and this concludes the proof of part (b). □

Exercises

7.2.1. Use Exercise 1.4.12(a) and (b) to deduce that

$$\big\|b\big\|_{BMO} \approx \sup_{\|f\|_{H^1}\le 1} \big|L_b(f)\big|,$$
$$\big\|f\big\|_{H^1} \approx \sup_{\|b\|_{BMO}\le 1} \big|L_b(f)\big|.$$

7.2.2. Suppose that a locally integrable function u is supported in a cube Q in $\mathbf{R}^n$ and satisfies

$$\int_Q u(x)g(x)\,dx = 0$$

for all square integrable functions g on Q with mean value zero. Show that u is almost everywhere equal to a constant.

7.3 Nontangential Maximal Functions and Carleson Measures

Many properties of functions defined on $\mathbf{R}^n$ are related to corresponding properties of associated functions defined on $\mathbf{R}^{n+1}_+$ in a natural way. A typical example of this situation is the relation between an $L^p(\mathbf{R}^n)$ function f and its Poisson integral $f * P_t$ or more generally $f * \Phi_t$, where $\{\Phi_t\}_{t>0}$ is an approximate identity. Here Φ is a Schwartz function on $\mathbf{R}^n$ with integral 1. A maximal operator associated to the approximate identity $\{f * \Phi_t\}_{t>0}$ is

$$f \to \sup_{t>0} |f * \Phi_t|,$$

which we know is pointwise controlled by a multiple of the Hardy–Littlewood maximal function $M(f)$. Another example of a maximal operator associated to the previous approximate identity is the nontangential maximal function

$$f \to M^*(f;\Phi)(x) = \sup_{t>0} \sup_{|y-x|<t} |(f * \Phi_t)(y)|.$$

To study nontangential behavior we consider general functions F defined on $\mathbf{R}^{n+1}_+$ that are not necessarily given as an average of functions defined on $\mathbf{R}^n$. Throughout this section we use capital letters to denote functions defined on $\mathbf{R}^{n+1}_+$. When we write $F(x,t)$ we mean that $x \in \mathbf{R}^n$ and $t > 0$.

7.3.1 Definition and Basic Properties of Carleson Measures

Definition 7.3.1. Let F be a measurable function on $\mathbf{R}^{n+1}_+$. For x in $\mathbf{R}^n$ let $\Gamma(x)$ be the cone with vertex x defined by

$$\Gamma(x) = \{(y,t) \in \mathbf{R}^n \times \mathbf{R}^+ : \ |y-x| < t\}.$$

A picture of this cone is shown in Figure 7.1. The *nontangential maximal function* of F is the function

$$F^*(x) = \sup_{(y,t)\in\Gamma(x)} |F(y,t)|$$

defined on $\mathbf{R}^n$. This function is obtained by taking the supremum of the values of F inside the cone $\Gamma(x)$.

We remark that if $F^*(x) = 0$ for almost all $x \in \mathbf{R}^n$, then F is identically equal to zero on $\mathbf{R}^{n+1}_+$.

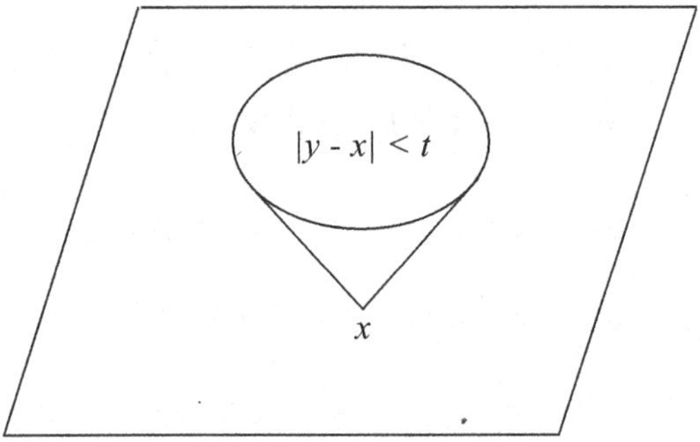

Fig. 7.1 The cone $\Gamma(x)$ truncated at height t.

Definition 7.3.2. Given a ball $B = B(x_0, r)$ in $\mathbf{R}^n$ we define the *tent* or *cylindrical tent* over B to be the "cylindrical set"

$$T(B) = \{(x,t) \in \mathbf{R}^{n+1}_+ : \ x \in B, \quad 0 < t \le r\}.$$

For a cube Q in $\mathbf{R}^n$ we define the tent over Q to be the cube

$$T(Q) = Q \times (0, \ell(Q)].$$

A tent over a ball and over a cube are shown in Figure 7.2. A positive measure μ on $\mathbf{R}^{n+1}_+$ is called a *Carleson measure* if

$$\left\|\mu\right\|_{\mathscr{C}} = \sup_{Q} \frac{1}{|Q|}\mu(T(Q)) < \infty, \tag{7.3.1}$$

where the supremum in (7.3.1) is taken over all cubes Q in $\mathbf{R}^n$. The *Carleson function* of the measure μ is defined as

$$\mathscr{C}(\mu)(x) = \sup_{Q \ni x} \frac{1}{|Q|}\mu(T(Q)), \tag{7.3.2}$$

where the supremum in (7.3.2) is taken over all cubes in $\mathbf{R}^n$ containing the point x. Observe that $\left\|\mathscr{C}(\mu)\right\|_{L^\infty} = \left\|\mu\right\|_{\mathscr{C}}$.

We also define

$$\left\|\mu\right\|_{\mathscr{C}}^{\text{cylinder}} = \sup_{B} \frac{1}{|B|}\mu(T(B)), \tag{7.3.3}$$

where the supremum is taken over all balls B in $\mathbf{R}^n$. One should verify that there exist dimensional constants c_n and C_n such that

$$c_n\left\|\mu\right\|_{\mathscr{C}} \le \left\|\mu\right\|_{\mathscr{C}}^{\text{cylinder}} \le C_n\left\|\mu\right\|_{\mathscr{C}}$$

for all measures μ on $\mathbf{R}^{n+1}_+$, that is, a measure satisfies the Carleson condition (7.3.1) with respect to cubes if and only if it satisfies the analogous condition (7.3.3) with respect to balls. Likewise, the Carleson function $\mathscr{C}(\mu)$ defined with respect to cubes is comparable to $\mathscr{C}^{\text{cylinder}}(\mu)$ defined with respect to cylinders over balls.

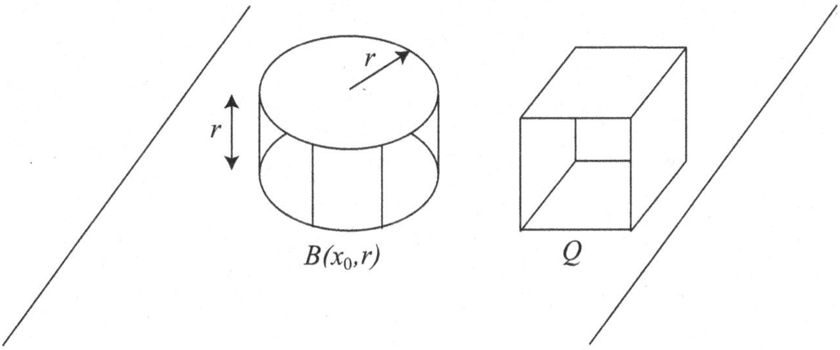

Fig. 7.2 The tents over the ball $B(x_0,r)$ and over a cube Q in $\mathbf{R}^2$.

Examples 7.3.3. The Lebesgue measure on $\mathbf{R}^{n+1}_+$ is not a Carleson measure. Indeed, it is not difficult to see that condition (7.3.1) cannot hold for large balls.

Let L be a line in $\mathbf{R}^2$. For A measurable subsets of $\mathbf{R}^2_+$ define $\mu(A)$ to be the linear Lebesgue measure of the set $L \cap A$. Then μ is a Carleson measure on $\mathbf{R}^2_+$. Indeed, the linear measure of the part of a line inside the box $[x_0 - r, x_0 + r] \times (0, r]$ is at most equal to the diagonal of the box, that is, $\sqrt{5}r$.

Likewise, let P be an affine plane in $\mathbf{R}^{n+1}$ and define a measure ν by setting $\nu(A)$ to be the n-dimensional Lebesgue measure of the set $A \cap P$ for any $A \subseteq \mathbf{R}^{n+1}_+$. A similar idea shows that ν is a Carleson measure on $\mathbf{R}^{n+1}_+$.

We now turn to the study of some interesting boundedness properties of functions on $\mathbf{R}^{n+1}_+$ with respect to Carleson measures.

A useful tool in this study is the *Whitney decomposition* of an open set in $\mathbf{R}^n$. This is a decomposition of a general open set Ω in $\mathbf{R}^n$ as a union of disjoint cubes whose lengths are proportional to their distance from the boundary of the open set. For a given cube Q in $\mathbf{R}^n$, we denote by $\ell(Q)$ its length.

Proposition 7.3.4. ***(Whitney decomposition)*** *Let Ω be an open nonempty proper subset of $\mathbf{R}^n$. Then there exists a family of closed cubes $\{Q_j\}_j$ such that*

(a) $\bigcup_j Q_j = \Omega$ and the Q_j's have disjoint interiors;

(b) $\sqrt{n}\,\ell(Q_j) \le dist(Q_j, \Omega^c) \le 4\sqrt{n}\,\ell(Q_j)$;

(c) if the boundaries of two cubes Q_j and Q_k touch, then

$$\frac{1}{4} \le \frac{\ell(Q_j)}{\ell(Q_k)} \le 4;$$

(d) for a given Q_j there exist at most 12^n Q_k's that touch it.

The proof of Proposition 7.3.4 is given in Appendix J.

Theorem 7.3.5. *There exists a dimensional constant C_n such that for all $\alpha > 0$, all measures $\mu \ge 0$ on $\mathbf{R}^{n+1}_+$, and all μ-measurable functions F on $\mathbf{R}^{n+1}_+$, we have*

$$\mu(\{(x,t) \in \mathbf{R}^{n+1}_+ : |F(x,t)| > \alpha\}) \le C_n \int_{\{F^* > \alpha\}} \mathscr{C}(\mu)(x)\,dx. \tag{7.3.4}$$

In particular, if μ is a Carleson measure, then

$$\mu(\{|F| > \alpha\}) \le C_n \big\|\mu\big\|_{\mathscr{C}} |\{F^* > \alpha\}|.$$

Proof. We prove this theorem by working with the equivalent definition of Carleson measures and Carleson functions using balls and cylinders over balls. As observed earlier, these quantities are comparable to the corresponding quantities using cubes.

We begin by observing that for any function F the set $\Omega_\alpha = \{F^* > \alpha\}$ is open, and in particular, F^* is Lebesgue measurable. Indeed, if $x_0 \in \Omega_\alpha$, then there is a $(y_0, t_0) \in \mathbf{R}^{n+1}_+$ such that $|F(y_0, t_0)| > \alpha$. If d_0 is the distance from y_0 to the sphere formed by the intersection of the hyperplane $t_0 + \mathbf{R}^n$ with the boundary of the cone $\Gamma(x_0)$, then $|x_0 - y_0| \le t_0 - d_0$. It follows that the open ball $B(x_0, d_0)$ is contained in Ω_α, since for $z \in B(x_0, d_0)$ we have $|z - y_0| < t_0$, hence $F^*(z) \ge |F(y_0, t_0)| > \alpha$.

Let $\{Q_k\}$ be the Whitney decomposition of the set Ω_α. For each $x \in \Omega_\alpha$, set $\delta_\alpha(x) = \operatorname{dist}(x, \Omega_\alpha^c)$. Then for $z \in Q_k$ we have

$$\delta_\alpha(z) \le \sqrt{n}\,\ell(Q_k) + \operatorname{dist}(Q_k, \Omega_\alpha^c) \le 5\sqrt{n}\,\ell(Q_k) \tag{7.3.5}$$

in view of Proposition 7.3.4 (b). For each Q_k, let B_k be the smallest ball that contains Q_k. Then the radius of B_k is $\sqrt{n}\,\ell(Q_k)/2$. Combine this observation with (7.3.5) to obtain that

$$z \in Q_k \qquad \Longrightarrow \qquad B(z, \delta_\alpha(z)) \subseteq 12 B_k.$$

This implies that

$$\bigcup_{z \in \Omega_\alpha} T\big(B(z, \delta_\alpha(z))\big) \subseteq \bigcup_k T(12 B_k). \tag{7.3.6}$$

Next we claim that

$$\{|F| > \alpha\} \subseteq \bigcup_{z \in \Omega_\alpha} T\big(B(z, \delta_\alpha(z))\big). \tag{7.3.7}$$

Indeed, let $(x,t) \in \mathbf{R}^{n+1}_+$ such that $|F(x,t)| > \alpha$. Then by the definition of F^* we have that $F^*(y) > \alpha$ for all $y \in \mathbf{R}^n$ satisfying $|x-y| < t$. Thus $B(x,t) \subseteq \Omega_\alpha$ and so $\delta_\alpha(x) \ge t$. This gives that $(x,t) \in T\big(B(x, \delta_\alpha(x))\big)$, which proves (7.3.7).

Combining (7.3.6) and (7.3.7) we obtain

$$\{|F| > \alpha\} \subseteq \bigcup_k T(12 B_k).$$

Applying the measure μ and using the definition of the Carleson function, we obtain

$$\begin{aligned}
\mu(\{|F| > \alpha\}) &\le \sum_k \mu(T(12 B_k)) \\
&\le \sum_k |12 B_k| \inf_{x \in Q_k} \mathscr{C}^{\text{cylinder}}(\mu)(x) \\
&\le 12^n \sum_k \frac{|B_k|}{|Q_k|} \int_{Q_k} \mathscr{C}^{\text{cylinder}}(\mu)(x)\, dx \\
&\le C_n \int_{\Omega_\alpha} \mathscr{C}(\mu)(x)\, dx,
\end{aligned}$$

since $|B_k| = 2^{-n} n^{n/2} v_n |Q_k|$. This proves (7.3.4). □

Corollary 7.3.6. *For any Carleson measure μ and every μ-measurable function F on $\mathbf{R}^{n+1}_+$ we have*

$$\int_{\mathbf{R}^{n+1}_+} |F(x,t)|^p \, d\mu(x,t) \le C_n \big\|\mu\big\|_{\mathscr{C}} \int_{\mathbf{R}^n} (F^*(x))^p \, dx$$

for all $0 < p < \infty$.

Proof. Simply use Proposition 1.1.4 and the previous theorem. □

A particular example of this situation arises when $F(x,t) = f * \Phi_t(x)$ for some nice integrable function Φ. Here and in the sequel, $\Phi_t(x) = t^{-n}\Phi(t^{-1}x)$. For instance one may take Φ_t to be the Poisson kernel P_t.

Theorem 7.3.7. *Let Φ be a function on $\mathbf{R}^n$ that satisfies for some $0 < C, \delta < \infty$,*

$$|\Phi(x)| \leq \frac{C}{(1+|x|)^{n+\delta}}. \tag{7.3.8}$$

Let μ be a Carleson measure on $\mathbf{R}^{n+1}_+$. Then for every $1 < p < \infty$ there is a constant $C_{p,n}(\mu)$ such that for all $f \in L^p(\mathbf{R}^n)$ we have

$$\int_{\mathbf{R}^{n+1}_+} |(\Phi_t * f)(x)|^p \, d\mu(x,t) \leq C_{p,n}(\mu) \int_{\mathbf{R}^n} |f(x)|^p \, dx, \tag{7.3.9}$$

where $C_{p,n}(\mu) \leq C(p,n)\big\|\mu\big\|_{\mathscr{C}}$. Conversely, suppose that Φ is nonnegative and satisfies (7.3.8) and $\int_{|x|\leq 1} \Phi(x)\, dx > 0$. If μ is a measure on $\mathbf{R}^{n+1}_+$ such that for some $1 < p < \infty$ there is a constant $C_{p,n}(\mu)$ such that (7.3.9) holds for all $f \in L^p(\mathbf{R}^n)$, then μ is a Carleson measure with norm at most a multiple of $C_{p,n}(\mu)$.

Proof. If μ is a Carleson measure, we may obtain (7.3.9) as a consequence of Corollary 7.3.6. Indeed, for $F(x,t) = (\Phi_t * f)(x)$ we have

$$F^*(x) = \sup_{t>0} \sup_{\substack{y \in \mathbf{R}^n \\ |y-x|<t}} |(\Phi_t * f)(y)|.$$

Using (7.3.8) and Corollary 2.1.12, this is easily seen to be pointwise controlled by the Hardy–Littlewood maximal operator, which is L^p bounded. See also Exercise 7.3.4.

Conversely, if (7.3.9) holds, then we fix a ball $B = B(x_0, r)$ in $\mathbf{R}^n$ with center x_0 and radius $r > 0$. Then for (x,t) in $T(B)$ we have

$$(\Phi_t * \chi_{2B})(x) = \int\limits_{x-2B} \Phi_t(y)\, dy \geq \int\limits_{B(0,t)} \Phi_t(y)\, dy = \int\limits_{B(0,1)} \Phi(y)\, dy = c_n > 0,$$

since $B(0,t) \subseteq x - 2B(x_0, r)$ whenever $t \leq r$. Therefore, we have

$$\begin{aligned} \mu(T(B)) &\leq \frac{1}{c_n^p} \int_{\mathbf{R}^{n+1}_+} |(\Phi_t * \chi_{2B})(x)|^p \, d\mu(x,t) \\ &\leq \frac{C_{p,n}(\mu)}{c_n^p} \int_{\mathbf{R}^n} |\chi_{2B}(x)|^p \, dx \\ &= \frac{2^n C_{p,n}(\mu)}{c_n^p} |B|. \end{aligned}$$

This proves that μ is a Carleson measure with $\big\|\mu\big\|_{\mathscr{C}} \leq 2^n c_n^{-p} C_{p,n}(\mu)$. □

7.3.2 BMO Functions and Carleson Measures

We now turn to an interesting connection between *BMO* functions and Carleson measures. We have the following.

Theorem 7.3.8. *Let b be a BMO function on $\mathbf{R}^n$ and let Ψ be an integrable function with mean value zero on $\mathbf{R}^n$ that satisfies*

$$|\Psi(x)| \le A(1+|x|)^{-n-\delta} \tag{7.3.10}$$

for some $0 < A, \delta < \infty$. Consider the dilations $\Psi_t = t^{-n}\Psi(t^{-1}x)$ and define the Littlewood–Paley operators $\Delta_j(f) = f * \Psi_{2^{-j}}$.
(a) Suppose that

$$\sup_{\xi\in\mathbf{R}^n} \sum_{j\in\mathbf{Z}} |\widehat{\Psi}(2^{-j}\xi)|^2 \le B^2 < \infty \tag{7.3.11}$$

and let $\delta_{2^{-j}}(t)$ be Dirac mass at the point $t = 2^{-j}$. Then there is a constant $C_{n,\delta}$ such that

$$d\mu(x,t) = \sum_{j\in\mathbf{Z}} |(\Psi_{2^{-j}} * b)(x)|^2\, dx\, \delta_{2^{-j}}(t)$$

is a Carleson measure on $\mathbf{R}^{n+1}_+$ with norm at most $C_{n,\delta}(A+B)^2\big\|b\big\|^2_{BMO}$.
(b) Suppose that

$$\sup_{\xi\in\mathbf{R}^n} \int_0^\infty |\widehat{\Psi}(t\xi)|^2 \frac{dt}{t} \le B^2 < \infty. \tag{7.3.12}$$

Then the continuous version $d\nu(x,t)$ of $d\mu(x,t)$ defined by

$$d\nu(x,t) = |(\Psi_t * b)(x)|^2\, dx\, \frac{dt}{t}$$

is a Carleson measure on $\mathbf{R}^{n+1}_+$ with norm at most $C_{n,\delta}(A+B)^2\big\|b\big\|^2_{BMO}$ for some constant $C_{n,\delta}$.
(c) Let $\delta, A > 0$. Suppose that $\{K_t\}_{t>0}$ are functions on $\mathbf{R}^n \times \mathbf{R}^n$ that satisfy

$$|K_t(x,y)| \le \frac{A\,t^\delta}{(t+|x-y|)^{n+\delta}} \tag{7.3.13}$$

for all $t > 0$ and all $x, y \in \mathbf{R}^n$. Let R_t be the linear operator

$$R_t(f)(x) = \int_{\mathbf{R}^n} K_t(x,y)\, f(y)\, dy,$$

which is well defined for all $f \in \bigcup_{1\le p\le\infty} L^p(\mathbf{R}^n)$. Suppose that $R_t(1) = 0$ for all $t > 0$ and that there is a constant $B > 0$ such that

$$\int_0^\infty \int_{\mathbf{R}^n} \big|R_t(f)(x)\big|^2 \frac{dx\,dt}{t} \le B\big\|f\big\|^2_{L^2(\mathbf{R}^n)} \tag{7.3.14}$$

for all $f \in L^2(\mathbf{R}^n)$. Then for all b in BMO the measure

$$|R_t(b)(x)|^2 \frac{dx\,dt}{t}$$

is Carleson with norm at most a constant multiple of $(A+B)^2\|b\|_{BMO}^2$.

We note that if in addition to (7.3.10), the function Ψ has mean value zero and satisfies $|\nabla\Psi(x)| \le A(1+|x|)^{-n-\delta}$, then (7.3.11) and (7.3.12) hold and therefore conclusions (a) and (b) of Theorem 7.3.8 follow.

Proof. We prove (a). The measure μ is defined so that for every μ-integrable function F on $\mathbf{R}^{n+1}_+$ we have

$$\int_{\mathbf{R}^{n+1}_+} F(x,t)\,d\mu(x,t) = \sum_{j\in\mathbf{Z}} \int_{\mathbf{R}^n} |(\Psi_{2^{-j}} * b)(x)|^2 F(x,2^{-j})\,dx. \tag{7.3.15}$$

For a cube Q in $\mathbf{R}^n$ we let Q^* be the cube with the same center and orientation whose side length is $3\sqrt{n}\,\ell(Q)$, where $\ell(Q)$ is the side length of Q. Fix a cube Q in $\mathbf{R}^n$, take F to be the characteristic function of the tent of Q, and split b as

$$b = \big(b - \operatorname*{Avg}_Q b\big)\chi_{Q^*} + \big(b - \operatorname*{Avg}_Q b\big)\chi_{(Q^*)^c} + \operatorname*{Avg}_Q b.$$

Since Ψ has mean value zero, $\Psi_{2^{-j}} * \operatorname{Avg}_Q b = 0$. Then (7.3.15) gives

$$\mu(T(Q)) = \sum_{2^{-j}\le\ell(Q)} \int_Q |\Delta_j(b)(x)|^2\,dx \le 2\Sigma_1 + 2\Sigma_2,$$

where

$$\begin{aligned}
\Sigma_1 &= \sum_{j\in\mathbf{Z}} \int_{\mathbf{R}^n} \big|\Delta_j\big((b - \operatorname*{Avg}_Q b)\chi_{Q^*}\big)(x)\big|^2\,dx,\\
\Sigma_2 &= \sum_{2^{-j}\le\ell(Q)} \int_Q \big|\Delta_j\big((b - \operatorname*{Avg}_Q b)\chi_{(Q^*)^c}\big)(x)\big|^2\,dx.
\end{aligned}$$

Using Plancherel's theorem and (7.3.11), we obtain

$$\begin{aligned}
\Sigma_1 &\le \sup_\xi \sum_{j\in\mathbf{Z}} |\widehat{\Psi}(2^{-j}\xi)|^2 \int_{\mathbf{R}^n} \big|\big((b - \operatorname*{Avg}_Q b)\chi_{Q^*}\big)\widehat{\ }(\xi)\big|^2\,d\xi\\
&\le B^2 \int_{Q^*} \big|b(x) - \operatorname*{Avg}_Q b\big|^2\,dx\\
&\le 2B^2 \int_{Q^*} \big|b(x) - \operatorname*{Avg}_{Q*} b\big|^2\,dx + 2A^2|Q^*|\,\big|\operatorname*{Avg}_{Q*} b - \operatorname*{Avg}_Q b\big|^2\\
&\le B^2 \int_{Q^*} \big|b(x) - \operatorname*{Avg}_{Q^*} b\big|^2\,dx + c_n B^2\|b\|_{BMO}^2\,|Q|\\
&\le C_n B^2\|b\|_{BMO}^2\,|Q|
\end{aligned}$$

in view of Proposition 7.1.5 (i) and Corollary 7.1.8. To estimate Σ_2, we use the size estimate of the function Ψ. We obtain

$$\left|\left(\Psi_{2^{-j}} * (b - \operatorname*{Avg}_{Q} b)\chi_{(Q^*)^c}\right)(x)\right| \le \int_{(Q^*)^c} \frac{A\, 2^{-j\delta}\left|b(y) - \operatorname{Avg}_Q b\right|}{(2^{-j} + |x-y|)^{n+\delta}}\, dy. \tag{7.3.16}$$

But note that if c_Q is the center of Q, then

$$\begin{aligned}
2^{-j} + |x-y| &\ge |y-x| \\
&\ge |y-c_Q| - |c_Q - x| \\
&\ge \frac{1}{2}|c_Q - y| + \frac{3\sqrt{n}}{4}\ell(Q) - |c_Q - x| \\
&\ge \frac{1}{2}|c_Q - y| + \frac{3\sqrt{n}}{4}\ell(Q) - \frac{\sqrt{n}}{2} l(Q) \\
&= \frac{1}{2}\Big(|c_Q - y| + \frac{\sqrt{n}}{2}\ell(Q)\Big)
\end{aligned}$$

when $y \in (Q^*)^c$ and $x \in Q$. Inserting this estimate in (7.3.16), integrating over Q, and summing over j with $2^{-j} \le \ell(Q)$, we obtain

$$\begin{aligned}
\Sigma_2 &\le C_n \sum_{2^{-j\delta} \le \ell(Q)} 2^{-2j} \int_Q \left(A \int_{\mathbf{R}^n} \frac{\left|b(y) - \operatorname{Avg}_Q b\right|}{(\ell(Q) + |c_Q - y|)^{n+\delta}}\, dy\right)^2 dx \\
&\le C_n A^2 |Q| \left(\int_{\mathbf{R}^n} \frac{\ell(Q)^\delta \left|b(y) - \operatorname{Avg}_Q b\right|}{(\ell(Q) + |y - c_Q|)^{n+\delta}}\, dy\right)^2 \\
&\le C'_{n,\delta} |Q| \left\|b\right\|_{BMO}^2
\end{aligned}$$

in view of (7.1.5). This proves that

$$\Sigma_1 + \Sigma_2 \le C_{n,\delta}(A^2 + B^2)|Q| \left\|b\right\|_{BMO}^2,$$

which implies that $\mu(T(Q)) \le C_{n,\delta}(A+B)^2 \left\|b\right\|_{BMO}^2 |Q|$.

The proof of part (b) of the theorem is obtained in a similar fashion. Finally, part (c) is a generalization of part (b) and is proved likewise. We sketch its proof. Write

$$b = (b - \operatorname*{Avg}_{Q} b)\chi_{Q^*} + (b - \operatorname*{Avg}_{Q} b)\chi_{(Q^*)^c} + \operatorname*{Avg}_{Q} b$$

and note that $R_t(\operatorname{Avg}_Q b) = 0$. We handle the term containing $R_t\big((b - \operatorname{Avg}_Q b)\chi_{Q^*}\big)$ using an L^2 estimate over Q^* and condition (7.3.14), while for the term containing $R_t\big((b - \operatorname{Avg}_Q b)\chi_{(Q^*)^c}\big)$ we use an L^1 estimate and condition (7.3.13). In both cases we obtain the required conclusion in a way analogous to that in part (a). □

Exercises

7.3.1. Let a_j, b_j be sequences of positive real numbers such that $\sum_j b_j < \infty$. Define a measure μ on $\mathbf{R}^{n+1}_+$ by setting

$$\mu(E) = \sum_j b_j |E \cap \{(x, a_j) : x \in \mathbf{R}^n\}|,$$

where E is a subset of $\mathbf{R}^{n+1}_+$ and $|\ |$ denotes n-dimensional Lebesgue measure on the affine planes $t = a_j$. Show that μ is a Carleson measure with norm

$$\|\mu\|_{\mathscr{C}}^{\text{cylinder}} = \|\mu\|_{\mathscr{C}} = \sum_j b_j .$$

7.3.2. Let $x_0 \in \mathbf{R}^n$ and $\mu = \delta_{(x_0,1)}$ be the Dirac mass at the point $(x_0, 1)$. Show that μ is Carleson measure and compute $\|\mu\|_{\mathscr{C}}^{\text{cylinder}}$ and $\|\mu\|_{\mathscr{C}}$. Which of these norms is larger?

7.3.3. Define *conical* and *hemispherical* tents over balls in $\mathbf{R}^n$ as well as *pyramidal* tents over cubes in $\mathbf{R}^n$ and define the expressions $\|\mu\|_{\mathscr{C}}^{\text{cone}}$, $\|\mu\|_{\mathscr{C}}^{\text{hemisphere}}$, and $\|\mu\|_{\mathscr{C}}^{\text{pyramid}}$. Show that

$$\|\mu\|_{\mathscr{C}}^{\text{cone}} \approx \|\mu\|_{\mathscr{C}}^{\text{hemisphere}} \approx \|\mu\|_{\mathscr{C}}^{\text{pyramid}} \approx \|\mu\|_{\mathscr{C}} ,$$

where all the implicit constants in the previous estimates depend only on the dimension.

7.3.4. Suppose that Φ has a radial, bounded, symmetrically decreasing integrable majorant. Set $F(x,t) = (f * \Phi_t)(x)$, where f is a locally integrable function on $\mathbf{R}^n$. Prove that

$$F^*(x) \le CM(f)(x) ,$$

where M is the Hardy–Littlewood maximal operator and C is a constant that depends only on the dimension.
[*Hint:* If $\varphi(|x|)$ is the claimed majorant of $\Phi(x)$, then the function $\psi(|x|) = \varphi(0)$ for $|x| \le 1$ and $\psi(|x|) = \varphi(|x| - 1)$ for $|x| \ge 1$ is a majorant for the function $\Psi(x) = \sup_{|u| \le 1} |\Phi(x-u)|$.]

7.3.5. Let F be a function on $\mathbf{R}^{n+1}_+$, let F^* be the nontangential maximal function derived from F, and let $\mu \ge 0$ be a measure on $\mathbf{R}^{n+1}_+$. Prove that

$$\|F\|_{L^r(\mathbf{R}^{n+1}_+, \mu)} \le C_n^{1/r} \left(\int_{\mathbf{R}^n} \mathscr{C}(\mu)(x) F^*(x)^r \, dx \right)^{1/r} ,$$

where C_n is the constant of Theorem 7.3.5 and $0 < r < \infty$.

7.3.6. (a) Given A a closed subset of $\mathbf{R}^n$ and $0<\gamma<1$, define

$$A_\gamma^* = \left\{ x \in \mathbf{R}^n : \inf_{r>0} \frac{|A\cap B(x,r)|}{|B(x,r)|} \ge \gamma \right\}.$$

Show that A^* is a closed subset of A and that it satisfies

$$|(A_\gamma^*)^c| \le \frac{3^n}{1-\gamma}|A^c|.$$

[*Hint:* Consider the Hardy–Littlewood maximal function of χ_{A^c}.]
(b) For a function F on $\mathbf{R}_+^{n+1}$ and $0<a<\infty$, set

$$F_a^*(x) = \sup_{t>0} \sup_{|y-x|<at} |F(y,t)|.$$

Let $0<a<b<\infty$ be given. Prove that for all $\lambda>0$ we have

$$|\{F_a^*>\lambda\}| \le |\{F_b^*>\lambda\}| \le 3^n a^{-n}(a+b)^n |\{F_a^*>\lambda\}|.$$

7.3.7. Let μ be a Carleson measure on $\mathbf{R}_+^{n+1}$. Show that for any $z_0\in\mathbf{R}^n$ and $t>0$ we have

$$\iint_{\mathbf{R}^n\times(0,t)} \frac{t}{(|z-z_0|^2+t^2+s^2)^{\frac{n+1}{2}}}\, d\mu(z,s) \le \big\|\mu\big\|_{\mathscr{C}}^{\text{cylinder}} \frac{\pi^{\frac{n+1}{2}}}{\Gamma(\frac{n+1}{2})}.$$

[*Hint:* Begin by writing

$$\frac{t}{(|z-z_0|^2+t^2+s^2)^{\frac{n+1}{2}}} = (n+1)t\int_Q^\infty \frac{dr}{r^{n+2}},$$

where $Q=\sqrt{|z-z_0|^2+t^2+s^2}$. Apply Fubini's theorem to estimate the required expression by

$$t(n+1)\int_t^\infty \int_{T\left(B(z_0,\sqrt{r^2-t^2})\right)} d\mu(z,s)\,\frac{dr}{r^{n+2}} \le t(n+1)v_n\big\|\mu\big\|_{\mathscr{C}}^{\text{cylinder}}\int_t^\infty (r^2-t^2)^{\frac{n}{2}}\frac{dr}{r^{n+2}},$$

where v_n is the volume of the unit ball in $\mathbf{R}^n$. Reduce the last integral to a beta function.]

7.3.8. (*Verbitsky [325]*) Let μ be a Carleson measure on $\mathbf{R}_+^{n+1}$. Show that for all $p>2$ there exists a dimensionless constant C_p such that

$$\int_{\mathbf{R}_+^{n+1}} |(P_t*f)(x)|^p\, d\mu(x,t) \le C_p\big\|\mu\big\|_{\mathscr{C}}^{\text{cylinder}} \int_{\mathbf{R}^n} |f(x)|^p\, dx.$$

[*Hint:* It suffices to prove that the operator $f \mapsto P_t * f$ maps $L^2(\mathbf{R}^n)$ to $L^{2,\infty}(\mathbf{R}^{n+1}_+, d\mu)$ with a dimensionless constant C, since then the conclusion follows by interpolation with the corresponding L^∞ estimate, which holds with constant 1. By duality and Exercise 1.4.7 this is equivalent to showing that

$$\int_{\mathbf{R}^n} \left[\iint\limits_E P_t(x-y)\, d\mu(y,t) \iint\limits_E P_s(x-z)\, d\mu(z,s) \right] dx \le C\mu(E)$$

for any set E in $\mathbf{R}^{n+1}_+$ with $\mu(E) < \infty$. Apply Fubini's theorem, use the identity

$$\int_{\mathbf{R}^n} P_t(x-y) P_s(x-z)\, dx = P_{t+s}(y-z),$$

and consider the cases $t \le s$ and $s \le t$.]

7.4 The Sharp Maximal Function

In Section 7.1 we defined *BMO* as the space of all locally integrable functions on $\mathbf{R}^n$ whose mean oscillation is at most a finite constant. In this section we introduce a quantitative way to measure the mean oscillation of a function near any point.

7.4.1 Definition and Basic Properties of the Sharp Maximal Function

The local behavior of the mean oscillation of a function is captured to a certain extent by the sharp maximal function. This is a device that enables us to relate integrability properties of a function to those of its mean oscillations.

Definition 7.4.1. Given a locally integrable function f on $\mathbf{R}^n$, we define its *sharp maximal function* $M^\#(f)$ as

$$M^\#(f)(x) = \sup_{Q \ni x} \frac{1}{|Q|} \int_Q \left| f(t) - \operatorname*{Avg}_Q f \right| dt,$$

where the supremum is taken over all cubes Q in $\mathbf{R}^n$ that contain the given point x.

The sharp maximal function is an analogue of the Hardy–Littlewood maximal function, but it has some advantages over it, especially in dealing with the endpoint space L^∞. The very definition of $M^\#(f)$ brings up a connection with *BMO* that is crucial in interpolation. Precisely, we have

$$BMO(\mathbf{R}^n) = \{f \in L^1_{\mathrm{loc}}(\mathbf{R}^n) : M^\#(f) \in L^\infty(\mathbf{R}^n)\},$$

and in this case

$$\|f\|_{BMO} = \|M^{\#}(f)\|_{L^\infty}.$$

We summarize some properties of the sharp maximal function.

Proposition 7.4.2. *Let f,g be a locally integrable functions on $\mathbf{R}^n$. Then*

(1) $M^{\#}(f) \le 2M_c(f)$, where M_c is the Hardy–Littlewood maximal operator with respect to cubes in $\mathbf{R}^n$.

(2) For all cubes Q in $\mathbf{R}^n$ we have

$$\frac{1}{2}M^{\#}(f)(x) \le \sup_{x\in Q}\inf_{a\in\mathbf{C}} \frac{1}{|Q|}\int_Q |f(y)-a|\,dy \le M^{\#}(f)(x).$$

(3) $M^{\#}(|f|) \le 2M^{\#}(f)$.

(4) We have $M^{\#}(f+g) \le M^{\#}(f) + M^{\#}(g)$.

Proof. The proof of (1) is trivial. To prove (2) we fix $\varepsilon > 0$ and for any cube Q we pick a constant a_Q such that

$$\frac{1}{|Q|}\int_Q |f(y)-a_Q|\,dy \le \inf_{a\in Q}\frac{1}{|Q|}\int_Q |f(y)-a|\,dy + \varepsilon.$$

Then

$$\begin{aligned}
\frac{1}{|Q|}\int_Q \Big|f(y)-\operatorname*{Avg}_Q f\Big|\,dy &\le \frac{1}{|Q|}\int_Q |f(y)-a_Q|\,dy + \frac{1}{|Q|}\int_Q \Big|\operatorname*{Avg}_Q f - a_Q\Big|\,dy \\
&\le \frac{1}{|Q|}\int_Q |f(y)-a_Q|\,dy + \frac{1}{|Q|}\int_Q |f(y)-a_Q|\,dy \\
&\le 2\inf_{a\in Q}\frac{1}{|Q|}\int_Q |f(y)-a|\,dy + 2\varepsilon.
\end{aligned}$$

Taking the supremum over all cubes Q in $\mathbf{R}^n$, we obtain the first inequality in (2), since $\varepsilon > 0$ was arbitrary. The other inequality in (2) is simple. The proofs of (3) and (4) are immediate. □

We saw that $M^{\#}(f) \le 2M_c(f)$, which implies that

$$\|M^{\#}(f)\|_{L^p} \le C_n p(p-1)^{-1}\|f\|_{L^p} \tag{7.4.1}$$

for $1 < p < \infty$. Thus the sharp function of an L^p function is also in L^p whenever $1 < p < \infty$. The fact that the converse inequality is also valid is one of the main results in this section. We obtain this estimate via a distributional inequality for the sharp function called a *good lambda* inequality.

7.4.2 A Good Lambda Estimate for the Sharp Function

A useful tool in obtaining the converse inequality to (7.4.1) is the dyadic maximal function.

Definition 7.4.3. Given a locally integrable function f on $\mathbf{R}^n$, we define its *dyadic maximal function* $M_d(f)$ by

$$M_d(f)(x) = \sup_{\substack{Q \ni x \\ Q \text{ dyadic cube}}} \frac{1}{|Q|} \int_Q |f(t)|\,dt\,.$$

The supremum is taken over all dyadic cubes in $\mathbf{R}^n$ that contain a given point x. Recalling the expectation operators E_k from Section 5.4, we have

$$M_d(f)(x) = \sup_{k \in \mathbf{Z}} E_k(f)(x)\,.$$

Obviously, one has the pointwise estimate

$$M_d(f) \le M_c(f) \tag{7.4.2}$$

for all locally integrable functions. This yields the boundedness of M_d on L^p for $1 < p \le \infty$ and the weak type $(1,1)$ property of M_d. More precise estimates on the norm of M_d can derived. In fact, in view of the result of Exercise 2.1.12, M_d is of weak type $(1,1)$ with norm at most 1. By interpolation (precisely Exercise 1.3.3(a)), it follows that M_d maps $L^p(\mathbf{R}^n)$ to itself with norm at most

$$\big\|M_d\big\|_{L^p(\mathbf{R}^n)\to L^p(\mathbf{R}^n)} \le \frac{p}{p-1}$$

when $1 < p < \infty$.

One may wonder whether an estimate converse to (7.4.2) holds. But a quick observation shows that for a locally integrable function f that vanishes on certain open sets, $M_d(f)$ could have zeros, but $M_c(f)$ never vanishes. Therefore, there is no hope for $M_d(f)$ and $M_c(f)$ to be pointwise comparable. Although the functions $M_d(f)$ and $M(f)$ are not pointwise comparable, we will show that they are comparable in norm.

The next result provides an example of a *good lambda distributional inequality.*

Theorem 7.4.4. *For all $\gamma > 0$, all $\lambda > 0$, and all locally integrable functions f on $\mathbf{R}^n$, we have the estimate*

$$\big|\{x \in \mathbf{R}^n : M_d(f)(x) > 2\lambda,\, M^{\#}(f)(x) \le \gamma\lambda\}\big| \le 2^n\gamma\big|\{x \in \mathbf{R}^n : M_d(f)(x) > \lambda\}\big|.$$

Proof. We may suppose that the set $\Omega_\lambda = \{x \in \mathbf{R}^n : M_d(f)(x) > \lambda\}$ has finite measure; otherwise, there is nothing to prove. Then for each $x \in \Omega_\lambda$ there is a maximal dyadic cube Q^x that contains x such that

$$\frac{1}{|Q^x|}\int_{Q^x}|f(y)|\,dy>\lambda\,; \tag{7.4.3}$$

otherwise, Ω_λ would have infinite measure. Let Q_j be the collection of all such maximal dyadic cubes containing all x in Ω_λ, i.e., $\{Q_j\}_j=\{Q^x:\,x\in\Omega_\lambda\}$. Maximal dyadic cubes are disjoint; hence any two different Q_j's are disjoint; Moreover, we note that if $x,y\in Q_j$, then $Q_j=Q^x=Q^y$. It follows that $\Omega_\lambda=\bigcup_j Q_j$. To prove the required estimate, it suffices to show that for all Q_j we have

$$\big|\{x\in Q_j:\,M_d(f)(x)>2\lambda,\,M^\#(f)(x)\le\gamma\lambda\}\big|\le 2^n\gamma\big|Q_j\big|, \tag{7.4.4}$$

for once (7.4.4) is established, the conclusion follows by summing on j.

We fix j and $x\in Q_j$ such that $M_d(f)(x)>2\lambda$. Then the supremum

$$M_d(f)(x)=\sup_{R\ni x}\frac{1}{|R|}\int_R|f(y)|\,dy \tag{7.4.5}$$

is taken over all dyadic cubes R that either contain Q_j or are contained in Q_j (since $Q_j\cap R\neq\emptyset$). If $R\supsetneq Q_j$, the maximality of Q_j implies that (7.4.3) does not hold for R; thus the average of $|f|$ over R is at most λ. Thus if $M_d(f)(x)>2\lambda$, then the supremum in (7.4.5) is attained for some dyadic cube R contained (not properly) in Q_j. Therefore, if $x\in Q_j$ and $M_d(f)(x)>2\lambda$, then we can replace f by $f\chi_{Q_j}$ in (7.4.5) and we must have $M_d(f\chi_{Q_j})(x)>2\lambda$. We let Q_j' be the unique dyadic cube of twice the side length of Q_j. Therefore, for $x\in Q_j$ we have

$$M_d\Big(\big(f-\operatorname*{Avg}_{Q_j'}f\big)\chi_{Q_j}\Big)(x)\ge M_d(f\chi_{Q_j})(x)-\big|\operatorname*{Avg}_{Q_j'}f\big|>2\lambda-\lambda=\lambda\,,$$

since $|\operatorname{Avg}_{Q_j'}f|\le\operatorname{Avg}_{Q_j'}|f|\le\lambda$ because of the maximality of Q_j. We conclude that

$$\big|\{x\in Q_j:M_d(f)(x)>2\lambda\}\big|\le\Big|\Big\{x\in Q_j:M_d\big((f-\operatorname*{Avg}_{Q_j'}f)\chi_{Q_j}\big)(x)>\lambda\Big\}\Big|, \tag{7.4.6}$$

and using the fact that M_d is of weak type $(1,1)$ with constant 1, we control the last expression in (7.4.6) by

$$\begin{aligned}\frac{1}{\lambda}\int_{Q_j}\big|f(y)-\operatorname*{Avg}_{Q_j'}f\big|\,dy&\le\frac{2^n|Q_j|}{\lambda}\frac{1}{|Q_j'|}\int_{Q_j'}\big|f(y)-\operatorname*{Avg}_{Q_j'}f\big|\,dy\\&\le\frac{2^n|Q_j|}{\lambda}M^\#(f)(\xi_j)\end{aligned} \tag{7.4.7}$$

for all $\xi_j\in Q_j$. In proving (7.4.4) we may assume that for some $\xi_j\in Q_j$ we have $M^\#(f)(\xi_j)\le\gamma\lambda$; otherwise, there is nothing to prove. For this ξ_j, using (7.4.6) and (7.4.7) we obtain (7.4.4). □

Good lambda inequalities can be used to obtain L^p bounds for quantities they contain. For example, we use Theorem 7.4.4 to obtain the equivalence of the L^p

norms of $M_d(f)$ and $M^\#(f)$. Since $M^\#(f)$ is pointwise controlled by $2M_c(f)$ and

$$\big\|M_c(f)\big\|_{L^p} \le C(p,n)\big\|f\big\|_{L^p} \le C(p,n)\big\|M_d(f)\big\|_{L^p},$$

we have the estimate

$$\big\|M^\#(f)\big\|_{L^p(\mathbf{R}^n)} \le 2C(p,n)\big\|M_d(f)\big\|_{L^p(\mathbf{R}^n)}$$

for all f in $L^p(\mathbf{R}^n)$. The next theorem says that the converse estimate is valid.

Theorem 7.4.5. *Let $0 < p_0 < \infty$. Then for any p with $p_0 \le p < \infty$ there is a constant $C_n(p)$ such that for all functions f with $M_d(f) \in L^{p_0}(\mathbf{R}^n)$ we have*

$$\big\|M_d(f)\big\|_{L^p(\mathbf{R}^n)} \le C_n(p)\big\|M^\#(f)\big\|_{L^p(\mathbf{R}^n)}. \tag{7.4.8}$$

Proof. For a positive real number N we set

$$I_N = \int_0^N p\lambda^{p-1}\big|\{x \in \mathbf{R}^n : M_d(f)(x) > \lambda\}\big|\,d\lambda\,.$$

We note that I_N is finite, since $p \ge p_0$ and it is bounded by

$$\frac{pN^{p-p_0}}{p_0}\int_0^N p_0\lambda^{p_0-1}\big|\{x \in \mathbf{R}^n : M_d(f)(x) > \lambda\}\big|\,d\lambda \le \frac{pN^{p-p_0}}{p_0}\big\|M_d(f)\big\|_{L^{p_0}}^{p_0} < \infty\,.$$

We now write

$$I_N = 2^p\int_0^{\frac{N}{2}} p\lambda^{p-1}\big|\{x \in \mathbf{R}^n : M_d(f)(x) > 2\lambda\}\big|\,d\lambda$$

and we use Theorem 7.4.4 to obtain the following sequence of inequalities:

$$\begin{aligned}
I_N \le\ & 2^p\int_0^{\frac{N}{2}} p\lambda^{p-1}\big|\{x \in \mathbf{R}^n : M_d(f)(x) > 2\lambda,\, M^\#(f)(x) \le \gamma\lambda\}\big|\,d\lambda \\
&\qquad + 2^p\int_0^{\frac{N}{2}} p\lambda^{p-1}\big|\{x \in \mathbf{R}^n : M^\#(f)(x) > \gamma\lambda\}\big|\,d\lambda \\
\le\ & 2^p 2^n\gamma\int_0^{\frac{N}{2}} p\lambda^{p-1}\big|\{x \in \mathbf{R}^n : M_d(f)(x) > \lambda\}\big|\,d\lambda \\
&\qquad + 2^p\int_0^{\frac{N}{2}} p\lambda^{p-1}\big|\{x \in \mathbf{R}^n : M^\#(f)(x) > \gamma\lambda\}\big|\,d\lambda \\
\le\ & 2^p 2^n\gamma I_N + \frac{2^p}{\gamma^p}\int_0^{\frac{N\gamma}{2}} p\lambda^{p-1}\big|\{x \in \mathbf{R}^n : M^\#(f)(x) > \lambda\}\big|\,d\lambda\,.
\end{aligned}$$

At this point we pick a γ such that $2^p 2^n\gamma = 1/2$. Since I_N is finite, we can subtract from both sides of the inequality the quantity $\frac{1}{2}I_N$ to obtain

$$I_N \le 2^{p+1} 2^{p(n+p+1)} \int_0^{\frac{N\gamma}{2}} p\lambda^{p-1} \big|\{x \in \mathbf{R}^n : M^{\#}(f)(x) > \lambda\}\big| \, d\lambda \,,$$

from which we obtain (7.4.8) with $C_n(p) = 2^{n+p+2+\frac{1}{p}}$ letting $N \to \infty$. □

Corollary 7.4.6. *Let $0 < p_0 < \infty$. Then for any p with $p_0 \le p < \infty$ and for all locally integrable functions f with $M_d(f) \in L^{p_0}(\mathbf{R}^n)$ we have*

$$\big\|f\big\|_{L^p(\mathbf{R}^n)} \le C_n(p) \big\|M^{\#}(f)\big\|_{L^p(\mathbf{R}^n)} \,, \tag{7.4.9}$$

where $C_n(p)$ is the constant in Theorem 7.4.5.

Proof. Since for every point in $\mathbf{R}^n$ there is a sequence of dyadic cubes shrinking to it, the Lebesgue differentiation theorem yields that for almost every point x in $\mathbf{R}^n$ the averages of the locally integrable function f over the dyadic cubes containing x converge to $f(x)$. Consequently,

$$|f| \le M_d(f) \qquad \text{a.e.}$$

Using this fact, the proof of (7.4.9) is immediate, since

$$\big\|f\big\|_{L^p(\mathbf{R}^n)} \le \big\|M_d(f)\big\|_{L^p(\mathbf{R}^n)} \,,$$

and by Theorem 7.4.5 the latter is controlled by $C_n(p)\big\|M^{\#}(f)\big\|_{L^p(\mathbf{R}^n)}$. □

Estimate (7.4.9) provides the sought converse to (7.4.1).

7.4.3 Interpolation Using BMO

We continue this section by proving an interpolation result in which the space L^∞ is replaced by BMO. The sharp function plays a key role in the following theorem.

Theorem 7.4.7. *Let $1 \le p_0 < \infty$. Let T be a linear operator that maps $L^{p_0}(\mathbf{R}^n)$ to $L^{p_0}(\mathbf{R}^n)$ with bound A_0, and $L^\infty(\mathbf{R}^n)$ to $BMO(\mathbf{R}^n)$ with bound A_1. Then for all p with $p_0 < p < \infty$ there is a constant $C_{n,p}$ such that for all $f \in L^p$ we have*

$$\big\|T(f)\big\|_{L^p(\mathbf{R}^n)} \le C_{n,p,p_0} A_0^{\frac{p_0}{p}} A_1^{1-\frac{p_0}{p}} \big\|f\big\|_{L^p(\mathbf{R}^n)} \,. \tag{7.4.10}$$

Remark 7.4.8. In certain applications, the operator T may not be a priori defined on all of $L^{p_0} + L^\infty$ but only on some subspace of it. In this case one may state that the hypotheses and the conclusion of the preceding theorem hold for a subspace of these spaces.

Proof. We consider the operator

$$S(f) = M^{\#}(T(f))$$

defined for $f \in L^{p_0} + L^\infty$. It is easy to see that S is a sublinear operator. We prove that S maps L^{p_0} to itself and L^∞ to itself. For $f \in L^{p_0}$ we have

$$\begin{aligned} \big\|S(f)\big\|_{L^{p_0}} &= \big\|M^{\#}(T(f))\big\|_{L^{p_0}} \le 2\big\|M_c(T(f))\big\|_{L^{p_0}} \\ &\le C_{n,p_0}\big\|T(f)\big\|_{L^{p_0}} \le C_{n,p_0}A_0\big\|f\big\|_{L^{p_0}}, \end{aligned}$$

while for $f \in L^\infty$ one has

$$\big\|S(f)\big\|_{L^\infty} = \big\|M^{\#}(T(f))\big\|_{L^\infty} = \big\|T(f)\big\|_{BMO} \le A_1\big\|f\big\|_{L^\infty}.$$

Interpolating between these estimates using Theorem 1.3.2, we deduce

$$\big\|M^{\#}(T(f))\big\|_{L^p} = \big\|S(f)\big\|_{L^p} \le C_{p,p_0} A_0^{\frac{p_0}{p}} A_1^{1-\frac{p_0}{p}} \big\|f\big\|_{L^p}$$

for all $f \in L^p$, where $p_0 < p < \infty$. Consider now a function $h \in L^p \cap L^{p_0}$. In the case $p_0 > 1$, $M_d(T(h)) \in L^{p_0}$; hence Corollary 7.4.6 is applicable and gives

$$\big\|T(h)\big\|_{L^p} \le C_n(p)\, C_{p,p_0} A_0^{\frac{p_0}{p}} A_1^{1-\frac{p_0}{p}} \big\|h\big\|_{L^p}.$$

Density yields the same estimate for all $f \in L^p(\mathbf{R}^n)$. If $p_0 = 1$, one applies the same idea but needs the endpoint estimate of Exercise 7.4.6, since $M_d(T(h)) \in L^{1,\infty}$. □

7.4.4 Estimates for Singular Integrals Involving the Sharp Function

We use the sharp function to obtain pointwise estimates for singular integrals. These enable us to recover previously obtained estimates for singular integrals, but also to deduce a new endpoint boundedness result from L^∞ to BMO.

Let us recall some facts from Chapter 4. Suppose that K is defined on $\mathbf{R}^n \setminus \{0\}$ and satisfies

$$|K(x)| \le A_1|x|^{-n}, \tag{7.4.11}$$

$$|K(x-y) - K(x)| \le A_2|y|^{\delta}|x|^{-n-\delta} \quad \text{whenever } |x| \ge 2|y| > 0, \tag{7.4.12}$$

$$\sup_{r<R<\infty}\left|\int_{r\le|x|\le R} K(x)\,dx\right| \le A_3. \tag{7.4.13}$$

Let W be a tempered distribution that coincides with K on $\mathbf{R}^n \setminus \{0\}$ and let T be the linear operator given by convolution with W.

Under these assumptions we have that T is L^2 bounded with norm at most a constant multiple of $A_1 + A_2 + A_3$ (Theorem 4.4.1), and hence it is also L^p bounded with a similar norm on L^p for $1 < p < \infty$ (Theorem 4.3.3). Furthermore, under the preceding conditions, the maximal singular integral $T^{(*)}$ is also bounded from $L^p(\mathbf{R}^n)$ to itself for $1 < p < \infty$ (Theorem 4.3.4).

Theorem 7.4.9. *Let T be given by convolution with a distribution W that coincides with a function K on $\mathbf{R}^n \setminus \{0\}$ satisfying (7.4.12). Assume that T has an extension that is L^2 bounded with a norm B. Then there is a constant C_n such that for any $s > 1$ the estimate*

$$M^{\#}(T(f))(x) \le C_n(A_2+B)\max(s,(s-1)^{-1})M(|f|^s)^{\frac{1}{s}}(x) \tag{7.4.14}$$

is valid for all f in $\bigcup_{s\le p<\infty} L^p$ and almost all $x \in \mathbf{R}^n$.

Proof. In view of Proposition 7.4.2 (2), given any cube Q, it suffices to find a constant a such that

$$\frac{1}{|Q|}\int_Q |T(f)(y)-a|\,dy \le C_{s,n}(A_2+B)M(|f|^s)^{\frac{1}{s}}(x) \tag{7.4.15}$$

for all $x \in Q$. To prove this estimate we employ a theme that we have seen several times before. We write $f = f_Q^0 + f_Q^\infty$, where $f_Q^0 = f\chi_{6\sqrt{n}Q}$ and $f_Q^\infty = f\chi_{(6\sqrt{n}Q)^c}$. Here $6\sqrt{n}Q$ denotes the cube that is concentric with Q, has sides parallel to those of Q, and has side length $6\sqrt{n}\ell(Q)$, where $\ell(Q)$ is the side length of Q.

We now fix an f in $\bigcup_{s\le p<\infty} L^p$ and we select $a = T(f_Q^\infty)(x)$. Then a is finite for almost all $x \in Q$. It follows that

$$\begin{aligned}&\frac{1}{|Q|}\int_Q |T(f)(y)-a|\,dy\\ &\qquad \le \frac{1}{|Q|}\int_Q |T(f_Q^0)(y)|\,dy + \frac{1}{|Q|}\int_Q |T(f_Q^\infty)(y)-T(f_Q^\infty)(x)|\,dy.\end{aligned} \tag{7.4.16}$$

In view of Theorem 4.3.3, T maps L^s to L^s with norm at most a dimensional constant multiple of $\max(s,(s-1)^{-1})(B+A_2)$. The first term in (7.4.16) is controlled by

$$\begin{aligned}\left(\frac{1}{|Q|}\int_Q |T(f_Q^0)(y)|^s\,dy\right)^{\frac{1}{s}} &\le C_n\max(s,(s-1)^{-1})(B+A_2)\left(\frac{1}{|Q|}\int_{\mathbf{R}^n} |f_Q^0(y)|^s\,dy\right)^{\frac{1}{s}}\\ &\le C_n'\max(s,(s-1)^{-1})(B+A_2)M(|f|^s)^{\frac{1}{s}}(x).\end{aligned}$$

To estimate the second term in (7.4.16), we first note that

$$\int_Q |T(f_Q^\infty)(y)-T(f_Q^\infty)(x)|\,dy \le \int_Q \left| \int_{(6\sqrt{n}Q)^c} (K(y-z)-K(x-z))f(z)\,dz\right| dy.$$

We make a few geometric observations. Since both x and y are in Q, we have $|x-y| \le \sqrt{n}\,\ell(Q)$. Also (see Figure 7.3), since $z \notin 6\sqrt{n}Q$ and $x \in Q$, we must have

$$|x-z| \ge \operatorname{dist}\left(Q,(6\sqrt{n}Q)^c\right) \ge (3\sqrt{n}-\tfrac{1}{2})\ell(Q) \ge 2\sqrt{n}\,\ell(Q) \ge 2|x-y|.$$

Therefore, we have $|x-z| \geq 2|x-y|$, and this allows us to conclude that

$$\left|K(y-z)-K(x-z)\right| = \left|K((x-z)-(x-y))-K(x-z)\right| \leq A_2 \frac{|x-y|^\delta}{|x-z|^{n+\delta}}$$

using condition (7.4.12). Using these observations, we bound the second term in (7.4.16) by

$$\begin{aligned}
\frac{1}{|Q|}\int_Q\int_{(6\sqrt{n}Q)^c} \frac{A_2|x-y|^\delta}{|x-z|^{n+\delta}}|f(z)|\,dz\,dy &\leq C_n\frac{A_2}{|Q|}\int_{(6\sqrt{n}Q)^c} \frac{\ell(Q)^{n+\delta}}{|x-z|^{n+\delta}}|f(z)|\,dz \\
&\leq C_nA_2\int_{\mathbf{R}^n} \frac{\ell(Q)^\delta}{(\ell(Q)+|x-z|)^{n+\delta}}|f(z)|\,dz \\
&\leq C_nA_2M(f)(x) \\
&\leq C_nA_2(M(|f|^s)(x))^{\frac{1}{s}},
\end{aligned}$$

where we used the fact that $|x-z|$ is at least $\ell(Q)$ and Theorem 2.1.10. This proves (7.4.15) and hence (7.4.14). □

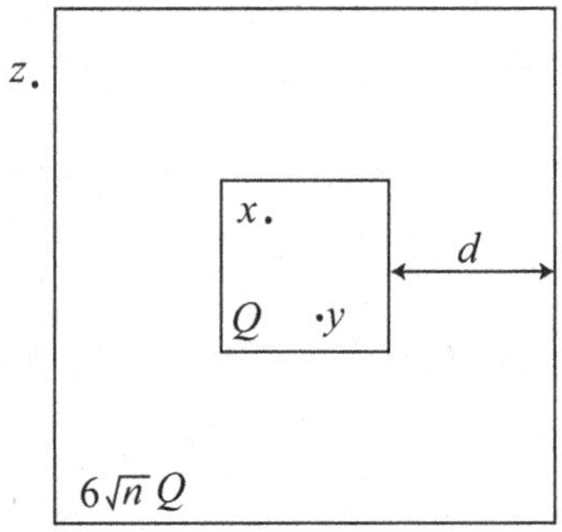

Fig. 7.3 The cubes Q and $6\sqrt{n}Q$. The distance d is equal to $(3\sqrt{n}-\frac{1}{2})\ell(Q)$.

The inequality (7.4.14) in Theorem 7.4.9 is noteworthy, since it provides a pointwise estimate for $T(f)$ in terms of a maximal function. This clearly strengthens the L^p boundedness of T. As a consequence of this estimate, we deduce the following result.

Corollary 7.4.10. *Let T be given by convolution with a distribution W that coincides with a function K on $\mathbf{R}^n \setminus \{0\}$ that satisfies (7.4.12). Assume that T has an extension that is L^2 bounded with a norm B. Then there is a constant C_n such that the estimate*

$$\left\|T(f)\right\|_{BMO} \leq C_n(A_2+B)\left\|f\right\|_{L^\infty} \tag{7.4.17}$$

is valid for all $f \in L^\infty \bigcap \left(\bigcup_{1\leq p<\infty} L^p\right)$.

Proof. We take $s = 2$ in Theorem 7.4.9 and we observe that

$$\|T(f)\|_{BMO} = \|M^{\#}(T(f))\|_{L^\infty} \le C_n(A_2 + B)\|M(|f|^2)^{\frac{1}{2}}\|_{L^\infty},$$

and the last expression is easily controlled by $C_n(A_2 + B)\|f\|_{L^\infty}$. □

At this point we have not defined the action of $T(f)$ when f lies merely in L^∞; and for this reason we restricted the functions f in Corollary 7.4.10 to be also in some L^p. There is, however, a way to define T on L^∞ abstractly via duality. Theorem 6.7.1 gives that T and thus also its adjoint T^* map H^1 to L^1. Then the adjoint operator of T^* (i.e., T) maps L^∞ to BMO and is therefore well defined on L^∞. In this way, however, $T(f)$ is not defined explicitly when f is in L^∞. Such an explicit definition is given in the next chapter in a slightly more general setting.

Remark 7.4.11. In the hypotheses of Theorem 7.4.9 we could have replaced the condition that T maps L^2 to L^2 by the condition that T maps L^r to $L^{r,\infty}$ with norm B for some $1 < r < \infty$.

Exercises

7.4.1. Let $0 < q < \infty$. Prove that for every p with $q < p < \infty$ there is a constant $C_{n,p,q}$ such that for all functions f on $\mathbf{R}^n$ with $M_d(f) \in L^q(\mathbf{R}^n)$ we have

$$\|f\|_{L^p} \le C_{n,p,q}\|f\|_{L^q}^{1-\theta}\|f\|_{BMO}^{\theta},$$

where $\frac{1}{p} = \frac{1-\theta}{q}$.

7.4.2. Let μ be a positive Borel measure on $\mathbf{R}^n$.
(a) Show that the maximal operator

$$M_\mu^d(f)(x) = \sup_{\substack{Q \ni x \\ Q \text{ dyadic cube}}} \frac{1}{\mu(Q)} \int_Q |f(t)|\, d\mu(t)$$

maps $L^1(\mathbf{R}^n, d\mu)$ to $L^{1,\infty}(\mathbf{R}^n, d\mu)$ with constant 1.
(b) For a μ-locally integrable function f, define the *sharp maximal function with respect to* μ,

$$M_\mu^{\#}(f)(x) = \sup_{Q \ni x} \frac{1}{\mu(Q)} \int_Q \Big| f(t) - \operatorname*{Avg}_{Q,\mu} f \Big|\, d\mu(t),$$

where $\operatorname{Avg}_{Q,\mu} f$ denotes the average of f over Q with respect to μ. Assume that μ is a doubling measure with doubling constant $C(\mu)$ [this means that $\mu(3Q) \le C(\mu)\mu(Q)$ for all cubes Q]. Prove that for all $\gamma > 0$, all $\lambda > 0$, and all μ-locally integrable functions f on $\mathbf{R}^n$ we have the estimate

$$\mu\big(\{x : M_\mu^d(f)(x) > 2\lambda,\ M_\mu^{\#}(f)(x) \le \gamma\lambda\}\big) \le C(\mu)\gamma\mu\big(\{x : M_\mu^d(f)(x) > \lambda\}\big).$$

[*Hint:* Part (a): For any x in the set $\{x \in \mathbf{R}^n : M_\mu^d(f)(x) > \lambda\}$, choose a maximal dyadic cube $Q = Q(x)$ such that $\int_Q |f(t)|\,d\mu(t) > \lambda\mu(Q)$. Part (b): Mimic the proof of Theorem 7.4.4.]

7.4.3. Let $0 < p_0 < \infty$ and let M_μ^d and $M_\mu^\#$ be as in Exercise 7.4.2. Prove that for any p with $p_0 \le p < \infty$ there is a constant $C_n(p,\mu)$ such that for all locally integrable functions f with $M_\mu^d(f) \in L^{p_0}(\mathbf{R}^n)$ we have

$$\big\|M_\mu^d(f)\big\|_{L^p(\mathbf{R}^n,d\mu)} \le C_n(p,\mu)\big\|M_\mu^\#(f)\big\|_{L^p(\mathbf{R}^n,d\mu)}.$$

7.4.4. We say that a function f on $\mathbf{R}^n$ is in BMO_d (or dyadic BMO) if

$$\big\|f\big\|_{BMO_d} = \sup_{Q\,\text{dyadic cube}} \frac{1}{|Q|}\int_Q \big|f(x) - \operatorname*{Avg}_Q f\big|\,dx < \infty.$$

(a) Show that BMO is a proper subset of BMO_d.
(b) Suppose that A is a finite constant and that a function f in BMO_d satisfies

$$\big|\operatorname*{Avg}_{Q_1} f - \operatorname*{Avg}_{Q_2} f\big| \le A$$

for all adjacent dyadic cubes of the same length. Show that f is in BMO.
[*Hint:* Consider first the case $n = 1$. Given an interval I, find adjacent dyadic intervals of the same length I_1 and I_2 such that $I \subsetneqq I_1 \bigcup I_2$ and $|I_1| \le |I| < 2|I_1|$.]

7.4.5. Suppose that K is a function on $\mathbf{R}^n \setminus \{0\}$ that satisfies (7.4.11), (7.4.12), and (7.4.13). Let η be a smooth function that vanishes in a neighborhood of the origin and is equal to 1 in a neighborhood of infinity. For $\varepsilon > 0$ let $K_\eta^{(\varepsilon)}(x) = K(x)\eta(x/\varepsilon)$ and let $T_\eta^{(\varepsilon)}$ be the operator given by convolution with $K_\eta^{(\varepsilon)}$. Prove that for any $1 < s < \infty$ there is a constant $C_{n,s}$ such that for all p with $s < p < \infty$ and f in L^p we have

$$\Big\|\sup_{\varepsilon>0} M^\#(T_\eta^{(\varepsilon)}(f))\Big\|_{L^p(\mathbf{R}^n)} \le C_{n,s}(A_1 + A_2 + A_3)\big\|f\big\|_{L^p(\mathbf{R}^n)}.$$

[*Hint:* Observe that the kernels $K_\eta^{(\varepsilon)}$ satisfy (7.4.11), (7.4.12), and (7.4.13) uniformly in $\varepsilon > 0$ and use Theorems 4.4.1 and 7.4.9.]

7.4.6. Let $0 < p_0 < \infty$ and suppose that for some locally integrable function f we have that $M_d(f)$ lies in $L^{p_0,\infty}(\mathbf{R}^n)$. Show that for any p in (p_0,∞) there exists a constant $C_n(p)$ such that

$$\big\|f\big\|_{L^p(\mathbf{R}^n)} \le \big\|M_d(f)\big\|_{L^p(\mathbf{R}^n)} \le C_n(p)\big\|M^\#(f)\big\|_{L^p(\mathbf{R}^n)},$$

where $C_n(p)$ depends only on n and p.
[*Hint:* With the same notation as in the proof of Theorem 7.4.5, use the hypothesis $\big\|M_d(f)\big\|_{L^{p_0,\infty}} < \infty$ to prove that $I_N < \infty$ whenever $p > p_0$. Then the arguments in the proofs of Theorem 7.4.5 and Corollary 7.4.6 remain unchanged.]

7.4.7. Prove that the expressions

$$\Sigma_N(x) = \sum_{k=1}^{N} \frac{\sin(2\pi kx)}{k}$$

are uniformly bounded in N and x. Then use Corollary 7.4.10 to prove that

$$\sup_{N\ge 1} \left\| \sum_{k=1}^{N} \frac{e^{2\pi ikx}}{k} \right\|_{BMO} \le C < \infty .$$

Deduce that the limit of $\Sigma_N(x)$ as $N \to \infty$ can be defined as an element of *BMO*. [*Hint:* Use that the Hilbert transform of $\sin(2\pi kx)$ is $\cos(2\pi kx)$. Also note that the series $\sum_{k=1}^{\infty} \frac{\sin(2\pi kx)}{k}$ coincides with the periodic extension of the (bounded) function $= \pi(\frac{1}{2} - x)$ on $[0,1)$.]

7.5 Commutators of Singular Integrals with *BMO* Functions

The mean value zero property of $H^1(\mathbf{R}^n)$ is often manifested when its elements are paired with functions in *BMO*. It is therefore natural to expect that *BMO* can be utilized to express and quantify the cancellation of expressions in H^1. Let us be specific through an example. We saw in Exercise 6.7.3 that the bilinear operator

$$(f,g) \mapsto fH(g) + H(f)g$$

maps $L^2(\mathbf{R}^n) \times L^2(\mathbf{R}^n)$ to $H^1(\mathbf{R}^n)$; here H is the Hilbert transform. Pairing with a *BMO* function b and using that $H^t = -H$, we obtain that

$$\langle fH(g) + H(f)g\,, b\rangle = \langle f\,, H(g)b - H(gb)\rangle ,$$

and hence the operator $g \mapsto H(g)b - H(gb)$ should be L^2 bounded. This expression $H(g)b - H(gb)$ is called the *commutator* of H with the *BMO* function b. More generally, we give the following definition.

Definition 7.5.1. The *commutator* of a singular integral operator T with a function b is defined as

$$[b,T](f) = bT(f) - T(bf) .$$

If the function b is locally integrable and has at most polynomial growth at infinity, then the operation $[b,T]$ is well defined when acting on Schwartz functions f.

In view of the preceding remarks, the L^p boundedness of the commutator $[b,T]$ for b in *BMO* exactly captures the cancellation property of the bilinear expression

$$(f,g) \mapsto T(f)g - fT^t(g) .$$

As in the case with the Hilbert transform, it is natural to expect that the commutator $[b,T]$ of a general singular integral T is L^p bounded for all $1 < p < \infty$. This fact is proved in this section. Since BMO functions are unbounded in general, one may surmise that the presence of the negative sign in the definition of the commutator plays a crucial cancellation role.

We introduce some material needed in the study of the boundedness of the commutator.

7.5.1 An Orlicz-Type Maximal Function

We can express the L^p norm ($1 \le p < \infty$) of a function f on a measure space X by

$$\big\|f\big\|_{L^p(X)} = \left(\int_X |f|^p \, d\mu\right)^{\frac{1}{p}} = \inf\left\{\lambda > 0 : \int_X \left|\frac{|f|}{\lambda}\right|^p d\mu \le 1\right\}.$$

Motivated by the second expression, we may replace the function t^p by a general increasing convex function $\Phi(t)$. We give the following definition.

Definition 7.5.2. A *Young's function* is a continuous increasing convex function Φ on $[0,\infty)$ that satisfies $\Phi(0) = 0$ and $\lim_{t\to\infty} \Phi(t) = \infty$. The *Orlicz norm* of a measurable function f on a measure space (X,μ) with respect to a Young's function Φ is defined as

$$\big\|f\big\|_{\Phi(L)(X,\mu)} = \inf\left\{\lambda > 0 : \int_X \Phi(|f|/\lambda)\, d\mu \le 1\right\}.$$

The *Orlicz space* $\Phi(L)(X,\mu)$ is then defined as the space of all measurable functions f on X such that $\big\|f\big\|_{\Phi(L)(X,\mu)} < \infty$.

We are mostly concerned with the case in which the measure space X is a cube in $\mathbf{R}^n$ with normalized Lebesgue measure $|Q|^{-1}dx$. For a measurable function f on a cube Q in $\mathbf{R}^n$, the Orlicz norm of f is therefore

$$\big\|f\big\|_{\Phi(L)(Q,\frac{dx}{|Q|})} = \inf\left\{\lambda > 0 : \frac{1}{|Q|}\int_Q \Phi(|f|/\lambda)\, dx \le 1\right\},$$

which is simply denoted by $\big\|f\big\|_{\Phi(L)(Q)}$, since the measure is understood to be normalized Lebesgue whenever the ambient space is a cube.

Since for $C > 1$ convexity gives $\Phi(t/C) \le \Phi(t)/C$ for all $t \ge 0$, it follows that

$$\big\|f\big\|_{C\Phi(Q)} \le C\big\|f\big\|_{\Phi(Q)}, \tag{7.5.1}$$

which implies that the norms with respect to Φ and $C\Phi$ are comparable.

A case of particular interest arises when $\Phi(t) = t\log(e+t)$. This function is pointwise comparable to $t(1+\log^+ t)$ for $t \ge 0$. We make use in the sequel of a certain maximal operator defined in terms of the corresponding Orlicz norm.

Definition 7.5.3. We define the *Orlicz maximal operator*

$$M_{L\log(e+L)}(f)(x) = \sup_{Q\ni x} \|f\|_{L\log(e+L)(Q)},$$

where the supremum is taken over all cubes Q with sides parallel to the axes that contain the given point x.

The boundedness properties of this maximal operator are a consequence of the following lemma.

Lemma 7.5.4. *There is a positive constant $c(n)$ such that for any cube Q in $\mathbf{R}^n$ and any nonnegative locally integrable function w, we have*

$$\|w\|_{L\log(e+L)(Q)} \le \frac{c(n)}{|Q|}\int_Q M_c(w)\,dx, \tag{7.5.2}$$

where M_c is the Hardy–Littlewood maximal operator with respect to cubes. Hence, for some other dimensional constant $c'(n)$ and all nonnegative w in $L^1_{\text{loc}}(\mathbf{R}^n)$ the inequality

$$M_{L\log(e+L)}(w)(x) \le c'(n)M^2(w)(x) \tag{7.5.3}$$

is valid, where $M^2 = M\circ M$ and M is the Hardy–Littlewood maximal operator.

Proof. Fix a cube Q in $\mathbf{R}^n$ with sides parallel to the axes. We introduce a *maximal operator associated with Q* as follows:

$$M_c^Q(f)(x) = \sup_{\substack{R\ni x\\ R\subseteq Q}} \frac{1}{|R|}\int_R |f(y)|\,dy,$$

where the supremum is taken over cubes R in $\mathbf{R}^n$ with sides parallel to the axes. The key estimate follows from the following local version of the reverse weak type $(1,1)$ estimate of Exercise 2.1.4(b). For each nonnegative function f on $\mathbf{R}^n$ and $\alpha \ge \text{Avg}_Q f$, we have

$$\frac{1}{\alpha}\int_{Q\cap\{f>\alpha\}} f\,dx \le 2^n \big|\{x\in Q : M_c^Q(f)(x) > \alpha\}\big|. \tag{7.5.4}$$

Indeed, to prove (7.5.4), we apply Corollary 2.1.21 to the function f and the number α. With the notation of that corollary, we have $Q\setminus(\bigcup_j Q_j) \subseteq \{f\le\alpha\}$. This implies that $Q\cap\{f>\alpha\}\subseteq\bigcup_j Q_j$, which is contained in $\{x\in Q:\ M_c^Q(f)(x)>\alpha\}$. Multiplying both sides of (2.1.20) by $|Q_j|$, summing over all j, and using these observations, we obtain (7.5.4).

Using the definition of $M_{L\log(e+L)}$, (7.5.2) follows from the fact that for some constant $c>1$ independent of w we have

$$\frac{1}{|Q|}\int_Q \frac{w}{\lambda_Q}\log\Big(e+\frac{w}{\lambda_Q}\Big)\,d\mu \le 1, \tag{7.5.5}$$

where

$$\lambda_Q = \frac{c}{|Q|}\int_Q M_c(w)\,dx = c\,\operatorname*{Avg}_Q M_c(w).$$

We let $f = w/\lambda_Q$; by the Lebesgue differentiation theorem we have that $0 \le \operatorname{Avg}_Q f \le 1/c$. Recall identity (1.1.7),

$$\int_X \phi(f)\,d\nu = \int_0^\infty \phi'(t)\,\nu(\{x\in X : f(x) > t\})\,dt,$$

where $\nu \ge 0$, (X,ν) is a measure space, and ϕ is an increasing continuously differentiable function with $\phi(0) = 0$. We take $X = Q$, $d\nu = |Q|^{-1} f\chi_Q\,dx$, and $\phi(t) = \log(e+t) - 1$ to deduce

$$\begin{aligned}\frac{1}{|Q|}\int_Q f\log(e+f)\,dx &= \frac{1}{|Q|}\int_Q f\,dx + \frac{1}{|Q|}\int_0^\infty \frac{1}{e+t}\left(\int_{Q\cap\{f>t\}} f\,dx\right)dt\\ &= I_0 + I_1 + I_2,\end{aligned}$$

where

$$\begin{aligned} I_0 &= \frac{1}{|Q|}\int_Q f\,dx,\\ I_1 &= \frac{1}{|Q|}\int_0^{\operatorname{Avg}_Q f} \frac{1}{e+t}\left(\int_{Q\cap\{f>t\}} f\,dx\right)dt,\\ I_2 &= \frac{1}{|Q|}\int_{\operatorname{Avg}_Q f}^\infty \frac{1}{e+t}\left(\int_{Q\cap\{f>t\}} f\,dx\right)dt.\end{aligned}$$

We now clearly have that $I_0 = \operatorname{Avg}_Q f \le 1/c$, while $I_1 \le (\operatorname{Avg}_Q f)^2 \le 1/c^2$. For I_2 we use estimate (7.5.4). Indeed, one has

$$\begin{aligned} I_2 &= \frac{1}{|Q|}\int_{\operatorname{Avg}_Q f}^\infty \frac{1}{e+t}\left(\int_{Q\cap\{f>t\}} f\,dx\right)dt\\ &\le \frac{2^n}{|Q|}\int_{\operatorname{Avg}_Q f}^\infty \frac{t}{e+t}\big|\{x\in Q : M_c^Q(f)(x) > t\}\big|\,dt\\ &\le \frac{2^n}{|Q|}\int_0^\infty \big|\{x\in Q : M_c^Q(f)(x) > \lambda\}\big|\,d\lambda\\ &= \frac{2^n}{|Q|}\int_Q M_c^Q(f)\,dx\\ &= \frac{2^n}{|Q|}\int_Q M_c(w)\,dx\,\frac{1}{\lambda_Q} = \frac{2^n}{c}\end{aligned}$$

using the definition of λ_Q. Combining all the estimates obtained, we deduce that

$$I_0 + I_1 + I_2 \le \frac{1}{c} + \frac{1}{c^2} + \frac{2^n}{c} \le 1,$$

provided c is large enough. □

7.5.2 A Pointwise Estimate for the Commutator

For $\delta > 0$, $M^\#_\delta$ denotes the following modification of the sharp maximal operator introduced in Section 7.4:

$$M^\#_\delta(f) = M^\#(|f|^\delta)^{1/\delta}.$$

It is often useful to work with the following characterization of $M^\#$ [see Proposition 7.4.2 (2)]:

$$M^\#(f)(x) \approx \sup_{Q \ni x} \inf_c \frac{1}{|Q|} \int_Q |f(y) - c| \, dy.$$

We also need the following version of the Hardy–Littlewood maximal operator:

$$M_\varepsilon(f) = M(|f|^\varepsilon)^{1/\varepsilon}.$$

The next lemma expresses the fact that commutators of singular integral operators with *BMO* functions are pointwise controlled by the maximal function $M^2 = M \circ M$.

Lemma 7.5.5. *Let T be a linear operator given by convolution with a tempered distribution on $\mathbf{R}^n$ that coincides with a function $K(x)$ on $\mathbf{R}^n \setminus \{0\}$ satisfying (7.4.11), (7.4.12), and (7.4.13). Let b be in $BMO(\mathbf{R}^n)$, and let $0 < \delta < \varepsilon$. Then there exists a positive constant $C = C_{\delta,\varepsilon,n}$ such that for any smooth function f with compact support we have*

$$M^\#_\delta([b,T](f)) \le C \|b\|_{BMO} \{M_\varepsilon(T(f)) + M^2(f)\}. \tag{7.5.6}$$

Proof. Fix a cube Q in $\mathbf{R}^n$ with sides parallel to the axes centered at the point x. Since for $0 < \delta < 1$ we have $\big||\alpha|^\delta - |\beta|^\delta\big| \le |\alpha - \beta|^\delta$ for $\alpha, \beta \in \mathbf{R}$, it is enough to show for some complex constant $c = c_Q$ that there exists $C = C_\delta > 0$ such that

$$\left(\frac{1}{|Q|} \int_Q \big|[b,T](f)(y) - c\big|^\delta dy\right)^{\frac{1}{\delta}} \le C \|b\|_{BMO} \{M_\varepsilon(T(f))(x) + M^2(f)(x)\}. \tag{7.5.7}$$

Denote by Q^* the cube $5\sqrt{n}Q$ that has side length $5\sqrt{n}$ times the side length of Q and the same center x as Q. Let $f = f_1 + f_2$, where $f_1 = f\chi_{Q^*}$. For an arbitrary constant a we write

$$[b,T](f) = (b-a)T(f) - T((b-a)f_1) - T((b-a)f_2).$$

Selecting

$$c = \operatorname*{Avg}_Q T((b-a)f_2) \qquad \text{and} \qquad a = \operatorname*{Avg}_{Q^*} b,$$

we can estimate the left-hand side of (7.5.7) by a multiple of $L_1 + L_2 + L_3$, where

$$L_1 = \left(\frac{1}{|Q|}\int_Q \left|(b(y)-\operatorname*{Avg}_{Q^*} b)\,T(f)(y)\right|^{\delta} dy\right)^{\frac{1}{\delta}},$$

$$L_2 = \left(\frac{1}{|Q|}\int_Q \left|T\big((b-\operatorname*{Avg}_{Q^*} b)f_1\big)(y)\right|^{\delta} dy\right)^{\frac{1}{\delta}},$$

$$L_3 = \left(\frac{1}{|Q|}\int_Q \left|T\big((b-\operatorname*{Avg}_{Q^*} b)f_2\big)-\operatorname*{Avg}_{Q} T\big((b-\operatorname*{Avg}_{Q^*} b)f_2\big)\right|^{\delta} dy\right)^{\frac{1}{\delta}}.$$

To estimate L_1, we use Hölder's inequality with exponents r and r' for some $1<r<\varepsilon/\delta$:

$$\begin{aligned} L_1 &\le \left(\frac{1}{|Q|}\int_Q \left|b(y)-\operatorname*{Avg}_{Q^*} b\right|^{\delta r'} dy\right)^{\frac{1}{\delta r'}} \left(\frac{1}{|Q|}\int_Q |T(f)(y)|^{\delta r} dy\right)^{\frac{1}{\delta r}} \\ &\le C\left\|b\right\|_{BMO} M_{\delta r}(T(f))(x) \\ &\le C\left\|b\right\|_{BMO} M_{\varepsilon}(T(f))(x), \end{aligned}$$

recalling that x is the center of Q. Since $T : L^1(\mathbf{R}^n) \to L^{1,\infty}(\mathbf{R}^n)$ and $0 < \delta < 1$, Kolmogorov's inequality (Exercise 2.1.5) yields

$$\begin{aligned} L_2 &\le \frac{C}{|Q|}\int_Q \left|(b(y)-\operatorname*{Avg}_{Q^*} b)f_1(y)\right| dy \\ &= \frac{C'}{|Q^*|}\int_{Q^*} \left|(b(y)-\operatorname*{Avg}_{Q^*} b)f(y)\right| dy \\ &\le 2C'\left\|b-\operatorname*{Avg}_{Q^*} b\right\|_{(e^L-1)(Q^*)}\left\|f\right\|_{L\log(1+L)(Q^*)}, \end{aligned}$$

using Exercise 7.5.2(c).

For some $0<\gamma<(2^n e)^{-1}$, let $C_{n,\gamma}>2$ be a constant larger than that appearing on the right-hand side of the inequality in Corollary 7.1.7. We set $c_0 = C_{n,\gamma}-1>1$. We use (7.5.1) and we claim that

$$\left\|b-\operatorname*{Avg}_{Q^*} b\right\|_{(e^L-1)(Q^*)} \le c_0\left\|b-\operatorname*{Avg}_{Q^*} b\right\|_{c_0^{-1}(e^L-1)(Q^*)} \le \frac{c_0}{\gamma}\left\|b\right\|_{BMO}. \tag{7.5.8}$$

Indeed, the last inequality is equivalent to

$$\frac{1}{|Q^*|}\int_{Q^*} c_0^{-1}\left[e^{\gamma|b(y)-\operatorname{Avg}_{Q^*} b|/\|b\|_{BMO}}-1\right] dy \le 1,$$

which is a restatement of Corollary 7.1.7. We therefore conclude that

$$L_2 \le C\left\|b\right\|_{BMO} M_{L\log(1+L)}(f)(x).$$

Finally, we turn our attention to the term L_3. Note that if $z,y \in Q$ and $w \notin Q^*$, then $|z-w| \geq 2|z-y|$. Using Fubini's theorem and property (7.4.12) succesively, we control L_3 pointwise by

$$\begin{aligned}
&\frac{1}{|Q|}\int_Q \Big|T\big((b-\operatorname*{Avg}_{Q^*} b)f_2\big)(y) - \operatorname*{Avg}_{Q} T\big((b-\operatorname*{Avg}_{Q^*} b)f_2\big)\Big|\,dy \\
&\leq \frac{1}{|Q|^2}\int_Q\int_Q\int_{\mathbf{R}^n\setminus Q^*} |K(y-w)-K(z-w)|\,\big|(b(w)-\operatorname*{Avg}_{Q^*} b)f(w)\big|\,dw\,dz\,dy \\
&\leq \frac{1}{|Q|^2}\int_Q\int_Q\sum_{j=0}^{\infty}\int_{2^{j+1}Q^*\setminus 2^jQ^*} \frac{A_2|y-z|^\delta}{|z-w|^{n+\delta}}\big|b(w)-\operatorname*{Avg}_{Q^*} b\big|\,|f(w)|\,dw\,dz\,dy \\
&\leq CA_2\sum_{j=0}^{\infty}\frac{\ell(Q)^\delta}{(2^j\ell(Q))^{n+\delta}}\int_{2^{j+1}Q^*}\big|b(w)-\operatorname*{Avg}_{Q^*} b\big|\,|f(w)|\,dw \\
&\leq CA_2\Bigg(\sum_{j=0}^{\infty}\frac{2^{-j\delta}}{(2^j\ell(Q))^n}\int_{2^{j+1}Q^*}\big|b(w)-\operatorname*{Avg}_{2^{j+1}Q^*} b\big|\,|f(w)|\,dw \\
&\qquad\qquad +\sum_{j=0}^{\infty}2^{-j\delta}\Big|\operatorname*{Avg}_{2^{j+1}Q^*} b-\operatorname*{Avg}_{Q^*} b\Big|\frac{1}{(2^j\ell(Q))^n}\int_{2^{j+1}Q^*}|f(w)|\,dw\Bigg) \\
&\leq C'A_2\sum_{j=0}^{\infty}2^{-j\delta}\big\|b-\operatorname*{Avg}_{2^{j+1}Q^*} b\big\|_{(e^L-1)(2^{j+1}Q^*)}\big\|f\big\|_{L\log(1+L)(2^{j+1}Q^*)} \\
&\qquad\qquad +C'A_2\big\|b\big\|_{BMO}\sum_{j=1}^{\infty}\frac{j}{2^{j\delta}}M(f)(x) \\
&\leq C''A_2\big\|b\big\|_{BMO}M_{L\log(1+L)}(f)(x)+C''A_2\big\|b\big\|_{BMO}M(f)(x) \\
&\leq C'''A_2\big\|b\big\|_{BMO}M^2(f)(x),
\end{aligned}$$

where we have used inequality (7.5.8), Lemma 7.5.4, and the simple estimate

$$\Big|\operatorname*{Avg}_{2^{j+1}Q^*} b-\operatorname*{Avg}_{Q^*} b\Big| \leq C_n j\big\|b\big\|_{BMO}$$

of Exercise 7.1.5. □

7.5.3 L^p Boundedness of the Commutator

We note that if f has compact support and b is in BMO, then bf lies in $L^q(\mathbf{R}^n)$ for all $q < \infty$ and therefore $T(bf)$ is well defined whenever T is a singular integral operator. Likewise, $[b,T]$ is a well defined operator on $\mathscr{C}_0^\infty$ for all b in BMO.

Having obtained the crucial Lemma 7.5.5, we now pass to an important result concerning its L^p boundedness.

Theorem 7.5.6. *Let T be as in Lemma 7.5.5. Then for any $1 < p < \infty$ there exists a constant $C = C_{p,n}$ such that for all smooth functions with compact support f and all BMO functions b, the following estimate is valid:*

$$\big\|[b,T](f)\big\|_{L^p(\mathbf{R}^n)} \le C\big\|b\big\|_{BMO}\big\|f\big\|_{L^p(\mathbf{R}^n)}. \tag{7.5.9}$$

Consequently, the linear operator

$$f \mapsto [b,T](f)$$

admits a bounded extension from $L^p(\mathbf{R}^n)$ to $L^p(\mathbf{R}^n)$ for all $1 < p < \infty$ with norm at most a multiple of $\big\|b\big\|_{BMO}$.

Proof. Using the inequality of Theorem 7.4.4, we obtain for functions g, with $|g|^\delta$ locally integrable,

$$\big|\{M_d(|g|^\delta)^{\frac{1}{\delta}} > 2^{\frac{1}{\delta}}\lambda\} \cap \{M^\#_\delta(g) \le \gamma\lambda\}\big| \le 2^n\gamma^\delta\big|\{M_d(|g|^\delta)^{\frac{1}{\delta}} > \lambda\}\big| \tag{7.5.10}$$

for all $\lambda, \gamma, \delta > 0$. Then a repetition of the proof of Theorem 7.4.5 yields the second inequality:

$$\big\|M(|g|^\delta)^{\frac{1}{\delta}}\big\|_{L^p} \le C_n\big\|M_d(|g|^\delta)^{\frac{1}{\delta}}\big\|_{L^p} \le C_n(p)\big\|M^\#_\delta(g)\big\|_{L^p} \tag{7.5.11}$$

for all $p \in (p_0, \infty)$, provided $M_d(|g|^\delta)^{\frac{1}{\delta}} \in L^{p_0}(\mathbf{R}^n)$ for some $p_0 > 0$.

For the following argument, it is convenient to replace b by the bounded function

$$b_k(x) = \begin{cases} k & \text{if } b(x) < k, \\ b(x) & \text{if } -k \le b(x) \le k, \\ -k & \text{if } b(x) > -k, \end{cases}$$

which satisfies $\big\|b_k\big\|_{BMO} \le \big\|b\big\|_{BMO}$ for any $k > 0$; see Exercise 7.1.4.

For given $1 < p < \infty$, select p_0 such that $1 < p_0 < p$. Given a smooth function with compact support f, we note that the function $b_k f$ lies in L^{p_0}; thus $T(b_k f)$ also lies in L^{p_0}. Likewise, $b_k T(f)$ also lies in L^{p_0}. Since M_δ is bounded on L^{p_0} for $0 < \delta < 1$, we conclude that

$$\big\|M_\delta([b_k,T](f))\big\|_{L^{p_0}} \le C_\delta\big(\big\|M_\delta(b_k T(f))\big\|_{L^{p_0}} + \big\|M_\delta(T(b_k f))\big\|_{L^{p_0}}\big) < \infty.$$

This allows us to obtain (7.5.11) with $g = [b_k,T](f)$. We now turn to Lemma 7.5.5, in which we pick $0 < \delta < \varepsilon < 1$. Taking L^p norms on both sides of (7.5.6) and using (7.5.11) with $g = [b_k,T](f)$ and the boundedness of M_ε, T, and M^2 on $L^p(\mathbf{R}^n)$, we deduce the a priori estimate (7.5.9) for smooth functions with compact support f and the truncated *BMO* functions b_k.

The Lebesgue dominated convergence theorem gives that $b_k \to b$ in L^2 of every compact set and, in particular, in $L^2(\text{supp} f)$. It follows that $b_k f \to bf$ in L^2 and therefore $T(b_k f) \to T(bf)$ in L^2 by the boundedness of T on L^2. We deduce that for some subsequence of integers k_j, $T(b_{k_j}f) \to T(bf)$ a.e. For this subsequence we have $[b_{k_j},T](f) \to [b,T](f)$ a.e. Letting $j \to \infty$ and using Fatou's lemma, we deduce that (7.5.9) holds for all *BMO* functions b and smooth functions f with compact support.

Since smooth functions with compact support are dense in L^p, it follows that the commutator admits a bounded extension on L^p that satisfies (7.5.9). □

We refer to Exercise 7.5.4 for an analogue of Theorem 7.5.6 when $p = 1$.

Exercises

7.5.1. Use Jensen's inequality to show that M is pointwise controlled by $M_{L\log(1+L)}$.

7.5.2. (a) (*Young's inequality for Orlicz spaces*) Let φ be a continuous, real-valued, strictly increasing function defined on $[0,\infty)$ such that $\varphi(0) = 0$ and $\lim_{t\to\infty} \varphi(t) = \infty$. Let $\psi = \varphi^{-1}$ and for $x \in [0,\infty)$ define

$$\Phi(x) = \int_0^x \varphi(t)\,dt\,, \qquad \Psi(x) = \int_0^x \psi(t)\,dt\,.$$

Show that for $s,t \in [0,\infty)$ we have

$$st \le \Phi(s) + \Psi(t)\,.$$

(b) (cf. Exercise 4.2.3) Choose a suitable function φ in part (a) to deduce for s,t in $[0,\infty)$ the inequality

$$st \le (t+1)\log(t+1) - t + e^s - s - 1 \le t\log(t+1) + e^s - 1\,.$$

(c) (*Hölder's inequality for Orlicz spaces*) Deduce the inequality

$$\big|\langle f,g\rangle\big| \le 2\big\|f\big\|_{\Phi(L)}\big\|g\big\|_{\Psi(L)}\,.$$

[*Hint:* Give a geometric proof distinguishing the cases $t > \varphi(s)$ and $t \le \varphi(s)$. Use that for $u \ge 0$ we have $\int_0^u \varphi(t)\,dt + \int_0^{\varphi(u)} \psi(s)\,ds = u\varphi(u)$.]

7.5.3. Let T be as in Lemma 7.5.5. Show that there is a constant $C_n < \infty$ such that for all $f \in L^p(\mathbf{R}^n)$ and $g \in L^{p'}(\mathbf{R}^n)$ we have

$$\big\|T(f)g - fT^t(g)\big\|_{H^1(\mathbf{R}^n)} \le C\big\|f\big\|_{L^p(\mathbf{R}^n)}\big\|g\big\|_{L^{p'}(\mathbf{R}^n)}\,.$$

In other words, show that the bilinear operator $(f,g) \mapsto T(f)g - fT^t(g)$ maps $L^p(\mathbf{R}^n) \times L^{p'}(\mathbf{R}^n)$ to $H^1(\mathbf{R}^n)$.

7.5.4. (*Pérez [260]*) Let $\Phi(t) = t\log(1+t)$. Then there exists a positive constant C, depending on the *BMO* constant of b, such that for any smooth function with compact support f the following is valid:

$$\sup_{\alpha>0}\frac{1}{\Phi(\frac{1}{\alpha})}\big|\{\big|[b,T](f)\big|>\alpha\}\big|\le C\sup_{\alpha>0}\frac{1}{\Phi(\frac{1}{\alpha})}\big|\{M^2(f)>\alpha\}\big|.$$

7.5.5. Let R_1, R_2 be the Riesz transforms in $\mathbf{R}^2$. Show that there is a constant $C<\infty$ such that for all square integrable functions g_1, g_2 on $\mathbf{R}^2$ the following is valid:

$$\big\|R_1(g_1)R_2(g_2)-R_1(g_2)R_2(g_1)\big\|_{H^1}\le C_p\big\|g_1\big\|_{L^2}\big\|g_2\big\|_{L^2}.$$

[*Hint:* Consider the pairing $\langle g_1,R_2([b,R_1](g_2))-R_1([b,R_2](g_2))\rangle$ with $b\in BMO$.]

7.5.6. (*Coifman, Lions, Meyer, and Semmes [78]*) Use Exercise 7.5.5 to prove that the Jacobian J_f of a map $f=(f_1,f_2):\mathbf{R}^2\to\mathbf{R}^2$,

$$J_f=\det\begin{pmatrix}\partial_1 f_1 & \partial_2 f_1\\ \partial_1 f_2 & \partial_2 f_2\end{pmatrix},$$

lies in $H^1(\mathbf{R}^2)$ whenever $f_1,f_2\in \dot L^2_1(\mathbf{R}^2)$.
[*Hint:* Set $g_j=\Delta^{1/2}(f_j)$.]

7.5.7. Let $\Phi(t)=t(1+\log^+ t)^\alpha$, where $0\le\alpha<\infty$. Let T be a linear (or sublinear) operator that maps $L^{p_0}(\mathbf{R}^n)$ to $L^{p_0,\infty}(\mathbf{R}^n)$ with norm B for some $1<p_0\le\infty$ and also satisfies the following *weak type Orlicz estimate*: for all functions f in $\Phi(L)$,

$$|\{x\in\mathbf{R}^n:\ |T(f)(x)|>\lambda\}|\le A\int_{\mathbf{R}^n}\Phi\Big(\frac{|f(x)|}{\lambda}\Big)\,dx,$$

for some $A<\infty$ and all $\lambda>0$. Prove that T is bounded from $L^p(\mathbf{R}^n)$ to itself, whenever $1<p<p_0$.
[*Hint:* Set $f^\lambda=f\chi_{|f|>\lambda}$ and $f_\lambda=f-f^\lambda$. When $p_0<\infty$, estimate $|\{|T(f)|>2\lambda\}|$ by $|\{|T(f^\lambda)|>\lambda\}|+|\{|T(f_\lambda)|>\lambda\}|\le A\int_{|f|>\lambda}\Phi(\frac{|f(x)|}{\lambda})dx+B^{p_0}\int_{|f|\le\lambda}\frac{|f(x)|^{p_0}}{\lambda^{p_0}}dx$. Multiply by p, integrate with respect to the measure $\lambda^{p-1}d\lambda$ from 0 to infinity, apply Fubini's theorem, and use that $\int_0^1\Phi(1/\lambda)\lambda^{p-1}\,d\lambda<\infty$ to deduce that T maps L^p to $L^{p,\infty}$. When $p_0=\infty$, use that $|\{|T(f)|>2B\lambda\}|\le|\{|T(f^\lambda)|>B\lambda\}|$ and argue as in the case $p_0<\infty$. Boundedness from L^p to L^p follows by applying Theorem 1.3.2.]

HISTORICAL NOTES

The space of functions of bounded mean oscillation first appeared in the work of John and Nirenberg [177] in the context of nonlinear partial differential equations that arise in the study of minimal surfaces. Theorem 7.1.6 was obtained by John and Nirenberg [177]. The relationship of *BMO* functions and Carleson measures is due to Fefferman and Stein [130]. For a variety of issues relating *BMO* to complex function theory one may consult the book of Garnett [142]. The duality of H^1 and *BMO* (Theorem 7.2.2) was announced by Fefferman in [124], but its first proof appeared in the article of Fefferman and Stein [130]. This article actually contains two proofs of this result. The proof of Theorem 7.2.2 is based on the atomic decomposition of H^1, which was obtained subsequently. An alternative proof of the duality between H^1 and *BMO* was given by Carleson [57]. Dyadic *BMO* (Exercise 7.4.4) in relation to *BMO* is studied in Garnett and Jones [144]. The same authors studied the distance in *BMO* to L^∞ in [143].

Carleson measures first appeared in the work of Carleson [53] and [54]. Corollary 7.3.6 was first proved by Carleson, but the proof given here is due to Stein. The characterization of Carleson measures in Theorem 7.3.8 was obtained by Carleson [53]. A theory of balayage for studying *BMO* was developed by Varopoulos [323]. The space *BMO* can also be characterized in terms Carleson measures via Theorem 7.3.8. The converse of Theorem 7.3.8 (see Fefferman and Stein [130]) states that if the function Ψ satisfies a nondegeneracy condition and $|f * \Psi_t|^2 \frac{dx\,dt}{t}$ is a Carleson measure, then f must be a *BMO* function. We refer to Stein [292] (page 159) for a proof of this fact, which uses a duality idea related to tent spaces. The latter were introduced by Coifman, Meyer, and Stein [83] to systematically study the connection between square functions and Carleson measures.

The sharp maximal function was introduced by Fefferman and Stein [130], who first used it to prove Theorem 7.4.5 and derive interpolation for analytic families of operators when one endpoint space is *BMO*. Theorem 7.4.7 provides the main idea why L^∞ can be replaced by *BMO* in this context. The fact that L^2-bounded singular integrals also map L^∞ to *BMO* was independently obtained by Peetre [254], Spanne [286], and Stein [290]. Peetre [254] also observed that translation-invariant singular integrals (such as the ones in Corollary 7.4.10) actually map *BMO* to itself. Another interesting property of *BMO* is that it is preserved under the action of the Hardy–Littlewood maximal operator. This was proved by Bennett, DeVore, and Sharpley [19]; see also the almost simultaneous proof of Chiarenza and Frasca [60]. The decomposition of open sets given in Proposition 7.3.4 is due to Whitney [331].

An alternative characterization of *BMO* can be obtained in terms of commutators of singular integrals. Precisely, we have that the commutator $[b,T](f)$ is L^p bounded for $1 < p < \infty$ if and only if the function b is in *BMO*. The sufficiency of this result (Theorem 7.5.6) is due to Coifman, Rochberg, and Weiss [85], who used it to extend the classical theory of H^p spaces to higher dimensions. The necessity was obtained by Janson [176], who also obtained a simpler proof of the sufficiency. The exposition in Section 7.5 is based on the article of Pérez [260]. This approach is not the shortest available, but the information derived in Lemma 7.5.5 is often useful; for instance, it is used in the substitute of the weak type $(1,1)$ estimate of Exercise 7.5.4. The inequality (7.5.3) in Lemma 7.5.4 can be reversed as shown by Pérez and Wheeden [263]. Weighted L^p estimates for the commutator in terms of the double iteration of the Hardy–Littlewood maximal operator can be deduced as a consequence of Lemma 7.5.5; see the article of Pérez [261].

Orlicz spaces were introduced by Birbaum and Orlicz [26] and furher elaborated by Orlicz [251], [252]. For a modern treatment one may consult the book of Rao and Ren [269]. Bounded mean oscillation with Orlicz norms was considered by Strömberg [297].

The space of functions of vanishing mean oscillation (*VMO*) was introduced by Sarason [277] as the set of integrable functions f on $\mathbf{T}^1$ satisfying $\lim_{\delta\to 0}\sup_{I:\,|I|\le\delta}|I|^{-1}\int_I|f-\mathrm{Avg}_I f|\,dx=0$. This space is the closure in the *BMO* norm of the subspace of $BMO(\mathbf{T}^1)$ consisting of all uniformly continuous functions on $\mathbf{T}^1$. One may define $VMO(\mathbf{R}^n)$ as the space of functions on $\mathbf{R}^n$ that satisfy $\lim_{\delta\to 0}\sup_{Q:\,|Q|\le\delta}|Q|^{-1}\int_Q|f-\mathrm{Avg}_Q f|\,dx=0$, $\lim_{N\to\infty}\sup_{Q:\,\ell(Q)\ge N}|Q|^{-1}\int_Q|f-\mathrm{Avg}_Q f|\,dx=0$, and $\lim_{R\to\infty}\sup_{Q:\,Q\cap B(0,R)=\emptyset}|Q|^{-1}\int_Q|f-\mathrm{Avg}_Q f|\,dx=0$; here I denotes intervals in $\mathbf{T}^1$ and Q cubes in $\mathbf{R}^n$. Then $VMO(\mathbf{R}^n)$ is the closure of the the space of continuous functions that vanish at infinity in the $BMO(\mathbf{R}^n)$ norm. One of the imporant features of $VMO(\mathbf{R}^n)$ is that it is the predual of $H^1(\mathbf{R}^n)$, as was shown by Coifman and Weiss [86]. As a companion to Corollary 7.4.10, singular integral operators can be shown to map the space of continuous functions that vanish at infinity into *VMO*. We refer to the article of Dafni [101] for a short and elegant exposition of these results as well as for a local version of the *VMO*-H^1 duality.

Chapter 8
Singular Integrals of Nonconvolution Type

Up to this point we have studied singular integrals given by convolution with certain tempered distributions. These operators commute with translations. We are now ready to broaden our perspective and study a class of more general singular integrals that are not necessarily translation invariant. Such operators appear in many places in harmonic analysis and partial differential equations. For instance, a large class of pseudodifferential operators falls under the scope of this theory.

This broader point of view does not necessarily bring additional complications in the development of the subject except at one point, the study of L^2 boundedness, where Fourier transform techniques are lacking. The L^2 boundedness of convolution operators is easily understood via a careful examination of the Fourier transform of the kernel, but for nonconvolution operators different tools are required in this study. The main result of this chapter is the derivation of a set of necessary and sufficient conditions for nonconvolution singular integrals to be L^2 bounded. This result is referred to as the $T(1)$ theorem and owes its name to a condition expressed in terms of the action of the operator T on the function 1.

An extension of the $T(1)$ theorem, called the $T(b)$ theorem, is obtained in Section 8.6 and is used to deduce the L^2 boundedness of the Cauchy integral along Lipschitz curves. A variant of the $T(b)$ theorem is also used in the boundedness of the square root of a divergence form elliptic operator discussed in Section 8.7.

8.1 General Background and the Role of *BMO*

We begin by recalling the notion of the adjoint and transpose operator. One may choose to work with either a real or a complex inner product on pairs of functions. For f,g complex-valued functions with integrable product, we denote the real inner product by

$$\langle f,g\rangle = \int_{\mathbf{R}^n} f(x)g(x)\,dx.$$

L. Grafakos, *Modern Fourier Analysis*, DOI: 10.1007/978-0-387-09434-2_8,

This notation is suitable when we think of f as a distribution acting on a test function g. We also have the complex inner product

$$\langle f \mid g \rangle = \int_{\mathbf{R}^n} f(x)\overline{g(x)}\,dx,$$

which is an appropriate notation when we think of f and g as elements of a Hilbert space over the complex numbers. Now suppose that T is a linear operator bounded on L^p. Then the *adjoint* operator T^* of T is uniquely defined via the identity

$$\langle T(f) \mid g \rangle = \langle f \mid T^*(g) \rangle$$

for all f in L^p and g in $L^{p'}$. The *transpose* operator T^t of T is uniquely defined via the identity

$$\langle T(f), g \rangle = \langle f, T^t(g) \rangle = \langle T^t(g), f \rangle$$

for all functions f in L^p and g in $L^{p'}$. The name *transpose* comes from matrix theory, where if A^t denotes the transpose of a complex $n \times n$ matrix A, then we have the identity

$$\langle Ax, y \rangle = \sum_{j=1}^{n} (Ax)_j y_j = Ax \cdot y = x \cdot A^t y = \sum_{j=1}^{n} x_j (A^t y)_j = \langle x, A^t y \rangle$$

for all column vectors $x = (x_1, \dots, x_n)$, $y = (y_1, \dots, y_n)$ in $\mathbf{C}^n$. We may easily check the following intimate relationship between the transpose and the adjoint of a linear operator T:

$$T^*(f) = \overline{T^t(\overline{f})},$$

indicating that they have almost interchangeable use. However, in many cases, it is convenient to avoid complex conjugates and work with the transpose operator for simplicity. Observe that if a linear operator T has kernel $K(x,y)$, that is,

$$T(f)(x) = \int K(x,y) f(y)\,dy,$$

then the kernel of T^t is $K^t(x,y) = K(y,x)$ and that of T^* is $K^*(x,y) = \overline{K(y,x)}$.

An operator is called *self-adjoint* if $T = T^*$ and *self-transpose* if $T = T^t$. For example, the operator iH, where H is the Hilbert transform, is self-adjoint but not self-transpose, and the operator with kernel $i(x+y)^{-1}$ is self-transpose but not self-adjoint.

8.1.1 Standard Kernels

The singular integrals we study in this chapter have kernels that satisfy size and regularity properties similar to those encountered in Chapter 4 for convolution-type

Calderón–Zygmund operators. Let us be specific and introduce the relevant background. We consider functions $K(x,y)$ defined on $\mathbf{R}^n \times \mathbf{R}^n \setminus \{(x,x) : x \in \mathbf{R}^n\}$ that satisfy for some $A > 0$ the size condition

$$|K(x,y)| \le \frac{A}{|x-y|^n} \tag{8.1.1}$$

and for some $\delta > 0$ the regularity conditions

$$|K(x,y) - K(x',y)| \le \frac{A\,|x-x'|^\delta}{(|x-y|+|x'-y|)^{n+\delta}}, \tag{8.1.2}$$

whenever $|x-x'| \le \frac{1}{2}\max\big(|x-y|,|x'-y|\big)$ and

$$|K(x,y) - K(x,y')| \le \frac{A\,|y-y'|^\delta}{(|x-y|+|x-y'|)^{n+\delta}}, \tag{8.1.3}$$

whenever $|y-y'| \le \frac{1}{2}\max\big(|x-y|,|x-y'|\big)$.

Remark 8.1.1. Observe that if

$$|x-x'| \le \frac{1}{2}\max\big(|x-y|,|x'-y|\big),$$

then

$$\max\big(|x-y|,|x'-y|\big) \le 2\min\big(|x-y|,|x'-y|\big),$$

implying that the numbers $|x-y|$ and $|x'-y|$ are comparable. This fact is useful in specific calculations.

Another important observation is that if (8.1.1) holds and we have

$$|\nabla_x K(x,y)| + |\nabla_y K(x,y)| \le \frac{A}{|x-y|^{n+1}}$$

for all $x \neq y$, then K is in $SK(1,4^{n+1}A)$.

Definition 8.1.2. Functions on $\mathbf{R}^n \times \mathbf{R}^n \setminus \{(x,x) : x \in \mathbf{R}^n\}$ that satisfy (8.1.1), (8.1.2), and (8.1.3) are called *standard kernels* with constants δ, A. The class of all standard kernels with constants δ, A is denoted by $SK(\delta,A)$. Given a kernel $K(x,y)$ in $SK(\delta,A)$, we observe that the functions $K(y,x)$ and $\overline{K(y,x)}$ are also in $SK(\delta,A)$. These functions have special names. The function

$$K^t(x,y) = K(y,x)$$

is called the *transpose kernel* of K, and the function

$$K^*(x,y) = \overline{K(y,x)}$$

is called the *adjoint kernel* of K.

Example 8.1.3. The function $K(x,y) = |x-y|^{-n}$ defined away from the diagonal of $\mathbf{R}^n \times \mathbf{R}^n$ is in $SK(1, n4^{n+1})$. Indeed, for

$$|x-x'| \le \frac{1}{2}\max\big(|x-y|, |x'-y|\big)$$

the mean value theorem gives

$$\big|\,|x-y|^{-n} - |x'-y|^{-n}\big| \le \frac{n|x-x'|}{|\theta-y|^{n+1}}$$

for some θ that lies on the line segment joining x and x'. But then we have $|\theta - y| \ge \frac{1}{2}\max\big(|x-y|, |x'-y|\big)$, which gives (8.1.2) with $A = n4^{n+1}$.

Remark 8.1.4. The previous example can be modified to give that if $K(x,y)$ satisfies

$$|\nabla_x K(x,y)| \le A'|x-y|^{-n-1}$$

for all $x \neq y$ in $\mathbf{R}^n$, then $K(x,y)$ also satisfies (8.1.2) with $\delta = 1$ and A controlled by a constant multiple of A'. Likewise, if

$$|\nabla_y K(x,y)| \le A'|x-y|^{-n-1}$$

for all $x \neq y$ in $\mathbf{R}^n$, then $K(x,y)$ satisfies (8.1.3) with with $\delta = 1$ and A bounded by a multiple of A'.

We are interested in standard kernels K that can be extended to tempered distributions on $\mathbf{R}^n \times \mathbf{R}^n$. We begin by observing that given a standard kernel $K(x,y)$, there may not exist a tempered distribution W on $\mathbf{R}^n \times \mathbf{R}^n$ that coincides with the given $K(x,y)$ on $\mathbf{R}^n \times \mathbf{R}^n \setminus \{(x,x) : x \in \mathbf{R}^n\}$. For example, the function $K(x,y) = |x-y|^{-n}$ does not admit such an extension; see Exercise 8.1.2.

We are concerned with kernels $K(x,y)$ in $SK(\delta, A)$ for which there are tempered distributions W on $\mathbf{R}^n \times \mathbf{R}^n$ that coincide with K on $\mathbf{R}^n \times \mathbf{R}^n \setminus \{(x,x) : x \in \mathbf{R}^n\}$. This means that the convergent integral representation

$$\langle W, F\rangle = \int_{\mathbf{R}^n}\int_{\mathbf{R}^n} K(x,y)F(x,y)\,dx\,dy \tag{8.1.4}$$

is valid whenever the Schwartz function F on $\mathbf{R}^n \times \mathbf{R}^n$ is supported away from the diagonal $\{(x,x) : x \in \mathbf{R}^n\}$. Note that the integral in (8.1.4) is well defined and absolutely convergent whenever F is a Schwartz function that vanishes in a neighborhood of the set $\{(x,x) : x \in \mathbf{R}^n\}$. Also observe that there may be several distributions W coinciding with a fixed function $K(x,y)$. In fact, if W is such a distribution, then so is $W + \delta_{x=y}$, where $\delta_{x=y}$ denotes Lebesgue measure on the diagonal of $\mathbf{R}^{2n}$. (This is some sort of a Dirac distribution.)

We now consider continuous linear operators

$$T : \mathscr{S}(\mathbf{R}^n) \to \mathscr{S}'(\mathbf{R}^n)$$

from the space of Schwartz functions $\mathscr{S}(\mathbf{R}^n)$ to the space of all tempered distributions $\mathscr{S}'(\mathbf{R}^n)$. By the *Schwartz kernel theorem* (see Hörmander [168, p. 129]), for such an operator T there is a distribution W in $\mathscr{S}'(\mathbf{R}^{2n})$ that satisfies

$$\langle T(f),\varphi\rangle = \langle W, f\otimes\varphi\rangle \qquad \text{when} \quad f,\varphi\in\mathscr{S}(\mathbf{R}^n), \tag{8.1.5}$$

where $(f\otimes\varphi)(x,y)=f(x)\varphi(y)$. Furthermore, as a consequence of the same theorem, there exist constants C,N,M such that for all $f,g\in\mathscr{S}(\mathbf{R}^n)$ we have

$$|\langle T(f),g\rangle| = |\langle W,f\otimes g\rangle| \le C\Bigg[\sum_{|\alpha|,|\beta|\le N}\rho_{\alpha,\beta}(f)\Bigg]\Bigg[\sum_{|\alpha|,|\beta|\le M}\rho_{\alpha,\beta}(g)\Bigg], \tag{8.1.6}$$

where $\rho_{\alpha,\beta}(\varphi)=\sup_{x\in\mathbf{R}^n}|\partial_x^\alpha(x^\beta\varphi)(x)|$ is the set of seminorms for the topology in $\mathscr{S}$. A distribution W that satisfies (8.1.5) and (8.1.6) is called a *Schwartz kernel.*

We study continuous linear operators $T:\mathscr{S}(\mathbf{R}^n)\to\mathscr{S}'(\mathbf{R}^n)$ whose Schwartz kernels coincide with standard kernels $K(x,y)$ on $\mathbf{R}^n\times\mathbf{R}^n\setminus\{(x,x):\,x\in\mathbf{R}^n\}$. This means that (8.1.5) admits the absolutely convergent integral representation

$$\langle T(f),\varphi\rangle = \int_{\mathbf{R}^n}\int_{\mathbf{R}^n}K(x,y)f(y)\varphi(x)\,dx\,dy \tag{8.1.7}$$

whenever f and φ are Schwartz functions whose supports do not intersect.

We make some remarks concerning duality in this context. Given a continuous linear operator $T:\ \mathscr{S}(\mathbf{R}^n)\to\mathscr{S}'(\mathbf{R}^n)$ with a Schwartz kernel W, we can define another distribution W^t as follows:

$$\langle W^t,F\rangle = \langle W,F^t\rangle,$$

where $F^t(x,y)=F(y,x)$. This means that for all $f,\varphi\in\mathscr{S}(\mathbf{R}^n)$ we have

$$\langle W,f\otimes\varphi\rangle = \langle W^t,\varphi\otimes f\rangle.$$

It is a simple fact that the transpose operator T^t of T, which satisfies

$$\langle T(\varphi),f\rangle = \langle T^t(f),\varphi\rangle \tag{8.1.8}$$

for all f,φ in $\mathscr{S}(\mathbf{R}^n)$, is the unique continuous linear operator from $\mathscr{S}(\mathbf{R}^n)$ to $\mathscr{S}'(\mathbf{R}^n)$ whose Schwartz kernel is the distribution W^t, that is, we have

$$\langle T^t(f),\varphi\rangle = \langle T(\varphi),f\rangle = \langle W,\varphi\otimes f\rangle = \langle W^t,f\otimes\varphi\rangle. \tag{8.1.9}$$

We now observe that a large class of standard kernels admits extensions to tempered distributions W on $\mathbf{R}^{2n}$.

Example 8.1.5. Suppose that $K(x,y)$ satisfies (8.1.1) and (8.1.2) and is *antisymmetric*, in the sense that

$$K(x,y) = -K(y,x)$$

for all $x \neq y$ in $\mathbf{R}^n$. Then K also satisfies (8.1.3), and moreover, there is a distribution W on $\mathbf{R}^{2n}$ that extends K on $\mathbf{R}^n \times \mathbf{R}^n$.

Indeed, define

$$\langle W, F\rangle = \lim_{\varepsilon \to 0} \iint\limits_{|x-y|>\varepsilon} K(x,y)F(x,y)\,dy\,dx \tag{8.1.10}$$

for all F in the Schwartz class of $\mathbf{R}^{2n}$. In view of antisymmetry, we may write

$$\iint\limits_{|x-y|>\varepsilon} K(x,y)F(x,y)\,dy\,dx = \frac{1}{2}\iint\limits_{|x-y|>\varepsilon} K(x,y)\big(F(x,y)-F(y,x)\big)\,dy\,dx.$$

Using (8.1.1), the observation that

$$|F(x,y)-F(y,x)| \le \frac{2\,|x-y|}{(1+|x|^2+|y|^2)^{n+1}} \sup_{(x,y)\in\mathbf{R}^{2n}} \left|\nabla_{x,y}\Big((1+|x|^2+|y|^2)^{n+1}F(x,y)\Big)\right|,$$

and the fact that the preceding supremum is controlled by a finite sum of Schwartz seminorms of F, it follows that the limit in (8.1.10) exists and gives a tempered distribution on $\mathbf{R}^{2n}$. We can therefore define an operator $T : \mathscr{S}(\mathbf{R}^n) \to \mathscr{S}'(\mathbf{R}^n)$ with kernel W as follows:

$$\langle T(f),\varphi\rangle = \lim_{\varepsilon \to 0} \iint\limits_{|x-y|>\varepsilon} K(x,y)f(x)\varphi(y)\,dy\,dx.$$

Example 8.1.6. Let A be a Lipschitz function on $\mathbf{R}$. This means that it satisfies the estimate $|A(x)-A(y)| \le L|x-y|$ for some $L<\infty$ and all $x,y\in\mathbf{R}$. For $x,y\in\mathbf{R}$, $x\neq y$, we let

$$K(x,y) = \frac{1}{x-y+i(A(x)-A(y))} \tag{8.1.11}$$

and we observe that $K(x,y)$ is a standard kernel in $SK(1,4+4L)$. The details are left to the reader. Note that the kernel K defined in (8.1.11) is antisymmetric.

Example 8.1.7. Let the function A be as in the previous example. For each integer $m\ge 1$ we set

$$K_m(x,y) = \left(\frac{A(x)-A(y)}{x-y}\right)^m \frac{1}{x-y}, \qquad x,y\in\mathbf{R}. \tag{8.1.12}$$

Clearly, K_m is an antisymmetric function. To see that each K_m is a standard kernel, we use the simple fact that

$$\max\big(|\nabla_x K_m(x,y)|, |\nabla_y K_m(x,y)|\big) \le \frac{(2m+1)L^m}{|x-y|^2}$$

and the observation made in Remark 8.1.1. It follows that K_m lies in $SK(\delta,C)$ with $\delta = 1$ and $C = 16(2m+1)L^m$. The linear operator with kernel $(\pi i)^{-1}K_m$ is called the mth *Calderón commutator*.

8.1.2 Operators Associated with Standard Kernels

Having introduced standard kernels, we are in a position to define linear operators associated with them.

Definition 8.1.8. Let $0 < \delta, A < \infty$ and K in $SK(\delta,A)$. A continuous linear operator T from $\mathscr{S}(\mathbf{R}^n)$ to $\mathscr{S}'(\mathbf{R}^n)$ is said to be *associated with* K if it satisfies

$$T(f)(x) = \int_{\mathbf{R}^n} K(x,y)f(y)\,dy \tag{8.1.13}$$

for all $f \in \mathscr{C}_0^\infty$ and x not in the support of f. If T is associated with K, then the Schwartz kernel W of T coincides with K on $\mathbf{R}^n \times \mathbf{R}^n \setminus \{(x,x) : x \in \mathbf{R}^n\}$.

If T is associated with K and admits a bounded extension on $L^2(\mathbf{R}^n)$, that is, it satisfies

$$\big\|T(f)\big\|_{L^2} \le B\big\|f\big\|_{L^2} \tag{8.1.14}$$

for all $f \in \mathscr{S}(\mathbf{R}^n)$, then T is called a *Calderón–Zygmund operator* associated with the standard kernel K. In this case we use the same notation for the L^2 extension.

In the sequel we denote by $CZO(\delta,A,B)$ the class of all Calderón–Zygmund operators associated with standard kernels in $SK(\delta,A)$ that admit L^2 bounded extensions with norm at most B.

We make the point that there may be several Calderón–Zygmund operators associated with a given standard kernel K. For instance, we may check that the zero operator and the identity operator have the same kernel $K(x,y) = 0$. We investigate connections between any two such operators in Proposition 8.1.11. Next we discuss the important fact that once an operator T admits an extension that is L^2 bounded, then (8.1.13) holds for all f that are bounded and compactly supported whenever x does not lie in its support.

Proposition 8.1.9. *Let T be an element of $CZO(\delta,A,B)$ associated with a standard kernel K. Then for all f in L^∞ with compact support and every $x \notin \operatorname{supp} f$ we have the absolutely convergent integral representation*

$$T(f)(x) = \int_{\mathbf{R}^n} K(x,y)f(y)\,dy. \tag{8.1.15}$$

Proof. Identity (8.1.15) can be deduced from the fact that whenever f and φ are bounded and compactly supported functions that satisfy

$$\operatorname{dist}(\operatorname{supp}\varphi, \operatorname{supp} f) > 0, \tag{8.1.16}$$

then we have the integral representation

$$\int_{\mathbf{R}^n} T(f)(x)\,\varphi(x)\,dx = \int_{\mathbf{R}^n}\int_{\mathbf{R}^n} K(x,y)f(y)\varphi(x)\,dy\,dx. \tag{8.1.17}$$

To see this, given f and φ as previously, select $f_j, \varphi_j \in \mathscr{C}_0^\infty$ such that φ_j are uniformly bounded and supported in a small neighborhood of the support of φ, $\varphi_j \to \varphi$ in L^2 and almost everywhere, $f_j \to f$ in L^2 and almost everywhere, and

$$\operatorname{dist}(\operatorname{supp}\varphi_j, \operatorname{supp} f_j) \geq \frac{1}{2}\operatorname{dist}(\operatorname{supp}\varphi, \operatorname{supp} f) > 0$$

for all j. Because of (8.1.7), identity (8.1.17) is valid for the functions f_j and φ_j in place of f and φ. By the boundedness of T, it follows that $T(f_j)$ converges to $T(f)$ in L^2 and thus

$$\int_{\mathbf{R}^n} T(f_j)(x)\varphi_j(x)\,dx \to \int_{\mathbf{R}^n} T(f)(x)\varphi(x)\,dx.$$

Now write $f_j\varphi_j - f\varphi = (f_j - f)\varphi_j + f(\varphi_j - \varphi)$ and observe that

$$\int_{\mathbf{R}^n}\int_{\mathbf{R}^n} K(x,y)f(y)(\varphi_j(x) - \varphi(x))\,dy\,dx \to 0,$$

since it is controlled by a multiple of $\big\|T(f)\big\|_{L^2}\big\|\varphi_j - \varphi\big\|_{L^2}$, while

$$\int_{\mathbf{R}^n}\int_{\mathbf{R}^n} K(x,y)(f_j(y) - f(y))\varphi_j(x)\,dy\,dx \to 0,$$

since it is controlled by a multiple of $\sup_j \big\|T^t(\varphi_j)\big\|_{L^2}\big\|f_j - f\big\|_{L^2}$. This gives that

$$\int_{\mathbf{R}^n}\int_{\mathbf{R}^n} K(x,y)f_j(y)\varphi_j(x)\,dy\,dx \to \int_{\mathbf{R}^n}\int_{\mathbf{R}^n} K(x,y)f(y)\varphi(x)\,dy\,dx$$

as $j \to \infty$, which proves the validity of (8.1.17). □

We now define truncated kernels and operators.

Definition 8.1.10. Given a kernel K in $SK(\delta,A)$ and $\varepsilon > 0$, we define the *truncated kernel*

$$K^{(\varepsilon)}(x,y) = K(x,y)\chi_{|x-y|>\varepsilon}\,.$$

Given a continuous linear operator T from $\mathscr{S}(\mathbf{R}^n)$ to $\mathscr{S}'(\mathbf{R}^n)$ and $\varepsilon > 0$, we define the *truncated operator* $T^{(\varepsilon)}$ by

$$T^{(\varepsilon)}(f)(x) = \int_{\mathbf{R}^n} K^{(\varepsilon)}(x,y)\,f(y)\,dy$$

and the *maximal singular operator* associated with T as follows:

$$T^{(*)}(f)(x) = \sup_{\varepsilon>0}\big|T^{(\varepsilon)}(f)(x)\big|\,.$$

Note that both $T^{(\varepsilon)}$ and $T^{(*)}$ are well defined for f in $\bigcup_{1\le p<\infty} L^p(\mathbf{R}^n)$.

We investigate a certain connection between the boundedness of T and the boundedness of the family $\{T^{(\varepsilon)}\}_{\varepsilon>0}$ uniformly in $\varepsilon>0$.

Proposition 8.1.11. *Let K be a kernel in $SK(\delta,A)$ and let T in $CZO(\delta,A,B)$ be associated with K. For $\varepsilon>0$, let $T^{(\varepsilon)}$ be the truncated operators obtained from T. Assume that there exists a constant $B'<\infty$ such that*

$$\sup_{\varepsilon>0} \big\|T^{(\varepsilon)}\big\|_{L^2\to L^2} \le B'. \tag{8.1.18}$$

Then there exists a linear operator T_0 defined on $L^2(\mathbf{R}^n)$ such that

(1) The Schwartz kernel of T_0 coincides with K on

$$\mathbf{R}^n\times\mathbf{R}^n\setminus\{(x,x):\, x\in\mathbf{R}^n\}.$$

(2) For some subsequence $\varepsilon_j\downarrow 0$, we have

$$\int_{\mathbf{R}^n} T^{(\varepsilon_j)}(f)(x)g(x)\,dx \to \int_{\mathbf{R}^n} (T_0 f)(x)g(x)\,dx$$

as $j\to\infty$ for all f,g in $L^2(\mathbf{R}^n)$.
(3) T_0 is bounded on $L^2(\mathbf{R}^n)$ with norm

$$\big\|T_0\big\|_{L^2\to L^2}\le B'.$$

(4) There exists a measurable function b on $\mathbf{R}^n$ with $\big\|b\big\|_{L^\infty}\le B+B'$ such that

$$T(f)-T_0(f)=bf,$$

for all $f\in L^2(\mathbf{R}^n)$.

Proof. Consider the Banach space $X=\mathscr{B}(L^2,L^2)$ of all bounded linear operators from $L^2(\mathbf{R}^n)$ to itself. Then X is isomorphic to $\mathscr{B}((L^2)^*,(L^2)^*)^*$, which is a dual space. Since the unit ball of a dual space is weak* compact, and the operators $T^{(\varepsilon)}$ lie in a multiple of this unit ball, the Banach–Alaoglu theorem gives the existence of a sequence $\varepsilon_j\downarrow 0$ such that $T^{(\varepsilon_j)}$ converges to some T_0 in the weak* topology of $\mathscr{B}(L^2,L^2)$ as $j\to\infty$. This means that

$$\int_{\mathbf{R}^n} T^{(\varepsilon_j)}(f)(x)g(x)\,dx \to \int_{\mathbf{R}^n} T_0(f)(x)g(x)\,dx \tag{8.1.19}$$

for all f,g in $L^2(\mathbf{R}^n)$ as $j\to\infty$. This proves (2). The L^2 boundedness of T_0 is a consequence of (8.1.19), hypothesis (8.1.18), and duality, since

$$\big\|T_0(f)\big\|_{L^2}\le \sup_{\|g\|_{L^2}\le 1}\ \limsup_{j\to\infty}\left|\int_{\mathbf{R}^n} T^{(\varepsilon_j)}(f)(x)g(x)\,dx\right| \le B'\big\|f\big\|_{L^2}.$$

This proves (3). Finally, (1) is a consequence of the integral representation

$$\int_{\mathbf{R}^n} T^{(\varepsilon_j)}(f)(x)g(x)\,dx = \int_{\mathbf{R}^n}\int_{\mathbf{R}^n} K^{(\varepsilon_j)}(x,y)f(y)\,dy\,g(x)\,dx,$$

whenever f, g are Schwartz functions with disjoint supports, by letting $j \to \infty$.

We finally prove (4). We first observe that if g is a bounded function with compact support and Q is an open cube in $\mathbf{R}^n$, we have

$$(T^{(\varepsilon)} - T)(g\chi_Q)(x) = \chi_Q(x)\,(T^{(\varepsilon)} - T)(g)(x)\,, \tag{8.1.20}$$

whenever $x \notin \partial Q$ and ε is small enough. Indeed, take first $x \notin \overline{Q}$; then x is not in the support of $g\chi_Q$. Note that since $g\chi_Q$ is bounded and has compact support, we can use the integral representation formula (8.1.15) obtained in Proposition 8.1.9. Then we have that for $\varepsilon < \operatorname{dist}(x, \operatorname{supp} g\chi_Q)$, the left-hand side in (8.1.20) is zero. Moreover, for $x \in Q$, we have that x does not lie in the support of $g\chi_{Q^c}$, and again because of (8.1.15) we obtain $(T^{(\varepsilon)} - T)(g\chi_{Q^c})(x) = 0$ whenever $\varepsilon < \operatorname{dist}(x, \operatorname{supp} g\chi_{Q^c})$. This proves (8.1.20) for all x not in the boundary ∂Q of Q. Taking weak limits in (8.1.20) as $\varepsilon \to 0$, we obtain that

$$(T_0 - T)(g\chi_Q) = \chi_Q\,(T_0 - T)(g) \qquad \text{a.e.} \tag{8.1.21}$$

for all open cubes Q in $\mathbf{R}^n$. By linearity we extend (8.1.21) to simple functions. Using the fact that $T_0 - T$ is L^2 bounded and a simple density argument, we obtain

$$(T_0 - T)(gf) = f\,(T_0 - T)(g) \qquad \text{a.e.} \tag{8.1.22}$$

whenever f is in L^2 and g is bounded and has compact support. If $B(0, j)$ is the open ball with center 0 and radius j on $\mathbf{R}^n$, when $j \le j'$ we have

$$(T_0 - T)(\chi_{B(0,j)}) = (T_0 - T)(\chi_{B(0,j)}\chi_{B(0,j')}) = \chi_{B(0,j)}\,(T_0 - T)(\chi_{B(0,j')})\,.$$

Therefore, the sequence of functions $(T_0 - T)(\chi_{B(0,j)})$ satisfies the "consistency" property

$$(T_0 - T)(\chi_{B(0,j)}) = (T_0 - T)(\chi_{B(0,j')}) \quad \text{in } B(0, j)$$

when $j \le j'$. It follows that there exists a well defined function b such that

$$b = (T_0 - T)(\chi_{B(0,j)}) \quad \text{a.e. in } B(0, j)\,.$$

Applying (8.1.22) with f supported in $B(0, j)$ and $g = \chi_{B(0,j)}$, we obtain

$$(T_0 - T)(f) = (T_0 - T)(f\chi_{B(0,j)}) = f\,(T_0 - T)(\chi_{B(0,j)}) = f\,b \qquad \text{a.e.},$$

from which it follows that $(T_0 - T)(f) = bf$ for all $f \in L^2$. Since the norm of $T - T_0$ on L^2 is at most $B + B'$, it follows that the norm of the linear map $f \mapsto bf$ from L^2 to itself is at most $B + B'$. From this we obtain that $\big\|b\big\|_{L^\infty} \le B + B'$. □

Remark 8.1.12. We show in the next section (cf. Corollary 8.2.4) that if a Calderón–Zygmund operator maps L^2 to L^2, then so do all of its truncations $T^{(\varepsilon)}$ uniformly in $\varepsilon > 0$. By Proposition 8.1.11, there exists a linear operator T_0 that has the form

$$T_0(f)(x) = \lim_{j\to\infty} \int_{|x-y|>\varepsilon_j} K(x,y)f(y)\,dy,$$

where the limit is taken in the weak topology of L^2, so that T is equal to T_0 plus a bounded function times the identity operator.

We give a special name to operators of this form.

Definition 8.1.13. Suppose that for a given T in $CZO(\delta,A,B)$ there is a sequence ε_j of positive numbers that tends to zero as $j \to \infty$ such that for all $f \in L^2(\mathbf{R}^n)$,

$$T^{(\varepsilon_j)}(f) \to T(f)$$

weakly in L^2. Then T is called a *Calderón–Zygmund singular integral operator.* Thus Calderón–Zygmund singular integral operators are special kinds of Calderón–Zygmund operators. The subclass of $CZO(\delta,A,B)$ consisting of all Calderón–Zygmund singular integral operators is denoted by $CZSIO(\delta,A,B)$.

In view of Proposition 8.1.11 and Remark 8.1.12, a Calderón–Zygmund operator is equal to a Calderón–Zygmund singular integral operator plus a bounded function times the identity operator. For this reason, the study of Calderón–Zygmund operators is equivalent to the study of Calderón–Zygmund singular integral operators, and in almost all situations it suffices to restrict attention to the latter.

8.1.3 Calderón–Zygmund Operators Acting on Bounded Functions

We are now interested in defining the action of a Calderón–Zygmund operator T on bounded and smooth functions. To achieve this we first need to define the space of special test functions $\mathscr{D}_0$.

Definition 8.1.14. Recall the space $\mathscr{D}(\mathbf{R}^n) = \mathscr{C}_0^\infty(\mathbf{R}^n)$ of all smooth functions with compact support on $\mathbf{R}^n$. We define $\mathscr{D}_0(\mathbf{R}^n)$ to be the space of all smooth functions with compact support and integral zero. We equip $\mathscr{D}_0(\mathbf{R}^n)$ with the same topology as the space $\mathscr{D}(\mathbf{R}^n)$ (cf. Definition 2.3.1). The dual space of $\mathscr{D}_0(\mathbf{R}^n)$ under this topology is denoted by $\mathscr{D}_0'(\mathbf{R}^n)$. This is a space of distributions larger than $\mathscr{D}'(\mathbf{R}^n)$.

Example 8.1.15. *BMO* functions are examples of elements of $\mathscr{D}_0'(\mathbf{R}^n)$. Indeed, given $b \in BMO(\mathbf{R}^n)$, for any compact set K there is a constant $C_K = \big\|b\big\|_{L^1(K)}$ such that

$$\left|\int_{\mathbf{R}^n} b(x)\varphi(x)\,dx\right| \le C_K \big\|\varphi\big\|_{L^\infty}$$

for any $\varphi \in \mathscr{D}_0(\mathbf{R}^n)$. Moreover, observe that the preceding integral remains unchanged if the BMO function b is replaced by $b+c$, where c is a constant.

Definition 8.1.16. Let T be a continuous linear operator from $\mathscr{S}(\mathbf{R}^n)$ to $\mathscr{S}'(\mathbf{R}^n)$ that satisfies (8.1.5) for some distribution W that coincides with a standard kernel $K(x,y)$ satisfying (8.1.1), (8.1.2), and (8.1.3). Given f bounded and smooth, we define an element $T(f)$ of $\mathscr{D}_0'(\mathbf{R}^n)$ as follows: For a given φ in $\mathscr{D}_0(\mathbf{R}^n)$, select η in $\mathscr{C}_0^\infty$ with $0 \le \eta \le 1$ and equal to 1 in a neighborhood of the support of φ. Since T maps $\mathscr{S}$ to $\mathscr{S}'$, the expression $T(f\eta)$ is a tempered distribution, and its action on φ is well defined. We define the action of $T(f)$ on φ via

$$\langle T(f),\varphi\rangle = \langle T(f\eta),\varphi\rangle + \int_{\mathbf{R}^n}\left[\int_{\mathbf{R}^n} K(x,y)\varphi(x)\,dx\right] f(y)(1-\eta(y))\,dy, \quad (8.1.23)$$

provided we make sense of the double integral as an absolutely convergent integral. To do this, we pick x_0 in the support of φ and we split the y-integral in (8.1.23) into the sum of integrals over the regions $I_0 = \{y \in \mathbf{R}^n : |x-x_0| > \frac{1}{2}|x_0-y|\}$ and $I_\infty = \{y \in \mathbf{R}^n : |x-x_0| \le \frac{1}{2}|x_0-y|\}$. By the choice of η we must necessarily have $\operatorname{dist}(\operatorname{supp}\eta, \operatorname{supp}\varphi) > 0$, and hence the part of the double integral in (8.1.23) when y is restricted to I_0 is absolutely convergent in view of (8.1.1). For $y \in I_\infty$ we use the mean value property of φ to write the expression inside the square brackets in (8.1.23) as

$$\int_{\mathbf{R}^n} \big(K(x,y) - K(x_0,y)\big)\,\varphi(x)\,dx.$$

With the aid of (8.1.2) we deduce the absolute convergence of the double integral in (8.1.23) as follows:

$$\begin{aligned}
&\iint_{|y-x_0|\ge 2|x-x_0|} |K(x,y)-K(x_0,y)|\,|\varphi(x)|\,(1-\eta(y))\,|f(y)|\,dx\,dy \\
&\qquad \le \int_{\mathbf{R}^n} A|x-x_0|^\delta \int_{|y-x_0|\ge 2|x-x_0|} |x_0-y|^{-n-\delta}\,|f(y)|\,dy\,|\varphi(x)|\,dx \\
&\qquad \le A\,\frac{\omega_{n-1}}{\delta 2^\delta}\,\big\|\varphi\big\|_{L^1}\,\big\|f\big\|_{L^\infty} < \infty.
\end{aligned}$$

This completes the definition of $T(f)$ as an element of $\mathscr{D}_0'$ when $f \in \mathscr{C}^\infty \cap L^\infty$ but leaves two points open. We need to show that this definition is independent of η and secondly that whenever f is a Schwartz function, the distribution $T(f)$ defined in (8.1.23) coincides with the original element of $\mathscr{S}'(\mathbf{R}^n)$ given in Definition 8.1.8.

Remark 8.1.17. We show that the definition of $T(f)$ is independent of the choice of the function η. Indeed, if ζ is another function satisfying $0 \le \zeta \le 1$ that is also equal to 1 in a neighborhood of the support of φ, then $f(\eta-\zeta)$ and φ have disjoint supports, and by (8.1.7) we have the absolutely convergent integral realization

$$\langle T(f(\eta-\zeta)),\varphi\rangle = \int_{\mathbf{R}^n}\int_{\mathbf{R}^n} K(x,y)f(y)(\eta-\zeta)(y)\,dy\,\varphi(x)\,dx.$$

It follows that the expression in (8.1.23) coincides with the corresponding expression obtained when η is replaced by ζ.

Next, if f is a Schwartz function, then both ηf and $(1-\eta)f$ are Schwartz functions; by the linearity of T one has $\langle T(f),\varphi\rangle = \langle T(\eta f),\varphi\rangle + \langle T((1-\eta)f),\varphi\rangle$, and by (8.1.7) the second expression can be written as the double absolutely convergent integral in (8.1.23), since φ and $(1-\eta)f$ have disjoint supports. Thus the distribution $T(f)$ defined in (8.1.23) coincides with the original element of $\mathscr{S}'(\mathbf{R}^n)$ given in Definition 8.1.8.

Remark 8.1.18. When T has a bounded extension that maps L^2 to itself, we may define $T(f)$ for all $f\in L^\infty(\mathbf{R}^n)$, not necessarily smooth. Simply observe that under this assumption, the expression $T(f\eta)$ is a well defined L^2 function and thus

$$\langle T(f\eta),\varphi\rangle = \int_{\mathbf{R}^n} T(f\eta)(x)\varphi(x)\,dx$$

is given by an absolutely convergent integral for all $\varphi\in\mathscr{D}_0$.

Finally, observe that although $\langle T(f),\varphi\rangle$ is defined for f in L^∞ and φ in $\mathscr{D}_0$, this definition is valid for all square integrable functions φ with compact support and integral zero; indeed, the smoothness of φ was never an issue in the definition of $\langle T(f),\varphi\rangle$.

In summary, if T is a Calderón–Zygmund operator and f lies in $L^\infty(\mathbf{R}^n)$, then $T(f)$ has a well defined *action* $\langle T(f),\varphi\rangle$ on square integrable functions φ with compact support and integral zero. This action satisfies

$$\left|\langle T(f),\varphi\rangle\right| \le \left\|T(f\eta)\right\|_{L^2}\left\|\varphi\right\|_{L^2} + C_{n,\delta}A\left\|\varphi\right\|_{L^1}\left\|f\right\|_{L^\infty} < \infty. \tag{8.1.24}$$

In the next section we show that in this case, $T(f)$ is in fact an element of *BMO*.

Exercises

8.1.1. Suppose that K is a function defined away from the diagonal on $\mathbf{R}^n\times\mathbf{R}^n$ that satisfies for some $\delta>0$ the condition

$$|K(x,y)-K(x',y)| \le A'\frac{|x-x'|^\delta}{|x-y|^{n+\delta}}$$

whenever $|x-x'|\le\frac{1}{2}|x-y|$. Prove that K satisfies (8.1.2) with constant $A = (\frac{5}{2})^{n+\delta}A'$. Obtain an analogous statement for condition (8.1.3).

8.1.2. Prove that there does not exist a tempered distribution W on $\mathbf{R}^{2n}$ that extends the function $|x-y|^{-n}$ defined on $\mathbf{R}^{2n}\setminus\{(x,x):\ x\in\mathbf{R}^n\}$.

[*Hint:* Apply such a distribution to a positive smooth bump that does not vanish at the origin.]

8.1.3. Let $\varphi(x)$ be a smooth radial function that is equal to 1 when $|x| \ge 1$ and vanishes when $|x| \le \frac{1}{2}$. Prove that if K lies in $SK(\delta,A)$, then all the smooth truncations $K_\varphi^{(\varepsilon)}(x,y) = K(x,y)\varphi(\frac{x-y}{\varepsilon})$ lie in $SK(\delta,cA)$ for some $c>0$ independent of $\varepsilon>0$.

8.1.4. Suppose that A is a Lipschitz map from $\mathbf{R}^n$ to $\mathbf{R}^m$. This means that there exists a constant L such that $|A(x)-A(y)| \le L|x-y|$ for all $x,y \in \mathbf{R}^n$. Suppose that F is a $\mathscr{C}^\infty$ odd function defined on $\mathbf{R}^m$. Show that the kernel

$$K(x,y) = \frac{1}{|x-y|^n} F\left(\frac{A(x)-A(y)}{|x-y|}\right)$$

is in $SK(1,C)$ for some $C>0$.

8.1.5. Extend the result of Proposition 8.1.11 to the case that the space L^2 is replaced by L^q for some $1<q<\infty$.

8.1.6. Observe that for an operator T as in Definition 8.1.16, the condition $T(1)=0$ is equivalent to the statement that for all φ smooth with compact support and integral zero we have $\int_{\mathbf{R}^n} T^t(\varphi)(x)\,dx = 0$. A similar statement holds for T^t.

8.1.7. Suppose that $K(x,y)$ is continuous, bounded, and nonnegative on $\mathbf{R}^n \times \mathbf{R}^n$ and satisfies $\int_{\mathbf{R}^n} K(x,y)\,dy = 1$ for all $x \in \mathbf{R}^n$. Define a linear operator T by setting $T(f)(x) = \int_{\mathbf{R}^n} K(x,y) f(y)\,dy$ for $f \in L^1(\mathbf{R}^n)$.
(a) Suppose that h is a continuous and integrable function on $\mathbf{R}^n$ that has a global minimum [i.e., there exists $x_0 \in \mathbf{R}^n$ such that $h(x_0) \le h(x)$ for all $x \in \mathbf{R}^n$]. If we have

$$T(h)(x) = h(x)$$

for all $x \in \mathbf{R}^n$, prove that h is a constant function.
(b) Show that T preserves the set of integrable functions that are bounded below by a fixed constant.
(c) Suppose that $T(T(f)) = f$ for some everywhere positive and continuous function f on $\mathbf{R}^n$. Show that $T(f) = f$.
[*Hint:* Part (c): Let $L(x,y)$ be the kernel of $T \circ T$. Show that

$$\int_{\mathbf{R}^n} L(x,y) \frac{f(y)}{f(x)} \frac{T(f)(y)}{f(y)}\,dy = \frac{T(f)(x)}{f(x)}$$

and conclude by part (a) that $\frac{T(f)(y)}{f(y)}$ is a constant.]

8.2 Consequences of L^2 Boundedness

Calderón–Zygmund singular integral operators admit L^2 bounded extensions. As in the case of convolution operators, L^2 boundedness has several consequences. In this section we are concerned with consequences of the L^2 boundedness of Calderón–Zygmund singular integral operators. Throughout the entire discussion, we assume that $K(x,y)$ is a kernel defined away from the diagonal in $\mathbf{R}^{2n}$ that satisfies the standard size and regularity conditions (8.1.1), (8.1.2), and (8.1.3). These conditions may be relaxed; see the exercises at the end of this section.

8.2.1 Weak Type $(1,1)$ and L^p Boundedness of Singular Integrals

We begin by proving that operators in $CZO(\delta,A,B)$ are bounded from L^1 to weak L^1. This result is completely analogous to that in Theorem 4.3.3.

Theorem 8.2.1. *Assume that $K(x,y)$ is in $SK(\delta,A)$ and let T be an element of $CZO(\delta,A,B)$ associated with the kernel K. Then T has a bounded extension that maps $L^1(\mathbf{R}^n)$ to $L^{1,\infty}(\mathbf{R}^n)$ with norm*

$$\big\|T\big\|_{L^1\to L^{1,\infty}} \le C_n(A+B),$$

and also maps $L^p(\mathbf{R}^n)$ to itself for $1<p<\infty$ with norm

$$\big\|T\big\|_{L^p\to L^p} \le C_n \max(p,(p-1)^{-1})(A+B),$$

where C_n is a dimensional constant.

Proof. The proof of this theorem is a reprise of the argument of the proof of Theorem (4.3.3). Fix $\alpha>0$ and let f be in $L^1(\mathbf{R}^n)$. Since $T(f)$ may not be defined when f is a general integrable function, we take f to be a Schwartz class function. Once we obtain a weak type $(1,1)$ estimate for Schwartz functions, it is only a matter of density to extend it to all f in L^1.

We apply the Calderón–Zygmund decomposition to f at height $\gamma\alpha$, where γ is a positive constant to be chosen later. Write $f=g+b$, where $b=\sum_j b_j$ and conditions (1)–(6) of Theorem 4.3.1 are satisfied with the constant α replaced by $\gamma\alpha$. Since we are assuming that f is Schwartz function, it follows that each bad function b_j is bounded and compactly supported. Thus $T(b_j)$ is an L^2 function, and when x is not in the support of b_j we have the integral representation

$$T(b_j)(x) = \int_{Q_j} b_j(y)K(x,y)\,dy$$

in view of Proposition 8.1.9.

As usual, we denote by $\ell(Q)$ the side length of a cube Q. Let Q_j^* be the unique cube with sides parallel to the axes having the same center as Q_j and having side length

$$\ell(Q_j^*) = 2\sqrt{n}\,\ell(Q_j)\,.$$

We have

$$
\begin{aligned}
&|\{x\in\mathbf{R}^n : |T(f)(x)|>\alpha\}| \\
&\quad\le \left|\left\{x\in\mathbf{R}^n :\ |T(g)(x)|>\frac{\alpha}{2}\right\}\right| + \left|\left\{x\in\mathbf{R}^n :\ |T(b)(x)|>\frac{\alpha}{2}\right\}\right| \\
&\quad\le \frac{2^2}{\alpha^2}\big\|T(g)\big\|_{L^2}^2 + \Big|\bigcup_j Q_j^*\Big| + \left|\left\{x\notin\bigcup_j Q_j^* :\ |T(b)(x)|>\frac{\alpha}{2}\right\}\right| \\
&\quad\le \frac{2^2}{\alpha^2}B^2\big\|g\big\|_{L^2}^2 + \sum_j |Q_j^*| + \frac{2}{\alpha}\int_{(\bigcup_j Q_j^*)^c} |T(b)(x)|\,dx \\
&\quad\le \frac{2^2}{\alpha^2}2^nB^2(\gamma\alpha)\big\|f\big\|_{L^1} + (2\sqrt{n})^n\frac{\big\|f\big\|_{L^1}}{\gamma\alpha} + \frac{2}{\alpha}\sum_j\int_{(Q_j^*)^c}|T(b_j)(x)|\,dx \\
&\quad\le \left(\frac{(2^{n+1}B\gamma)^2}{2^n\gamma}+\frac{(2\sqrt{n})^n}{\gamma}\right)\frac{\big\|f\big\|_{L^1}}{\alpha} + \frac{2}{\alpha}\sum_j\int_{(Q_j^*)^c}|T(b_j)(x)|\,dx.
\end{aligned}
$$

It suffices to show that the last sum is bounded by some constant multiple of $\big\|f\big\|_{L^1}$. Let y_j be the center of the cube Q_j. For $x\in(Q_j^*)^c$, we have $|x-y_j|\ge\frac{1}{2}\ell(Q_j^*)=\sqrt{n}\ell(Q_j)$. But if $y\in Q_j$ we have $|y-y_j|\le\sqrt{n}\ell(Q_j)/2$; thus $|y-y_j|\le\frac{1}{2}|x-y_j|$, since the diameter of a cube is equal to $\sqrt{n}$ times its side length. We now estimate the last displayed sum as follows:

$$
\begin{aligned}
\sum_j\int_{(Q_j^*)^c}|T(b_j)(x)|\,dx &= \sum_j\int_{(Q_j^*)^c}\left|\int_{Q_j} b_j(y)K(x,y)\,dy\right|dx \\
&= \sum_j\int_{(Q_j^*)^c}\left|\int_{Q_j} b_j(y)\big(K(x,y)-K(x,y_j)\big)\,dy\right|dx \\
&\le \sum_j\int_{Q_j}|b_j(y)|\int_{(Q_j^*)^c}|K(x,y)-K(x,y_j)|\,dx\,dy \\
&\le \sum_j\int_{Q_j}|b_j(y)|\int_{|x-y_j|\ge 2|y-y_j|}|K(x,y)-K(x,y_j)|\,dx\,dy \\
&\le A_2\sum_j\int_{Q_j}|b_j(y)|\,dy \\
&= A_2\sum_j\big\|b_j\big\|_{L^1} \\
&\le A_2 2^{n+1}\big\|f\big\|_{L^1}.
\end{aligned}
$$

Combining the facts proved and choosing $\gamma=B^{-1}$, we deduce a weak type $(1,1)$ estimate for $T(f)$ when f is in the Schwartz class. We obtain that T has a bounded extension from L^1 to $L^{1,\infty}$ with bound at most $C_n(A+B)$. The L^p result for $1<p<2$ follows by interpolation and Exercise 1.3.2. The result for $2<p<\infty$ follows by duality; one uses here that the dual operator T^t has a kernel $K^t(x,y)=K(y,x)$ that satisfies the same estimates as K, and by the result just proved, it is also bounded on

L^p for $1 < p < 2$ with norm at most $C_n(A+B)$. Thus T must be bounded on L^p for $2 < p < \infty$ with norm at most a constant multiple of $A+B$. $\square$

Consequently, for operators T in $CZO(\delta,A,B)$ and L^p functions f, $1 \le p < \infty$, the expressions $T(f)$ make sense as L^p (or $L^{1,\infty}$ when $p = 1$) functions. The following result addresses the question whether these functions can be expressed as integrals.

Proposition 8.2.2. *Let T be an operator in $CZO(\delta,A,B)$ associated with a kernel K. Then for $g \in L^p(\mathbf{R}^n)$, $1 \le p < \infty$, the following absolutely convergent integral representation is valid:*

$$T(g)(x) = \int_{\mathbf{R}^n} K(x,y)\,g(y)\,dy \tag{8.2.1}$$

for almost all $x \in \mathbf{R}^n \setminus$ supp g, *provided that* supp $g \subsetneq \mathbf{R}^n$.

Proof. Set $g_k(x) = g(x)\chi_{|g(x)|\le k}\chi_{|x|\le k}$. These are L^p functions with compact support that is contained in the support of g. Also, the g_k converge to g in L^p as $k \to \infty$. In view of Proposition 8.1.9, for every k we have

$$T(g_k)(x) = \int_{\mathbf{R}^n} K(x,y)\,g_k(y)\,dy$$

for all $x \in \mathbf{R}^n \setminus \operatorname{supp} g$. Since T maps L^p to L^p (or to weak L^1 when $p = 1$), it follows that $T(g_k)$ converges to $T(g)$ in weak L^p and hence in measure. By Proposition 1.1.9, a subsequence of $T(g_k)$ converges to $T(g)$ almost everywhere. On the other hand, for $x \in \mathbf{R}^n \setminus \operatorname{supp} g$ we have

$$\int_{\mathbf{R}^n} K(x,y)\,g_k(y)\,dy \to \int_{\mathbf{R}^n} K(x,y)\,g(y)\,dy$$

when $k \to \infty$, since the absolute value of the difference is bounded by $B\big\|g_k - g\big\|_{L^p}$, which tends to zero. The constant B is the $L^{p'}$ norm of the function $|x-y|^{-n-\delta}$ on the support of g; one has $|x-y| \ge c > 0$ for all y in the support of g and thus $B < \infty$. Therefore $T(g_k)(x)$ converges a.e. to both sides of the identity (8.2.1) for x not in the support of g. This concludes the proof of this identity. $\square$

8.2.2 Boundedness of Maximal Singular Integrals

We pose the question whether there is an analogous boundedness result to Theorem 8.2.1 concerning the maximal singular integral operator $T^{(*)}$. We note that given f in $L^p(\mathbf{R}^n)$ for some $1 \le p < \infty$, the expression $T^{(*)}(f)(x)$ is well defined for all $x \in \mathbf{R}^n$. This is a simple consequence of estimate (8.1.1) and Hölder's inequality.

Theorem 8.2.3. *Let K be in $SK(\delta,A)$ and T in $CZO(\delta,A,B)$ be associated with K. Let $r \in (0,1)$. Then there is a constant $C(n,r)$ such that*

$$|T^{(*)}(f)(x)| \le C(n,r)\Big[M(|T(f)|^r)(x)^{\frac{1}{r}} + (A+B)M(f)(x)\Big] \tag{8.2.2}$$

is valid for all functions in $\bigcup_{1\le p<\infty} L^p(\mathbf{R}^n)$. *Also, there exist dimensional constants* C_n, C_n' *such that*

$$\big\|T^{(*)}(f)\big\|_{L^{1,\infty}(\mathbf{R}^n)} \le C_n'(A+B)\big\|f\big\|_{L^1(\mathbf{R}^n)}, \tag{8.2.3}$$

$$\big\|T^{(*)}(f)\big\|_{L^p(\mathbf{R}^n)} \le C_n(A+B)\max(p,(p-1)^{-1})\big\|f\big\|_{L^p(\mathbf{R}^n)}, \tag{8.2.4}$$

for all $1\le p<\infty$ *and all* f *in* $L^p(\mathbf{R}^n)$.

Estimate (8.2.2) is referred to as *Cotlar's inequality*.

Proof. We fix r so that $0<r<1$ and $f\in L^p(\mathbf{R}^n)$ for some p satisfying $1\le p<\infty$. To prove (8.2.2), we also fix $\varepsilon>0$ and we set $f_0^{\varepsilon,x} = f\chi_{B(x,\varepsilon)}$ and $f_\infty^{\varepsilon,x} = f\chi_{B(x,\varepsilon)^c}$. Since $x\notin \operatorname{supp} f_\infty^{\varepsilon,x}$ whenever $|x-y|\ge\varepsilon$, using Proposition 8.2.2 we can write

$$T(f_\infty^{\varepsilon,x})(x) = \int_{\mathbf{R}^n} K(x,y)\,f_\infty^{\varepsilon,x}(y)\,dy = \int_{|x-y|\ge\varepsilon} K(x,y)\,f(y)\,dy = T^{(\varepsilon)}(f)(x)\,.$$

In view of (8.1.2), for $z\in B(x,\frac{\varepsilon}{2})$ we have $|z-x|\le\frac12|x-y|$ whenever $|x-y|\ge\varepsilon$ and thus

$$\begin{aligned}
|T(f_\infty^{\varepsilon,x})(x)-T(f_\infty^{\varepsilon,x})(z)| &= \bigg|\int_{|x-y|\ge\varepsilon}\big(K(z,y)-K(x,y)\big)f(y)\,dy\bigg| \\
&\le |z-x|^\delta \int_{|x-y|\ge\varepsilon}\frac{A\,|f(y)|}{(|x-y|+|y-z|)^{n+\delta}}\,dy \\
&\le \Big(\frac{\varepsilon}{2}\Big)^\delta \int_{|x-y|\ge\varepsilon}\frac{A\,|f(y)|}{(|x-y|+\varepsilon/2)^{n+\delta}}\,dy \\
&\le C_{n,\delta}\,A\,M(f)(x),
\end{aligned}$$

where the last estimate is a consequence of Theorem 2.1.10. We conclude that for all $z\in B(x,\frac{\varepsilon}{2})$ we have

$$\begin{aligned}
|T^{(\varepsilon)}(f)(x)| &= |T(f_\infty^{\varepsilon,x})(x)| \\
&\le |T(f_\infty^{\varepsilon})(x)-T(f_\infty^{\varepsilon,x})(z)| + |T(f_\infty^{\varepsilon,x})(z)| \\
&\le C_{n,\delta}\,A\,M(f)(x) + |T(f_0^{\varepsilon,x})(z)| + |T(f)(z)|\,.
\end{aligned} \tag{8.2.5}$$

For $0<r<1$ it follows from (8.2.5) that for $z\in B(x,\frac{\varepsilon}{2})$ we have

$$|T^{(\varepsilon)}(f)(x)|^r \le C_{n,\delta}^r A^r M(f)(x)^r + |T(f_0^{\varepsilon,x})(z)|^r + |T(f)(z)|^r\,. \tag{8.2.6}$$

Integrating over $z\in B(x,\frac{\varepsilon}{2})$, dividing by $|B(x,\frac{\varepsilon}{2})|$, and raising to the power $\frac1r$, we obtain

$$|T^{(\varepsilon)}(f)(x)| \le 3^{\frac{1}{r}} \bigg[C_{n,\delta} A M(f)(x) + \bigg(\frac{1}{|B(x,\frac{\varepsilon}{2})|} \int_{B(x,\frac{\varepsilon}{2})} |T(f_0^{\varepsilon,x})(z)|^r dz \bigg)^{\frac{1}{r}} + M(|T(f)|^r)(x)^{\frac{1}{r}} \bigg].$$

Using Exercise 2.1.5, we estimate the middle term on the right-hand side of the preceding equation by

$$\bigg(\frac{1}{|B(x,\frac{\varepsilon}{2})|} \frac{\|T\|^r_{L^1 \to L^{1,\infty}}}{1-r} |B(x,\tfrac{\varepsilon}{2})|^{1-r} \|f_0^{\varepsilon,x}\|^r_{L^1} \bigg)^{\frac{1}{r}} \le C_{n,r}(A+B)M(f)(x).$$

This proves (8.2.2).

We now use estimate (8.2.2) to show that T is L^p bounded and of weak type $(1,1)$. To obtain the weak type $(1,1)$ estimate for $T^{(*)}$ we need to use that the Hardy–Littlewood maximal operator maps $L^{p,\infty}$ to $L^{p,\infty}$ for all $1 < p < \infty$. See Exercise 2.1.13. We also use the trivial fact that for all $0 < p, q < \infty$ we have

$$\big\| |f|^q \big\|_{L^{p,\infty}} = \|f\|^q_{L^{pq,\infty}}.$$

Take any $r < 1$ in (8.2.2). Then we have

$$\begin{aligned} \big\| M(|T(f)|^r)^{\frac{1}{r}} \big\|_{L^{1,\infty}} &= \big\| M(|T(f)|^r) \big\|^{\frac{1}{r}}_{L^{\frac{1}{r},\infty}} \\ &\le C_{n,r} \big\| |T(f)|^r \big\|^{\frac{1}{r}}_{L^{\frac{1}{r},\infty}} \\ &= C_{n,r} \|T(f)\|_{L^{1,\infty}} \\ &\le \widetilde{C}_{n,r}(A+B)\|f\|_{L^1}, \end{aligned}$$

where we used the weak type $(1,1)$ bound for T in the last estimate.

To obtain the L^p boundedness of $T^{(*)}$ for $1 < p < \infty$, we use the same argument as before. We fix $r = \frac{1}{2}$. Recall that the maximal function is bounded on L^{2p} with norm at most $3^{\frac{n}{2p}} \frac{2p}{2p-1} \le 2 \cdot 3^{\frac{n}{2}}$ [see (2.1.5)]. We have

$$\begin{aligned} \big\| M(|T(f)|^{\frac{1}{2}})^2 \big\|_{L^p} &= \big\| M(|T(f)|^{\frac{1}{2}}) \big\|^2_{L^{2p}} \\ &\le \big(3^{\frac{n}{2p}} \tfrac{2p}{2p-1}\big)^2 \big\| |T(f)|^{\frac{1}{2}} \big\|^2_{L^{2p}} \\ &\le 4 \cdot 3^n \|T(f)\|_{L^p} \\ &\le C_n \max(\tfrac{1}{p-1}, p)(A+B)\|f\|_{L^p}, \end{aligned}$$

where we used the L^p boundedness of T in the last estimate. □

We end this section with two corollaries, the first of which confirms a fact mentioned in Remark 8.1.12.

Corollary 8.2.4. *Let K be in $SK(\delta,A)$ and T in $CZO(\delta,A,B)$ be associated with K. Then there exists a dimensional constant C_n such that*

$$\sup_{\varepsilon>0} \|T^{(\varepsilon)}\|_{L^2\to L^2} \le C_n\big(A+\|T\|_{L^2\to L^2}\big).$$

Corollary 8.2.5. *Let K be in $SK(\delta,A)$ and let $T=\lim_{\varepsilon_j\to 0} T^{(\varepsilon_j)}$ be an element of $CZSIO(\delta,A,B)$ associated with K. Then for $1\le p<\infty$ and all $f\in L^p(\mathbf{R}^n)$ we have that*

$$T^{(\varepsilon_j)}(f)\to T(f)$$

almost everywhere.

Proof. Using (8.1.1), (8.1.2), and (8.1.3), we see that the alleged convergence holds (everywhere) for smooth functions with compact support. The general case follows from Theorem 8.2.3 and Theorem 2.1.14. □

8.2.3 $H^1\to L^1$ and $L^\infty\to BMO$ Boundedness of Singular Integrals

Theorem 8.2.6. *Let T be an element of $CZO(\delta,A,B)$. Then T has an extension that maps $H^1(\mathbf{R}^n)$ to $L^1(\mathbf{R}^n)$. Precisely, there is a constant $C_{n,\delta}$ such that*

$$\|T\|_{H^1\to L^1} \le C_{n,\delta}\big(A+\|T\|_{L^2\to L^2}\big).$$

Proof. The proof is analogous to that of Theorem 6.7.1. Let $B=\|T\|_{L^2\to L^2}$. We start by examining the action of T on L^2 atoms for H^1. Let $f=a$ be such an atom, supported in a cube Q. Let c_Q be the center of Q and let $Q^*=2\sqrt{n}\,Q$. We write

$$\int_{\mathbf{R}^n}|T(a)(x)|\,dx=\int_{Q^*}|T(a)(x)|\,dx+\int_{(Q^*)^c}|T(a)(x)|\,dx \tag{8.2.7}$$

and we estimate each term separately. We have

$$\begin{aligned}
\int_{Q^*}|T(a)(x)|\,dx &\le |Q^*|^{\frac12}\Big(\int_{Q^*}|T(a)(x)|^2\,dx\Big)^{\frac12}\\
&\le B|Q^*|^{\frac12}\Big(\int_{Q}|a(x)|^2\,dx\Big)^{\frac12}\\
&\le B|Q^*|^{\frac12}|Q|^{-\frac12}\\
&= C_nB,
\end{aligned}$$

where we used property (b) of atoms in Definition 6.6.8. Now observe that if $x\notin Q^*$ and $y\in Q$, then

$$|y-c_Q|\le\frac12|x-c_Q|;$$

hence $x-y$ stays away from zero and $T(a)(x)$ can be expressed as a convergent integral by Proposition 8.2.2. We have

$$\begin{aligned}
\int_{(Q^*)^c} |T(a)(x)|\,dx &= \int_{(Q^*)^c} \left| \int_Q K(x,y)a(y)\,dy \right| dx \\
&= \int_{(Q^*)^c} \left| \int_Q \big(K(x,y)-K(x,c_Q)\big)a(y)\,dy \right| dx \\
&\le \int_Q \int_{(Q^*)^c} |K(x,y)-K(x,c_Q)|\,dx\,|a(y)|\,dy \\
&\le \int_Q \int_{(Q^*)^c} \frac{A|y-c_Q|^\delta}{|x-c_Q|^{n+\delta}}\,dx\,|a(y)|\,dy \\
&\le C'_{n,\delta} A \int_Q |a(y)|\,dy \\
&\le C'_{n,\delta} A |Q|^{\frac{1}{2}} \big\|a\big\|_{L^2} \\
&\le C'_{n,\delta} A |Q|^{\frac{1}{2}} |Q|^{-\frac{1}{2}} \\
&= C'_{n,\delta} A\,.
\end{aligned}$$

Combining this calculation with the previous one and inserting the final conclusions in (8.2.7), we deduce that L^2 atoms for H^1 satisfy

$$\big\|T(a)\big\|_{L^1} \le C_{n,\delta}\,(A+B)\,. \tag{8.2.8}$$

To pass to general functions in H^1, we use Theorem 6.6.10 to write an $f \in H^1$ as

$$f = \sum_{j=1}^{\infty} \lambda_j a_j\,,$$

where the series converges in H^1, the a_j are L^2 atoms for H^1, and

$$\big\|f\big\|_{H^1} \approx \sum_{j=1}^{\infty} |\lambda_j|\,. \tag{8.2.9}$$

Since T maps L^1 to weak L^1 by Theorem 8.2.1, $T(f)$ is already a well defined $L^{1,\infty}$ function. We plan to prove that

$$T(f) = \sum_{j=1}^{\infty} \lambda_j T(a_j) \qquad \text{a.e.} \tag{8.2.10}$$

Note that the series in (8.2.10) converges in L^1 and defines an integrable function almost everywhere. Once (8.2.10) is established, the required conclusion (6.7.5) follows easily by taking L^1 norms in (8.2.10) and using (8.2.8) and (8.2.9).

To prove (8.2.10), we use that T is of weak type $(1,1)$. For a given $\mu > 0$ we have

$$
\begin{aligned}
&\Big|\Big\{\Big|T(f)-\sum_{j=1}^{\infty}\lambda_j T(a_j)\Big|>\mu\Big\}\Big| \\
&\leq \Big|\Big\{\Big|T(f)-\sum_{j=1}^{N}\lambda_j T(a_j)\Big|>\mu/2\Big\}\Big| + \Big|\Big\{\Big|\sum_{j=N+1}^{\infty}\lambda_j T(a_j)\Big|>\mu/2\Big\}\Big| \\
&\leq \frac{2}{\mu}\big\|T\big\|_{L^1\to L^{1,\infty}}\Big\|f-\sum_{j=1}^{N}\lambda_j a_j\Big\|_{L^1} + \frac{2}{\mu}\Big\|\sum_{j=N+1}^{\infty}\lambda_j T(a_j)\Big\|_{L^1} \\
&\leq \frac{2}{\mu}\big\|T\big\|_{L^1\to L^{1,\infty}}\Big\|f-\sum_{j=1}^{N}\lambda_j a_j\Big\|_{H^1} + \frac{2}{\mu}C_{n,\delta}(A+B)\sum_{j=N+1}^{\infty}|\lambda_j|\,.
\end{aligned}
$$

Since $\sum_{j=1}^{N}\lambda_j a_j$ converges to f in H^1 and $\sum_{j=1}^{\infty}|\lambda_j|<\infty$, both terms in the sum converge to zero as $N\to\infty$. We conclude that

$$
\Big|\Big\{\Big|T(f)-\sum_{j=1}^{\infty}\lambda_j T(a_j)\Big|>\mu\Big\}\Big| = 0
$$

for all $\mu>0$, which implies (8.2.10). □

Theorem 8.2.7. *Let T be in $CZO(\delta,A,B)$. Then for any bounded function f, the distribution $T(f)$ can be identified with a BMO function that satisfies*

$$
\big\|T(f)\big\|_{BMO}\leq C'_{n,\delta}(A+B)\big\|f\big\|_{L^\infty}, \tag{8.2.11}
$$

where $C_{n,\delta}$ is a constant.

Proof. Let $L^2_{0,c}$ be the space of all square integrable functions with compact support and integral zero on $\mathbf{R}^n$. This space is contained in $H^1(\mathbf{R}^n)$ (cf. Exercise 6.4.3) and contains the set of finite sums of L^2 atoms for H^1, which is dense in H^1 (cf. Exercise 6.6.5); thus $L^2_{0,c}$ is dense in H^1. Recall that for $f\in L^\infty$, $T(f)$ has a well defined action $\langle T(f),\varphi\rangle$ on functions φ in $L^2_{0,c}$ that satisfies (8.1.24).

Suppose we have proved the identity

$$
\big\langle T(f),\varphi\big\rangle = \int_{\mathbf{R}^n} T^t(\varphi)(x)f(x)\,dx, \tag{8.2.12}
$$

for all bounded functions f and all φ in $L^2_{0,c}$. Since such a φ is in H^1, Theorem 8.2.6 yields that $T^t(\varphi)$ is in L^1, and consequently, the integral in (8.2.12) converges absolutely. Assuming (8.2.12) and using Theorem 8.2.6 we obtain that

$$
\big|\big\langle T(f),\varphi\big\rangle\big| \leq \big\|T^t(\varphi)\big\|_{L^1}\big\|f\big\|_{L^\infty} \leq C_{n,\delta}(A+B)\,\big\|\varphi\big\|_{H^1}\,\big\|f\big\|_{L^\infty}.
$$

We conclude that $L(\varphi)=\langle T(f),\varphi\rangle$ is a bounded linear functional on $L^2_{0,c}$ with norm at most $C_{n,\delta}(A+B)\,\|f\|_{L^\infty}$. Obviously, L has a bounded extension on H^1 with the same norm. By Theorem 7.2.2 there exists a BMO function b_f that satisfies $\|b_f\|_{BMO}\leq C'_n\|L\|_{H^1\to\mathbf{C}}$ such that the linear functional L has the form L_{b_f} (using the notation of Theorem 7.2.2). In other words, the distribution $T(f)$ can be identified with a *BMO* function that satisfies (8.2.11) with $C_{n,\delta} = C_n' C_{n,\delta}$, i.e.,

$$\|T(f)\|_{BMO} \le C_n' C_{n,\delta}(A+B)\,\|f\|_{L^\infty}.$$

We return to the proof of identity (8.2.12). Pick a smooth function with compact support η that satisfies $0 \le \eta \le 1$ and is equal to 1 in a neighborhood of the support of φ. We write the right-hand side of (8.2.12) as

$$\int_{\mathbf{R}^n} T^t(\varphi)\eta f\,dx + \int_{\mathbf{R}^n} T^t(\varphi)(1-\eta) f\,dx = \langle T(\eta f),\varphi\rangle + \int_{\mathbf{R}^n} T^t(\varphi)(1-\eta) f\,dx.$$

In view of Definition 8.1.16, to prove (8.2.12) it will suffice to show that

$$\int_{\mathbf{R}^n} T^t(\varphi)(1-\eta) f\,dx = \int_{\mathbf{R}^n}\int_{\mathbf{R}^n} \big(K(x,y)-K(x_0,y)\big)\varphi(x)\,dx(1-\eta(y))f(y)\,dy,$$

where x_0 lies in the support of φ. But the inner integral above is absolutely convergent and equal to

$$\int_{\mathbf{R}^n} \big(K(x,y)-K(x_0,y)\big)\varphi(x)\,dx = \int_{\mathbf{R}^n} K^t(y,x)\varphi(x)\,dx = T^t(\varphi)(y),$$

since $y \notin \operatorname{supp} \varphi$, by Proposition 8.1.9. Thus (8.2.12) is valid. □

Exercises

8.2.1. Let $T : \mathscr{S}(\mathbf{R}^n) \to \mathscr{S}'(\mathbf{R}^n)$ be a continuous linear operator whose Schwartz kernel coincides with a function $K(x,y)$ on $\mathbf{R}^n \times \mathbf{R}^n$ minus its diagonal. Suppose that the function $K(x,y)$ satisfies

$$\sup_{R>0}\int_{R\le|x-y|\le 2R} |K(x,y)|\,dy \le A < \infty.$$

(a) Show that the previous condition is equivalent to

$$\sup_{R>0}\frac{1}{R}\int_{|x-y|\le R} |x-y|\,|K(x,y)|\,dy \le A' < \infty$$

by proving that $A' \le A \le 2A'$.
(b) For $\varepsilon > 0$, let $T^{(\varepsilon)}$ be the truncated linear operators with kernels $K^{(\varepsilon)}(x,y) = K(x,y)\chi_{|x-y|>\varepsilon}$. Show that $T^{(\varepsilon)}(f)$ is well defined for Schwartz functions.
[*Hint:* Consider the annuli $\varepsilon 2^j \le |x| \le \varepsilon 2^{j+1}$ for $j \ge 0$.]

8.2.2. Let T be as in Exercise 8.2.1. Prove that the limit $T^{(\varepsilon)}(f)(x)$ exists for all f in the Schwartz class and for almost all $x \in \mathbf{R}^n$ as $\varepsilon \to 0$ if and only if the limit

$$\lim_{\varepsilon\to 0}\int_{\varepsilon<|x-y|<1} K(x,y)\,dy$$

exists for almost all $x\in\mathbf{R}^n$.

8.2.3. Let $K(x,y)$ be a function defined away from the diagonal in $\mathbf{R}^{2n}$ that satisfies

$$\sup_{R>0}\int_{R\le|x-y|\le 2R}|K(x,y)|\,dy\le A<\infty$$

and also *Hörmander's condition*

$$\sup_{\substack{y,y'\in\mathbf{R}^n\\ y\neq y'}}\int_{|x-y|\ge 2|y-y'|}|K(x,y)-K(x,y')|\,dx\le A''<\infty.$$

Show that all the truncations $K^{(\varepsilon)}(x,y)$ also satisfy Hörmander's condition uniformly in $\varepsilon>0$ with a constant $A+A$".

8.2.4. Let T be as in Exercise 8.2.1 and assume that T maps $L^r(\mathbf{R}^n)$ to itself for some $1<r\le\infty$.
(a) Assume that $K(x,y)$ satisfies Hörmander's condition, Then T has an extension that maps $L^1(\mathbf{R}^n)$ to $L^{1,\infty}(\mathbf{R}^n)$ with norm

$$\big\|T\big\|_{L^1\to L^{1,\infty}}\le C_n(A+B),$$

and therefore T maps $L^p(\mathbf{R}^n)$ to itself for $1<p<r$ with norm

$$\big\|T\big\|_{L^p\to L^p}\le C_n(p-1)^{-1}(A+B),$$

where C_n is a dimensional constant.
(b) Assuming that $K^t(x,y)=K(y,x)$ satisfies Hörmander's condition, prove that T maps $L^p(\mathbf{R}^n)$ to itself for $r<p<\infty$ with norm

$$\big\|T\big\|_{L^p\to L^p}\le C_n\,p\,(A+B),$$

where C_n is independent of p.

8.2.5. Show that estimate (8.2.2) also holds when $r=1$.
[*Hint:* Estimate (8.2.6) holds when $r=1$. For fixed $\varepsilon>0$, take $0<b<|T^{(\varepsilon)}(f)(x)|$ and define $B_1^\varepsilon(x)=B(x,\frac{\varepsilon}{2})\cap\{|T(f)|>\frac{b}{3}\}$, $B_2^\varepsilon(x)=B(x,\frac{\varepsilon}{2})\cap\{|T(f_0^{\varepsilon,x})|>\frac{b}{3}\}$, and $B_3^\varepsilon(x)=B(x,\frac{\varepsilon}{2})$ if $C_{n,\delta}M(f)(x)>\frac{b}{3}$ and empty otherwise. Then $|B(x,\frac{\varepsilon}{2})|\le|B_1^\varepsilon(x)|+|B_2^\varepsilon(x)|+|B_3^\varepsilon(x)|$. Use the weak type $(1,1)$ property of T to show that $b\le C(n)\big(M(|T(f)|)(x)+M(f)(x)\big)$, and take the supremum over all $b<|T^{(\varepsilon)}(f)(x)|$.]

8.2.6. Prove that if $|f|\log^+|f|$ is integrable over a ball, then $T^{(*)}(f)$ is integrable over the same ball.
[*Hint:* Use the behavior of the norm of $T^{(*)}$ on L^p as $p\to 1$ and use Exercise 1.3.7.]

8.3 The $T(1)$ Theorem

We now turn to one of the main results of this chapter, the so-called $T(1)$ theorem. This theorem gives necessary and sufficient conditions for linear operators T with standard kernels to be bounded on $L^2(\mathbf{R}^n)$. In this section we obtain several such equivalent conditions. The name of theorem $T(1)$ is due to the fact that one of the conditions that we derive is expressed in terms of properties of the distribution $T(1)$, which was introduced in Definition 8.1.16.

8.3.1 Preliminaries and Statement of the Theorem

We begin with some preliminary facts and definitions.

Definition 8.3.1. A *normalized bump* is a smooth function φ supported in the ball $B(0,10)$ that satisfies

$$|(\partial_x^\alpha \varphi)(x)| \le 1$$

for all multi-indices $|\alpha| \le 2\left[\frac{n}{2}\right]+2$, where $[x]$ denotes here the integer part of x.

Observe that every smooth function supported inside the ball $B(0,10)$ is a constant multiple of a normalized bump. Also note that if a normalized bump is supported in a compact subset of $B(0,10)$, then small translations of it are also normalized bumps.

Given a function f on $\mathbf{R}^n$, $R>0$, and $x_0 \in \mathbf{R}^n$, we use the notation f_R to denote the function $f_R(x) = R^{-n} f(R^{-1}x)$ and $\tau^{x_0}(f)$ to denote the function $\tau^{x_0}(f)(x) = f(x-x_0)$. Thus

$$\tau^{x_0}(f_R)(y) = f_R(y-x_0) = R^{-n} f\big(R^{-1}(y-x_0)\big).$$

Set $N = \left[\frac{n}{2}\right]+1$. Using that all derivatives up to order $2N$ of normalized bumps are bounded by 1, we easily deduce that for all $x_0 \in \mathbf{R}^n$, all $R>0$, and all normalized bumps φ we have the estimate

$$\begin{aligned} R^n \int_{\mathbf{R}^n} \big|\widehat{\tau^{x_0}(\varphi_R)}(\xi)\big|\,d\xi & \\ = \int_{\mathbf{R}^n} |\widehat{\varphi}(\xi)|\,d\xi & \\ = \int_{\mathbf{R}^n} \left| \int_{\mathbf{R}^n} \varphi(y) e^{-2\pi i y\cdot\xi}\,dy \right| d\xi & \qquad (8.3.1)\\ = \int_{\mathbf{R}^n} \left| \int_{\mathbf{R}^n} (I-\Delta)^N(\varphi)(y) e^{-2\pi i y\cdot\xi}\,dy \right| \frac{d\xi}{(1+4\pi^2|\xi|^2)^N} & \\ \le C_n, & \end{aligned}$$

since $|(\partial_x^\alpha \varphi)(x)| \le 1$ for all multi-indices α with $|\alpha| \le \left[\frac{n}{2}\right]+1$, and C_n is independent of the bump φ. Here $I-\Delta$ denotes the operator

$$(I-\Delta)(\varphi)=\varphi+\sum_{j=1}^{n}\frac{\partial^2\varphi}{\partial x_j^2}.$$

Definition 8.3.2. We say that a continuous linear operator

$$T:\ \mathscr{S}(\mathbf{R}^n)\to\mathscr{S}'(\mathbf{R}^n)$$

satisfies the *weak boundedness property* (WBP) if there is a constant C such that for all f and g normalized bumps and for all $x_0\in\mathbf{R}^n$ and $R>0$ we have

$$|\langle T(\tau^{x_0}(f_R)),\tau^{x_0}(g_R)\rangle|\le CR^{-n}. \tag{8.3.2}$$

The smallest constant C in (8.3.2) is denoted by $\|T\|_{WB}$.

Note that $\|\tau^{x_0}(f_R)\|_{L^2}=\|f\|_{L^2}R^{-n/2}$ and thus if T has a bounded extension from $L^2(\mathbf{R}^n)$ to itself, then T satisfies the weak boundedness property with bound

$$\|T\|_{WB}\le 10^n v_n\|T\|_{L^2\to L^2},$$

where v_n is the volume of the unit ball in $\mathbf{R}^n$.

We now state one of the main theorems in this chapter.

Theorem 8.3.3. *Let T be a continuous linear operator from $\mathscr{S}(\mathbf{R}^n)$ to $\mathscr{S}'(\mathbf{R}^n)$ whose Schwartz kernel coincides with a function K on $\mathbf{R}^n\times\mathbf{R}^n\setminus\{(x,x):\,x\in\mathbf{R}^n\}$ that satisfies (8.1.1), (8.1.2), and (8.1.3) for some $0<\delta,A<\infty$. Let $K^{(\varepsilon)}$ and $T^{(\varepsilon)}$ be the usual truncated kernel and operator for $\varepsilon>0$. Assume that there exists a sequence $\varepsilon_j\downarrow 0$ such that for all $f,g\in\mathscr{S}(\mathbf{R}^n)$ we have*

$$\langle T^{(\varepsilon_j)}(f),g\rangle\to\langle T(f),g\rangle. \tag{8.3.3}$$

Consider the assertions:

(i) The following statement is valid:

$$B_1=\sup_B\sup_{\varepsilon>0}\left[\frac{\|T^{(\varepsilon)}(\chi_B)\|_{L^2}}{|B|^{\frac12}}+\frac{\|(T^{(\varepsilon)})^t(\chi_B)\|_{L^2}}{|B|^{\frac12}}\right]<\infty,$$

where the first supremum is taken over all balls B in $\mathbf{R}^n$.

(ii) The following statement is valid:

$$B_2=\sup_{\varepsilon,N,x_0}\left[\frac{1}{N^n}\int_{B(x_0,N)}\left|\int_{|x-y|<N}K^{(\varepsilon)}(x,y)\,dy\right|^2dx\right.$$
$$\left.+\frac{1}{N^n}\int_{B(x_0,N)}\left|\int_{|x-y|<N}K^{(\varepsilon)}(y,x)\,dy\right|^2dx\right]^{\frac12}<\infty,$$

where the supremum is taken over all $0 < \varepsilon < N < \infty$ *and all* $x_0 \in \mathbf{R}^n$.

(iii) The following statement is valid:

$$B_3 = \sup_{\varphi} \sup_{x_0 \in \mathbf{R}^n} \sup_{R>0} R^{\frac{n}{2}} \Big[\big\| T(\tau^{x_0}(\varphi_R)) \big\|_{L^2} + \big\| T^t(\tau^{x_0}(\varphi_R)) \big\|_{L^2} \Big] < \infty,$$

where the first supremum is taken over all normalized bumps φ.

(iv) The operator T *satisfies the weak boundedness property and the distributions* $T(1)$ *and* $T^t(1)$ *coincide with BMO functions, that is,*

$$B_4 = \big\| T(1) \big\|_{BMO} + \big\| T^t(1) \big\|_{BMO} + \big\| T \big\|_{WB} < \infty.$$

(v) For every $\xi \in \mathbf{R}^n$ *the distributions* $T(e^{2\pi i(\cdot)\cdot\xi})$ *and* $T^t(e^{2\pi i(\cdot)\cdot\xi})$ *coincide with BMO functions such that*

$$B_5 = \sup_{\xi \in \mathbf{R}^n} \big\| T(e^{2\pi i(\cdot)\cdot\xi}) \big\|_{BMO} + \sup_{\xi \in \mathbf{R}^n} \big\| T^t(e^{2\pi i(\cdot)\cdot\xi}) \big\|_{BMO} < \infty.$$

(vi) The following statement is valid:

$$B_6 = \sup_{\varphi} \sup_{x_0 \in \mathbf{R}^n} \sup_{R>0} R^{n} \Big[\big\| T(\tau^{x_0}(\varphi_R)) \big\|_{BMO} + \big\| T^t(\tau^{x_0}(\varphi_R)) \big\|_{BMO} \Big] < \infty,$$

where the first supremum is taken over all normalized bumps φ.

Then assertions (i)–(vi) *are all equivalent to each other and to the* L^2 *boundedness of* T, *and we have the following equivalence of the previous quantities:*

$$c_{n,\delta}(A + B_j) \le \big\| T \big\|_{L^2 \to L^2} \le C_{n,\delta}(A + B_j),$$

for all $j \in \{1,2,3,4,5,6\}$, *for some constants* $c_{n,\delta}, C_{n,\delta}$ *that depend only on the dimension* n *and on the parameter* $\delta > 0$.

Remark 8.3.4. Condition (8.3.3) says that the operator T is the weak limit of a sequence of its truncations. We already know that if T is bounded on L^2, then it must be equal to an operator that satisfies (8.3.3) plus a bounded function times the identity operator. (See Proposition 8.1.11.) Therefore, it is not a serious restriction to assume this. See Remark 8.3.6 for a version of Theorem 8.3.3 in which this assumption is not imposed. However, the reader should always keep in mind the following pathological situation: Let K be a function on $\mathbf{R}^n \times \mathbf{R}^n \setminus \{(x,x) : x \in \mathbf{R}^n\}$ that satisfies condition (ii) of the theorem. Then nothing prevents the Schwartz kernel W of T from having the form

$$W = K(x,y) + h(x)\delta_{x=y},$$

where $h(x)$ is an unbounded function and $\delta_{x=y}$ is Lebesgue measure on the subspace $x = y$. In this case, although the $T^{(\varepsilon)}$'s are uniformly bounded on L^2, T cannot be L^2 bounded, since h is not a bounded function.

Before we begin the lengthy proof of this theorem, we state a lemma that we need.

Lemma 8.3.5. *Under assumptions (8.1.1), (8.1.2), and (8.1.3), there is a constant C_n such that for all normalized bumps φ we have*

$$\sup_{x_0\in\mathbf{R}^n}\int_{|x-x_0|\geq 20R}\left|\int_{\mathbf{R}^n}K(x,y)\tau^{x_0}(\varphi_R)(y)\,dy\right|^2 dx \leq \frac{C_nA^2}{R^n}. \tag{8.3.4}$$

Proof. Note that the interior integral in (8.3.4) is absolutely convergent, since $\tau^{x_0}(\varphi_R)$ is supported in the ball $B(x_0,10R)$ and x lies in the complement of the double of this ball. To prove (8.3.4), simply observe that since $|K(x,y)|\leq A|x-y|^{-n}$, we have that

$$|T(\tau^{x_0}(\varphi_R))(x)|\leq\frac{C_nA}{|x-x_0|^n}$$

whenever $|x-x_0|\geq 20R$. The estimate follows easily. □

8.3.2 The Proof of Theorem 8.3.3

This subsection is dedicated to the proof of Theorem 8.3.3.

Proof. The proof is based on a series of steps. We begin by showing that condition (iii) implies condition (iv).

(iii) $\Longrightarrow$ (iv)

Fix a $\mathscr{C}_0^\infty$ function ϕ with $0\leq\phi\leq 1$, supported in the ball $B(0,4)$, and equal to 1 on the ball $B(0,2)$. We consider the functions $\phi(\cdot/R)$ that tend to 1 as $R\to\infty$ and we show that $T(1)$ is the weak limit of the functions $T(\phi(\cdot/R))$. This means that for all $g\in\mathscr{D}_0'$ (smooth functions with compact support and integral zero) one has

$$\big\langle T(\phi(\cdot/R)),g\big\rangle\to\big\langle T(1),g\big\rangle \tag{8.3.5}$$

as $R\to\infty$. To prove (8.3.5) we fix a $\mathscr{C}_0^\infty$ function η that is equal to one on the support of g. Then we write

$$\begin{aligned}\big\langle T(\phi(\cdot/R)),g\big\rangle &= \big\langle T(\eta\phi(\cdot/R)),g\big\rangle+\big\langle T((1-\eta)\phi(\cdot/R)),g\big\rangle\\ &= \big\langle T(\eta\phi(\cdot/R)),g\big\rangle\\ &\quad+\int_{\mathbf{R}^n}\int_{\mathbf{R}^n}\big(K(x,y)-K(x_0,y)\big)g(x)(1-\eta(y))\phi(y/R)\,dy\,dx,\end{aligned}$$

where x_0 is a point in the support of g. There exists an $R_0>0$ such that for $R\geq R_0$, $\phi(\cdot/R)$ is equal to 1 on the support of η, and moreover the expressions

$$\int_{\mathbf{R}^n}\int_{\mathbf{R}^n}\big(K(x,y)-K(x_0,y)\big)g(x)(1-\eta(y))\phi(y/R)\,dy\,dx$$

converge to

$$\int_{\mathbf{R}^n}\int_{\mathbf{R}^n}\big(K(x,y)-K(x_0,y)\big)g(x)(1-\eta(y))\,dy\,dx$$

as $R\to\infty$ by the Lebesgue dominated convergence theorem. Using Definition 8.1.16, we obtain the validity of (8.3.5).

Next we observe that the functions $\phi(\cdot/R)$ are in L^2, since $\phi(x/R)=R^{-n}\phi_R(x)$, and by hypothesis (iii), ϕ_R are in L^2. We show that

$$\big\|T(\phi(\cdot/R))\big\|_{BMO}\le C_{n,\delta}(A+B_3) \tag{8.3.6}$$

uniformly in $R>0$. Once (8.3.6) is established, then the sequence $\{T(\phi(\cdot/j))\}_{j=1}^{\infty}$ lies in a multiple of the unit ball of $BMO=(H^1)^*$, and by the Banach–Alaoglou theorem, there is a subsequence of the positive integers R_j such that $T(\phi(\cdot/R_j))$ converges weakly to an element b in BMO. This means that

$$\big\langle T(\phi(\cdot/R_j)),g\big\rangle\to\big\langle b,g\big\rangle \tag{8.3.7}$$

as $j\to\infty$ for all $g\in\mathscr{D}_0$. Using (8.3.5), we conclude that $T(1)$ can be identified with the BMO function b, and as a consequence of (8.3.6) it satisfies

$$\big\|T(1)\big\|_{BMO}\le C_{n,\delta}(A+B_3)\,.$$

In a similar fashion, we identify $T^t(1)$ with a BMO function with norm satisfying

$$\big\|T^t(1)\big\|_{BMO}\le C_{n,\delta}(A+B_3)\,.$$

We return to the proof of (8.3.6). We fix a ball $B=B(x_0,r)$ with radius $r>0$ centered at $x_0\in\mathbf{R}^n$. If we had a constant c_B such that

$$\frac{1}{|B|}\int_B|T(\phi(\cdot/R))(x)-c_B|\,dx\le c_{n,\delta}\,B_3 \tag{8.3.8}$$

for all $R>0$, then property (3) in Proposition 7.1.2 (adapted to balls) would yield (8.3.6). Obviously, (8.3.8) is a consequence of the two estimates

$$\frac{1}{|B|}\int_B\big|T\big[\phi(\tfrac{\cdot-x_0}{r})\phi(\tfrac{\cdot}{R})\big](x)\big|\,dx\le c_n\,B_3\,, \tag{8.3.9}$$

$$\frac{1}{|B|}\int_B\big|T\big[(1-\phi(\tfrac{\cdot-x_0}{r}))\phi(\tfrac{\cdot}{R})\big](x)-T\big[(1-\phi(\tfrac{\cdot-x_0}{r}))\phi(\tfrac{\cdot}{R})\big](x_0)\big|\,dx\le\frac{c_n}{\delta}A\,. \tag{8.3.10}$$

We bound the double integral in (8.3.10) by

$$\frac{1}{|B|}\int_B\int_{|y-x_0|\ge 2r}|K(x,y)-K(x_0,y)|\,\phi(y/R)\,dy\,dx\,, \tag{8.3.11}$$

since $1-\phi((y-x_0)/r)=0$ when $|y-x_0|\le 2r$. Since $|x-x_0|\le r\le \frac{1}{2}|y-x_0|$, condition (8.1.2) gives that (8.3.11) holds with $c_n=\omega_{n-1}=|\mathbf{S}^{n-1}|$.

It remains to prove (8.3.9). It is easy to verify that there is a constant $C_0=C_0(n,\phi)$ such that for $0<\varepsilon\le 1$ and for all $a\in\mathbf{R}^n$ the functions

$$C_0^{-1}\phi(\varepsilon(x+a))\phi(x),\qquad C_0^{-1}\phi(x)\phi(-a+\varepsilon x) \tag{8.3.12}$$

are normalized bumps. The important observation is that with $a=x_0/r$ we have

$$\begin{aligned}\phi(\tfrac{x}{R})\phi(\tfrac{x-x_0}{r}) &= r^n\tau^{x_0}\Big[\Big(\phi(\tfrac{r}{R}(\cdot+a))\phi(\cdot)\Big)_r\Big](x) && (8.3.13)\\ &= R^n\Big(\phi(\cdot)\phi\big(-a+\tfrac{R}{r}(\cdot)\big)\Big)_R(x), && (8.3.14)\end{aligned}$$

and thus in either case $r\le R$ or $R\le r$, one may express the product $\phi(\frac{x}{R})\phi(\frac{x-x_0}{r})$ as a multiple of a translation of an L^1-dilation of a normalized bump.

Let us suppose that $r\le R$. In view of (8.3.13) we write

$$T\big[\phi(\tfrac{\cdot-x_0}{r})\phi(\tfrac{\cdot}{R})\big](x)=C_0\,r^nT\big[\tau^{x_0}(\varphi_r)\big](x)$$

for some normalized bump φ. Using this fact and the Cauchy–Schwarz inequality, we estimate the expression on the left in (8.3.9) by

$$\frac{C_0\,r^{n/2}}{|B|^{\frac12}}r^{n/2}\bigg(\int_B\big|T\big[\tau^{x_0}(\varphi_r)\big](x)\big|^2dx\bigg)^{\frac12}\le\frac{C_0\,r^{n/2}}{|B|^{\frac12}}B_3=c_n\,B_3\,,$$

where the first inequality follows by applying hypothesis (iii).

We now consider the case $R\le r$. In view of (8.3.14) we write

$$T\big[\phi(\tfrac{\cdot-x_0}{r})\phi(\tfrac{\cdot}{R})\big](x)=C_0\,R^nT\big(\varphi_R\big)(x)$$

for some other normalized bump φ. Using this fact and the Cauchy–Schwarz inequality, we estimate the expression on the left in (8.3.9) by

$$\frac{C_0\,R^{n/2}}{|B|^{\frac12}}R^{n/2}\bigg(\int_B|T(\zeta_R)(x)|^2dx\bigg)^{\frac12}\le\frac{C_0\,R^{n/2}}{|B|^{\frac12}}B_3\le c_n\,B_3$$

by applying hypothesis (iii) and recalling that $R\le r$. This proves (8.3.9).

To finish the proof of (iv), we need to prove that T satisfies the weak boundedness property. But this is elementary, since for all normalized bumps φ and ψ and all $x\in\mathbf{R}^n$ and $R>0$ we have

$$\begin{aligned}\big|\big\langle T(\tau^x(\psi_R)),\tau^x(\varphi_R)\big\rangle\big| &\le \big\|T(\tau^x(\psi_R))\big\|_{L^2}\big\|\tau^x(\varphi_R)\big\|_{L^2}\\ &\le B_3R^{-\frac{n}{2}}\big\|\tau^x(\varphi_R)\big\|_{L^2}\\ &\le C_nB_3R^{-n}.\end{aligned}$$

This gives $\|T\|_{WB} \le C_n B_3$, which implies the estimate $B_4 \le C_{n,\delta}(A+B_3)$ and concludes the proof of the fact that condition (iii) implies (iv).

(iv) $\Longrightarrow$ (L^2 boundedness of T)

We now assume condition (iv) and we present the most important step of the proof, establishing the fact that T has an extension that maps $L^2(\mathbf{R}^n)$ to itself. The assumption that the distributions $T(1)$ and $T^t(1)$ coincide with BMO functions leads to the construction of Carleson measures that provide the key tool in the boundedness of T.

We pick a smooth radial function Φ with compact support that is supported in the ball $B(0,\frac{1}{2})$ and that satisfies $\int_{\mathbf{R}^n}\Phi(x)\,dx = 1$. For $t>0$ we define $\Phi_t(x) = t^{-n}\Phi(\frac{x}{t})$. Since Φ is a radial function, the operator

$$P_t(f) = f * \Phi_t \tag{8.3.15}$$

is self-transpose. The operator P_t is a continuous analogue of $S_j = \sum_{k\le j}\Delta_k$, where the Δ_j's are the Littlewood–Paley operators.

We now fix a Schwartz function f whose Fourier transform is supported away from a neighborhood of the origin. We discuss an integral representation for $T(f)$. We begin with the facts, which can be found in Exercises 8.3.1 and 8.3.2, that

$$\begin{aligned} T(f) &= \lim_{s\to 0} P_s^2 T P_s^2(f)\,, \\ 0 &= \lim_{s\to\infty} P_s^2 T P_s^2(f)\,, \end{aligned}$$

where the limits are interpreted in the topology of $\mathscr{S}'(\mathbf{R}^n)$. Thus, with the use of the fundamental theorem of calculus and the product rule, we are able to write

$$\begin{aligned} T(f) &= \lim_{s\to 0} P_s^2 T P_s^2(f) - \lim_{s\to\infty} P_s^2 T P_s^2(f) \\ &= -\lim_{\varepsilon\to 0}\int_{\varepsilon}^{\frac{1}{\varepsilon}} s\frac{d}{ds}(P_s^2 T P_s^2)(f)\,\frac{ds}{s} \\ &= -\lim_{\varepsilon\to 0}\int_{\varepsilon}^{\frac{1}{\varepsilon}} \left[s\left(\frac{d}{ds}P_s^2\right)TP_s^2(f) + P_s^2\left(Ts\frac{d}{ds}P_s^2\right)(f)\right]\frac{ds}{s}\,. \end{aligned} \tag{8.3.16}$$

For a Schwartz function g we have

$$\begin{aligned} \left(s\frac{d}{ds}P_s^2(g)\right)^{\widehat{}}(\xi) &= \widehat{g}(\xi)\, s\frac{d}{ds}\widehat{\Phi}(s\xi)^2 \\ &= \widehat{g}(\xi)\,\widehat{\Phi}(s\xi)\,(2s\xi\cdot\nabla\widehat{\Phi}(s\xi)) \\ &= \widehat{g}(\xi)\sum_{k=1}^{n}\widehat{\Psi_k}(s\xi)\widehat{\Theta_k}(s\xi) \\ &= \sum_{k=1}^{n}\left(\widetilde{Q}_{k,s}Q_{k,s}(g)\right)^{\widehat{}}(\xi) = \sum_{k=1}^{n}\left(Q_{k,s}\widetilde{Q}_{k,s}(g)\right)^{\widehat{}}(\xi)\,, \end{aligned}$$

where for $1 \le k \le n$, $\widehat{\Psi_k}(\xi) = 2\xi_k \widehat{\Phi}(\xi)$, $\widehat{\Theta_k}(\xi) = \partial_k \widehat{\Phi}(\xi)$ and $Q_{k,s}$, $\widetilde{Q}_{k,s}$ are operators defined by

$$Q_{k,s}(g) = g * (\Psi_k)_s, \qquad \widetilde{Q}_{k,s}(g) = g * (\Theta_k)_s;$$

here $(\Theta_k)_s(x) = s^{-n}\Theta_k(s^{-1}x)$ and $(\Psi_k)_s$ are defined similarly. Observe that Ψ_k and Θ_k are smooth odd bumps supported in $B(0, \frac{1}{2})$ and have integral zero. Since Ψ_k and Θ_k are odd, they are anti-self-transpose, meaning that $(Q_{k,s})^t = -Q_{k,s}$ and $(\widetilde{Q}_{k,s})^t = -\widetilde{Q}_{k,s}$. We now write the expression in (8.3.16) as

$$-\lim_{\varepsilon \to 0} \sum_{k=1}^{n} \left[\int_{\varepsilon}^{\frac{1}{\varepsilon}} \widetilde{Q}_{k,s} Q_{k,s} T P_s P_s(f) \frac{ds}{s} + \int_{\varepsilon}^{\frac{1}{\varepsilon}} P_s P_s T Q_{k,s} \widetilde{Q}_{k,s}(f) \frac{ds}{s} \right], \tag{8.3.17}$$

where the limit converges in $\mathscr{S}'(\mathbf{R}^n)$. We set

$$T_{k,s} = Q_{k,s} T P_s,$$

and we observe that the operator $P_s T Q_{k,s}$ is equal to $-((T^t)_{k,s})^t$.

Recall the notation $\tau^x(h)(z) = h(z-x)$. In view of identity (2.3.21) and the convergence of the Riemann sums to the integral defining $f * \Phi_s$ in the topology of $\mathscr{S}$ (see the proof of Theorem 2.3.20), we deduce that the operator $T_{k,s}$ has kernel

$$K_{k,s}(x,y) = -\left\langle T(\tau^y(\Phi_s)), \tau^x((\Psi_k)_s) \right\rangle = -\left\langle T^t(\tau^x((\Psi_k)_s)), \tau^y(\Phi_s) \right\rangle. \tag{8.3.18}$$

Likewise, the operator $-(T^t)^t_{k,s}$ has kernel

$$\left\langle T^t(\tau^x(\Phi_s)), \tau^y((\Psi_k)_s) \right\rangle = \left\langle T(\tau^y((\Psi_k)_s)), \tau^x(\Phi_s) \right\rangle.$$

For $1 \le k \le n$ we need the following facts regarding the kernels of these operators:

$$\left|\left\langle T(\tau^x((\Psi_k)_s)), \tau^y(\Phi_s) \right\rangle\right| \le C_{n,\delta}\left(\|T\|_{WB} + A\right) p_s(x-y), \tag{8.3.19}$$

$$\left|\left\langle T^t(\tau^x((\Psi_k)_s)), \tau^y(\Phi_s) \right\rangle\right| \le C_{n,\delta}\left(\|T\|_{WB} + A\right) p_s(x-y), \tag{8.3.20}$$

where

$$p_t(u) = \frac{1}{t^n} \frac{1}{(1+|\frac{u}{t}|)^{n+\delta}}$$

is the L^1 dilation of the function $p(u) = (1+|u|)^{-n-\delta}$.

To prove (8.3.20), we consider the following two cases: If $|x-y| \le 5s$, then the weak boundedness property gives

$$\left|\left\langle T(\tau^y(\Phi_s)), \tau^x((\Psi_k)_s) \right\rangle\right| = \left|\left\langle T(\tau^x((\tau^{\frac{y-x}{s}}(\Phi))_s)), \tau^x((\Psi_k)_s) \right\rangle\right| \le \frac{C_n \|T\|_{WB}}{s^n},$$

since both Ψ_k and $\tau^{\frac{y-x}{s}}(\Phi)$ are multiples of normalized bumps. Notice here that both of these functions are supported in $B(0,10)$, since $\frac{1}{s}|x-y| \le 5$. This estimate proves (8.3.20) when $|x-y| \le 5s$.

We now turn to the case $|x-y| \geq 5s$. Then the functions $\tau^y(\Phi_s)$ and $\tau^x((\Psi_k)_s)$ have disjoint supports and so we have the integral representation

$$\langle T^t(\tau^x((\Psi_k)_s)), \tau^y(\Phi_s)\rangle = \int_{\mathbf{R}^n}\int_{\mathbf{R}^n} \Phi_s(v-y)K(u,v)(\Psi_k)_s(u-x)\,du\,dv\,.$$

Using that Ψ_k has mean value zero, we can write the previous expression as

$$\int_{\mathbf{R}^n}\int_{\mathbf{R}^n} \Phi_s(v-y)\big(K(u,v)-K(x,v)\big)(\Psi_k)_s(u-x)\,du\,dv\,.$$

We observe that $|u-x| \leq s$ and $|v-y| \leq s$ in the preceding double integral. Since $|x-y| \geq 5s$, this makes $|u-v| \geq |x-y| - 2s \geq 3s$, which implies that $|u-x| \leq \frac{1}{2}|u-v|$. Using (8.1.2), we obtain

$$|K(u,v)-K(x,v)| \leq \frac{A|x-u|^{\delta}}{(|u-v|+|x-v|)^{n+\delta}} \leq C_{n,\delta} A \frac{s^{\delta}}{|x-y|^{n+\delta}}\,,$$

where we used the fact that $|u-v| \approx |x-y|$. Inserting this estimate in the double integral, we obtain (8.3.20). Estimate (8.3.19) is proved similarly.

At this point we drop the dependence of $Q_{k,s}$ and $\widetilde{Q}_{k,s}$ on the index k, since we can concentrate on one term of the sum in (8.3.17). We have managed to express $T(f)$ as a finite sum of operators of the form

$$\int_0^{\infty} \widetilde{Q}_s T_s P_s(f)\,\frac{ds}{s} \tag{8.3.21}$$

and of the form

$$\int_0^{\infty} P_s T_s \widetilde{Q}_s(f)\,\frac{ds}{s}\,, \tag{8.3.22}$$

where the preceding integrals converge in $\mathscr{S}'(\mathbf{R}^n)$ and the T_s's have kernels $K_s(x,y)$, which are pointwise dominated by a constant multiple of

$$(A+B_4)p_s(x-y)\,.$$

It suffices to obtain L^2 bounds for an operator of the form (8.3.21) with constant at most a multiple of $A+B_4$. Then by duality the same estimate also holds for the operators of the form (8.3.22). We make one more observation. Using (8.3.18) (recall that we have dropped the indices k), we obtain

$$T_s(1)(x) = \int_{\mathbf{R}^n} K_s(x,y)\,dy = -\langle T^t(\tau^x(\Psi_s)), 1\rangle = -(\Psi_s * T(1))(x)\,, \tag{8.3.23}$$

where all integrals converge absolutely.

We can therefore concentrate on the L^2 boundedness of the operator in (8.3.21). We pair this operator with a Schwartz function g and we use the convergence of the integral in $\mathscr{S}'$ and the property $(\widetilde{Q}_s)^t = -\widetilde{Q}_s$ to obtain

$$\left\langle \int_0^\infty \widetilde{Q}_s T_s P_s(f) \frac{ds}{s}, g \right\rangle = \int_0^\infty \langle \widetilde{Q}_s T_s P_s(f), g \rangle \frac{ds}{s} = -\int_0^\infty \langle T_s P_s(f), \widetilde{Q}_s(g) \rangle \frac{ds}{s}.$$

The intuition here is as follows: T_s is an averaging operator at scale s and $P_s(f)$ is essentially constant on that scale. Therefore, the expression $T_s P_s(f)$ must look like $T_s(1)P_s(f)$. To be precise, we introduce this term and try to estimate the error that occurs. We have

$$T_s P_s(f) = T_s(1)P_s(f) + \left[T_s P_s(f) - T_s(1)P_s(f)\right]. \tag{8.3.24}$$

We estimate the terms that arise from this splitting. Recalling (8.3.23), we write

$$\left|\int_0^\infty \left\langle (\Psi_s * T(1))P_s(f), \widetilde{Q}_s(g) \right\rangle \frac{ds}{s}\right| \tag{8.3.25}$$

$$\leq \left(\int_0^\infty \big\|P_s(f)\,(\Psi_s * T(1))\big\|_{L^2}^2 \frac{ds}{s}\right)^{\frac{1}{2}} \left(\int_0^\infty \big\|\widetilde{Q}_s(g)\big\|_{L^2}^2 \frac{ds}{s}\right)^{\frac{1}{2}}$$

$$= \left\|\left(\int_0^\infty |P_s(f)\,(\Psi_s * T(1))|^2 \frac{ds}{s}\right)^{\frac{1}{2}}\right\|_{L^2} \left\|\left(\int_0^\infty |\widetilde{Q}_s(g)|^2 \frac{ds}{s}\right)^{\frac{1}{2}}\right\|_{L^2}. \tag{8.3.26}$$

Since $T(1)$ is a BMO function, $|(\Psi_s * T(1))(x)|^2 dx\frac{ds}{s}$ is a Carleson measure on $\mathbf{R}^{n+1}_+$. Using Theorem 7.3.8 and the Littlewood–Paley theorem (Exercise 5.1.4), we obtain that (8.3.26) is controlled by

$$C_n \big\|T(1)\big\|_{BMO} \big\|f\big\|_{L^2} \big\|g\big\|_{L^2} \leq C_n B_4 \big\|f\big\|_{L^2} \big\|g\big\|_{L^2}.$$

This gives the sought estimate for the first term in (8.3.24). For the second term in (8.3.24) we have

$$\left|\int_0^\infty \int_{\mathbf{R}^n} \widetilde{Q}_s(g)(x) \left[T_s P_s(f) - T_s(1)\, P_s(f)\right](x)\, dx \frac{ds}{s}\right|$$

$$\leq \left(\int_0^\infty \int_{\mathbf{R}^n} |\widetilde{Q}_s(g)(x)|^2\, dx \frac{ds}{s}\right)^{\frac{1}{2}} \left(\int_0^\infty \int_{\mathbf{R}^n} |(T_s P_s(f) - T_s(1)P_s(f))(x)|^2\, dx \frac{ds}{s}\right)^{\frac{1}{2}}$$

$$\leq C_n \big\|g\big\|_{L^2} \left(\int_0^\infty \int_{\mathbf{R}^n} \left|\int_{\mathbf{R}^n} K_s(x,y)[P_s(f)(y) - P_s(f)(x)]\, dy\right|^2 dx \frac{ds}{s}\right)^{\frac{1}{2}}$$

$$\leq C_n(A + B_4) \big\|g\big\|_{L^2} \left(\int_0^\infty \int_{\mathbf{R}^n} \int_{\mathbf{R}^n} p_s(x-y) \left|P_s(f)(y) - P_s(f)(x)\right|^2 dy\, dx \frac{ds}{s}\right)^{\frac{1}{2}},$$

where in the last estimate we used the fact that the measure $p_t(x-y)\,dy$ is a multiple of a probability measure. It suffices to estimate the last displayed square root. Changing variables $u = x - y$ and applying Plancherel's theorem, we express this square root as

$$\begin{aligned}
&\left(\int_0^\infty \int_{\mathbf{R}^n}\int_{\mathbf{R}^n} p_s(u)\big|P_s(f)(y)-P_s(f)(y+u)\big|^2\,du\,dy\,\frac{ds}{s}\right)^{\frac{1}{2}}\\
&=\left(\int_0^\infty \int_{\mathbf{R}^n}\int_{\mathbf{R}^n} p_s(u)\big|\widehat{\Phi}(s\xi)-\widehat{\Phi}(s\xi)e^{2\pi i u\cdot\xi}\big|^2\big|\widehat{f}(\xi)\big|^2\,du\,d\xi\,\frac{ds}{s}\right)^{\frac{1}{2}}\\
&\le\left(\int_0^\infty \int_{\mathbf{R}^n}\int_{\mathbf{R}^n} p_s(u)\big|\widehat{\Phi}(s\xi)\big|^2 4\pi^{\frac{\delta}{2}}|u|^{\frac{\delta}{2}}|\xi|^{\frac{\delta}{2}}\big|\widehat{f}(\xi)\big|^2\,du\,d\xi\,\frac{ds}{s}\right)^{\frac{1}{2}}\\
&=2\pi^{\frac{\delta}{4}}\left(\int_{\mathbf{R}^n}\int_0^\infty \left(\int_{\mathbf{R}^n} p_s(u)\big|\tfrac{u}{s}\big|^{\frac{\delta}{2}}du\right)\big|\widehat{\Phi}(s\xi)\big|^2|s\xi|^{\frac{\delta}{2}}\frac{ds}{s}\big|\widehat{f}(\xi)\big|^2\,d\xi\right)^{\frac{1}{2}},
\end{aligned}$$

and we claim that this last expression is bounded by $C_{n,\delta}\big\|f\big\|_{L^2}$. Indeed, we first bound the quantity $\int_{\mathbf{R}^n} p_s(u)\big|\frac{u}{s}\big|^{\delta/2}du$ by a constant, and then we use the estimate

$$\int_0^\infty \big|\widehat{\Phi}(s\xi)\big|^2|s\xi|^{\frac{\delta}{2}}\frac{ds}{s}=\int_0^\infty \big|\widehat{\Phi}(se_1)\big|^2 s^{\frac{\delta}{2}}\frac{ds}{s}\le C'_{n,\delta}<\infty$$

and Plancherel's theorem to obtain the claim. [Here $e_1=(1,0,\dots,0)$.] Taking g to be an arbitrary Schwartz function with L^2 norm at most 1 and using duality, we deduce the estimate $\big\|T(f)\big\|_{L^2}\le C_{n,\delta}(A+B_4)\big\|f\big\|_{L^2}$ for all Schwartz functions f whose Fourier transform does not contain a neighborhood of the origin. Such functions are dense in $L^2(\mathbf{R}^n)$ (cf. Exercise 5.2.9) and thus T admits an extension on L^2 that satisfies $\big\|T\big\|_{L^2\to L^2}\le C_{n,\delta}(A+B_4)$.

(L^2 boundedness of T) $\Longrightarrow$ (v)

If T has an extension that maps L^2 to itself, then by Theorem 8.2.7 we have

$$B_5\le C_{n,\delta}\big(A+\big\|T\big\|_{L^2\to L^2}\big)<\infty.$$

Thus the boundedness of T on L^2 implies condition (v).

(v) $\Longrightarrow$ (vi)

At a formal level the proof of this fact is clear, since we can write a normalized bump as the inverse Fourier transform of its Fourier transform and interchange the integrations with the action of T to obtain

$$T(\tau^{x_0}(\varphi_R))=\int_{\mathbf{R}^n}\widehat{\tau^{x_0}(\varphi_R)}(\xi)T(e^{2\pi i\xi\cdot(\cdot)})\,d\xi. \tag{8.3.27}$$

The conclusion follows by taking BMO norms. To make identity (8.3.27) precise we provide the following argument.

Let us fix a normalized bump φ and a smooth and compactly supported function g with mean value zero. We pick a smooth function η with compact support that is equal to 1 on the double of a ball containing the support of g and vanishes off the triple of that ball. Define $\eta_k(\xi)=\eta(\xi/k)$ and note that η_k tends pointwise to 1 as

$k \to \infty$. Observe that $\eta_k \tau^{x_0}(\varphi_R)$ converges to $\tau^{x_0}(\varphi_R)$ in $\mathscr{S}(\mathbf{R}^n)$ as $k \to \infty$, and by the continuity of T we obtain

$$\lim_{k\to\infty} \langle T(\eta_k \tau^{x_0}(\varphi_R)), g\rangle = \langle T(\tau^{x_0}(\varphi_R)), g\rangle.$$

The continuity and linearity of T also allow us to write

$$\langle T(\tau^{x_0}(\varphi_R)), g\rangle = \lim_{k\to\infty} \int_{\mathbf{R}^n} \widehat{\tau^{x_0}(\varphi_R)}(\xi) \left\langle T\left(\eta_k e^{2\pi i \xi\cdot(\cdot)}\right), g\right\rangle d\xi. \tag{8.3.28}$$

Let W be the Schwartz kernel of T. By (8.1.5) we have

$$\left\langle T\left(\eta_k e^{2\pi i \xi\cdot(\cdot)}\right), g\right\rangle = \left\langle W, g\otimes \eta_k e^{2\pi i \xi\cdot(\cdot)}\right\rangle. \tag{8.3.29}$$

Using (8.1.6), we obtain that the expression in (8.3.29) is controlled by a finite sum of L^∞ norms of derivatives of the function

$$g(x)\,\eta_k(y) e^{2\pi i \xi\cdot y}$$

on some compact set (that depends on g). Then for some $M > 0$ and some constant $C(g)$ depending on g, we have that this sum of L^∞ norms of derivatives is controlled by

$$C(g)\,(1+|\xi|)^M$$

uniformly in $k \ge 1$. Since $\widehat{\tau^{x_0}(\varphi_R)}$ is integrable, the Lebesgue dominated convergence theorem allows us to pass the limit inside the integrals in (8.3.28) to obtain

$$\langle T(\tau^{x_0}(\varphi_R)), g\rangle = \int_{\mathbf{R}^n} \widehat{\tau^{x_0}(\varphi_R)}(\xi) \left\langle T\left(e^{2\pi i \xi\cdot(\cdot)}\right), g\right\rangle d\xi.$$

We now use assumption (v). The distributions $T\left(e^{2\pi i \xi\cdot(\cdot)}\right)$ coincide with BMO functions whose norm is at most B_5. It follows that

$$\begin{aligned} \left|\langle T(\tau^{x_0}(\varphi_R)), g\rangle\right| &\le \left\|\widehat{\tau^{x_0}(\varphi_R)}\right\|_{L^1} \sup_{\xi\in\mathbf{R}^n} \left\|T\left(e^{2\pi i \xi\cdot(\cdot)}\right)\right\|_{BMO} \left\|g\right\|_{H^1} \\ &\le C_n B_5 R^{-n} \left\|g\right\|_{H^1}, \end{aligned} \tag{8.3.30}$$

where the constant C_n is independent of the normalized bump φ in view of (8.3.1). It follows from (8.3.30) that

$$g \mapsto \langle T(\tau^{x_0}(\varphi_R)), g\rangle$$

is a bounded linear functional on BMO with norm at most a multiple of $B_5 R^{-n}$. It follows from Theorem 7.2.2 that $T(\tau^{x_0}(\varphi_R))$ coincides with a BMO function that satisfies

$$R^n \left\|T(\tau^{x_0}(\varphi_R))\right\|_{BMO} \le C_n B_5.$$

The same argument is valid for T^t, and this shows that

$$B_6 \le C_{n,\delta}(A + B_5)$$

and concludes the proof that (v) implies (vi).

(vi) $\Longrightarrow$ (iii)

We fix $x_0 \in \mathbf{R}^n$ and $R > 0$. Pick z_0 in $\mathbf{R}^n$ such that $|x_0 - z_0| = 40R$. Then if $|y - x_0| \le 10R$ and $|x - z_0| \le 20R$ we have

$$\begin{aligned} 10R &\le |z_0 - x_0| - |x - z_0| - |y - x_0| \\ &\le |x - y| \\ &\le |x - z_0| + |z_0 - x_0| + |x_0 - y| \le 70R. \end{aligned}$$

From this it follows that when $|x - z_0| \le 20R$ we have

$$\left| \int_{|y-x_0|\le 10R} K(x,y)\tau^{x_0}(\varphi_R)(y)\,dy \right| \le \int_{10R\le|x-y|\le 70R} |K(x,y)| \frac{dy}{R^n} \le \frac{C_{n,\delta}A}{R^n}$$

and thus

$$\left| \operatorname*{Avg}_{B(z_0,20R)} T(\tau^{x_0}(\varphi_R)) \right| \le \frac{C_{n,\delta}A}{R^n}, \tag{8.3.31}$$

where $\operatorname{Avg}_B g$ denotes the average of g over B. Because of assumption (vi), the *BMO* norm of the function $T(\tau^{x_0}(\varphi_R))$ is bounded by a multiple of $B_6 R^{-n}$, a fact used in the following sequence of implications. We have

$$\begin{aligned} &\big\|T(\tau^{x_0}(\varphi_R))\big\|_{L^2(B(x_0,20R))} \\ &\quad\le \Big\| T(\tau^{x_0}(\varphi_R)) - \operatorname*{Avg}_{B(x_0,20R)} T(\tau^{x_0}(\varphi_R)) \Big\|_{L^2(B(x_0,20R))} \\ &\qquad + v_n^{\frac12}(20R)^{\frac n2} \Big| \operatorname*{Avg}_{B(x_0,20R)} T(\tau^{x_0}(\varphi_R)) - \operatorname*{Avg}_{B(z_0,20R)} T(\tau^{x_0}(\varphi_R)) \Big| \\ &\qquad + v_n^{\frac12}(20R)^{\frac n2} \Big| \operatorname*{Avg}_{B(z_0,20R)} T(\tau^{x_0}(\varphi_R)) \Big| \\ &\quad\le C_{n,\delta}\Big(R^{\frac n2} \big\|T(\tau^{x_0}(\varphi_R))\big\|_{BMO} + R^{\frac n2} \big\|T(\tau^{x_0}(\varphi_R))\big\|_{BMO} + R^{-\frac n2}A \Big) \\ &\quad\le C_{n,\delta} R^{-\frac n2}(B_6 + A), \end{aligned}$$

where we used (8.3.31) and Exercise 7.1.6. Now we have that

$$\big\|T(\tau^{x_0}(\varphi_R))\big\|_{L^2(B(x_0,20R)^c)} \le C_{n,\delta} A R^{-\frac n2}$$

in view of Lemma 8.3.5. Since the same computations apply to T^t, it follows that

$$R^{\frac n2}\left(\big\|T(\tau^{x_0}(\varphi_R))\big\|_{L^2} + \big\|T^t(\tau^{x_0}(\varphi_R))\big\|_{L^2} \right) \le C_{n,\delta}(A + B_6), \tag{8.3.32}$$

which proves that $B_3 \le C_{n,\delta}(A+B_6)$ and hence (iii). This concludes the proof of the fact that (vi) implies (iii)

We have now completed the proof of the following equivalence of statements:

$$\big(L^2 \text{ boundedness of } T\big) \iff \text{(iii)} \iff \text{(iv)} \iff \text{(v)} \iff \text{(vi)}. \tag{8.3.33}$$

(i) $\iff$ (ii)

We show that the quantities $A+B_1$ and $A+B_2$ are controlled by constant multiples of each other. Let us set

$$I_{\varepsilon,N}(x) = \int_{\varepsilon<|x-y|<N} K(x,y)\,dy \qquad \text{and} \qquad I^t_{\varepsilon,N}(x) = \int_{\varepsilon<|x-y|<N} K^t(x,y)\,dy.$$

We work with a ball $B(x_0,N)$. Observe that

$$\begin{aligned} I_{\varepsilon,N}(x) - T^{(\varepsilon)}(\chi_{B(x_0,N)})(x) &= \int_{\varepsilon<|x-y|<N} K(x,y)\,dy - \int_{\substack{\varepsilon<|x-y|\\ |x_0-y|<N}} K(x,y)\,dy \\ &= -\int_{S_{\varepsilon,N}(x,x_0)} K(x,y)\,dy, \end{aligned} \tag{8.3.34}$$

where $S_{\varepsilon,N}(x,x_0)$ is the set of all $y\in\mathbf{R}^n$ that satisfy $\varepsilon<|x-y|$ and $|x_0-y|<N$ but do not satisfy $\varepsilon<|x-y|<N$. But observe that when $|x_0-x|<N$, then

$$S_{\varepsilon,N}(x,x_0) \subseteq \{y\in\mathbf{R}^n:\ N\le|x-y|<2N\}. \tag{8.3.35}$$

Using (8.3.34), (8.3.35), and (8.1.1), we obtain

$$\big|I_{\varepsilon,N}(x) - T^{(\varepsilon)}(\chi_{B(x_0,N)})(x)\big| \le \int_{N\le|x-y|\le 2N} |K(x,y)|\,dy \le (\omega_{n-1}\log 2)A \tag{8.3.36}$$

whenever $|x_0-x|<N$. It follows that

$$\big\|I_{\varepsilon,N} - T^{(\varepsilon)}(\chi_{B(x_0,N)})\big\|_{L^2(B(x_0,N))} \le C_n A N^{\frac{n}{2}},$$

and similarly, it follows that

$$\big\|I^t_{\varepsilon,N} - (T^{(\varepsilon)})^t(\chi_{B(x_0,N)})\big\|_{L^2(B(x_0,N))} \le C_n A N^{\frac{n}{2}}.$$

These two estimates easily imply the equivalence of conditions (i) and (ii).

We now consider the following condition analogous to (iii):

$$\text{(iii)}' \quad B_3' = \sup_{\varphi} \sup_{x_0 \in \mathbf{R}^n} \sup_{\substack{\varepsilon>0 \\ R>0}} R^{\frac{n}{2}} \Big[\big\| T^{(\varepsilon)}(\tau^{x_0}(\varphi_R)) \big\|_{L^2} + \big\| (T^{(\varepsilon)})^t(\tau^{x_0}(\varphi_R)) \big\|_{L^2} \Big] < \infty,$$

where the first supremum is taken over all normalized bumps φ. We continue by showing that this condition is a consequence of (ii).

<u>(ii) $\Longrightarrow$ (iii)′</u>

More precisely, we prove that $B_3' \le C_{n,\delta}(A+B_2)$. To prove (iii)′, fix a normalized bump φ, a point $x_0 \in \mathbf{R}^n$, and $R > 0$. Also fix $x \in \mathbf{R}^n$ with $|x - x_0| \le 20R$. Then we have

$$T^{(\varepsilon)}(\tau^{x_0}(\varphi_R))(x) = \int_{\varepsilon<|x-y|\le 30R} K^{(\varepsilon)}(x,y)\tau^{x_0}(\varphi_R)(y)\,dx = U_1(x) + U_2(x),$$

where

$$\begin{aligned} U_1(x) &= \int_{\varepsilon<|x-y|\le 30R} K(x,y)\big(\tau^{x_0}(\varphi_R)(y) - \tau^{x_0}(\varphi_R)(x)\big)\,dy, \\ U_2(x) &= \tau^{x_0}(\varphi_R)(x) \int_{\varepsilon<|x-y|\le 30R} K(x,y)\,dy. \end{aligned}$$

But we have that $|\tau^{x_0}(\varphi_R)(y) - \tau^{x_0}(\varphi_R)(x)| \le C_n R^{-1-n}|x-y|$; thus we obtain

$$|U_1(x)| \le C_n A R^{-n}$$

on $B(x_0, 20R)$; hence $\big\|U_1\big\|_{L^2(B(x_0,20R))} \le C_n A R^{-\frac{n}{2}}$. Condition (ii) gives that

$$\big\|U_2\big\|_{L^2(B(x_0,20R))} \le R^{-n}\big\|I_{\varepsilon,30R}\big\|_{L^2(B(x_0,30R))} \le B_2(30R)^{\frac{n}{2}}R^{-n}.$$

Combining these two, we obtain

$$\big\|T^{(\varepsilon)}(\tau^{x_0}(\varphi_R))\big\|_{L^2(B(x_0,20R))} \le C_n(A+B_2)R^{-\frac{n}{2}} \tag{8.3.37}$$

and likewise for $(T^{(\varepsilon)})^t$. It follows from Lemma 8.3.5 that

$$\big\|T^{(\varepsilon)}(\tau^{x_0}(\varphi_R))\big\|_{L^2(B(x_0,20R)^c)} \le C_{n,\delta} A R^{-\frac{n}{2}},$$

which combined with (8.3.37) gives condition (iii)′ with constant

$$B_3' \le C_{n,\delta}(A+B_2).$$

This concludes the proof that condition (ii) implies (iii)′.

<u>(iii)′ $\Longrightarrow$ [$T^{(\varepsilon)} : L^2 \to L^2$ uniformly in $\varepsilon > 0$]</u>

For $\varepsilon > 0$ we introduce the smooth truncations $T_\zeta^{(\varepsilon)}$ of T by setting

$$T_\zeta^{(\varepsilon)}(f)(x) = \int_{\mathbf{R}^n} K(x,y)\zeta(\tfrac{x-y}{\varepsilon})f(y)\,dy,$$

where $\zeta(x)$ is a smooth function that is equal to 1 for $|x| \ge 1$ and vanishes for $|x| \le \frac{1}{2}$. We observe that

$$\big|T_\zeta^{(\varepsilon)}(f) - T^{(\varepsilon)}(f)\big| \le C_n A M(f); \tag{8.3.38}$$

thus the uniform boundedness of $T^{(\varepsilon)}$ on L^2 is equivalent to the uniform boundedness of $T_\zeta^{(\varepsilon)}$. In view of Exercise 8.1.3, the kernels of the operators $T_\zeta^{(\varepsilon)}$ lie in $SK(\delta, cA)$ uniformly in $\varepsilon > 0$ (for some constant c). Moreover, because of (8.3.38), we see that the operators $T_\zeta^{(\varepsilon)}$ satisfy (iii)$'$ with constant $C_nA + B_3'$. The point to be noted here is that condition (iii) for T (with constant B_3) is identical to condition (iii)$'$ for the operators $T_\zeta^{(\varepsilon)}$ uniformly in $\varepsilon > 0$ (with constant $C_nA + B_3'$).

A careful examination of the proof of the implications

$$\text{(iii)} \implies \text{(iv)} \implies (L^2 \text{ boundedness of } T)$$

reveals that all the estimates obtained depend only on the constants B_3, B_4, and A, but not on the specific operator T. Therefore, these estimates are valid for the operators $T_\zeta^{(\varepsilon)}$ that satisfy condition (iii)$'$. This gives the uniform boundedness of the $T_\zeta^{(\varepsilon)}$ on $L^2(\mathbf{R}^n)$ with bounds at most a constant multiple of $A + B_3'$. The same conclusion also holds for the operators $T^{(\varepsilon)}$.

$[T^{(\varepsilon)} : L^2 \to L^2$ uniformly in $\varepsilon > 0] \implies$ (i)

This implication holds trivially.

We have now established the following equivalence of statements:

$$\text{(i)} \iff \text{(ii)} \iff \text{(iii)}' \iff [T^{(\varepsilon)} : L^2 \to L^2 \text{ uniformly in } \varepsilon > 0] \tag{8.3.39}$$

(iii) $\iff$ (iii)$'$

Finally, we link the sets of equivalent conditions (8.3.33) and (8.3.39). We first observe that (iii)$'$ implies (iii). Indeed, using duality and (8.3.3), we obtain

$$\begin{aligned}
\big\|T(\tau^{x_0}(\varphi_R))\big\|_{L^2} &= \sup_{\substack{h\in\mathscr{S}\\ \|h\|_{L^2}\le 1}} \left|\int_{\mathbf{R}^n} T(\tau^{x_0}(\varphi_R))(x)\,h(x)\,dx\right| \\
&\le \sup_{\substack{h\in\mathscr{S}\\ \|h\|_{L^2}\le 1}} \limsup_{j\to\infty} \left|\int_{\mathbf{R}^n} T^{(\varepsilon_j)}(\tau^{x_0}(\varphi_R))(x)\,h(x)\,dx\right| \\
&\le B_3' R^{-\frac{n}{2}},
\end{aligned}$$

which gives $B_3 \le B_3'$. Thus under assumption (8.3.3), (ii) implies (iii) and as we have shown, (iii) implies the boundedness of T on L^2. But in view of Corollary 8.2.4, the boundedness of T on L^2 implies the boundedness of $T^{(\varepsilon)}$ on L^2 uniformly in $\varepsilon > 0$, which implies (iii)$'$.

This completes the proof of the equivalence of the six statements (i)–(vi) in such a way that

$$\big\|T\big\|_{L^2\to L^2} \approx (A + B_j)$$

for all $j \in \{1,2,3,4,5,6\}$. The proof of the theorem is now complete. □

Remark 8.3.6. Suppose that condition (8.3.3) is removed from the hypothesis of Theorem 8.3.3. Then the given proof of Theorem 8.3.3 actually shows that (i) and (ii) are equivalent to each other and to the statement that the $T^{(\varepsilon)}$'s have bounded extensions on $L^2(\mathbf{R}^n)$ that satisfy

$$\sup_{\varepsilon>0} \big\|T^{(\varepsilon)}\big\|_{L^2\to L^2} < \infty .$$

Also, without hypothesis (8.3.3), conditions (iii), (iv), (v), and (vi) are equivalent to each other and to the statement that T has an extension that maps $L^2(\mathbf{R}^n)$ to $L^2(\mathbf{R}^n)$.

8.3.3 An Application

We end this section with one application of the $T(1)$ theorem. We begin with the following observation.

Corollary 8.3.7. *Let K be a standard kernel that is antisymmetric, i.e., it satisfies $K(x,y) = -K(y,x)$ for all $x \neq y$. Then a linear continuous operator T associated with K is L^2 bounded if and only if $T(1)$ is in BMO.*

Proof. In view of Exercise 8.3.3, T automatically satisfies the weak boundedness property. Moreover, $T^t = -T$. Therefore, the three conditions of Theorem 8.3.3 (iv) reduce to the single condition $T(1) \in BMO$. □

Example 8.3.8. Let us recall the kernels K_m of Example 8.1.7. These arise in the expansion of the kernel in Example 8.1.6 in geometric series

$$\frac{1}{x-y+i(A(x)-A(y))} = \frac{1}{x-y}\sum_{m=0}^{\infty}\left(i\,\frac{A(x)-A(y)}{x-y}\right)^m \tag{8.3.40}$$

when $L = \sup_{x\neq y} \frac{|A(x)-A(y)|}{|x-y|} < 1$. The operator with kernel $(i\pi)^{-1}K_m(x,y)$, i.e.,

$$\mathscr{C}_m(f)(x) = \frac{1}{\pi i}\lim_{\varepsilon\to 0}\int_{|x-y|>\varepsilon}\left(\frac{A(x)-A(y)}{x-y}\right)^m \frac{1}{x-y} f(y)\,dy, \tag{8.3.41}$$

is called the mth *Calderón commutator*. We use the $T(1)$ theorem to show that the operators $\mathscr{C}_m$ are L^2 bounded.

We show that there exists a constant $R > 0$ such that for all $m \geq 0$ we have

$$\big\|\mathscr{C}_m\big\|_{L^2 \to L^2} \leq R^m L^m . \tag{8.3.42}$$

We prove (8.3.42) by induction. We note that (8.3.42) is trivially true when $m = 0$, since $\mathscr{C}_0 = -iH$, where H is the Hilbert transform.

Assume that (8.3.42) holds for a certain m. We show its validity for $m+1$. Recall that K_m is a kernel in $SK(1, 16(2m+1)L^m)$ by the discussion in Example 8.1.7. We need the following estimate proved in Theorem 8.2.7:

$$\big\|\mathscr{C}_m\big\|_{L^\infty \to BMO} \leq C_2\big[16(2m+1)L^m + \big\|\mathscr{C}_m\big\|_{L^2 \to L^2}\big], \tag{8.3.43}$$

which holds for some absolute constant C_2.

We start with the following consequence of Theorem 8.3.3:

$$\big\|\mathscr{C}_{m+1}\big\|_{L^2 \to L^2} \leq C_1\big[\big\|\mathscr{C}_{m+1}(1)\big\|_{BMO} + \big\|(C_{m+1})^t(1)\big\|_{BMO} + \big\|\mathscr{C}_{m+1}\big\|_{WB}\big], \tag{8.3.44}$$

valid for some absolute constant C_1. The key observation is that

$$\mathscr{C}_{m+1}(1) = \mathscr{C}_m(A'), \tag{8.3.45}$$

for which we refer to Exercise 8.3.4. Here A' denotes the derivative of A, which exists almost everywhere, since Lipschitz functions are differentiable almost everywhere. Note that the kernel of $\mathscr{C}_m$ is antisymmetric; consequently, $(\mathscr{C}_m)^t = -\mathscr{C}_m$ and Exercise 8.3.3 gives that $\big\|\mathscr{C}_m\big\|_{WB} \leq C_3\, 16\,(2m+1)\, L^m$ for some other absolute constant C_3. Using all these facts we deduce from (8.3.44) that

$$\big\|\mathscr{C}_{m+1}\big\|_{L^2 \to L^2} \leq C_1\big[2\big\|\mathscr{C}_m(A')\big\|_{BMO} + C_3\, 16\,(2m+3)\, L^{m+1}\big].$$

Using (8.3.43) and the fact that $\big\|A'\big\|_{L^\infty} \leq L$ we obtain that

$$\big\|\mathscr{C}_{m+1}\big\|_{L^2 \to L^2} \leq C_1\Big[2C_2 L\big\{16\,(2m+1)L^m + \big\|\mathscr{C}_m\big\|_{L^2 \to L^2}\big\} + C_3\, 16\,(2m+3)\, L^{m+1}\Big].$$

Combining this estimate with the induction hypothesis (8.3.42), we obtain

$$\big\|\mathscr{C}_{m+1}(1)\big\|_{BMO} \leq R^{m+1} L^{m+1},$$

provided that R is chosen so that

$$\begin{aligned} R^{m+1} &> 96 C_1 C_2 (2m+1), \\ R &> 6 C_1 C_2, \\ R^{m+1} &> 48 C_1 C_3 (2m+3) \end{aligned}$$

for all $m \geq 0$. Such an R exists independent of m. This completes the proof of (8.3.42) by induction.

Exercises

8.3.1. Let T be a continuous linear operator from $\mathscr{S}(\mathbf{R}^n)$ to $\mathscr{S}'(\mathbf{R}^n)$ and let f be in $\mathscr{S}(\mathbf{R}^n)$. Let P_t be as in (8.3.15).
(a) Show that $P_t(f)$ converges to f in $\mathscr{S}(\mathbf{R}^n)$ as $t \to 0$.
(b) Conclude that $TP_t(f) \to T(f)$ in $\mathscr{S}'(\mathbf{R}^n)$ as $t \to 0$.
(c) Conclude that $P_tTP_t(f) \to T(f)$ in $\mathscr{S}'(\mathbf{R}^n)$ as $t \to 0$.
(d) Observe that (a)–(c) are also valid if P_t is replaced by P_t^2.
[*Hint:* Part (a): Use that $g_k \to g$ in $\mathscr{S}$ if and only if $\widehat{g_k} \to \widehat{g}$ in $\mathscr{S}$.]

8.3.2. Let T and P_t be as in Exercise 8.3.1 and let f be a Schwartz function whose Fourier transform vanishes in a neighborhood of the origin.
(a) Show that $P_t(f)$ converges to 0 in $\mathscr{S}(\mathbf{R}^n)$ as $t \to \infty$.
(b) Conclude that $TP_t(f) \to 0$ in $\mathscr{S}'(\mathbf{R}^n)$ as $t \to \infty$.
(c) Conclude that $P_tTP_t(f) \to 0$ in $\mathscr{S}'(\mathbf{R}^n)$ as $t \to \infty$.
(d) Observe that (a)–(c) are also valid if P_t is replaced by P_t^2.
[*Hint:* Part (a): Use the hint in Exercise 8.3.1 and the observation that $|\widehat{\Phi}(t\xi)\widehat{f}(\xi)| \le C(1+tc_0)^{-1}|\widehat{f}(\xi)|$ if $\widehat{f}$ is supported outside the ball $B(0,c_0)$. Part (c): Pair with a Schwartz function g and use part (a) and the fact that all Schwartz seminorms of $P_t(g)$ are bounded uniformly in $t > 0$. To prove the latter you may need that all Schwartz seminorms of $P_t(g)$ are bounded uniformly in $t > 0$ if and only if all Schwartz seminorms of $\widehat{P_t(g)}$ are bounded uniformly in $t > 0$.]

8.3.3. (a) Prove that every linear operator T from $\mathscr{S}(\mathbf{R}^n)$ to $\mathscr{S}'(\mathbf{R}^n)$ associated with an antisymmetric kernel in $SK(\delta,A)$ satisfies the weak boundedness property. Precisely, for some dimensional constant C_n we have

$$\big\|T\big\|_{WB} \le C_n A\,.$$

(b) Conclude that for some $c < \infty$, the Calderón commutators satisfy

$$\big\|\mathscr{C}_m\big\|_{WB} \le c\,16(2m+1)L^m\,.$$

[*Hint:* Write $\big\langle T(\tau^{x_0}(f_R)), \tau^{x_0}(g_R)\big\rangle$ as

$$\frac{1}{2}\int_{\mathbf{R}^n}\int_{\mathbf{R}^n} K(x,y)\big(\tau^{x_0}(f_R)(y)\tau^{x_0}(g_R)(x) - \tau^{x_0}(f_R)(x)\tau^{x_0}(g_R)(y)\big)\,dy\,dx$$

and use the mean value theorem.]

8.3.4. Prove identity (8.3.45). This identity is obvious by a formal integration by parts, but to prove it properly, one should interpret things in the sense of distributions.

8.3.5. Suppose that a standard kernel $K(x,y)$ has the form $k(x-y)$ for some function k on $\mathbf{R}^n \setminus \{0\}$. Suppose that k extends to a tempered distribution on $\mathbf{R}^n$ whose Fourier transform is a bounded function. Let T be a continuous linear operator from $\mathscr{S}(\mathbf{R}^n)$

to $\mathscr{S}'(\mathbf{R}^n)$ associated with K.
(a) Identify the functions $T(e^{2\pi i\xi\cdot()})$ and $T^t(e^{2\pi i\xi\cdot()})$ and restrict to $\xi = 0$ to obtain $T(1)$ and $T^t(1)$.
(b) Use Theorem 8.3.3 to obtain the L^2 boundedness of T.
(c) What are $H(1)$ and $H^t(1)$ equal to when H is the Hilbert transform?

8.3.6. (*A. Calderón*) Let A be a Lipschitz function on $\mathbf{R}$. Use expansion (8.3.40) and estimate (8.3.42) to show that the operator

$$\mathscr{C}_A(f)(x) = \frac{1}{\pi i}\lim_{\varepsilon\to 0}\int_{|x-y|>\varepsilon}\frac{f(y)\,dy}{x-y+i(A(x)-A(y))}$$

is bounded on $L^2(\mathbf{R})$ when $\|A'\|_{L^\infty} < R^{-1}$, where R satisfies (8.3.43).

8.3.7. Prove that condition (i) of Theorem 8.3.3 is equivalent to the statement that

$$\sup_Q\sup_{\varepsilon>0}\left(\frac{\|T^{(\varepsilon)}(\chi_Q)\|_{L^2}}{|Q|^{\frac{1}{2}}}+\frac{\|(T^{(\varepsilon)})^t(\chi_Q)\|_{L^2}}{|Q|^{\frac{1}{2}}}\right)=B_1'<\infty,$$

where the first supremum is taken over all cubes Q in $\mathbf{R}^n$.
[*Hint:* You may repeat the argument in the equivalence (i) $\Longleftrightarrow$ (ii) replacing the ball $B(x_0,N)$ by a cube centered at x_0 with side length N.]

8.4 Paraproducts

In this section we study a useful class of operators called paraproducts. Their name suggests they are related to products; in fact, they are "half products" in some sense that needs to be made precise. Paraproducts provide interesting examples of nonconvolution operators with standard kernels whose L^2 boundedness was discussed in the Section 8.3. They have found use in many situations, including a proof of the main implication in Theorem 8.3.3. This proof is discussed in the present section.

8.4.1 Introduction to Paraproducts

Throughout this section we fix a Schwartz radial function Ψ whose Fourier transform is supported in the annulus $\frac{1}{2}\le|\xi|\le 2$ and that satisfies

$$\sum_{j\in\mathbf{Z}}\widehat{\Psi}(2^{-j}\xi)=1,\qquad\text{when}\quad \xi\in\mathbf{R}^n\setminus\{0\}.\tag{8.4.1}$$

Associated with this Ψ we define the Littlewood–Paley operator $\Delta_j(f) = f*\Psi_{2^{-j}}$, where $\Psi_t(x) = t^{-n}\Psi(t^{-1}x)$. Using (8.4.1), we easily obtain

$$\sum_{j\in\mathbf{Z}} \Delta_j = I, \tag{8.4.2}$$

where (8.4.2) is interpreted as an identity on Schwartz functions with mean value zero. See Exercise 8.4.1. Note that by construction, the function Ψ is radial and thus even. This makes the operator Δ_j equal to its transpose.

We now observe that in view of the properties of Ψ, the function

$$\xi \mapsto \sum_{j\le 0} \widehat{\Psi}(2^{-j}\xi) \tag{8.4.3}$$

is supported in $|\xi| \le 2$, and is equal to 1 when $0 < |\xi| \le \frac{1}{2}$. But $\widehat{\Psi}(0) = 0$, which implies that the function in (8.4.3) also vanishes at the origin. We can easily fix this discontinuity by introducing the Schwartz function whose Fourier transform is equal to

$$\widehat{\Phi}(\xi) = \begin{cases} \sum_{j\le 0} \widehat{\Psi}(2^{-j}\xi) & \text{when } \xi \neq 0, \\ 1 & \text{when } \xi = 0. \end{cases}$$

Definition 8.4.1. We define the *partial sum operator* S_j as

$$S_j = \sum_{k\le j} \Delta_k. \tag{8.4.4}$$

In view of the preceding discussion, S_j is given by convolution with $\Phi_{2^{-j}}$, that is,

$$S_j(f)(x) = (f * \Phi_{2^{-j}})(x), \tag{8.4.5}$$

and the expression in (8.4.5) is well defined for all f in $\bigcup_{1\le p\le\infty} L^p(\mathbf{R}^n)$. Since Φ is a radial function by construction, the operator S_j is self-transpose.

Similarly, $\Delta_j(g)$ is also well defined for all g in $\bigcup_{1\le p\le\infty} L^p(\mathbf{R}^n)$. Moreover, since Δ_j is given by convolution with a function with mean value zero, it also follows that $\Delta_j(b)$ is well defined when $b \in BMO(\mathbf{R}^n)$. See Exercise 8.4.2 for details.

Definition 8.4.2. Given a function g on $\mathbf{R}^n$, we define the *paraproduct operator* P_g as follows:

$$P_g(f) = \sum_{j\in\mathbf{Z}} \Delta_j(g) S_{j-3}(f) = \sum_{j\in\mathbf{Z}} \sum_{k\le j-3} \Delta_j(g)\Delta_k(f), \tag{8.4.6}$$

for f in $L^1_{\text{loc}}(\mathbf{R}^n)$. It is not clear for which functions g and in what sense the series in (8.4.6) converges even when f is a Schwartz function. One may verify that the series in (8.4.6) converges absolutely almost everywhere when g is a Schwartz function with mean value zero; in this case, by Exercise 8.4.1 the series $\sum_j \Delta_j(g)$ converges absolutely (everywhere) and $S_j(f)$ is uniformly bounded by the Hardy–Littlewood maximal function $M(f)$, which is finite almost everywhere.

One of the main goals of this section is to show that the series in (8.4.6) converges in L^2 when f is in $L^2(\mathbf{R}^n)$ and g is a BMO function.

The name *paraproduct* is derived from the fact that $P_g(f)$ is essentially "half" the product of fg. Namely, in view of the identity in (8.4.2) the product fg can be written as

$$fg = \sum_j \sum_k \Delta_j(f)\,\Delta_k(g).$$

Restricting the summation of the indices to $k < j$ defines an operator that corresponds to "half" the product of fg. It is only for minor technical reasons that we take $k \le j-3$ in (8.4.6).

The main feature of the paraproduct operator P_g is that it is essentially a sum of orthogonal L^2 functions. Indeed, the Fourier transform of the function $\widehat{\Delta_j(g)}$ is supported in the set

$$\{\xi \in \mathbf{R}^n : \, 2^{j-1} \le |\xi| \le 2^{j+1}\},$$

while the Fourier transform of the function $\widehat{S_{j-3}(f)}$ is supported in the set

$$\bigcup_{k \le j-3} \{\xi \in \mathbf{R}^n : \, 2^{k-1} \le |\xi| \le 2^{k+1}\}.$$

This implies that the Fourier transform of the function $\Delta_j(g)\,S_{j-3}(f)$ is supported in the algebraic sum

$$\{\xi \in \mathbf{R}^n : \, 2^{j-1} \le |\xi| \le 2^{j+1}\} + \{\xi \in \mathbf{R}^n : \, |\xi| \le 2^{j-2}\}.$$

But this sum is contained in the set

$$\{\xi \in \mathbf{R}^n : \, 2^{j-2} \le |\xi| \le 2^{j+2}\}, \tag{8.4.7}$$

and the family of sets in (8.4.7) is "almost disjoint" as j varies. This means that every point in $\mathbf{R}^n$ belongs to at most four annuli of the form (8.4.7). Therefore, the paraproduct $P_g(f)$ can be written as a sum of functions h_j such that the families $\{h_j : \, j \in 4\mathbf{Z}+r\}$ are mutually orthogonal in L^2, for all $r \in \{0,1,2,3\}$. This orthogonal decomposition of the paraproduct has as an immediate consequence its L^2 boundedness when g is an element of BMO.

8.4.2 L^2 Boundedness of Paraproducts

The following theorem is the main result of this subsection.

Theorem 8.4.3. *For fixed $b \in BMO(\mathbf{R}^n)$ and $f \in L^2(\mathbf{R}^n)$ the series*

$$\sum_{|j| \le M} \Delta_j(b)\,S_{j-3}(f)$$

converges in L^2 as $M \to \infty$ to a function that we denote by $P_b(f)$. The operator P_b thus defined is bounded on $L^2(\mathbf{R}^n)$, and there is a dimensional constant C_n such that *for all $b \in BMO(\mathbf{R}^n)$ we have*

$$\big\|P_b\big\|_{L^2\to L^2} \le C_n\big\|b\big\|_{BMO}.$$

Proof. The proof of this result follows by putting together some of the powerful ideas developed in Chapter 7. First we define a measure on $\mathbf{R}^{n+1}_+$ by setting

$$d\mu(x,t) = \sum_{j\in\mathbf{Z}} |\Delta_j(b)(x)|^2\,dx\,\delta_{2^{-(j-3)}}(t).$$

By Theorem 7.3.8 we have that μ is a Carleson measure on $\mathbf{R}^{n+1}_+$ whose norm is controlled by a constant multiple of $\big\|b\big\|^2_{BMO}$. Now fix $f \in L^2(\mathbf{R}^n)$ and recall that $\Phi(x) = \sum_{r\le 0}\Psi_{2^{-r}}(x)$. We define a function $F(x,t)$ on $\mathbf{R}^{n+1}_+$ by setting

$$F(x,t) = (\Phi_t * f)(x).$$

Observe that $F(x,2^{-k}) = S_k(f)(x)$ for all $k \in \mathbf{Z}$. We estimate the L^2 norm of a finite sum of terms of the form $\Delta_j(b)\,S_{j-3}(f)$. For $M,N \in \mathbf{Z}^+$ with $M \ge N$ we have

$$\begin{aligned}\int_{\mathbf{R}^n}\Big|\sum_{N\le|j|\le M}\Delta_j(b)(x)\,S_{j-3}(f)(x)\Big|^2dx \\ = \int_{\mathbf{R}^n}\Big|\sum_{N\le|j|\le M}\big(\Delta_j(b)\,S_{j-3}(f)\big)\widehat{\ }(\xi)\Big|^2 d\xi\,.\end{aligned}\tag{8.4.8}$$

It is a simple fact that every $\xi \in \mathbf{R}^n$ belongs to at most four annuli of the form (8.4.7). It follows that at most four terms in the last sum in (8.4.8) are nonzero. Thus

$$\begin{aligned}\int_{\mathbf{R}^n}\Big|\sum_{N\le|j|\le M}&\big(\Delta_j(b)S_{j-3}(f)\big)\widehat{\ }(\xi)\Big|^2 d\xi \\ &\le 4\sum_{N\le|j|\le M}\int_{\mathbf{R}^n}\big|\big(\Delta_j(b)\,S_{j-3}(f)\big)\widehat{\ }(\xi)\big|^2 d\xi \\ &\le 4\sum_{j\in\mathbf{Z}}\int_{\mathbf{R}^n}\big|\Delta_j(b)(x)S_{j-3}(f)(x)\big|^2 dx \\ &= 4\int_{\mathbf{R}^n}\big|F(x,t)\big|^2 d\mu(x,t) \\ &\le C_n\big\|b\big\|^2_{BMO}\int_{\mathbf{R}^n}F^*(x)^2\,dx,\end{aligned}\tag{8.4.9}$$

where we used Corollary 7.3.6 in the last inequality.

Next we note that the nontangential maximal function F^* of F is controlled by the Hardy–Littlewood maximal function of f. Indeed, since Φ is a Schwartz function, we have

$$F^*(x) \le C_n \sup_{t>0} \sup_{|y-x|<t} \int_{\mathbf{R}^n} \frac{1}{t^n} \frac{|f(z)|}{(1+\frac{|z-y|}{t})^{n+1}}\,dz. \tag{8.4.10}$$

Now break the previous integral into parts such that $|z-y| \ge 3t$ and $|z-y| \le 3t$. In the first case we have $|z-y| \ge |z-x| - t \ge \frac{1}{2}|z-x|$, and the last inequality is valid, since $|z-x| \ge |z-y| - t \ge 2t$. Using this estimate together with Theorem 2.1.10 we obtain that this part of the integral is controlled by a constant multiple of $M(f)(x)$. The part of the integral in (8.4.10) where $|z-y| \le 3t$ is controlled by the integral over the larger set $|z-x| \le 4t$, and since the denominator in (8.4.10) is always bounded by 1, we also obtain that this part of the integral is controlled by a constant multiple of $M(f)(x)$. We conclude that

$$\int_{\mathbf{R}^n} F^*(x)^2\,dx \le C_n \int_{\mathbf{R}^n} M(f)(x)^2\,dx \le C_n \int_{\mathbf{R}^n} |f(x)|^2\,dx. \tag{8.4.11}$$

Combining (8.4.9) and (8.4.11), we obtain the estimate

$$4\sum_{j\in\mathbf{Z}} \int_{\mathbf{R}^n} |(\Delta_j(b)\,S_{j-3}(f))\widehat{\ }(\xi)|^2\,d\xi \le C_n \|b\|_{BMO}^2 \|f\|_{L^2}^2 < \infty.$$

This implies that given $\varepsilon > 0$, we can find an $N_0 > 0$ such that

$$M \ge N \ge N_0 \implies \sum_{N\le|j|\le M} \int_{\mathbf{R}^n} \big|(\Delta_j(b)\,S_{j-3}(f))\widehat{\ }(\xi)\big|^2\,d\xi < \varepsilon.$$

But recall from (8.4.8) and (8.4.9) that

$$\int_{\mathbf{R}^n} \bigg| \sum_{N\le|j|\le M} \Delta_j(b)(x)\,S_{j-3}(f)(x)\bigg|^2\,dx \le 4 \sum_{N\le|j|\le M} \int_{\mathbf{R}^n} \big|(\Delta_j(b)S_{j-3}(f))\widehat{\ }(\xi)\big|^2\,d\xi.$$

We conclude that the sequence

$$\bigg\{ \sum_{|j|\le M} \Delta_j(b) S_{j-3}(f) \bigg\}_M$$

is Cauchy in $L^2(\mathbf{R}^n)$, and therefore it converges in L^2 to a function $P_b(f)$. The boundedness of P_b on L^2 follows from the sequence of inequalities already proved. □

8.4.3 Fundamental Properties of Paraproducts

Having established the L^2 boundedness of paraproducts, we turn to some properties that they possess. First we study their kernels. Paraproducts are not operators of convolution type but are more general integral operators of the form discussed in Section 8.1. We show that the kernel of P_b is a tempered distribution L_b that coincides with a standard kernel on $\mathbf{R}^n \times \mathbf{R}^n \setminus \{(x,x) : x \in \mathbf{R}^n\}$.

First we study the kernel of the operator $f \mapsto \Delta_j(b)S_{j-3}(f)$ for any $j \in \mathbf{Z}$. We have that

$$\Delta_j(b)(x)S_{j-3}(f)(x) = \int_{\mathbf{R}^n} L_j(x,y)f(y)\,dy,$$

where L_j is the integrable function

$$L_j(x,y) = (b * \Psi_{2^{-j}})(x)2^{(j-3)n}\Phi(2^{j-3}(x-y)).$$

Next we can easily verify the following size and regularity estimates for L_j:

$$|L_j(x,y)| \le C_n \|b\|_{BMO} \frac{2^{nj}}{(1+2^j|x-y|)^{n+1}}, \tag{8.4.12}$$

$$|\partial_x^\alpha \partial_y^\beta L_j(x,y)| \le C_{n,\alpha,\beta,N} \|b\|_{BMO} \frac{2^{j(n+|\alpha|+|\beta|)}}{(1+2^j|x-y|)^{n+1+N}}, \tag{8.4.13}$$

for all multi-indices α and β and all $N \ge |\alpha| + |\beta|$.

It follows from (8.4.12) that when $x \ne y$ the series

$$\sum_{j\in\mathbf{Z}} L_j(x,y) \tag{8.4.14}$$

converges absolutely and is controlled in absolute value by

$$C_n \|b\|_{BMO} \sum_{j\in\mathbf{Z}} \frac{2^{nj}}{(1+2^j|x-y|)^{n+1}} \le \frac{C_n' \|b\|_{BMO}}{|x-y|^n}.$$

Similarly, by taking $N \ge |\alpha| + |\beta|$, it can be shown that the series

$$\sum_{j\in\mathbf{Z}} \partial_x^\alpha \partial_y^\beta L_j(x,y) \tag{8.4.15}$$

converges absolutely when $x \ne y$ and is controlled in absolute value by

$$C_{n,\alpha,\beta,N} \|b\|_{BMO} \sum_{j\in\mathbf{Z}} \frac{2^{j(n+|\alpha|+|\beta|)}}{(1+2^j|x-y|)^{n+1+N}} \le \frac{C'_{n,\alpha,\beta} \|b\|_{BMO}}{|x-y|^{n+|\alpha|+|\beta|}}$$

for all multi-indices α and β.

The Schwartz kernel of P_b is a distribution W_b on $\mathbf{R}^{2n}$. It follows from the preceding discussion that the distribution W_b coincides with the function

$$L_b(x,y) = \sum_{j\in\mathbf{Z}} L_j(x,y)$$

on $\mathbf{R}^n \times \mathbf{R}^n \setminus \{(x,x) : x \in \mathbf{R}^n\}$, and also that the function L_b satisfies the estimates

$$|\partial_x^\alpha \partial_y^\beta L_b(x,y)| \le \frac{C'_{n,\alpha,\beta}\|b\|_{BMO}}{|x-y|^{n+|\alpha|+|\beta|}} \tag{8.4.16}$$

away from the diagonal $x = y$.

We note that the transpose of the operator P_b is formally given by the identity

$$P_b^t(f) = \sum_{j\in\mathbf{Z}} S_{j-3}(f\Delta_j(b)).$$

As remarked in the previous section, the kernel of the operator P_b^t is a distribution W_b^t that coincides with the function

$$L_b^t(x,y) = L_b(y,x)$$

away from the diagonal of $\mathbf{R}^{2n}$. It is trivial to observe that L_b^t satisfies the same size and regularity estimates (8.4.16) as L_b. Moreover, it follows from Theorem 8.4.3 that the operator P_b^t is bounded on $L^2(\mathbf{R}^n)$ with norm at most a multiple of the BMO norm of b.

We now turn to two important properties of paraproducts. In view of Definition 8.1.16, we have a meaning for $P_b(1)$ and $P_b^t(1)$, where P_b is the paraproduct operator. The first property we prove is that $P_b(1) = b$. Observe that this statement is trivially valid at a formal level, since $S_j(1) = 1$ for all j and $\sum_j \Delta_j(b) = b$. The second property is that $P_b^t(1) = 0$. This is also trivially checked at a formal level, since $S_{j-3}(\Delta_j(b)) = 0$ for all j, as a Fourier transform calculation shows. We make both of these statements precise in the following proposition.

Proposition 8.4.4. *Given $b \in BMO(\mathbf{R}^n)$, let P_b be the paraproduct operator defined as in (8.4.6). Then the distributions $P_b(1)$ and $P_b^t(1)$ coincide with elements of BMO. Precisely, we have*

$$P_b(1) = b \qquad \textit{and} \qquad P_b^t(1) = 0. \tag{8.4.17}$$

Proof. Let φ be an element of $\mathscr{D}_0(\mathbf{R}^n)$. Find a uniformly bounded sequence of smooth functions with compact support $\{\eta_N\}_{N=1}^\infty$ that converges to the function 1 as $N \to \infty$. Without loss of generality assume that all the functions η_N are equal to 1 on the ball $B(y_0, 3R)$, where $B(y_0,R)$ is a ball that contains the support of φ. As we observed in Remark 8.1.17, the definition of $P_b(1)$ is independent of the choice of sequence η_N, so we have the following identity for all $N \ge 1$:

$$\begin{aligned}\langle P_b(1),\varphi\rangle = &\int_{\mathbf{R}^n} \sum_{j\in\mathbf{Z}} \Delta_j(b)(x)\, S_{j-3}(\eta_N)(x)\,\varphi(x)\,dx \\ &+ \int_{\mathbf{R}^n}\left[\int_{\mathbf{R}^n} L_b(x,y)\,\varphi(x)\,dx\right](1-\eta_N(y))\,dy.\end{aligned} \tag{8.4.18}$$

Since φ has mean value zero, we can subtract the constant $L_b(y_0,y)$ from $L_b(x,y)$ in the integral inside the square brackets in (8.4.18). Then we estimate the absolute value of the double integral in (8.4.18) by

$$\int_{|y-y_0|\ge 3R}\int_{|x-y_0|\le R} A\frac{|y_0-x|}{|y_0-y|^{n+1}}|1-\eta_N(y)|\,|\varphi(x)|\,dx\,dy,$$

which tends to zero as $N\to\infty$ by the Lebesgue dominated convergence theorem.

It suffices to prove that the first integral in (8.4.18) tends to $\int_{\mathbf{R}^n} b(x)\varphi(x)\,dx$ as $N\to\infty$. Let us make some preliminary observations. Since the Fourier transform of the product $\Delta_j(b)S_{j-3}(\eta_N)$ is supported in the annulus

$$\{\xi\in\mathbf{R}^n:\ 2^{j-2}\le|\xi|\le 2^{j+2}\},\tag{8.4.19}$$

we may introduce a smooth and compactly supported function $\widehat{Z}(\xi)$ such that for all $j\in\mathbf{Z}$ the function $\widehat{Z}(2^{-j}\xi)$ is equal to 1 on the annulus (8.4.19) and vanishes outside the annulus $\{\xi\in\mathbf{R}^n:\ 2^{j-3}\le|\xi|\le 2^{j+3}\}$. Let us denote by Q_j the operator given by multiplication on the Fourier transform by the function $\widehat{Z}(2^{-j}\xi)$.

Note that $S_j(1)$ is well defined and equal to 1 for all j. This is because Φ has integral equal to 1. Also, the duality identity

$$\int f\,S_j(g)\,dx=\int g\,S_j(f)\,dx\tag{8.4.20}$$

holds for all $f\in L^1$ and $g\in L^\infty$. For φ in $\mathscr{D}_0(\mathbf{R}^n)$ we have

$$\begin{aligned}
&\int_{\mathbf{R}^n}\sum_{j\in\mathbf{Z}}\Delta_j(b)\,S_{j-3}(\eta_N)\,\varphi\,dx\\
&=\sum_{j\in\mathbf{Z}}\int_{\mathbf{R}^n}\Delta_j(b)\,S_{j-3}(\eta_N)\,\varphi\,dx && \text{(series converges in } L^2 \text{ and } \varphi\in L^2)\\
&=\sum_{j\in\mathbf{Z}}\int_{\mathbf{R}^n}\Delta_j(b)\,S_{j-3}(\eta_N)\,Q_j(\varphi)\,dx && \big[\widehat{Q_j(\varphi)}=\widehat{\varphi}\text{ on the}\\
& && \text{support of }\big((\Delta_j(b)\,S_{j-3}(\eta_N))^\wedge\big]\\
&=\sum_{j\in\mathbf{Z}}\int_{\mathbf{R}^n}\eta_N\,S_{j-3}\big(\Delta_j(b)Q_j(\varphi)\big)\,dx && \text{(duality)}\\
&=\int_{\mathbf{R}^n}\eta_N\sum_{j\in\mathbf{Z}}S_{j-3}\big(\Delta_j(b)Q_j(\varphi)\big)\,dx && \text{(series converges in } L^1 \text{ and } \eta_N\in L^\infty).
\end{aligned}$$

We now explain why the last series of the foregoing converges in L^1. Since φ is in $\mathscr{D}_0(\mathbf{R}^n)$, Exercise 8.4.1 gives that the series $\sum_{j\in\mathbf{Z}}Q_j(\varphi)$ converges in L^1. Since S_j preserves L^1 and

$$\sup_j\big\|\Delta_j(b)\big\|_{L^\infty}\le C_n\big\|b\big\|_{BMO}$$

by Exercise 8.4.2, it follows that the series $\sum_{j\in\mathbf{Z}}S_{j-3}\big(\Delta_j(b)Q_j(\varphi)\big)$ also converges in L^1.

We now use the Lebesgue dominated convergence theorem to obtain that the expression

$$\int_{\mathbf{R}^n} \eta_N \sum_{j\in\mathbf{Z}} S_{j-3}\big(\Delta_j(b)Q_j(\varphi)\big)\,dx$$

converges as $N \to \infty$ to

$$\begin{aligned}
&\int_{\mathbf{R}^n} \sum_{j\in\mathbf{Z}} S_{j-3}\big(\Delta_j(b)Q_j(\varphi)\big)\,dx \\
&\qquad = \sum_{j\in\mathbf{Z}} \int_{\mathbf{R}^n} S_{j-3}\big(\Delta_j(b)Q_j(\varphi)\big)\,dx && \text{(series converges in } L^1) \\
&\qquad = \sum_{j\in\mathbf{Z}} \int_{\mathbf{R}^n} S_{j-3}(1)\,\Delta_j(b)\,Q_j(\varphi)\,dx && \text{(in view of (8.4.20))} \\
&\qquad = \sum_{j\in\mathbf{Z}} \int_{\mathbf{R}^n} \Delta_j(b)\,Q_j(\varphi)\,dx && \text{(since } S_{j-3}(1)=1) \\
&\qquad = \sum_{j\in\mathbf{Z}} \int_{\mathbf{R}^n} \Delta_j(b)\,\varphi\,dx && \big(\widehat{Q_j(\varphi)} = \widehat{\varphi} \text{ on support } \widehat{\Delta_j(b)}\big) \\
&\qquad = \sum_{j\in\mathbf{Z}} \big\langle b, \Delta_j(\varphi)\big\rangle && \text{(duality)} \\
&\qquad = \big\langle b, \sum_{j\in\mathbf{Z}} \Delta_j(\varphi)\big\rangle && \text{(series converges in } H^1,\ b\in BMO) \\
&\qquad = \big\langle b, \varphi\big\rangle && \text{(Exercise 8.4.1(a)).}
\end{aligned}$$

Regarding the fact that the series $\sum_j \Delta_j(\varphi)$ converges in H^1, we refer to Exercise 8.4.1. We now obtain that the first integral in (8.4.18) tends to $\langle b, \varphi\rangle$ as $N \to \infty$. We have therefore proved that

$$\big\langle P_b(1), \varphi\big\rangle = \big\langle b, \varphi\big\rangle$$

for all φ in $\mathscr{D}_0(\mathbf{R}^n)$. In other words, we have now identified $P_b(1)$ as an element of $\mathscr{D}_0'$ with the BMO function b.

For the transpose operator P_b^t we observe that we have the identity

$$\begin{aligned}
\big\langle P_b^t(1), \varphi\big\rangle = &\int_{\mathbf{R}^n} \sum_{j\in\mathbf{Z}} S_{j-3}^t\big(\Delta_j(b)\,\eta_N\big)(x)\,\varphi(x)\,dx \\
&+ \int_{\mathbf{R}^n}\int_{\mathbf{R}^n} L_b^t(x,y)(1-\eta_N(y))\,\varphi(x)\,dy\,dx.
\end{aligned} \tag{8.4.21}$$

As before, we can use the Lebesgue dominated convergence theorem to show that the double integral in (8.4.21) tends to zero. As for the first integral in (8.4.21), we have the identity

$$\int_{\mathbf{R}^n} P_b^t(\eta_N)\,\varphi\,dx = \int_{\mathbf{R}^n} \eta_N\,P_b(\varphi)\,dx.$$

Since φ is a multiple of an L^2-atom for H^1, Theorem 8.2.6 gives that $P_b(\varphi)$ is an L^1 function. The Lebesgue dominated convergence theorem now implies that

$$\int_{\mathbf{R}^n} \eta_N P_b(\varphi)\,dx \to \int_{\mathbf{R}^n} P_b(\varphi)\,dx = \int_{\mathbf{R}^n} \sum_{j\in\mathbf{Z}} \Delta_j(b)\,S_{j-3}(\varphi)\,dx$$

as $N \to \infty$. The required conclusion would follow if we could prove that the function $P_b(\varphi)$ has integral zero. Since $\Delta_j(b)$ and $S_{j-3}(\varphi)$ have disjoint Fourier transforms, it follows that

$$\int_{\mathbf{R}^n} \Delta_j(b)\,S_{j-3}(\varphi)\,dx = 0$$

for all j in $\mathbf{Z}$. But the series

$$\sum_{j\in\mathbf{Z}} \Delta_j(b)\,S_{j-3}(\varphi) \tag{8.4.22}$$

defining $P_b(\varphi)$ converges in L^2 and not necessarily in L^1, and for this reason we need to justify the interchange of the following integrals:

$$\int_{\mathbf{R}^n} \sum_{j\in\mathbf{Z}} \Delta_j(b)\,S_{j-3}(\varphi)\,dx = \sum_{j\in\mathbf{Z}} \int_{\mathbf{R}^n} \Delta_j(b)\,S_{j-3}(\varphi)\,dx. \tag{8.4.23}$$

To complete the proof, it suffices to show that when φ is in $\mathscr{D}_0(\mathbf{R}^n)$, the series in (8.4.22) converges in L^1. To prove this, pick a ball $B(y_0,R)$ that contains the support of φ. The series in (8.4.22) converges in $L^2(3B)$ and hence converges in $L^1(3B)$. It remains to prove that it converges in $L^1((3B)^c)$. For a fixed $x \in (3B)^c$ and a finite subset F of $\mathbf{Z}$, we have

$$\sum_{j\in F} \int_{\mathbf{R}^n} L_j(x,y)\varphi(y)\,dy = \sum_{j\in F} \int_B \big(L_j(x,y) - L_j(x,y_0)\big)\,\varphi(y)\,dy. \tag{8.4.24}$$

Using estimates (8.4.13), we obtain that the expression in (8.4.24) is controlled by a constant multiple of

$$\int_B \sum_{j\in F} \frac{|y-y_0|2^{nj}2^j}{(1+2^j|x-y_0|)^{n+2}}|\varphi(y)|\,dy \le c\,\frac{1}{|x-y_0|^{n+1}} \int_{\mathbf{R}^n} |y-y_0|\,|\varphi(y)|\,dy.$$

Integrating this estimate with respect to $x \in (3B)^c$, we obtain that

$$\sum_{j\in F} \big\|\Delta_j(b)S_{j-3}(\varphi)\big\|_{L^1((3B)^c)} \le C_n \big\|\varphi\big\|_{L^1} < \infty \;.$$

for all finite subsets F of $\mathbf{Z}$. This proves that the series in (8.4.22) converges in L^1.

We have now proved that $\langle P_b^t(1), \varphi\rangle = 0$ for all $\varphi \in \mathscr{D}_0(\mathbf{R}^n)$. This shows that the distribution $P_b^t(1)$ is a constant function, which is of course identified with zero if considered as an element of BMO. □

Remark 8.4.5. The boundedness of P_b on L^2 is a consequence of Theorem 8.3.3, since hypothesis (iv) is satisfied. Indeed, $P_b(1) = b$, $P_b^t(1) = 0$ are both BMO functions, and see Exercise 8.4.4 for a sketch of a proof of the estimate $\big\|P_b\big\|_{WB} \le C_n\big\|b\big\|_{BMO}$. This provides another proof of the fact that $\big\|P_b\big\|_{L^2\to L^2} \le C_n\big\|b\big\|_{BMO}$, bypassing Theorem 8.3.3. We use this result to obtain a different proof of the main direction in the $T(1)$ theorem in the next section.

Exercises

8.4.1. Let $f \in \mathscr{S}(\mathbf{R}^n)$ have mean value zero, and consider the series

$$\sum_{j\in\mathbf{Z}} \Delta_j(f).$$

(a) Show that this series converges to f absolutely everywhere.
(b) Show that this series converges in L^1.
(b) Show that this series converges in H^1.
[*Hint:* To obtain convergence in L^1 for $j \ge 0$ use the estimate $\big\|\Delta_j(f)\big\|_{L^1} \le 2^{-j}\int_{\mathbf{R}^n}\int_{\mathbf{R}^n} 2^{jn}|\Psi(2^j y)|\,|2^j y|\,|(\nabla f)(x-\theta y)|\,dy\,dx$ for some θ in $[0,1]$ and consider the cases $|x| \ge 2|y|$ and $|x| \le 2|y|$. When $j \le 0$ use the simple identity $f * \Psi_{2^{-j}} = (f_{2^j} * \Psi)_{2^{-j}}$ and reverse the roles of f and Ψ. To show convergence in H^1, use that $\big\|\Delta_j(\varphi)\big\|_{H^1} \approx \big\|(\sum_k |\Delta_k\Delta_j(\varphi)|^2)^{\frac{1}{2}}\big\|_{L^1}$ and that only at most three terms in the square function are nonzero.]

8.4.2. Without appealing to the H^1-BMO duality theorem, prove that there is a dimensional constant C_n such that for all $b \in BMO(\mathbf{R}^n)$ we have

$$\sup_{j\in\mathbf{Z}} \big\|\Delta_j(b)\big\|_{L^\infty} \le C_n \big\|b\big\|_{BMO}.$$

8.4.3. (a) Show that for all $1 < p,q,r < \infty$ with $\frac{1}{p}+\frac{1}{q}=\frac{1}{r}$ there is a constant C_{pqr} such that for all Schwartz functions f,g on $\mathbf{R}^n$ we have

$$\big\|P_g(f)\big\|_{L^r} \le C_{pqr}\big\|f\big\|_{L^p}\big\|g\big\|_{L^q}.$$

(b) Obtain the same conclusion for the bilinear operator

$$\widetilde{P}_g(f) = \sum_j \sum_{k\le j} \Delta_j(g)\,\Delta_k(f).$$

[*Hint:* Part (a): Estimate the L^r norm using duality. Part (b): Use part (a).]

8.4.4. (a) Let f be a normalized bump (see Definition 8.3.1). Prove that

$$\big\|\Delta_j(f_R)\big\|_{L^\infty} \le C(n,\Psi)\min\big(2^{-j}R^{-(n+1)}, 2^{nj}\big)$$

for all $R > 0$. Then interpolate between L^1 and L^∞ to obtain

$$\big\|\Delta_j(f_R)\big\|_{L^2} \le C(n,\Psi)\min\Big(2^{-\frac{j}{2}}R^{-\frac{n+1}{2}}, 2^{\frac{nj}{2}}\Big).$$

(b) Observe that the same result is valid for the operators Q_j as defined in Proposition 8.4.4. Conclude that for some constant C_n we have

$$\sum_{j\in\mathbf{Z}} \big\|Q_j(g_R)\big\|_{L^2} \le C_n R^{-\frac{n}{2}}.$$

(c) Show that there is a constant C_n such that for all normalized bumps f and g we have

$$\big|\big\langle P_b(\tau^{x_0}(f_R)), \tau^{x_0}(g_R)\big\rangle\big| \le C_n R^{-n}\big\|b\big\|_{BMO}.$$

[*Hint:* Part (a): Use the cancellation of the functions f and Ψ. Part (c): Write

$$\big\langle P_b(\tau^{x_0}(f_R)), \tau^{x_0}(g_R)\big\rangle = \sum_j \int_{\mathbf{R}^n} S_{j-3}[\Delta_j(\tau^{-x_0}(b))Q_j(g_R)]f_R\,dx.$$

Apply the Cauchy–Schwarz inequality, and use the boundedness of S_{j-3} on L^2, Exercise 8.4.2, and part (b).]

8.4.5. (*Continuous paraproducts*) (a) Let Φ and Ψ be Schwartz functions on $\mathbf{R}^n$ with $\int_{\mathbf{R}^n}\Phi(x)\,dx = 1$ and $\int_{\mathbf{R}^n}\Psi(x)\,dx = 0$. For $t>0$ define operators $P_t(f) = \Phi_t * f$ and $Q_t(f) = \Psi_t * f$. Let $b\in BMO(\mathbf{R}^n)$ and $f\in L^2(\mathbf{R}^n)$. Show that the limit

$$\lim_{\substack{\varepsilon\to 0\\ N\to\infty}} \int_\varepsilon^N Q_t\big(Q_t(b)\,P_t(f)\big)\,\frac{dt}{t}$$

converges in $L^2(\mathbf{R}^n)$ and defines an operator $\Pi_b(f)$ that satisfies

$$\big\|\Pi_b\big\|_{L^2\to L^2} \le C_n\big\|b\big\|_{BMO}$$

for some dimensional constant C_n.
(b) Under the additional assumption that

$$\lim_{\substack{\varepsilon\to 0\\ N\to\infty}} \int_\varepsilon^N Q_t^2\,\frac{dt}{t} = I,$$

identify $\Pi_b(1)$ and $\Pi_b(b)$.
[*Hint:* Suitably adapt the proofs of Theorem 8.4.3 and Proposition 8.4.4.]

8.5 An Almost Orthogonality Lemma and Applications

In this section we discuss an important lemma regarding orthogonality of operators and some of its applications.

It is often the case that a linear operator T is given as an infinite sum of other linear operators T_j such that the T_j's are uniformly bounded on L^2. This sole condition is not enough to imply that the sum of the T_j's is also L^2 bounded, although this is often the case. Let us consider, for instance, the linear operators $\{T_j\}_{j\in\mathbf{Z}}$ given by convolution with the smooth functions $e^{2\pi i j t}$ on the circle $\mathbf{T}^1$. Each T_j can be written as $T_j(f) = (\widehat{f} \otimes \delta_j)^\vee$, where $\widehat{f}$ is the sequence of Fourier coefficients of f; here δ_j is the infinite sequence consisting of zeros everywhere except at the jth entry, in which it has 1, and $\otimes$ denotes term-by-term multiplication of infinite sequences. It follows that each operator T_j is bounded on $L^2(\mathbf{T}^1)$ with norm 1. Moreover, the sum of the T_j's is the identity operator, which is also L^2 bounded with norm 1.

It is apparent from the preceding discussion that the crucial property of the T_j's that makes their sum bounded is their orthogonality. In the preceding example we have $T_jT_k = 0$ unless $j = k$. It turns out that this orthogonality condition is a bit too strong, and it can be weakened significantly.

8.5.1 The Cotlar–Knapp–Stein Almost Orthogonality Lemma

The next result provides a sufficient orthogonality criterion for boundedness of sums of linear operators on a Hilbert space.

Lemma 8.5.1. *Let $\{T_j\}_{j\in\mathbf{Z}}$ be a family of operators mapping a Hilbert space H to itself. Assume that there is a a function $\gamma : \mathbf{Z} \to \mathbf{R}^+$ such that*

$$\big\|T_j^*T_k\big\|_{H\to H} + \big\|T_jT_k^*\big\|_{H\to H} \le \gamma(j-k) \tag{8.5.1}$$

for all j,k in $\mathbf{Z}$. Suppose that

$$A = \sum_{j\in\mathbf{Z}} \sqrt{\gamma(j)} < \infty.$$

Then the following three conclusions are valid:

(i) For all finite subsets Λ of $\mathbf{Z}$ we have

$$\Big\|\sum_{j\in\Lambda} T_j\Big\|_{H\to H} \le A.$$

(ii) For all $x \in H$ we have

$$\sum_{j\in\mathbf{Z}} \big\|T_j(x)\big\|_H^2 \le A^2\big\|x\big\|_H^2.$$

(iii) For all $x \in H$ the sequence $\sum_{|j|\le N} T_j(x)$ converges to some $T(x)$ as $N \to \infty$ in the norm topology of H. The linear operator T defined in this way is bounded from H to H with norm

$$\big\|T\big\|_{H\to H} \le A.$$

Proof. As usual we denote by S^* the adjoint of a linear operator S. It is a simple fact that any bounded linear operator $S:\ H \to H$ satisfies

$$\big\|S\big\|_{H\to H}^2 = \big\|SS^*\big\|_{H\to H}. \tag{8.5.2}$$

See Exercise 8.5.1. By taking $j = k$ in (8.5.1) and using (8.5.2), we obtain

$$\big\|T_j\big\|_{H\to H} \le \sqrt{\gamma(0)} \tag{8.5.3}$$

for all $j \in \mathbf{Z}$. It also follows from (8.5.2) that if an operator S is self-adjoint, then $\big\|S\big\|^2_{H\to H} = \big\|S^2\big\|_{H\to H}$, and more generally,

$$\big\|S\big\|^m_{H\to H} = \big\|S^m\big\|_{H\to H} \tag{8.5.4}$$

for m that are powers of 2. Now observe that the linear operator

$$\Big(\sum_{j\in\Lambda} T_j\Big)\Big(\sum_{j\in\Lambda} T_j^*\Big)$$

is self-adjoint. Applying (8.5.2) and (8.5.4) to this operator, we obtain

$$\Big\|\sum_{j\in\Lambda} T_j\Big\|^2_{H\to H} = \Big\|\Big[\Big(\sum_{j\in\Lambda} T_j\Big)\Big(\sum_{j\in\Lambda} T_j^*\Big)\Big]^m\Big\|^{\frac{1}{m}}_{H\to H}, \tag{8.5.5}$$

where m is a power of 2. We now expand the mth power of the expression in (8.5.5). So we write the right side of the identity in (8.5.5) as

$$\Big\|\sum_{j_1,\cdots,j_{2m}\in\Lambda} T_{j_1}T_{j_2}^*\cdots T_{j_{2m-1}}T_{j_{2m}}^*\Big\|^{\frac{1}{m}}_{H\to H}, \tag{8.5.6}$$

which is controlled by

$$\Big(\sum_{j_1,\cdots,j_{2m}\in\Lambda} \big\|T_{j_1}T_{j_2}^*\cdots T_{j_{2m-1}}T_{j_{2m}}^*\big\|_{H\to H}\Big)^{\frac{1}{m}}. \tag{8.5.7}$$

We estimate the expression inside the sum in (8.5.7) in two different ways. First we group j_1 with j_2, j_3 with j_4, ..., j_{2m-1} with j_{2m} and we apply (8.5.3) and (8.5.1) to control this expression by

$$\gamma(j_1-j_2)\gamma(j_3-j_4)\cdots\gamma(j_{2m-1}-j_{2m}).$$

Grouping j_2 with j_3, j_4 with j_5, ..., j_{2m-2} with j_{2m-1} and leaving j_1 and j_{2m} alone, we also control the expression inside the sum in (8.5.7) by

$$\sqrt{\gamma(0)}\gamma(j_2-j_3)\gamma(j_4-j_5)\cdots\gamma(j_{2m-2}-j_{2m-1})\sqrt{\gamma(0)}.$$

Taking the geometric mean of these two estimates, we obtain the following bound for (8.5.7):

$$\Big(\sum_{j_1,\dots,j_{2m}\in\Lambda} \sqrt{\gamma(0)}\sqrt{\gamma(j_1-j_2)}\sqrt{\gamma(j_2-j_3)}\cdots\sqrt{\gamma(j_{2m-1}-j_{2m})}\Big)^{\frac{1}{m}}.$$

Summing first over j_1, then over j_2, and finally over j_{2m-1}, we obtain the estimate

$$\gamma(0)^{\frac{1}{2m}} A^{\frac{2m-1}{m}} \Big(\sum_{j_{2m}\in\Lambda} 1 \Big)^{\frac{1}{m}}$$

for (8.5.7). Using (8.5.5), we conclude that

$$\Big\| \sum_{j\in\Lambda} T_j \Big\|_{H\to H}^2 \le \gamma(0)^{\frac{1}{2m}} A^{\frac{2m-1}{m}} |\Lambda|^{\frac{1}{m}},$$

and letting $m \to \infty$, we obtain conclusion (i) of the proposition.

To prove (ii) we use the Rademacher functions r_j of Appendix C.1. These functions are defined for nonnegative integers j, but we can reindex them so that the subscript j runs through the integers. The fundamental property of these functions is their orthogonality, that is, $\int_0^1 r_j(\omega) r_k(\omega)\, d\omega = 0$ when $j \neq k$. Using the fact that the norm $\|\cdot\|_H$ comes from an inner product, for every finite subset Λ of $\mathbf{Z}$ and x in H we obtain

$$\begin{aligned} &\int_0^1 \Big\| \sum_{j\in\Lambda} r_j(\omega) T_j(x) \Big\|_H^2 \, d\omega \\ &\quad = \sum_{j\in\Lambda} \big\| T_j(x) \big\|_H^2 + \int_0^1 \sum_{\substack{j,k\in\Lambda \\ j\neq k}} r_j(\omega) r_k(\omega) \big\langle T_j(x), T_k(x) \big\rangle_H \, d\omega \qquad (8.5.8) \\ &\quad = \sum_{j\in\Lambda} \big\| T_j(x) \big\|_H^2 . \end{aligned}$$

For any fixed $\omega \in [0,1]$ we now use conclusion (i) of the proposition for the operators $r_j(\omega) T_j$, which also satisfy assumption (8.5.1), since $r_j(\omega) = \pm 1$. We obtain that

$$\Big\| \sum_{j\in\Lambda} r_j(\omega) T_j(x) \Big\|_H^2 \le A^2 \|x\|_H^2 ,$$

which, combined with (8.5.8), gives conclusion (ii).

We now prove (iii). First we show that given $x \in H$ the sequence

$$\Big\{ \sum_{j=-N}^{N} T_j(x) \Big\}_N$$

is Cauchy in H. Suppose that this is not the case. This means that there is some $\varepsilon > 0$ and a subsequence of integers $1 \le N_1 < N_2 < N_3 < \cdots$ such that

$$\big\| \widetilde{T}_k(x) \big\|_H \ge \varepsilon , \qquad (8.5.9)$$

where we set

$$\widetilde{T}_k(x) = \sum_{N_k \le |j| < N_{k+1}} T_j(x).$$

For any fixed $\omega \in [0,1]$, apply conclusion (i) to the operators $S_j = r_k(\omega)T_j$ whenever $N_k \le |j| < N_{k+1}$, since these operators clearly satisfy hypothesis (8.5.1). Taking $N_1 \le |j| \le N_{K+1}$, we obtain

$$\Big\| \sum_{k=1}^{K} r_k(\omega) \sum_{N_k \le |j| < N_{k+1}} T_j(x) \Big\|_H = \Big\| \sum_{k=1}^{K} r_k(\omega) \widetilde{T}_k(x) \Big\|_H \le A \|x\|_H .$$

Squaring and integrating this inequality with respect to ω in $[0,1]$, and using (8.5.8) with $\widetilde{T}_k$ in the place of T_k and $\{1,2,\dots,K\}$ in the place of Λ, we obtain

$$\sum_{k=1}^{K} \|\widetilde{T}_k(x)\|_H^2 \le A^2 \|x\|_H^2 .$$

But this clearly contradicts (8.5.9) as $K \to \infty$.

We conclude that every sequence $\big\{ \sum_{j=-N}^{N} T_j(x) \big\}_N$ is Cauchy in H and thus it converges to Tx for some linear operator T. In view of conclusion (i), it follows that T is a bounded operator on H with norm at most A. □

Remark 8.5.2. At first sight, it appears strange that the norm of the operator T is independent of the norm of every piece T_j and depends only on the quantity A in (8.5.1). But as observed in the proof, if we take $j = k$ in (8.5.1), we obtain

$$\|T_j\|_{H\to H}^2 = \|T_j T_j^*\|_{H\to H} \le \gamma(0) \le A^2 ;$$

thus the norm of each individual T_j is also controlled by the constant A.

We also note that there wasn't anything special about the role of the index set $\mathbf{Z}$ in Lemma 8.5.1. Indeed, the set $\mathbf{Z}$ can be replaced by any countable group, such as $\mathbf{Z}^k$ for some k. For instance, see Theorem 8.5.7, in which the index set is $\mathbf{Z}^{2n}$. See also Exercises 8.5.7 and 8.5.8, in which versions of Lemma 8.5.1 are given with no group structure on the set of indices.

8.5.2 An Application

We now discuss an application of the almost orthogonality lemma just proved concerning sums of nonconvolution operators on $L^2(\mathbf{R}^n)$. We begin with the following version of Theorem 8.3.3, in which it is assumed that $T(1) = T^t(1) = 0$.

Proposition 8.5.3. *Suppose that $K_j(x,y)$ are functions on $\mathbf{R}^n \times \mathbf{R}^n$ indexed by $j \in \mathbf{Z}$ that satisfy*

$$|K_j(x,y)| \le \frac{A 2^{nj}}{(1+2^j|x-y|)^{n+\delta}}, \tag{8.5.10}$$

$$|K_j(x,y) - K_j(x,y')| \le A 2^{\gamma j} 2^{nj} |y-y'|^{\gamma}, \tag{8.5.11}$$

$$|K_j(x,y) - K_j(x',y)| \le A 2^{\gamma j} 2^{nj} |x-x'|^{\gamma}, \tag{8.5.12}$$

for some $0 < A, \gamma, \delta < \infty$ *and all* $x, y, x', y' \in \mathbf{R}^n$. *Suppose also that*

$$\int_{\mathbf{R}^n} K_j(z,y)\,dz = 0 = \int_{\mathbf{R}^n} K_j(x,z)\,dz, \tag{8.5.13}$$

for all $x, y \in \mathbf{R}^n$ *and all* $j \in \mathbf{Z}$. *For* $j \in \mathbf{Z}$ *define integral operators*

$$T_j(f)(x) = \int_{\mathbf{R}^n} K_j(x,y) f(y)\,dy$$

for $f \in L^2(\mathbf{R}^n)$. *Then the series*

$$\sum_{j\in\mathbf{Z}} T_j(f)$$

converges in L^2 *to some* $T(f)$ *for all* $f \in L^2(\mathbf{R}^n)$, *and the linear operator* T *defined in this way is* L^2 *bounded.*

Proof. It is a consequence of (8.5.10) that the kernels K_j are in $L^1(dy)$ uniformly in $x \in \mathbf{R}^n$ and $j \in \mathbf{Z}$ and hence the operators T_j map $L^2(\mathbf{R}^n)$ to $L^2(\mathbf{R}^n)$ uniformly in j. Our goal is to show that the sum of the T_j's is also bounded on $L^2(\mathbf{R}^n)$. We achieve this using the orthogonality considerations of Lemma 8.5.1. To be able to use Lemma 8.5.1, we need to prove (8.5.1). Indeed, we show that for all $k, j \in \mathbf{Z}$ we have

$$\big\|T_j T_k^*\big\|_{L^2\to L^2} \le C A^2 2^{-\frac{1}{4}\frac{\delta}{n+\delta}\min(\gamma,\delta)|j-k|}, \tag{8.5.14}$$

$$\big\|T_j^* T_k\big\|_{L^2\to L^2} \le C A^2 2^{-\frac{1}{4}\frac{\delta}{n+\delta}\min(\gamma,\delta)|j-k|}, \tag{8.5.15}$$

for some $0 < C = C_{n,\gamma,\delta} < \infty$. We prove only (8.5.15), since the proof of (8.5.14) is similar. In fact, since the kernels of T_j and T_j^* satisfy similar size, regularity, and cancellation estimates, (8.5.15) is directly obtained from (8.5.14) when T_j are replaced by T_j^*.

It suffices to prove (8.5.15) under the extra assumption that $k \le j$. Once (8.5.15) is established under this assumption, taking $j \le k$ yields

$$\big\|T_j^* T_k\big\|_{L^2\to L^2} = \big\|(T_k^* T_j)^*\big\|_{L^2\to L^2} = \big\|T_k^* T_j\big\|_{L^2\to L^2} \le C A^2 2^{-\frac{1}{2}\min(\gamma,\delta)|j-k|},$$

thus proving (8.5.15) also under the assumption $j \le k$.

We therefore take $k \le j$ in the proof of (8.5.15). Note that the kernel of $T_j^* T_k$ is

$$L_{jk}(x,y) = \int_{\mathbf{R}^n} \overline{K_j(z,x)} K_k(z,y)\,dz.$$

We prove that

$$\sup_{x\in\mathbf{R}^n} \int_{\mathbf{R}^n} |L_{kj}(x,y)|\,dy \le C A^2 2^{-\frac{1}{4}\frac{\delta}{n+\delta}\min(\gamma,\delta)|k-j|}, \tag{8.5.16}$$

$$\sup_{y\in\mathbf{R}^n} \int_{\mathbf{R}^n} |L_{kj}(x,y)|\,dx \le C A^2 2^{-\frac{1}{4}\frac{\delta}{n+\delta}\min(\gamma,\delta)|k-j|}. \tag{8.5.17}$$

Once (8.5.16) and (8.5.17) are established, (8.5.15) follows directly from the classical Schur lemma in Appendix I.1.

We need to use the following estimate, valid for $k \le j$:

$$\int_{\mathbf{R}^n} \frac{2^{nj}\min(1,(2^k|u|)^\gamma)}{(1+2^j|u|)^{n+\delta}}\,du \le C_{n,\delta} 2^{-\frac{1}{2}\min(\gamma,\delta)(j-k)}. \tag{8.5.18}$$

Indeed, to prove (8.5.18), we observe that by changing variables we may assume that $j=0$ and $k\le 0$. Taking $r=k-j\le 0$, we establish (8.5.18) as follows:

$$\begin{aligned}
\int_{\mathbf{R}^n} \frac{\min(1,(2^r|u|)^\gamma)}{(1+|u|)^{n+\delta}}\,du &\le \int_{\mathbf{R}^n} \frac{\min\left(1,(2^r|u|)^{\frac{1}{2}\min(\gamma,\delta)}\right)}{(1+|u|)^{n+\delta}}\,du \\
&\le \int_{|u|\le 2^{-r}} \frac{(2^r|u|)^{\frac{1}{2}\min(\gamma,\delta)}}{(1+|u|)^{n+\delta}}\,du + \int_{|u|\ge 2^{-r}} \frac{1}{(1+|u|)^{n+\delta}}\,du \\
&\le 2^{\frac{1}{2}\min(\gamma,\delta)r} \int_{\mathbf{R}^n} \frac{1}{(1+|u|)^{n+\frac{\delta}{2}}}\,du + \int_{|u|\ge 2^{-r}} \frac{1}{|u|^{n+\delta}}\,du \\
&\le C_{n,\delta}\left[2^{\frac{1}{2}\min(\gamma,\delta)r} + 2^{\delta r}\right] \\
&\le C_{n,\delta}\, 2^{-\frac{1}{2}\min(\gamma,\delta)|r|},
\end{aligned}$$

We now obtain estimates for L_{jk} in the case $k\le j$. Using (8.5.13), we write

$$\begin{aligned}
|L_{jk}(x,y)| &= \left|\int_{\mathbf{R}^n} K_k(z,y)\overline{K_j(z,x)}\,dz\right| \\
&= \left|\int_{\mathbf{R}^n} \left[K_k(z,y)-K_k(x,y)\right]\overline{K_j(z,x)}\,dz\right| \\
&\le A^2 \int_{\mathbf{R}^n} 2^{nk}\min(1,(2^k|x-z|)^\gamma)\frac{2^{nj}}{(1+2^j|z-x|)^{n+\delta}}\,dz \\
&\le CA^2\, 2^{kn}\, 2^{-\frac{1}{2}\min(\gamma,\delta)(j-k)}
\end{aligned}$$

using estimate (8.5.18). Combining this estimate with

$$|L_{jk}(x,y)| \le \int_{\mathbf{R}^n} |K_j(z,x)|\,|K_k(z,y)|\,dz \le \frac{CA^2 2^{kn}}{(1+2^k|x-y|)^{n+\delta}},$$

which follows from (8.5.10) and the result in Appendix K.1 (since $k\le j$), yields

$$|L_{jk}(x,y)| \le C_{n,\gamma,\delta} A^2\, 2^{-\frac{1}{2}\frac{\delta/2}{n+\delta}\min(\gamma,\delta)(j-k)} \frac{2^{kn}}{(1+2^k|x-y|)^{n+\frac{\delta}{2}}},$$

which easily implies (8.5.16) and (8.5.17). This concludes the proof of the proposition. □

8.5.3 Almost Orthogonality and the $T(1)$ Theorem

We now give an important application of the proposition just proved. We re-prove the difficult direction of the $T(1)$ theorem proved in Section 8.3. We have the following:

Theorem 8.5.4. *Let K be in $SK(\delta,A)$ and let T be a continuous linear operator from $\mathscr{S}(\mathbf{R}^n)$ to $\mathscr{S}'(\mathbf{R}^n)$ associated with K. Assume that*

$$\|T(1)\|_{BMO}+\|T^t(1)\|_{BMO}+\|T\|_{WB}=B_4<\infty.$$

Then T extends to bounded linear operator on $L^2(\mathbf{R}^n)$ with norm at most a constant multiple of $A+B_4$.

Proof. Consider the paraproduct operators $P_{T(1)}$ and $P_{T^t(1)}$ introduced in the previous section. Then, as we showed in Proposition 8.4.4, we have

$$\begin{aligned} P_{T(1)}(1) &= T(1), & (P_{T(1)})^t(1) &= 0,\\ P_{T^t(1)}(1) &= T^t(1), & (P_{T^t(1)})^t(1) &= 0. \end{aligned}$$

Let us define an operator

$$L=T-P_{T(1)}-(P_{T^t(1)})^t.$$

Using Proposition 8.4.4, we obtain that

$$L(1)=L^t(1)=0.$$

In view of (8.4.16), we have that L is an operator whose kernel satisfies the estimates (8.1.1), (8.1.2), and (8.1.3) with constants controlled by a dimensional constant multiple of

$$A+\|T(1)\|_{BMO}+\|T^t(1)\|_{BMO}.$$

Both of these numbers are controlled by $A+B_4$. We also have

$$\begin{aligned} \|L\|_{WB} &\le C_n\big(\|T\|_{WB}+\|P_{T(1)}\|_{L^2\to L^2}+\|(P_{T^t(1)})^t\|_{L^2\to L^2}\big)\\ &\le C_n\big(\|T\|_{WB}+\|T(1)\|_{BMO}+\|T^t(1)\|_{BMO}\big)\\ &\le C_n(A+B_4), \end{aligned}$$

which is a very useful fact.

Next we introduce operators Δ_j and S_j; one should be cautious as these are not the operators Δ_j and S_j introduced in Section 8.4 but rather discrete analogues of those introduced in the proof of Theorem 8.3.3. We pick a smooth radial real-valued function Φ with compact support contained in the unit ball $B(0,\frac{1}{2})$ that satisfies $\int_{\mathbf{R}^n}\Phi(x)\,dx=1$ and we define

$$\Psi(x)=\Phi(x)-2^{-n}\Phi(\tfrac{x}{2}). \tag{8.5.19}$$

Notice that Ψ has mean value zero. We define

$$\Phi_{2^{-j}}(x) = 2^{nj}\Phi(2^j x) \qquad \text{and} \qquad \Psi_{2^{-j}}(x) = 2^{nj}\Psi(2^j x)$$

and we observe that both Φ and Ψ are supported in $B(0,1)$ and are multiples of normalized bumps. We then define Δ_j to be the operator given by convolution with the function $\Psi_{2^{-j}}$ and S_j the operator given by convolution with the function $\Phi_{2^{-j}}$. In view of identity (8.5.19) we have that $\Delta_j = S_j - S_{j-1}$. Notice that

$$S_jLS_j = S_{j-1}LS_{j-1} + \Delta_jLS_j + S_{j-1}L\Delta_j,$$

which implies that for all integers $N < M$ we have

$$\begin{aligned} S_MLS_M - S_{N-1}LS_{N-1} &= \sum_{j=N}^{M} \big(S_jLS_j - S_{j-1}LS_{j-1}\big) \\ &= \sum_{j=N}^{M} \Delta_jLS_j - \sum_{j=N}^{M} S_{j-1}L\Delta_j. \end{aligned} \tag{8.5.20}$$

Until the end of the proof we fix a Schwartz function f whose Fourier transform vanishes in a neighborhood of the origin; such functions are dense in L^2; see Exercise 5.2.9. We would like to use Proposition 8.5.3 to conclude that

$$\sup_{M\in\mathbf{Z}}\sup_{N<M} \big\|S_MLS_M(f) - S_{N-1}LS_{N-1}(f)\big\|_{L^2} \le C_n(A_2+B_4)\big\|f\big\|_{L^2} \tag{8.5.21}$$

and that $S_MLS_M(f) - S_{N-1}LS_{N-1}(f) \to \widetilde{L}(f)$ in L^2 as $M\to\infty$ and $N\to-\infty$. Once these statements are proved, we deduce that $\widetilde{L}(f) = L(f)$. To see this, it suffices to prove that $S_MLS_M(f) - S_{N-1}LS_{N-1}(f)$ converges to $L(f)$ weakly in L^2. Indeed, let g be another Schwartz function. Then

$$\begin{aligned} &\big\langle S_MLS_M(f) - S_{N-1}LS_{N-1}(f), g\big\rangle - \big\langle L(f), g\big\rangle \\ &\qquad = \big\langle S_MLS_M(f) - L(f), g\big\rangle - \big\langle S_{N-1}LS_{N-1}(f), g\big\rangle. \end{aligned} \tag{8.5.22}$$

We first prove that the first term in (8.5.22) tends to zero as $M\to\infty$. Indeed,

$$\begin{aligned} \big\langle S_MLS_M(f) - L(f), g\big\rangle &= \big\langle LS_M(f), S_Mg\big\rangle - \big\langle L(f), g\big\rangle \\ &= \big\langle L(S_M(f) - f), S_M(g)\big\rangle + \big\langle L(f), S_M(g) - g\big\rangle, \end{aligned}$$

and both terms converge to zero, since $S_M(f) - f \to 0$ and $S_M(g) - g$ tend to zero in $\mathscr{S}$, L is continuous from $\mathscr{S}$ to $\mathscr{S}'$, and all Schwartz seminorms of $S_M(g)$ are bounded uniformly in M; see also Exercise 8.3.1.

The second term in (8.5.22) is $\big\langle S_{N-1}LS_{N-1}(f), g\big\rangle = \big\langle LS_{N-1}(f), S_{N-1}(g)\big\rangle$. Since $\widehat{f}$ is supported away from the origin, $S_N(f) \to 0$ in $\mathscr{S}$ as $N\to-\infty$; see Exercise 8.3.2. By the continuity of L, $LS_{N-1}(f) \to 0$ in $\mathscr{S}'$, and since all Schwartz seminorms of $S_{N-1}(g)$ are bounded uniformly in N, we conclude that the term $\langle LS_{N-1}(f), S_{N-1}(g)\rangle$ tends to zero as $N \to -\infty$. We thus deduce that $\widetilde{L}(f) = L(f)$.

It remains to prove (8.5.21). We now define

$$L_j = \Delta_j L S_j \qquad \text{and} \qquad L_j' = S_{j-1} L \Delta_j$$

for $j \in \mathbf{Z}$. In view of identity (2.3.21) and the convergence of the Riemann sums to the integral defining $f * \Phi_{2^{-j}}$ in the topology of $\mathscr{S}$ (see the proof of Theorem 2.3.20), we have

$$\big(L(f * \Phi_{2^{-j}}) * \Psi_{2^{-j}}\big)(x) = \int_{\mathbf{R}^n} \big\langle L(\tau^y(\Phi_{2^{-j}})), \tau^x(\Psi_{2^{-j}})\big\rangle f(y)\,dy,$$

where $\tau^y(g)(u) = g(u-y)$. Thus the kernel K_j of L_j is

$$K_j(x,y) = \big\langle L(\tau^y(\Phi_{2^{-j}})), \tau^x(\Psi_{2^{-j}})\big\rangle$$

and the kernel K_j' of L_j' is

$$K_j'(x,y) = \big\langle L(\tau^y(\Psi_{2^{-j}})), \tau^x(\Phi_{2^{-(j-1)}})\big\rangle .$$

We plan to prove that

$$|K_j(x,y)| + 2^{-j}|\nabla K_j(x,y)| \le C_n(A+B_4)2^{nj}(1+2^j|x-y|)^{-n-\delta}, \tag{8.5.23}$$

noting that an analogous estimate holds for $K_j'(x,y)$. Once (8.5.23) is established, Exercise 8.5.2 and the conditions

$$L_j(1) = \Delta_j L S_j(1) = \Delta_j L(1) = 0, \qquad L_j'(1) = S_{j-1} L \Delta_j(1) = 0,$$

yield the hypotheses of Proposition 8.5.3. Recalling (8.5.20), the conclusion of this proposition yields (8.5.21).

To prove (8.5.23) we quickly repeat the corresponding argument from the proof of Theorem 8.3.3. We consider the following two cases: If $|x-y| \le 5\cdot 2^{-j}$, then the weak boundedness property gives

$$\begin{aligned} \big|\big\langle L(\tau^y(\Phi_{2^{-j}})), \tau^x(\Psi_{2^{-j}})\big\rangle\big| &= \big|\big\langle L(\tau^x(\tau^{2^j(y-x)}(\Phi)_{2^{-j}})), \tau^x(\Psi_{2^{-j}})\big\rangle\big| \\ &\le C_n \big\|L\big\|_{WB} 2^{jn}, \end{aligned}$$

since Ψ and $\tau^{2^j(y-x)}(\Phi)$, whose support is contained in $B(0,\frac{1}{2}) + B(0,5) \subseteq B(0,10)$, are multiples of normalized bumps. This proves the first of the two estimates in (8.5.23) when $|x-y| \le 5\cdot 2^{-j}$.

We now turn to the case $|x-y| \ge 5\cdot 2^{-j}$. Then the functions $\tau^y(\Phi_{2^{-j}})$ and $\tau^x(\Psi_{2^{-j}})$ have disjoint supports, and so we have the integral representation

$$K_j(x,y) = \int_{\mathbf{R}^n}\int_{\mathbf{R}^n} \Phi_{2^{-j}}(v-y) K(u,v) \Psi_{2^{-j}}(u-x)\,du\,dv .$$

Using that Ψ has mean value zero, we can write the previous expression as

$$\int_{\mathbf{R}^n}\int_{\mathbf{R}^n} \Phi_{2^{-j}}(v-y)\big(K(u,v)-K(x,v)\big)\Psi_{2^{-j}}(u-x)\,du\,dv.$$

We observe that $|u-x|\le 2^{-j}$ and $|v-y|\le 2^{-j}$ in the preceding integral. Since $|x-y|\ge 5\cdot 2^{-j}$, this makes $|u-v|\ge |x-y|-2\cdot 2^{-j}\ge 2\cdot 2^{-j}$, which implies that $|u-x|\le \frac{1}{2}|u-v|$. Using the regularity condition (8.1.2), we deduce

$$|K(u,v)-K(x,v)|\le A\frac{|x-u|^{\delta}}{|u-v|^{n+\delta}}\le C_{n,\delta}A\,\frac{2^{-j\delta}}{|x-y|^{n+\delta}}\,.$$

Inserting this estimate in the preceding double integral, we obtain the first estimate in (8.5.23). The second estimate in (8.5.23) is proved similarly. □

8.5.4 Pseudodifferential Operators

We now turn to another elegant application of Lemma 8.5.1 regarding pseudodifferential operators. We first introduce pseudodifferential operators.

Definition 8.5.5. Let $m\in\mathbf{R}$ and $0<\rho,\delta\le 1$. A $\mathscr{C}^\infty$ function $\sigma(x,\xi)$ on $\mathbf{R}^n\times\mathbf{R}^n$ is called a *symbol of class* $S^m_{\rho,\delta}$ if for all multi-indices α and β there is a constant $B_{\alpha,\beta}$ such that

$$|\partial_x^\alpha\partial_\xi^\beta\sigma(x,\xi)|\le B_{\alpha,\beta}(1+|\xi|)^{m-\rho|\beta|+\delta|\alpha|}. \tag{8.5.24}$$

For $\sigma\in S^m_{\rho,\delta}$, the linear operator

$$T_\sigma(f)(x)=\int_{\mathbf{R}^n}\sigma(x,\xi)\widehat{f}(\xi)e^{2\pi i x\cdot\xi}\,d\xi$$

initially defined for f in $\mathscr{S}(\mathbf{R}^n)$ is called a *pseudodifferential operator* with symbol $\sigma(x,\xi)$.

Example 8.5.6. The paraproduct P_b introduced in the previous section is a pseudodifferential operator with symbol

$$\sigma_b(x,\xi)=\sum_{j\in\mathbf{Z}}\Delta_j(b)(x)\widehat{\Psi}(2^{-j}\xi). \tag{8.5.25}$$

It is not hard to see that the symbol σ_b satisfies

$$|\partial_x^\alpha\partial_\xi^\beta\sigma_b(x,\xi)|\le B_{\alpha,\beta}|\xi|^{-|\beta|+|\alpha|} \tag{8.5.26}$$

for all multi-indices α and β. Indeed, every differentiation in x produces a factor of 2^j, while every differentiation in ξ produces a factor of 2^{-j}. But since $\widehat{\Psi}$ is supported in $\frac{1}{2}\cdot 2^j\le|\xi|\le 2\cdot 2^j$, it follows that $|\xi|\approx 2^j$, which yields (8.5.26).

It follows that σ_b is not in any of the classes $S^m_{\rho,\delta}$ introduced in Definition 8.5.5. However, if we restrict the indices of summation in (8.5.25) to $j \ge 0$, then $|\xi| \approx 1+|\xi|$ and we obtain a symbol of class $S^0_{1,1}$. Note that not all symbols in $S^0_{1,1}$ give rise to bounded operators on L^2. See Exercise 8.5.6.

An example of a symbol in $S^m_{0,0}$ is $(1+|\xi|^2)^{\frac{1}{2}(m+it)}$ when $m,t \in \mathbf{R}$.

We do not plan to embark on a systematic study of pseudodifferential operators here, but we would like to study the L^2 boundedness of symbols of class $S^0_{0,0}$.

Theorem 8.5.7. *Suppose that a symbol σ belongs to the class $S^0_{0,0}$. Then the pseudodifferential operator T_σ with symbol σ, initially defined on $\mathscr{S}(\mathbf{R}^n)$, has a bounded extension on $L^2(\mathbf{R}^n)$.*

Proof. In view of Plancherel's theorem, it suffices to obtain the L^2 boundedness of the linear operator

$$\widetilde{T_\sigma}(f)(x) = \int_{\mathbf{R}^n} \sigma(x,\xi) f(\xi) e^{2\pi i x\cdot\xi}\, d\xi\,. \tag{8.5.27}$$

We fix a nonnegative smooth function $\varphi(\xi)$ supported in a small multiple of the unit cube $Q_0 = [0,1]^n$ (say in $[-\frac{1}{9},\frac{10}{9}]^n$) that satisfies

$$\sum_{j\in\mathbf{Z}^n} \varphi(x-j) = 1\,, \qquad x \in \mathbf{R}^n\,. \tag{8.5.28}$$

For $j,k \in \mathbf{Z}^n$ we define symbols

$$\sigma_{j,k}(x,\xi) = \varphi(x-j)\sigma(x,\xi)\varphi(\xi-k)$$

and corresponding operators $T_{j,k}$ given by (8.5.27) in which $\sigma(x,\xi)$ is replaced by $\sigma_{j,k}(x,\xi)$. Using (8.5.28), we obtain that

$$\widetilde{T_\sigma} = \sum_{j,k\in\mathbf{Z}^n} T_{j,k}\,,$$

where the double sum is easily shown to converge in the topology of $\mathscr{S}(\mathbf{R}^n)$. Our goal is to show that for all $N \in \mathbf{Z}^+$ we have

$$\big\|T^*_{j,k}T_{j',k'}\big\|_{L^2\to L^2} \le C_N(1+|j-j'|+|k-k'|)^{-2N}\,, \tag{8.5.29}$$

$$\big\|T_{j,k}T^*_{j',k'}\big\|_{L^2\to L^2} \le C_N(1+|j-j'|+|k-k'|)^{-2N}\,, \tag{8.5.30}$$

where C_N depends on N and n but is independent of j,j',k,k'.

We note that

$$T^*_{j,k}T_{j',k'}(f)(x) = \int_{\mathbf{R}^n} K_{j,k,j',k'}(x,y) f(y)\, dy\,,$$

where

$$K_{j,k,j',k'}(x,y) = \int_{\mathbf{R}^n} \overline{\sigma_{j,k}(z,x)}\sigma_{j',k'}(z,y)e^{2\pi i(y-x)\cdot z}\,dz. \tag{8.5.31}$$

We integrate by parts in (8.5.31) using the identity

$$e^{2\pi i z\cdot(y-x)} = \frac{(I-\Delta_z)^N(e^{2\pi i z\cdot(y-x)})}{(1+4\pi^2|x-y|^2)^N},$$

and we obtain the pointwise estimate

$$\frac{\varphi(x-k)\varphi(y-k')}{(1+4\pi^2|x-y|^2)^N}\left|(I-\Delta_z)^N(\varphi(z-j)\overline{\sigma(z,x)}\sigma(z,y)\varphi(z-j'))\right|$$

for the integrand in (8.5.31). The support property of φ forces $|j-j'|\le c_n$ for some dimensional constant c_n; indeed, $c_n = 2\sqrt{n}$ suffices. Moreover, all derivatives of σ and φ are controlled by constants, and φ is supported in a cube of finite measure. We also have $1+|x-y|\approx 1+|k-k'|$. It follows that

$$|K_{j,k,j',k'}(x,y)| \le \begin{cases} \dfrac{C_N\varphi(x-k)\varphi(y-k')}{(1+|k-k'|)^{2N}} & \text{when } |j-j'|\le c_n, \\ 0 & \text{otherwise.} \end{cases}$$

We can rewrite the preceding estimates in a more compact (and symmetric) form as

$$|K_{j,k,j',k'}(x,y)| \le \frac{C_{n,N}\varphi(x-k)\varphi(y-k')}{(1+|j-j'|+|k-k'|)^{2N}},$$

from which we easily obtain that

$$\sup_{x\in\mathbf{R}^n}\int_{\mathbf{R}^n}|K_{j,k,j',k'}(x,y)|\,dy \le \frac{C_{n,N}}{(1+|j-j'|+|k-k'|)^{2N}}, \tag{8.5.32}$$

$$\sup_{y\in\mathbf{R}^n}\int_{\mathbf{R}^n}|K_{j,k,j',k'}(x,y)|\,dx \le \frac{C_{n,N}}{(1+|j-j'|+|k-k'|)^{2N}}. \tag{8.5.33}$$

Using the classical Schur lemma in Appendix I.1, we obtain that

$$\left\|T^*_{j,k}T_{j',k'}\right\|_{L^2\to L^2} \le \frac{C_{n,N}}{(1+|j-j'|+|k-k'|)^{2N}},$$

which proves (8.5.29). Since $\rho=\delta=0$, the roles of the variables x and ξ are symmetric, and (8.5.30) can be proved in exactly the same way as (8.5.29). The almost orthogonality Lemma 8.5.1 now applies, since

$$\sum_{j,k\in\mathbf{Z}^n}\sqrt{\frac{1}{(1+|j|+|k|)^{2N}}} \le \sum_{j\in\mathbf{Z}^n}\sum_{k\in\mathbf{Z}^n}\frac{1}{(1+|j|)^{\frac{N}{2}}}\frac{1}{(1+|k|)^{\frac{N}{2}}} < \infty$$

for $N\ge 2n+2$, and the boundedness of $\widetilde{T_\sigma}$ on L^2 follows. □

Remark 8.5.8. The reader may want to check that the argument in Theorem 8.5.7 is also valid for symbols of the class $S^0_{\rho,\rho}$ whenever $0<\rho<1$.

Exercises

8.5.1. Prove that any bounded linear operator $S: H \to H$ satisfies

$$\|S\|^2_{H\to H} = \|SS^*\|_{H\to H}.$$

8.5.2. Show that if a family of kernels K_j satisfy (8.5.10) and

$$|\nabla_x K_j(x,y)| + |\nabla_y K_j(x,y)| \le \frac{A 2^{(n+1)j}}{(1+2^j|x-y|)^{n+\delta}}$$

for all $x,y \in \mathbf{R}^n$, then conditions (8.5.11) and (8.5.12) hold with $\gamma = 1$.

8.5.3. Prove the boundedness of the Hilbert transform using Lemma 8.5.1 and without using the Fourier transform.
[*Hint:* Pick a smooth function η supported in $[1/2,2]$ such that $\sum_{j\in\mathbf{Z}} \eta(2^{-j}x) = 1$ for $x \neq 0$ and set $K_j(x) = x^{-1}\eta(2^{-j}|x|)$ and $H_j(f) = f * K_j$. Note that $H_j^* = -H_j$. Estimate $\|H_k H_j\|_{L^2\to L^2}$ by $\|K_k * K_j\|_{L^1} \le \|K_k * K_j\|_{L^\infty} |\text{supp}\,(K_k * K_j)|$. When $j<k$, use the mean value property of K_j and that $\|K_k'\|_{L^\infty} \le C2^{-2k}$ to obtain that $\|K_k * K_j\|_{L^\infty} \le C2^{-2k+j}$. Conclude that $\|H_k H_j\|_{L^2\to L^2} \le C2^{-|j-k|}$.]

8.5.4. For a symbol $\sigma(x,\xi)$ in $S^0_{1,0}$, let $k(x,z)$ denote the inverse Fourier transform (evaluated at z) of the function $\sigma(x,\cdot)$ with x fixed. Show that for all $x\in\mathbf{R}^n$, the distribution $k(x,\cdot)$ coincides with a smooth function away from the origin in $\mathbf{R}^n$ that satisfies the estimates

$$|\partial_x^\alpha \partial_z^\beta k(x,z)| \le C_{\alpha,\beta} |z|^{-n-|\beta|},$$

and conclude that the kernels $K(x,y) = k(x,x-y)$ are well defined and smooth functions away from the diagonal in $\mathbf{R}^{2n}$ that belong to $SK(1,A)$ for some $A>0$. Conclude that pseudodifferential operators with symbols in $S^0_{1,0}$ are associated with standard kernels.
[*Hint:* Consider the distribution $(\partial^\gamma \sigma(x,\cdot))^\vee = (-2\pi i z)^\gamma k(x,\cdot)$. Since $\partial_\xi^\gamma \sigma(x,\xi)$ is integrable in ξ when $|\gamma| \ge n+1$, it follows that $k(x,\cdot)$ coincides with a smooth function on $\mathbf{R}^n \setminus \{0\}$. Next, set $\sigma_j(x,\xi) = \sigma(x,\xi)\widehat{\Psi}(2^{-j}\xi)$, where Ψ is as in Section 8.4 and k_j the inverse Fourier transform of σ_j in z. For $|\gamma| = M$ use that

$$(-2\pi i z)^\gamma \partial_x^\alpha \partial_\xi^\beta k_j(x,z) = \int_{\mathbf{R}^n} \partial_\xi^\gamma \big((2\pi i \xi)^\beta \partial_x^\alpha \sigma_j(x,\xi)\big) 2^{2\pi i \xi\cdot z}\, d\xi$$

to obtain $|\partial_x^\alpha \partial_\xi^\beta k_j(x,z)| \le B_{M,\alpha,\beta} 2^{jn} 2^{j|\alpha|} (2^j n|z|)^{-M}$ and sum over $j\in\mathbf{Z}$.]

8.5.5. Prove that pseudodifferential operators with symbols in $S^0_{1,0}$ that have compact support in x are elements of $CZO(1,A,B)$ for some $A,B>0$.
[*Hint:* Write

$$T_\sigma(f)(x) = \int_{\mathbf{R}^n} \left(\int_{\mathbf{R}^n} \widehat{\sigma}(a,\xi)\widehat{f}(\xi)e^{2\pi i x\cdot\xi}\,d\xi \right) e^{2\pi i x\cdot a}\,da,$$

where $\widehat{\sigma}(a,\xi)$ denotes the Fourier transform of $\sigma(x,\xi)$ in the variable x. Use integration by parts to obtain $\sup_\xi |\widehat{\sigma}(a,\xi)| \le C_N(1+|a|)^{-N}$ and pass the L^2 norm inside the integral in a to obtain the required conclusion using the translation-invariant case.]

8.5.6. Let $\widehat{\eta}(\xi)$ be a smooth bump on $\mathbf{R}$ that is supported in $2^{-\frac{1}{2}} \le |\xi| \le 2^{\frac{1}{2}}$ and is equal to 1 on $2^{-\frac{1}{4}} \le |\xi| \le 2^{\frac{1}{4}}$. Let

$$\sigma(x,\xi) = \sum_{k=1}^{\infty} e^{-2\pi i 2^k x}\widehat{\eta}(2^{-k}\xi).$$

Show that σ is an element of $S^0_{1,1}$ on the line but the corresponding pseudodifferential operator T_σ is not L^2 bounded.
[*Hint:* To see the latter statement, consider the sequence of functions $f_N(x) = \sum_{k=5}^{N} \frac{1}{k} e^{2\pi i 2^k x}h(x)$, where $h(x)$ is a Schwartz function whose Fourier transform is supported in the set $|\xi| \le \frac{1}{4}$. Show that $\|f_N\|_{L^2} \le C\|h\|_{L^2}$ but $\|T_\sigma(f_N)\|_{L^2} \ge c\log N \|h\|_{L^2}$ for some positive constants c, C.]

8.5.7. Prove conclusions (i) and (ii) of Lemma 8.5.1 if hypothesis (8.5.1) is replaced by

$$\|T_j^* T_k\|_{H\to H} + \|T_j T_k^*\|_{H\to H} \le \Gamma(j,k),$$

where Γ is a nonnegative function on $\mathbf{Z}\times\mathbf{Z}$ such that

$$\sup_j \sum_{k\in\mathbf{Z}} \sqrt{\Gamma(j,k)} = A < \infty.$$

8.5.8. Let $\{T_t\}_{t\in\mathbf{R}^+}$ be a family of operators mapping a Hilbert space H to itself. Assume that there is a function $\gamma: \mathbf{R}^+\times\mathbf{R}^+ \to \mathbf{R}^+\cup\{0\}$ satisfying

$$A_\gamma = \sup_{t>0} \int_0^\infty \sqrt{\gamma(t,s)}\,\frac{ds}{s} < \infty$$

such that

$$\|T_t^* T_s\|_{H\to H} + \|T_t T_s^*\|_{H\to H} \le \gamma(t,s)$$

for all t,s in $\mathbf{R}^+$. [An example of a function with $A_\gamma<\infty$ is $\gamma(t,s) = \min\left(\frac{s}{t},\frac{t}{s}\right)^\varepsilon$ for some $\varepsilon>0$.] Then prove that for all $0<\varepsilon<N$ we have

$$\left\| \int_\varepsilon^N T_t\,\frac{dt}{t} \right\|_{H\to H} \le A_\gamma.$$

8.6 The Cauchy Integral of Calderón and the $T(b)$ Theorem

The Cauchy integral is almost as old as complex analysis itself. In the classical theory of complex analysis, if Γ is a curve in $\mathbf{C}$ and f is a function on the curve, the Cauchy integral of f is given by

$$\frac{1}{2\pi i}\int_\Gamma \frac{f(\zeta)}{\zeta - z}\,d\zeta\,.$$

One situation in which this operator appears is the following: If Γ is a closed simple curve (i.e., a Jordan curve), Ω_+ is the interior connected component of $\mathbf{C}\setminus\Gamma$, Ω_- is the exterior connected component of $\mathbf{C}\setminus\Gamma$, and f is a smooth complex function on Γ, is it possible to find analytic functions F_+ on Ω_+ and F_- on Ω_-, respectively, that have continuous extensions on Γ such that their difference is equal to the given f on Γ? It turns out that a solution of this problem is given by

$$F_+(w) = \frac{1}{2\pi i}\int_\Gamma \frac{f(\zeta)}{\zeta - w}\,d\zeta\,, \qquad w \in \Omega_+\,,$$

and

$$F_-(w) = \frac{1}{2\pi i}\int_\Gamma \frac{f(\zeta)}{\zeta - w}\,d\zeta\,, \qquad w \in \Omega_-\,.$$

We are would like to study the case in which the Jordan curve Γ passes through infinity, in particular, when it is the graph of a Lipschitz function on $\mathbf{R}$. In this case we compute the boundary limits of F_+ and F_- and we see that they give rise to a very interesting operator on the curve Γ. To fix notation we let

$$A:\ \mathbf{R}\to\mathbf{R}$$

be a Lipschitz function. This means that there is a constant $L > 0$ such that for all $x, y \in \mathbf{R}$ we have $|A(x) - A(y)| \le L|x - y|$. We define a curve

$$\gamma:\ \mathbf{R}\to\mathbf{C}$$

by setting

$$\gamma(x) = x + iA(x)$$

and we denote by

$$\Gamma = \{\gamma(x):\ x \in \mathbf{R}\} \tag{8.6.1}$$

the graph of γ. Given a smooth function f on Γ we set

$$F(w) = \frac{1}{2\pi i}\int_\Gamma \frac{f(\zeta)}{\zeta - w}\,d\zeta\,, \qquad w \in \mathbf{C}\setminus\Gamma\,. \tag{8.6.2}$$

We now show that for $z \in \Gamma$, both $F(z + i\delta)$ and $F(z - i\delta)$ have limits as $\delta \downarrow 0$, and these limits give rise to an operator on the curve Γ that we would like to study.

8.6.1 Introduction of the Cauchy Integral Operator along a Lipschitz Curve

For a smooth function f on the curve Γ and $z \in \Gamma$ we define the *Cauchy integral of f at z* as

$$\mathfrak{C}_\Gamma(f)(z) = \lim_{\varepsilon \to 0+} \frac{1}{\pi i} \int\limits_{\substack{\zeta \in \Gamma \\ |\mathrm{Re}\,\zeta - \mathrm{Re}\,z| > \varepsilon}} \frac{f(\zeta)}{\zeta - z}\, d\zeta\,, \tag{8.6.3}$$

assuming that $f(\zeta)$ has some decay as $|\zeta| \to \infty$. The latter assumption makes the integral in (8.6.3) converge when $|\mathrm{Re}\,\zeta - \mathrm{Re}\,z| \geq 1$. The fact that the limit in (8.6.3) exists as $\varepsilon \to 0$ for almost all $z \in \Gamma$ is shown in the next proposition.

Proposition 8.6.1. *Let Γ be as in (8.6.1). Let $f(\zeta)$ be a smooth function on Γ that has decay as $|\zeta| \to \infty$. Given f, we define a function F as in (8.6.2) related to f. Then the limit in (8.6.3) exists as $\varepsilon \to 0$ for almost all $z \in \Gamma$ and gives rise to a well defined operator $\mathfrak{C}_\Gamma(f)$ acting on such functions f. Moreover, for almost all $z \in \Gamma$ we have that*

$$\lim_{\delta \downarrow 0} F(z + i\delta) = \frac{1}{2}\mathfrak{C}_\Gamma(f)(z) - \frac{1}{2} f(z)\,, \tag{8.6.4}$$

$$\lim_{\delta \downarrow 0} F(z - i\delta) = \frac{1}{2}\mathfrak{C}_\Gamma(f)(z) + \frac{1}{2} f(z)\,. \tag{8.6.5}$$

Proof. We show first that the limit in (8.6.3) exists as $\varepsilon \to 0$. For $z \in \Gamma$ and $0 < \varepsilon < 1$ we write

$$\begin{aligned}\frac{1}{\pi i} \int\limits_{\substack{\zeta \in \Gamma \\ |\mathrm{Re}\,\zeta - \mathrm{Re}\,z| > \varepsilon}} \frac{f(\zeta)\,d\zeta}{\zeta - z} = & \frac{1}{\pi i} \int\limits_{\substack{\zeta \in \Gamma \\ |\mathrm{Re}\,\zeta - \mathrm{Re}\,z| > 1}} \frac{f(\zeta)\,d\zeta}{\zeta - z} \\ & + \frac{1}{\pi i} \int\limits_{\substack{\zeta \in \Gamma \\ \varepsilon \leq |\mathrm{Re}\,\zeta - \mathrm{Re}\,z| \leq 1}} \frac{(f(\zeta) - f(z))\,d\zeta}{\zeta - z} \\ & + \frac{f(z)}{\pi i} \int\limits_{\substack{\zeta \in \Gamma \\ \varepsilon \leq |\mathrm{Re}\,\zeta - \mathrm{Re}\,z| \leq 1}} \frac{d\zeta}{\zeta - z}\,.\end{aligned} \tag{8.6.6}$$

By the smoothness of f, the middle term of the sum in (8.6.6) has a limit as $\varepsilon \to 0$. We therefore study the third (last) term of this sum.

We consider two branches of the complex logarithm: first $\log_{upper}(z)$ defined for z in $\mathbf{C} \setminus \{0\}$ minus the negative imaginary axis normalized so that $\log_{upper}(1) = 0$; this logarithm satisfies $\log_{upper}(i) = \frac{\pi i}{2}$ and $\log_{upper}(-1) = \pi i$. Second, $\log_{lower}(z)$ defined for z in $\mathbf{C} \setminus \{0\}$ minus the positive imaginary axis normalized so that $\log_{lower}(1) = 0$; this logarithm satisfies $\log_{lower}(-i) = -\frac{\pi i}{2}$ and $\log_{lower}(-1) = -\pi i$.

Let $\tau = \operatorname{Re} z$ and $t = \operatorname{Re} \zeta$; then $z = \gamma(\tau) = \tau + iA(\tau)$ and $\zeta = \gamma(t)$. The function A is Lipschitz and thus differentiable almost everywhere; consequently, the function $\gamma(\tau) = \tau + iA(\tau)$ is differentiable a.e. in $\tau \in \mathbf{R}$. Moreover, $\gamma'(\tau) = 1 + iA'(\tau) \neq 0$ whenever γ is differentiable at τ. Fix a $\tau = \operatorname{Re} z$ at which γ is differentiable.

We rewrite the last term in the sum in (8.6.6) as

$$\int_\varepsilon^1 \frac{\gamma'(t)}{\gamma(t+\tau) - \gamma(\tau)}\, dt + \int_{-1}^{-\varepsilon} \frac{\gamma'(t)}{\gamma(t+\tau) - \gamma(\tau)}\, dt\,. \tag{8.6.7}$$

The curve $t \mapsto \gamma(t+\tau) - \gamma(\tau) = t + i(A(t+\tau) - A(\tau))$ lies in the complex plane minus a small angle centered at the origin that does not contain the negative imaginary axis. Using the upper branch of the logarithm, we evaluate (8.6.7) as

$$\begin{aligned}
\frac{f(z)}{\pi i}\Big[&\log_{upper}\big(\gamma(1+\tau) - \gamma(\tau)\big) - \log_{upper}\big(\gamma(\varepsilon+\tau) - \gamma(\tau)\big) \\
&\qquad - \log_{upper}\big(\gamma(-1+\tau) - \gamma(\tau)\big) + \log_{upper}\big(\gamma(-\varepsilon+\tau) - \gamma(\tau)\big)\Big] \\
&= \log_{upper}\big(\gamma(\tau-\varepsilon) - \gamma(\tau)\big) - \log_{upper}\big(\gamma(\varepsilon+\tau) - \gamma(\tau)\big) \\
&= \log_{upper} \frac{\dfrac{\gamma(\tau-\varepsilon) - \gamma(\tau)}{\varepsilon}}{\dfrac{\gamma(\varepsilon+\tau) - \gamma(\tau)}{\varepsilon}}\,.
\end{aligned}$$

This expression converges to $\log_{upper}\big(-\frac{\gamma'(\tau)}{\gamma'(\tau)}\big) = \log_{upper}(-1) = i\pi$ as $\varepsilon \to 0$. Thus the limit in (8.6.6), and hence in (8.6.3), exists as $\varepsilon \to 0$ for almost all z on the curve. Hence $\mathfrak{C}_\Gamma(f)$ is a well defined operator whenever f is a smooth function with decay at infinity.

We proceed with the proof of (8.6.4). For fixed $\delta > 0$ and $0 < \varepsilon < 1$ we write

$$\begin{aligned}
F(z+i\delta) = {} & \frac{1}{2\pi i} \int_{\substack{\zeta\in\Gamma \\ |\operatorname{Re}\zeta - \operatorname{Re} z| > \varepsilon}} \frac{f(\zeta)}{\zeta - z - i\delta}\, d\zeta \\
& + \frac{1}{2\pi i} \int_{\substack{\zeta\in\Gamma \\ |\operatorname{Re}\zeta - \operatorname{Re} z| \le \varepsilon}} \frac{f(\zeta) - f(z)}{\zeta - z - i\delta}\, d\zeta \\
& + f(z)\frac{1}{2\pi i} \int_{\substack{\zeta\in\Gamma \\ |\operatorname{Re}\zeta - \operatorname{Re} z| \le \varepsilon}} \frac{1}{\zeta - z - i\delta}\, d\zeta\,.
\end{aligned} \tag{8.6.8}$$

With $\tau = \operatorname{Re} z$, the last term in the sum in (8.6.8) is equal to

$$\int_\varepsilon^1 \frac{\gamma'(t)}{\gamma(t+\tau) - (\gamma(\tau) + i\delta)}\, dt + \int_{-1}^{-\varepsilon} \frac{\gamma'(t)}{\gamma(t+\tau) - (\gamma(\tau) + i\delta)}\, dt\,. \tag{8.6.9}$$

Since $\delta > 0$, the curve $\gamma(t+\tau)-(\gamma(\tau)+i\delta)$ lies below the curve $t \mapsto \gamma(t+\tau)-\gamma(\tau)$ and therefore outside a small angle centered at the origin that does not contain the positive imaginary axis. In this region, $\log_{lower}$ is an analytic branch of the logarithm. Evaluation of (8.6.9) yields

$$\frac{f(z)}{2\pi i}\log_{lower}\frac{\gamma(\varepsilon+\tau)-\gamma(\tau)-i\delta}{\gamma(-\varepsilon+\tau)-\gamma(\tau)-i\delta}.$$

So, taking limits as $\delta \downarrow 0$ in (8.6.8), we obtain that

$$\begin{aligned}\lim_{\delta\downarrow 0} F(z+i\delta) = & \frac{1}{2\pi i}\int\limits_{\substack{\zeta\in\Gamma \\ |\mathrm{Re}\,\zeta-\mathrm{Re}\,z|>\varepsilon}} \frac{f(\zeta)}{\zeta-z}\,d\zeta \\ & + \frac{1}{2\pi i}\int\limits_{\substack{\zeta\in\Gamma \\ |\mathrm{Re}\,\zeta-\mathrm{Re}\,z|\le\varepsilon}} \frac{f(\zeta)-f(z)}{\zeta-z}\,d\zeta + \frac{f(z)}{2\pi i}\log_{lower}\frac{\gamma(\tau+\varepsilon)-\gamma(\tau)}{\gamma(\tau-\varepsilon)-\gamma(\tau)},\end{aligned} \tag{8.6.10}$$

in which $z=\gamma(\tau)=\tau+iA(\tau)$ and both integrals converge absolutely.

Up until this point, $\varepsilon\in(0,1)$ was arbitrary and we may let it tend to zero. In doing so we first observe that the middle integral in (8.6.10) tends to zero because of the smoothness of f. But for almost all $\tau\in\mathbf{R}$, the limit as $\varepsilon\to 0$ of the logarithm in (8.6.10) is equal to $\log_{lower}(-\frac{\gamma'(\tau)}{\gamma'(\tau)}) = \log_{lower}(-1) = -\pi i$. From this we conclude that for almost all $z\in\Gamma$ we have

$$\lim_{\delta\downarrow 0} F(z+i\delta) = \lim_{\varepsilon\to 0}\frac{1}{2\pi i}\int\limits_{\substack{\zeta\in\Gamma \\ |\mathrm{Re}\,\zeta-\mathrm{Re}\,z|>\varepsilon}} \frac{f(\zeta)}{\zeta-z}\,d\zeta + f(z)\frac{1}{2\pi i}(-\pi i), \tag{8.6.11}$$

which proves (8.6.4).

The only difference in the proof of (8.6.5) is that $\log_{upper}$ is replaced by $\log_{lower}$, and for this reason $(-\pi i)$ should be replaced by πi in (8.6.11). □

Remark 8.6.2. If we let F_+ be the restriction of F on the region above the graph Γ and let F_- be the restriction of F on the region below the graph Γ, we have that F_+ and F_- have continuous extensions on Γ, and moreover,

$$F_+ - F_- = -f,$$

where f is the given smooth function on the curve. We also note that the argument given in Proposition 8.6.1 does not require f to be smoother than $\mathscr{C}^1$.

8.6.2 Resolution of the Cauchy Integral and Reduction of Its L^2 Boundedness to a Quadratic Estimate

Having introduced the Cauchy integral $\mathfrak{C}_\Gamma$ as an operator defined on smooth functions on the graph Γ of a Lipschitz function A, we turn to some of its properties. We are mostly interested in obtaining an a priori L^2 estimate for $\mathfrak{C}_\Gamma$. Before we achieve this goal, we make some observations. First we can write $\mathfrak{C}_\Gamma$ as

$$\mathfrak{C}_\Gamma(H)(x+iA(x)) = \lim_{\varepsilon\to 0}\frac{1}{\pi i}\int_{|x-y|>\varepsilon}\frac{H(y+iA(y))(1+iA'(y))}{y+iA(y)-x-iA(x)}\,dy, \tag{8.6.12}$$

where the integral is over the real line and H is a function on the curve Γ. (Recall that Lipschitz functions are differentiable almost everywhere.) To any function H on Γ we can associate a function h on the line $\mathbf{R}$ by setting

$$h(y) = H(y+iA(y)).$$

We have that

$$\int_\Gamma |H(y)|^2\,dy = \int_{\mathbf{R}} |h(y)|^2(1+|A'(y)|^2)^{\frac{1}{2}}\,dy \approx \int_{\mathbf{R}} |h(y)|^2\,dy$$

for some constants that depend on the Lipschitz constant L of A. Therefore, the boundedness of the operator in (8.6.12) is equivalent to that of the operator

$$\mathcal{C}_\Gamma(h)(x) = \lim_{\varepsilon\to 0}\frac{1}{\pi i}\int_{|x-y|>\varepsilon}\frac{h(y)(1+iA'(y))}{y-x+i(A(y)-A(x))}\,dy \tag{8.6.13}$$

acting on Schwartz functions h on the line. It is this operator that we concentrate on in the remainder of this section. We recall that (see Example 8.1.6) the function

$$\frac{1}{y-x+i(A(y)-A(x))}$$

defined on $\mathbf{R}\times\mathbf{R}\setminus\{(x,x):\, x\in\mathbf{R}\}$ is a standard kernel in $SK(1,cL)$ for some $c>0$. We note that this is not the case with the kernel

$$\frac{1+iA'(y)}{y-x+i(A(y)-A(x))}, \tag{8.6.14}$$

for conditions (8.1.2) and (8.1.3) fail for this kernel, since the function $1+iA'$ does not possess any smoothness. [Condition (8.1.1) trivially holds for the function in (8.6.14).] We note, however, that the L^p boundedness of the operator in (8.6.13) is equivalent to that of

$$\widetilde{\mathcal{C}}_{\Gamma}(h)(x) = \lim_{\varepsilon\to 0}\frac{1}{\pi i}\int\limits_{|x-y|>\varepsilon}\frac{h(y)}{y-x+i(A(y)-A(x))}\,dy, \tag{8.6.15}$$

since the function $1+iA'$ is bounded above and below and can be absorbed in h. Therefore, the L^2 boundedness of $\mathcal{C}_\Gamma$ is equivalent to that of $\widetilde{\mathcal{C}}_\Gamma$, which has a kernel that satisfies standard estimates. This equivalence, however, is not as useful in the approach we take in the sequel. We choose to work with the operator $\mathcal{C}_\Gamma$, in which the appearance of the term $1+iA'(y)$ plays a crucial cancellation role.

In the proof of Theorem 8.3.3 we used a *resolution* of an operator T with standard kernel of the form

$$\int_0^\infty P_sT_sQ_s\,\frac{ds}{s},$$

where P_s and Q_s are nice averaging operators that approximate the identity and the zero operator, respectively. Our goal is to achieve a similar resolution for the operator $\mathcal{C}_\Gamma$ defined in (8.6.13). To achieve this, for every $s>0$ we introduce the auxiliary operator

$$\mathcal{C}_\Gamma(h)(x;s) = \frac{1}{\pi i}\int_{\mathbf{R}}\frac{h(y)(1+iA'(y))}{y-x+i(A(y)-A(x))+is}\,dy \tag{8.6.16}$$

defined for Schwartz functions h on the line. We make two preliminary observations regarding this operator: For almost all $x\in\mathbf{R}$ we have

$$\lim_{s\to\infty}\mathcal{C}_\Gamma(h)(x;s) = 0, \tag{8.6.17}$$

$$\lim_{s\to 0}\mathcal{C}_\Gamma(h)(x;s) = \mathcal{C}_\Gamma(h)(x)+h(x). \tag{8.6.18}$$

Identity (8.6.17) is trivial. To obtain (8.6.18), for a fixed $\varepsilon>0$ we write

$$\begin{aligned}\mathcal{C}_\Gamma(h)(x;s) &= \frac{1}{\pi i}\int\limits_{|x-y|>\varepsilon}\frac{h(y)(1+iA'(y))}{y-x+i(A(y)-A(x))+is}\,dy\\ &\quad+\frac{1}{\pi i}\int\limits_{|x-y|\le\varepsilon}\frac{(h(y)-h(x))(1+iA'(y))}{y-x+i(A(y)-A(x))+is}\,dy\\ &\quad+h(x)\frac{1}{\pi i}\log_{upper}\frac{\varepsilon+i(A(x+\varepsilon)-A(x))+is}{-\varepsilon+i(A(x-\varepsilon)-A(x))+is},\end{aligned} \tag{8.6.19}$$

where $\log_{upper}$ denotes the analytic branch of the complex logarithm defined in the proof of Proposition 8.6.1. We used this branch of the logarithm, since for $s>0$, the graph of the function $y\mapsto y+i(A(y+x)-A(x))+is$ lies outside a small angle centered at the origin that contains the negative imaginary axis.

We now take successive limits first as $s\to 0$ and then as $\varepsilon\to 0$ in (8.6.19). We obtain that

$$\lim_{s\to 0} \mathcal{C}_\Gamma(h)(x;s) = \lim_{\varepsilon\to 0} \frac{1}{\pi i} \int_{|x-y|>\varepsilon} \frac{h(y)(1+iA'(y))}{y-x+i(A(y)-A(x))}\,dy$$
$$+h(x) \lim_{\varepsilon\to 0} \frac{1}{\pi i} \log_{upper} \frac{\varepsilon+i(A(x+\varepsilon)-A(x))}{-\varepsilon+i(A(x-\varepsilon)-A(x))}.$$

Since this expression inside the logarithm tends to -1 as $\varepsilon \to 0$, this logarithm tends to πi, and this concludes the proof of (8.6.18).

We now consider the second derivative in s of the auxiliary operator $\mathcal{C}_\Gamma(h)(x;s)$.

$$\begin{aligned}
&\int_0^\infty s^2 \frac{d^2}{ds^2}\mathcal{C}_\Gamma(h)(x;s)\,\frac{ds}{s}\\
&= \int_0^\infty s\frac{d^2}{ds^2}\mathcal{C}_\Gamma(h)(x;s)\,ds\\
&= \lim_{s\to\infty} s\frac{d}{ds}\mathcal{C}_\Gamma(h)(x;s) - \lim_{s\to 0} s\frac{d}{ds}\mathcal{C}_\Gamma(h)(x;s) - \int_0^\infty \frac{d}{ds}\mathcal{C}_\Gamma(h)(x;s)\,ds\\
&= 0-0+\lim_{s\to 0}\mathcal{C}_\Gamma(h)(x;s) - \lim_{s\to\infty}\mathcal{C}_\Gamma(h)(x;s)\\
&= \mathcal{C}_\Gamma(h)(x)+h(x),
\end{aligned}$$

where we used integration by parts, the fact that for almost all $x \in \mathbf{R}$ we have

$$\lim_{s\to\infty} s\frac{d}{ds}\mathcal{C}_\Gamma(h)(x;s) = \lim_{s\to 0} s\frac{d}{ds}\mathcal{C}_\Gamma(h)(x;s) = 0, \tag{8.6.20}$$

and identities (8.6.17) and (8.6.18) whenever h is a Schwartz function. One may consult Exercise 8.6.2 for a proof of the identities in (8.6.20). So we have succeeded in writing the operator $\mathcal{C}_\Gamma(h)+h$ as an average of smoother operators. Precisely, we have shown that for $h \in \mathscr{S}(\mathbf{R})$ we have

$$\mathcal{C}_\Gamma(h)(x)+h(x) = \int_0^\infty s^2\frac{d^2}{ds^2}\mathcal{C}_\Gamma(h)(x;s)\,\frac{ds}{s}, \tag{8.6.21}$$

and it remains to understand what the operator

$$\frac{d^2}{ds^2}\mathcal{C}_\Gamma(h)(x;s) = \mathcal{C}_\Gamma(h)''(x;s)$$

really is. Differentiating (8.6.16) twice, we obtain

$$\begin{aligned}
\mathcal{C}_\Gamma(h)(x)+h(x) &= \int_0^\infty s^2\mathcal{C}_\Gamma(h)''(x;s)\,\frac{ds}{s}\\
&= 4\int_0^\infty s^2\mathcal{C}_\Gamma(h)''(x;2s)\,\frac{ds}{s}\\
&= -\frac{8}{\pi i}\int_0^\infty\int_{\mathbf{R}} \frac{s^2h(y)(1+iA'(y))}{(y-x+i(A(y)-A(x))+2is)^3}\,dy\,\frac{ds}{s}\\
&= -\frac{8}{\pi i}\int_0^\infty\int_\Gamma \frac{s^2H(\zeta)}{(\zeta-z+2is)^3}\,d\zeta\,\frac{ds}{s},
\end{aligned}$$

where in the last step we set $z = x + iA(x)$, $H(z) = h(x)$, and we switched to complex integration over the curve Γ. We now use the following identity from complex analysis. For $\zeta, z \in \Gamma$ we have

$$\frac{1}{(\zeta - z + 2is)^3} = -\frac{1}{4\pi i}\int_\Gamma \frac{1}{(\zeta - w + is)^2}\frac{1}{(w - z + is)^2}\,dw, \tag{8.6.22}$$

for which we refer to Exercise 8.6.3. Inserting this identity in the preceding expression for $\mathcal{C}_\Gamma(h)(x) + h(x)$, we obtain

$$\mathcal{C}_\Gamma(h)(x) + h(x) = -\frac{2}{\pi^2}\int_0^\infty \left[\int_\Gamma \frac{s}{(w - z + is)^2}\left(\int_\Gamma \frac{s\,H(\zeta)}{(\zeta - w + is)^2}\,d\zeta\right)dw\right]\frac{ds}{s},$$

recalling that $z = x + iA(x)$. Introducing the linear operator

$$\Theta_s(h)(x) = \int_{\mathbf{R}} \theta_s(x,y)\,h(y)\,dy, \tag{8.6.23}$$

where

$$\theta_s(x,y) = \frac{s}{(y - x + i(A(y) - A(x)) + is)^2}, \tag{8.6.24}$$

we may therefore write

$$\mathcal{C}_\Gamma(h)(x) + h(x) = -\frac{2}{\pi^2}\int_0^\infty \Theta_s\big((1 + iA')\Theta_s((1 + iA')h)\big)(x)\,\frac{ds}{s}. \tag{8.6.25}$$

We also introduce the multiplication operator

$$M_b(h) = bh,$$

which enables us to write (8.6.25) in a more compact form as

$$\mathcal{C}_\Gamma(h) = -h - \frac{2}{\pi^2}\int_0^\infty \Theta_s M_{1+iA'}\Theta_s M_{1+iA'}(h)\,\frac{ds}{s}. \tag{8.6.26}$$

This gives us the desired resolution of the operator $\mathcal{C}_\Gamma$. It suffices to obtain an L^2 estimate for the integral expression in (8.6.26). Using duality, we write

$$\left\langle \int_0^\infty \Theta_s M_{1+iA'}\Theta_s M_{1+iA'}(h)\,\frac{ds}{s}, g\right\rangle = \int_0^\infty \left\langle M_{1+iA'}\Theta_s M_{1+iA'}(h), \Theta_s^t(g)\right\rangle \frac{ds}{s},$$

which is easily bounded by

$$\begin{aligned}&\sqrt{1+L^2}\int_0^\infty \big\|\Theta_s M_{1+iA'}(h)\big\|_{L^2}\big\|\Theta_s^t(g)\big\|_{L^2}\,\frac{ds}{s}\\ &\quad\le \sqrt{1+L^2}\left(\int_0^\infty \big\|\Theta_s M_{1+iA'}(h)\big\|_{L^2}^2\,\frac{ds}{s}\right)^{\frac12}\left(\int_0^\infty \big\|\Theta_s(g)\big\|_{L^2}^2\,\frac{ds}{s}\right)^{\frac12}.\end{aligned}$$

We have now reduced matters to the following estimate:

$$\left(\int_0^\infty \big\|\Theta_s(h)\big\|_{L^2}^2 \frac{ds}{s}\right)^{\frac{1}{2}} \le C\big\|h\big\|_{L^2}. \tag{8.6.27}$$

We derive (8.6.27) as a consequence of Theorem 8.6.6 discussed in Section 8.6.4.

8.6.3 A Quadratic $T(1)$ Type Theorem

We review what we have achieved so far and we introduce definitions that place matters into a new framework.

For the purposes of the subsequent exposition we can switch to $\mathbf{R}^n$, since there are no differences from the one-dimensional argument. Suppose that for all $s > 0$, there is a family of functions θ_s defined on $\mathbf{R}^n \times \mathbf{R}^n$ such that

$$|\theta_s(x,y)| \le \frac{1}{s^n} \frac{A}{\left(1+\frac{|x-y|}{s}\right)^{n+\delta}} \tag{8.6.28}$$

and

$$|\theta_s(x,y) - \theta_s(x,y')| \le \frac{A}{s^n} \frac{|y-y'|^\gamma}{s^\gamma} \tag{8.6.29}$$

for all $x,y,y' \in \mathbf{R}^n$ and some $0 < \gamma, \delta, A < \infty$. Let Θ_s be the operator with kernel θ_s, that is,

$$\Theta_s(h)(x) = \int_{\mathbf{R}^n} \theta_s(x,y)\, h(y)\, dy, \tag{8.6.30}$$

which is well defined for all h in $\bigcup_{1\le p\le\infty} L^p(\mathbf{R}^n)$ in view of (8.6.28).

At this point we observe that both (8.6.28) and (8.6.29) hold for the θ_s defined in (8.6.24) with $\gamma = \delta = 1$ and A a constant multiple of L. We leave the details of this calculation to the reader but we note that (8.6.29) can be obtained quickly using the mean value theorem. Our goal is to figure out under what additional conditions on Θ_s the quadratic estimate (8.6.27) holds. If we can find such a condition that is easily verifiable for the Θ_s associated with the Cauchy integral, this will conclude the proof of its L^2 boundedness.

We first consider a simple condition that implies the quadratic estimate (8.6.27).

Theorem 8.6.3. *For $s > 0$, let θ_s be a family of kernels satisfying (8.6.28) and (8.6.29) and let Θ_s be the linear operator whose kernel is θ_s. Suppose that for all $s > 0$ we have*

$$\Theta_s(1) = 0. \tag{8.6.31}$$

Then there is a constant $C_{n,\delta}$ such that for all $f \in L^2$ we have

$$\left(\int_0^\infty \big\|\Theta_s(f)\big\|_{L^2}^2 \frac{ds}{s}\right)^{\frac{1}{2}} \le C_{n,\delta}A\big\|f\big\|_{L^2}. \tag{8.6.32}$$

We note that condition (8.6.31) is not satisfied for the operators Θ_s associated with the Cauchy integral as defined in (8.6.23). However, Theorem 8.6.3 gives us an idea of what we are looking for, something like the action of Θ_s on a specific function. We also observe that condition (8.6.31) is "basically" saying that $\Theta(1) = 0$, where

$$\Theta = \int_0^\infty \Theta_s \frac{ds}{s}.$$

Proof. We introduce Littlewood–Paley operators Q_s given by convolution with a smooth function $\Psi_s = \frac{1}{s^n}\Psi(\frac{\cdot}{s})$ whose Fourier transform is supported in the annulus $s/2 \le |\xi| \le 2s$ that satisfies

$$\int_0^\infty Q_s^2 \frac{ds}{s} = \lim_{\substack{\varepsilon\to 0\\ N\to\infty}} \int_\varepsilon^N Q_s^2 \frac{ds}{s} = I, \tag{8.6.33}$$

where the limit is taken in the sense of distributions and the identity holds in $\mathscr{S}'(\mathbf{R}^n)/\mathscr{P}$. This identity and properties of Θ_t imply the operator identity

$$\Theta_t = \Theta_t \int_0^\infty Q_s^2 \frac{ds}{s} = \int_0^\infty \Theta_t Q_s^2 \frac{ds}{s}.$$

The key fact is the following estimate:

$$\big\|\Theta_t Q_s\big\|_{L^2\to L^2} \le A\, C_{n,\Psi} \min\Big(\frac{s}{t}, \frac{t}{s}\Big)^{\varepsilon}, \tag{8.6.34}$$

which holds for some $\varepsilon = \varepsilon(\gamma, \delta, n) > 0$. [Recall that A, γ, and δ are as in (8.6.28) and (8.6.29).] Assuming momentarily estimate (8.6.34), we can quickly prove Theorem 8.6.3 using duality. Indeed, let us take a function $G(x,t)$ such that

$$\int_0^\infty \int_{\mathbf{R}^n} |G(x,t)|^2\, dx \frac{dt}{t} \le 1. \tag{8.6.35}$$

Then we have

$$\begin{aligned}
&\int_0^\infty \int_{\mathbf{R}^n} G(x,t)\,\Theta_t(f)(x)\, dx \frac{dt}{t}\\
&= \int_0^\infty \int_{\mathbf{R}^n} G(x,t) \int_0^\infty \Theta_t Q_s^2(f)(x) \frac{ds}{s}\, dx \frac{dt}{t}\\
&= \int_0^\infty \int_0^\infty \int_{\mathbf{R}^n} G(x,t)\,\Theta_t Q_s^2(f)(x)\, dx \frac{dt}{t}\frac{ds}{s}\\
&\le \bigg(\int_0^\infty \int_0^\infty \int_{\mathbf{R}^n} |G(x,t)|^2 dx \min\Big(\frac{s}{t}, \frac{t}{s}\Big)^{\varepsilon} \frac{dt}{t}\frac{ds}{s}\bigg)^{\frac{1}{2}}\\
&\quad\times \bigg(\int_0^\infty \int_0^\infty \int_{\mathbf{R}^n} |\Theta_t Q_s(Q_s(f))(x)|^2\, dx \min\Big(\frac{s}{t}, \frac{t}{s}\Big)^{-\varepsilon} \frac{dt}{t}\frac{ds}{s}\bigg)^{\frac{1}{2}}.
\end{aligned}$$

But we have the estimate

$$\sup_{t>0}\int_0^\infty \min\left(\frac{s}{t},\frac{t}{s}\right)^{\varepsilon}\frac{ds}{s}\le C_\varepsilon\,,$$

which, combined with (8.6.35), yields that the first term in the product of the two preceding square functions is controlled by $\sqrt{C_\varepsilon}$. Using this fact and (8.6.34), we write

$$\begin{aligned}
&\int_0^\infty\int_{\mathbf{R}^n} G(x,t)\,\Theta_t(f)(x)\,dx\,\frac{dt}{t}\\
&\quad\le \sqrt{C_\varepsilon}\left(\int_0^\infty\int_0^\infty\int_{\mathbf{R}^n}|\Theta_t Q_s(Q_s(f))(x)|^2\,dx\,\min\left(\frac{s}{t},\frac{t}{s}\right)^{-\varepsilon}\frac{dt}{t}\frac{ds}{s}\right)^{\frac12}\\
&\quad\le A\sqrt{C_\varepsilon}\left(\int_0^\infty\int_0^\infty\int_{\mathbf{R}^n}|Q_s(f)(x)|^2\,dx\,\min\left(\frac{s}{t},\frac{t}{s}\right)^{2\varepsilon}\min\left(\frac{s}{t},\frac{t}{s}\right)^{-\varepsilon}\frac{dt}{t}\frac{ds}{s}\right)^{\frac12}\\
&\quad\le A\sqrt{C_\varepsilon}\left(\int_0^\infty\int_0^\infty\int_{\mathbf{R}^n}|Q_s(f)(x)|^2\,dx\,\min\left(\frac{s}{t},\frac{t}{s}\right)^{\varepsilon}\frac{dt}{t}\frac{ds}{s}\right)^{\frac12}\\
&\quad\le C_\varepsilon A\left(\int_0^\infty\int_{\mathbf{R}^n}|Q_s(f)(x)|^2\,dx\,\frac{ds}{s}\right)^{\frac12}\\
&\quad\le C_{n,\varepsilon}A\big\|f\big\|_{L^2}\,,
\end{aligned}$$

where in the last step we used the continuous version of Theorem 5.1.2 (cf. Exercise 5.1.4). Taking the supremum over all functions $G(x,t)$ that satisfy (8.6.35) yields estimate (8.6.32).

It remains to prove (8.6.34). What is crucial here is that both Θ_t and Q_s satisfy the cancellation conditions $\Theta_t(1)=0$ and $Q_s(1)=0$. The proof of estimate (8.6.34) is similar to that of estimates (8.5.14) and (8.5.15) in Proposition 8.5.3. Using ideas from the proof of Proposition 8.5.3, we quickly dispose of the proof of (8.6.34).

The kernel of $\Theta_t Q_s$ is seen easily to be

$$L_{t,s}(x,y)=\int_{\mathbf{R}^n}\theta_t(x,z)\Psi_s(z-y)\,dz\,.$$

Notice that the function $(y,z)\mapsto\Psi_s(z-y)$ satisfies (8.6.28) with $\delta=1$ and $A=C_\Psi$ and satisfies

$$|\Psi_s(z-y)-\Psi_s(z'-y)|\le\frac{C_\Psi}{s^n}\frac{|z-z'|}{s}$$

for all $z,z',y\in\mathbf{R}^n$ for some $C_\Psi<\infty$. We prove that

$$\sup_{x\in\mathbf{R}^n}\int_{\mathbf{R}^n}|L_{t,s}(x,y)|\,dy\;\le\;C_\Psi A\min\left(\frac{t}{s},\frac{s}{t}\right)^{\frac14\frac{\min(\delta,1)}{n+\min(\delta,1)}\min(\gamma,\delta,1)}\,,\tag{8.6.36}$$

$$\sup_{y\in\mathbf{R}^n}\int_{\mathbf{R}^n}|L_{t,s}(x,y)|\,dx\;\le\;C_\Psi A\min\left(\frac{t}{s},\frac{s}{t}\right)^{\frac14\frac{\min(\delta,1)}{n+\min(\delta,1)}\min(\gamma,\delta,1)}\,.\tag{8.6.37}$$

Once (8.6.36) and (8.6.37) are established, (8.6.34) follows directly from the lemma in Appendix I.1 with $\varepsilon = \frac{1}{4}\frac{\min(\delta,1)}{n+\min(\delta,1)}\min(\gamma,\delta,1)$.

We begin by observing that when $s \le t$ we have the estimate

$$\int_{\mathbf{R}^n} \frac{s^{-n}\min(2,(t^{-1}|u|)^{\gamma})}{(1+s^{-1}|u|)^{n+1}}\,du \le C_n\Big(\frac{s}{t}\Big)^{\frac{1}{2}\min(\gamma,1)}. \tag{8.6.38}$$

Also when $t \le s$ we have the analogous estimate

$$\int_{\mathbf{R}^n} \frac{t^{-n}\min(2,s^{-1}|u|)}{(1+t^{-1}|u|)^{n+\delta}}\,du \le C_n\Big(\frac{t}{s}\Big)^{\frac{1}{2}\min(\delta,1)}. \tag{8.6.39}$$

Both (8.6.38) and (8.6.39) are trivial reformulations or consequences of (8.5.18).

We now take $s \le t$ and we use that $Q_s(1) = 0$ for all $s > 0$ to obtain

$$\begin{aligned}
|L_{t,s}(x,y)| &= \Big|\int_{\mathbf{R}^n} \theta_t(x,z)\Psi_s(z-y)\,dz\Big| \\
&= \Big|\int_{\mathbf{R}^n} \big[\theta_t(x,z)-\theta_t(x,y)\big]\Psi_s(z-y)\,dz\Big| \\
&\le CA\int_{\mathbf{R}^n} \frac{\min(2,(t^{-1}|z-y|)^{\gamma})}{t^n}\,\frac{s^{-n}}{(1+s^{-1}|z-y|)^{n+1}}\,dz \\
&\le C_n' A\,\frac{1}{t^n}\Big(\frac{s}{t}\Big)^{\frac{1}{2}\min(\gamma,1)} \\
&\le C_n' A\min\Big(\frac{1}{t},\frac{1}{s}\Big)^n \min\Big(\frac{t}{s},\frac{s}{t}\Big)^{\frac{1}{2}\min(\gamma,\delta,1)}
\end{aligned}$$

using estimate (8.6.38). Similarly, using (8.6.39) and the hypothesis that $\Theta_t(1) = 0$ for all $t > 0$, we obtain for $t \le s$,

$$\begin{aligned}
|L_{t,s}(x,y)| &= \Big|\int_{\mathbf{R}^n} \theta_t(x,z)\Psi_s(z-y)\,dz\Big| \\
&= \Big|\int_{\mathbf{R}^n} \theta_t(x,z)\big[\Psi_s(z-y)-\Psi_s(x-y)\big]\,dz\Big| \\
&\le xCA\int_{\mathbf{R}^n} \frac{t^{-n}}{(1+t^{-1}|x-z|)^{n+\delta}}\,\frac{\min(2,s^{-1}|x-z|)}{s^n}\,dz \\
&\le C_n' A\,\frac{1}{s^n}\Big(\frac{t}{s}\Big)^{\frac{1}{2}\min(\delta,1)} \\
&\le C_n' A\min\Big(\frac{1}{t},\frac{1}{s}\Big)^n \min\Big(\frac{t}{s},\frac{s}{t}\Big)^{\frac{1}{2}\min(\gamma,\delta,1)}.
\end{aligned}$$

Combining the estimates for $|L_{t,s}(x,y)|$ in the preceding cases $t \le s$ and $s \le t$ with the estimate

$$|L_{t,s}(x,y)| \le \int_{\mathbf{R}^n} |\Theta_t(x,z)|\,|\Psi_s(z-y)|\,dz \le \frac{CA\min(\frac{1}{t},\frac{1}{s})^n}{\big(1+\min(\frac{1}{t},\frac{1}{s})|x-y|\big)^{n+\min(\delta,1)}},$$

which is a consequence of the result in Appendix K.1, gives

$$|L_{t,s}(x,y)| \le \frac{C\min(\frac{t}{s},\frac{s}{t})^{\frac{1}{2}\min(\gamma,\delta,1)(1-\beta)}A\min(\frac{1}{t},\frac{1}{s})^n}{\Big(\big(1+\min(\frac{1}{t},\frac{1}{s})|x-y|\big)^{n+\min(\delta,1)}\Big)^{\beta}}$$

for any $0<\beta<1$. Choosing $\beta=(n+\frac{1}{2}\min(\delta,1))(n+\min(\delta,1))^{-1}$ and integrating over x or y yields (8.6.36) and (8.6.37), respectively, and thus concludes the proof of estimate (8.6.34). □

We end this subsection with a small generalization of the previous theorem that follows by an examination of its proof. The simple details are left to the reader.

Corollary 8.6.4. *For $s>0$ let Θ_s be linear operators that are uniformly bounded on $L^2(\mathbf{R}^n)$ by a constant B. Let Ψ be a Schwartz function whose Fourier transform is supported in the annulus $1/2\le|x|\le 2$ such that the Littlewood–Paley operator Q_s given by convolution with $\Psi_s(x)=s^{-n}\Psi(s^{-1}x)$ satisfies* (8.6.33)*. Suppose that for some $C_{n,\Psi},A,\varepsilon<\infty$,*

$$\big\|\Theta_t Q_s\big\|_{L^2\to L^2} \le A\,C_{n,\Psi}\min\Big(\frac{s}{t},\frac{t}{s}\Big)^{\varepsilon} \tag{8.6.40}$$

is satisfied for all $t,s>0$. Then there is a constant $C_{n,\Psi,\varepsilon}$ such that for all $f\in L^2(\mathbf{R}^n)$ we have

$$\Big(\int_0^\infty \big\|\Theta_s(f)\big\|_{L^2}^2\frac{ds}{s}\Big)^{\frac{1}{2}} \le C_{n,\Psi,\varepsilon}(A+B)\big\|f\big\|_{L^2}.$$

8.6.4 A $T(b)$ Theorem and the L^2 Boundedness of the Cauchy Integral

The operators Θ_s defined in (8.6.23) and (8.6.24) that appear in the resolution of the Cauchy integral operator $\mathcal{C}_\Gamma$ do not satisfy the condition $\Theta_s(1)=0$ of Theorem 8.6.3. It turns out that a certain variant of this theorem is needed for the purposes of the application we have in mind, the L^2 boundedness of the Cauchy integral operator. This variant is a quadratic type $T(b)$ theorem discussed in this subsection. Before we state the main theorem, we need a definition.

Definition 8.6.5. A bounded complex-valued function b on $\mathbf{R}^n$ is said to be *accretive* if there is a constant $c_0>0$ such that $\operatorname{Re} b(x)\ge c_0$ for almost all $x\in\mathbf{R}^n$.

The following theorem is the main result of this section.

Theorem 8.6.6. *Let θ_s be a complex-valued function on $\mathbf{R}^n \times \mathbf{R}^n$ that satisfies (8.6.28) and (8.6.29), and let Θ_s be the linear operator in (8.6.30) whose kernel is θ_s. If there is an accretive function b such that*

$$\Theta_s(b) = 0 \tag{8.6.41}$$

for all $s > 0$, then there is a constant $C_n(b)$ such that the estimate

$$\left(\int_0^\infty \|\Theta_s(f)\|_{L^2}^2 \frac{ds}{s}\right)^{\frac{1}{2}} \le C_n(b)\|f\|_{L^2} \tag{8.6.42}$$

holds for all $f \in L^2$.

Corollary 8.6.7. *The Cauchy integral operator $\mathcal{C}_\Gamma$ maps $L^2(\mathbf{R})$ to itself.*

The corollary is a consequence of Theorem 8.6.6. Indeed, the crucial and important cancellation property

$$\Theta_s(1+iA') = 0 \tag{8.6.43}$$

is valid for the accretive function $1+iA'$, when Θ_s and θ_s are as in (8.6.23) and (8.6.24). To prove (8.6.43) we simply note that

$$\begin{aligned}
\Theta_s(1+iA')(x) &= \int_{\mathbf{R}} \frac{s(1+iA'(y))\,dy}{(y-x+i(A(y)-A(x))+is)^2} \\
&= \left[\frac{-s}{y-x+i(A(y)-A(x))+is}\right]_{y=-\infty}^{y=+\infty} \\
&= 0-0 = 0.
\end{aligned}$$

This condition plays exactly the role of (8.6.31), which may fail in general. The necessary "internal cancellation" of the family of operators Θ_s is exactly captured by the single condition (8.6.43).

It remains to prove Theorem 8.6.6.

Proof. We fix an approximation of the identity operator, such as

$$P_s(f)(x) = \int_{\mathbf{R}^n} \Phi_s(x-y)\,f(y)\,dy,$$

where $\Phi_s(x) = s^{-n}\Phi(s^{-1}x)$, and Φ is a nonnegative Schwartz function with integral 1. Then P_s is a nice positive averaging operator that satisfies $P_s(1) = 1$ for all $s > 0$. The key idea is to decompose the operator Θ_s as

$$\Theta_s = \left(\Theta_s - M_{\Theta_s(1)}P_s\right) + M_{\Theta_s(1)}P_s\,, \tag{8.6.44}$$

where $M_{\Theta_s(1)}$ is the operator given by multiplication by $\Theta_s(1)$. We begin with the first term in (8.6.44), which is essentially an error term. We simply observe that

$$\left(\Theta_s - M_{\Theta_s(1)}P_s\right)(1) = \Theta_s(1) - \Theta_s(1)P_s(1) = \Theta_s(1) - \Theta_s(1) = 0\,.$$

Therefore, Theorem 8.6.3 is applicable once we check that the kernel of the operator $\Theta_s - M_{\Theta_s(1)}P_s$ satisfies (8.6.28) and (8.6.29). But these are verified easily, since the kernels of both Θ_s and P_s satisfy these estimates and $\Theta_s(1)$ is a bounded function uniformly in s. The latter statement is a consequence of condition (8.6.28).

We now need to obtain the required quadratic estimate for the term $M_{\Theta_s(1)}P_s$. With the use of Theorem 7.3.7, this follows once we prove that the measure

$$\left|\Theta_s(1)(x)\right|^2 \frac{dx\,ds}{s}$$

is Carleson. It is here that we use condition (8.6.41). Since $\Theta_s(b) = 0$ we have

$$P_s(b)\,\Theta_s(1) = \big(P_s(b)\,\Theta_s(1) - \Theta_s P_s(b)\big) + \big(\Theta_s P_s(b) - \Theta_s(b)\big)\,. \tag{8.6.45}$$

Suppose we could show that the measures

$$\left|\Theta_s(b)(x) - \Theta_s P_s(b)(x)\right|^2 \frac{dx\,ds}{s}, \tag{8.6.46}$$

$$\left|\Theta_s P_s(b)(x) - P_s(b)(x)\,\Theta_s(1)(x)\right|^2 \frac{dxds}{s}, \tag{8.6.47}$$

are Carleson. Then it would follow from (8.6.45) that the measure

$$\left|P_s(b)(x)\,\Theta_s(1)(x)\right|^2 \frac{dx\,ds}{s}$$

is also Carleson. Using the accretivity condition on b and the positivity of P_s we obtain

$$\left|P_s(b)\right| \ge \operatorname{Re} P_s(b) = P_s(\operatorname{Re} b) \ge P_s(c_0) = c_0,$$

from which it follows that $|\Theta_s(1)(x)|^2 \le c_0^{-2}|P_s(b)(x)\,\Theta_s(1)(x)|^2$. Thus the measure $|\Theta_s(1)(x)|^2 dx\,ds/s$ must be Carleson.

Therefore, the proof will be complete if we can show that both measures (8.6.46) and (8.6.47) are Carleson. Theorem 7.3.8 plays a key role here.

We begin with the measure in (8.6.46). First we observe that the kernel

$$L_s(x,y) = \int_{\mathbf{R}^n} \theta_s(x,z)\Phi_s(z-y)\,dz$$

of $\Theta_s P_s$ satisfies (8.6.28) and (8.6.29). The verification of (8.6.28) is a straightforward consequence of the estimate in Appendix K.1, while (8.6.29) follows easily from the mean value theorem. It follows that the kernel of

$$R_s = \Theta_s - \Theta_s P_s$$

satisfies the same estimates. Moreover, it is easy to see that $R_s(1) = 0$ and thus the quadratic estimate (8.6.32) holds for R_s in view of Theorem 8.6.3. Therefore, the hypotheses of Theorem 7.3.8(c) are satisfied, and this gives that the measure in (8.6.46) is Carleson.

We now continue with the measure in (8.6.47). Here we set

$$T_s(f)(x) = \Theta_s P_s(f)(x) - P_s(f)(x)\Theta_s(1)(x).$$

The kernel of T_s is $L_s(x,y) - \Theta_s(1)(x)\Phi_s(x-y)$, which clearly satisfies (8.6.28) and (8.6.29), since $\Theta_s(1)(x)$ is a bounded function uniformly in $s > 0$. We also observe that $T_s(1) = 0$. Using Theorem 8.6.3, we conclude that the quadratic estimate (8.6.32) holds for T_s. Therefore, the hypotheses of Theorem 7.3.8(c) are satisfied; hence the measure in (8.6.46) is Carleson. □

We conclude by observing that if we attempt to replace Θ_s with $\widetilde{\Theta}_s = \Theta_s M_{1+iA'}$ in the resolution identity (8.6.26), then $\widetilde{\Theta}_s(1) = 0$ would hold, but the kernel of $\widetilde{\Theta}_s$ would not satisfy the regularity estimate (8.6.29). The whole purpose of Theorem 8.6.6 was to find a certain balance between regularity and cancellation.

Exercises

8.6.1. Given a function H on a Lipschitz graph Γ, we associate a function h on the line by setting $h(t) = H(t+iA(t))$. Prove that for all $0 < p < \infty$ we have

$$\|h\|_{L^p(\mathbf{R})}^p \le \|H\|_{L^p(\Gamma)}^p \le \sqrt{1+L^2}\,\|h\|_{L^p(\mathbf{R})}^p,$$

where L is the Lipschitz constant of the defining function A of the graph Γ.

8.6.2. Let $A : \mathbf{R} \to \mathbf{R}$ satisfy $|A(x)-A(y)| \le L|x-y|$ for all $x,y \in \mathbf{R}$ for some $L > 0$. Also, let h be a Schwartz function on $\mathbf{R}$.
(a) Show that for all $s > 0$ and $x,y \in \mathbf{R}$ we have

$$\frac{s^2+|x-y|^2}{|x-y|^2+|A(x)-A(y)+s|^2} \le 4L^2+2.$$

(b) Use the Lebesgue dominated convergence theorem to prove that

$$\int_{|x-y|>\sqrt{s}} \frac{s(1+iA'(y))h(y)}{(y-x+i(A(y)-A(x))+is)^2}\,dy \to 0$$

as $s \to 0$.
(c) Integrate directly to show that as $s \to 0$,

$$\int_{|x-y|\le\sqrt{s}} \frac{s(1+iA'(y))}{(y-x+i(A(y)-A(x))+is)^2}\,dy \to 0$$

for every point x at which A is differentiable.
(d) Use part (a) and the Lebesgue dominated convergence theorem to show that as

$s \to 0$,

$$\int_{|x-y|\le\sqrt{s}} \frac{s(1+iA'(y))(h(y)-h(x))}{(y-x+i(A(y)-A(x))+is)^2}\,dy \to 0.$$

(e) Use part (a) and the Lebesgue dominated convergence theorem to show that as $s \to \infty$,

$$\int_{\mathbf{R}} \frac{s(1+iA'(y))h(y)}{(y-x+i(A(y)-A(x))+is)^2}\,dy \to 0.$$

Conclude the validity of the statements in (8.6.20) for almost all $x \in \mathbf{R}$.

8.6.3. Prove identity (8.6.22).
[*Hint:* Write the identity in (8.6.22) as

$$\frac{-2}{((\zeta+is)-(z-is))^3} = \frac{1}{2\pi i}\int_\Gamma \frac{\frac{1}{(w-(z-is))^2}}{(w-(\zeta+is))^2}\,dw$$

and interpret it as Cauchy's integral formula for the derivative of the analytic function $w \mapsto (w-(z-is))^{-2}$ defined on the region above Γ. If Γ were a closed curve containing $\zeta+is$ but not $z-is$, then the previous assertion would be immediate. In general, consider a circle of radius R centered at the point $\zeta+is$ and the region U_R inside this circle and above Γ. See Figure 8.1. Integrate over the boundary of U_R and let $R \to \infty$.]

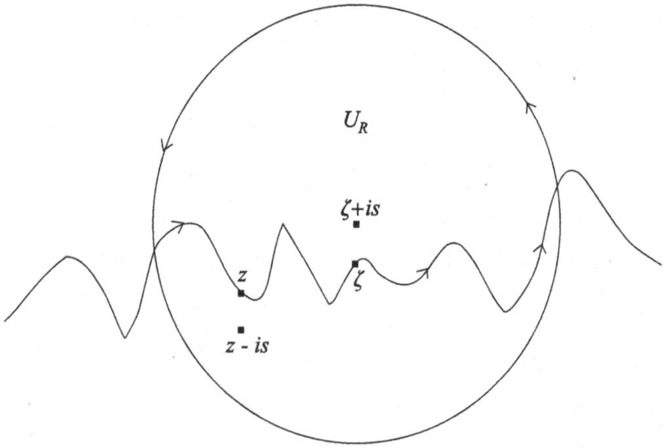

Fig. 8.1 The region U_R inside the circle and above the curve.

8.6.4. Given an accretive function b, define a pseudo-inner product

$$\langle f,g\rangle_b = \int_{\mathbf{R}^n} f(x)\,g(x)\,b(x)\,dx$$

on L^2. For an interval I, set $b_I = \int_I b(x)\,dx$. Let I_L denote the left half of a dyadic interval I and let I_R denote its right half. For a complex number z, let $z^{\frac{1}{2}} = e^{\frac{1}{2}\log_{right} z}$, where $\log_{right}$ is the branch of the logarithm defined on the complex plane minus the negative real axis normalized so that $\log_{right} 1 = 0$ [and $\log_{right}(\pm i) = \pm\frac{\pi}{2}i$]. Show that the family of functions

$$h_I = \frac{-1}{b(I)^{\frac{1}{2}}}\left(\frac{b(I_R)^{\frac{1}{2}}}{b(I_L)^{\frac{1}{2}}}\chi_{I_L} - \frac{b(I_L)^{\frac{1}{2}}}{b(I_R)^{\frac{1}{2}}}\chi_{I_R}\right),$$

where I runs over all dyadic intervals, is an orthonormal family on $L^2(\mathbf{R})$ with respect to the preceding inner product. (This family of functions is called a *pseudo-Haar basis associated with b*.)

8.6.5. Let $I = (a,b)$ be a dyadic interval and let $3I$ be its triple. For a given $x \in \mathbf{R}$, let

$$d_I(x) = \min\left(|x-a|, |x-b|, |x-\tfrac{a+b}{2}|\right).$$

Show that there exists a constant C such that

$$\left|\mathcal{C}_\Gamma(h_I)(x)\right| \le C\,|I|^{-\frac{1}{2}} \log\frac{10|I|}{|x-d_I(x)|}$$

whenever $x \in 3I$ and also

$$\left|\mathcal{C}_\Gamma(h_I)(x)\right| \le \frac{C\,|I|^{\frac{3}{2}}}{|x-d_I(x)|^2}$$

for $x \notin 3I$. In the latter case, $d_I(x)$ can be any of $a, b, \frac{a+b}{2}$.

8.6.6. (*Semmes [281]*) We say that a bounded function b is *para-accretive* if for all $s > 0$ there is a linear operator R_s with kernel satisfying (8.6.28) and (8.6.29) such that $|R_s(b)| \ge c_0$ for all $s > 0$. Let Θ_s and P_s be as in Theorem 8.6.6.
(a) Prove that

$$\left|R_s(b)(x) - R_s(1)(x)P_s(b)(x)\right|^2 \frac{dx\,ds}{s}$$

is a Carleson measure.
(b) Use the result in part (a) and the fact that $\sup_{s>0}|R_s(1)| \le C$ to obtain that $\chi_\Omega(x,s)\,dx\,ds/s$ is a Carleson measure, where

$$\Omega = \left\{(x,s) :\ |P_s(b)(x)| \le \frac{c_0}{2}\left(\sup_{s>0}|R_s(1)|\right)^{-1}\right\}.$$

(c) Conclude that the measure $\left|\Theta_s(1)(x)\right|^2 dx\,ds/s$ is Carleson, thus obtaining a generalization of Theorem 8.6.6 for para-accretive functions.

8.6.7. Using the operator $\widetilde{\mathcal{C}}_\gamma$ defined in (8.6.15), obtain that $\mathcal{C}_\Gamma$ is of weak type $(1,1)$ and bounded on $L^p(\mathbf{R})$ for all $1 < p < \infty$.

8.7 Square Roots of Elliptic Operators

In this section we prove an L^2 estimate for the square root of a divergence form second-order elliptic operator on $\mathbf{R}^n$. This estimate is based on an approach in the spirit of the $T(b)$ theorem discussed in the previous section. However, matters here are significantly more complicated for two main reasons: the roughness of the variable coefficients of the aforementioned elliptic operator and the higher-dimensional nature of the problem.

8.7.1 Preliminaries and Statement of the Main Result

For $\xi = (\xi_1, \dots, \xi_n) \in \mathbf{C}^n$ we denote its complex conjugate $(\overline{\xi_1}, \dots, \overline{\xi_n})$ by $\overline{\xi}$. Moreover, for $\xi, \zeta \in \mathbf{C}^n$ we use the inner product notation

$$\xi \cdot \zeta = \sum_{k=1}^{n} \xi_k \zeta_k .$$

Throughout this section, $A = A(x)$ is an $n \times n$ matrix of complex-valued L^∞ functions, defined on $\mathbf{R}^n$, that satisfies the *ellipticity* (or *accretivity*) conditions for some $0 < \lambda \le \Lambda < \infty$, that

$$\begin{aligned} \lambda |\xi|^2 &\le \operatorname{Re}\,(A(x)\,\xi \cdot \overline{\xi}), \\ |A(x)\,\xi \cdot \overline{\zeta}| &\le \Lambda |\xi||\zeta|, \end{aligned} \tag{8.7.1}$$

for all $x \in \mathbf{R}^n$ and $\xi, \zeta \in \mathbf{C}^n$. We interpret an element ξ of $\mathbf{C}^n$ as a column vector in $\mathbf{C}^n$ when the matrix A acts on it.

Associated with such a matrix A, we define a second-order *divergence form operator*

$$L(f) = -\operatorname{div}(A\nabla f) = \sum_{j=1}^{n} \partial_j\big((A\nabla f)_j\big), \tag{8.7.2}$$

which we interpret in the weak sense whenever f is a distribution.

The accretivity condition (8.7.1) enables us to define a square root operator $L^{1/2} = \sqrt{L}$ so that the operator identity $L = \sqrt{L}\sqrt{L}$ holds. The *square root operator* can be written in several ways, one of which is

$$\sqrt{L}(f) = \frac{16}{\pi} \int_0^{+\infty} (I + t^2 L)^{-3} t^3 L^2(f)\,\frac{dt}{t}. \tag{8.7.3}$$

We refer the reader to Exercise 8.7.3 for the existence of the square root operator and the validity of identity (8.7.3).

An important problem in the subject is to determine whether the estimate

$$\big\|\sqrt{L}(f)\big\|_{L^2} \le C_{n,\lambda,\Lambda} \big\|\nabla f\big\|_{L^2} \tag{8.7.4}$$

holds for functions f in a dense subspace of the homogeneous Sobolev space $\dot{L}_1^2(\mathbf{R}^n)$, where $C_{n,\lambda,\Lambda}$ is a constant depending only on n, λ, and Λ. Once (8.7.4) is known for a dense subspace of $\dot{L}_1^2(\mathbf{R}^n)$, then it can be extended to the entire space by density. The main purpose of this section is to discuss a detailed proof of the following result.

Theorem 8.7.1. *Let L be as in (8.7.2). Then there is a constant $C_{n,\lambda,\Lambda}$ such that for all smooth functions f with compact support, estimate (8.7.4) is valid.*

The proof of this theorem requires certain estimates concerning elliptic operators. These are presented in the next subsection, while the proof of the theorem follows in the remaining four subsections.

8.7.2 Estimates for Elliptic Operators on $\mathbf{R}^n$

The following lemma provides a quantitative expression for the mean decay of the resolvent kernel.

Lemma 8.7.2. *Let E and F be two closed sets of $\mathbf{R}^n$ and set*

$$d = \operatorname{dist}(E,F),$$

the distance between E and F. Then for all complex-valued functions f supported in E and all vector-valued functions $\vec{f}$ supported in E, we have

$$\int_F |(I+t^2L)^{-1}(f)(x)|^2\,dx \le Ce^{-c\frac{d}{t}}\int_E |f(x)|^2\,dx, \tag{8.7.5}$$

$$\int_F |t\nabla(I+t^2L)^{-1}(f)(x)|^2\,dx \le Ce^{-c\frac{d}{t}}\int_E |f(x)|^2\,dx, \tag{8.7.6}$$

$$\int_F |(I+t^2L)^{-1}(t\operatorname{div}\vec{f}\,)(x)|^2\,dx \le Ce^{-c\frac{d}{t}}\int_E |\vec{f}(x)|^2\,dx, \tag{8.7.7}$$

where $c = c(\lambda,\Lambda)$, $C = C(n,\lambda,\Lambda)$ are finite constants.

Proof. It suffices to obtain these inequalities whenever $d \ge t > 0$. Let us set $u_t = (I+t^2L)^{-1}(f)$. For all $v \in L_1^2(\mathbf{R}^n)$ we have

$$\int_{\mathbf{R}^n} u_t v\,dx + t^2\int_{\mathbf{R}^n} A\nabla u_t\cdot\nabla v\,dx = \int_{\mathbf{R}^n} fv\,dx.$$

Let η be a nonnegative smooth function with compact support that does not meet E and that satisfies $\|\eta\|_{L^\infty} = 1$. Taking $v = \overline{u_t}\,\eta^2$ and using that f is supported in E, we obtain

$$\int_{\mathbf{R}^n} |u_t|^2\eta^2\,dx + t^2\int_{\mathbf{R}^n} A\nabla u_t\cdot\overline{\nabla u_t}\,\eta^2\,dx = -2t^2\int_{\mathbf{R}^n} A(\eta\nabla u_t)\cdot\overline{u_t\nabla\eta}\,dx.$$

Using (8.7.1) and the inequality $2ab \le \varepsilon|a|^2 + \varepsilon^{-1}|b|^2$, we obtain for all $\varepsilon > 0$,

$$\begin{aligned}\int_{\mathbf{R}^n} |u_t|^2 \eta^2\,dx + \lambda t^2 \int_{\mathbf{R}^n} |\nabla u_t|^2\,\eta^2\,dx \\ \le \Lambda \varepsilon t^2 \int_{\mathbf{R}^n} |\nabla u_t|^2\,\eta^2\,dx + \Lambda \varepsilon^{-1} t^2 \int_{\mathbf{R}^n} |u_t|^2 |\nabla \eta|^2\,dx,\end{aligned}$$

and this reduces to

$$\int_{\mathbf{R}^n} |u_t|^2 |\eta|^2\,dx \le \frac{\Lambda^2 t^2}{\lambda} \int_{\mathbf{R}^n} |u_t|^2 |\nabla \eta|^2\,dx \tag{8.7.8}$$

by choosing $\varepsilon = \frac{\lambda}{\Lambda}$. Replacing η by $e^{k\eta} - 1$ in (8.7.8), where

$$k = \frac{\sqrt{\lambda}}{2\Lambda t \|\nabla \eta\|_{L^\infty}},$$

yields

$$\int_{\mathbf{R}^n} |u_t|^2 |e^{k\eta} - 1|^2\,dx \le \frac{1}{4} \int_{\mathbf{R}^n} |u_t|^2 |e^{k\eta}|^2\,dx. \tag{8.7.9}$$

Using that $|e^{k\eta} - 1|^2 \ge \frac{1}{2}|e^{k\eta}|^2 - 1$, we obtain

$$\int_{\mathbf{R}^n} |u_t|^2 |e^{k\eta}|^2\,dx \le 4 \int_{\mathbf{R}^n} |u_t|^2\,dx \le 4C \int_E |f|^2\,dx,$$

where in the last estimate we use the uniform boundedness of $(I + t^2 L)^{-1}$ on $L^2(\mathbf{R}^n)$ (Exercise 8.7.2). If, in addition, we have $\eta = 1$ on F, then

$$|e^k|^2 \int_F |u_t|^2\,dx \le \int_{\mathbf{R}^n} |u_t|^2 |e^{k\eta}|^2\,dx,$$

and picking η so that $\|\nabla \eta\|_{L^\infty} \approx 1/d$, we conclude (8.7.5).

Next, choose $\varepsilon = \lambda/2\Lambda$ and η as before to obtain

$$\begin{aligned}\int_F |t\nabla u_t|^2\,dx &\le \int_{\mathbf{R}^n} |t\nabla u_t|^2 \eta^2\,dx \\ &\le \frac{2\Lambda^2 t^2}{\lambda} \int_{\mathbf{R}^n} |u_t|^2 |\nabla \eta|^2\,dx \\ &\le C t^2 d^{-2} e^{-c\frac{d}{t}} \int_E |f|^2\,dx,\end{aligned}$$

which gives (8.7.6). Finally, (8.7.7) is obtained by duality from (8.7.6) applied to $L^* = -\mathrm{div}\,(A^*\nabla)$ when the roles of E and F are interchanged. □

Lemma 8.7.3. *Let M_f be the operator given by multiplication by a Lipschitz function f. Then there is a constant C that depends only on n, λ, and Λ such that*

$$\big\| \big[(I + t^2 L)^{-1}, M_f\big] \big\|_{L^2 \to L^2} \le C t \, \|\nabla f\|_{L^\infty} \tag{8.7.10}$$

and

$$\big\|\nabla\big[(I+t^2L)^{-1},M_f\big]\big\|_{L^2\to L^2}\le C\big\|\nabla f\big\|_{L^\infty} \tag{8.7.11}$$

for all $t>0$. *Here* $[T,S]=TS-ST$ *is the* commutator *of the operators* T *and* S.

Proof. Set $\vec{b}=A\nabla f$, $\vec{d}=A^t\nabla f$ and note that the operators given by pointwise multiplication by these vectors are L^2 bounded with norms at most a multiple of $C\big\|\nabla f\big\|_{L^\infty}$. Write

$$\begin{aligned}\big[(I+t^2L)^{-1},M_f\big] &= -(I+t^2L)^{-1}\big[(I+t^2L),M_f\big](I+t^2L)^{-1}\\ &= -(I+t^2L)^{-1}t^2(\operatorname{div}\vec{b}+\vec{d}\cdot\nabla)(1+t^2L)^{-1}.\end{aligned}$$

The uniform L^2 boundedness of $(I+t^2L)^{-1}\ t\nabla(I+t^2L)^{-1}$ and $(I+t^2L)^{-1}t\operatorname{div}$ on L^2 (see Exercise 8.7.2) implies (8.7.10). Finally, using the L^2 boundedness of the operator $t^2\nabla(I+t^2L)^{-1}\operatorname{div}$ yields (8.7.11). □

Next we have a technical lemma concerning the mean square deviation of f from $(I+t^2L)^{-1}$.

Lemma 8.7.4. *There exists a constant* C *depending only on* n, λ, *and* Λ *such that for all* Q *cubes in* $\mathbf{R}^n$ *with sides parallel to the axes, for all* $t\le\ell(Q)$, *and all Lipschitz functions* f *on* $\mathbf{R}^n$ *we have*

$$\frac{1}{|Q|}\int_Q|(I+t^2L)^{-1}(f)-f|^2\,dx \le Ct^2\big\|\nabla f\big\|_{L^\infty}^2, \tag{8.7.12}$$

$$\frac{1}{|Q|}\int_Q|\nabla((I+t^2L)^{-1}(f)-f)|^2\,dxx \le C\big\|\nabla f\big\|_{L^\infty}^2. \tag{8.7.13}$$

Proof. We begin by proving (8.7.12), while we omit the proof of (8.7.13), since it is similar. By a simple rescaling, we may assume that $\ell(Q)=1$ and that $\|\nabla f\|_{L^\infty}=1$. Set $Q_0=2Q$ (i.e., the cube with the same center as Q with twice its side length) and write $\mathbf{R}^n$ as a union of cubes Q_k of side length 2 with disjoint interiors and sides parallel to the axes. Lemma 8.7.2 implies that

$$(I+t^2L)^{-1}(1)=1$$

in the sense that

$$\lim_{R\to\infty}(I+t^2L)^{-1}(\eta_R)=1$$

in $L^2_{\text{loc}}(\mathbf{R}^n)$, where $\eta_R(x)=\eta(x/R)$ and η is a smooth bump function with $\eta\equiv 1$ near 0. Hence, we may write

$$(I+t^2L)^{-1}(f)(x)-f(x)=\sum_{k\in\mathbf{Z}^n}(I+t^2L)^{-1}((f-f(x))\chi_{Q_k})(x)=\sum_{k\in\mathbf{Z}^n}g_k(x).$$

The term for $k=0$ in the sum is $[(I+t^2L)^{-1},M_f](\chi_{Q_0})(x)$. Hence, its $L^2(Q)$ norm is controlled by $Ct\big\|\chi_{Q_0}\big\|_{L^2}$ by (8.7.10). The terms for $k\neq 0$ are dealt with using the further decomposition

$$g_k(x) = (I+t^2L)^{-1}((f-f(x_k))\chi_{Q_k})(x) + (f(x_k)-f(x))(I+t^2L)^{-1}(\chi_{Q_k})(x),$$

where x_k is the center of Q_k. Applying Lemma 8.7.2 for $(I+t^2L)^{-1}$ on the sets $E = Q_k$ and $F = Q$ and using that f is a Lipschitz function, we obtain

$$\int_Q |g_k|^2\,dx \le Ct^2 e^{-c\frac{|x_k|}{t}} \|\chi_{Q_k}\|_{L^2}^2 = Ct^2 e^{-c\frac{|x_k|}{t}} 2^n |Q|.$$

The desired bound on the $L^2(Q)$ norm of $(I+t^2L)^{-1}(f) - f$ follows from these estimates, Minkowski's inequality, and the fact that $t \le 1 = \ell(Q)$. □

8.7.3 Reduction to a Quadratic Estimate

We are given a divergence form elliptic operator as in (8.7.2) with ellipticity constants λ and Λ in (8.7.1). Our goal is to obtain the a priori estimate (8.7.4) for functions f in some dense subspace of $\dot{L}^2_1(\mathbf{R}^n)$.

To obtain this estimate we need to resolve the operator $\sqrt{L}$ as an average of simpler operators that are uniformly bounded from $\dot{L}^2_1(\mathbf{R}^n)$ to $L^2(\mathbf{R}^n)$. In the sequel we use the following resolution of the square root:

$$\sqrt{L}(f) = \frac{16}{\pi}\int_0^\infty (I+t^2L)^{-3}t^3L^2(f)\,\frac{dt}{t},$$

in which the integral converges in $L^2(\mathbf{R}^n)$ for $f \in \mathscr{C}_0^\infty(\mathbf{R}^n)$. Take $g \in \mathscr{C}_0^\infty(\mathbf{R}^n)$ with $\|g\|_{L^2} = 1$. Using duality and the Cauchy–Schwarz inequality, we can control the quantity $\big|\langle\sqrt{L}(f)\,|\,g\rangle\big|^2$ by

$$\frac{256}{\pi^2}\bigg(\int_0^\infty \|(I+t^2L)^{-1}tL(f)\|_2^2\,\frac{dt}{t}\bigg)\bigg(\int_0^\infty \|V_t(g)\|_{L^2}^2\,\frac{dt}{t}\bigg), \tag{8.7.14}$$

where we set

$$V_t = t^2L^*(I+t^2L^*)^{-2}.$$

Here L^* is the adjoint operator to L and note that the matrix corresponding to L^* is the conjugate-transpose matrix A^* of A (i.e., the transpose of the matrix whose entries are the complex conjugates of the matrix A). We explain why the estimate

$$\int_0^\infty \|V_t(g)\|_{L^2}^2\,\frac{dt}{t} \le C\|g\|_{L^2}^2 \tag{8.7.15}$$

is valid. Fix a real-valued function $\Psi \in \mathscr{C}_0^\infty(\mathbf{R}^n)$ with mean value zero normalized so that

$$\int_0^\infty |\widehat{\Psi}(s\xi)|^2\,\frac{ds}{s} = 1$$

for all $\xi \in \mathbf{R}^n$ and define $\Psi_s(x) = \frac{1}{s^n}\Psi(\frac{x}{s})$. Throughout the proof, Q_s denotes the operator

$$Q_s(h) = h * \Psi_s. \tag{8.7.16}$$

Obviously we have

$$\int_0^\infty \big\|Q_s(g)\big\|_{L^2}^2 \frac{ds}{s} = \big\|g\big\|_{L^2}^2$$

for all L^2 functions g.

We obtain estimate (8.7.15) as a consequence of Corollary 8.6.4 applied to the operators V_t that have uniform (in t) bounded extensions on $L^2(\mathbf{R}^n)$. To apply Corollary 8.6.4, we need to check that condition (8.6.40) holds for $\Theta_t = V_t$. Since

$$V_tQ_s = -(I+t^2L^*)^{-2}t^2\operatorname{div} A^*\nabla Q_s\,,$$

we have

$$\big\|V_tQ_s\big\|_{L^2\to L^2} \le \big\|(I+t^2L^*)^{-2}t^2\operatorname{div} A^*\big\|_{L^2\to L^2}\big\|\nabla Q_s\big\|_{L^2\to L^2} \le c\,\frac{t}{s}\,, \tag{8.7.17}$$

with C depending only on n, λ, and Λ. Choose $\Psi = \Delta\varphi$ with $\varphi \in \mathscr{C}_0^\infty(\mathbf{R}^n)$ radial so that in particular, $\Psi = \operatorname{div}\vec{h}$. This yields $Q_s = s\operatorname{div}\vec{R}_s$ with $\vec{R}_s$ uniformly bounded; hence

$$\big\|V_tQ_s\big\|_{L^2\to L^2} \le \big\|t^2L^*(I+t^2L^*)^{-2}\operatorname{div}\big\|_{L^2\to L^2}\big\|s\vec{R}_s\big\|_{L^2\to L^2} \le c\,\frac{s}{t}\,, \tag{8.7.18}$$

with C depending only on n, λ, and Λ.

Combining (8.7.17) and (8.7.18) proves (8.6.40) with $\Theta_t = V_t$. Hence Corollary 8.6.4 is applicable and (8.7.15) follows.

Therefore, the second integral on the right-hand side of (8.7.14) is bounded, and estimate (8.7.4) is reduced to proving

$$\int_0^\infty \big\|(I+t^2L)^{-1}tL(f)\big\|_2^2\frac{dt}{t} \le C\int_{\mathbf{R}^n}|\nabla f|^2\,dx \tag{8.7.19}$$

for all $f \in \mathscr{C}_0^\infty(\mathbf{R}^n)$.

8.7.4 Reduction to a Carleson Measure Estimate

Our next goal is to reduce matters to a Carleson measure estimate. We first introduce some notation to be used throughout. For $\mathbf{C}^n$-valued functions $\vec{f} = (f_1,\dots,f_n)$ define

$$Z_t(\vec{f}) = -\sum_{k=1}^n\sum_{j=1}^n (I+t^2L)^{-1}t\partial_j(a_{j,k}f_k)\,.$$

In short, we write $Z_t = -(I+t^2L)^{-1}t\,\mathrm{div}A$. With this notation, we reformulate (8.7.19) as

$$\int_0^\infty \|Z_t(\nabla f)\|_2^2 \frac{dt}{t} \le C\int_{\mathbf{R}^n} |\nabla f|^2\,dx. \tag{8.7.20}$$

Also, define

$$\gamma_t(x) = Z_t(\mathbf{1})(x) = \Big(-\sum_{j=1}^n (I+t^2L)^{-1}t\,\partial_j(a_{j,k})(x)\Big)_{1\le k\le n},$$

where $\mathbf{1}$ is the $n\times n$ identity matrix and the action of Z_t on $\mathbf{1}$ is columnwise.

The reduction to a Carleson measure estimate and to a $T(b)$ argument requires the following inequality:

$$\int_{\mathbf{R}^n}\int_0^\infty |\gamma_t(x)\cdot P_t^2(\nabla g)(x) - Z_t(\nabla g)(x)|^2 \frac{dx\,dt}{t} \le C\int_{\mathbf{R}^n} |\nabla g|^2\,dx, \tag{8.7.21}$$

where C depends only on n, λ, and Λ. Here, P_t denotes the operator

$$P_t(h) = h * p_t, \tag{8.7.22}$$

where $p_t(x) = t^{-n}p(t^{-1}x)$ and p denotes a nonnegative smooth function supported in the unit ball of $\mathbf{R}^n$ with integral equal to 1. To prove this, we need to handle Littlewood–Paley theory in a setting a bit more general than the one encountered in the previous section.

Lemma 8.7.5. *For $t>0$, let U_t be integral operators defined on $L^2(\mathbf{R}^n)$ with measurable kernels $L_t(x,y)$. Suppose that for some $m>n$ and for all $y\in\mathbf{R}^n$ and $t>0$ we have*

$$\int_{\mathbf{R}^n}\Big(1+\frac{|x-y|}{t}\Big)^{2m} |L_t(x,y)|^2\,dx \le t^{-n}. \tag{8.7.23}$$

Assume that for any ball $B(y,t)$, U_t has a bounded extension from $L^\infty(\mathbf{R}^n)$ to $L^2(B(y,t))$ such that for all f in $L^\infty(\mathbf{R}^n)$ and $y\in\mathbf{R}^n$ we have

$$\frac{1}{t^n}\int_{B(y,t)} |U_t(f)(x)|^2\,dx \le \|f\|_{L^\infty}^2. \tag{8.7.24}$$

Finally, assume that $U_t(1)=0$ in the sense that

$$U_t(\chi_{B(0,R)}) \to 0 \qquad \text{in} \quad L^2(B(y,t)) \tag{8.7.25}$$

as $R\to\infty$ for all $y\in\mathbf{R}^n$ and $t>0$.

Let Q_s and P_t be as in (8.7.16) and (8.7.22), respectively. Then for some $\alpha>0$ and C depending on n and m we have

$$\|U_tP_tQ_s\|_{L^2\to L^2} \le C\min\Big(\frac{t}{s},\frac{s}{t}\Big)^\alpha \tag{8.7.26}$$

and also

$$\big\|U_tQ_s\big\|_{L^2\to L^2} \le C\Big(\frac{t}{s}\Big)^{\alpha}, \qquad t \le s. \tag{8.7.27}$$

Proof. We begin by observing that $U_t^*U_t$ has a kernel $K_t(x,y)$ given by

$$K_t(x,y) = \int_{\mathbf{R}^n} \overline{L_t(z,x)}L_t(z,y)\,dz.$$

The simple inequality $(1+a+b) \le (1+a)(1+b)$ for $a,b>0$ combined with the Cauchy–Schwarz inequality and (8.7.23) yield that $\left(1+\frac{|x-y|}{t}\right)^m |K_t(x,y)|$ is bounded by

$$\int_{\mathbf{R}^n}\Big(1+\frac{|x-z|}{t}\Big)^m |L_t(z,x)|\,|L_t(z,y)|\,\Big(1+\frac{|z-y|}{t}\Big)^m\,dy \le t^{-n}.$$

We conclude that

$$|K_t(x,y)| \le \frac{1}{t^n}\Big(1+\frac{|x-y|}{t}\Big)^{-m}. \tag{8.7.28}$$

Hence $U_t^*U_t$ is bounded on all L^p, $1\le p\le +\infty$, and in particular, for $p=2$. Since L^2 is a Hilbert space, it follows that U_t is bounded on $L^2(\mathbf{R}^n)$ uniformly in $t>0$.

For $s\le t$ we use that $\big\|U_t\big\|_{L^2\to L^2}\le B<\infty$ and basic estimates to deduce that

$$\big\|U_tP_tQ_s\big\|_{L^2\to L^2} \le B\big\|P_tQ_s\big\|_{L^2\to L^2} \le CB\Big(\frac{s}{t}\Big)^{\alpha}.$$

Next, we consider the case $t\le s$. Since P_t has an integrable kernel, and the kernel of $U_t^*U_t$ satisfies (8.7.28), it follows that $W_t=U_t^*U_tP_t$ has a kernel that satisfies a similar estimate. If we prove that $W_t(1)=0$, then we can deduce from standard arguments that when $t\le s$ we have

$$\big\|W_tQ_s\big\|_{L^2\to L^2} \le C\Big(\frac{t}{s}\Big)^{2\alpha} \tag{8.7.29}$$

for $0<\alpha<m-n$. This would imply the required estimate (8.7.26), since

$$\big\|U_tP_tQ_s\big\|^2_{L^2\to L^2} = \big\|Q_s^*P_tU_t^*U_tP_tQ_s\big\|_{L^2\to L^2} \le C\big\|U_t^*U_tP_tQ_s\big\|_{L^2\to L^2}.$$

We have that $W_t(1)=U_t^*U_t(1)$. Suppose that a function φ in $L^2(\mathbf{R}^n)$ is compactly supported. Then φ is integrable over $\mathbf{R}^n$ and we have

$$\big\langle U_t^*U_t(1)\,|\,\varphi\big\rangle = \lim_{R\to\infty}\big\langle U_t^*U_t(\chi_{B(0,R)})\,|\,\varphi\big\rangle = \lim_{R\to\infty}\big\langle U_t(\chi_{B(0,R)})\,|\,U_t(\varphi)\big\rangle.$$

We have

$$\big\langle U_t(\chi_{B(0,R)})\,|\,U_t(\varphi)\big\rangle = \int_{\mathbf{R}^n}\int_{\mathbf{R}^n} U_t(\chi_{B(0,R)})(x)\overline{U_t(x,y)\varphi(y)}\,dy\,dx,$$

and this is in absolute value at most a constant multiple of

$$\left(t^{-n}\int_{\mathbf{R}^n}\int_{\mathbf{R}^n}\left(1+\frac{|x-y|}{t}\right)^{-2m}|U_t(\chi_{B(0,R)})(x)|^2|\varphi(y)|\,dy\,dx\right)^{\frac{1}{2}}\|\varphi\|_{L^1}^{\frac{1}{2}}$$

by (8.7.23) and the Cauchy–Schwarz inequality for the measure $|\varphi(y)|\,dy\,dx$. Using a covering in the x variable by a family of balls $B(y+ckt,t)$, $k\in\mathbf{Z}^n$, we deduce easily that the last displayed expression is at most

$$C_\varphi\left(\sum_{k\in\mathbf{Z}^n}\int_{\mathbf{R}^n}(1+|k|)^{-2m}c_R(y,k)|\varphi(y)|\,dy\right)^{\frac{1}{2}},$$

where C_φ is a constant that depends on φ and

$$c_R(y,k)=t^{-n}\int_{B(y+ckt,t)}|U_t(\chi_{B(0,R)})(x)|^2\,dx.$$

Applying the dominated convergence theorem and invoking (8.7.24) and (8.7.25) as $R\to\infty$, we conclude that $\langle U_t^*U_t(1)\,|\,\varphi\rangle=0$. The latter implies that $U_t^*U_t(1)=0$. The same conclusion follows for W_t, since $P_t(1)=1$.

To prove (8.7.27) when $t\le s$ we repeat the previous argument with $W_t=U_t^*U_t$. Since $W_t(1)=0$ and W_t has a nice kernel, it follows that (8.7.29) holds. Thus

$$\|U_tQ_s\|_{L^2\to L^2}^2=\|Q_s^*U_t^*U_tQ_s\|_{L^2\to L^2}\le C\|U_t^*U_tQ_s\|_{L^2\to L^2}\le C\left(\frac{t}{s}\right)^{2\alpha}.$$

This concludes the proof of the lemma. □

Lemma 8.7.6. *Let P_t be as in Lemma 8.7.5. Then the operator U_t defined by $U_t(\vec{f})(x)=\gamma_t(x)\cdot P_t(\vec{f})(x)-Z_tP_t(\vec{f})(x)$ satisfies*

$$\int_0^\infty\|U_tP_t(\vec{f})\|_{L^2}^2\frac{dt}{t}\le C\|\vec{f}\|_{L^2}^2,$$

where C depends only on n, λ, and Λ. Here the action of P_t on $\vec{f}$ is componentwise.

Proof. By the off-diagonal estimates of Lemma 8.7.2 for Z_t and the fact that p has support in the unit ball, it is simple to show that there is a constant C depending on n, λ, and Λ such that for all $y\in\mathbf{R}^n$,

$$\frac{1}{t^n}\int_{B(y,t)}|\gamma_t(x)|^2\,dx\le C \tag{8.7.30}$$

and that the kernel of $C^{-1}U_t$ satisfies the hypotheses in Lemma 8.7.5. The conclusion follows from Corollary 8.6.4 applied to U_tP_t. □

We now return to (8.7.21). We begin by writing

$$\gamma_t(x)\cdot P_t^2(\nabla g)(x)-Z_t(\nabla g)(x)=U_tP_t(\nabla g)(x)+Z_t(P_t^2-I)(\nabla g)(x),$$

and we prove (8.7.21) for each term that appears on the right. For the first term we apply Lemma 8.7.6. Since P_t commutes with partial derivatives, we may use that

$$\|Z_t \nabla\|_{L^2 \to L^2} = \|(I+t^2L)^{-1} tL\|_{L^2 \to L^2} \le Ct^{-1},$$

and therefore we obtain for the second term

$$\begin{aligned}\int_{\mathbf{R}^n}\int_0^\infty |Z_t(P_t^2 - I)(\nabla g)(x)|^2 \frac{dx\,dt}{t} &\le C^2 \int_{\mathbf{R}^n}\int_0^\infty |(P_t^2 - I)(g)(x)|^2 \frac{dt}{t^3}\,dx \\ &\le C^2 c(p)\|\nabla g\|_2^2\end{aligned}$$

by Plancherel's theorem, where C depends only on n, λ, and Λ. This concludes the proof of (8.7.21).

Lemma 8.7.7. *The required estimate (8.7.4) follows from the Carleson measure estimate*

$$\sup_Q \frac{1}{|Q|}\int_Q\int_0^{\ell(Q)} |\gamma_t(x)|^2 \frac{dx\,dt}{t} < \infty, \tag{8.7.31}$$

where the supremum is taken over all cubes in $\mathbf{R}^n$ with sides parallel to the axes.

Proof. Indeed, (8.7.31) and Theorem 7.3.7 imply

$$\int_{\mathbf{R}^n}\int_0^\infty |P_t^2(\nabla g)(x)\cdot\gamma_t(x)|^2 \frac{dx\,dt}{t} \le C\int_{\mathbf{R}^n} |\nabla g|^2\,dx,$$

and together with (8.7.21) we deduce that (8.7.20) holds. □

Next we introduce an auxiliary averaging operator. We define a dyadic averaging operator S_t^Q as follows:

$$S_t^Q(\vec f)(x) = \left(\frac{1}{|Q'_x|}\int_{Q'_x} \vec f(y)\,dy\right)\chi_{Q'_x}(x),$$

where Q'_x is the unique dyadic cube contained in Q that contains x and satisfies $\frac12 \ell(Q'_x) < t \le \ell(Q'_x)$. Notice that S_t^Q is a projection, i.e., it satisfies $S_t^Q S_t^Q = S_t^Q$. We have the following technical lemma concerning S_t^Q.

Lemma 8.7.8. *For some C depending only on n, λ, and Λ, we have*

$$\int_Q\int_0^{\ell(Q)} |\gamma_t(x)\cdot(S_t^Q - P_t^2)(\vec f)(x)|^2 \frac{dx\,dt}{t} \le C\int_{\mathbf{R}^n} |\vec f|^2\,dx. \tag{8.7.32}$$

Proof. We actually obtain a stronger version of (8.7.32) in which the t-integration on the left is taken over $(0,+\infty)$. Let Q_s be as in (8.7.16). Set $\Theta_t = \gamma_t \cdot (S_t^Q - P_t^2)$. The proof of (8.7.32) is based on Corollary 8.6.4 provided we show that for some $\alpha > 0$,

$$\|\Theta_t Q_s\|_{L^2 \to L^2} \le C \min\left(\frac{t}{s}, \frac{s}{t}\right)^\alpha.$$

Suppose first that $t \le s$. Notice that $\Theta_t(1) = 0$, and thus (8.7.25) holds. With the aid of (8.7.30), we observe that Θ_t satisfies the hypotheses (8.7.23) and (8.7.24) of Lemma 8.7.5. Conclusion (8.7.27) of this lemma yields that for some $\alpha > 0$ we have

$$\big\|\Theta_t Q_s\big\|_{L^2\to L^2} \le C\Big(\frac{t}{s}\Big)^{\alpha}.$$

We now turn to the case $s \le t$. Since the kernel of P_t is bounded by $ct^{-n}\chi_{|x-y|\le t}$, condition (8.7.30) yields that $\gamma_t P_t$ is uniformly bounded on L^2 and thus

$$\big\|\gamma_t P_t^2 Q_s\big\|_{L^2\to L^2} \le C\big\|P_t Q_s\big\|_{L^2\to L^2} \le C'\frac{s}{t}.$$

It remains to consider the case $s \le t$ for the operator $U_t = \gamma_t \cdot S_t^Q$. We begin by observing that U_t is L^2 bounded uniformly in $t > 0$; this follows from a standard $U_t^* U_t$ argument using condition (8.7.23). Secondly, as already observed, S_t^Q is an orthogonal projection. Therefore, we have

$$\begin{aligned}
\big\|(\gamma_t \cdot S_t^Q) Q_s\big\|_{L^2\to L^2} &\le \big\|(\gamma_t \cdot S_t^Q) S_t^Q Q_s\big\|_{L^2\to L^2}\\
&\le \big\|S_t^Q Q_s\big\|_{L^2\to L^2}\\
&\le \big\|S_t^Q\big\|_{L^2\to \dot L^2_\alpha}\big\|Q_s\big\|_{\dot L^2_\alpha\to L^2}\\
&\le C s^{\alpha} t^{-\alpha}.
\end{aligned}$$

The last inequality follows from the facts that for any α in $(0,\frac{1}{2})$, Q_s maps the homogeneous Sobolev space $\dot L^2_\alpha$ to L^2 with norm at most a multiple of Cs^α and that the dyadic averaging operator S_t^Q maps $L^2(\mathbf{R}^n)$ to $\dot L^2_\alpha(\mathbf{R}^n)$ with norm $Ct^{-\alpha}$. The former of these statements is trivially verified by taking the Fourier transform, while the latter statement requires some explanation.

Fix an $\alpha \in (0,\frac{1}{2})$ and take $h, g \in L^2(\mathbf{R}^n)$. Also fix $j \in \mathbf{Z}$ such that $2^{-j-1} \le t < 2^{-j}$. We then have

$$\big\langle S_t^Q(-\Delta)^{\frac{\alpha}{2}}(h), g\big\rangle = \sum_{J_{j,k}\subseteq Q}\Big\langle (-\Delta)^{\frac{\alpha}{2}}(h), \chi_{J_{j,k}}(x)\big(\operatorname*{Avg}_{J_{j,k}} \overline{g}\big)\Big\rangle,$$

where $J_{j,k} = \prod_{r=1}^{n}[2^{-j}k_r, 2^{-j}(k_r+1))$ and $k = (k_1,\dots,k_n)$. It follows that

$$\begin{aligned}
\big\langle S_t^Q(-\Delta)^{\frac{\alpha}{2}}(h), g\big\rangle &= \sum_{J_{j,k}\subseteq Q}\Big\langle h, \big(\operatorname*{Avg}_{J_{j,k}} \overline{g}\big)(-\Delta)^{\frac{\alpha}{2}}(\chi_{J_{j,k}})(x)\Big\rangle\\
&= \Big\langle h, \sum_{J_{j,k}\subseteq Q} 2^{\alpha j}\big(\operatorname*{Avg}_{J_{j,k}} \overline{g}\big)(-\Delta)^{\frac{\alpha}{2}}(\chi_{[0,1)^n})(2^j(\cdot)-k)\Big\rangle.
\end{aligned}$$

Set $\chi_\alpha = (-\Delta)^{\frac{\alpha}{2}}(\chi_{[0,1)^n})$. We estimate the L^2 norm of the preceding sum. We have

$$
\begin{aligned}
&\int_{\mathbf{R}^n} \Big| \sum_{J_{j,k} \subseteq Q} 2^{\alpha j} \big(\operatorname*{Avg}_{J_{j,k}} \overline{g} \big) \chi_\alpha(2^j x - k) \Big|^2 dx \\
&= 2^{2\alpha j - nj} \int_{\mathbf{R}^n} \Big| \sum_{J_{j,k} \subseteq Q} \big(\operatorname*{Avg}_{J_{j,k}} \overline{g} \big) \chi_\alpha(x - k) \Big|^2 dx \\
&= 2^{2\alpha j - nj} \int_{\mathbf{R}^n} \Big| \sum_{J_{j,k} \subseteq Q} e^{-2\pi i k \cdot \xi} \big(\operatorname*{Avg}_{J_{j,k}} \overline{g} \big) \Big|^2 |\widehat{\chi_\alpha}(\xi)|^2 d\xi \\
&= 2^{2\alpha j - nj} \int_{[0,1]^n} \Big| \sum_{J_{j,k} \subseteq Q} e^{-2\pi i k \cdot \xi} \big(\operatorname*{Avg}_{J_{j,k}} \overline{g} \big) \Big|^2 \sum_{l \in \mathbf{Z}^n} |\widehat{\chi_\alpha}(\xi + l)|^2 d\xi \\
&\le 2^{2\alpha j - nj} \int_{[0,1]^n} \Big| \sum_{J_{j,k} \subseteq Q} e^{-2\pi i k \cdot \xi} \big(\operatorname*{Avg}_{J_{j,k}} \overline{g} \big) \Big|^2 d\xi \sup_{\xi \in [0,1]^n} \sum_{l \in \mathbf{Z}^n} |\widehat{\chi_\alpha}(\xi + l)|^2 \\
&= 2^{2\alpha j - nj} \sum_{k \in \mathbf{Z}^n} \Big| \operatorname*{Avg}_{J_{j,k}} \overline{g} \Big|^2 C(n,\alpha)^2,
\end{aligned}
$$

where we used Plancherel's identity on the torus (Proposition 3.1.16) and we set

$$
C(n,\alpha)^2 = \sup_{\xi \in [0,1]^n} \sum_{l \in \mathbf{Z}^n} |\widehat{\chi_\alpha}(\xi + l)|^2 .
$$

Since

$$
\widehat{\chi_\alpha}(\xi) = |\xi|^\alpha \prod_{r=1}^n \frac{1 - e^{-2\pi i \xi_r}}{2\pi i \xi_r},
$$

it follows that $C(n,\alpha) < \infty$ when $0 < \alpha < \frac{1}{2}$. In this case we conclude that

$$
\begin{aligned}
\big| \langle S_t^Q (-\Delta)^{\frac{\alpha}{2}}(h), g \rangle \big| &\le C(n,\alpha) \big\| h \big\|_{L^2} 2^{j\alpha} \Big(2^{-nj} \sum_{k \in \mathbf{Z}^n} \Big| \operatorname*{Avg}_{J_{j,k}} \overline{g} \Big|^2 \Big)^{\frac{1}{2}} \\
&\le C' \big\| h \big\|_{L^2} t^{-\alpha} \big\| g \big\|_{L^2},
\end{aligned}
$$

and this implies that $\big\| S_t^Q \big\|_{L^2 \to L^2_\alpha} \le C t^{-\alpha}$ and hence the required conclusion. □

8.7.5 The $T(b)$ Argument

To obtain (8.7.31), we adapt the $T(b)$ theorem of the previous section for square roots of divergence form elliptic operators. We fix a cube Q with center c_Q, an $\varepsilon \in (0,1)$, and a unit vector w in $\mathbf{C}^n$. We define a scalar-valued function

$$
f_{Q,w}^\varepsilon = (1 + (\varepsilon \ell(Q))^2 L)^{-1} (\Phi_Q \cdot \overline{w}), \tag{8.7.33}
$$

where

$$
\Phi_Q(x) = x - c_Q .
$$

We begin by observing that the following estimates are consequences of Lemma 8.7.4:

$$\int_{5Q} |f^{\varepsilon}_{Q,w} - \Phi_Q \cdot \overline{w}|^2 \, dx \le C_1 \varepsilon^2 \ell(Q)^2 |Q| \tag{8.7.34}$$

and

$$\int_{5Q} |\nabla(f^{\varepsilon}_{Q,w} - \Phi_Q \cdot \overline{w})|^2 \, dx \le C_2 |Q| , \tag{8.7.35}$$

where C_1, C_2 depend on n, λ, Λ and not on ε, Q, and w. It is important to observe that the constants C_1, C_2 are independent of ε.

The proof of (8.7.31) follows by combining the next two lemmas. The rest of this section is devoted to their proofs.

Lemma 8.7.9. *There exists an $\varepsilon > 0$ depending on n, λ, Λ, and a finite set $\mathscr{F}$ of unit vectors in $\mathbf{C}^n$ whose cardinality depends on ε and n, such that*

$$\sup_Q \frac{1}{|Q|} \int_Q \int_0^{\ell(Q)} |\gamma_t(x)|^2 \, \frac{dx\,dt}{t}$$
$$\le C \sum_{w \in \mathscr{F}} \sup_Q \frac{1}{|Q|} \int_Q \int_0^{\ell(Q)} |\gamma_t(x) \cdot (S_t^Q \nabla f^{\varepsilon}_{Q,w})(x)|^2 \, \frac{dx\,dt}{t} ,$$

where C depends only on ε, n, λ, and Λ. The suprema are taken over all cubes Q in $\mathbf{R}^n$ with sides parallel to the axes.

Lemma 8.7.10. *For C depending only on n, λ, Λ, and $\varepsilon > 0$, we have*

$$\int_Q \int_0^{\ell(Q)} \left|\gamma_t(x) \cdot (S_t^Q \nabla f^{\varepsilon}_{Q,w})(x)\right|^2 \frac{dx\,dt}{t} \le C|Q| . \tag{8.7.36}$$

We begin with the proof of Lemma 8.7.10, which is the easiest of the two.

Proof of Lemma 8.7.10. Pick a smooth bump function $\mathscr{X}_Q$ localized on $4Q$ and equal to 1 on $2Q$ with $\left\|\mathscr{X}_Q\right\|_{L^\infty} + \ell(Q)\left\|\nabla \mathscr{X}_Q\right\|_{L^\infty} \le c_n$. By Lemma 8.7.5 and estimate (8.7.21), the left-hand side of (8.7.36) is bounded by

$$C \int_{\mathbf{R}^n} \left|\nabla(\mathscr{X}_Q f^{\varepsilon}_{Q,w})\right|^2 dx + 2 \int_Q \int_0^{\ell(Q)} \left|\gamma_t(x) \cdot (P_t^2 \nabla(\mathscr{X}_Q f^{\varepsilon}_{Q,w}))(x)\right|^2 \frac{dx\,dt}{t}$$
$$\le C \int_{\mathbf{R}^n} \left|\nabla(\mathscr{X}_Q f^{\varepsilon}_{Q,w})\right|^2 dx + 4 \int_Q \int_0^{\ell(Q)} \left|(Z_t \nabla(\mathscr{X}_Q f^{\varepsilon}_{Q,w}))(x)\right|^2 \frac{dx\,dt}{t} .$$

It remains to control the last displayed expression by $C|Q|$.

First, it follows easily from (8.7.34) and (8.7.35) that

$$\int_{\mathbf{R}^n} |\nabla(\mathscr{X}_Q f^{\varepsilon}_{Q,w})|^2 \, dx \le C|Q| ,$$

where C is independent of Q and w (but it may depend on ε). Next, we write

$$Z_t\nabla(\mathscr{X}_Q f^{\varepsilon}_{Q,w}) = W_t^1 + W_t^2 + W_t^3,$$

where

$$\begin{aligned} W_t^1 &= (I+t^2L)^{-1}t\left(\mathscr{X}_Q L(f^{\varepsilon}_{Q,w})\right), \\ W_t^2 &= -(I+t^2L)^{-1}t\left(\operatorname{div}(Af^{\varepsilon}_{Q,w}\nabla\mathscr{X}_Q)\right), \\ W_t^3 &= -(I+t^2L)^{-1}t\left(A\nabla f^{\varepsilon}_{Q,w}\cdot\nabla\mathscr{X}_Q\right), \end{aligned}$$

and we use different arguments to treat each term W_t^j.

To handle W_t^1, observe that

$$L(f^{\varepsilon}_{Q,w}) = \frac{f^{\varepsilon}_{Q,w} - \Phi_Q\cdot\overline{w}}{\varepsilon^2\ell(Q)^2},$$

and therefore it follows from (8.7.34) that

$$\int_{\mathbf{R}^n} |\mathscr{X}_Q L(f^{\varepsilon}_{Q,w})|^2 \le C|Q|(\varepsilon\ell(Q))^{-2},$$

where C is independent of Q and w. Using the (uniform in t) boundedness of the operator $(I+t^2L)^{-1}$ on $L^2(\mathbf{R}^n)$, we obtain

$$\int_Q\int_0^{\ell(Q)} \left|W_t^1(x)\right|^2 \frac{dx\,dt}{t} \le \int_0^{\ell(Q)} \frac{C|Q|t^2}{(\varepsilon\ell(Q))^2}\frac{dt}{t} \le \frac{C|Q|}{\varepsilon^2},$$

which establishes the required quadratic estimate for W_t^1.

To obtain a similar quadratic estimate for W_t^2, we apply Lemma 8.7.2 for the operator $(I+t^2L)^{-1}t\operatorname{div}$ with sets $F=Q$ and $E=\operatorname{supp}(f^{\varepsilon}_{Q,w}\nabla\mathscr{X}_Q)\subseteq 4Q\setminus 2Q$. We obtain that

$$\int_Q\int_0^{\ell(Q)} |W_t^2(x)|^2\frac{dx\,dt}{t} \le C\int_0^{\ell(Q)} e^{-\frac{\ell(Q)}{ct}}\frac{dt}{t}\int_{4Q\setminus 2Q}|Af^{\varepsilon}_{Q,w}\nabla\mathscr{X}_Q|^2\,dx.$$

The first integral on the right provides at most a constant factor, while we handle the second integral by writing

$$f^{\varepsilon}_{Q,w} = (f^{\varepsilon}_{Q,w} - \Phi_Q\cdot\overline{w}) + \Phi_Q\cdot\overline{w}.$$

Using (8.7.34) and the facts that $\left\|\nabla\mathscr{X}_Q\right\|_{L^\infty} \le c_n\ell(Q)^{-1}$ and that $|\Phi_Q| \le c_n\ell(Q)$ on the support of $\mathscr{X}_Q$, we obtain that

$$\int_{4Q\setminus 2Q}|Af^{\varepsilon}_{Q,w}\nabla\mathscr{X}_Q|^2\,dx \le C|Q|,$$

where C depends only on n, λ, and Λ. This yields the required result for W_t^2.

To obtain a similar estimate for W_t^3, we use the (uniform in t) boundedness of $(I+t^2L)^{-1}$ on $L^2(\mathbf{R}^n)$ (Exercise 8.7.2) to obtain that

$$\int_Q \int_0^{\ell(Q)} |W_t^3(x)|^2 \frac{dx\,dt}{t} \le C \int_0^{\ell(Q)} t^2 \frac{dt}{t} \int_{4Q\setminus 2Q} |A\nabla f_{Q,w}^{\varepsilon} \cdot \nabla \mathscr{X}_Q|^2 \, dx.$$

But the last integral is shown easily to be bounded by $C|Q|$ by writing $f_{Q,w}^{\varepsilon}$, as in the previous case, and using (8.7.35) and the properties of $\mathscr{X}_Q$ and Φ_Q. Note that C here depends only on n, λ, and Λ. This concludes the proof of Lemma 8.7.10. □

8.7.6 The Proof of Lemma 8.7.9

It remains to prove Lemma 8.7.9. The main ingredient in the proof of Lemma 8.7.9 is the following proposition, which we state and prove first.

Proposition 8.7.11. *There exists an $\varepsilon > 0$ depending on n, λ, and Λ, and $\eta = \eta(\varepsilon) > 0$ such that for each unit vector w in $\mathbf{C}^n$ and each cube Q with sides parallel to the axes, there exists a collection $\mathscr{S}_w' = \{Q'\}$ of nonoverlapping dyadic subcubes of Q such that*

$$\Big|\bigcup_{Q' \in \mathscr{S}_w'} Q'\Big| \le (1-\eta)|Q|, \tag{8.7.37}$$

and moreover, if $\mathscr{S}_w''$ is the collection of all dyadic subcubes of Q not contained in any $Q' \in \mathscr{S}_w'$, then for any $Q'' \in \mathscr{S}_w''$ we have

$$\frac{1}{|Q''|}\int_{Q''} \operatorname{Re}(\nabla f_{Q,w}^{\varepsilon}(y)\cdot w)\,dy \ge \frac{3}{4} \tag{8.7.38}$$

and

$$\frac{1}{|Q''|}\int_{Q''} |\nabla f_{Q,w}^{\varepsilon}(y)|^2\,dy \le (4\varepsilon)^{-2}. \tag{8.7.39}$$

Proof. We begin by proving the following crucial estimate:

$$\left|\int_Q (1-\nabla f_{Q,w}^{\varepsilon}(x)\cdot w)\,dx\right| \le C\varepsilon^{\frac{1}{2}}|Q|, \tag{8.7.40}$$

where C depends on n, λ, and Λ, but not on ε, Q, and w. Indeed, we observe that

$$\nabla(\Phi_Q\cdot\overline{w})(x)\cdot w = |w|^2 = 1,$$

so that

$$1-\nabla f_{Q,w}^{\varepsilon}(x)\cdot w = \nabla g_{Q,w}^{\varepsilon}(x)\cdot w,$$

where we set

$$g_{Q,w}^{\varepsilon}(x) = \Phi_Q(x)\cdot\overline{w} - f_{Q,w}^{\varepsilon}(x).$$

Next we state another useful lemma, whose proof is postponed until the end of this subsection.

Lemma 8.7.12. *There exists a constant* $C = C_n$ *such that for all* $h \in \dot{L}_1^2$ *we have*

$$\left| \int_Q \nabla h(x)\, dx \right| \le C\ell(Q)^{\frac{n-1}{2}} \left(\int_Q |h(x)|^2\, dx \right)^{\frac{1}{4}} \left(\int_Q |\nabla h(x)|^2\, dx \right)^{\frac{1}{4}}.$$

Applying Lemma 8.7.12 to the function $g^{\varepsilon}_{Q,w}$, we deduce (8.7.40) as a consequence of (8.7.34) and (8.7.35).

We now proceed with the proof of Proposition 8.7.11. First we deduce from (8.7.40) that

$$\frac{1}{|Q|} \int_Q \mathrm{Re}\,(\nabla f^{\varepsilon}_{Q,w}(x) \cdot w)\, dx \ge \frac{7}{8},$$

provided that ε is small enough. We also observe that as a consequence of (8.7.35) we have

$$\frac{1}{|Q|} \int_Q |\nabla f^{\varepsilon}_{Q,w}(x)|^2\, dx \le C_3,$$

where C_3 is independent of ε. Now we perform a stopping-time decomposition to select a collection $\mathscr{S}'_w$ of dyadic subcubes of Q that are maximal with respect to either one of the following conditions:

$$\frac{1}{|Q'|} \int_{Q'} \mathrm{Re}\,(\nabla f^{\varepsilon}_{Q,w}(x) \cdot w)\, dx, \le \frac{3}{4} \tag{8.7.41}$$

$$\frac{1}{|Q'|} \int_{Q'} |\nabla f^{\varepsilon}_{Q,w}(x)|^2\, dx \ge (4\varepsilon)^{-2}. \tag{8.7.42}$$

This is achieved by subdividing Q dyadically and by selecting those cubes Q' for which either (8.7.41) or (8.7.42) holds, subdividing all the nonselected cubes, and repeating the procedure. The validity of (8.7.38) and (8.7.39) now follows from the construction and (8.7.41) and (8.7.42).

It remains to establish (8.7.37). Let B_1 be the union of the cubes in $\mathscr{S}'_w$ for which (8.7.41) holds. Also, let B_2 be the union of those cubes in $\mathscr{S}'_w$ for which (8.7.42) holds. We then have

$$\Big| \bigcup_{Q' \in \mathscr{S}'_w} Q' \Big| \le |B_1| + |B_2|.$$

The fact that the cubes in $\mathscr{S}'_w$ do not overlap yields

$$|B_2| \le (4\varepsilon)^2 \int_Q |\nabla f^{\varepsilon}_{Q,w}(x)|^2\, dx \le (4\varepsilon)^2 C_3 |Q|.$$

Setting $b^{\varepsilon}_{Q,w}(x) = 1 - \mathrm{Re}\,(\nabla f^{\varepsilon}_{Q,w}(x) \cdot w)$, we also have

$$|B_1| \le 4 \sum \int_{Q'} b^{\varepsilon}_{Q,w}\, dx = 4 \int_Q b^{\varepsilon}_{Q,w}\, dx - 4 \int_{Q \setminus B_1} b^{\varepsilon}_{Q,w}\, dx, \tag{8.7.43}$$

where the sum is taken over all cubes Q' that comprise B_1. The first term on the right in (8.7.43) is bounded above by $C\varepsilon^{\frac{1}{2}}|Q|$ in view of (8.7.40). The second term on the right in (8.7.43) is controlled in absolute value by

$$4|Q\setminus B_1|+4|Q\setminus B_1|^{\frac{1}{2}}(C_3|Q|)^{\frac{1}{2}} \le 4|Q\setminus B_1|+4C_3\varepsilon^{\frac{1}{2}}|Q|+\varepsilon^{-\frac{1}{2}}|Q\setminus B_1|.$$

Since $|Q\setminus B_1| = |Q|-|B_1|$, we obtain

$$(5+\varepsilon^{-\frac{1}{2}})|B_1| \le (4+C\varepsilon^{\frac{1}{2}}+\varepsilon^{-\frac{1}{2}})|Q|,$$

which yields $|B_1| \le (1-\varepsilon^{\frac{1}{2}}+o(\varepsilon^{\frac{1}{2}}))|Q|$ if ε is small enough. Hence

$$|B| \le (1-\eta(\varepsilon))|Q|$$

with $\eta(\varepsilon)\approx\varepsilon^{\frac{1}{2}}$ for small ε. This concludes the proof of Proposition 8.7.11. □

Next, we need the following simple geometric fact.

Lemma 8.7.13. *Let w,u,v be in $\mathbf{C}^n$ such that $|w|=1$ and let $0<\varepsilon\le 1$ be such that*

$$|u-(u\cdot\overline{w})w| \le \varepsilon\,|u\cdot\overline{w}|, \tag{8.7.44}$$

$$\mathrm{Re}\,(v\cdot w) \ge \frac{3}{4}, \tag{8.7.45}$$

$$|v| \le (4\varepsilon)^{-1}. \tag{8.7.46}$$

Then we have $|u|\le 4\,|u\cdot v|$.

Proof. It follows from (8.7.45) that

$$\tfrac{3}{4}\,|u\cdot\overline{w}| \le |(u\cdot\overline{w})(v\cdot w)|. \tag{8.7.47}$$

Moreover, (8.7.44) and the triangle inequality imply that

$$|u| \le (1+\varepsilon)|u\cdot\overline{w}| \le 2\,|u\cdot\overline{w}|. \tag{8.7.48}$$

Also, as a consequence of (8.7.44) and (8.7.46), we obtain

$$|(u-(u\cdot\overline{w})w)\cdot v| \le \tfrac{1}{4}\,|u\cdot\overline{w}|. \tag{8.7.49}$$

Finally, using (8.7.47) and (8.7.49) together with the triangle inequality, we deduce that

$$|u\cdot v| \ge |(u\cdot\overline{w})(v\cdot w)| - |(u-(u\cdot\overline{w})w)\cdot v| \ge (\tfrac{3}{4}-\tfrac{1}{4})\,|u\cdot\overline{w}| \ge \tfrac{1}{4}\,|u|,$$

where in the last inequality we used (8.7.48). □

We now proceed with the proof of Lemma 8.7.9. We fix an $\varepsilon>0$ to be chosen later and we choose a finite number of cones $\mathscr{C}_w$ indexed by a finite set $\mathscr{F}$ of unit vectors w in $\mathbf{C}^n$ defined by

$$\mathscr{C}_w = \big\{u\in\mathbf{C}^n:\ |u-(u\cdot\overline{w})w| \le \varepsilon\,|u\cdot\overline{w}|\big\}, \tag{8.7.50}$$

so that

$$\mathbf{C}^n = \bigcup_{w \in \mathscr{F}} \mathscr{C}_w .$$

Note that the size of the set $\mathscr{F}$ can be chosen to depend only on ε and the dimension n.

It suffices to show that for each fixed $w \in \mathscr{F}$ we have a Carleson measure estimate for $\gamma_{t,w}(x) \equiv \chi_{\mathscr{C}_w}(\gamma_t(x))\gamma_t(x)$, where $\chi_{\mathscr{C}_w}$ denotes the characteristic function of $\mathscr{C}_w$. To achieve this we define

$$A_w \equiv \sup_Q \frac{1}{|Q|} \int_Q \int_0^{\ell(Q)} |\gamma_{t,w}(x)|^2 \frac{dx\,dt}{t}, \tag{8.7.51}$$

where the supremum is taken over all cubes Q in $\mathbf{R}^n$ with sides parallel to the axes. By truncating $\gamma_{t,w}(x)$ for t small and t large, we may assume that this quantity is finite. Once an a priori bound independent of these truncations is obtained, we can pass to the limit by monotone convergence to deduce the same bound for $\gamma_{t,w}(x)$.

We now fix a cube Q and let $\mathscr{S}_w''$ be as in Proposition 8.7.11. We pick Q'' in $\mathscr{S}_w''$ and we set

$$v = \frac{1}{|Q''|} \int_{Q''} \nabla f_{Q,w}^{\varepsilon}(y)\, dy \in \mathbf{C}^n .$$

It is obvious that statements (8.7.38) and (8.7.39) in Proposition 8.7.11 yield conditions (8.7.45) and (8.7.46) of Lemma 8.7.13. Set $u = \gamma_{t,w}(x)$ and note that if $x \in Q''$ and $\frac{1}{2}\ell(Q'') < t \le \ell(Q'')$, then $v = S_t^Q(\nabla f_{Q,w}^{\varepsilon})(x)$; hence

$$\big|\gamma_{t,w}(x)\big| \le 4\,\big|\gamma_{t,w}(x) \cdot S_t^Q(\nabla f_{Q,w}^{\varepsilon})(x)\big| \le 4\,\big|\gamma_t(x) \cdot S_t^Q(\nabla f_{Q,w}^{\varepsilon})(x)\big| \tag{8.7.52}$$

from Lemma 8.7.13 and the definition of $\gamma_{t,w}(x)$.

We partition the Carleson region $Q \times (0, \ell(Q)]$ as a union of boxes $Q' \times (0, \ell(Q')]$ for Q' in $\mathscr{S}_w'$ and Whitney rectangles $Q'' \times (\frac{1}{2}\ell(Q''), \ell(Q'')]$ for Q'' in $\mathscr{S}_w''$. This allows us to write

$$\begin{aligned}\int_Q \int_0^{\ell(Q)} |\gamma_{t,w}(x)|^2 \frac{dx\,dt}{t} = &\sum_{Q' \in \mathscr{S}_w'} \int_{Q'} \int_0^{\ell(Q')} |\gamma_{t,w}(x)|^2 \frac{dx\,dt}{t} \\ &+ \sum_{Q'' \in \mathscr{S}_w''} \int_{Q''} \int_{\frac{1}{2}\ell(Q'')}^{\ell(Q'')} |\gamma_{t,w}(x)|^2 \frac{dx\,dt}{t} .\end{aligned}$$

First observe that

$$\sum_{Q' \in \mathscr{S}_w'} \int_{Q'} \int_0^{\ell(Q')} |\gamma_{t,w}(x)|^2 \frac{dx\,dt}{t} \le \sum_{Q' \in \mathscr{S}_w'} A_w |Q'| A_w (1-\eta)|Q| .$$

Second, using (8.7.52), we obtain

$$\sum_{Q''\in\mathscr{S}''_w}\int_{Q''}\int_{\frac{1}{2}\ell(Q'')}^{\ell(Q'')}|\gamma_{t,w}(x)|^2\frac{dx\,dt}{t}$$
$$\leq 16\sum_{Q''\in\mathscr{S}''_w}\int_{Q''}\int_{\frac{1}{2}\ell(Q'')}^{\ell(Q'')}|\gamma_t(x)\cdot S_t^Q(\nabla f^\varepsilon_{Q,w})(x)|^2\frac{dx\,dt}{t}$$
$$\leq 16\int_Q\int_0^{\ell(Q)}|\gamma_t(x)\cdot S_t^Q(\nabla f^\varepsilon_{Q,w})(x)|^2\frac{dx\,dt}{t}.$$

Altogether, we obtain the bound

$$\int_Q\int_0^{\ell(Q)}|\gamma_{t,w}(x)|^2\frac{dx\,dt}{t}$$
$$\leq A_w(1-\eta)|Q|+16\int_Q\int_0^{\ell(Q)}|\gamma_t(x)\cdot S_t^Q(\nabla f^\varepsilon_{Q,w})(x)|^2\frac{dx\,dt}{t}.$$

We divide by $|Q|$, we take the supremum over all cubes Q with sides parallel to the axes, and we use the definition and the finiteness of A_w to obtain the required estimate

$$A_w\leq 16\eta^{-1}\sup_Q\frac{1}{|Q|}\int_Q\int_0^{\ell(Q)}|\gamma_t(x)\cdot S_t^Q(\nabla f^\varepsilon_{Q,w})(x)|^2\frac{dx\,dt}{t},$$

thus concluding the proof of the lemma. □

We end by verifying the validity of Lemma 8.7.12 used earlier.

Proof of Lemma 8.7.12. For simplicity we may take Q to be the cube $[-1,1]^n$. Once this case is established, the case of a general cube follows by translation and rescaling. Set

$$M=\left(\int_Q|h(x)|^2\,dx\right)^{\frac{1}{2}},\qquad M'=\left(\int_Q|\nabla h(x)|^2\,dx\right)^{\frac{1}{2}}.$$

If $M\geq M'$, there is nothing to prove, so we may assume that $M<M'$. Take $t\in(0,1)$ and $\varphi\in\mathscr{C}_0^\infty(Q)$ with $\varphi(x)=1$ when $\operatorname{dist}(x,\partial Q)\geq t$ and $0\leq\varphi\leq 1$, $\|\nabla\varphi\|_{L^\infty}\leq C/t$, $C=C(n)$; here the distance is taken in the L^∞ norm of $\mathbf{R}^n$. Then

$$\int_Q\nabla h(x)\,dx=\int_Q(1-\varphi(x))\nabla h(x)\,dx-\int_Q h(x)\nabla\varphi(x)\,dx,$$

and the Cauchy–Schwarz inequality yields

$$\left|\int_Q\nabla h(x)\,dx\right|\leq C(M't^{\frac{1}{2}}+Mt^{-\frac{1}{2}}).$$

Choosing $t=M/M'$, we conclude the proof of the lemma. □

The proof of Theorem 8.7.1 is now complete. □

Exercises

8.7.1. Let A and L be as in the statement of Theorem 8.7.1.
(a) Consider the generalized heat equation

$$\frac{\partial u}{\partial t} - \operatorname{div}(A\nabla u) = 0$$

on $\mathbf{R}^{n+1}_+$ with initial condition $u(0,x) = u_0$. Assume a uniqueness theorem for solutions of these equations to obtain that the solution of the equation in part (a) is

$$u(t,x) = e^{-tL}(u_0).$$

(b) Take $u_0 = 1$ to deduce the identity

$$e^{-tL}(1) = 1$$

for all $t > 0$. Conclude that the family of $\{e^{-tL}\}_{t>0}$ is an approximate identity, in the sense that

$$\lim_{t\to 0} e^{-tL} = I.$$

8.7.2. Let L be as in (8.7.2). Show that the operators

$$\begin{aligned} L_1 &= (I+t^2L)^{-1}, \\ L_2 &= t\nabla(I+t^2L)^{-1}, \\ L_3 &= (I+t^2L)^{-1}t\operatorname{div} \end{aligned}$$

are bounded on $L^2(\mathbf{R}^n)$ uniformly in t with bounds depending only on n, λ, and Λ. [*Hint:* The L^2 boundedness of L_3 follows from that of L_2 via duality and integration by parts. To prove the L^2 boundedness of L_1 and L_2, let $u_t = (I+t^2L)^{-1}(f)$. Then $u_t + t^2L(u_t) = f$, which implies $\int_{\mathbf{R}^n} |u_t|^2\,dx + t^2 \int_{\mathbf{R}^n} \overline{u_t}\,L(u_t)\,dx = \int_{\mathbf{R}^n} \overline{u_t}\,f\,dx$. The definition of L and integration by parts yield $\int_{\mathbf{R}^n} |u_t|^2\,dx + t^2 \int_{\mathbf{R}^n} A\nabla u_t\,\nabla\overline{u_t}\,dx = \int_{\mathbf{R}^n} \overline{u_t}\,f\,dx$. Apply the ellipticity condition to bound the left side of this identity from below by $\int_{\mathbf{R}^n} |u_t|^2\,dx + \lambda \int_{\mathbf{R}^n} |t\nabla u_t|^2\,dx$. Also $\int_{\mathbf{R}^n} \overline{u_t}\,f\,dx$ is at most $\varepsilon^{-1}\int_{\mathbf{R}^n} |f|^2\,dx + \varepsilon \int_{\mathbf{R}^n} |u_t|^2\,dx$ by the Cauchy–Schwarz inequality. Choose ε small enough to complete the proof when $\|u_t\|_{L^2} < \infty$. In the case $\|u_t\|_{L^2} = \infty$, multiply the identity $u_t + t^2L(u_t) = f$ by $\overline{u_t}\eta_R$, where η_R is a suitable cutoff localized in a ball $B(0,R)$, and use the idea of Lemma 8.7.2. Then let $R \to \infty$.]

8.7.3. Let L be as in the proof of Theorem 8.7.1.
(a) Show that for all $t > 0$ we have

$$(I+t^2L^2)^{-2} = \int_0^\infty e^{-u(I+t^2L)}\,u\,du$$

by checking the identities

$$\int_0^\infty (I+t^2L)^2 e^{-u(I+t^2L)}u\,du = \int_0^\infty e^{-u(I+t^2L)}(I+t^2L)^2 u\,du = I\,.$$

(b) Prove that the operator

$$T = \frac{4}{\pi}\int_0^\infty L(I+t^2L)^{-2}\,dt$$

satisfies $TT = L$.
(c) Conclude that the operator

$$S = \frac{16}{\pi}\int_0^{+\infty} t^3L^2(I+t^2L)^{-3}\,\frac{dt}{t}$$

satisfies $SS = L$, that is, S is the square root of L. Moreover, all the integrals converge in $L^2(\mathbf{R}^n)$ when restricted to functions in $f\in\mathscr{C}_0^\infty(\mathbf{R}^n)$.
[*Hint:* Part (a): Write $(I+t^2L)e^{-u(I+t^2L)} = -\frac{d}{du}(e^{-u(I+t^2L)})$, apply integration by parts twice, and use Exercise 8.7.1. Part (b): Write the integrand as in part (a) and use the identity

$$\int_0^\infty\int_0^\infty e^{-(ut^2+vs^2)L}L^2\,dt\,ds = \frac{\pi}{4}(uv)^{-\frac{1}{2}}\int_0^\infty e^{-r^2L}L^2\,2r\,dr.$$

Set $\rho = r^2$ and use $e^{-\rho L}L = \frac{d}{d\rho}(e^{-\rho L})$. Part (c): Show that $T = S$ using an integration by parts starting with the identity $L = \frac{d}{dt}(tL)$.]

8.7.4. Suppose that μ is a measure on $\mathbf{R}^{n+1}_+$. For a cube Q in $\mathbf{R}^n$ we define the tent $T(Q)$ of Q as the set $Q\times(0,\ell(Q))$. Suppose that there exist two positive constants $\alpha<1$ and β such that for all cubes Q in $\mathbf{R}^n$ there exist subcubes Q_j of Q with disjoint interiors such that

1. $\left|Q\setminus\bigcup_j Q_j\right| > \alpha\,|Q|$,
2. $\mu\Big(T(Q)\setminus\bigcup_j T(Q_j)\Big) \le \beta\,|Q|$.

Then μ is a Carleson measure with constant

$$\big\|\mu\big\|_{\mathscr{C}} \le \frac{\beta}{\alpha}.$$

[*Hint:* We have

$$\begin{aligned}\mu(T(Q)) &\le \mu\Big(T(Q)\setminus\bigcup_j T(Q_j)\Big) + \sum_j \mu(T(Q_j))\\ &\le \beta|Q| + \big\|\mu\big\|_{\mathscr{C}}\sum_j |Q_j|,\end{aligned}$$

and the last expression is at most $(\beta + (1-\alpha)\|\mu\|_{\mathscr{C}})|Q|$. Assuming that $\|\mu\|_{\mathscr{C}} < \infty$, we obtain the required conclusion. In general, approximate the measure by a sequence of truncated measures.]

HISTORICAL NOTES

Most of the material in Sections 8.1 and 8.2 has been in the literature since the early development of the subject. Theorem 8.2.7 was independently obtained by Peetre [254], Spanne [286], and Stein [290].

The original proof of the $T(1)$ theorem obtained by David and Journé [103] stated that if $T(1)$, $T^t(1)$ are in BMO and T satisfies the weak boundedness property, then T is L^2 bounded. This proof is based on the boundedness of paraproducts and is given in Theorem 8.5.4. Paraproducts were first exploited by Bony [28] and Coifman and Meyer [81]. The proof of L^2 boundedness using condition (iv) given in the proof of Theorem 8.3.3 was later obtained by Coifman and Meyer [82]. The equivalent conditions (ii), (iii), and (vi) first appeared in Stein [292], while condition (iv) is also due to David and Journé [103]. Condition (i) appears in the article of Nazarov, Volberg, and Treil [245] in the context of nondoubling measures. The same authors [246] obtained a proof of Theorems 8.2.1 and 8.2.3 for Calderón–Zygmund operators on nonhomogeneous spaces. Multilinear versions of the $T(1)$ theorem were obtained by Christ and Journé [70], Grafakos and Torres [154], and Bényi, Demeter, Nahmod, Thiele, Torres, and Villaroya [20]. The article [70] also contains a proof of the quadratic $T(1)$ type Theorem 8.6.3. Smooth paraproducts viewed as bilinear operators have been studied by Bényi, Maldonado, Nahmod, and Torres [21] and Dini-continuous versions of them by Maldonado and Naibo [225].

The orthogonality Lemma 8.5.1 was first proved by Cotlar [94] for self-adjoint and mutually commuting operators T_j. The case of general noncommuting operators was obtained by Knapp and Stein [190]. Theorem 8.5.7 is due to Calderón and Vaillancourt [49] and is also valid for symbols of class $S^0_{\rho,\rho}$ when $0 \le \rho < 1$. For additional topics on pseudodifferential operators we refer to the books of Coifman and Meyer [81], Journé [180], Stein [292], Taylor [309], Torres [315], and the references therein. The last reference presents a careful study of the action of linear operators with standard kernels on general function spaces. The continuous version of the orthogonality Lemma 8.5.1 given in Exercise 8.5.8 is due to Calderón and Vaillancourt [49]. Conclusion (iii) in the orthogonality Lemma 8.5.1 follows from a general principle saying that if $\sum x_j$ is a series in a Hilbert space such that $\|\sum_{j\in F} x_j\| \le M$ for all finite sets F, then the series $\sum x_j$ converges in norm. This is a consequence of the Orlicz–Pettis theorem, which states that in any Banach space, if $\sum x_{n_j}$ converges weakly for every subsequence of integers n_j, then $\sum x_j$ converges in norm.

A nice exposition on the Cauchy integral that presents several historical aspects of its study is the book of Muskhelishvili [243]. See also the book of Journé [180]. Proposition 8.6.1 is due to Plemelj [265] when Γ is a closed Jordan curve. The L^2 boundedness of the first commutator $\mathscr{C}_1$ in Example 8.3.8 is due to Calderón [42]. The L^2 boundedness of the remaining commutators $\mathscr{C}_m$, $m \ge 2$, is due to Coifman and Meyer [80], but with bounds of order $m!\,\|A'\|^m_{L^\infty}$. These bounds are not as good as those obtained in Example 8.3.8 and do not suffice in obtaining the boundedness of the Cauchy integral by summing the series of commutators. The L^2 boundedness of the Cauchy integral when $\|A'\|_{L^\infty}$ is small enough is due to Calderón [43]. The first proof of the boundedness of the Cauchy integral with arbitrary $\|A'\|_{L^\infty}$ was obtained by Coifman, McIntosh, and Meyer [79]. This proof is based on an improved operator norm for the commutators $\|\mathscr{C}_m\|_{L^2\to L^2} \le C_0 m^4 \|A'\|^m_{L^\infty}$. The quantity m^4 was improved by Christ and Journé [70] to $m^{1+\delta}$ for any $\delta > 0$; it is announced in Verdera [326] that Mateu and Verdera have improved this result by taking $\delta = 0$. Another proof of the L^2 boundedness of the Cauchy integral was given by David [102] by employing the following bootstrapping argument: If the Cauchy integral is L^2 bounded whenever $\|A'\|_{L^\infty} \le \varepsilon$, then it is also L^2 bounded whenever $\|A'\|_{L^\infty} \le \frac{10}{9}\varepsilon$. A refinement of this bootstrapping technique was independently obtained by Murai [241], who was also able to obtain the best possible bound for the operator norm $\|\widetilde{\mathcal{C}}_\Gamma\|_{L^2\to L^2} \le C\big(1+\|A'\|_{L^\infty}\big)^{1/2}$ in terms of $\|A'\|_{L^\infty}$. Here $\widetilde{\mathcal{C}}_\Gamma$ is the operator defined in (8.6.15). Note that the corresponding estimate for $\mathcal{C}_\Gamma$ involves the power $3/2$ instead of $1/2$. See the book of Murai [242] for this result and a variety of topics related to the commutators and the Cauchy integral. Two elementary proofs of the L^2 boundedness of the Cauchy integral were given by Coifman, Jones, and Semmes [77]. The first of these proofs uses complex variables and the second a pseudo-Haar basis of L^2 adapted to the accretive function $1+iA'$. A geometric proof was given by Melnikov and Verdera [231]. Other proofs were obtained by Verdera [326] and Tchamitchian [310]. The proof of boundedness of the Cauchy integral given in Section 8.6 is taken from Semmes [281]. The book of Christ [67] contains an insightful exposition of many of the preceding results and discusses connections between the Cauchy integral and analytic capacity. The book of David and Semmes [105] presents several extensions of the results in this chapter to singular integrals along higher-dimensional surfaces.

The $T(1)$ theorem is applicable to many problems only after a considerable amount of work; see, for instance, Christ [67] for the case of the Cauchy integral. A more direct approach to many problems was given by McIntosh and Meyer [224], who replaced the function 1 by an accretive function b and showed that any operator T with standard kernel that satisfies $T(b) = T^t(b) = 0$ and $\|M_bTM_b\|_{WB} < \infty$ must be L^2 bounded. (M_b here is the operator given by multiplication by b.) This theorem easily implies the boundedness of the Cauchy integral. David, Journé, and Semmes [104] generalized this theorem even further as follows: If b_1 and b_2 are para-accretive functions such that T maps $b_1\mathscr{C}_0^\infty \to (b_2\mathscr{C}_0^\infty)'$ and is associated with a standard kernel, then T is L^2 bounded if and only if $T(b_1) \in BMO$, $T^t(b_2) \in BMO$, and $\|M_{b_1}TM_{b_2}\|_{WB} < \infty$. This is called the $T(b)$ theorem. The article of Semmes [281] contains a different proof of this theorem in the special case $T(b) = 0$ and $T^t(1) = 0$ (Exercise 8.6.6). Our proof of Theorem 8.6.6 is based on ideas from [281]. An alternative proof of the $T(b)$ theorem was given by Fabes, Mitrea, and Mitrea [121] based on a lemma due to Krein [200]. Another version of the $T(b)$ theorem that is applicable to spaces with no Euclidean structure was obtained by Christ [66].

Theorem 8.7.1 was posed as a problem by Kato [181] for maximal accretive operators and reformulated by McIntosh [222], [223] for square roots of elliptic operators. The reformulation was motivated by counterexamples found to Kato's original abstract formulation, first by Lions [215] for maximal accretive operators, and later by McIntosh [220] for regularly accretive ones. The one-dimensional Kato problem and the boundeness of the Cauchy integral along Lipschitz curves are equivalent problems as shown by Kenig and Meyer [188]. See also Auscher, McIntosh, and Nahmod [8]. Coifman, Deng, and Meyer [73] and independently Fabes, Jerison, and Kenig [119], [120] solved the square root problem for small peturbations of the identity matrix. This method used multilinear expansions and can be extended to operators with smooth coefficients. McIntosh [221] considered coefficients in Sobolev spaces, Escauriaza in VMO (unpublished), and Alexopoulos [3] real Hölder coefficients using homogenization techniques. Peturbations of real symmetric matrices with L^∞ coefficients were treated in Auscher, Hofmann, Lewis, and Tchamitchian [10]. The solution of the two-dimensional Kato problem was obtained by Hofmann and McIntosh [164] using a previously derived $T(b)$ type reduction due to Auscher and Tchamitchian [9]. Hofmann, Lacey, and McIntosh [165] extended this theorem to the case in which the heat kernel of e^{-tL} satisfies Gaussian bounds. Theorem 8.7.1 was obtained by Auscher, Hofmann, Lacey, McIntosh, and Tchamitchian [11]; the exposition in the text is based on this reference. Combining Theorem 8.7.1 with a theorem of Lions [215], it follows that the domain of $\sqrt{L}$ is $\dot{L}_1^2(\mathbf{R}^n)$ and that for functions f in this space the equivalence of norms $\|\sqrt{L}(f)\|_{L^2} \approx \|\nabla f\|_{L^2}$ is valid.

Chapter 9
Weighted Inequalities

Weighted inequalities arise naturally in Fourier analysis, but their use is best justified by the variety of applications in which they appear. For example, the theory of weights plays an important role in the study of boundary value problems for Laplace's equation on Lipschitz domains. Other applications of weighted inequalities include extrapolation theory, vector-valued inequalities, and estimates for certain classes of nonlinear partial differential equations.

The theory of weighted inequalities is a natural development of the principles and methods we have acquainted ourselves with in earlier chapters. Although a variety of ideas related to weighted inequalities appeared almost simultaneously with the birth of singular integrals, it was only in the 1970s that a better understanding of the subject was obtained. This was spurred by Muckenhoupt's characterization of positive functions w for which the Hardy–Littlewood maximal operator M maps $L^p(\mathbf{R}^n, w(x)\,dx)$ to itself. This characterization led to the introduction of the class A_p and the development of weighted inequalities. We pursue exactly this approach in the next section to motivate the introduction of the A_p classes.

9.1 The A_p Condition

A *weight* is a nonnegative locally integrable function on $\mathbf{R}^n$ that takes values in $(0,\infty)$ almost everywhere. Therefore, weights are allowed to be zero or infinite only on a set of Lebesgue measure zero. Hence, if w is a weight and $1/w$ is locally integrable, then $1/w$ is also a weight.

Given a weight w and a measurable set E, we use the notation

$$w(E) = \int_E w(x)\,dx$$

to denote the w-measure of the set E. Since weights are locally integrable functions, $w(E) < \infty$ for all sets E contained in some ball. The weighted L^p spaces are denoted by $L^p(\mathbf{R}^n, w)$ or simply $L^p(w)$. Recall the uncentered Hardy–Littlewood maximal

L. Grafakos, *Modern Fourier Analysis*, DOI: 10.1007/978-0-387-09434-2_9,

operators on $\mathbf{R}^n$ over balls

$$M(f)(x) = \sup_{B \ni x} \operatorname{Avg}_B |f| = \sup_{B \ni x} \frac{1}{|B|} \int_B |f(y)|\, dy,$$

and over cubes

$$M_c(f)(x) = \sup_{Q \ni x} \operatorname{Avg}_Q |f| = \sup_{Q \ni x} \frac{1}{|Q|} \int_Q |f(y)|\, dy,$$

where the suprema are taken over all balls B and cubes Q (with sides parallel to the axes) that contain the given point x. It is a classical result proved in Section 2.1 that for all $1 < p < \infty$ there is a constant $C_p(n) > 0$ such that

$$\int_{\mathbf{R}^n} M(f)(x)^p\, dx \le C_p(n)^p \int_{\mathbf{R}^n} |f(x)|^p\, dx \tag{9.1.1}$$

for all functions $f \in L^p(\mathbf{R}^n)$. We are concerned with the situation in which the measure dx in (9.1.1) is replaced by $w(x)\,dx$ for some weight $w(x)$.

9.1.1 Motivation for the A_p Condition

The question we raise is whether there is a characterization of all weights $w(x)$ such that the strong type (p,p) inequality

$$\int_{\mathbf{R}^n} M(f)(x)^p\, w(x)\, dx \le C_p^p \int_{\mathbf{R}^n} |f(x)|^p\, w(x)\, dx \tag{9.1.2}$$

is valid for all $f \in L^p(w)$.

Suppose that (9.1.2) is valid for some weight w and all $f \in L^p(w)$ for some $1 < p < \infty$. Apply (9.1.2) to the function $f\chi_B$ supported in a ball B and use that $\operatorname{Avg}_B |f| \le M(f\chi_B)(x)$ for all $x \in B$ to obtain

$$w(B)\big(\operatorname{Avg}_B |f|\big)^p \le \int_B M(f\chi_B)^p\, w\, dx \le C_p^p \int_B |f|^p\, w\, dx. \tag{9.1.3}$$

It follows that

$$\left(\frac{1}{|B|}\int_B |f(t)|\, dt\right)^p \le \frac{C_p^p}{w(B)} \int_B |f(x)|^p\, w(x)\, dx \tag{9.1.4}$$

for all balls B and all functions f. At this point, it is tempting to choose a function such that the two integrands are equal. We do so by setting $f = w^{-p'/p}$, which gives $f^p w = w^{-p'/p}$. Under the assumption that $\inf_B w > 0$ for all balls B, it would follow from (9.1.4) that

$$\sup_{B \text{ balls}} \left(\frac{1}{|B|}\int_B w(x)\,dx\right)\left(\frac{1}{|B|}\int_B w(x)^{-\frac{1}{p-1}}\,dx\right)^{p-1} \le C_p^p. \tag{9.1.5}$$

If $\inf_B w = 0$ for some balls B, we take $f = (w+\varepsilon)^{-p'/p}$ to obtain

$$\left(\frac{1}{|B|}\int_B w(x)dx\right)\left(\frac{1}{|B|}\int_B (w(x)+\varepsilon)^{-\frac{p'}{p}}dx\right)^{p}\left(\frac{1}{|B|}\int_B \frac{w(x)\,dx}{(w(x)+\varepsilon)^{p'}}\right)^{-1} \le C_p^p \tag{9.1.6}$$

for all $\varepsilon > 0$. Replacing $w(x)\,dx$ by $(w(x)+\varepsilon)\,dx$ in the last integral in (9.1.6) we obtain a smaller expression, which is also bounded by C_p^p. Since $-p'/p = -p'+1$, (9.1.6) implies that

$$\left(\frac{1}{|B|}\int_B w(x)dx\right)\left(\frac{1}{|B|}\int_B (w(x)+\varepsilon)^{-\frac{p'}{p}}dx\right)^{p-1} \le C_p^p, \tag{9.1.7}$$

from which we can still deduce (9.1.5) via the Lebesgue monotone convergence theorem by letting $\varepsilon \to 0$. We have now obtained that every weight w that satisfies (9.1.2) must also satisfy the rather strange-looking condition (9.1.5), which we refer to in the sequel as the A_p condition. It is a remarkable fact, to be proved in this chapter, that the implication obtained can be reversed, that is, (9.1.2) is a consequence of (9.1.5). This is the first significant achievement of the theory of weights [i.e., a characterization of all functions w for which (9.1.2) holds]. This characterization is based on some deep principles discussed in the next section and provides a solid motivation for the introduction and careful examination of condition (9.1.5).

Before we study the converse statements, we consider the case $p = 1$. Assume that for some weight w the weak type $(1,1)$ inequality

$$w\big(\{x \in \mathbf{R}^n : M(f)(x) > \alpha\}\big) \le \frac{C_1}{\alpha}\int_{\mathbf{R}^n} |f(x)|w(x)\,dx \tag{9.1.8}$$

holds for all functions $f \in L^1(\mathbf{R}^n)$. Since $M(f)(x) \ge \operatorname{Avg}_B |f|$ for all $x \in B$, it follows from (9.1.8) that for all $\alpha < \operatorname{Avg}_B |f|$ we have

$$w(B) \le w\big(\{x \in \mathbf{R}^n : M(f)(x) > \alpha\}\big) \le \frac{C_1}{\alpha}\int_{\mathbf{R}^n} |f(x)|w(x)\,dx. \tag{9.1.9}$$

Taking $f\chi_B$ instead of f in (9.1.9), we deduce that

$$\operatorname*{Avg}_B |f| = \frac{1}{|B|}\int_B |f(t)|\,dt \le \frac{C_1}{w(B)}\int_B |f(x)|\,w(x)\,dx \tag{9.1.10}$$

for all functions f and balls B. Taking $f = \chi_S$, we obtain

$$\frac{|S|}{|B|} \le C_1 \frac{w(S)}{w(B)}, \tag{9.1.11}$$

where S is any measurable subset of the ball B.

Recall that the *essential infimum* of a function w over a set E is defined as

$$\operatorname*{ess.inf}_{E}(w) = \inf\{b>0:\ |\{x\in E:\ w(x)<b\}|>0\}.$$

Then for every $a > \text{ess.inf}_B(w)$ there exists a subset S_a of B with positive measure such that $w(x) < a$ for all $x \in S_a$. Applying (9.1.11) to the set S_a, we obtain

$$\frac{1}{|B|}\int_B w(t)\,dt \le \frac{C_1}{|S_a|}\int_{S_a} w(t)\,dt \le C_1 a, \tag{9.1.12}$$

which implies

$$\frac{1}{|B|}\int_B w(t)\,dt \le C_1 w(x) \qquad \text{for all balls } B \text{ and almost all } x\in B. \tag{9.1.13}$$

It remains to understand what condition (9.1.13) really means. For every ball B, there exists a null set $N(B)$ such that (9.1.13) holds for all x in $B\setminus N(B)$. Let N be the union of all the null sets $N(B)$ for all balls B with centers in $\mathbf{Q}^n$ and rational radii. Then N is a null set and for every x in $B\setminus N$, (9.1.13) holds for all balls B with centers in $\mathbf{Q}^n$ and rational radii. By density, (9.1.13) must also hold for all balls B that contain a fixed x in $\mathbf{R}^n\setminus N$. It follows that for $x\in\mathbf{R}^n\setminus N$ we have

$$M(w)(x) = \sup_{B\ni x}\frac{1}{|B|}\int_B w(t)\,dt \le C_1 w(x). \tag{9.1.14}$$

Therefore, assuming (9.1.8), we have arrived at the condition

$$M(w)(x) \le C_1 w(x) \qquad \text{for almost all } x\in\mathbf{R}^n, \tag{9.1.15}$$

where C_1 is the same constant as in (9.1.13).

We later see that this deduction can be reversed and we can obtain (9.1.8) as a consequence of (9.1.15). This motivates a careful study of condition (9.1.15), which we refer to as the A_1 condition. Since in all the previous arguments we could have replaced cubes with balls, we give the following definitions in terms of cubes.

Definition 9.1.1. A function $w(x)\ge 0$ is called an A_1 *weight* if

$$M(w)(x) \le C_1 w(x) \qquad \text{for almost all } x\in\mathbf{R}^n \tag{9.1.16}$$

for some constant C_1. If w is an A_1 weight, then the (finite) quantity

$$[w]_{A_1} = \sup_{Q \text{ cubes in } \mathbf{R}^n}\left(\frac{1}{|Q|}\int_Q w(t)\,dt\right)\big\|w^{-1}\big\|_{L^\infty(Q)} \tag{9.1.17}$$

is called the A_1 *Muckenhoupt characteristic constant* of w, or simply the A_1 *characteristic constant* of w. Note that A_1 weights w satisfy

$$\frac{1}{|Q|}\int_Q w(t)\,dt \le [w]_{A_1}\ \operatorname*{ess.inf}_{y\in Q} w(y) \tag{9.1.18}$$

for all cubes Q in $\mathbf{R}^n$.

Remark 9.1.2. We also define

$$[w]_{A_1}^{\text{balls}} = \sup_{B \text{ balls in } \mathbf{R}^n} \left(\frac{1}{|B|}\int_B w(t)\,dt\right)\big\|w^{-1}\big\|_{L^\infty(B)}. \tag{9.1.19}$$

Using (9.1.13), we see that the smallest constant C_1 that appears in (9.1.16) is equal to the A_1 characteristic constant of w as defined in (9.1.19). This is also equal to the smallest constant that appears in (9.1.13). All these constants are bounded above and below by dimensional multiples of $[w]_{A_1}$.

We now recall condition (9.1.5), which motivates the following definition of A_p weights for $1 < p < \infty$.

Definition 9.1.3. Let $1 < p < \infty$. A weight w is said to be *of class* A_p if

$$\sup_{Q \text{ cubes in } \mathbf{R}^n} \left(\frac{1}{|Q|}\int_Q w(x)\,dx\right)\left(\frac{1}{|Q|}\int_Q w(x)^{-\frac{1}{p-1}}\,dx\right)^{p-1} < \infty. \tag{9.1.20}$$

The expression in (9.1.20) is called the A_p Muckenhoupt characteristic constant of w (or simply the A_p characteristic constant of w) and is denoted by $[w]_{A_p}$.

Remark 9.1.4. Note that Definitions 9.1.1 and 9.1.3 could have been given with the set of all cubes in $\mathbf{R}^n$ replaced by the set of all balls in $\mathbf{R}^n$. Defining $[w]_{A_p}^{\text{balls}}$ as in (9.1.20) except that cubes are replaced by balls, we see that

$$(v_n 2^{-n})^p \le \frac{[w]_{A_p}}{[w]_{A_p}^{\text{balls}}} \le (n^{n/2} v_n 2^{-n})^p. \tag{9.1.21}$$

9.1.2 Properties of A_p Weights

It is straightforward that translations, isotropic dilations, and scalar multiples of A_p weights are also A_p weights with the same A_p characteristic. We summarize some basic properties of A_p weights in the following proposition.

Proposition 9.1.5. *Let $w \in A_p$ for some $1 \le p < \infty$. Then*

(1) $[\delta^\lambda(w)]_{A_p} = [w]_{A_p}$, where $\delta^\lambda(w)(x) = w(\lambda x_1, \dots, \lambda x_n)$.

(2) $[\tau^z(w)]_{A_p} = [w]_{A_p}$, where $\tau^z(w)(x) = w(x - z)$, $z \in \mathbf{R}^n$.

(3) $[\lambda w]_{A_p} = [w]_{A_p}$ for all $\lambda > 0$.

(4) When $1<p<\infty$, *the function* $w^{-\frac{1}{p-1}}$ *is in* $A_{p'}$ *with characteristic constant*

$$\big[w^{-\frac{1}{p-1}}\big]_{A_{p'}} = [w]_{A_p}^{\frac{1}{p-1}}.$$

Therefore, $w \in A_2$ *if and only if* $w^{-1} \in A_2$ *and both weights have the same* A_2 *characteristic constant.*

(5) $[w]_{A_p} \geq 1$ *for all* $w \in A_p$. *Equality holds if and only if* w *is a constant.*

(6) The classes A_p *are increasing as* p *increases; precisely, for* $1 \leq p < q < \infty$ *we have*

$$[w]_{A_q} \leq [w]_{A_p}.$$

(7) $\lim_{q\to 1+} [w]_{A_q} = [w]_{A_1}$ *if* $w \in A_1$.

(8) The following is an equivalent characterization of the A_p *characteristic constant of* w:

$$[w]_{A_p} = \sup_{\substack{Q \text{ cubes} \\ \text{in } \mathbf{R}^n}} \sup_{\substack{f \text{ in } L^p(Q, w\,dx) \\ |f|>0 \text{ a.e. on } Q}} \left\{ \frac{\left(\frac{1}{|Q|}\int_Q |f(t)|\,dt\right)^p}{\frac{1}{w(Q)}\int_Q |f(t)|^p w(t)\,dt} \right\}.$$

(9) The measure $w(x)\,dx$ *is doubling: precisely, for all* $\lambda > 1$ *and all cubes* Q *we have*

$$w(\lambda Q) \leq \lambda^{np}[w]_{A_p} w(Q).$$

(λQ *denotes the cube with the same center as* Q *and side length* λ *times the side length of* Q.*)*

Proof. The simple proofs of (1), (2), and (3) are left as an exercise. Property (4) is also easy to check and plays the role of duality in this context. To prove (5) we use Hölder's inequality with exponents p and p' to obtain

$$1 = \frac{1}{|Q|}\int_Q dx = \frac{1}{|Q|}\int_Q w(x)^{\frac{1}{p}} w(x)^{-\frac{1}{p}}\,dx \leq [w]_{A_p}^{\frac{1}{p}},$$

with equality holding only when $w(x)^{\frac{1}{p}} = c\,w(x)^{-\frac{1}{p}}$ for some $c>0$ (i.e., when w is a constant). To prove (6), observe that $0 < q'-1 < p'-1 \leq \infty$ and that the statement

$$[w]_{A_q} \leq [w]_{A_p}$$

is equivalent to the fact

$$\big\|w^{-1}\big\|_{L^{q'-1}(Q,\frac{dx}{|Q|})} \leq \big\|w^{-1}\big\|_{L^{p'-1}(Q,\frac{dx}{|Q|})}.$$

Property (7) is a consequence of part (a) of Exercise 1.1.3.

To prove (8), apply Hölder's inequality with exponents p and p' to get

$$
\begin{aligned}
(\operatorname*{Avg}_Q |f|)^p &= \left(\frac{1}{|Q|}\int_Q |f(x)|\,dx\right)^p \\
&= \left(\frac{1}{|Q|}\int_Q |f(x)| w(x)^{\frac{1}{p}} w(x)^{-\frac{1}{p}}\,dx\right)^p \\
&\le \frac{1}{|Q|^p}\left(\int_Q |f(x)|^p w(x)\,dx\right)\left(\int_Q w(x)^{-\frac{p'}{p}}\,dx\right)^{\frac{p}{p'}} \\
&= \left(\frac{1}{\omega(Q)}\int_Q |f(x)|^p w(x)\,dx\right)\left(\frac{1}{|Q|}\int_Q w(x)\,dx\right)\left(\frac{1}{|Q|}\int_Q w(x)^{-\frac{1}{p-1}}\,dx\right)^{p-1} \\
&\le [w]_{A_p}\left(\frac{1}{\omega(Q)}\int_Q |f(x)|^p w(x)\,dx\right).
\end{aligned}
$$

This argument proves the inequality $\geq$ in (8) when $p > 1$. In the case $p = 1$ the obvious modification yields the same inequality. The reverse inequality follows by taking $f = (w+\varepsilon)^{-p'/p}$ as in (9.1.6) and letting $\varepsilon \to 0$.

Applying (8) to the function $f = \chi_Q$ and putting λQ in the place of Q in (8), we obtain

$$w(\lambda Q) \le \lambda^{np}[w]_{A_p} w(Q),$$

which says that $w(x)\,dx$ is a doubling measure. This proves (9). □

Example 9.1.6. A positive measure $d\mu$ is called doubling if for some $C < \infty$,

$$\mu(2B) \le C\mu(B) \tag{9.1.22}$$

for all balls B. We show that the measures $|x|^a\,dx$ are doubling when $a > -n$. We divide all balls $B(x_0,R)$ in $\mathbf{R}^n$ into two categories: balls of type I that satisfy $|x_0| \ge 3R$ and type II that satisfy $|x_0| < 3R$. For balls of type I we observe that

$$\int_{B(x_0,2R)} |x|^a\,dx \le v_n(2R)^n \begin{cases} (|x_0|+2R)^a & \text{when } a \ge 0, \\ (|x_0|-2R)^a & \text{when } a < 0, \end{cases}$$

$$\int_{B(x_0,R)} |x|^a\,dx \ge v_n R^n \begin{cases} (|x_0|-R)^a & \text{when } a \ge 0, \\ (|x_0|+R)^a & \text{when } a < 0. \end{cases}$$

Since $|x_0| \ge 3R$, we have $|x_0|+2R \le 4(|x_0|-R)$ and $|x_0|-2R \ge \frac{1}{4}(|x_0|+R)$, from which (9.1.22) follows with $C = 2^{3n}4^{|a|}$.

For balls of type II, we have $|x_0| \le 3R$ and we note two things: first

$$\int_{B(x_0,2R)} |x|^a\,dx \le \int_{|x|\le 5R} |x|^a\,dx = c_n R^{n+a},$$

and second, since $|x|^a$ is radially decreasing for $a < 0$ and radially increasing for $a \ge 0$, we have

$$\int_{B(x_0,R)} |x|^a\,dx \geq \begin{cases} \displaystyle\int_{B(0,R)} |x|^a\,dx & \text{when } a \geq 0, \\ \displaystyle\int_{B(3R\frac{x_0}{|x_0|},R)} |x|^a\,dx & \text{when } a < 0. \end{cases}$$

For $x \in B(3R\frac{x_0}{|x_0|},R)$ we must have $|x| \geq 2R$, and hence both integrals on the right are at least a multiple of R^{n+a}. This establishes (9.1.22) for balls of type II.

Example 9.1.7. We investigate for which real numbers a the power function $|x|^a$ is an A_p weight on $\mathbf{R}^n$. For $1 < p < \infty$, we need to examine for which a the following expression is finite:

$$\sup_{B \text{ balls}} \left(\frac{1}{|B|}\int_B |x|^a\,dx\right)\left(\frac{1}{|B|}\int_B |x|^{-a\frac{p'}{p}}\,dx\right)^{\frac{p}{p'}}. \tag{9.1.23}$$

As in the previous example we split the balls in $\mathbf{R}^n$ into those of type I and those of type II. If $B = B(x_0,R)$ is of type I, then the presence of the origin does not affect the behavior of either integral in (9.1.23), and we see that the expression inside the supremum in (9.1.23) is comparable to

$$|x_0|^a\left(|x_0|^{-a\frac{p'}{p}}\right)^{\frac{p}{p'}} = 1.$$

If $B(x_0,R)$ is a ball of type II, then $B(0,5R)$ has size comparable to $B(x_0,R)$ and contains it. Since the measure $|x|^a\,dx$ is doubling, the integrals of the function $|x|^a$ over $B(x_0,R)$ and over $B(0,5R)$ are comparable. It suffices therefore to estimate the expression inside the supremum in (9.1.23), in which we have replaced $B(x_0,R)$ by $B(0,5R)$. But this is

$$\begin{aligned}&\left(\frac{1}{v_n(5R)^n}\int_{B(0,5R)} |x|^a\,dx\right)\left(\frac{1}{v_n(5R)^n}\int_{B(0,5R)} |x|^{-a\frac{p'}{p}}\,dx\right)^{\frac{p}{p'}} \\ &\qquad = \left(\frac{n}{(5R)^n}\int_0^{5R} r^{a+n-1}\,dr\right)\left(\frac{n}{(5R)^n}\int_0^{5R} r^{-a\frac{p'}{p}+n-1}\,dr\right)^{\frac{p}{p'}},\end{aligned}$$

which is seen easily to be finite and independent of R exactly when $-n < a < n\frac{p}{p'}$. We conclude that $|x|^a$ is an A_p weight, $1 < p < \infty$, if and only if $-n < a < n(p-1)$.

The previous proof can be suitably modified to include the case $p = 1$. In this case we obtain that $|x|^a$ is an A_1 weight if and only if $-n < a \leq 0$. As we have seen, the measure $|x|^a\,dx$ is doubling on the larger range $-n < a < \infty$. Thus for $a > n(p-1)$, the function $|x|^a$ provides an example of a doubling measure that is not in A_p.

Example 9.1.8. On $\mathbf{R}^n$ the function

$$u(x) = \begin{cases} \log\frac{1}{|x|} & \text{when } |x| < \frac{1}{e}, \\ 1 & \text{otherwise,} \end{cases}$$

is an A_1 weight. Indeed, to check condition (9.1.19) it suffices to consider balls of type I and type II as defined in Example 9.1.6. In either case the required estimate follows easily.

We now return to a point alluded to earlier, that the A_p condition implies the boundedness of the Hardy–Littlewood maximal function M on the space $L^p(w)$. To this end we introduce four maximal functions acting on functions f that are locally integrable with respect to w:

$$M^w(f)(x) = \sup_{B \ni x} \frac{1}{w(B)} \int_B |f|\, w\, dy,$$

where the supremum is taken over open balls B that contain the point x and

$$\begin{aligned} \mathcal{M}^w(f)(x) &= \sup_{\delta>0} \frac{1}{w(B(x,\delta))} \int_{B(x,\delta)} |f|\, w\, dy, \\ M_c^w(f)(x) &= \sup_{Q \ni x} \frac{1}{w(Q)} \int_Q |f|\, w\, dy, \end{aligned}$$

where Q is an open cube containing the point x, and

$$\mathcal{M}_c^w(f)(x) = \sup_{\delta>0} \frac{1}{w(Q(x,\delta))} \int_{Q(x,\delta)} |f|\, w\, dy,$$

where $Q(x,\delta) = \prod_{j=1}^n (x_j - \delta, x_j + \delta)$ is a cube of side length 2δ centered at $x = (x_1, \dots, x_n)$. When $w = 1$, these maximal functions reduce to the standard ones $M(f)$, $\mathcal{M}(f)$, $M_c(f)$, and $\mathcal{M}_c(f)$, the uncentered and centered Hardy–Littlewood maximal functions with respect to balls and cubes, respectively.

Theorem 9.1.9. *Let $w \in A_p(\mathbf{R}^n)$ for some $1 < p < \infty$. Then there is a constant $C_{n,p}$ such that*

$$\big\| \mathcal{M}_c \big\|_{L^p(w) \to L^p(w)} \le C_{n,p} [w]_{A_p}^{\frac{1}{p-1}}. \tag{9.1.24}$$

Since the operators $\mathcal{M}_c$, M_c, $\mathcal{M}$, and M are pointwise comparable, a similar conclusion holds for the other three as well.

Proof. Fix a weight w and let $\sigma = w^{-\frac{1}{p-1}}$ be the dual weight. Fix an open cube $Q = Q(x_0, r)$ in $\mathbf{R}^n$ with center x_0 and side length $2r$ and write

$$\frac{1}{|Q|} \int_Q |f|\, dy = \frac{w(Q)^{\frac{1}{p-1}} \sigma(3Q)}{|Q|^{\frac{p}{p-1}}} \left\{ \frac{|Q|}{w(Q)} \left(\frac{1}{\sigma(3Q)} \int_Q |f|\, dy \right)^{p-1} \right\}^{\frac{1}{p-1}}. \tag{9.1.25}$$

For any $x \in Q$, consider the cube $Q(x, 2r)$. Then $Q \subseteq Q(x,2r) \subseteq 3Q = Q(x_0, 3r)$ and thus

$$\frac{1}{\sigma(3Q)} \int_Q |f|\, dy \le \frac{1}{\sigma(Q(x,2r))} \int_{Q(x,2r)} |f|\, dy \le \mathcal{M}_c^\sigma(|f|\sigma^{-1})(x)$$

for any $x \in Q$. Inserting this expression in (9.1.25), we obtain

$$\frac{1}{|Q|}\int_Q |f|\,dy \leq \frac{w(Q)^{\frac{1}{p-1}}\sigma(3Q)}{|Q|^{\frac{p}{p-1}}}\left\{\frac{1}{w(Q)}\int_Q \mathcal{M}_c^\sigma(|f|\sigma^{-1})^{p-1}\,dy\right\}^{\frac{1}{p-1}}. \tag{9.1.26}$$

Since one may easily verify that

$$\frac{w(Q)\sigma(3Q)^{p-1}}{|Q|^p} \leq 3^{np}[w]_{A_p},$$

it follows that

$$\frac{1}{|Q|}\int_Q |f|\,dy \leq 3^{\frac{np}{p-1}}[w]_{A_p}^{\frac{1}{p-1}}\Big(\mathcal{M}_c^w\big[(\mathcal{M}_c^\sigma(|f|\sigma^{-1}))^{p-1}w^{-1}\big](x_0)\Big)^{\frac{1}{p-1}},$$

since x_0 is the center of Q. Hence, we have

$$\mathcal{M}_c(f) \leq 3^{\frac{np}{p-1}}[w]_{A_p}^{\frac{1}{p-1}}\Big(\mathcal{M}_c^w\big[(\mathcal{M}_c^\sigma(|f|\sigma^{-1}))^{p-1}w^{-1}\big]\Big)^{\frac{1}{p-1}}.$$

Applying $L^p(w)$ norms, we deduce

$$\begin{aligned}
\big\|\mathcal{M}_c(f)\big\|_{L^p(w)} &\leq 3^{\frac{np}{p-1}}[w]_{A_p}^{\frac{1}{p-1}}\big\|\mathcal{M}_c^w\big[(\mathcal{M}_c^\sigma(|f|\sigma^{-1}))^{p-1}w^{-1}\big]\big\|_{L^{p'}(w)}^{\frac{1}{p-1}}\\
&\leq 3^{\frac{np}{p-1}}[w]_{A_p}^{\frac{1}{p-1}}\big\|\mathcal{M}_c^w\big\|_{L^{p'}(w)\to L^{p'}(w)}^{\frac{1}{p-1}}\big\|(\mathcal{M}_c^\sigma(|f|\sigma^{-1}))^{p-1}w^{-1}\big\|_{L^{p'}(w)}^{\frac{1}{p-1}}\\
&= 3^{\frac{np}{p-1}}[w]_{A_p}^{\frac{1}{p-1}}\big\|\mathcal{M}_c^w\big\|_{L^{p'}(w)\to L^{p'}(w)}^{\frac{1}{p-1}}\big\|\mathcal{M}_c^\sigma(|f|\sigma^{-1})\big\|_{L^p(\sigma)}\\
&\leq 3^{\frac{np}{p-1}}[w]_{A_p}^{\frac{1}{p-1}}\big\|\mathcal{M}_c^w\big\|_{L^{p'}(w)\to L^{p'}(w)}^{\frac{1}{p-1}}\big\|\mathcal{M}_c^\sigma\big\|_{L^p(\sigma)\to L^p(\sigma)}\big\|f\big\|_{L^p(w)},
\end{aligned}$$

and conclusion (9.1.24) follows, provided we show that

$$\big\|\mathcal{M}_c^w\big\|_{L^q(w)\to L^q(w)} \leq C(q,n) < \infty \tag{9.1.27}$$

for any $1 < q < \infty$ and any weight w.

We obtain this estimate by interpolation. Obviously (9.1.27) is valid when $q = \infty$ with $C(\infty, n) = 1$. If we prove that

$$\big\|\mathcal{M}_c^w\big\|_{L^1(w)\to L^{1,\infty}(w)} \leq C(1,n) < \infty, \tag{9.1.28}$$

then (9.1.27) will follow from Theorem 1.3.2.

To prove (9.1.28) we fix $f \in L^1(\mathbf{R}^n, w\,dx)$. We first show that the set

$$E_\lambda = \{\mathcal{M}_c^w(f) > \lambda\}$$

is open. For any $r > 0$, let $Q(x,r)$ denote an open cube of side length $2r$ with center $x \in \mathbf{R}^n$. If we show that for any $r > 0$ and $x \in \mathbf{R}^n$ the function

$$x \mapsto \frac{1}{w(Q(x,r))} \int_{Q(x,r)} |f| \, w \, dy \tag{9.1.29}$$

is continuous, then $\mathcal{M}_c^w(f)$ is the supremum of continuous functions; hence it is lower semicontinuous and thus the set E_λ is open. But this is straightforward. If $x_n \to x_0$, then $w(Q(x_n,r)) \to w(Q(x_0,r))$ and also $\int_{Q(x_n,r)} |f| \, w \, dy \to \int_{Q(x_0,r)} |f| \, w \, dy$ by the Lebesgue dominated convergence theorem. Since $w(Q(x_0,r)) \neq 0$, it follows that the function in (9.1.29) is continuous.

Given K a compact subset of E_λ, for any $x \in K$ select an open cube Q_x centered at x such that

$$\frac{1}{w(Q_x)} \int_{Q_x} |f| \, w \, dy > \lambda \, .$$

Applying Lemma 9.1.10 (proved immediately afterward) we find a subfamily $\{Q_{x_j}\}_{j=1}^m$ of the family of the balls $\{Q_x : x \in K\}$ such that (9.1.30) and (9.1.31) hold. Then

$$w(K) \le \sum_{j=1}^m w(Q_{x_j}) \le \sum_{j=1}^m \frac{1}{\lambda} \int_{Q_{x_j}} |f| \, w \, dy \le \frac{24^n}{\lambda} \int_{\mathbf{R}^n} |f| \, w \, dy \, ,$$

where the last inequality follows by multiplying (9.1.31) by $|f| w$ and integrating over $\mathbf{R}^n$. Taking the supremum over all compact subsets K of E_λ and using the inner regularity of $w\,dx$, which is a consequence of the Lebesgue monotone convergence theorem, we deduce that $\mathcal{M}_c^w$ maps $L^1(w)$ to $L^{1,\infty}(w)$ with constant at most 24^n. Thus (9.1.28) holds with $C(1,n) = 24^n$. □

Lemma 9.1.10. *Let K be a bounded set in $\mathbf{R}^n$ and for every $x \in K$, let Q_x be an open cube with center x and sides parallel to the axes. Then there are an $m \in \mathbf{Z}^+ \cup \{\infty\}$ and a sequence of points $\{x_j\}_{j=1}^m$ in K such that*

$$K \subseteq \bigcup_{j=1}^m Q_{x_j} \tag{9.1.30}$$

and for almost all $y \in \mathbf{R}^n$ one has

$$\sum_{j=1}^m \chi_{Q_{x_j}}(y) \le 24^n \, . \tag{9.1.31}$$

Proof. Let

$$s_0 = \sup\{\ell(Q_x) : x \in K\} \, .$$

If $s_0 = \infty$, then there exists $x_1 \in K$ such that $\ell(Q_{x_1}) > 4L$, where $[-L,L]^n$ contains K. Then K is contained in Q_{x_1} and the statement of the lemma is valid with $m = 1$.

Suppose now that $s_0 < \infty$. Select $x_1 \in K$ such that $\ell(Q_{x_1}) > s_0/2$. Then define

$$K_1 = K \setminus Q_{x_1} \, , \qquad s_1 = \sup\{\ell(Q_x) : x \in K_1\} \, ,$$

and select $x_2 \in K_1$ such that $\ell(Q_{x_2}) > s_1/2$. Next define

$$K_2 = K \setminus (Q_{x_1} \cup Q_{x_2}), \qquad s_2 = \sup\{\ell(Q_x) : x \in K_2\},$$

and select $x_3 \in K_2$ such that $\ell(Q_{x_3}) > s_2/2$. Continue until the first integer m is found such that K_m is an empty set. If no such integer exists, continue this process indefinitely and set $m = \infty$.

We claim that for all $i \neq j$ we have $\frac{1}{3}Q_{x_i} \cap \frac{1}{3}Q_{x_j} = \emptyset$. Indeed, suppose that $i > j$. Then $x_i \in K_{i-1} = K \setminus (Q_{x_1} \cup \cdots \cup Q_{x_{i-1}})$; thus $x_i \notin Q_j$. Also $x_i \in K_{i-1} \subseteqq K_{j-1}$, which implies that $\ell(Q_{x_i}) \le s_{j-1} < 2\ell(Q_{x_j})$. If $x_i \notin Q_j$ and $\ell(Q_{x_j}) > \frac{1}{2}\ell(Q_{x_i})$, it easily follows that $\frac{1}{3}Q_{x_i} \cap \frac{1}{3}Q_{x_j} = \emptyset$.

We now prove (9.1.30). If $m < \infty$, then $K_m = \emptyset$ and therefore $K \subseteqq \bigcup_{j=1}^m Q_{x_j}$. If $m = \infty$, then there is an infinite number of selected cubes Q_{x_j}. Since the cubes $\frac{1}{3}Q_{x_j}$ are pairwise disjoint and have centers in a bounded set, it must be the case that some subsequence of the sequence of their lengths converges to zero. If there exists a $y \in K \setminus \bigcup_{j=1}^\infty Q_{x_j}$, this y would belong to all K_j, $j = 1, 2, \dots$, and then $s_j \ge \ell(Q_y)$ for all j. Since some subsequence of the s_j's tends to zero, it would follow that $\ell(Q_y) = 0$, which would force the open cube Q_y to be the empty set, a contradiction. Thus (9.1.30) holds.

Finally, we show that $\sum_{j=1}^m \chi_{Q_{x_j}}(y) \le 24^n$ for almost every point $y \in \mathbf{R}^n$. To prove this we consider the n hyperplanes H_i that are parallel to the coordinate hyperplanes and pass through the point y. Then we may write $\mathbf{R}^n$ as a union of 2^n higher-dimensional open "octants" O_r and n hyperplanes H_i of n-dimensional Lebesgue measure zero. We show that there are only finitely many points x_j in a given such open "octant" O_r. Indeed, let us fix an O_r and pick an $x_{k_0} \in K \cap O_r$ such that $Q_{x_{k_0}}$ contains y and the distance from x_{k_0} to y is largest possible. If x_j is another point in $K \cap O_r$ such that Q_{x_j} contains y, then $\ell(Q_{x_{k_0}}) \ge \ell(Q_{x_j})$, which yields that $x_j \in Q_{x_{k_0}}$. As previously observed, one must then have $j < k_0$, which implies that $\frac{1}{2}\ell(Q_{x_{k_0}}) < \ell(Q_{x_j})$. Thus all cubes Q_{x_j} with centers in $K \cap O_r$ that contain the fixed point y have side lengths comparable to that of $Q_{x_{k_0}}$. A simple geometric argument now gives that there are at most finitely many cubes Q_{x_j} of side length between α and 2α that contain the given point y such that $\frac{1}{3}Q_{x_j}$ are pairwise disjoint. Indeed, let $\alpha = \frac{1}{2}\ell(Q_{x_{k_0}})$ and let $\{Q_{x_r}\}_{r\in I}$ be the cubes with these properties. Then we have

$$\frac{\alpha^n |I|}{3^n} \le \sum_{r\in I} \left|\tfrac{1}{3}Q_{x_r}\right| = \Big|\bigcup_{r\in I} \tfrac{1}{3}Q_{x_r}\Big| \le \Big|\bigcup_{r\in I} Q_{x_r}\Big| \le (4\alpha)^n;$$

since all the cubes Q_{x_r} contain the point y and have length at most 2α and they must therefore be contained in a cube of side length 4α centered at y. This observation shows that $|I| \le 12^n$, and since there are 2^n sets O_r, we conclude the proof of (9.1.31). □

Remark 9.1.11. Without use of the covering Lemma 9.1.10, (9.1.28) can be proved via the doubling property of w (cf. Exercise 2.1.1(a)), but then the resulting constant $C(q,n)$ would depend on the doubling constant of the measure $w\,dx$ and thus on $[w]_{A_p}$; this would yield a worse dependence on $[w]_{A_p}$ in the constant in (9.1.24).

Exercises

9.1.1. Let k be a nonnegative measurable function such that k, k^{-1} are in $L^\infty(\mathbf{R}^n)$. Prove that if w is an A_p weight for some $1 \le p < \infty$, then so is kw.

9.1.2. Let w_1, w_2 be two A_1 weights and let $1 < p < \infty$. Prove that $w_1 w_2^{1-p}$ is an A_p weight by showing that

$$[w_1 w_2^{1-p}]_{A_p} \le [w_1]_{A_1} [w_2]_{A_1}^{p-1}.$$

9.1.3. Suppose that $w \in A_p$ for some $p \in [1,\infty)$ and $0 < \delta < 1$. Prove that $w^\delta \in A_q$, where $q = \delta p + 1 - \delta$, by showing that

$$[w^\delta]_{A_q} \le [w]_{A_p}^\delta .$$

9.1.4. Show that if the A_p characteristic constants of a weight w are uniformly bounded for all $p > 1$, then $w \in A_1$.

9.1.5. Let $w_0 \in A_{p_0}$ and $w_1 \in A_{p_1}$ for some $1 \le p_0, p_1 < \infty$. Let $0 \le \theta \le 1$ and define

$$\frac{1}{p} = \frac{1-\theta}{p_0} + \frac{\theta}{p_1} \qquad \text{and} \qquad w^{\frac{1}{p}} = w_0^{\frac{1-\theta}{p_0}} w_1^{\frac{\theta}{p_1}} .$$

Prove that

$$[w]_{A_p} \le [w_0]_{A_{p_0}}^{(1-\theta)\frac{p}{p_0}} [w_1]_{A_{p_1}}^{\theta \frac{p}{p_1}} ;$$

thus w is in A_p.

9.1.6. Let $1 < p < \infty$. A pair of weights (u,w) that satisfies

$$[u,w]_{(A_p,A_p)} = \sup_{\substack{Q \text{ cubes} \\ \text{in } \mathbf{R}^n}} \left(\frac{1}{|Q|}\int_Q u\,dx\right)\left(\frac{1}{|Q|}\int_Q w^{-\frac{1}{p-1}}\,dx\right)^{p-1} < \infty$$

is said to be of class (A_p,A_p). The quantity $[u,w]_{(A_p,A_p)}$ is called the (A_p,A_p) characteristic constant of the pair.
(a) Show that for any $f \in L^1_{\text{loc}}(\mathbf{R}^n)$ with $0 < f < \infty$ a.e., the pair $(f, M(f))$ is of *class* (A_p,A_p) for every $1<p<\infty$ with characteristic constant independent of f.
(b) If (u,w) is of class (A_p,A_p), then the Hardy–Littlewood maximal operator M may not map $L^p(w)$ to $L^p(u)$.
[*Hint:* Try the pair $(M(g)^{1-p}, |g|^{1-p})$ for a suitable g.]

9.1.7. In contrast to part (b) of Exercise 9.1.6, show that if the pair of weights (u,w) is of class (A_p,A_p) for some $1 < p < \infty$, then M must map $L^p(w)$ to $L^{p,\infty}(u)$ with norm at most $C(n,p)[u,w]_{(A_p,A_p)}^{\frac{1}{p}}$.
[*Hint:* Show first using Hölder's inequality that for all functions f and all cubes Q' we have

$$\left(\frac{1}{|Q'|}\int_{Q'}|f|\,dx\right)^p u(Q') \le [u,w]_{(A_p,A_p)}\int_{Q'}|f|^p w\,dx.$$

Replacing f by $f\chi_Q$, where $Q \subseteq Q'$, obtain that

$$u(Q') \le [u,w]_{(A_p,A_p)}|Q'|^p \frac{\int_Q |f|^p w\,dx}{\left(\int_Q |f|\,dx\right)^p}.$$

Then use Exercise 4.3.9 to find disjoint cubes Q_j such that the set $E_\alpha = \{x \in \mathbf{R}^n : M_c(f)(x) > \alpha\}$ is contained in the union of $3Q_j$ and $\frac{\alpha}{4^n} < \frac{1}{|Q_j|}\int_{Q_j}|f(t)|\,dt \le \frac{\alpha}{2^n}$. Then $u(E_\alpha) \le \sum_j u(3Q_j)$, and bound each $u(3Q_j)$ by taking $Q' = 3Q_j$ and $Q = Q_j$ in the preceding estimate.]

9.1.8. Use Exercise 9.1.7 to prove that for all $1 < q < \infty$ there is a constant $C_q < \infty$ such that for all $f, g \ge 0$ locally integrable functions on $\mathbf{R}^n$ we have

$$\int_{\mathbf{R}^n} M(f)(x)^q\, g(x)\,dx \le C_q \int_{\mathbf{R}^n} f(x)^q\, M(g)(x)\,dx.$$

[*Hint:* Take $1 < p < q$ and interpolate between L^p and L^∞.]

9.1.9. Let $w \in A_p$ for some $1 \le p < \infty$ and $k \ge 1$. Show that $\min(w,k)$ is in A_p and satisfies

$$[\min(w,k)]_{A_p} \le c_p\left([w]_{A_p} + 1\right),$$

where $c_p = 1$ when $p \le 2$ and $c_p = 2^{p-2}$ when $p > 2$.
[*Hint:* Use that $\frac{1}{|Q|}\int_Q \min(w,k)^{-\frac{1}{p-1}}dx \le \frac{1}{|Q|}\int_Q w^{-\frac{1}{p-1}}dx + k^{-\frac{1}{p-1}}$, raise to the power $p-1$, and multiply by $\min\left(k, \frac{1}{|Q|}\int_Q w\,dx\right)$.]

9.1.10. Suppose that $w_j \in A_{p_j}$ with $1 \le j \le m$ for some $1 \le p_1, \dots, p_m < \infty$ and let $0 < \theta_1, \dots, \theta_m < 1$ be such that $\theta_1 + \cdots + \theta_m = 1$. Show that

$$w_1^{\theta_1} \cdots w_m^{\theta_m} \in A_{\max\{p_1,\dots,p_m\}}.$$

[*Hint:* First note that each weight w_j lies in $A_{\max\{p_1,\dots,p_m\}}$ and then apply Hölder's inequality.]

9.1.11. Let $w_1 \in A_{p_1}$ and $w_2 \in A_{p_2}$ for some $1 \le p_1, p_2 < \infty$. Prove that

$$[w_1 + w_2]_{A_p} \le [w_1]_{A_{p_1}} + [w_2]_{A_{p_2}},$$

where $p = \max(p_1, p_2)$.

9.1.12. Prove the claim of Example 9.1.8.

9.2 Reverse Hölder Inequality for A_p Weights and Consequences

An essential property of A_p weights is that they assign to subsets of balls mass proportional to the percentage of the Lebesgue measure of the subset within the ball. The following lemma provides a way to quantify this statement.

Lemma 9.2.1. *Let $w \in A_p$ for some $1 \le p < \infty$ and let $0 < \alpha < 1$. Then there exists $\beta < 1$ such that whenever S is a measurable subset of a cube Q that satisfies $|S| \le \alpha|Q|$, we have $w(S) \le \beta\, w(Q)$.*

Proof. Taking $f = \chi_A$ in property (8) of Proposition 9.1.5, we obtain

$$\left(\frac{|A|}{|Q|}\right)^p \le [w]_{A_p}\frac{w(A)}{w(Q)}. \tag{9.2.1}$$

We write $S = Q \setminus A$ to get

$$\left(1-\frac{|S|}{|Q|}\right)^p \le [w]_{A_p}\left(1-\frac{w(S)}{w(Q)}\right). \tag{9.2.2}$$

Given $0 < \alpha < 1$, set

$$\beta = 1 - \frac{(1-\alpha)^p}{[w]_{A_p}} \tag{9.2.3}$$

and use (9.2.2) to obtain the required conclusion. □

9.2.1 The Reverse Hölder Property of A_p Weights

We are now ready to state and prove one of the main results of the theory of weights, the reverse Hölder inequality for A_p weights.

Theorem 9.2.2. *Let $w \in A_p$ for some $1 \le p < \infty$. Then there exist constants C and $\gamma > 0$ that depend only on the dimension n, on p, and on $[w]_{A_p}$ such that for every cube Q we have*

$$\left(\frac{1}{|Q|}\int_Q w(t)^{1+\gamma}\,dt\right)^{\frac{1}{1+\gamma}} \le \frac{C}{|Q|}\int_Q w(t)\,dt. \tag{9.2.4}$$

Proof. Let us fix a cube Q and set

$$\alpha_0 = \frac{1}{|Q|}\int_Q w(x)\,dx.$$

We also fix $0 < \alpha < 1$. We define an increasing sequence of scalars

$$\alpha_0 < \alpha_1 < \alpha_2 < \cdots < \alpha_k < \cdots$$

for $k \geq 0$ by setting

$$\alpha_{k+1} = 2^n \alpha^{-1} \alpha_k \qquad \text{or} \qquad \alpha_k = (2^n \alpha^{-1})^k \alpha_0,$$

and for each $k \geq 1$ we apply a Calderón–Zygmund decomposition to w at height α_k. Precisely, for dyadic subcubes R of Q, we let

$$\frac{1}{|R|} \int_R w(x)\,dx > \alpha_k \tag{9.2.5}$$

be the selection criterion. Since Q does not satisfy the selection criterion, it is not selected. We divide the cube Q into a mesh of 2^n subcubes of equal side length, and among these cubes we select those that satisfy (9.2.5). We subdivide each unselected subcube into 2^n cubes of equal side length and we continue in this way indefinitely. We denote by $\{Q_{k,j}\}_j$ the collection of all selected subcubes of Q. We observe that the following properties are satisfied:

(1) $\alpha_k < \dfrac{1}{|Q_{k,j}|} \displaystyle\int_{Q_{k,j}} w(t)\,dt \leq 2^n \alpha_k$.

(2) For almost all $x \notin U_k$ we have $w(x) \leq \alpha_k$, where $U_k = \bigcup\limits_j Q_{k,j}$.

(3) Each $Q_{k+1,j}$ is contained in some $Q_{k,l}$.

Property (1) is satisfied since the unique dyadic parent of $Q_{k,j}$ was not chosen in the selection procedure. Property (2) follows from the Lebesgue differentiation theorem using the fact that for almost all $x \notin U_k$ there exists a sequence of unselected cubes of decreasing lengths whose closures' intersection is the singleton $\{x\}$. Property (3) is satisfied since each $Q_{k,j}$ is the maximal subcube of Q satisfying (9.2.5). And since the average of w over $Q_{k+1,j}$ is also bigger than α_k, it follows that $Q_{k+1,j}$ must be contained in some maximal cube that possesses this property.

We now compute the portion of $Q_{k,l}$ that is covered by cubes of the form $Q_{k+1,j}$ for some j. We have

$$\begin{aligned}
2^n \alpha_k &\geq \frac{1}{|Q_{k,l}|} \int_{Q_{k,l} \cap U_{k+1}} w(t)\,dt \\
&= \frac{1}{|Q_{k,l}|} \sum_{j:\, Q_{k+1,j} \subseteq Q_{k,l}} |Q_{k+1,j}| \frac{1}{|Q_{k+1,j}|} \int_{Q_{k+1,j}} w(t)\,dt \\
&> \frac{|Q_{k,l} \cap U_{k+1}|}{|Q_{k,l}|} \alpha_{k+1} \\
&= \frac{|Q_{k,l} \cap U_{k+1}|}{|Q_{k,l}|} 2^n \alpha^{-1} \alpha_k .
\end{aligned}$$

It follows that $|Q_{k,l} \cap U_{k+1}| \leq \alpha |Q_{k,l}|$; thus, applying Lemma 9.2.1, we obtain

$$\frac{w(Q_{k,l} \cap U_{k+1})}{w(Q_{k,l})} < \beta = 1 - \frac{(1-\alpha)^p}{[w]_{A_p}},$$

from which, summing over all l, we obtain

$$w(U_{k+1}) \le \beta w(U_k).$$

The latter gives $w(U_k) \le \beta^k w(U_0)$. We also have $|U_{k+1}| \le \alpha |U_k|$; hence $|U_k| \to 0$ as $k \to \infty$. Therefore, the intersection of the U_k's is a set of Lebesgue measure zero. We can therefore write

$$Q = (Q \setminus U_0) \bigcup \Big(\bigcup_{k=0}^{\infty} U_k \setminus U_{k+1} \Big)$$

modulo a set of Lebesgue measure zero. Let us now find a $\gamma > 0$ such that the reverse Hölder inequality (9.2.4) holds. We have $w(x) \le \alpha_k$ for almost all x in $Q \setminus U_k$ and therefore

$$\begin{aligned}
\int_Q w(t)^{1+\gamma}\,dt &= \int_{Q\setminus U_0} w(t)^\gamma w(t)\,dt + \sum_{k=0}^{\infty} \int_{U_k \setminus U_{k+1}} w(t)^\gamma w(t)\,dt \\
&\le \alpha_0^\gamma w(Q\setminus U_0) + \sum_{k=0}^{\infty} \alpha_{k+1}^\gamma w(U_k) \\
&\le \alpha_0^\gamma w(Q\setminus U_0) + \sum_{k=0}^{\infty} ((2^n\alpha^{-1})^{k+1}\alpha_0)^\gamma \beta^k w(U_0) \\
&\le \alpha_0^\gamma \Big(1 + (2^n\alpha^{-1})^\gamma \sum_{k=0}^{\infty} (2^n\alpha^{-1})^{\gamma k}\beta^k \Big) w(Q) \\
&= \Big(\frac{1}{|Q|} \int_Q w(t)\,dt \Big)^\gamma \Big(1 + \frac{(2^n\alpha^{-1})^\gamma}{1-(2^n\alpha^{-1})^\gamma \beta} \Big) \int_Q w(t)\,dt,
\end{aligned}$$

provided $\gamma > 0$ is chosen small enough that $(2^n\alpha^{-1})^\gamma \beta < 1$. Keeping track of the constants, we conclude the proof of the theorem with

$$\gamma < \frac{-\log \beta}{\log 2^n - \log \alpha} = \frac{\log\big([w]_{A_p}\big) - \log\big([w]_{A_p} - (1-\alpha)^p\big)}{\log 2^n - \log \alpha} \tag{9.2.6}$$

and

$$C = 1 + \frac{(2^n\alpha^{-1})^\gamma}{1-(2^n\alpha^{-1})^\gamma \beta} = 1 + \frac{(2^n\alpha^{-1})^\gamma}{1-(2^n\alpha^{-1})^\gamma \big(1 - \frac{(1-\alpha)^p}{[w]_{A_p}}\big)}. \tag{9.2.7}$$

Note that up to this point, α was an arbitrary number in $(0,1)$, and it may be chosen to maximize (9.2.6). □

Remark 9.2.3. It is worth observing that for any fixed $0 < \alpha < 1$, the constant in (9.2.6) decreases as $[w]_{A_p}$ increases, while the constant in (9.2.7) increases as $[w]_{A_p}$ increases. This allows us to obtain the following stronger version of Theorem 9.2.2:

For any $1 \le p < \infty$ and $B > 1$, there exist positive constants $C = C(n,p,B)$ and $\gamma = \gamma(n,p,B)$ such that for all $w \in A_p$ satisfying $[w]_{A_p} \le B$ the reverse Hölder condition (9.2.4) holds for every cube Q. See Exercise 9.2.4(a) for details.

Observe that in the proof of Theorem 9.2.2 it was crucial to know that for some $0 < \alpha, \beta < 1$ we have

$$|S| \le \alpha |Q| \implies w(S) \le \beta w(Q) \tag{9.2.8}$$

whenever S is a subset of the cube Q. No special property of Lebesgue measure was used in the proof of Theorem 9.2.2 other than its doubling property. Therefore, it is reasonable to ask whether Lebesgue measure in (9.2.8) can be replaced by a general measure μ satisfying the doubling property

$$\mu(3Q) \le C_n \mu(Q) < \infty \tag{9.2.9}$$

for all cubes Q in $\mathbf{R}^n$. A straightforward adjustment of the proof of the previous theorem indicates that this is indeed the case.

Corollary 9.2.4. *Let w be a weight and let μ be a measure on $\mathbf{R}^n$ satisfying (9.2.9). Suppose that there exist $0 < \alpha, \beta < 1$, such that*

$$\mu(S) \le \alpha \mu(Q) \implies \int_S w(t)\,d\mu(t) \le \beta \int_Q w(t)\,d\mu(t)$$

whenever S is a μ-measurable subset of a cube Q. Then there exist $0 < C, \gamma < \infty$ [which depend only on the dimension n, the constant C_n in (9.2.9), α, and β] such that for every cube Q in $\mathbf{R}^n$ we have

$$\left(\frac{1}{\mu(Q)}\int_Q w(t)^{1+\gamma}\,d\mu(t)\right)^{\frac{1}{1+\gamma}} \le \frac{C}{\mu(Q)}\int_Q w(t)\,d\mu(t). \tag{9.2.10}$$

Proof. The proof of the corollary can be obtained almost verbatim from that of Theorem 9.2.2 by replacing Lebesgue measure with the doubling measure $d\mu$ and the constant 2^n by C_n.

Precisely, we define $\alpha_k = (C_n \alpha^{-1})^k \alpha_0$, where α_0 is the μ-average of w over Q; then properties (1), (2), (3) concerning the selected cubes $\{Q_{k,j}\}_j$ are replaced by

(1_μ) $\alpha_k < \dfrac{1}{\mu(Q_{k,j})}\displaystyle\int_{Q_{k,j}} w(t)\,d\mu(t) \le C_n \alpha_k$.

(2_μ) On $Q \setminus U_k$ we have $w \le \alpha_k$ μ-almost everywhere, where $U_k = \bigcup_j Q_{k,j}$.

(3_μ) Each $Q_{k+1,j}$ is contained in some $Q_{k,l}$.

To prove the upper inequality in (1_μ) we use that the dyadic parent of each selected cube $Q_{k,j}$ was not selected and is contained in $3Q_{k,j}$. To prove (2_μ) we need a differentiation theorem for doubling measures, analogous to that in Corollary 2.1.16. This can be found in Exercise 2.1.1. The remaining details of the proof are trivially adapted to the new setting. The conclusion is that for

$$0 < \gamma < \frac{-\log \beta}{\log C_n - \log \alpha} \tag{9.2.11}$$

and

$$C = 1 + \frac{(2\alpha^{-1})^{\gamma}}{1-(C_n\alpha^{-1})^{\gamma}\beta}, \tag{9.2.12}$$

(9.2.10) is satisfied. □

9.2.2 Consequences of the Reverse Hölder Property

Having established the crucial reverse Hölder inequality for A_p weights, we now pass to some very important applications. Among them, the first result of this section yields that an A_p weight that lies a priori in $L^1_{\mathrm{loc}}(\mathbf{R}^n)$ must actually lie in the better space $L^{1+\sigma}_{\mathrm{loc}}(\mathbf{R}^n)$ for some $\sigma > 0$ depending on the weight.

Theorem 9.2.5. *If $w \in A_p$ for some $1 \le p < \infty$, then there exists a number $\gamma > 0$ (that depends on $[w]_{A_p}$, p, and n) such that $w^{1+\gamma} \in A_p$.*

Proof. The proof is simple. When $p = 1$, we apply the reverse Hölder inequality of Theorem 9.2.2 to the weight w to obtain

$$\frac{1}{|Q|}\int_Q w(t)^{1+\gamma}\,dt \le \left(\frac{C}{|Q|}\int_Q w(t)\,dt\right)^{1+\gamma} \le C^{1+\gamma}[w]_{A_1}^{1+\gamma} w(x)^{1+\gamma}$$

for almost all x in the cube Q. Therefore, $w^{1+\gamma}$ is an A_1 weight with characteristic constant at most $C^{1+\gamma}[w]_{A_1}^{1+\gamma}$. ($C$ is here the constant of Theorem 9.2.2.) When $p > 1$, there exist $\gamma_1, \gamma_2 > 0$ and $C_1, C_2 > 0$ such that the reverse Hölder inequality of Theorem 9.2.2 holds for the weights $w \in A_p$ and $w^{-\frac{1}{p-1}} \in A_{p'}$, that is,

$$\begin{aligned}
\left(\frac{1}{|Q|}\int_Q w(t)^{1+\gamma_1}\,dt\right)^{\frac{1}{1+\gamma_1}} &\le \frac{C_1}{|Q|}\int_Q w(t)\,dt, \\
\left(\frac{1}{|Q|}\int_Q w(t)^{-\frac{1}{p-1}(1+\gamma_2)}\,dt\right)^{\frac{1}{1+\gamma_2}} &\le \frac{C_2}{|Q|}\int_Q w(t)^{-\frac{1}{p-1}}\,dt.
\end{aligned}$$

Taking $\gamma = \min(\gamma_1, \gamma_2)$, both inequalities are satisfied with γ in the place of γ_1, γ_2. It follows that $w^{1+\gamma}$ is in A_p and satisfies

$$[w^{1+\gamma}]_{A_p} \le (C_1 C_2^{p-1})^{1+\gamma}[w]_{A_p}^{1+\gamma}. \tag{9.2.13}$$

This concludes the proof of the theorem. □

Corollary 9.2.6. *For any $1 < p < \infty$ and for every $w \in A_p$ there is a $q = q([w]_{A_p}, p, n)$ with $q < p$ such that $w \in A_q$. In other words, we have*

$$A_p = \bigcup_{q \in (1,p)} A_q.$$

Proof. Given $w \in A_p$, let γ, C_1, C_2 be as in the proof of Theorem 9.2.5. In view of the result in Exercise 9.1.3 (with $\delta = 1/(1+\gamma)$, if $w^{1+\gamma} \in A_p$ and

$$q = p\frac{1}{1+\gamma} + 1 - \frac{1}{1+\gamma} = \frac{p+\gamma}{1+\gamma},$$

then $w \in A_q$ and

$$[w]_{A_q} = [(w^{1+\gamma})^{\frac{1}{1+\gamma}}]_{A_q} \le [w^{1+\gamma}]_{A_p}^{\frac{1}{1+\gamma}} \le C_1 C_2^{p-1} [w]_{A_p},$$

where the last estimate comes from (9.2.13). Since $1 < q = \frac{p+\gamma}{1+\gamma} < p$, the required conclusion follows. Observe that the constants $C_1 C_2^{p-1}$, q, and $\frac{1}{\gamma}$ increase as $[w]_{A_p}$ increases. □

Another powerful consequence of the reverse Hölder property of A_p weights is the following characterization of all A_1 weights.

Theorem 9.2.7. *Let w be an A_1 weight. Then there exist $0 < \varepsilon < 1$, a nonnegative function k such that $k, k^{-1} \in L^\infty$, and a nonnegative locally integrable function f that satisfies $M(f) < \infty$ a.e. such that*

$$w(x) = k(x) M(f)(x)^{\varepsilon}. \tag{9.2.14}$$

Conversely, every weight w of the form (9.2.14) for some k, f as previously is in A_1 with

$$[w]_{A_1} \le \frac{C_n}{1-\varepsilon} \|k\|_{L^\infty} \|k^{-1}\|_{L^\infty},$$

where C_n is a universal dimensional constant.

Proof. In view of Theorem 9.2.2, there exist $0 < \gamma, C < \infty$ such that the reverse Hölder condition

$$\left(\frac{1}{|Q|}\int_Q w(t)^{1+\gamma}\,dt\right)^{\frac{1}{1+\gamma}} \le \frac{C}{|Q|}\int_Q w(t)\,dt \le C[w]_{A_1} w(x) \tag{9.2.15}$$

holds for all cubes Q for all x in $Q \setminus E_Q$, where E_Q is a null subset of Q. We set

$$\varepsilon = \frac{1}{1+\gamma} \qquad \text{and} \qquad f(x) = w(x)^{1+\gamma} = w(x)^{\frac{1}{\varepsilon}}.$$

Letting N be the union of E_Q over all Q with rational radii and centers in $\mathbf{Q}^n$, it follows from (9.2.15) that the uncentered Hardy–Littlewood maximal function $M_c(f)$ with respect to cubes satisfies

$$M_c(f)(x) \le C^{1+\gamma}[w]_{A_1}^{1+\gamma} f(x) \qquad \text{for } x \in \mathbf{R}^n \setminus N.$$

This implies that $M(f) \le C_n C^{1+\gamma}[w]_{A_1}^{1+\gamma} f$ a.e. for some constant C_n that depends only on the dimension. We now set

$$k(x) = \frac{f(x)^\varepsilon}{M(f)(x)^\varepsilon},$$

and we observe that $C^{-1}C_n^{-\varepsilon}[w]_{A_1}^{-1} \le k \le 1$ a.e.

It remains to prove the converse. Given a weight $w = kM(f)^\varepsilon$ in the form (9.2.14) and a cube Q, it suffices to show that

$$\frac{1}{|Q|}\int_Q M(f)(t)^\varepsilon\,dt \le \frac{C_n}{1-\varepsilon}M(f)^\varepsilon(x) \qquad \text{for almost all } x \in Q, \tag{9.2.16}$$

since the corresponding statement for $kM(f)^\varepsilon$ follows trivially from (9.2.16) using that $k, k^{-1} \in L^\infty$. To prove (9.2.16), we write

$$f = f\chi_{3Q} + f\chi_{(3Q)^c}.$$

Then

$$\frac{1}{|Q|}\int_Q M(f\chi_{3Q})(t)^\varepsilon\,dt \le \frac{C_n'}{1-\varepsilon}\left(\frac{1}{|Q|}\int_{\mathbf{R}^n}(f\chi_{3Q})(t)\,dt\right)^\varepsilon \tag{9.2.17}$$

in view of Kolmogorov's inequality (Exercise 2.1.5). But the last expression in (9.2.17) is at most a dimensional multiple of $M(f)(x)^\varepsilon$ for almost all $x \in Q$, which proves (9.2.16) when f is replaced by $f\chi_{3Q}$ on the left-hand side of the inequality. And for $f\chi_{(3Q)^c}$ we only need to notice that

$$M(f\chi_{(3Q)^c})(t) \le 2^n\mathcal{M}(f\chi_{(3Q)^c})(t) \le 2^n n^{\frac{n}{2}} M(f)(x)$$

for all x,t in Q, since any ball B centered at t that gives a nonzero average for $f\chi_{(3Q)^c}$ must have radius at least the side length of Q, and thus $\sqrt{n}B$ must also contain x. (Here $\mathcal{M}$ is the centered Hardy–Littlewood maximal operator introduced in Definition 2.1.1.) Hence (9.2.16) also holds when f is replaced by $f\chi_{(3Q)^c}$ on the left-hand side. Combining these two estimates and using the subbaditivity property $M(f_1+f_2)^\varepsilon \le M(f_1)^\varepsilon + M(f_2)^\varepsilon$, we obtain (9.2.16). □

Exercises

9.2.1. Let $w \in A_p$ for some $1 < p < \infty$ and let $1 \le q < \infty$. Prove that the sublinear operator

$$S(f) = \left(M(|f|^q w)w^{-1}\right)^{\frac{1}{q}}$$

is bounded on $L^{p'q}(w)$.

9.2.2. Let v be a real-valued locally integrable function on $\mathbf{R}^n$ and let $1 < p < \infty$. Prove that e^v is an A_p weight if and only if the following two conditions are satisfied for some constant $C < \infty$:

$$\sup_{Q \text{ cubes}} \frac{1}{|Q|} \int_Q e^{v(t)-v_Q}\, dt \leq C,$$
$$\sup_{Q \text{ cubes}} \frac{1}{|Q|} \int_Q e^{-(v(t)-v_Q)\frac{1}{p-1}}\, dt \leq C.$$

[*Hint:* If $e^v \in A_p$, use that

$$\frac{1}{|Q|} \int_Q e^{v(t)-v_Q}\, dt \leq \Big(\operatorname*{Avg}_Q e^{-\frac{v}{p-1}}\Big)^{p-1} \Big(\operatorname*{Avg}_Q e^{v}\Big)$$

and obtain a similar estimate for the second quantity.]

9.2.3. Let v be a real-valued locally integrable function on $\mathbf{R}^n$ and let $1 < p < \infty$.
(a) Use the result of Exercise 9.2.2 to show that e^v is in A_2 if and only if for some constant $C < \infty$, we have

$$\sup_{Q \text{ cubes}} \frac{1}{|Q|} \int_Q e^{|v(t)-v_Q|}\, dt \leq C.$$

Conclude that $\big\|\log \varphi\big\|_{BMO} \leq [\varphi]_{A_2}$; thus if $\varphi \in A_2$, then $\log \varphi \in BMO$.
(b) Use part (a) and Theorem 7.1.6 to prove the converse, namely that every BMO function is equal to a constant multiple of the logarithm of an A_2 weight.
(c) Prove that if φ is in A_p for some $1 < p < \infty$, then $\log \varphi$ is in BMO by showing that

$$\big\|\log \varphi\big\|_{BMO} \leq \begin{cases} [\varphi]_{A_p} & \text{when } 1 < p \leq 2, \\ (p-1)[\varphi]_{A_p}^{\frac{1}{p-1}} & \text{when } 2 < p < \infty\,. \end{cases}$$

[*Hint:* Use that $\varphi^{-\frac{1}{p-1}} \in A_{p'}$ when $p > 2$.]

9.2.4. Prove the following quantitative versions of Theorem 9.2.2 and Corollary 9.2.6.
(a) For any $1 \leq p < \infty$ and $B > 1$, there exist positive constants $C = C_1(n,p,B)$ and $\gamma = \gamma(n,p,B)$ such that for all $w \in A_p$ satisfying $[w]_{A_p} \leq B$, (9.2.4) holds for every cube Q.
(b) Given any $1 < p < \infty$ and $B > 1$ there exist constants $C = C_2(n,p,B)$ and $\delta = \delta(n,p,B)$ such that for all $w \in A_p$ we have

$$[w]_{A_p} \leq B \implies [w]_{A_{p-\delta}} \leq C.$$

9.2.5. Given a positive doubling measure μ on $\mathbf{R}^n$, define the characteristic constant $[w]_{A_p(\mu)}$ and the class $A_p(\mu)$ for $1 < p < \infty$.
(a) Show that statement (8) of Proposition 9.1.5 remains valid if Lebesgue measure is replaced by μ.
(b) Obtain as a consequence that if $w \in A_p(\mu)$, then for all cubes Q and all μ-measurable subsets A of Q we have

$$\left(\frac{\mu(A)}{\mu(Q)}\right)^p \le [w]_{A_p(\mu)} \frac{w(A)}{w(Q)}.$$

Conclude that if Lebesgue measure is replaced by μ in Lemma 9.2.1, then the lemma is valid for $w \in A_p(\mu)$.
(c) Use Corollary 9.2.4 to obtain that weights in $A_p(\mu)$ satisfy a reverse Hölder condition.
(d) Prove that given a weight $w \in A_p(\mu)$, there exists $1 < q < p$ [which depends on $[w]_{A_p(\mu)}$] such that $w \in A_q(\mu)$.

9.2.6. Let $1 < q < \infty$ and μ a positive measure on $\mathbf{R}^n$. We say that a positive function K on $\mathbf{R}^n$ satisfies a *reverse Hölder condition* of order q with respect to μ [symbolically $K \in RH_q(\mu)$] if

$$[K]_{RH_q(\mu)} = \sup_{Q \text{cubes in } \mathbf{R}^n} \frac{\left(\frac{1}{\mu(Q)}\int_Q K^q \, d\mu\right)^{\frac{1}{q}}}{\frac{1}{\mu(Q)}\int_Q K \, d\mu} < \infty.$$

For positive functions u, v on $\mathbf{R}^n$ and $1 < p < \infty$, show that

$$[vu^{-1}]_{RH_{p'}(u\,dx)} = [uv^{-1}]_{A_p(v\,dx)}^{\frac{1}{p}},$$

that is, vu^{-1} satisfies a reverse Hölder condition of order p' with respect to $u\,dx$ if and only if uv^{-1} is in $A_p(v\,dx)$. Conclude that

$$w \in RH_{p'}(dx) \iff w^{-1} \in A_p(w\,dx),$$
$$w \in A_p(dx) \iff w^{-1} \in RH_{p'}(w\,dx).$$

9.2.7. (*Gehring [145]*) Suppose that a positive function K on $\mathbf{R}^n$ lies in $RH_p(dx)$ for some $1 < p < \infty$. Show that there exists a $\delta > 0$ such that K lies in $RH_{p+\delta}(dx)$.
[*Hint:* By Exercise 9.2.6, $K \in RH_p(dx)$ is equivalent to the fact that $K^{-1} \in A_{p'}(K\,dx)$, and the index p' can be improved by Exercise 9.2.5(d).]

9.2.8. (a) Show that for any $w \in A_1$ and any cube Q in $\mathbf{R}^n$ and $a > 1$ we have

$$\operatorname*{ess.inf}_{Q} w \le a^n [w]_{A_1} \operatorname*{ess.inf}_{aQ} w.$$

(b) Prove that there is a constant C_n such that for all locally integrable functions f on $\mathbf{R}^n$ and all cubes Q in $\mathbf{R}^n$ we have

$$\operatorname*{ess.inf}_{Q} M(f) \le C_n \operatorname*{ess.inf}_{3Q} M(f),$$

and an analogous statement is valid for M_c.
[*Hint:* Part (a): Use (9.1.18). Part (b): Apply part (a) to $M(f)^{\frac{1}{2}}$, which is an A_1 weight in view of Theorem 9.2.7.]

9.2.9. *(Lerner, Pérez, and Ombrosi [211])* For a weight $w \in A_1(\mathbf{R}^n)$ define a quantity $r_w = 1 + \frac{1}{2^{n+1}[w]_{A_1}}$. Show that

$$M_c(w^{r_w})^{\frac{1}{r_w}} \leq 2\,[w]_{A_1}\, w\,.$$

[*Hint:* Fix a cube Q and consider the family $\mathscr{F}_Q$ of all cubes obtained by subdividing Q into a mesh of $(2^n)^m$ subcubes of side length $2^{-m}\ell(Q)$ for all $m = 1,2,\dots$. Define $M_Q^d(f)(x) = \sup_{R\in\mathscr{F}_Q, R\ni x} |R|^{-1}\int_R |f|\,dy$. Using Corollary 2.1.21, adapt the result of Exercise 2.1.4(b) to the maximal function M_Q^d; i.e., obtain $\int_{Q\cap\{M_Q^d(w)>\lambda\}} w(x)\,dx \leq 2^n\lambda|\{x\in Q:\, M_Q^d(w)(x)>\lambda\}|$ for $\lambda > w_Q = \frac{1}{|Q|}\int_Q w\,dt$. Multiply by $\lambda^{\delta-1}$ and integrate to obtain $\int_Q M_Q^d(w)^\delta w\,dx \leq (w_Q)^\delta \int_Q w\,dx + \frac{2^n\delta}{\delta+1}\int_Q M_Q^d(w)^{\delta+1}dx$. Setting $\delta = \frac{1}{2^{n+1}\|w\|_{A_1}}$, deduce that $\frac{1}{|Q|}\int_Q w^{\delta+1}dx \leq \frac{1}{|Q|}\int_Q M_Q^d(w)^\delta w\,dx \leq 2(w_Q)^{\delta+1}$.]

9.3 The A_∞ Condition

In this section we examine more closely the class of all A_p weights. It turns out that A_p weights possess properties that are p-independent but delicate enough to characterize them without reference to a specific value of p. The A_p classes increase as p increases, and it is only natural to consider their limit as $p \to \infty$. Not surprisingly, a condition obtained as a limit of the A_p conditions as $p \to \infty$ provides some unexpected but insightful characterizations of the class of all A_p weights.

9.3.1 The Class of A_∞ Weights

Let us start by recalling a simple consequence of Jensen's inequality:

$$\left(\int_X |h(t)|^q\,d\mu(t)\right)^{\frac{1}{q}} \geq \exp\left(\int_X \log|h(t)|\,d\mu(t)\right), \tag{9.3.1}$$

which holds for all measurable functions h on a probability space (X,μ) and all $0 < q < \infty$. See Exercise 1.1.3(b). Moreover, part (c) of the same exercise says that the limit of the expressions on the left in (9.3.1) as $q \to 0$ is equal to the expression on the right in (9.3.1).

We apply (9.3.1) to the function $h = w^{-1}$ for some weight w in A_p with $q = 1/(p-1)$. We obtain

$$\frac{w(Q)}{|Q|}\left(\frac{1}{|Q|}\int_Q w(t)^{-\frac{1}{p-1}}\,dt\right)^{p-1} \geq \frac{w(Q)}{|Q|}\exp\left(\frac{1}{|Q|}\int_Q \log w(t)^{-1}\,dt\right), \tag{9.3.2}$$

and the limit of the expressions on the left in (9.3.2) as $p \to \infty$ is equal to the expression on the right in (9.3.2). This observation provides the motivation for the following definition.

Definition 9.3.1. A weight w is called an A_∞ weight if

$$[w]_{A_\infty} = \sup_{Q \text{ cubes in } \mathbf{R}^n} \left\{ \left(\frac{1}{|Q|} \int_Q w(t)\, dt \right) \exp\left(\frac{1}{|Q|} \int_Q \log w(t)^{-1}\, dt \right) \right\} < \infty .$$

The quantity $[w]_{A_\infty}$ is called the A_∞ *characteristic constant* of w.

It follows from the previous definition and (9.3.2) that for all $1 \le p < \infty$ we have

$$[w]_{A_\infty} \le [w]_{A_p} .$$

This means that

$$\bigcup_{1 \le p < \infty} A_p \subseteq A_\infty , \tag{9.3.3}$$

but the remarkable thing is that equality actually holds in (9.3.3), a deep property that requires some work.

Before we examine this and other characterizations of A_∞ weights, we discuss some of their elementary properties.

Proposition 9.3.2. *Let $w \in A_\infty$. Then*

(1) $[\delta^\lambda(w)]_{A_\infty} = [w]_{A_\infty}$, where $\delta^\lambda(w)(x) = w(\lambda x_1, \dots, \lambda x_n)$ and $\lambda > 0$.

(2) $[\tau^z(w)]_{A_\infty} = [w]_{A_\infty}$, where $\tau^z(w)(x) = w(x - z)$, $z \in \mathbf{R}^n$.

(3) $[\lambda w]_{A_\infty} = [w]_{A_\infty}$ for all $\lambda > 0$.

(4) $[w]_{A_\infty} \ge 1$.

(5) The following is an equivalent characterization of the A_∞ characteristic constant of w:

$$[w]_{A_\infty} = \sup_{\substack{Q \text{ cubes} \\ \text{in } \mathbf{R}^n}} \sup_{\substack{\log|f| \in L^1(Q) \\ |f| > 0 \text{ a.e. on } Q}} \left\{ \frac{w(Q)}{\int_Q |f(t)| w(t)\, dt} \exp\left(\frac{1}{|Q|} \int_Q \log|f(t)|\, dt \right) \right\} .$$

(6) The measure $w(x)\,dx$ is doubling; precisely, for all $\lambda > 1$ and all cubes Q we have

$$w(\lambda Q) \le 2^{\lambda^n} [w]_{A_\infty}^{\lambda^n} w(Q) .$$

As usual, λQ here denotes the cube with the same center as Q and side length λ times that of Q.

We note that estimate (6) is not as good as $\lambda \to \infty$ but it can be substantially improved using the case $\lambda = 2$. We refer to Exercise 9.3.1 for an improvement.

Proof. Properties (1)–(3) are elementary, while property (4) is a consequence of Exercise 1.1.3(b). To show (5), first observe that by taking $f = w^{-1}$, the expression on the right in (5) is at least as big as $[w]_{A_\infty}$. Conversely, (9.3.1) gives

$$\exp\left(\frac{1}{|Q|}\int_Q \log\big(|f(t)|w(t)\big)\,dt\right) \le \frac{1}{|Q|}\int_Q |f(t)|w(t)\,dt\,,$$

which, after a simple algebraic manipulation, can be written as

$$\frac{w(Q)}{\int_Q |f|w\,dt}\exp\left(\frac{1}{|Q|}\int_Q \log|f|\,dt\right) \le \frac{w(Q)}{|Q|}\exp\left(-\frac{1}{|Q|}\int_Q \log|w|\,dt\right),$$

whenever f does not vanish almost everywhere on Q. Taking the supremum over all such f and all cubes Q in $\mathbf{R}^n$, we obtain that the expression on the right in (5) is at most $[w]_{A_\infty}$.

To prove the doubling property for A_∞ weights, we fix $\lambda > 1$ and we apply property (5) to the cube λQ in place of Q and to the function

$$f = \begin{cases} c & \text{on } Q, \\ 1 & \text{on } \mathbf{R}^n \setminus Q, \end{cases} \tag{9.3.4}$$

where c is chosen so that $c^{1/\lambda^n} = 2[w]_{A_\infty}$. We obtain

$$\frac{w(\lambda Q)}{w(\lambda Q \setminus Q) + c\,w(Q)}\exp\left(\frac{\log c}{\lambda^n}\right) \le [w]_{A_\infty}\,,$$

which implies (6) if we take into account the chosen value of c. □

9.3.2 Characterizations of A_∞ Weights

Having established some elementary properties of A_∞ weights, we now turn to some of their deeper properties, one of which is that every A_∞ weight lies in some A_p for $p < \infty$. It also turns out that A_∞ weights are characterized by the reverse Hölder property, which as we saw is a fundamental property of A_p weights. The following is the main theorem of this section.

Theorem 9.3.3. *Suppose that w is a weight. Then w is in A_∞ if and only if any one of the following conditions holds:*
(a) There exist $0 < \gamma, \delta < 1$ such that for all cubes Q in $\mathbf{R}^n$ we have

$$\big|\{x \in Q:\ w(x) \le \gamma\,\mathrm{Avg}_Q w\}\big| \le \delta\,|Q|\,.$$

(b) There exist $0 < \alpha, \beta < 1$ such that for all cubes Q and all measurable subsets A of Q we have

$$|A| \le \alpha\,|Q| \implies w(A) \le \beta\,w(Q)\,.$$

(c) The reverse Hölder condition holds for w, that is, there exist $0 < C_1, \varepsilon < \infty$ *such that for all cubes Q we have*

$$\left(\frac{1}{|Q|}\int_Q w(t)^{1+\varepsilon}\,dt\right)^{\frac{1}{1+\varepsilon}} \le \frac{C_1}{|Q|}\int_Q w(t)\,dt\,.$$

(d) There exist $0 < C_2, \varepsilon_0 < \infty$ *such that for all cubes Q and all measurable subsets A of Q we have*

$$\frac{w(A)}{w(Q)} \le C_2\left(\frac{|A|}{|Q|}\right)^{\varepsilon_0}.$$

(e) There exist $0 < \alpha', \beta' < 1$ *such that for all cubes Q and all measurable subsets A of Q we have*

$$w(A) < \alpha' w(Q) \implies |A| < \beta' |Q|\,.$$

(f) There exist $p, C_3 < \infty$ *such that* $[w]_{A_p} \le C_3$. *In other words, w lies in* A_p *for some* $p \in [1,\infty)$.

All the constants $C_1, C_2, C_3, \alpha, \beta, \gamma, \delta, \alpha', \beta', \varepsilon, \varepsilon_0$, *and p in (a)–(f) depend only on the dimension n and on* $[w]_{A_\infty}$. *Moreover, if any of the statements in (a)–(f) is valid, then so is any other statement in (a)–(f) with constants that depend only on the dimension n and the constants that appear in the assumed statement.*

Proof. The proof follows from the sequence of implications

$$w \in A_\infty \implies (a) \implies (b) \implies (c) \implies (d) \implies (e) \implies (f) \implies w \in A_\infty\,.$$

At each step we keep track of the way the constants depend on the constants of the previous step. This is needed to validate the last assertion of the theorem.

$w \in A_\infty \implies (a)$

Fix a cube Q. Since multiplication of an A_∞ weight with a positive scalar does not alter its A_∞ characteristic, we may assume that $\int_Q \log w(t)\,dt = 0$. This implies that $\mathrm{Avg}_Q\, w \le [w]_{A_\infty}$. Then we have

$$\begin{aligned}
\left|\{x\in Q:\ w(x)\le \gamma \operatorname*{Avg}_Q w\}\right| &\le \left|\{x\in Q:\ w(x)\le \gamma [w]_{A_\infty}\}\right| \\
&= \left|\{x\in Q:\ \log(1+w(x)^{-1})\ge \log(1+(\gamma[w]_{A_\infty})^{-1})\}\right| \\
&\le \frac{1}{\log(1+(\gamma[w]_{A_\infty})^{-1})}\int_Q \log\frac{1+w(t)}{w(t)}\,dt \\
&= \frac{1}{\log(1+(\gamma[w]_{A_\infty})^{-1})}\int_Q \log(1+w(t))\,dt \\
&\le \frac{1}{\log(1+(\gamma[w]_{A_\infty})^{-1})}\int_Q w(t)\,dt \\
&\le \frac{[w]_{A_\infty}|Q|}{\log(1+(\gamma[w]_{A_\infty})^{-1})} \\
&= \frac{1}{2}|Q|,
\end{aligned}$$

which proves (a) with $\gamma = [w]_{A_\infty}^{-1}(e^{2[w]_{A_\infty}} - 1)^{-1}$ and $\delta = \frac{1}{2}$.

$(a) \implies (b)$

Let Q be fixed and let A be a subset of Q with $w(A) > \beta w(Q)$ for some β to be chosen later. Setting $S = Q \setminus A$, we have $w(S) < (1-\beta)w(Q)$. We write $S = S_1 \cup S_2$, where

$$S_1 = \{x \in S :\ w(x) > \gamma \operatorname{Avg}_Q w\} \quad \text{and} \quad S_2 = \{x \in S :\ w(x) \le \gamma \operatorname{Avg}_Q w\}.$$

For S_2 we have $|S_2| \le \delta |Q|$ by assumption (a). For S_1 we use Chebyshev's inequality to obtain

$$|S_1| \le \frac{1}{\gamma \operatorname{Avg}_Q w} \int_S w(t)\,dt = \frac{|Q|}{\gamma}\frac{w(S)}{w(Q)} \le \frac{1-\beta}{\gamma}|Q|.$$

Adding the estimates for $|S_1|$ and $|S_2|$, we obtain

$$|S| \le |S_1| + |S_2| \le \frac{1-\beta}{\gamma}|Q| + \delta|Q| = \left(\delta + \frac{1-\beta}{\gamma}\right)|Q|.$$

Choosing numbers α, β in $(0,1)$ such that $\delta + \frac{1-\beta}{\gamma} = 1 - \alpha$, for example $\alpha = \frac{1-\delta}{2}$ and $\beta = 1 - \frac{(1-\delta)\gamma}{2}$, we obtain $|S| \le (1-\alpha)|Q|$, that is, $|A| > \alpha|Q|$.

$(b) \implies (c)$

This was proved in Corollary 9.2.4. To keep track of the constants, we note that the choices

$$\varepsilon = \frac{-\frac{1}{2}\log\beta}{\log 2^n - \log\alpha} \quad \text{and} \quad C_1 = 1 + \frac{(2^n\alpha^{-1})^\varepsilon}{1-(2^n\alpha^{-1})^\varepsilon\beta}$$

as given in (9.2.6) and (9.2.7) serve our purposes.

$(c) \implies (d)$

We apply first Hölder's inequality with exponents $1+\varepsilon$ and $(1+\varepsilon)/\varepsilon$ and then the reverse Hölder estimate to obtain

$$\begin{aligned}
\int_A w(x)\,dx &\le \left(\int_A w(x)^{1+\varepsilon}\,dx\right)^{\frac{1}{1+\varepsilon}} |A|^{\frac{\varepsilon}{1+\varepsilon}} \\
&\le \left(\frac{1}{|Q|}\int_Q w(x)^{1+\varepsilon}\,dx\right)^{\frac{1}{1+\varepsilon}} |Q|^{\frac{1}{1+\varepsilon}} |A|^{\frac{\varepsilon}{1+\varepsilon}} \\
&\le \frac{C_1}{|Q|}\int_Q w(x)\,dx\, |Q|^{\frac{1}{1+\varepsilon}} |A|^{\frac{\varepsilon}{1+\varepsilon}},
\end{aligned}$$

which gives

$$\frac{w(A)}{w(Q)} \le C_1\left(\frac{|A|}{|Q|}\right)^{\frac{\varepsilon}{1+\varepsilon}}.$$

This proves (d) with $\varepsilon_0 = \frac{\varepsilon}{1+\varepsilon}$ and $C_2 = C_1$.

$(d) \implies (e)$

Pick an $0 < \alpha'' < 1$ small enough that $\beta'' = C_2(\alpha'')^{\varepsilon_0} < 1$. It follows from (d) that

$$|A| < \alpha''|Q| \implies w(A) < \beta'' w(Q) \tag{9.3.5}$$

for all cubes Q and all A measurable subsets of Q. Replacing A by $Q \setminus A$, the implication in (9.3.5) can be equivalently written as

$$|A| \ge (1-\alpha'')|Q| \implies w(A) \ge (1-\beta'')w(Q).$$

In other words, for measurable subsets A of Q we have

$$w(A) < (1-\beta'')w(Q) \implies |A| < (1-\alpha'')|Q|, \tag{9.3.6}$$

which is the statement in (e) if we set $\alpha' = (1-\beta'')$ and $\beta' = 1-\alpha''$. Note that (9.3.5) and (9.3.6) are indeed equivalent.

$(e) \implies (f)$

We begin by examining condition (e), which can be written as

$$\int_A w(t)\,dt \le \alpha' \int_Q w(t)\,dt \implies \int_A w(t)^{-1} w(t)\,dt \le \beta' \int_Q w(t)^{-1} w(t)\,dt,$$

or, equivalently, as

$$\mu(A) \le \alpha' \mu(Q) \implies \int_A w(t)^{-1}\,d\mu(t) \le \beta' \int_Q w(t)^{-1}\,d\mu(t)$$

after defining the measure $d\mu(t) = w(t)\,dt$. As we have already seen, the assertions in (9.3.5) and (9.3.6) are equivalent. Therefore, we may use Exercise 9.3.2 to deduce that the measure μ is doubling [i.e., it satisfies property (9.2.9) for some constant $C_n = C_n(\alpha', \beta')$] and hence the hypotheses of Corollary 9.2.4 are satisfied. We conclude that the weight w^{-1} satisfies a reverse Hölder estimate with respect to the measure μ, that is, if γ, C are defined as in (9.2.11) and (9.2.12) [in which α is replaced by α', β by β', and C_n is the doubling constant of $w(x)\,dx$], then we have

$$\left(\frac{1}{\mu(Q)} \int_Q w(t)^{-1-\gamma}\,d\mu(t)\right)^{\frac{1}{1+\gamma}} \le \frac{C}{\mu(Q)} \int_Q w(t)^{-1}\,d\mu(t) \tag{9.3.7}$$

for all cubes Q in $\mathbf{R}^n$. Setting $p = 1 + \frac{1}{\gamma}$ and raising to the pth power, we can rewrite (9.3.7) as the A_p condition for w. We can therefore take $C_3 = C^p$ to conclude the proof of (f).

$(f) \implies w \in A_\infty$

This is trivial, since $[w]_{A_\infty} \le [w]_{A_p}$. □

An immediate consequence of the preceding theorem is the following result relating A_∞ to A_p.

Corollary 9.3.4. *The following equality is valid:*

$$A_\infty = \bigcup_{1\le p<\infty} A_p.$$

Exercises

9.3.1. (a) Show that property (6) in Proposition 9.3.2 can be improved to

$$w(\lambda Q) \le \min_{\varepsilon>0} \frac{(1+\varepsilon)^{\lambda^n}[w]_{A_\infty}^{\lambda^n} - 1}{\varepsilon}\, w(Q).$$

(b) Take $\lambda = 2$ in property (6) of Proposition 9.3.2 and iterate the estimate obtained to deduce that

$$w(\lambda Q) \le (2\lambda)^{2^n(1+\log_2[w]_{A_\infty})} w(Q).$$

[*Hint:* Part (a): Take c in (9.3.4) such that $c^{1/\lambda^n} = (1+\varepsilon)[w]_{A_\infty}$.]

9.3.2. Suppose that μ is a positive Borel measure on $\mathbf{R}^n$ with the property that for all cubes Q and all measurable subsets A of Q we have

$$|A| < \alpha|Q| \implies \mu(A) < \beta\mu(Q)$$

for some fixed $0 < \alpha, \beta < 1$. Show that μ is doubling [i.e., it satisfies (9.2.9)].
[*Hint:* Choose $\lambda = (1+\alpha/2)^{1/n}$ such that $|\lambda Q \setminus Q| < \alpha|Q|$. Write λQ as the union of c_n shifts Q_s of Q, for some dimensional constant c_n. Then $\mu((\lambda Q\setminus Q)\cap Q_s)| < \beta\mu(Q_s) \le \beta\mu(Q) + \beta\mu((\lambda Q\setminus Q)\cap Q_s)$. Conclude that $\mu(\lambda Q) \le c_n \frac{\beta}{1-\beta}\mu(Q)$ and from this derive an estimate for $\mu(3Q)$.]

9.3.3. Prove that a weight w is in A_p if and only if both w and $w^{-\frac{1}{p-1}}$ are in A_∞.
[*Hint:* You may want to use the result of Exercise 9.2.2.]

9.3.4. (*Stein [291]*) Prove that if $P(x)$ is a polynomial of degree k in $\mathbf{R}^n$, then $\log|P(x)|$ is in *BMO* with norm depending only on k and n and not on the coefficients of the polynomial.
[*Hint:* Use that all norms on the finite-dimensional space of polynomials of degree at most k are equivalent to show that $|P(x)|$ satisfies a reverse Hölder inequality. Therefore, $|P(x)|$ is an A_∞ weight and thus Exercise 9.2.3(c) is applicable.]

9.3.5. Show that the product of two A_1 weights may not be an A_∞ weight.

9.3.6. Let g be in $L^p(w)$ for some $1 \le p \le \infty$ and $w \in A_p$. Prove that $g \in L^1_{\mathrm{loc}}(\mathbf{R}^n)$.
[*Hint:* Let B be a ball. In the case $p < \infty$, write $\int_B |g|\,dx = \int_B (|g|w^{-\frac{1}{p}})w^{\frac{1}{p}}\,dx$ and apply Hölder's inequality. In the case $p = \infty$, use that $w \in A_{p_0}$ for some $p_0 < \infty$.]

9.3.7. (*Pérez [262]*) Show that a weight w lies in A_∞ if and only if there exist $\gamma, C > 0$ such that for all cubes Q we have

$$w\big(\{x\in Q:\, w(x)>\lambda\}\big) \le C\lambda\big|\{x\in Q:\, w(x)>\gamma\lambda\}\big|$$

for all $\lambda > \mathrm{Avg}_Q\, w$.
[*Hint:* The displayed condition easily implies that

$$\frac{1}{|Q|}\int_Q w_k^{1+\varepsilon}\,dx \le \Big(\frac{w(Q)}{|Q|}\Big)^{\varepsilon+1} + \frac{C'\delta}{\gamma^{1+\varepsilon}}\frac{1}{|Q|}\int_Q w_k^{1+\varepsilon}\,dx,$$

where $k>0$, $w_k=\min(w,k)$ and $\delta=\varepsilon/(1+\varepsilon)$. Take $\varepsilon>0$ small enough to obtain the reverse Hölder condition (c) in Theorem 9.3.3 for w_k. Let $k\to\infty$ to obtain the same conclusion for w. Conversely, find constants $\gamma,\delta\in(0,1)$ as in condition (a) of Theorem 9.3.3 and for $\lambda > \mathrm{Avg}_Q\, w$ write the set $\{w>\lambda\}\cap Q$ as a union of maximal dyadic cubes Q_j such that $\lambda < \mathrm{Avg}_{Q_j} w \le 2^n\lambda$ for all j. Then $w(Q_j)\le 2^n\lambda|Q_j| \le \frac{2^n\lambda}{1-\delta}|Q_j\cap\{w>\gamma\lambda\}|$ and the required conclusion follows by summing on j.]

9.4 Weighted Norm Inequalities for Singular Integrals

We now address a topic of great interest in the theory of singular integrals, their boundedness properties on weighted L^p spaces. It turns out that a certain amount of regularity must be imposed on the kernels of these operators to obtain the aforementioned weighted estimates.

9.4.1 A Review of Singular Integrals

We begin by recalling some definitions from Chapter 8.

Definition 9.4.1. Let $0<\delta, A<\infty$. A function $K(x,y)$ defined for $x,y\in\mathbf{R}^n$ with $x\neq y$ is called a standard kernel (with constants δ and A) if

$$|K(x,y)| \le \frac{A}{|x-y|^n}, \qquad x\neq y, \tag{9.4.1}$$

and whenever $|x-x'|\le\frac{1}{2}\max\big(|x-y|,|x'-y|\big)$ we have

$$|K(x,y)-K(x',y)| \le \frac{A|x-x'|^\delta}{(|x-y|+|x'-y|)^{n+\delta}} \tag{9.4.2}$$

and also when $|y-y'|\le\frac{1}{2}\max\big(|x-y|,|x-y'|\big)$ we have

$$|K(x,y)-K(x,y')| \le \frac{A|y-y'|^{\delta}}{(|x-y|+|x-y'|)^{n+\delta}}. \tag{9.4.3}$$

The class of all kernels that satisfy (9.4.1), (9.4.2), and (9.4.3) are denoted by $SK(\delta,A)$.

Definition 9.4.2. Let $0<\delta,A<\infty$ and K in $SK(\delta,A)$. A Calderón–Zygmund operator associated with K is a linear operator T defined on $\mathscr{S}(\mathbf{R}^n)$ that admits a bounded extension on $L^2(\mathbf{R}^n)$,

$$\big\|T(f)\big\|_{L^2} \le B\big\|f\big\|_{L^2}, \tag{9.4.4}$$

and that satisfies

$$T(f)(x) = \int_{\mathbf{R}^n} K(x,y)f(y)\,dy \tag{9.4.5}$$

for all $f\in\mathscr{C}_0^\infty$ and x not in the support of f. The class of all Calderón–Zygmund operators associated with kernels in $SK(\delta,A)$ that are bounded on L^2 with norm at most B is denoted by $CZO(\delta,A,B)$. Given a Calderón–Zygmund operator T in $CZO(\delta,A,B)$, we define the truncated operator $T^{(\varepsilon)}$ as

$$T^{(\varepsilon)}(f)(x) = \int_{|x-y|>\varepsilon} K(x,y)\,f(y)\,dy$$

and the maximal operator associated with T as follows:

$$T^{(*)}(f)(x) = \sup_{\varepsilon>0}\big|T^{(\varepsilon)}(f)(x)\big|.$$

We note that if T is in $CZO(\delta,A,B)$, then $T^{(\varepsilon)}f$ and $T^{(*)}(f)$ is well defined for all f in $\bigcup_{1\le p<\infty}L^p(\mathbf{R}^n)$. It is also well defined whenever f is locally integrable and satisfies $\int_{|x-y|\ge\varepsilon}|f(y)|\,|x-y|^{-n}dy<\infty$ for all $x\in\mathbf{R}^n$ and $\varepsilon>0$.

9.4.2 A Good Lambda Estimate for Singular Integrals

The following theorem is the main result of this section.

Theorem 9.4.3. *Let $w\in A_\infty$ and T in $CZO(\delta,A,B)$. Then there exist positive constants $C_0=C_0(n,[w]_{A_\infty})$, $\varepsilon_0=\varepsilon_0(n,[w]_{A_\infty})$, and $\gamma_0=\gamma_0(n,\delta,A)$ such that for all $0<\gamma<\gamma_0$ we have*

$$w\big(\{T^{(*)}(f)>3\lambda\}\cap\{M(f)\le\gamma\lambda\}\big)\le C_0\gamma^{\varepsilon_0}(A+B)^{\varepsilon_0}w\big(\{T^{(*)}(f)>\lambda\}\big), \tag{9.4.6}$$

for all locally integrable functions f for which $\int_{|x-y|\ge\varepsilon}|f(y)|\,|x-y|^{-n}dy<\infty$ for all $x\in\mathbf{R}^n$ and $\varepsilon>0$. Here M denotes the Hardy–Littlewood maximal operator.

Proof. We write the open set

$$\Omega=\{T^{(*)}(f)>\lambda\}=\bigcup_j Q_j,$$

where Q_j are the Whitney cubes of Proposition 7.3.4. We set

$$Q_j^* = 10\sqrt{n}\,Q_j\,,$$
$$Q_j^{**} = 10\sqrt{n}\,Q_j^*\,,$$

where aQ denotes the cube with the same center as Q whose side length is $a\ell(Q)$, where $\ell(Q)$ is the side length of Q. We note that in view of Proposition 7.3.4, the distance from Q_j to Ω^c is at most $4\sqrt{n}\,\ell(Q_j)$. But the distance from Q_j to the boundary of Q_j^* is $(5\sqrt{n}-\frac{1}{2})\,\ell(Q_j)$, which is bigger than $4\sqrt{n}\,\ell(Q_j)$. Therefore, Q_j^* must meet Ω^c and for every cube Q_j we fix a point $y_j \in \Omega^c \cap Q_j^*$. See Figure 9.1.

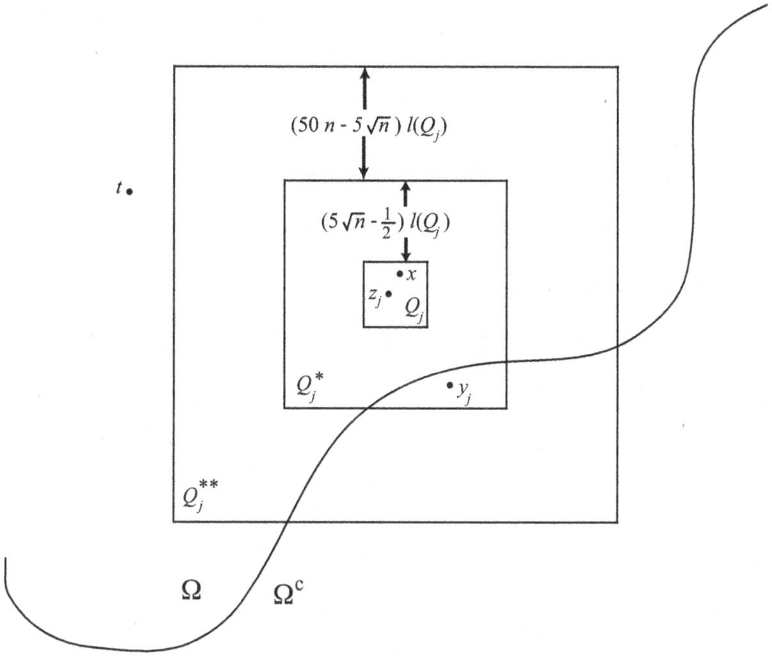

Fig. 9.1 A picture of the proof.

We also fix f in $\bigcup_{1\le p<\infty} L^p(\mathbf{R}^n)$, and for each j we write

$$f = f_0^j + f_\infty^j\,,$$

where f_0^j is the part of f near Q_j and f_∞^j is the part of f away from Q_j defined as follows:

$$f_0^j = f\chi_{Q_j^{**}}\,,$$
$$f_\infty^j = f\chi_{(Q_j^{**})^c}\,.$$

We now claim that the following estimate is true:

$$\big|Q_j \cap \{T^{(*)}(f) > 3\lambda\} \cap \{M(f) \le \gamma\lambda\}\big| \le C_n \gamma (A+B)\big|Q_j\big|. \tag{9.4.7}$$

Once the validity of (9.4.7) is established, we apply Theorem 9.3.3 part (d) to obtain constants $\varepsilon_0, C_2 > 0$ (which depend on $[w]_{A_\infty}$ and the dimension n) such that

$$w\big(Q_j \cap \{T^{(*)}(f) > 3\lambda\} \cap \{M(f) \le \gamma\lambda\}\big) \le C_2\,(C_n)^{\varepsilon_0}\,\gamma^{\varepsilon_0}\,(A+B)^{\varepsilon_0}\,w(Q_j).$$

Then a simple summation on j gives (9.4.6) with $C_0 = C_2 (C_n)^{\varepsilon_0}$, and recall that C_2 and ε_0 depend on n and $[w]_{A_\infty}$.

In proving estimate (9.4.7), we may assume that for each cube Q_j there exists a $z_j \in Q_j$ such that $M(f)(z_j) \le \gamma\lambda$; otherwise, the set on the left in (9.4.7) is empty.

We invoke Theorem 8.2.3, which states that $T^{(*)}$ maps $L^1(\mathbf{R}^n)$ to $L^{1,\infty}(\mathbf{R}^n)$ with norm at most $C(n)(A+B)$. We have the estimate

$$\big|Q_j \cap \{T^{(*)}(f) > 3\lambda\} \cap \{M(f) \le \gamma\lambda\}\big| \le I_0^\lambda + I_\infty^\lambda, \tag{9.4.8}$$

where

$$\begin{aligned} I_0^\lambda &= \big|Q_j \cap \{T^{(*)}(f_0^j) > \lambda\} \cap \{M(f) \le \gamma\lambda\}\big|, \\ I_\infty^\lambda &= \big|Q_j \cap \{T^{(*)}(f_\infty^j) > 2\lambda\} \cap \{M(f) \le \gamma\lambda\}\big|. \end{aligned}$$

To control I_0^λ we note that f_0^j is in $L^1(\mathbf{R}^n)$ and we argue as follows:

$$\begin{aligned} I_0^\lambda &\le \big|\{T^{(*)}(f_0^j) > \lambda\}\big| \\ &\le \frac{\big\|T^{(*)}\big\|_{L^1\to L^{1,\infty}}}{\lambda} \int_{\mathbf{R}^n} |f_0^j(x)|\,dx \\ &\le C(n)\,(A+B)\,\frac{|Q_j^{**}|}{\lambda}\,\frac{1}{|Q_j^{**}|}\int_{Q_j^{**}} |f(x)|\,dx \\ &\le C(n)\,(A+B)\,\frac{|Q_j^{**}|}{\lambda} M_c(f)(z_j) \\ &\le \widetilde{C}(n)\,(A+B)\,\frac{|Q_j^{**}|}{\lambda} M(f)(z_j) \\ &\le \widetilde{C}(n)\,(A+B)\,\frac{|Q_j^{**}|}{\lambda}\,\lambda\gamma \\ &= C_n\,(A+B)\,\gamma|Q_j|. \end{aligned} \tag{9.4.9}$$

Next we claim that $I_\infty^\lambda = 0$ if we take γ sufficiently small. We first show that for all $x \in Q_j$ we have

$$\sup_{\varepsilon>0}\big|T^{(\varepsilon)}(f_\infty^j)(x) - T^{(\varepsilon)}(f_\infty^j)(y_j)\big| \le C_{n,\delta}^{(1)}\,A\,M(f)(z_j). \tag{9.4.10}$$

Indeed, let us fix an $\varepsilon > 0$. We have

$$\begin{aligned}\left|T^{(\varepsilon)}(f_\infty^j)(x) - T^{(\varepsilon)}(f_\infty^j)(y_j)\right| &= \left| \int_{|t-x|>\varepsilon} K(x,t) f_\infty^j(t)\,dt - \int_{|t-y_j|>\varepsilon} K(y_j,t) f_\infty^j(t)\,dt \right| \\ &\leq L_1 + L_2 + L_3 ,\end{aligned}$$

where

$$\begin{aligned} L_1 &= \left| \int_{|t-y_j|>\varepsilon} \big[K(x,t) - K(y_j,t)\big] f_\infty^j(t)\,dt \right| , \\ L_2 &= \left| \int_{\substack{|t-x|>\varepsilon \\ |t-y_j|\leq\varepsilon}} K(x,t) f_\infty^j(t)\,dt \right| , \\ L_3 &= \left| \int_{\substack{|t-x|\leq\varepsilon \\ |t-y_j|>\varepsilon}} K(x,t) f_\infty^j(t)\,dt \right| , \end{aligned}$$

in view of identity (4.4.6).

We now make a couple of observations. For $t \notin Q_j^{**}$, $x, z_j \in Q_j$, and $y_j \in Q_j^*$ we have

$$\frac{3}{4} \leq \frac{|t-x|}{|t-y_j|} \leq \frac{5}{4}, \qquad\qquad \frac{48}{49} \leq \frac{|t-x|}{|t-z_j|} \leq \frac{50}{49}. \tag{9.4.11}$$

Indeed,

$$|t-y_j| \geq (50n - 5\sqrt{n})\,\ell(Q_j) \geq 44\,n\,\ell(Q_j)$$

and

$$|x-y_j| \leq \frac{1}{2}\sqrt{n}\,\ell(Q_j) + \sqrt{n}\,10\sqrt{n}\,\ell(Q_j) \leq 11\,n\,\ell(Q_j) \leq \frac{1}{4}\,|t-y_j| .$$

Using this estimate and the inequalities

$$\frac{3}{4}\,|t-y_j| \leq |t-y_j| - |x-y_j| \leq |t-x| \leq |t-y_j| + |x-y_j| \leq \frac{5}{4}\,|t-y_j| ,$$

we obtain the first estimate in (9.4.11). Likewise, we have

$$|x-z_j| \leq \sqrt{n}\,\ell(Q_j) \leq n\,\ell(Q_j)$$

and

$$|t-z_j| \geq (50n - \tfrac{1}{2})\ell(Q_j) \geq 49\,n\,\ell(Q_j) ,$$

and these give

$$\frac{48}{49}\,|t-z_j| \leq |t-z_j| - |x-z_j| \leq |t-x| \leq |t-z_j| + |x-z_j| \leq \frac{50}{49}\,|t-z_j| ,$$

yielding the second estimate in (9.4.11).

Since $|x-y_j| \le \frac{1}{2}|t-y_j| \le \frac{1}{2}\max\big(|t-x|,|t-y_j|\big)$, we have

$$|K(x,t)-K(y_j,t)| \le \frac{A|x-y_j|^\delta}{(|t-x|+|t-y_j|)^{n+\delta}} \le C'_{n,\delta}A\,\frac{\ell(Q_j)^\delta}{|t-z_j|^{n+\delta}}\,;$$

hence, we obtain

$$L_1 \le \int\limits_{|t-z_j|\ge 49n\ell(Q_j)} C'_{n,\delta}A\,\frac{\ell(Q_j)^\delta}{|t-z_j|^{n+\delta}}|f(t)|\,dt \le C''_{n,\delta}AM(f)(z_j)$$

using Theorem 2.1.10. Using (9.4.11) we deduce

$$L_2 \le \int\limits_{|t-z_j|\le\frac{5}{4}\cdot\frac{49}{48}\varepsilon} \frac{A}{|x-t|^n}\chi_{|t-x|\ge\varepsilon}\,|f^j_\infty(t)|\,dt \le C'_nAM(f)(z_j)\,.$$

Again using (9.4.11), we obtain

$$L_3 \le \int\limits_{|t-z_j|\le\frac{49}{48}\varepsilon} \frac{A}{|x-t|^n}\chi_{|t-x|\ge\frac{3}{4}\varepsilon}\,|f^j_\infty(t)|\,dt \le C''_nAM(f)(z_j)\,.$$

This proves (9.4.10) with constant $C^{(1)}_{n,\delta} = C''_{n,\delta}+C'_n+C''_n$.

Having established (9.4.10), we next claim that

$$\sup_{\varepsilon>0}\big|T^{(\varepsilon)}(f^j_\infty)(y_j)\big| \le T^{(*)}(f)(y_j)+C^{(2)}_nAM(f)(z_j)\,. \tag{9.4.12}$$

To prove (9.4.12) we fix a cube Q_j and $\varepsilon>0$. We let R_j be the smallest number such that

$$Q^{**}_j \subseteq B(y_j,R_j)\,.$$

See Figure 9.2. We consider the following two cases.

Case (1): $\varepsilon \ge R_j$. Since $Q^{**}_j \subseteq B(y_j,\varepsilon)$, we have $B(y_j,\varepsilon)^c \subseteq (Q^{**}_j)^c$ and therefore

$$T^{(\varepsilon)}(f^j_\infty)(y_j) = T^{(\varepsilon)}(f)(y_j)\,,$$

so (9.4.12) holds easily in this case.

Case (2): $0<\varepsilon<R_j$. Note that if $t\in(Q^{**}_j)^c$, then $|t-y_j|\ge 40n\ell(Q_j)$. On the other hand, $R_j \le \operatorname{diam}(Q^{**}_j) = 100n^{\frac{3}{2}}\,\ell(Q_j)$. This implies that

$$R_j \le \tfrac{5\sqrt{n}}{2}|t-y_j|\,, \qquad \text{when} \quad t\in(Q^{**}_j)^c\,.$$

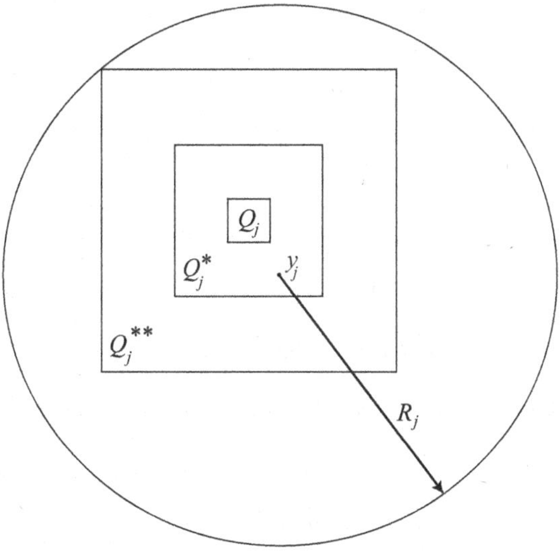

Fig. 9.2 The ball $B(y_j,R_j)$.

Notice also that in this case we have $B(y_j,R_j)^c \subseteq (Q_j^{**})^c$, hence

$$T^{(R_j)}(f_\infty^j)(y_j) = T^{(R_j)}(f)(y_j).$$

Therefore, we have

$$\begin{aligned}
\big|T^{(\varepsilon)}(f_\infty^j)(y_j)\big| &\le \big|T^{(\varepsilon)}(f_\infty^j)(y_j) - T^{(R_j)}(f_\infty^j)(y_j)\big| + \big|T^{(R_j)}(f)(y_j)\big| \\
&\le \int_{\varepsilon\le|y_j-t|\le R_j} |K(y_j,t)|\,|f_\infty^j(t)|\,dt + T^{(*)}(f)(y_j) \\
&\le \int_{\frac{2}{5\sqrt{n}}R_j\le|y_j-t|\le R_j} |K(y_j,t)||f_\infty^j(t)|\,dt + T^{(*)}(f)(y_j) \\
&\le \frac{A(\frac{2}{5\sqrt{n}})^{-n}}{R_j^n} \int_{|z_j-t|\le\frac{5}{4}\cdot\frac{49}{48}R_j} |f(t)|\,dt + T^{(*)}(f)(y_j) \\
&\le C_n^{(2)} A M(f)(z_j) + T^{(*)}(f)(y_j),
\end{aligned}$$

where in the penultimate estimate we used (9.4.11). The proof of (9.4.12) follows with the required bound $C_n^{(2)}A$.

Combining (9.4.10) and (9.4.12), we obtain

$$T^{(*)}(f_\infty^j)(x) \le T^{(*)}(f)(y_j) + (C_{n,\delta}^{(1)} + C_n^{(2)})\,A\,M(f)(z_j)\,.$$

Recalling that $y_j \notin \Omega$ and that $M(f)(z_j) \le \gamma\lambda$, we deduce

$$T^{(*)}(f_\infty^j)(x) \le \lambda + (C_{n,\delta}^{(1)} + C_n^{(2)})\,A\,\gamma\lambda\,.$$

Setting $\gamma_0 = \left(C_{n,\delta}^{(1)} + C_n^{(2)}\right)^{-1} A^{-1}$, for $0 < \gamma < \gamma_0$, we have that the set

$$Q_j \cap \{T^{(*)}(f_\infty^j) > 2\lambda\} \cap \{M(f) \le \gamma\lambda\}$$

is empty. This shows that the quantity I_∞^γ vanishes if γ is smaller than γ_0. Returning to (9.4.8) and using the estimate (9.4.9) proved earlier, we conclude the proof of (9.4.7), which, as indicated earlier, implies the theorem. □

Remark 9.4.4. We observe that for any $\delta > 0$, estimate (9.4.6) also holds for the operator

$$T_\delta^{(*)}(f)(x) = \sup_{\varepsilon \ge \delta} |T^{(\varepsilon)}(f)(x)|$$

with the same constant (which is independent of δ).

To see the validity of (9.4.6) for $T_\delta^{(*)}$, it suffices to prove

$$\left|T_\delta^{(*)}(f_\infty^j)(y_j)\right| \le T_\delta^{(*)}(f)(y_j) + C_n^{(2)} A\,M(f)(z_j)\,, \tag{9.4.13}$$

which is a version of (9.4.12) with $T^{(*)}$ replaced by $T_\delta^{(*)}$. The following cases arise:
Case (1′): $R_j \le \delta \le \varepsilon$ or $\delta \le R_j \le \varepsilon$. Here, as in Case (1) we have

$$|T^{(\varepsilon)}(f_\infty^j)(y_j)| = |T^{(\varepsilon)}(f)(y_j)| \le T_\delta^{(*)}(f)(y_j)\,.$$

Case (2′): $\delta \le \varepsilon < R_j$. As in Case (2) we have $T^{(R_j)}(f_\infty^j)(y_j) = T^{(R_j)}(f)(y_j)$, thus

$$\left|T^{(\varepsilon)}(f_\infty^j)(y_j)\right| \le \left|T^{(\varepsilon)}(f_\infty^j)(y_j) - T^{(R_j)}(f_\infty^j)(y_j)\right| + \left|T^{(R_j)}(f)(y_j)\right|.$$

As in the proof of Case (2), we bound the first term on the right of the last displayed expression by $C_n^{(2)} A\,M(f)(z_j)$ while the second term is at most $T_\delta^{(*)}(f)(y_j)$.

9.4.3 Consequences of the Good Lambda Estimate

Having obtained the important good lambda weighted estimate for singular integrals, we now pass to some of its consequences. We begin with the following lemma:

Lemma 9.4.5. *Let* $1 \le p < \infty$, $\delta > 0$, $w \in A_p$, $x \in \mathbf{R}^n$, *and* $f \in L^p(w)$. *Then we have*

$$\int_{|x-y|\geq\delta}\frac{|f(y)|}{|x-y|^n}\,dy \leq C_{00}([w]_{A_\infty},n,p,x,\delta)\,[w]_{A_p}^{\frac{1}{p}}\big\|f\big\|_{L^p(w)}$$

for some constant C_{00} depending on the stated parameters. In particular, $T^{(\delta)}(f)$ and $T^{(*)}(f)$ are defined for $f\in L^p(w)$.

Proof. For each $\delta>0$ and x pick a cube $Q_0=Q_0(x,\delta)$ of side length $c_n\delta$ (for some constant c_n) such that $Q_0\subsetneqq B(x,\delta)$. Set $Q_j=2^jQ_0$ for $j\geq 0$. We have

$$\begin{aligned}\int_{|y-x|\geq\delta}\frac{|f(y)|}{|x-y|^n}\,dy &\leq C_n\sum_{j=0}^{\infty}(2^j\delta)^{-n}\int_{Q_{j+1}\setminus Q_j}|f(y)|\,dy\\ &\leq C_n\sum_{j=1}^{\infty}\bigg(\frac{1}{|Q_j|}\int_{Q_j}|f(y)|^p w\,dy\bigg)^{\frac{1}{p}}\bigg(\frac{1}{|Q_j|}\int_{Q_j}w^{-\frac{p'}{p}}\,dy\bigg)^{\frac{1}{p'}}\\ &\leq C_n\,[w]_{A_p}^{\frac{1}{p}}\sum_{j=1}^{\infty}\bigg(\int_{Q_j}|f(y)|^p w\,dy\bigg)^{\frac{1}{p}}\bigg(\frac{1}{w(Q_j)}\bigg)^{\frac{1}{p}}\\ &\leq C_n\,[w]_{A_p}^{\frac{1}{p}}\big\|f\big\|_{L^p(w)}\sum_{j=1}^{\infty}\big(w(Q_j)\big)^{-\frac{1}{p}}.\end{aligned}$$

But Theorem 9.3.3 (d) gives for some $\varepsilon_0=\varepsilon_0(n,[w]_{A_\infty})$ that

$$\frac{w(Q_0)}{w(Q_j)}\leq C(n,[w]_{A_\infty})\frac{|Q_0|^{\varepsilon_0}}{|Q_j|^{\varepsilon_0}},$$

from which it follows that

$$w(Q_j)^{-1}\leq C'\,2^{-jn\varepsilon_0}.$$

In view of this estimate, the previous series converges. Note that C' and hence C_{00} depend on $[w]_{A_\infty},n,p,x,\delta$.

This argument is also valid in the case $p=1$ by an obvious modification. □

Theorem 9.4.6. *Let T be a $CZO(\delta,A,B)$, $1\leq p<\infty$, and $w\in A_p$. Then there is a constant $C_p=C_p(n,\delta,A+B,[w]_{A_\infty},[w]_{A_p})$ such that*

$$\big\|T^{(*)}(f)\big\|_{L^{1,\infty}(w)}\leq C_1\big\|f\big\|_{L^1(w)} \tag{9.4.14}$$

whenever $w\in A_1$ and $f\in L^1(w)$; and also

$$\big\|T^{(*)}(f)\big\|_{L^p(w)}\leq C_p\big\|f\big\|_{L^p(w)} \tag{9.4.15}$$

whenever $w\in A_p$ and $f\in L^p(w)$.

Proof. This theorem is a consequence of the estimate proved in the previous theorem. For technical reasons, it is useful to fix a $\delta>0$ and work with the auxiliary maximal operator $T_\delta^{(*)}$ instead of $T^{(*)}$. We begin by taking $1<p<\infty$ and $f\in L^p(w)$ for some $w\in A_p$. We write

$$\begin{aligned}\big\|T_\delta^{(*)}(f)\big\|_{L^p(w)}^p &= \int_0^\infty p\lambda^{p-1}w(\{T_\delta^{(*)}(f)>\lambda\})\,d\lambda\\ &= 3^p\int_0^\infty p\lambda^{p-1}w(\{T_\delta^{(*)}(f)>3\lambda\})\,d\lambda\,,\end{aligned}$$

which we control by

$$\begin{aligned}3^p\int_0^\infty p\lambda^{p-1}w(\{T_\delta^{(*)}(f)>3\lambda\}\cap\{M(f)\le\gamma\lambda\})\,d\lambda&\\ +\,3^p\int_0^\infty p\lambda^{p-1}w(\{M(f)>\gamma\lambda\})\,d\lambda\,.&\end{aligned}$$

Using Theorem 9.4.3 (or rather Remark 9.4.4), we estimate the last terms by

$$\begin{aligned}3^pC_0\gamma^{\varepsilon_0}(A+B)^{\varepsilon_0}\int_0^\infty p\lambda^{p-1}w(\{T_\delta^{(*)}(f)>\lambda\})\,d\lambda&\\ +\,\frac{3^p}{\gamma^p}\int_0^\infty p\lambda^{p-1}w(\{M(f)>\lambda\})\,d\lambda\,,&\end{aligned}$$

which is equal to

$$3^pC_0\gamma^{\varepsilon_0}(A+B)^{\varepsilon_0}\big\|T_\delta^{(*)}(f)\big\|_{L^p(w)}^p+\frac{3^p}{\gamma^p}\big\|M(f)\big\|_{L^p(w)}^p\,.$$

Taking $\gamma=\min\big(\frac{1}{2}\gamma_0,\frac{1}{2}(2C_03^p)^{-\frac{1}{\varepsilon_0}}(A+B)^{-1}\big)$, we conclude that

$$\begin{aligned}&\big\|T_\delta^{(*)}(f)\big\|_{L^p(w)}^p\\ &\quad\le\frac{1}{2}\big\|T_\delta^{(*)}(f)\big\|_{L^p(w)}^p+\widetilde{C}_p(n,\delta,A+B,[w]_{A_\infty})\big\|M(f)\big\|_{L^p(w)}^p\,.\end{aligned}\tag{9.4.16}$$

We now prove a similar estimate when $p=1$. For $f\in L^1(w)$ and $w\in A_1$ we have

$$\begin{aligned}&3\lambda w(\{T_\delta^{(*)}(f)>3\lambda\})\\ &\quad\le 3\lambda w(\{T_\delta^{(*)}(f)>3\lambda\}\cap\{M(f)\le\gamma\lambda)+3\lambda w(\{M(f)>\gamma\lambda\})\,,\end{aligned}$$

and this expression is controlled by

$$3\lambda C_0\gamma^{\varepsilon_0}(A+B)^{\varepsilon_0}w(\{T_\delta^{(*)}(f)>\lambda\})+\frac{3}{\gamma}\big\|M(f)\big\|_{L^{1,\infty}(w)}\,.$$

It follows that

$$\begin{aligned}&\big\|T_\delta^{(*)}(f)\big\|_{L^{1,\infty}(w)}\\&\le\frac{1}{2}\big\|T_\delta^{(*)}(f)\big\|_{L^{1,\infty}(w)}+\widetilde{C}_1(n,\delta,A+B,[w]_{A_\infty})\big\|M(f)\big\|_{L^{1,\infty}(w)}.\end{aligned}\tag{9.4.17}$$

Estimate (9.4.15) would follow from (9.4.16) if we knew that $\big\|T_\delta^{(*)}(f)\big\|_{L^p(w)}<\infty$ whenever $1<p<\infty$, $w\in A_p$ and $f\in L^p(w)$, while (9.4.14) would follow from (9.4.17) if we had $\big\|T_\delta^{(*)}(f)\big\|_{L^{1,\infty}(w)}<\infty$ whenever $w\in A_1$ and $f\in L^1(w)$. Since we do not know that these quantities are finite, a certain amount of work is needed.

To deal with this problem we momentarily restrict attention to a special class of functions on $\mathbf{R}^n$, the class of bounded functions with compact support. Note that in view of Exercise 9.4.1, such functions are dense in $L^p(w)$ when $w\in A_\infty$ and $1\le p<\infty$. Let h be a bounded function with compact support on $\mathbf{R}^n$. Then $T_\delta^{(*)}(h)\le C_1\delta^{-n}\big\|h\big\|_{L^1}$ and $T_\delta^{(*)}(h)(x)\le C_2(h)|x|^{-n}$ for x away from the support of h. It follows that

$$T_\delta^{(*)}(h)(x)\le C_3(h,\delta)(1+|x|)^{-n}$$

for all $x\in\mathbf{R}^n$. Furthermore, if h is nonzero, then

$$M(h)(x)\ge\frac{C_4(h)}{(1+|x|)^n},$$

and therefore for $w\in A_1$,

$$\big\|T_\delta^{(*)}(h)\big\|_{L^{1,\infty}(w\,dx)}\le C_5(h,\delta)\big\|M(h)\big\|_{L^{1,\infty}(w\,dx)}<\infty,$$

while for $1<p<\infty$ and $w\in A_p$,

$$\int_{\mathbf{R}^n}(T_\delta^{(*)}(h)(x))^p w(x)\,dx\le C_5(h,p,\delta)\int_{\mathbf{R}^n}M(h)(x)^p w(x)\,dx<\infty$$

in view of Theorem 9.1.9. Using these facts, (9.4.16), (9.4.17), and Theorem 9.1.9 once more, we conclude that for all $\delta>0$ and $1<p<\infty$ we have

$$\begin{aligned}\big\|T_\delta^{(*)}(h)\big\|_{L^p(w)}^p&\le 2\widetilde{C}_p\big\|M(h)\big\|_{L^p(w)}^p\le C_p^p\big\|h\big\|_{L^p(w)}^p,\\\big\|T_\delta^{(*)}(h)\big\|_{L^{1,\infty}(w)}&\le 2\widetilde{C}_1\big\|M(h)\big\|_{L^{1,\infty}(w)}\le C_1\big\|h\big\|_{L^1(w)},\end{aligned}\tag{9.4.18}$$

whenever h a bounded function with compact support. The constants $\widetilde{C}_p$ and $\widetilde{C}_1$ depend only on the parameters $A+B$, n, δ, and $[w]_{A_\infty}$, while C_p also depends on $[w]_{A_p}$, $1\le p<\infty$.

We now extend estimates (9.4.14) and (9.4.15) to functions in $L^p(\mathbf{R}^n,w\,dx)$. Given $1\le p<\infty$, $w\in A_p$, and $f\in L^p(w)$, let

$$f_N(x)=f(x)\chi_{|f|\le N}\chi_{|x|\le N}.$$

Then f_N is a bounded function with compact support that converges to f in $L^p(w)$ (i.e., $\|f_N - f\|_{L^p(w)} \to 0$ as $N \to \infty$) by the Lebesgue dominated convergence theorem. Also $|f_N| \le |f|$ for all N. Sublinearity and Lemma 9.4.5 give for all $x \in \mathbf{R}^n$,

$$\begin{aligned}|T_\delta^{(*)}(f_N)(x) - T_\delta^{(*)}(f)(x)| &\le T_\delta^{(*)}(f - f_N)(x)\\ &\le A\,C_{00}([w]_{A_\infty}, n, p, x, \delta)\,[w]_{A_p}^{\frac{1}{p}}\,\|f_N - f\|_{L^p(w)},\end{aligned}$$

and this converges to zero as $N \to \infty$. Therefore $T_\delta^{(*)}(f) = \lim_{N\to\infty} T_\delta^{(*)}(f_N)$ pointwise, and Fatou's lemma for weak type spaces [Exercise 1.1.12(d)]gives for $w \in A_1$ and $f \in L^1(w)$,

$$\begin{aligned}\|T_\delta^{(*)}(f)\|_{L^{1,\infty}(w)} &= \|\liminf_{N\to\infty} T_\delta^{(*)}(f_N)\|_{L^{1,\infty}(w)}\\ &\le \liminf_{N\to\infty} \|T_\delta^{(*)}(f_N)\|_{L^{1,\infty}(w)}\\ &\le C_1 \liminf_{N\to\infty} \|M(f_N)\|_{L^{1,\infty}(w)}\\ &\le C_1 \|M(f)\|_{L^{1,\infty}(w)},\end{aligned}$$

since $|f_N| \le |f|$ for all N. An analogous argument gives the estimate

$$\|T_\delta^{(*)}(f)\|_{L^p(w)} \le C_p \|f\|_{L^p(w)}$$

for $w \in A_p$ and $f \in L^p(w)$ when $1 < p < \infty$.

It remains to prove (9.4.15) and (9.4.14) for $T^{(*)}$. But this is also an easy consequence of Fatou's lemma, since the constants C_p and C_1 are independent of δ and

$$\lim_{\delta\to 0} T_\delta^{(*)}(f) = T^{(*)}(f)$$

for all $f \in L^p(w)$. □

Corollary 9.4.7. *Let T be a $CZO(\delta, A, B)$. Then for all $1 \le p < \infty$ and for every weight $w \in A_p$ there is a constant $C_p = C_p(n, [w]_{A_\infty}, \delta, A+B)$ such that*

$$\|T(f)\|_{L^{1,\infty}(w)} \le C_1 \|f\|_{L^1(w)}$$

and

$$\|T(f)\|_{L^p(w)} \le C_p \|f\|_{L^p(w)}$$

for all smooth functions f with compact support.

Proof. We use the fact that any element of $CZO(\delta, A, B)$ is a weak limit of a sequence of its truncations plus a bounded function times the identity operator, that is, $T = T_0 + aI$, where $\|a\|_{L^\infty} \le C_n(A+B)$ (cf. Proposition 8.1.11). Then $T^{(\varepsilon_j)}(f) \to T_0(f)$ weakly for some sequence $\varepsilon_j \to 0+$ and we have $|T_0(f)| \le T^{(*)}(f)$. Therefore,

$|T(f)| \le T^{(*)}(f) + C_n(A+B)|f|$, and this estimate implies the required result in view of the previous theorem. □

9.4.4 Necessity of the A_p Condition

We have established the main theorems relating Calderón–Zygmund operators and A_p weights, namely that such operators are bounded on $L^p(w)$ whenever w lies in A_p. It is natural to ask whether the A_p condition is necessary for the boundedness of singular integrals on L^p. We end this section by indicating the necessity of the A_p condition for the boundedness of the Riesz transforms on weighted L^p spaces.

Theorem 9.4.8. *Let w be a weight in $\mathbf{R}^n$ and let $1 \le p < \infty$. Suppose that each of the Riesz transforms R_j is of weak type (p,p) with respect to w. Then w must be an A_p weight. Similarly, let w be a weight in $\mathbf{R}$. If the Hilbert transform H is of weak type (p,p) with respect to w, then w must be an A_p weight.*

Proof. We prove the n-dimensional case, $n \ge 2$. The one-dimensional case is essentially contained in following argument, suitably adjusted.

Let Q be a cube and let f be a nonnegative function on $\mathbf{R}^n$ supported in Q that satisfies $\operatorname{Avg}_Q f > 0$. Let Q' be the cube that shares a corner with Q, has the same length as Q, and satisfies $x_j \ge y_j$ for all $1 \le j \le n$ whenever $x \in Q'$ and $y \in Q$. Then for $x \in Q'$ we have

$$\left|\sum_{j=1}^n R_j(f)(x)\right| = \frac{\Gamma(\frac{n+1}{2})}{\pi^{\frac{n+1}{2}}} \sum_{j=1}^n \int_Q \frac{x_j - y_j}{|x-y|^{n+1}} f(y)\,dy \ge \frac{\Gamma(\frac{n+1}{2})}{\pi^{\frac{n+1}{2}}} \int_Q \frac{f(y)}{|x-y|^n}\,dy.$$

But if $x \in Q'$ and $y \in Q$ we must have that $|x-y| \le 2\sqrt{n}\,\ell(Q)$, which implies that $|x-y|^{-n} \ge (2\sqrt{n})^{-n}|Q|^{-1}$. Let $C_n = \Gamma(\frac{n+1}{2})(2\sqrt{n})^{-n}\pi^{-\frac{n+1}{2}}$. It follows that for all $0 < \alpha < C_n \operatorname{Avg}_Q f$ we have

$$Q' \subseteq \Big\{x \in \mathbf{R}^n : \Big|\sum_{j=1}^n R_j(f)(x)\Big| > \alpha\Big\}.$$

Since the operator $\sum_{j=1}^n R_j$ is of weak type (p,p) with respect to w (with constant C), we must have

$$w(Q') \le \frac{C^p}{\alpha^p} \int_Q f(x)^p w(x)\,dx$$

for all $\alpha < C_n \operatorname{Avg}_Q f$, which implies that

$$\big(\operatorname*{Avg}_Q f\big)^p \le \frac{C_n^{-p} C^p}{w(Q')} \int_Q f(x)^p w(x)\,dx. \tag{9.4.19}$$

We observe that we can reverse the roles of Q and Q' and obtain

$$\left(\operatorname*{Avg}_{Q'} g\right)^p \le \frac{C_n^{-p} C^p}{w(Q)} \int_{Q'} g(x)^p w(x)\,dx \tag{9.4.20}$$

for all g supported in Q'. In particular, taking $g = \chi_{Q'}$ in (9.4.20) gives that $w(Q) \le C_n^{-p} C^p w(Q')$. Using this estimate and (9.4.19), we obtain

$$\left(\operatorname*{Avg}_{Q} f\right)^p \le \frac{(C_n^{-p} C^p)^2}{w(Q)} \int_{Q} f(x)^p w(x)\,dx. \tag{9.4.21}$$

Using the characterization of the A_p characteristic constant in Proposition 9.1.5 (8), it follows that $[w]_{A_p} \le (C_n^{-p} C^p)^2 < \infty$; hence $w \in A_p$. □

Exercises

9.4.1. Show that $\mathscr{C}_0^\infty$ is dense in $L^p(w)$ for all $w \in A_\infty$.

9.4.2. (*Córdoba and Fefferman [92]*) Let T be in $CZO(\delta, A, B)$. Show that for all $\varepsilon > 0$ and all $1 < p < \infty$ there exists a constant $C_{\varepsilon,p,n}$ such that for all functions u and f on $\mathbf{R}^n$ we have

$$\int_{\mathbf{R}^n} |T^{(*)}(f)|^p u\,dx \le C_{\varepsilon,p} \int_{\mathbf{R}^n} |f|^p M(u^{1+\varepsilon})^{\frac{1}{1+\varepsilon}}\,dx$$

whenever the right-hand side is finite.
[*Hint:* Obtain this result as a consequence of Theorem 9.4.6.]

9.4.3. Use the idea of the proof of Theorem 9.4.6 to prove the following result. Suppose that for some fixed $A, B > 0$ the nonnegative μ-measurable functions f and $T(f)$ satisfy the distributional inequality

$$\mu(\{T(f) > \alpha\} \cap \{f \le c\alpha\}) \le A\mu(\{T(f) > B\alpha\})$$

for all $\alpha > 0$. Given $0 < p < \infty$ and $A < B^p$, if $\|T(f)\|_{L^p(\mu)} < \infty$, then the following is valid:

$$\|T(f)\|_{L^p(\mu)} \le C(c, p, A, B) \|f\|_{L^p(\mu)},$$

for some constant $C(c, p, A, B)$ that depends only on the indicated parameters.

9.4.4. Let f be in $L^1(\mathbf{R}^n, w)$, where $w \in A_1$. Apply the Calderón–Zygmund decomposition to f at height $\alpha > 0$ to write $f = g + b$ as in Theorem 4.3.1. Prove that

$$\|g\|_{L^1(w)} \le [w]_{A_1} \|f\|_{L^1(w)}, \qquad \|b\|_{L^1(w)} \le 2[w]_{A_1} \|f\|_{L^1(w)}.$$

9.4.5. Assume that T has a kernel in $SK(\delta, A)$ and suppose that T maps $L^2(w)$ to $L^2(w)$ for every $w \in A_1$. Prove that T maps $L^1(w)$ to $L^{1,\infty}(w)$ for every $w \in A_1$.

[*Hint:* Use Theorem 4.3.1 to write $f = g + b$, where $b = \sum_j b_j$ and each b_j is supported in a cube Q_j with center c_j. To estimate $T(g)$ use an $L^2(w)$ estimate and Exercise 9.4.4. To estimate $T(b)$ use the mean value property, the fact that

$$\int_{\mathbf{R}^n \setminus Q_j^*} \frac{|y - c_j|^\delta}{|x - c_j|^{n+\delta}} w(x)\, dx \le C_{\delta,n} M(w)(y) \le C'_{\delta,n} [w]_{A_1} w(y),$$

and Exercise 9.4.4 to obtain the required estimate.]

9.4.6. Recall that the transpose T^t of a linear operator T is defined by

$$\langle T(f), g \rangle = \langle f, T^t(g) \rangle$$

for all suitable f and g. Suppose that T is a linear operator that maps $L^p(v)$ to itself for some $1 < p < \infty$ and some $v \in A_p$. Show that the transpose operator T^t maps $L^{p'}(w)$ to $L^{p'}(w)$ with the same norm, where $w = v^{1-p'} \in A_{p'}$.

9.4.7. Suppose that T is a linear operator that maps $L^2(v)$ to itself for all v such that $v^{-1} \in A_1$. Show that the transpose operator T^t of T maps $L^2(w)$ to $L^2(w)$ for all $w \in A_1$.

9.4.8. Let $1 < p < \infty$. Suppose that T is a linear operator that maps $L^p(v)$ to itself for all v satisfying $v^{-1} \in A_p$. Show that the transpose operator T^t of T maps $L^{p'}(w)$ to itself for all w satisfying $w^{-1} \in A_{p'}$.

9.4.9. Let $w \in A_\infty$ and assume that for some locally integrable function f we have $M(f) \in L^{p_0}(w)$ for some $0 < p_0 < \infty$. Show that for all p with $p_0 \le p < \infty$ there is a constant $C(p, n, [w]_{A_\infty})$ such that

$$\big\| M_d(f) \big\|_{L^p(w)} \le C(p, n, [w]_{A_\infty}) \big\| M^\#(f) \big\|_{L^p(w)},$$

where M_d is the dyadic maximal operator given in Definition 7.4.3. Conclude the same estimate for M.
[*Hint:* Let Q_j be as in the proof of Theorem 7.4.4. Combine estimate (7.4.4) with property (d) of Theorem 9.3.3,

$$w\big(Q_j \cap \{M_d(f) > 2\lambda,\ M^\#(f) \le \gamma\lambda\}\big) \le C_2 (2^n \gamma)^{\varepsilon_0} w(Q_j),$$

where both C_2 and ε_0 depend on the dimension n and $[w]_{A_\infty}$. Obtain the result of Theorem 7.4.4 in which the Lebesgue measure is replaced by w in A_∞ and the quantity $2^n\gamma$ is replaced by $C_2(2^n\gamma)^{\varepsilon_0}$. Finally, observe that Theorem 7.4.5 can be adapted to a general weight w in A_∞.]

9.5 Further Properties of A_p Weights

In this section we discuss other properties of A_p weights. Many of these properties indicate certain deep connections with other branches of analysis. We focus attention on three such properties: factorization, extrapolation, and relations of weighted inequalities to vector-valued inequalities.

9.5.1 Factorization of Weights

Recall the simple fact that if w_1, w_2 are A_1 weights, then $w = w_1 w_2^{1-p}$ is an A_p weight (Exercise 9.1.2). The factorization theorem for weights says that the converse of this statement is true. This provides a surprising and striking representation of A_p weights.

Theorem 9.5.1. *Suppose that w is an A_p weight for some $1 < p < \infty$. Then there exist A_1 weights w_1 and w_2 such that*

$$w = w_1 w_2^{1-p}.$$

Proof. Let us fix a $p \geq 2$ and $w \in A_p$. We define an operator T as follows:

$$T(f) = \left(w^{-\frac{1}{p}} M(f^{p-1} w^{\frac{1}{p}})\right)^{\frac{1}{p-1}} + w^{\frac{1}{p}} M(f w^{-\frac{1}{p}}),$$

where M is the Hardy–Littlewood maximal operator. We observe that T is well defined and bounded on $L^p(\mathbf{R}^n)$. This is a consequence of the facts that $w^{-\frac{1}{p-1}}$ is an $A_{p'}$ weight and that M maps $L^{p'}(w^{-\frac{1}{p-1}})$ to itself and also $L^p(w)$ to itself. Thus the norm of T on L^p depends only on the A_p characteristic constant of w. Let $B(w) = \big\|T\big\|_{L^p \to L^p}$, the norm of T on L^p. Next, we observe that for $f, g \geq 0$ in $L^p(\mathbf{R}^n)$ and $\lambda \geq 0$ we have

$$T(f+g) \leq T(f) + T(g), \quad T(\lambda f) = \lambda T(f). \tag{9.5.1}$$

To see the first assertion, we need only note that for every ball B, the operator

$$f \to \left(\frac{1}{|B|}\int_B |f|^{p-1} w^{\frac{1}{p}}\, dx\right)^{\frac{1}{p-1}}$$

is sublinear as a consequence of Minkowski's integral inequality, since $p - 1 \geq 1$.

We now fix an L^p function f_0 with $\big\|f_0\big\|_{L^p} = 1$ and we define a function φ in $L^p(\mathbf{R}^n)$ as the sum of the L^p convergent series

$$\varphi = \sum_{j=1}^{\infty} (2B(w))^{-j} T^j(f_0). \tag{9.5.2}$$

We define

$$w_1 = w^{\frac{1}{p}} \varphi^{p-1}, \qquad w_2 = w^{-\frac{1}{p}} \varphi,$$

so that $w = w_1 w_2^{1-p}$. It remains to show that w_1, w_2 are A_1 weights. Applying T and using (9.5.1), we obtain

$$\begin{aligned} T(\varphi) &\leq 2B(w) \sum_{j=1}^{\infty} (2B(w))^{-j-1} T^{j+1}(f_0) \\ &= 2B(w) \Big(\varphi - \frac{T(f_0)}{2B(w)} \Big) \\ &\leq 2B(w)\, \varphi , \end{aligned}$$

that is,

$$\big(w^{-\frac{1}{p}} M(\varphi^{p-1} w^{\frac{1}{p}}) \big)^{\frac{1}{p-1}} + w^{\frac{1}{p}} M(\varphi w^{-\frac{1}{p}}) \leq 2B(w)\, \varphi .$$

Using that $\varphi = (w^{-\frac{1}{p}} w_1)^{\frac{1}{p-1}} = w^{\frac{1}{p}} w_2$, we obtain

$$M(w_1) \leq (2B(w))^{p-1} w_1 \qquad \text{and} \qquad M(w_2) \leq 2B(w) w_2 .$$

These show that w_1 and w_2 are A_1 weights whose characteristic constants depend on $[w]_{A_p}$ (and also the dimension n and p). This concludes the case $p \geq 2$.

We now turn to the case $p < 2$. Given a weight $w \in A_p$ for $1 < p < 2$, we consider the weight $w^{-1/(p-1)}$, which is in $A_{p'}$. Since $p' > 2$, using the result we obtained, we write $w^{-1/(p-1)} = v_1 v_2^{1-p'}$, where v_1, v_2 are A_1 weights. It follows that $w = v_1^{1-p} v_2$, and this completes the asserted factorization of A_p weights. □

Combining the result just obtained with Theorem 9.2.7, we obtain the following description of A_p weights.

Corollary 9.5.2. *Let w be an A_p weight for some $1 < p < \infty$. Then there exist locally integrable functions f_1 and f_2 with*

$$M(f_1) + M(f_2) < \infty \qquad \textit{a.e.},$$

constants $0 < \varepsilon_1, \varepsilon_2 < 1$, and a nonnegative function k satisfying $k, k^{-1} \in L^\infty$ such that

$$w = k M(f_1)^{\varepsilon_1} M(f_2)^{\varepsilon_2(1-p)} . \tag{9.5.3}$$

9.5.2 Extrapolation from Weighted Estimates on a Single L^{p_0}

Our next topic concerns a striking application of weighted norm inequalities. This says that an estimate on $L^{p_0}(v)$ for a single p_0 and all A_{p_0} weights v implies a similar L^p estimate for all p in $(1, \infty)$. This property is referred to as extrapolation.

Surprisingly the operator T is not needed to be linear or sublinear in the following extrapolation theorem. The only condition required is that T be well defined on $\bigcup_{1\le q<\infty}\bigcup_{w\in A_q} L^q(w)$. If T happens to be a linear operator, this condition can be relaxed to T being well defined on $\mathscr{C}_0^\infty(\mathbf{R}^n)$.

Theorem 9.5.3. *Suppose that T is defined on $\bigcup_{1\le q<\infty}\bigcup_{w\in A_q} L^q(w)$ and takes values in the space of measurable complex-valued functions. Let $1\le p_0<\infty$ and suppose that there exists a positive increasing function N on $[1,\infty)$ such that for all weights v in A_{p_0} we have*

$$\big\|T\big\|_{L^{p_0}(v)\to L^{p_0}(v)} \le N\big([v]_{A_{p_0}}\big). \tag{9.5.4}$$

Then for any $1<p<\infty$ and for all weights w in A_p we have

$$\big\|T\big\|_{L^{p}(w)\to L^{p}(w)} \le K\big(n,p,p_0,[w]_{A_p}\big), \tag{9.5.5}$$

where

$$K\big(n,p,p_0,[w]_{A_p}\big)=\begin{cases} 2N\Big(\kappa_1(n,p,p_0)\,[w]_{A_p}^{\frac{p_0-1}{p-1}}\Big) & \text{when } p<p_0,\\[2ex] 2^{\frac{p-p_0}{p_0(p-1)}}N\big(\kappa_2(n,p,p_0)\,[w]_{A_p}\big) & \text{when } p>p_0,\end{cases}$$

and $\kappa_1(n,p,p_0)$ and $\kappa_2(n,p,p_0)$ are constants that depend on n, p, and p_0.

Proof. Let $1<p<\infty$ and $w\in A_p$. We define an operator

$$M'(f)=\frac{M(fw)}{w},$$

where M is the Hardy–Littlewood maximal operator. We observe that since $w^{1-p'}$ is in $A_{p'}$, the operator M' maps $L^{p'}(w)$ to itself; indeed, we have

$$\begin{aligned}\big\|M'\big\|_{L^{p'}(w)\to L^{p'}(w)} &= \big\|M\big\|_{L^{p'}(w^{1-p'})\to L^{p'}(w^{1-p'})}\\ &\le C_{n,p}[w^{1-p'}]^{\frac{1}{p'-1}}\\ &= C_{n,p}[w]_{A_p}\end{aligned} \tag{9.5.6}$$

in view of Theorem 9.1.9 and property (4) of Proposition 9.1.5.

We introduce operators $M^0(f)=|f|$ and $M^k=M\circ M\circ\cdots\circ M$, where M is the Hardy–Littlewood maximal function and the composition is taken k times. Likewise, we introduce powers $(M')^k$ of M' for $k\in\mathbf{Z}^+\cup\{0\}$. The following lemma provides the main tool in the proof of Theorem 9.5.3. Its simple proof uses Theorem 9.1.9 and (9.5.6) and is omitted.

Lemma 9.5.4. *(a) Let $1<p<\infty$ and $w\in A_p$. Define operators R and R'*

$$R(f)=\sum_{k=0}^{\infty}\frac{M^k(f)}{\big(2\big\|M\big\|_{L^p(w)\to L^p(w)}\big)^k}$$

for functions f in $L^p(w)$ and also

$$R'(f)=\sum_{k=0}^{\infty}\frac{(M')^k(f)}{\left(2\big\|M'\big\|_{L^{p'}(w)\to L^{p'}(w)}\right)^k}$$

for functions f in $L^{p'}(w)$. Then there exist constants $C_1(n,p)$ and $C_2(n,p)$ that depend on n and p such that

$$|f|\le R(f), \tag{9.5.7}$$
$$\big\|R(f)\big\|_{L^p(w)}\le 2\big\|f\big\|_{L^p(w)}, \tag{9.5.8}$$
$$M(R(f))\le C_1(n,p)\,[w]_{A_p}^{\frac{1}{p-1}}\,R(f), \tag{9.5.9}$$

for all functions f in $L^p(w)$ and such that

$$|h|\le R'(h), \tag{9.5.10}$$
$$\big\|R'(h)\big\|_{L^{p'}(w)}\le 2\big\|h\big\|_{L^{p'}(w)}, \tag{9.5.11}$$
$$M'(R'(h))\le C_2(n,p)\,[w]_{A_p}\,R'(h), \tag{9.5.12}$$

for all functions h in $L^{p'}(w)$.

We now proceed with the proof of the theorem. It is natural to split the proof into the cases $p<p_0$ and $p>p_0$.

Case (1): $p<p_0$. Assume momentarily that $R(f)^{-\frac{p_0}{(p_0/p)'}}$ is an A_{p_0} weight. Then we have

$$
\begin{aligned}
&\big\|T(f)\big\|_{L^p(w)}^p\\
&=\int_{\mathbf{R}^n}|T(f)|^pR(f)^{-\frac{p}{(p_0/p)'}}R(f)^{\frac{p}{(p_0/p)'}}\,w\,dx\\
&\le\left(\int_{\mathbf{R}^n}|T(f)|^{p_0}R(f)^{-\frac{p_0}{(p_0/p)'}}\,w\,dx\right)^{\frac{p}{p_0}}\left(\int_{\mathbf{R}^n}R(f)^p\,w\,dx\right)^{\frac{1}{(p_0/p)'}}\\
&\le N\Big(\big[R(f)^{-\frac{p_0}{(p_0/p)'}}\big]_{A_{p_0}}\Big)^p\left(\int_{\mathbf{R}^n}|f|^{p_0}R(f)^{-\frac{p_0}{(p_0/p)'}}\,w\,dx\right)^{\frac{p}{p_0}}\left(\int_{\mathbf{R}^n}R(f)^p\,w\,dx\right)^{\frac{1}{(p_0/p)'}}\\
&\le N\Big(\big[R(f)^{-\frac{p_0}{(p_0/p)'}}\big]_{A_{p_0}}\Big)^p\left(\int_{\mathbf{R}^n}R(f)^{p_0}R(f)^{-\frac{p_0}{(p_0/p)'}}\,w\,dx\right)^{\frac{p}{p_0}}\left(\int_{\mathbf{R}^n}R(f)^p\,w\,dx\right)^{\frac{1}{(p_0/p)'}}\\
&=N\Big(\big[R(f)^{-\frac{p_0}{(p_0/p)'}}\big]_{A_{p_0}}\Big)^p\left(\int_{\mathbf{R}^n}R(f)^p\,w\,dx\right)^{\frac{p}{p_0}}\left(\int_{\mathbf{R}^n}R(f)^p\,w\,dx\right)^{\frac{1}{(p_0/p)'}}\\
&\le N\Big(\big[R(f)^{-\frac{p_0}{(p_0/p)'}}\big]_{A_{p_0}}\Big)^p\left(2\big\|f\big\|_{L^p(w)}\right)^p,
\end{aligned}
$$

where we used Hölder's inequality with exponents p_0/p and $(p_0/p)'$, the hypothesis of the theorem, (9.5.7), and (9.5.8). Thus, we have the estimate

$$\big\|T(f)\big\|_{L^p(w)} \le 2N\Big(\big[R(f)^{-\frac{p_0}{(p_0/p)'}}\big]_{A_{p_0}}\Big)\big\|f\big\|_{L^p(w)} \tag{9.5.13}$$

and it remains to obtain a bound for the A_{p_0} characteristic constant of $R(f)^{-\frac{p_0}{(p_0/p)'}}$. In view of (9.5.9), the function $R(f)$ is an A_1 weight with characteristic constant at most a constant multiple of $[w]_{A_p}^{\frac{1}{p-1}}$. Consequently, there is a constant C_1' such that

$$R(f)^{-1} \le C_1'\,[w]_{A_p}^{\frac{1}{p-1}}\bigg(\frac{1}{|Q|}\int_Q R(f)\,dx\bigg)^{-1}$$

for any cube Q in $\mathbf{R}^n$. Thus we have

$$\begin{aligned}
&\frac{1}{|Q|}\int_Q R(f)^{-\frac{p_0}{(p_0/p)'}}\,w\,dx \\
&\le \big(C_1'\,[w]_{A_p}^{\frac{1}{p-1}}\big)^{\frac{p_0}{(p_0/p)'}}\bigg(\frac{1}{|Q|}\int_Q R(f)\,dx\bigg)^{-\frac{p_0}{(p_0/p)'}}\bigg(\frac{1}{|Q|}\int_Q w\,dx\bigg).
\end{aligned} \tag{9.5.14}$$

Next we have

$$\begin{aligned}
&\bigg(\frac{1}{|Q|}\int_Q\Big(R(f)^{-\frac{p_0}{(p_0/p)'}}\,w\Big)^{1-p_0'}\,dx\bigg)^{p_0-1} \\
&\qquad = \bigg(\frac{1}{|Q|}\int_Q R(f)^{\frac{p_0(p_0'-1)}{(p_0/p)'}}\,w^{1-p_0'}\,dx\bigg)^{p_0-1} \\
&\qquad \le \bigg(\frac{1}{|Q|}\int_Q R(f)\,dx\bigg)^{\frac{p_0}{(p_0/p)'}}\bigg(\frac{1}{|Q|}\int_Q w^{1-p'}\bigg)^{p-1},
\end{aligned} \tag{9.5.15}$$

where we applied Hölder's inequality with exponents

$$\bigg(\frac{p'-1}{p_0'-1}\bigg)' \qquad \text{and} \qquad \frac{p'-1}{p_0'-1},$$

and we used that

$$\frac{p_0(p_0'-1)}{(p_0/p)'}\bigg(\frac{p'-1}{p_0'-1}\bigg)' = 1 \qquad \text{and} \qquad \frac{p_0-1}{\Big(\frac{p'-1}{p_0'-1}\Big)'} = \frac{p_0}{(p_0/p)'}.$$

Multiplying (9.5.14) by (9.5.15) and taking the supremum over all cubes Q in $\mathbf{R}^n$ we deduce that

$$\Big[R(f)^{-\frac{p_0}{(p_0/p)'}}\Big]_{A_{p_0}} \le \big(C_1'\,[w]_{A_p}^{\frac{1}{p-1}}\big)^{\frac{p_0}{(p_0/p)'}}\,[w]_{A_p} = \kappa_1(n,p,p_0)\,[w]_{A_p}^{\frac{p_0-1}{p-1}}.$$

Combining this estimate with (9.5.13) and using the fact that N is an increasing function, we obtain the validity of (9.5.5) in the case $p < p_0$.

Case (2): $p > p_0$. In this case we set $r = p/p_0 > 1$. Then we have

$$\big\|T(f)\big\|_{L^p(w)}^p = \big\|\,|T(f)|^{p_0}\big\|_{L^r(w)}^r = \left(\int_{\mathbf{R}^n} |T(f)|^{p_0} h\, w\, dx\right)^r \tag{9.5.16}$$

for some nonnegative function h with $L^{r'}(w)$ norm equal to 1. We define a function

$$H = \big[R'(h^{\frac{r'}{p'}})\big]^{\frac{p'}{r'}}.$$

Obviously, we have $0 \le h \le H$ and thus

$$\begin{aligned}
\int_{\mathbf{R}^n} |T(f)|^{p_0} h\, w\, dx &\le \int_{\mathbf{R}^n} |T(f)|^{p_0} H\, w\, dx \\
&\le N\big([H\,w]_{A_{p_0}}\big)^{p_0} \big\|f\big\|_{L^{p_0}(H\,w)}^{p_0} \\
&\le N\big([H\,w]_{A_{p_0}}\big)^{p_0} \big\|\,|f|^{p_0}\big\|_{L^r(w)} \big\|H\big\|_{L^{r'}(w)} \\
&\le 2^{\frac{p'}{r'}} N\big([H\,w]_{A_{p_0}}\big)^{p_0} \big\|f\big\|_{L^p(w)}^{p_0},
\end{aligned} \tag{9.5.17}$$

noting that

$$\big\|H\big\|_{L^{r'}(w)}^{r'} = \int_{\mathbf{R}^n} R'(h^{r'/p'})^{p'}\, w\, dx \le 2^{p'} \int_{\mathbf{R}^n} h^{r'}\, w\, dx = 2^{p'},$$

which is valid in view of (9.5.11). Moreover, this argument is based on the hypothesis of the theorem and requires that $H\,w$ be an A_{p_0} weight. To see this, we observe that condition (9.5.12) implies that $H^{r'/p'}w$ is an A_1 weight with characteristic constant at most a multiple of $[w]_{A_1}$. Thus, there is a constant C_2' that depends only on n and p such that

$$\frac{1}{|Q|}\int_Q H^{\frac{r'}{p'}} w\, dx \le C_2' \, [w]_{A_p} H^{\frac{r'}{p'}} w$$

for all cubes Q in $\mathbf{R}^n$. From this it follows that

$$(H\,w)^{-1} \le \kappa_2(n,p,p_0)\, [w]_{A_p}^{\frac{p'}{r'}} \left(\frac{1}{|Q|}\int_Q H^{\frac{r'}{p'}} w\, dx\right)^{-\frac{p'}{r'}} w^{\frac{p'}{r'}-1},$$

where we set $\kappa_2(n,p,p_0) = (C_2')^{p'/r'}$. We raise the preceding displayed expression to the power $p_0' - 1$, we average over the cube Q, and then we raise to the power $p_0 - 1$. We deduce the estimate

$$
\begin{aligned}
&\left(\frac{1}{|Q|}\int_Q (Hw)^{1-p_0'}\,dx\right)^{p_0-1} \\
&\quad\le \kappa_2(n,p,p_0)\,[w]_{A_p}^{\frac{p'}{r'}}\left(\frac{1}{|Q|}\int_Q H^{\frac{r'}{p'}}w\,dx\right)^{-\frac{p'}{r'}}\left(\frac{1}{|Q|}\int_Q w^{1-p'}\,dx\right)^{p_0-1},
\end{aligned}
\tag{9.5.18}
$$

where we use the fact that

$$
\left(\frac{p'}{r'}-1\right)(p_0'-1)=1-p'.
$$

Note that $r'/p'\ge 1$, since $p_0\ge 1$. Using Hölder's inequality with exponents r'/p' and $(r'/p')^{-1}$ we obtain that

$$
\frac{1}{|Q|}\int_Q Hw\,dx\le\left(\frac{1}{|Q|}\int_Q H^{\frac{r'}{p'}}w\,dx\right)^{\frac{p'}{r'}}\left(\frac{1}{|Q|}\int_Q w\,dx\right)^{\frac{p_0-1}{p-1}},
\tag{9.5.19}
$$

where we used that

$$
\frac{1}{(\frac{r'}{p'})'}=\frac{p_0-1}{p-1}.
$$

Multiplying (9.5.18) by (9.5.19), we deduce the estimate

$$
\big[Hw\big]_{A_{p_0}}\le \kappa_2(n,p,p_0)\,[w]_{A_p}^{\frac{p'}{r'}}\,[w]_{A_p}^{\frac{p_0-1}{p-1}}=\kappa_2(n,p,p_0)\,[w]_{A_p}.
$$

Inserting this estimate in (9.5.17) we obtain

$$
\int_{\mathbf{R}^n}|T(f)|^{p_0}hw\,dx\le 2^{\frac{p'}{r'}}N\big(\kappa_2(n,p,p_0)\,[w]_{A_p}\big)^{p_0}\|f\|_{L^p(w)}^{p_0},
$$

and combining this with (9.5.16) we conclude that

$$
\|T(f)\|_{L^p(w)}^p\le 2^{\frac{p'r}{r'}}N\big(\kappa_2(n,p,p_0)\,[w]_{A_p}\big)^{p_0r}\|f\|_{L^p(w)}^{p_0r}.
$$

This proves the required estimate (9.5.5) in the case $p>p_0$. □

There is a version of Theorem 9.5.3 in which the initial strong type assumption is replaced by a weak type estimate.

Theorem 9.5.5. *Suppose that T is a well defined operator on $\bigcup_{1<q<\infty}\bigcup_{w\in A_q}L^q(w)$ that takes values in the space of measurable complex-valued functions. Fix $1\le p_0<\infty$ and suppose that there is an increasing function N on $[1,\infty)$ such that for all weights v in A_{p_0} we have*

$$
\|T\|_{L^{p_0}(v)\to L^{p_0,\infty}(v)}\le N([v]_{A_{p_0}}).
\tag{9.5.20}
$$

Then for any $1<p<\infty$ and for all weights w in A_p we have

$$\left\|T\right\|_{L^p(w)\to L^{p,\infty}(w)} \le K\big(n,p,p_0,[w]_{A_p}\big), \tag{9.5.21}$$

where $K\big(n,p,p_0,[w]_{A_p}\big)$ is as in Theorem 9.5.3.

Proof. For every fixed $\lambda > 0$ we define

$$T_\lambda(f) = \lambda \chi_{|T(f)|>\lambda}\,.$$

The operator T_λ is not linear but is well defined on $\bigcup_{1<q<\infty}\bigcup_{w\in A_q} L^q(w)$, since T is well defined on this union. We show that T_λ maps $L^{p_0}(v)$ to $L^{p_0}(v)$ for every $v \in A_{p_0}$. Indeed, we have

$$\begin{aligned}\bigg(\int_{\mathbf{R}^n} |T_\lambda(f)|^{p_0}\, v\,dx\bigg)^{\frac{1}{p_0}} &= \bigg(\int_{\mathbf{R}^n} \lambda^{p_0}\chi_{|T(f)|>\lambda}\, v\,dx\bigg)^{\frac{1}{p_0}}\\ &= \Big(\lambda^{p_0} v(\{|T(f)|>\lambda\})\Big)^{\frac{1}{p_0}}\\ &\le N([v]_{A_{p_0}})\left\|f\right\|_{L^{p_0}(v)}\end{aligned}$$

using the hypothesis on T. Applying Theorem 9.5.3, we obtain that T_λ maps $L^p(w)$ to itself for all $1<p<\infty$ and all $w \in A_p$ with a constant independent of λ. Precisely, for any $w \in A_p$ and any $f \in L^p(w)$ we have

$$\left\|T_\lambda(f)\right\|_{L^p(w)} \le K\big(n,p,p_0,[w]_{A_p}\big)\left\|f\right\|_{L^p(w)}\,.$$

Since

$$\left\|T(f)\right\|_{L^{p,\infty}(w)} = \sup_{\lambda>0}\left\|T_\lambda(f)\right\|_{L^p(w)},$$

it follows that T maps $L^p(w)$ to $L^{p,\infty}(w)$ with the asserted norm. □

Assuming that the operator T in the preceding theorem is sublinear (or quasi-sublinear), we obtain the following result that contains a stronger conclusion.

Corollary 9.5.6. *Suppose that T is a sublinear operator on $\bigcup_{1<q<\infty}\bigcup_{w\in A_q} L^q(w)$ that takes values in the space of measurable complex-valued functions. Fix $1 \le p_0 < \infty$ and suppose that there is an increasing function N on $[1,\infty)$ such that for all weights v in A_{p_0} we have*

$$\left\|T\right\|_{L^{p_0}(v)\to L^{p_0,\infty}(v)} \le N([v]_{A_{p_0}})\,. \tag{9.5.22}$$

Then for any $1<p<\infty$ and any weight w in A_p there is a constant $K'(n,p,p_0,[w]_{A_p})$ such that

$$\left\|T(f)\right\|_{L^p(w)} \le K'\big(n,p,p_0,[w]_{A_p}\big)\left\|f\right\|_{L^p(w)}\,.$$

Proof. The proof follows from Theorem 9.5.5 and the Marcinkiewicz interpolation theorem. □

We end this subsection by observing that the conclusion of the extrapolation Theorem 9.5.3 can be strengthened to yield vector-valued estimates. This strengthening may be achieved by a simple adaptation of the proof discussed.

Corollary 9.5.7. *Suppose that T is defined on $\bigcup_{1\le q<\infty}\bigcup_{w\in A_q} L^q(w)$ and takes values in the space of all measurable complex-valued functions. Fix $1 \le p_0 < \infty$ and suppose that there is an increasing function N on $[1,\infty)$ such that for all weights v in A_{p_0} we have*

$$\big\|T\big\|_{L^{p_0}(v)\to L^{p_0}(v)} \le N\big([v]_{A_{p_0}}\big).$$

Then for every $1 < p < \infty$ and every weight $w \in A_p$ we have

$$\Big\|\Big(\sum_j |T(f_j)|^{p_0}\Big)^{\frac{1}{p_0}}\Big\|_{L^p(w)} \le K(n,p,p_0,[w]_{A_p})\Big\|\Big(\sum_j |f_j|^{p_0}\Big)^{\frac{1}{p_0}}\Big\|_{L^p(w)}$$

for all sequences of functions f_j in $L^p(w)$, where $K\big(n,p,p_0,[w]_{A_p}\big)$ is as in Theorem 9.5.3.

Proof. To derive the claimed vector-valued inequality follow the proof of Theorem 9.5.3 replacing the function f by $(\sum_j |f_j|^{p_0})^{\frac{1}{p_0}}$ and $T(f)$ by $(\sum_j |T(f_j)|^{p_0})^{\frac{1}{p_0}}$. □

9.5.3 Weighted Inequalities Versus Vector-Valued Inequalities

We now turn to the last topic we are going to discuss in relation to A_p weights: connections between weighted inequalities and vector-valued inequalities. The next result provides strong evidence that there is a nontrivial connection of this sort. The following is a general theorem saying that any vector-valued inequality is equivalent to some weighted inequality. The proof of the theorem is based on a minimax lemma whose precise formulation and proof can be found in Appendix H.

Theorem 9.5.8. *(a) Let $0 < p < q, r < \infty$. Let $\{T_j\}_j$ be a sequence of sublinear operators that map $L^q(\mu)$ to $L^r(\nu)$, where μ and ν are arbitrary measures. Then the vector-valued inequality*

$$\Big\|\big(\sum_j |T_j(f_j)|^p\big)^{\frac{1}{p}}\Big\|_{L^r} \le C\Big\|\big(\sum_j |f_j|^p\big)^{\frac{1}{p}}\Big\|_{L^q} \tag{9.5.23}$$

holds for all $f_j \in L^q(\mu)$ if and only if for every $u \ge 0$ in $L^{\frac{r}{r-p}}(\nu)$ there exists $U \ge 0$ in $L^{\frac{q}{q-p}}(\mu)$ with

$$\begin{aligned} \big\|U\big\|_{L^{\frac{q}{q-p}}} &\le \big\|u\big\|_{L^{\frac{r}{r-p}}}, \\ \sup_j \int |T_j(f)|^p\, u\, d\nu &\le C^p \int |f|^p\, U\, d\mu. \end{aligned} \tag{9.5.24}$$

(b) Let $0 < q, r < p < \infty$. Let $\{T_j\}_j$ be as before. Then the vector-valued inequality (9.5.23) holds for all $f_j \in L^q(\mu)$ if and only if for every $u \geq 0$ in $L^{\frac{q}{p-q}}(\mu)$ there exists $U \geq 0$ in $L^{\frac{r}{p-r}}(\nu)$ with

$$\begin{aligned} \|U\|_{L^{\frac{r}{p-r}}} &\leq \|u\|_{L^{\frac{q}{p-q}}}, \\ \sup_j \int |T_j(f)|^p U^{-1} d\nu &\leq C^p \int |f|^p u^{-1} d\mu. \end{aligned} \tag{9.5.25}$$

Proof. We begin with part (a). Given $f_j \in L^q(\mathbf{R}^n, \mu)$, we use (9.5.24) to obtain

$$\begin{aligned} \Big\| \Big(\sum_j |T_j(f_j)|^p \Big)^{\frac{1}{p}} \Big\|_{L^r(\nu)} &= \Big\| \sum_j |T_j(f_j)|^p \Big\|_{L^{\frac{r}{p}}(\nu)}^{\frac{1}{p}} \\ &= \sup_{\|u\|_{L^{\frac{r}{r-p}}} \leq 1} \Big(\int_{\mathbf{R}^n} \sum_j |T_j(f_j)|^p u \, d\nu \Big)^{\frac{1}{p}} \\ &\leq \sup_{\|u\|_{L^{\frac{r}{r-p}}} \leq 1} C \Big(\int_{\mathbf{R}^n} \sum_j |f_j|^p U \, d\mu \Big)^{\frac{1}{p}} \\ &\leq \sup_{\|u\|_{L^{\frac{r}{r-p}}} \leq 1} C \Big\| \sum_j |f_j|^p \Big\|_{L^{\frac{q}{p}}(\mu)}^{\frac{1}{p}} \|U\|_{L^{\frac{q}{q-p}}}^{\frac{1}{p}} \\ &\leq C \Big\| \Big(\sum_j |f_j|^p \Big)^{\frac{1}{p}} \Big\|_{L^q(\mu)}, \end{aligned}$$

which proves (9.5.23) with the same constant C as in (9.5.24). To prove the converse, given a nonnegative $u \in L^{\frac{r}{r-p}}(\nu)$ with $\|u\|_{L^{\frac{r}{r-p}}} = 1$, we define

$$A = \Big\{ a = (a_0, a_1) : \, a_0 = \sum_j |f_j|^p, \quad a_1 = \sum_j |T_j(f_j)|^p, \quad f_j \in L^q(\mu) \Big\}$$

and

$$B = \Big\{ b \in L^{\frac{q}{q-p}}(\mu) : \, b \geq 0, \quad \|b\|_{L^{\frac{q}{q-p}}} \leq 1 = \|u\|_{L^{\frac{r}{r-p}}} \Big\}.$$

Notice that A and B are convex sets and B is weakly compact. (The sublinearity of each T_j is used here.) We define the function Φ on $A \times B$ by setting

$$\Phi(a,b) = \int a_1 u \, d\nu - C^p \int a_0 b \, d\mu = \sum_j \Big(\int |T_j(f_j)|^p u \, d\nu - C^p \int |f_j|^p b \, d\mu \Big).$$

Then Φ is concave on A and weakly continuous and convex on B. Thus the *minimax lemma* in Appendix H is applicable. This gives

$$\min_{b \in B} \sup_{a \in A} \Phi(a,b) = \sup_{a \in A} \min_{b \in B} \Phi(a,b). \tag{9.5.26}$$

At this point observe that for a fixed $a = \big(\sum_j |f_j|^p, \sum_j |T_j(f_j)|^p\big)$ in A we have

$$\begin{aligned}\min_{b\in B} \Phi(a,b) &\le \Big\|\sum_j |T_j(f_j)|^p\Big\|_{L^{\frac{r}{p}}(v)} \big\|u\big\|_{L^{\frac{r}{r-p}}} - C^p \max_{b\in B} \int \sum_j |f_j|^p \, b\, d\mu \\ &\le \Big\|\sum_j |T_j(f_j)|^p\Big\|_{L^{\frac{r}{p}}(v)} - C^p \Big\|\sum_j |f_j|^p\Big\|_{L^{\frac{q}{p}}(\mu)} \le 0\end{aligned}$$

using the hypothesis (9.5.23). It follows that $\sup_{a\in A} \min_{b\in B} \Phi(a,b) \le 0$ and hence (9.5.26) yields $\min_{b\in B} \sup_{a\in A} \Phi(a,b) \le 0$. Thus there exists a $U \in B$ such that $\Phi(a,U) \le 0$ for every $a \in A$. This completes the proof of part (a).

The proof of part (b) is similar. Using the result of Exercise 9.5.1 and (9.5.25), given $f_j \in L^q(\mathbf{R}^n, \mu)$ we have

$$\begin{aligned}\Big\|\Big(\sum_j |f_j|^p\Big)^{\frac{1}{p}}\Big\|_{L^q(\mu)} &= \Big\|\sum_j |f_j|^p\Big\|_{L^{\frac{q}{p}}(\mu)}^{\frac{1}{p}} \\ &= \inf_{\|u\|_{L^{\frac{q}{p-q}}} \le 1} \Big(\int_{\mathbf{R}^n} \sum_j |f_j|^p u^{-1} \, d\mu\Big)^{\frac{1}{p}} \\ &\ge \frac{1}{C} \inf_{\|U\|_{L^{\frac{r}{p-r}}} \le 1} \Big(\int_{\mathbf{R}^n} \sum_j |T_j(f_j)|^p U^{-1} \, dv\Big)^{\frac{1}{p}} \\ &= \frac{1}{C} \Big\|\sum_j |T_j(f_j)|^p\Big\|_{L^{\frac{r}{p}}(v)}^{\frac{1}{p}} \\ &= \frac{1}{C} \Big\|\Big(\sum_j |T_j(f_j)|^p\Big)^{\frac{1}{p}}\Big\|_{L^r(v)}.\end{aligned}$$

To prove the converse direction in part (b), given a fixed $u \ge 0$ in $L^{\frac{q}{p-q}}(\mu)$ with $\|u\|_{L^{\frac{q}{p-q}}} = 1$, we define A as in part (a) and

$$B = \big\{ b \in L^{\frac{p}{p-r}}(v) : \, b \ge 0, \quad \|b\|_{L^{\frac{p}{p-r}}} \le 1 = \|u\|_{L^{\frac{q}{p-q}}} \big\}.$$

We also define the function Φ on $A \times B$ by setting

$$\begin{aligned}\Phi(a,b) &= \int a_1 b^{-1} dv - C^p \int a_0 u^{-1} \, d\mu \\ &= \sum_j \Big(\int |T_j(f_j)|^p b^{-1} \, dv - C^p \int |f_j|^p u^{-1} \, d\mu \Big).\end{aligned}$$

Then Φ is concave on A and weakly continuous and convex on B. Also, using Exercise 9.5.1, for any $a = \big(\sum_j |f_j|^p, \sum_j |T_j(f_j)|^p\big)$ in A, we have

$$\min_{b\in B} \Phi(a,b) \le \Big\|\sum_j |T_j(f_j)|^p\Big\|_{L^{\frac{r}{p}}(v)} - C^p \Big\|\sum_j |f_j|^p\Big\|_{L^{\frac{q}{p}}(\mu)} \le 0.$$

Thus $\sup_{a\in A}\min_{b\in B}\Phi(a,b)\le 0$. Using (9.5.26), yields $\min_{b\in B}\sup_{a\in A}\Phi(a,b)\le 0$, and the latter implies the existence of a U in B such that $\Phi(a,U)\le 0$ for all $a\in A$. This proves (9.5.25). □

Example 9.5.9. We use the previous theorem to obtain another proof of the vector-valued Hardy–Littlewood maximal inequality in Corollary 4.6.5. We take $T_j = M$ for all j. For given $1<p<q<\infty$ and u in $L^{\frac{q}{q-p}}$ we set $s=\frac{q}{q-p}$ and $U=\big\|M\big\|_{L^s\to L^s}^{-1}M(u)$. In view of Exercise 9.1.8 we have

$$\big\|U\big\|_{L^s}\le\big\|u\big\|_{L^s}\qquad\text{and}\qquad\int_{\mathbf{R}^n}M(f)^p\,u\,dx\le C^p\int_{\mathbf{R}^n}|f|^p\,U\,dx\,.$$

Using Theorem 9.5.8, we obtain

$$\Big\|\Big(\sum_j|M(f_j)|^p\Big)^{\frac{1}{p}}\Big\|_{L^q}\le C_{n,p,q}\Big\|\Big(\sum_j|f_j|^p\Big)^{\frac{1}{p}}\Big\|_{L^q}\tag{9.5.27}$$

whenever $1<p<q<\infty$, an inequality obtained earlier in (4.6.17).

It turns out that no specific properties of the Hardy–Littlewood maximal function are used in the preceding inequality, and we can obtain a general result along these lines. For simplicity we take the operators T_j in the next theorem to be linear.

Exercises

9.5.1. Let $0<s<1$ and f be in $L^s(X,\mu)$. Show that

$$\big\|f\big\|_{L^s}=\inf\Big\{\int_X|f|u^{-1}\,d\mu:\ \big\|u\big\|_{L^{\frac{s}{1-s}}}\le 1\Big\}$$

and that the infimum is attained.
[*Hint:* Try $u=c|f|^{1-s}$ for a suitable constant c.]

9.5.2. Use the same idea of the proof of Theorem 9.5.1 to prove the following general result: Let μ be a positive measure on a measure space X and let T be a bounded sublinear operator on $L^p(X,\mu)$ for some $1\le p<\infty$. Suppose that $T(f)\ge 0$ for all f in $L^p(X,\mu)$. Prove that for all $f_0\in L^p(X,\mu)$, there exists an $f\in L^p(X,\mu)$ such that

(a) $f_0(x)\le f(x)$ for μ-almost all $x\in X$.

(b) $\big\|f\big\|_{L^p(X)}\le 2\big\|f_0\big\|_{L^p(X)}$.

(c) $T(f)(x)\le 2\big\|T\big\|_{L^p\to L^p}f(x)$ for μ-almost all $x\in X$.

[*Hint:* Try the expression in (9.5.2) starting the sum at $j=0$.]

9.5.3. (*Duoandikoetxea [116]*) Suppose that T is a well defined operator on the union $\bigcup_{1<q<\infty}\bigcup_{w\in A_q} L^q(w)$ that satisfies $\big\|T\big\|_{L^r(v)\to L^r(v)} \le N([v]_{A_r})$ for some increasing function $N:[1,\infty)\to \mathbf{R}^+$. Without using Theorem 9.5.3 prove that for $1<q<r$ and all $v\in A_1$, T maps $L^q(v)$ to $L^q(v)$ with constant depending on q,r,n, and $[v]_{A_1}$.
[*Hint:* Hölder's inequality gives that

$$\big\|T(f)\big\|_{L^q(v)} \le \bigg(\int_{\mathbf{R}^n} |T(f)(x)|^r M(f)(x)^{q-r}\, v(x)\,dx\bigg)^{\frac{1}{r}} \bigg(\int_{\mathbf{R}^n} M(f)(x)^q v(x)\,dx\bigg)^{\frac{r-q}{rq}}.$$

Then use the fact that the weight $M(f)^{\frac{r-q}{r-1}}$ is in A_1 and Exercise 9.1.2.]

9.5.4. Let T be a sublinear operator defined on $\bigcup_{2\le q<\infty} L^q$. Suppose that for all functions f and u we have

$$\int_{\mathbf{R}^n} |T(f)|^2 u\,dx \le \int_{\mathbf{R}^n} |f|^2 M(u)\,dx.$$

Prove that T maps $L^p(\mathbf{R}^n)$ to itself for all $2<p<\infty$.
[*Hint:* Use that

$$\big\|T(f)\big\|_{L^p} = \sup_{\|u\|_{L^{(p/2)'}}\le 1} \bigg(\int_{\mathbf{R}^n} |T(f)|^2 u\,dx\bigg)^{\frac{1}{2}}$$

and Hölder's inequality.]

9.5.5. (*X. C. Li*) Let T be a sublinear operator defined on $\bigcup_{1<q\le 2}\bigcup_{w\in A_q} L^q(w)$. Suppose that T maps $L^2(w)$ to $L^2(w)$ for all weights w that satisfy $w^{-1}\in A_1$. Prove that T maps L^p to itself for all $1<p<2$.
[*Hint:* We have

$$\big\|T(f)\big\|_{L^p} \le \bigg(\int_{\mathbf{R}^n} |T(f)|^2 M(f)^{-(2-p)}\,dx\bigg)^{\frac{1}{2}} \bigg(\int_{\mathbf{R}^n} M(f)^p\,dx\bigg)^{\frac{2-p}{2p}}$$

by Hölder's inequality. Apply the hypothesis to the first term of the product.]

HISTORICAL NOTES

Weighted inequalities can probably be traced back to the beginning of integration, but the A_p condition first appeared in a paper of Rosenblum [272] in a somewhat different form. The characterization of A_p when $n=1$ in terms of the boundedness of the Hardy–Littlewood maximal operator was obtained by Muckenhoupt [237]. The estimate on the norm in (9.1.24) can also be reversed, as shown by Buckley [36]. The simple proof of Theorem 9.1.9 is contained in Lerner's article [209] and yields both the Muckenhoupt theorem and Buckley's optimal growth of the norm of the Hardy–Littlewood maximal operator in terms of the A_p characteristic constant of the weight. Another proof of this result is given by Christ and Fefferman [69]. Versions of Lemma 9.1.10 for balls were first obtained by Besicovitch [23] and independently by Morse [235]. The particular version of Lemma 9.1.10 that appears in the text is adapted from that in de Guzmán [109]. Another version of this lemma is contained in the book of Mattila [227]. The fact that A_∞ is the union of the A_p spaces was independently obtained by Muckenhoupt [238] and Coifman and Fefferman [74]. The latter paper also contains a proof that A_p weights satisfy the crucial reverse Hölder condition. This condition first appeared in the work of Gehring [145] in the following context: If F is a quasiconformal homeomorphism from $\mathbf{R}^n$ into itself, then $|\det(\nabla F)|$ satisfies a reverse Hölder inequality. The characterization of A_1 weights is due to Coifman and Rochberg [84]. The fact that $M(f)^\delta$ is in A_∞ when $\delta < 1$ was previously obtained by Córdoba and Fefferman [92]. The different characterizations of A_∞ (Theorem 9.3.3) are implicit in [237] and [74]. Another characterization of A_∞ in terms of the Gurov-Reshetnyak condition $\sup_Q \frac{1}{|Q|}\int_Q |f - \mathrm{Avg}_Q f|\,dx \le \varepsilon \,\mathrm{Avg}_Q f$ for $f \ge 0$ and $0 < \varepsilon < 2$ was obtained by Korenovskyy, Lerner, and Stokolos [196]. The definition of A_∞ using the reverse Jensen inequality herein was obtained as an equivalent characterization of that space by García-Cuerva and Rubio de Francia [141] (p. 405) and independently by Hruščev [169]. The reverse Hölder condition was extensively studied by Cruz-Uribe and Neugebauer [98].

Weighted inequalities with weights of the form $|x|^a$ for the Hilbert transform were first obtained by Hardy and Littlewood [159] and later by Stein [288] for other singular integrals. The necessity and sufficiency of the A_p condition for the boundedness of the Hilbert transform on weighted L^p spaces was obtained by Hunt, Muckenhoupt, and Wheeden [172]. Historically, the first result relating A_p weights and the Hilbert transform is the Helson-Szegő theorem [162], which says that the Hilbert transform is bounded on $L^2(w)$ if and only if $\log w = u + Hv$, where $u, v \in L^\infty(\mathbf{R})$ and $\|v\|_{L^\infty} < \frac{\pi}{2}$. The Helson-Szegő condition easily implies the A_2 condition, but the only known direct proof for the converse gives $\|v\|_{L^\infty} < \pi$; see Coifman, Jones, and Rubio de Francia [76]. A related result in higher dimensions was obtained by Garnett and Jones [143]. Weighted L^p estimates controlling Calderón–Zygmund operators by the Hardy–Littlewood maximal operator were obtained by Coifman [71]. Coifman and Fefferman [74] extended one-dimensional weighted norm inequalities to higher dimensions and also obtained good lambda inequalities for A_∞ weights for more general singular integrals and maximal singular integrals (Theorem 9.4.3). Bagby and Kurtz [14], and later Alvarez and Pérez [4], gave a sharper version of Theorem 9.4.3, by replacing the good lambda inequality by a rearrangement inequality. See also the related work of Lerner [208]. The result of Exercise 9.4.9 relating the weighted norms of f and $M^\#(f)$ is also valid under weaker assumptions on f; for instance, the condition $M(f) \in L^{p_0}$ can be replaced by the condition $w(\{|f| > t\}) < \infty$ for every $t > 0$; see Kurtz [201]. Using that $\min(M, w)$ is an A_∞ weight with constant independent of M and Fatou's lemma, this condition can be relaxed to $|\{|f| > t\}| < \infty$ for every $t > 0$. A rearrangement inequality relating f and $M^\#(f)$ is given in Bagby and Kurtz [13].

The factorization of A_p weights was conjectured by Muckenhoupt and proved by Jones [178]. The simple proof given in the text can be found in [76]. Extrapolation of operators (Theorem 9.5.3) is due to Rubio de Francia [274]. An alternative proof of this theorem was given later by García-Cuerva [140]. The value of the constant $K(n, p, p_0, [w]_{A_p})$ first appeared in Dragičević, Grafakos, Pereyra, and Petermichl [110]. The present treatment of Theorem 9.5.3, based on crucial Lemma 9.5.4, was communicated to the author by J. M. Martell. One may also consult the related work of Cruz-Uribe, Martell, and Pérez [97]. The simple proof of Theorem 9.5.5 was conceived by J. M. Martell and first appeared in the treatment of extrapolation of operators of many variables; see Grafakos and Martell [151]. The idea of extrapolation can be carried to general pairs of functions, see Cruz-Uribe, Martell, and Pérez [96]. The equivalence between vector-valued inequalities and weighted norm inequalities of Theorem 9.5.8 is also due to Rubio de Francia [275]. The difficult direction in this equivalence is obtained using a minimax principle (see Fan [122]). Alternatively, one can use the factorization theory of Maurey [228], which brings an interesting connection with Banach space theory. The book of García-Cuerva and Rubio de Francia [141] provides an excellent reference on this and other topics related to weighted norm inequalities.

A primordial double-weighted norm inequality is the observation of Fefferman and Stein [129] that the maximal function maps $L^p(M(w))$ to $L^p(w)$ for nonnegative measurable functions w (Exercise 9.1.8). Sawyer [278] obtained that the condition $\sup_Q \big(\int_Q v^{1-p'} dx \big)^{-1} \int_Q M(v^{1-p'} \chi_Q)^p w\, dx < \infty$ provides a characterization of all pairs of weights (v,w) for which the Hardy–Littlewood maximal operator M maps $L^p(v)$ to $L^p(w)$. Simpler proofs of this result were obtained by Cruz-Uribe [95] and Verbitsky [324]. The fact that Sawyer's condition reduces to the usual A_p condition when $v=w$ was shown by Hunt, Kurtz, and Neugebauer [171]. The two-weight problem for singular integrals is more delicate, since they are not necessarily bounded from $L^p(M(w))$ to $L^p(w)$. Known results in this direction are that singular integrals map $L^p(M^{[p]+1}(w))$ to $L^p(w)$, where M^r denotes the rth iterate of the maximal operator. See Wilson [333] (for $1<p<2$) and Pérez [259] for the remaining p's. A necessary condition for the boundedness of the Hilbert transform from $L^p(v)$ to $L^p(w)$ was obtained by Muckenhoupt and Wheeden [239]. A necessary and sufficient such condition is yet to be found. A class of multiple weights that satisfy a vector A_p condition has been introduced and studied in the article of Lerner, Ombrosi, Pérez, Torres, and Trujillo-González [210].

For an approach to two-weighted inequalities using Bellman functions, we refer to the article of Nazarov, Treil, and Volberg [247]. The notion of Bellman functions originated in control theory; the article [248] of the previous authors analyzes the connections between optimal control and harmonic analysis. Bellman functions have been used to derive estimates for the norms of classical operators on weighted Lebesgue spaces; for instance, Petermichl [264] showed that for $w \in A_2(\mathbf{R})$, the norm of the Hilbert transform from $L^2(\mathbf{R},w)$ to $L^2(\mathbf{R},w)$ is bounded by a constant times the characteristic constant $[w]_{A_2}$.

The theory of A_p weights in this chapter carries through to the situation in which Lebesgue measure is replaced by a general doubling measure. This theory also has a substantial analogue when the underlying measure is nondoubling but satisfies $\mu(\partial Q)=0$ for all cubes Q in $\mathbf{R}^n$ with sides parallel to the axes; see Orobitg and Pérez [253]. A thorough account of weighted Littlewood–Paley theory and exponential-square function integrability is contained in the book of Wilson [334].

Chapter 10
Boundedness and Convergence of Fourier Integrals

In this chapter we return to fundamental questions in Fourier analysis related to convergence of Fourier series and Fourier integrals. Our main goal is to understand in what sense the inversion property of the Fourier transform

$$f(x) = \int_{\mathbf{R}^n} \widehat{f}(\xi) e^{2\pi i x\cdot\xi}\, d\xi$$

holds when f is a function on $\mathbf{R}^n$. This question is equivalent to the corresponding question for the Fourier series

$$f(x) = \sum_{m\in\mathbf{Z}^n} \widehat{f}(m) e^{2\pi i x\cdot m}$$

when f is a function on $\mathbf{T}^n$. The main problem is that the function (or sequence) $\widehat{f}$ may not be integrable and the convergence of the preceding integral (or series) needs to be suitably interpreted. To address this issue, a summability method is employed. This is achieved by the introduction of a localizing factor $\Phi(\xi/R)$, leading to the study of the convergence of the expressions

$$\int_{\mathbf{R}^n} \Phi(\xi/R)\widehat{f}(\xi) e^{2\pi i x\cdot\xi}\, d\xi$$

as $R \to \infty$. Here Φ is a function on $\mathbf{R}^n$ that decays sufficiently rapidly at infinity and satisfies $\Phi(0) = 1$. For instance, we may take $\Phi = \chi_{B(0,1)}$, where $B(0,1)$ is the unit ball in $\mathbf{R}^n$. Analogous summability methods arise in the torus.

An interesting case arises when $\Phi(\xi) = (1-|\xi|^2)_+^\lambda$, $\lambda \geq 0$, in which we obtain the Bochner–Riesz means introduced by Riesz when $n = 1$ and $\lambda = 0$ and Bochner for $n \geq 2$ and general $\lambda > 0$. The question is whether the Bochner–Riesz means

$$\sum_{m_1^2+\cdots+m_n^2\leq R^2} \left(1 - \frac{m_1^2+\cdots+m_n^2}{R^2}\right)^\lambda \widehat{f}(m_1,\dots,m_n)\, e^{2\pi i(m_1x_1+\cdots+m_nx_n)}$$

L. Grafakos, *Modern Fourier Analysis*, DOI: 10.1007/978-0-387-09434-2_10,

converge in L^p. This question is equivalent to whether the function $(1-|\xi|^2)_+^\lambda$ is an L^p multiplier on $\mathbf{R}^n$ and is investigated in this chapter. Analogous questions concerning the almost everywhere convergence of these families are also studied.

10.1 The Multiplier Problem for the Ball

In this section we show that the characteristic function of the unit disk in $\mathbf{R}^2$ is not an L^p multiplier when $p \neq 2$. This implies the same conclusion in dimensions $n \geq 3$, since sections of higher-dimensional balls are disks and by Theorem 2.5.16 we have that if $\chi_{B(0,r)} \notin \mathscr{M}_p(\mathbf{R}^2)$ for all $r > 0$, then $\chi_{B(0,1)} \notin \mathscr{M}_p(\mathbf{R}^n)$ for any $n \geq 3$.

10.1.1 Sprouting of Triangles

We begin with a certain geometric construction that at first sight has no apparent relationship to the multiplier problem for the ball in $\mathbf{R}^n$. Given a triangle ABC with base $b = AB$ and height h_0 we let M be the midpoint of AB. We construct two other triangles AMF and BME from ABC as follows. We fix a height $h_1 > h_0$ and we extend the sides AC and BC in the direction away from its base until they reach a certain height h_1. We let E be the unique point on the line passing through the points B and C such that the triangle EMB has height h_1. Similarly, F is uniquely chosen on the line through A and C so that the triangle AMF has height h_1.

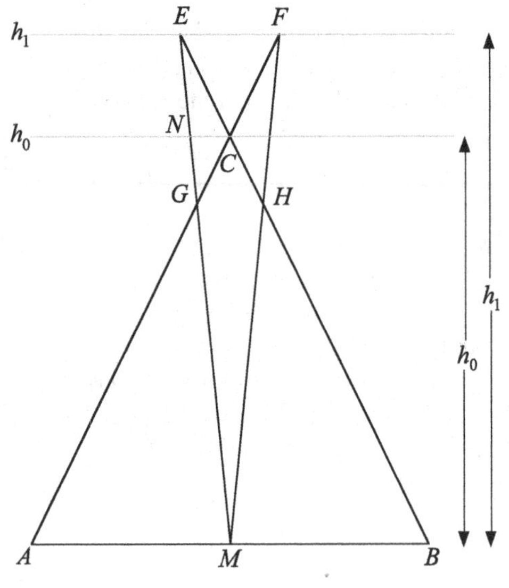

Fig. 10.1 The sprouting of the triangle ABC.

The triangle ABC now gives rise to two triangles AMF and BME called the *sprouts* of ABC. The union of the two sprouts AMF and BME is called the *sprouted figure* obtained from ABC and is denoted by $\mathrm{Spr}(ABC)$. Clearly $\mathrm{Spr}(ABC)$ contains ABC. We call the difference

$$\mathrm{Spr}(ABC) \setminus ABC$$

the *arms* of the sprouted figure. The sprouted figure $\mathrm{Spr}(ABC)$ has two arms of equal area, the triangles EGC and FCH as shown in Figure 10.1, and we can precisely compute the area of each arm. One may easily check (see Exercise 10.1.1) that

$$\text{Area (each arm of Spr}(ABC)) = \frac{b}{2}\frac{(h_1-h_0)^2}{2h_1-h_0}, \tag{10.1.1}$$

where $b = AB$.

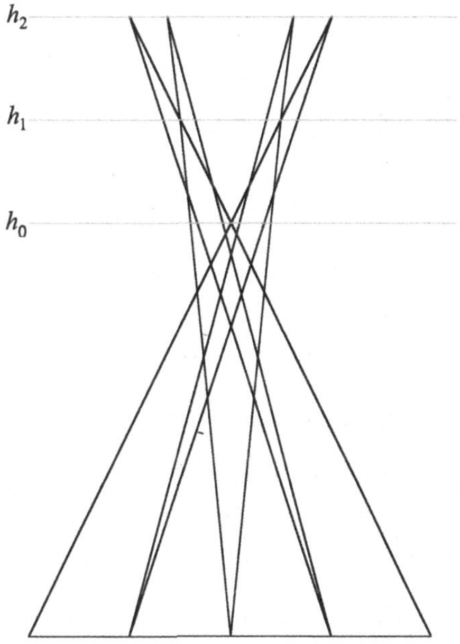

Fig. 10.2 The second step of the construction.

We start with an isosceles triangle $\Lambda = ABC$ in $\mathbf{R}^2$ with base AB of length $b_0 = \varepsilon$ and height $MC = h_0 = \varepsilon$, where M is the midpoint of AB. We define the heights

$$\begin{aligned} h_1 &= \left(1+\frac{1}{2}\right)\varepsilon, \\ h_2 &= \left(1+\frac{1}{2}+\frac{1}{3}\right)\varepsilon, \\ &\dots \\ h_j &= \left(1+\frac{1}{2}+\dots+\frac{1}{j+1}\right)\varepsilon. \end{aligned}$$

We apply the previously described sprouting procedure to Λ to obtain two sprouts $\Lambda_1 = AMF$ and $\Lambda_2 = EMB$, as in Figure 10.1, each with height h_1 and base length $b_0/2$. We now apply the same procedure to the triangles Λ_1 and Λ_2. We then obtain two sprouts Λ_{11} and Λ_{12} from Λ_1 and two sprouts Λ_{21} and Λ_{22} from Λ_2, a total of four sprouts with height h_2. See Figure 10.2. We continue this process, obtaining at the jth step 2^j sprouts $\Lambda_{r_1 \dots r_j}$, $r_1, \dots, r_j \in \{1,2\}$ each with base length $b_j = 2^{-j} b_0$ and height h_j. We stop this process when the kth step is completed.

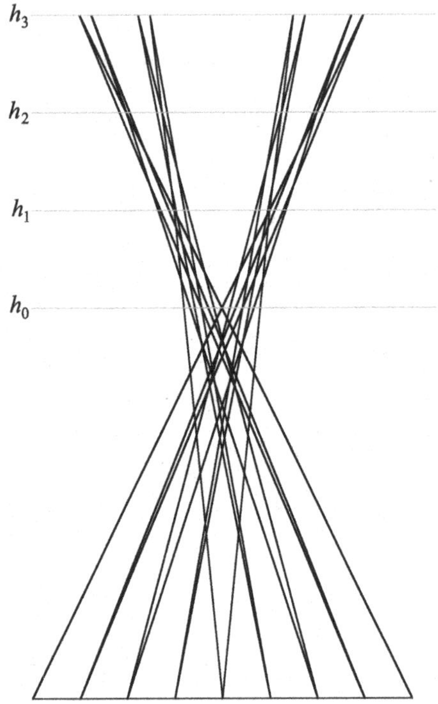

Fig. 10.3 The third step of the construction.

We let $E(\varepsilon, k)$ be the union of the triangles $\Lambda_{r_1 \dots r_k}$ over all sequences r_j of 1's and 2's. We obtain an estimate for the area of $E(\varepsilon, k)$ by adding to the area of Λ the areas of the arms of all the sprouted figures obtained during the construction. By (10.1.1) we have that each of the 2^j arms obtained at the jth step has area

$$\frac{b_{j-1}}{2} \frac{(h_j - h_{j-1})^2}{2h_j - h_{j-1}}.$$

Summing over all these areas and adding the area of the original triangle, we obtain the estimate

$$
\begin{aligned}
|E(\varepsilon,k)| &= \frac{1}{2}\varepsilon^2 + \sum_{j=1}^{k} 2^j \frac{b_{j-1}}{2}\frac{(h_j - h_{j-1})^2}{2h_j - h_{j-1}} \\
&\le \frac{1}{2}\varepsilon^2 + \sum_{j=1}^{k} 2^j \frac{2^{-(j-1)}b_0}{2}\frac{\varepsilon^2}{(j+1)^2\varepsilon} \\
&\le \frac{1}{2}\varepsilon^2 + \sum_{j=2}^{\infty} \frac{\varepsilon^2}{j^2} \le \Big(\frac{1}{2}+\frac{\pi^2}{6}-1\Big)\varepsilon^2 \\
&\le \frac{3}{2}\varepsilon^2,
\end{aligned}
$$

where we used the fact that $2h_j - h_{j-1} \ge \varepsilon$ for all $j \ge 1$.

Having completed the construction of the set $E(\varepsilon,k)$, we are now in a position to indicate some of the ideas that appear in the solution of the Kakeya problem. We first observe that no matter what k is, the measure of the set $E(\varepsilon,k)$ can be made as small as we wish if we take ε small enough. Our purpose is to make a needle of infinitesimal width and unit length move continuously from one side of this angle to the other utilizing each sprouted triangle in succession. To achieve this, we need to apply a similar construction to any of the 2^k triangles that make up the set $E(\varepsilon,k)$ and repeat the sprouting procedure a large enough number of times. We refer to [99] for details. An elaborate construction of this sort yields a set within which the needle can be turned only through a fixed angle. But adjoining a few such sets together allows us to rotate a needle through a half-turn within a set that still has arbitrarily small area. This is the idea used to solve the aforementioned needle problem.

10.1.2 The counterexample

We now return to the multiplier problem for the ball, which has an interesting connection with the Kakeya needle problem.

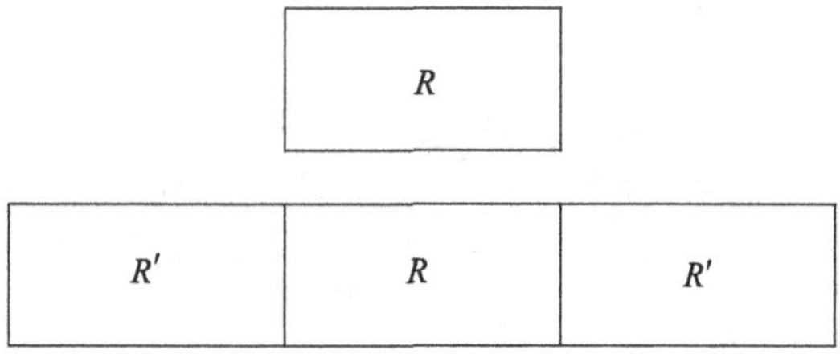

Fig. 10.4 A rectangle R and its adjacent rectangles R'.

In the discussion that follows we employ the following notation. Given a rectangle R in $\mathbf{R}^2$, we let R' be two copies of R adjacent to R along its shortest side so that $R \cup R'$ has the same width as R but three times its length. See Figure 10.4.

We need the following lemma.

Lemma 10.1.1. *Let $\delta > 0$ be a given number. Then there exists a measurable subset E of $\mathbf{R}^2$ and a finite collection of rectangles R_j in $\mathbf{R}^2$ such that*

(1) The R_j's are pairwise disjoint.
(2) We have $1/2 \le |E| \le 3/2$.
(3) We have $|E| \le \delta \sum_j |R_j|$.
(4) For all j we have $|R'_j \cap E| \ge \frac{1}{12}|R_j|$.

Proof. We start with an isosceles triangle ABC in the plane with height 1 and base AB, where $A = (0,0)$ and $B = (1,0)$. Given $\delta > 0$, we find a positive integer k such that $k+2 > e^{1/\delta}$. For this k we set $E = E(1,k)$, the set constructed earlier with $\varepsilon = 1$. We then have $1/2 \le |E| \le 3/2$; thus (2) is satisfied.

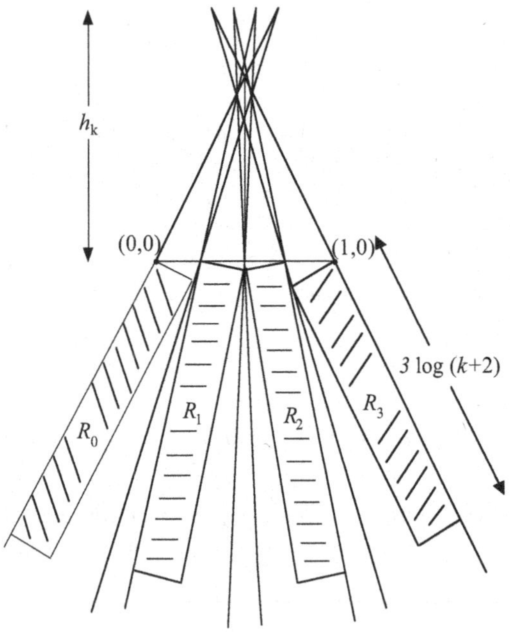

Fig. 10.5 The rectangles R_j.

Recall that each dyadic interval $[j2^{-k},(j+1)2^{-k}]$ in $[0,1]$ is the base of exactly one sprouted triangle $A_jB_jC_j$, where $j \in \{0,1,\dots,2^k-1\}$. Here we set $A_j = (j2^{-k},0)$, $B_j = ((j+1)2^{-k},0)$, and C_j the other vertex of the sprouted triangle. We define a rectangle R_j inside the angle $\angle A_jC_jB_j$ as in Figure 10.6. The rectangle R_j is defined so that one of its vertices is either A_j or B_j and the length of its longest side is $3\log(k+2)$.

We now make some calculations. First we observe that the longest possible length that either A_jC_j or B_jC_j can achieve is $\sqrt{5}h_k/2$. By symmetry we may assume that the length of A_jC_j is larger than that of B_jC_j as in Figure 10.6. We now have that

$$\frac{\sqrt{5}}{2}h_k < \frac{3}{2}\left(1+\frac{1}{2}+\cdots+\frac{1}{k+1}\right) < \frac{3}{2}\big(1+\log(k+1)\big) < 3\log(k+2),$$

since $k \geq 1$ and $e < 3$. Hence R'_j contains the triangle $A_jB_jC_j$. We also have that

$$h_k = 1+\frac{1}{2}+\cdots+\frac{1}{k+1} > \log(k+2).$$

Using these two facts, we obtain

$$|R'_j \cap E| \geq \text{Area}(A_jB_jC_j) = \frac{1}{2}2^{-k}h_k > 2^{-k-1}\log(k+2). \tag{10.1.2}$$

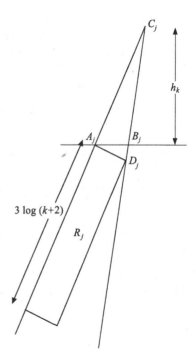

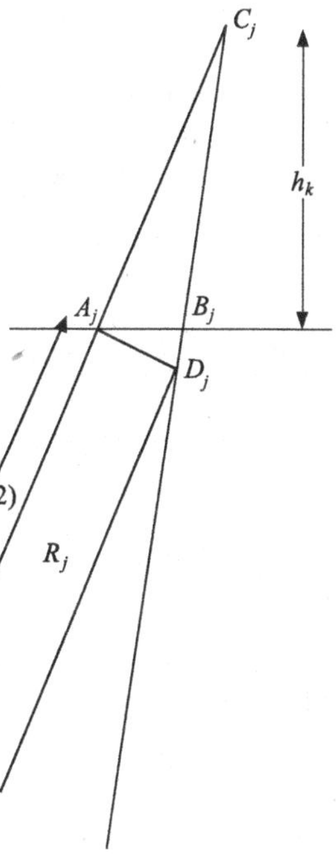

Fig. 10.6 A closer look at R_j.

Denote by $|XY|$ the length of the line segment through the points X and Y. The law of sines applied to the triangle $A_jB_jD_j$ gives

$$|A_jD_j| = 2^{-k}\frac{\sin(\angle A_jB_jD_j)}{\sin(\angle A_jD_jB_j)} \leq \frac{2^{-k}}{\cos(\angle A_jC_jB_j)}. \tag{10.1.3}$$

But the law of cosines applied to the triangle $A_jB_jC_j$ combined with the estimates $h_k \leq |A_jC_j|, |B_jC_j| \leq \sqrt{5}h_k/2$ give that

$$\cos(\angle A_jC_jB_j) \geq \frac{h_k^2 + h_k^2 - (2^{-k})^2}{2\frac{5}{4}h_k^2} \geq \frac{4}{5} - \frac{2}{5} \cdot \frac{1}{4} \geq \frac{1}{2}. \tag{10.1.4}$$

Combining (10.1.3) and (10.1.4), we obtain

$$|A_jD_j| \leq 2^{-k+1} = 2|A_jB_j|.$$

Using this fact and (10.1.2), we deduce

$$|R_j' \cap E| \geq 2^{-k-1}\log(k+2) = \frac{1}{12}2^{-k+1}3\log(k+2) \geq \frac{1}{12}|R_j|,$$

which proves the required conclusion (4).

Conclusion (1) in Lemma 10.1.1 follows from the fact that the regions inside the angles $\angle A_jC_jB_j$ and under the triangles $A_jC_jB_j$ are pairwise disjoint. This is shown in Figure 10.5. This can be proved rigorously by a careful examination of the construction of the sprouted triangles $A_jC_jB_j$, but the details are omitted.

It remains to prove (3). To achieve this we first estimate the length of the line segment A_jD_j from below. The law of sines gives

$$\frac{|A_jD_j|}{\sin(\angle A_jB_jD_j)} = \frac{2^{-k}}{\sin(\angle A_jD_jB_j)},$$

from which we obtain that

$$|A_jD_j| \geq 2^{-k}\sin(\angle A_jB_jD_j) \geq 2^{-k-1}\angle A_jB_jD_j \geq 2^{-k-1}\angle B_jA_jC_j.$$

(All angles are measured in radians.) But the smallest possible value of the angle $\angle B_jA_jC_j$ is attained when $j=0$, in which case $\angle B_0A_0C_0 = \arctan 2 > 1$. This gives that

$$|A_jD_j| \geq 2^{-k-1}.$$

It follows that each R_j has area at least $2^{-k-1}3\log(k+2)$. Therefore,

$$\left|\bigcup_{j=0}^{2^k-1} R_j\right| = \sum_{j=0}^{2^k-1}|R_j| \geq 2^k 2^{-k-1}3\log(k+2) \geq |E|\log(k+2) \geq \frac{|E|}{\delta},$$

since $|E| \leq 3/2$ and k was chosen so that $k+2 > e^{1/\delta}$. □

Next we have a calculation involving the Fourier transforms of characteristic functions of rectangles.

Proposition 10.1.2. *Let R be a rectangle whose center is the origin in $\mathbf{R}^2$ and let v be a unit vector parallel to its longest side. Consider the half-plane*

$$\mathscr{H} = \{x \in \mathbf{R}^2 : x \cdot v \geq 0\}$$

and the multiplier operator

$$S_{\mathscr{H}}(f) = (\widehat{f}\chi_{\mathscr{H}})^{\vee}.$$

Then we have $|S_{\mathscr{H}}(\chi_R)| \geq \frac{1}{10}\chi_{R'}$.

Remark 10.1.3. Applying a translation, we see that the same conclusion is valid for any rectangle in $\mathbf{R}^2$ whose longest side is parallel to v.

Proof. Applying a rotation, we reduce the problem to the case $R = [-a,a] \times [-b,b]$, where $0 < a \leq b < \infty$, and $v = e_2 = (0,1)$. Since the Fourier transform acts in each variable independently, we have the identity

$$\begin{aligned} S_{\mathscr{H}}(\chi_R)(x_1,x_2) &= \chi_{[-a,a]}(x_1)\big(\widehat{\chi_{[-b,b]}}\chi_{[0,\infty)}\big)^{\vee}(x_2) \\ &= \chi_{[-a,a]}(x_1)\,\frac{I+iH}{2}(\chi_{[-b,b]})(x_2). \end{aligned}$$

It follows that

$$\begin{aligned} |S_{\mathscr{H}}(\chi_R)(x_1,x_2)| &\geq \frac{1}{2}\chi_{[-a,a]}(x_1)|H(\chi_{[-b,b]})(x_2)| \\ &= \frac{1}{2\pi}\chi_{[-a,a]}(x_1)\left|\log\left|\frac{x_2+b}{x_2-b}\right|\right|. \end{aligned}$$

But for $(x_1,x_2) \in R'$ we have $\chi_{[-a,a]}(x_1) = 1$ and $b < |x_2| < 3b$. So we have two cases, $b < x_2 < 3b$ and $-3b < x_2 < -b$. When $b < x_2 < 3b$ we see that

$$\left|\frac{x_2+b}{x_2-b}\right| = \frac{x_2+b}{x_2-b} > 2,$$

and similarly, when $-3b < x_2 < -b$ we have

$$\left|\frac{x_2-b}{x_2+b}\right| = \frac{b-x_2}{-b-x_2} > 2.$$

It follows that for $(x_1,x_2) \in R'$ the lower estimate is valid:

$$|S_{\mathscr{H}}(\chi_R)(x_1,x_2)| \geq \frac{\log 2}{2\pi} \geq \frac{1}{10}.$$

□

Next we have a lemma regarding vector-valued inequalities of half-plane multipliers.

Lemma 10.1.4. *Let* $v_1, v_2, \ldots, v_j, \ldots$ *be a sequence of unit vectors in* $\mathbf{R}^2$. *Define the half-planes*

$$\mathscr{H}_j = \{x \in \mathbf{R}^2 : x \cdot v_j \geq 0\} \tag{10.1.5}$$

and linear operators

$$S_{\mathscr{H}_j}(f) = (\widehat{f}\,\chi_{\mathscr{H}_j})^{\vee}.$$

Assume that the disk multiplier operator

$$T(f) = (\widehat{f}\,\chi_{B(0,1)})^{\vee}$$

maps $L^p(\mathbf{R}^2)$ *to itself with norm* $B_p < \infty$. *Then we have the inequality*

$$\left\|\Big(\sum_j |S_{\mathscr{H}_j}(f_j)|^2\Big)^{\frac{1}{2}}\right\|_{L^p} \le B_p \left\|\Big(\sum_j |f_j|^2\Big)^{\frac{1}{2}}\right\|_{L^p} \tag{10.1.6}$$

for all bounded and compactly supported functions f_j.

Proof. We prove the lemma for Schwartz functions f_j and we obtain the general case by a simple limiting argument. We define disks $D_{j,R} = \{x \in \mathbf{R}^2 : |x - Rv_j| \le R\}$ and we let

$$T_{j,R}(f) = (\widehat{f}\,\chi_{D_{j,R}})^{\vee}$$

be the multiplier operator associated with the disk $D_{j,R}$. We observe that $\chi_{D_{j,R}} \to \chi_{\mathscr{H}_j}$ pointwise as $R \to \infty$, as shown in Figure 10.7.

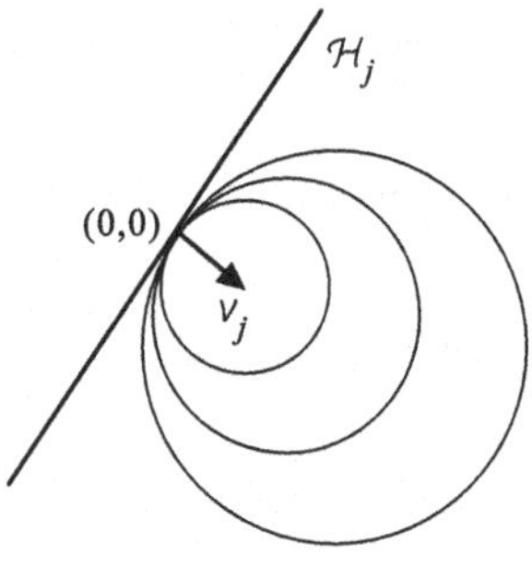

Fig. 10.7 A sequence of disks converging to a half-plane.

For $f \in \mathscr{S}(\mathbf{R}^2)$ and every $x \in \mathbf{R}^2$ we have

$$\lim_{R\to\infty} T_{j,R}(f)(x) = S_{\mathscr{H}_j}(f)(x)$$

by passing the limit inside the convergent integral. Fatou's lemma now yields

$$\left\|\Big(\sum_j |S_{\mathscr{H}_j}(f_j)|^2\Big)^{\frac{1}{2}}\right\|_{L^p} \le \liminf_{R\to\infty} \left\|\Big(\sum_j |T_{j,R}(f_j)|^2\Big)^{\frac{1}{2}}\right\|_{L^p}. \tag{10.1.7}$$

Next we observe that the following identity is valid:

$$T_{j,R}(f)(x) = e^{2\pi i R v_j \cdot x} T_R(e^{-2\pi i R v_j \cdot (\cdot)} f)(x), \tag{10.1.8}$$

where T_R is the multiplier operator $T_R(f) = (\widehat{f}\chi_{B(0,R)})^{\vee}$. Setting $g_j = e^{-2\pi i R v_j \cdot (\cdot)} f_j$ and using (10.1.7) and (10.1.8), we deduce

$$\Big\|\Big(\sum_j |S_{\mathscr{H}_j}(f_j)|^2\Big)^{\frac{1}{2}}\Big\|_{L^p} \le \liminf_{R\to\infty} \Big\|\Big(\sum_j |T_R(g_j)|^2\Big)^{\frac{1}{2}}\Big\|_{L^p}. \tag{10.1.9}$$

Observe that the operator T_R is L^p bounded with the same norm B_p as T in view of identity (2.5.15). Applying Theorem 4.5.1, we obtain that the last term in (10.1.9) is bounded by

$$\liminf_{R\to\infty} \big\|T_R\big\|_{L^p\to L^p} \Big\|\Big(\sum_j |g_j|^2\Big)^{\frac{1}{2}}\Big\|_{L^p} = B_p \Big\|\Big(\sum_j |f_j|^2\Big)^{\frac{1}{2}}\Big\|_{L^p}.$$

Combining this inequality with (10.1.9), we obtain (10.1.6). □

We have now completed all the preliminary material we need to prove that the characteristic function of the unit disk in $\mathbf{R}^2$ is not an L^p multiplier if $p \neq 2$.

Theorem 10.1.5. *The characteristic function of the unit ball in $\mathbf{R}^n$ is not an L^p multiplier when $1 < p \neq 2 < \infty$.*

Proof. As mentioned earlier, in view of Theorem 2.5.16, it suffices to prove the result in dimension $n = 2$. By duality it suffices to prove the result when $p > 2$. Suppose that $\chi_{B(0,1)} \in \mathscr{M}_p(\mathbf{R}^2)$ for some $p > 2$, say with norm $B_p < \infty$.

Suppose that $\delta > 0$ is given. Let E and R_j be as in Lemma 10.1.1. We let $f_j = \chi_{R_j}$. Let v_j be the unit vector parallel to the long side of R_j and let H_j be the half-plane defined as in (10.1.5). Using Proposition 10.1.2, we obtain

$$\begin{aligned}
\int_E \sum_j |S_{\mathscr{H}_j}(f_j)(x)|^2\,dx &= \sum_j \int_E |S_{\mathscr{H}_j}(f_j)(x)|^2\,dx \\
&\ge \sum_j \int_E \frac{1}{10^2}\chi_{R'_j}(x)\,dx \\
&= \frac{1}{100}\sum_j |E\cap R'_j| \\
&\ge \frac{1}{1200}\sum_j |R_j|,
\end{aligned} \tag{10.1.10}$$

where we used condition (4) of Lemma 10.1.1 in the last inequality. Hölder's inequality with exponents $p/2$ and $(p/2)' = p/(p-2)$ gives

$$\begin{aligned}
\int_E \sum_j |S_{\mathscr{H}_j}(f_j)(x)|^2\,dx &\le |E|^{\frac{p-2}{p}} \Big\|\Big(\sum_j |S_{\mathscr{H}_j}(f_j)|^2\Big)^{\frac{1}{2}}\Big\|_{L^p}^2 \\
&\le B_p^2 |E|^{\frac{p-2}{p}} \Big\|\Big(\sum_j |f_j|^2\Big)^{\frac{1}{2}}\Big\|_{L^p}^2 \\
&= B_p^2 |E|^{\frac{p-2}{p}} \Big(\sum_j |R_j|\Big)^{\frac{2}{p}} \\
&\le B_p^2 \delta^{\frac{p-2}{p}} \sum_j |R_j|,
\end{aligned} \tag{10.1.11}$$

where we used Lemma 10.1.4, the disjointness of the R_j's, and condition (3) of Lemma 10.1.1 successively. Combining (10.1.10) with (10.1.11), we obtain the inequality

$$\sum_j |R_j| \le 1200 B_p \delta^{\frac{p-2}{p}} \sum_j |R_j| ,$$

which provides a contradiction when δ is very small. □

Exercises

10.1.1. Prove identity (10.1.1).
[*Hint:* With the notation of Figure 10.1, first prove

$$\frac{h_1 - h_0}{h_1} = \frac{NC}{b/2}, \qquad \frac{\text{height}(NGC)}{h_0} = \frac{NC}{NC + b/2}$$

using similar triangles.]

10.1.2. Given a rectangle R, let R'' denote either of the two parts that make up R'. Prove that for any $k \in \mathbf{Z}^+$ and any $\delta > 0$, there exist rectangles S_j in $\mathbf{R}^2$, $0 \le j < 2^k$, with dimensions proportionate to $2^{-k} \times \log(k+1)$,

$$\left| \bigcup_{j=0}^{2^k-1} S_j \right| < \delta ,$$

such that for some choice of S_j'', the S_j'''s are disjoint.
[*Hint:* Consider the 2^k triangles that make up the set $E(\varepsilon, k)$ and choose each rectangle S_j inside a corresponding triangle. Then the parts of the S_j''s that point downward are disjoint. Choose ε depending on δ.]

10.1.3. Is the characteristic function of the cylinder

$$\{(\xi_1, \xi_2, \xi_3) \in \mathbf{R}^3 : \xi_1^2 + \xi_2^2 < 1\}$$

a Fourier multiplier on $L^p(\mathbf{R}^3)$ for $1 < p < \infty$ and $p \ne 2$?

10.1.4. Modify the ideas of the proof of Lemma 10.1.4 to show that the characteristic function of the set

$$\{(\xi_1, \xi_2) \in \mathbf{R}^2 : \xi_2 > \xi_1^2\}$$

is not in $\mathscr{M}_p(\mathbf{R}^2)$ when $p \ne 2$.
[*Hint:* Let $\mathscr{H}_j = \{(\xi_1, \xi_2) \in \mathbf{R}^2 : \xi_2 > s_j \xi_1^2\}$ for some $s_j > 0$. The parabolic regions $\{(\xi_1, \xi_2) \in \mathbf{R}^2 : \xi_2 + R\frac{s_j^2}{4} > \frac{1}{R}(\xi_1 + R\frac{s_j}{2})^2\}$ are contained in $\mathscr{H}_j$, are translates of the region $\{(\xi_1, \xi_2) \in \mathbf{R}^2 : \xi_2 > \frac{1}{R}\xi_1^2\}$, and tend to $\mathscr{H}_j$ as $R \to \infty$.]

10.1.5. Let $a_1,\dots,a_n > 0$. Show that the characteristic function of the ellipsoid

$$\Big\{(\xi,\dots,\xi_n)\in\mathbf{R}^n:\ \frac{\xi_1^2}{a_1^2}+\dots+\frac{\xi_n^2}{a_n^2}<1\Big\}$$

is not in $\mathscr{M}_p(\mathbf{R}^n)$ when $p\neq 2$.
[*Hint:* Think about dilations.]

10.2 Bochner–Riesz Means and the Carleson–Sjölin Theorem

We now address the problem of norm convergence for the Bochner–Riesz means. In this section we provide a satisfactory answer in dimension $n=2$, although a key ingredient required in the proof is left for the next section.

Definition 10.2.1. For a function f on $\mathbf{R}^n$ we define its *Bochner–Riesz means* of complex order λ with $\operatorname{Re}\lambda>0$ to be the family of operators

$$B_R^\lambda(f)(x)=\int_{\mathbf{R}^n}(1-|\xi/R|^2)_+^\lambda\widehat{f}(\xi)e^{2\pi i x\cdot\xi}\,d\xi,\qquad R>0.$$

We are interested in the convergence of the family $B_R^\lambda(f)$ as $R\to\infty$. Observe that when $R\to\infty$ and f is a Schwartz function, the sequence $B_R^\lambda(f)$ converges pointwise to f. Does it also converge in norm? Using Exercise 10.2.1, this question is equivalent to whether the function $(1-|\xi|^2)_+^\lambda$ is an L^p multiplier [it lies in $\mathscr{M}_p(\mathbf{R}^n)$], that is, whether the linear operator

$$B^\lambda(f)(x)=\int_{\mathbf{R}^n}(1-|\xi|^2)_+^\lambda\widehat{f}(\xi)e^{2\pi i x\cdot\xi}\,d\xi$$

maps $L^p(\mathbf{R}^n)$ to itself. The question that arises is given λ with $\operatorname{Re}\lambda>0$ find the range of p's for which $(1-|\xi|^2)_+^\lambda$ is an $L^p(\mathbf{R}^n)$ Fourier multiplier; this question is investigated in this section when $n=2$.

The analogous question for the operators B_R^λ on the n-torus introduced in Definition 3.4.1 is also equivalent to the fact that the function $(1-|\xi|^2)_+^\lambda$ is a Fourier multiplier in $\mathscr{M}_p(\mathbf{R}^n)$. This was shown in Corollary 3.6.10. Therefore the Bochner–Riesz problem for the torus $\mathbf{T}^n$ and the Euclidean space $\mathbf{R}^n$ are equivalent. Here we focus attention on the Euclidean case, and we start our investigation by studying the kernel of the operator B^λ.

10.2.1 The Bochner–Riesz Kernel and Simple Estimates

In view of the last identity in Appendix B.5, B^λ is a convolution operator with kernel

$$K_\lambda(x) = \frac{\Gamma(\lambda+1)}{\pi^\lambda} \frac{J_{\frac{n}{2}+\lambda}(2\pi|x|)}{|x|^{\frac{n}{2}+\lambda}}. \tag{10.2.1}$$

Following Appendix B.6, we have for $|x| \le 1$,

$$|K_\lambda(x)| = \frac{|\Gamma(\lambda+1)|}{|\pi^\lambda|} \frac{|J_{\frac{n}{2}+\lambda}(2\pi|x|)|}{|x|^{\frac{n}{2}+\operatorname{Re}\lambda}} \le \frac{\Gamma(\operatorname{Re}\lambda+1)}{\pi^{\operatorname{Re}\lambda}} C_0\, e^{\pi^2|\operatorname{Im}\lambda|^2},$$

where C_0 is a constant that depends only on $n/2+\operatorname{Re}\lambda$. Consequently, $K_\lambda(x)$ is bounded by a constant (that grows at most exponentially in $|\operatorname{Im}\lambda|^2$) in the unit ball of $\mathbf{R}^n$.

For $|x| \ge 1$, following Appendix B.7, we have

$$|K_\lambda(x)| = \frac{|\Gamma(\lambda+1)|}{|\pi^\lambda|} \frac{|J_{\frac{n}{2}+\lambda}(2\pi|x|)|}{|x|^{\frac{n}{2}+\operatorname{Re}\lambda}} \le C_0 \frac{e^{\pi|\operatorname{Im}\lambda|+\pi^2|\operatorname{Im}\lambda|^2}}{\pi^{\operatorname{Re}\lambda}(2\pi|x|)^{\frac{1}{2}}} \frac{\Gamma(\operatorname{Re}\lambda+1)}{|x|^{\frac{n}{2}+\operatorname{Re}\lambda}},$$

where C_0 depends only on $n/2+\operatorname{Re}\lambda$. Thus $K_\lambda(x)$ is pointwise bounded by a constant (that grows at most exponentially in $|\operatorname{Im}\lambda|$) times $|x|^{-\frac{n+1}{2}-\operatorname{Re}\lambda}$ for $|x| \ge 1$.

Combining these two observations, we obtain that for $\operatorname{Re}\lambda > \frac{n-1}{2}$, K_λ is a smooth integrable function on $\mathbf{R}^n$. Hence B^λ is a bounded operator on L^p for $1 \le p \le \infty$.

Proposition 10.2.2. *For all $1 \le p \le \infty$ and $\lambda > \frac{n-1}{2}$, B^λ is a bounded operator on $L^p(\mathbf{R}^n)$ with norm at most $C_1\, e^{c_1|\operatorname{Im}\lambda|^2}$, where C_1, c_1 depend only on $n, \operatorname{Re}\lambda$.*

Proof. The ingredients of the proof have already been discussed. □

We refer to Exercise 10.2.8 for an analogous result for the maximal Bochner–Riesz operator.

According to the asymptotics for Bessel functions in Appendix B.8, K_λ is a smooth function equal to

$$\frac{\Gamma(\lambda+1)}{\pi^{\lambda+1}} \frac{\cos(2\pi|x| - \frac{\pi(n+1)}{4} - \frac{\pi\lambda}{2})}{|x|^{\frac{n+1}{2}+\lambda}} + O(|x|^{-\frac{n+3}{2}-\lambda}) \tag{10.2.2}$$

for $|x| \ge 1$. It is natural to examine whether the operators B^λ are bounded on certain L^p spaces by testing them on specific functions. This may provide some indication as to the range of p's for which these operators may be bounded on L^p.

Proposition 10.2.3. *When $\lambda > 0$ and $p \le \frac{2n}{n+1+2\lambda}$ or $p \ge \frac{2n}{n-1-2\lambda}$, the operators B^λ are not bounded on $L^p(\mathbf{R}^n)$.*

Proof. Let h be a Schwartz function whose Fourier transform is equal to 1 on the ball $B(0,2)$ and vanishes off the ball $B(0,3)$. Then

$$B^\lambda(h)(x) = \int_{|\xi|\le 1} (1-|\xi|^2)^\lambda e^{2\pi i \xi\cdot x}\, dx = K_\lambda(x),$$

and it suffices to show that K_λ is not in $L^p(\mathbf{R}^n)$ for the claimed range of p's. Notice that

$$\sqrt{2}/2 \le \cos(2\pi|x| - \tfrac{\pi(n+1)}{4} - \tfrac{\pi\lambda}{2}) \le 1 \tag{10.2.3}$$

for all x lying in the annuli

$$A_k = \left\{x \in \mathbf{R}^n : k + \frac{n+2\lambda}{8} \le |x| \le k + \frac{n+2\lambda}{8} + \frac{1}{4}\right\}, \qquad k \in \mathbf{Z}^+.$$

Since in this range, the argument of the cosine in (10.2.2) lies in $[2\pi k, 2\pi k + \frac{\pi}{4}]$.

Consider the range of p's that satisfy

$$\frac{2n}{n+1+2\lambda} \ge p > \frac{2n}{n+3+2\lambda}. \tag{10.2.4}$$

If we can show that B^λ is unbounded in this range, it will also have to be unbounded in the bigger range $\frac{2n}{n+1+2\lambda} \ge p$. This follows by interpolation between the values $p = \frac{2n}{n+3+2\lambda} - \delta$ and $p = \frac{2n}{n+1+2\lambda} + \delta$, $\delta > 0$, for λ fixed.

In view of (10.2.2) and (10.2.3), we have that

$$\|K_\lambda\|_{L^p}^p \ge C' \sum_{k=n+2\lambda}^{\infty} \int_{A_k} |x|^{-p\frac{n+1}{2} - p\lambda} dx - C'' - C''' \int_{|x|\ge 1} |x|^{-p\frac{n+3}{2} - p\lambda} dx, \tag{10.2.5}$$

where C'' is the integral of K_λ in the unit ball. It is easy to see that for p in the range (10.2.4), the integral outside the unit ball converges, while the series diverges in (10.2.5).

The unboundedness of B^λ on $L^p(\mathbf{R}^n)$ in the range of $p \ge \frac{2n}{n-1-2\lambda}$ follows by duality. □

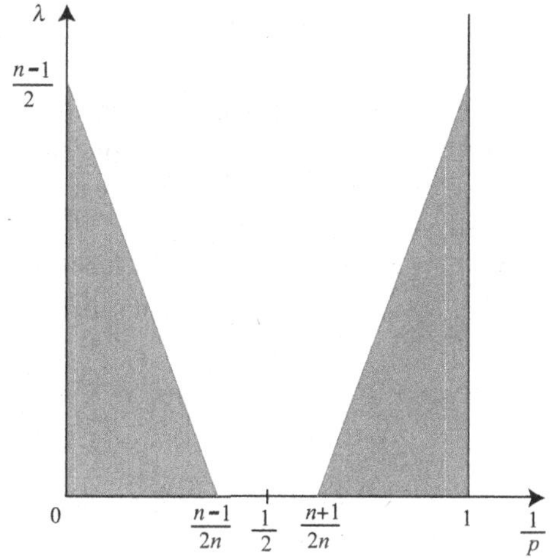

Fig. 10.8 The operator B^λ is unbounded on $L^p(\mathbf{R}^n)$ when $(1/p, \lambda)$ lies in the shaded region.

In Figure 10.8 the shaded region is the set of all pairs $(\frac{1}{p},\lambda)$ for which the operators B^λ are known to be unbounded on $L^p(\mathbf{R}^n)$.

10.2.2 The Carleson–Sjölin Theorem

We now pass to the main result in this section. We prove the boundedness of the operators B^λ, $\lambda>0$, in the range of p's not excluded by the previous proposition in dimension $n=2$.

Theorem 10.2.4. *Suppose that $0<\operatorname{Re}\lambda\le 1/2$. Then the Bochner–Riesz operator B^λ maps $L^p(\mathbf{R}^2)$ to itself when $\frac{4}{3+2\operatorname{Re}\lambda}<p<\frac{4}{1-2\operatorname{Re}\lambda}$. Moreover, for this range of p's and for all $f\in L^p(\mathbf{R}^2)$ we have that*

$$B_R^\lambda(f)\to f$$

in $L^p(\mathbf{R}^2)$ as $R\to\infty$.

Proof. Once the first assertion of the theorem is established, the second assertion will be a direct consequence of it and of the fact that the means $B_R^\lambda(h)$ converge to h in L^p for h in a dense subclass of L^p. Such a dense class is, for instance, the class of all Schwartz functions h whose Fourier transforms are compactly supported (Exercise 5.2.9). For a function h in this class, we see easily that $B_R^\lambda(h)\to h$ pointwise. But if $\widehat{h}$ is supported in $|\xi|\le c$, then for $R\ge 2c$, integration by parts gives that the functions $B_R^\lambda(h)(x)$ are pointwise controlled by the function $(1+|x|)^{-N}$ with N large; then the Lebesgue dominated convergence theorem gives that the $B_R^\lambda(h)$ converge to h in L^p. Finally, a standard $\varepsilon/3$ argument, using that

$$\sup_{R>0}\big\|B_R^\lambda\big\|_{L^p\to L^p}=\big\|(1-|\xi|^2)_+^\lambda\big\|_{\mathcal{M}_p}<\infty,$$

yields $B_R^\lambda(f)\to f$ in L^p for general L^p functions f.

It suffices to focus our attention on the first part of the theorem. We therefore fix a complex number λ with positive real part and we keep track of the growth of all involved constants in $\operatorname{Im}\lambda$.

We start by picking a smooth function φ supported in $[-\frac{1}{2},\frac{1}{2}]$ and a smooth function ψ supported in $[\frac{1}{8},\frac{5}{8}]$ that satisfy

$$\varphi(t)+\sum_{k=0}^{\infty}\psi\Big(\frac{1-t}{2^{-k}}\Big)=1$$

for all $t\in[0,1)$. We now decompose the multiplier $(1-|\xi|^2)_+^\lambda$ as

$$(1-|\xi|^2)_+^\lambda=m_{00}(\xi)+\sum_{k=0}^{\infty}2^{-k\lambda}m_k(\xi),\tag{10.2.6}$$

where $m_{00}(\xi) = \varphi(|\xi|)(1-|\xi|^2)^\lambda$ and for $k \ge 0$, m_k is defined by

$$m_k(\xi) = \Big(\frac{1-|\xi|}{2^{-k}}\Big)^\lambda \psi\Big(\frac{1-|\xi|}{2^{-k}}\Big)(1+|\xi|)^\lambda .$$

Note that m_{00} is a smooth function with compact support; hence the multiplier m_{00} lies in $\mathscr{M}_p(\mathbf{R}^2)$ for all $1 \le p \le \infty$. Each function m_k is also smooth, radial, and supported in the small annulus

$$1-\tfrac{5}{8}2^{-k} \le |\xi| \le 1-\tfrac{1}{8}2^{-k}$$

and therefore also lies in $\mathscr{M}_p$; nevertheless the $\mathscr{M}_p$ norms of the m_k's grow as k increases, and it is crucial to determine how this growth depends on k so that we can sum the series in (10.2.6).

Next we show that the Fourier multiplier norm of each m_k on $L^4(\mathbf{R}^2)$ is at most $C(1+|k|)^{1/2}(1+|\operatorname{Im}\lambda|)^3$. Summing on k implies that B^λ maps $L^4(\mathbf{R}^2)$ to itself with norm at most a multiple of $(1+|\operatorname{Im}\lambda|)^3$ when $\operatorname{Re}\lambda > 0$. Given this bound, we conclude the first (and main) statement of the theorem via Theorem 1.3.7 (precisely Exercise 1.3.4), which permits interpolation for the analytic family of operators $\lambda \mapsto B^\lambda$ between the estimates

$$\begin{aligned} \big\|B^\lambda\big\|_{L^4(\mathbf{R}^2)\to L^4(\mathbf{R}^2)} &\le C(1+|\operatorname{Im}\lambda|)^3 && \text{when } \operatorname{Re}\lambda > 0,\\ \big\|B^\lambda\big\|_{L^1(\mathbf{R}^2)\to L^1(\mathbf{R}^2)} &\le C_1\, e^{c_1|\operatorname{Im}\lambda|^2} && \text{when } \operatorname{Re}\lambda > \tfrac{1}{2}, \end{aligned}$$

where C, C_1, c_1 depend only on $\operatorname{Re}\lambda$. The second estimate above is proved in Proposition 10.2.2 while the set of points $(1/p,\lambda)$ obtained by interpolation can be seen in Figure 10.8.

To estimate the norm of each m_k in $\mathscr{M}_4(\mathbf{R}^2)$, we need an additional decomposition of the operator m_k that takes into account the radial nature of m_k. For each $k \ge 0$ we define the sectorial arcs (parts of a sector between two arcs)

$$\Gamma_{k,\ell} = \big\{re^{2\pi i\theta} \in \mathbf{R}^2 : \,|\theta - \ell 2^{-\frac{k}{2}}| < 2^{-\frac{k}{2}}, \quad 1-\tfrac{5}{8}2^{-k} \le r \le 1-\tfrac{1}{8}2^{-k}\big\}$$

for all $\ell \in \{0,1,2,\dots,[2^{k/2}]-1\}$. We now introduce a smooth function ω supported in $[-1,1]$ and equal to 1 on $[-1/4,1/4]$ such that for all $x \in \mathbf{R}$ we have

$$\sum_{\ell\in\mathbf{Z}} \omega(x-\ell) = 1 .$$

Then we define $m_{k,\ell}(re^{2\pi i\theta}) = m_k(re^{2\pi i\theta})\omega(2^{k/2}\theta - \ell)$ for integers ℓ in the set $\{0,1,2,\dots,[2^{k/2}]-1\}$. If k is an even integer, it follows from the construction that

$$m_k(\xi) = \sum_{\ell=0}^{[2^{k/2}]-1} m_{k,\ell}(\xi) \tag{10.2.7}$$

for all ξ in $\mathbf{R}^2$. If k is odd we replace the function $\theta \mapsto \omega(2^{k/2}\theta - ([2^{k/2}] - 1))$ by a function $\omega_k(\theta)$ supported in the bigger interval $\big[([2^{k/2}]-2)2^{-k/2}, 1\big]$ that satisfies $\omega_k(\theta) + \omega(2^{k/2}(\theta-1)) = 1$ on the interval $\big[([2^{k/2}]-1)2^{-k/2}, 1\big]$. This leads to a new definition of the function $m_{k,[2^{k/2}]-1}$ so that (10.2.7) is satisfied.

This provides the circular (angular) decomposition of m_k. Observe that for all positive integers α and β there exist constants $C_{\alpha,\beta}$ such that

$$|\partial_r^\alpha \partial_\theta^\beta m_{k,\ell}(re^{2\pi i\theta})| \le C_{\alpha,\beta}(1+|\lambda|)^{\alpha+\beta} 2^{k\alpha} 2^{\frac{k}{2}\beta}$$

and such that each $m_{k,\ell}$ is a smooth function supported in the sectorial arcs $\Gamma_{k,\ell}$.

We fix $k \ge 0$ and we group the set of all $\{m_{k,\ell}\}_\ell$ into five subsets: (a) those whose supports are contained in $Q = \{(x,y) \in \mathbf{R}^2 : x > 0, |y| < |x|\}$; (b) those $m_{k,\ell}$ whose supports are contained in the sector $Q' = \{(x,y) \in \mathbf{R}^2 : x < 0, |y| < |x|\}$; (c) those whose supports are contained in $Q'' = \{(x,y) \in \mathbf{R}^2 : y > 0, |y| > |x|\}$; (d) the $m_{k,\ell}$ with supports contained in $Q''' = \{(x,y) \in \mathbf{R}^2 : y < 0, |y| > |x|\}$; and finally (e) those $m_{k,\ell}$ whose supports intersect the lines $|y| = |x|$.

There are only at most eight $m_{k,\ell}$ in case (e), and their sum is easily shown to be an L^4 Fourier multilpier with a constant that grows like $(1+|\lambda|)^3$, as shown below. The remaining cases are symmetric, and we focus attention on case (a).

Let I be the set of all indices ℓ in the set $\{0,1,2,\dots,[2^{k/2}]-1\}$ corresponding to case (a), i.e., the sectorial arcs $\Gamma_{k,\ell}$ are contained in the quarter-plane Q. Let $T_{k,\ell}$ be the operator given on the Fourier transform by multiplication by the function $m_{k,\ell}$. We have

$$\begin{aligned} \Big\|\sum_{\ell\in I} T_{k,\ell}(f)\Big\|_{L^4}^4 &= \int_{\mathbf{R}^2} \Big|\sum_{\ell\in I} T_{k,\ell}(f)\Big|^4 dx \\ &= \int_{\mathbf{R}^2} \Big|\sum_{\ell\in I}\sum_{\ell'\in I} T_{k,\ell}(f) T_{k,\ell'}(f)\Big|^2 dx \\ &= \int_{\mathbf{R}^2} \Big|\sum_{\ell\in I}\sum_{\ell'\in I} \widehat{T_{k,\ell}(f)} * \widehat{T_{k,\ell'}(f)}\Big|^2 d\xi\,, \end{aligned} \tag{10.2.8}$$

where we used Plancherel's identity in the last equality. Each function $\widehat{T_{k,\ell}(f)}$ is supported in the sectorial arc $\Gamma_{k,\ell}$. Therefore, the function $\widehat{T_{k,\ell}(f)} * \widehat{T_{k,\ell'}(f)}$ is supported in $\Gamma_{k,\ell} + \Gamma_{k,\ell'}$ and we write the last integral as

$$\int_{\mathbf{R}^2} \Big|\sum_{\ell\in I}\sum_{\ell'\in I} \big(\widehat{T_{k,\ell}(f)} * \widehat{T_{k,\ell'}(f)}\big)\chi_{\Gamma_{k,\ell}+\Gamma_{k,\ell'}}\Big|^2 d\xi\,.$$

In view of the Cauchy–Schwarz inequality, the last expression is controlled by

$$\int_{\mathbf{R}^2} \Big(\sum_{\ell\in I}\sum_{\ell'\in I} \big|\widehat{T_{k,\ell}(f)} * \widehat{T_{k,\ell'}(f)}\big|^2\Big)\Big(\sum_{\ell\in I}\sum_{\ell'\in I} |\chi_{\Gamma_{k,\ell}+\Gamma_{k,\ell'}}|^2\Big) d\xi\,. \tag{10.2.9}$$

At this point we make use of the following lemma, in which the curvature of the circle is manifested.

Lemma 10.2.5. *There exists a constant C_0 such that for all $k \ge 0$ the following estimate holds:*

$$\sum_{\ell \in I} \sum_{\ell' \in I} \chi_{\Gamma_{k,\ell}+\Gamma_{k,\ell'}} \le C_0 .$$

We postpone the proof of this lemma until the end of this section. Using Lemma 10.2.5, we control the expression in (10.2.9) by

$$C_0 \int_{\mathbf{R}^2} \sum_{\ell \in I} \sum_{\ell' \in I} \big|\widehat{T_{k,\ell}(f)} * \widehat{T_{k,\ell'}(f)}\big|^2 \, d\xi = C_0 \bigg\| \Big(\sum_{\ell \in I} |T_{k,\ell}(f)|^2 \Big)^{\frac{1}{2}} \bigg\|_{L^4}^4 . \tag{10.2.10}$$

We examiine each $T_{k,\ell}$ a bit more carefully. We have that $m_{k,0}$ is supported in a rectangle with sides parallel to the axes and dimensions 2^{-k} (along the ξ_1-axis) and $2^{-\frac{k}{2}+1}$ (along the ξ_2-axis). Moreover, in that rectangle, $\partial_{\xi_1} \approx \partial_r$ and $\partial_{\xi_2} \approx \partial_\theta$, and it follows that the smooth function $m_{k,0}$ satisfies

$$|\partial_{\xi_1}^{\alpha} \partial_{\xi_2}^{\beta} m_{k,0}(\xi_1, \xi_2)| \le C_{\alpha,\beta}(1 + |\lambda|)^{\alpha+\beta} 2^{k\alpha} 2^{\frac{k}{2}\beta}$$

for all positive integers α and β. This estimate can also be written as

$$\big|\partial_{\xi_1}^{\alpha} \partial_{\xi_2}^{\beta} \big[m_{k,0}(2^{-k}\xi_1, 2^{-\frac{k}{2}}\xi_2)\big]\big| \le C_{\alpha,\beta}(1 + |\lambda|)^{\alpha+\beta} ,$$

which easily implies that

$$2^{\frac{3}{2}k} |m_{k,0}^{\vee}(2^k x_1, 2^{\frac{k}{2}} x_2)| \le C_{\alpha,\beta}(1 + |\lambda|)^3 (1 + |x_1| + |x_2|)^{-3} .$$

Let V_ℓ be the unit vector representing the point $e^{2\pi i \ell 2^{-k/2}}$ and $V_\ell^\perp$ the unit vector representing the point $ie^{2\pi i \ell 2^{-k/2}}$. Applying a rotation, we obtain that the functions $m_{k,\ell}^{\vee}$ satisfy

$$\big|m_{k,\ell}^{\vee}(x_1,x_2)\big| \le C(1 + |\lambda|)^3 3^{-\frac{3k}{2}} (1 + 2^{-k}|x \cdot V_\ell| + 2^{-\frac{k}{2}}|x \cdot V_\ell^\perp|)^{-3} \tag{10.2.11}$$

and hence

$$\sup_{k \ge 0} \sup_{\ell \in I} \big\|m_{k,\ell}^{\vee}\big\|_{L^1} \le C(1 + |\lambda|)^3 . \tag{10.2.12}$$

The crucial fact is that the constant C in (10.2.12) is independent of ℓ and k.

At this point, for each fixed $k \ge 0$ and $\ell \in I$ we let $J_{k,\ell}$ be the ξ_2-projection of the support of $m_{k,\ell}$. Based on the earlier definition of $m_{k,\ell}$, we easily see that when $\ell > 0$,

$$J_{k,\ell} = \big[(1 - \tfrac{5}{8} 2^{-k}) \sin(2\pi 2^{-\frac{k}{2}}(\ell - 1)), (1 - \tfrac{1}{8} 2^{-k}) \sin(2\pi 2^{-\frac{k}{2}}(\ell + 1))\big] .$$

A similar formula holds for $\ell < 0$ in I. The crucial observation is that for any fixed $k \geq 0$ the sets $J_{k,\ell}$ are "almost disjoint" for different $\ell \in I$. Indeed, the sets $J_{k,\ell}$ are contained in the intervals

$$\widetilde{J_{k,\ell}} = \left[(1-\tfrac{3}{8}2^{-k})\sin(2\pi 2^{-\frac{k}{2}}\ell) - 10\cdot 2^{-\frac{k}{2}}, (1-\tfrac{3}{8}2^{-k})\sin(2\pi 2^{-\frac{k}{2}}\ell) + 10\cdot 2^{-\frac{k}{2}}\right],$$

which have length $20\cdot 2^{-\frac{k}{2}}$ and are centered at the points $(1-\frac{3}{8}2^{-k})\sin(2\pi 2^{-\frac{k}{2}}\ell)$. For $\sigma \in \mathbf{Z}$ and $\tau \in \{0,1,\dots,39\}$ we define the strips

$$S_{k,\sigma,\tau} = \left\{(\xi_1,\xi_2):\ \xi_2 \in \left[40\sigma 2^{-\frac{k}{2}} + \tau 2^{-\frac{k}{2}}, 40(\sigma+1)2^{-\frac{k}{2}} + \tau 2^{-\frac{k}{2}}\right)\right\}.$$

These strips have length $40\cdot 2^{-\frac{k}{2}}$ and have the property that each $\widetilde{J_{k,\ell}}$ is contained in one of them; say $\widetilde{J_{k,\ell}}$ is contained in some $S_{k,\sigma_\ell,\tau_\ell}$, which we call $B_{k,\ell}$. Then we have

$$T_{k,\ell}(f) = T_{k,\ell}(f_{k,\ell}),$$

where we set

$$f_{k,\ell} = \left(\chi_{B_{k,\ell}}\widehat{f}\,\right)^{\vee} = \chi_{B_{k,\ell}}^{\vee} * f.$$

As a consequence of the Cauchy–Schwarz inequality (with respect to the measure $|m_{k,\ell}^{\vee}|\,dx$), we obtain

$$\begin{aligned} |T_{k,\ell}(f_{k,\ell})|^2 &\leq \left\|m_{k,\ell}^{\vee}\right\|_{L^1}\left(|m_{k,\ell}^{\vee}| * |f_{k,\ell}|^2\right) \\ &\leq C(1+|\lambda|)^3\left(|m_{k,\ell}^{\vee}| * |f_{k,\ell}|^2\right) \end{aligned}$$

in view of (10.2.12). We now return to (10.2.10), which controls (10.2.9) and hence (10.2.8). Using this estimate, we bound the term in (10.2.10) by

$$\begin{aligned} \left\|\Big(\sum_{\ell\in I}|T_{k,\ell}(f)|^2\Big)^{\frac{1}{2}}\right\|_{L^4}^4 &= \left\|\sum_{\ell\in I}|T_{k,\ell}(f_{k,\ell})|^2\right\|_{L^2}^2 \\ &\leq C^2(1+|\lambda|)^6\left\|\sum_{\ell\in I}|m_{k,\ell}^{\vee}| * |f_{k,\ell}|^2\right\|_{L^2}^2 \\ &= C^2(1+|\lambda|)^6\left(\int_{\mathbf{R}^2}\sum_{\ell\in I}\left(|m_{k,\ell}^{\vee}| * |f_{k,\ell}|^2\right)g\,dx\right)^2 \\ &= C^2(1+|\lambda|)^6\left(\sum_{\ell\in I}\int_{\mathbf{R}^2}\left(|\widetilde{m_{k,\ell}}| * g\right)|f_{k,\ell}|^2\,dx\right)^2 \\ &\leq C^2(1+|\lambda|)^6\left(\int_{\mathbf{R}^2}\sup_{\ell\in I}\left(|\widetilde{m_{k,\ell}}| * g|\right)\sum_{\ell\in I}|f_{k,\ell}|^2\,dx\right)^2 \\ &\leq C^2(1+|\lambda|)^6\left\|\sup_{\ell\in I}\left(|\widetilde{m_{k,\ell}}| * g\right)\right\|_{L^2}^2\left\|\Big(\sum_{\ell\in I}|f_{k,\ell}|^2\Big)^{\frac{1}{2}}\right\|_{L^4}^4, \end{aligned}$$

where g is an appropriate nonnegative function in $L^2(\mathbf{R}^2)$ of norm 1.

If we knew the validity of the estimates

$$\Big\| \sup_{\ell \in I} \big(|\widehat{m_{k,\ell}}| * g \big) \Big\|_{L^2} \le C(1+|\lambda|)^3 (1+k) \big\| g \big\|_{L^2} \tag{10.2.13}$$

and

$$\Big\| \Big(\sum_{\ell \in I} |f_{k,\ell}|^2 \Big)^{\frac{1}{2}} \Big\|_{L^4} \le C \big\| f \big\|_{L^4}, \tag{10.2.14}$$

then we would be able to conclude that

$$\big\| m_k \big\|_{\mathscr{M}_p} \le C(1+|\lambda|)^3 (1+k)^{\frac{1}{2}} \tag{10.2.15}$$

and hence we could sum the series in (10.2.6).

Estimates (10.2.13) and (10.2.14) are discussed in the next two subsections. □

10.2.3 The Kakeya Maximal Function

We showed in the previous subsection that $m_{k,0}^{\vee}$ is integrable over $\mathbf{R}^2$ and satisfies the estimate

$$2^{\frac{3}{2}k} |m_{k,0}^{\vee}(2^k x_1, 2^{\frac{k}{2}} x_2)| \le \frac{C(1+|\lambda|)^3}{(1+|x|)^3}.$$

Since

$$\frac{1}{(1+|x|)^3} \le C \sum_{s=0}^{\infty} \frac{2^{-s}}{2^{2s}} \chi_{[-2^s,2^s]\times[-2^s,2^s]}(x),$$

it follows that

$$|\widehat{m_{k,0}}(x)| \le C'(1+|\lambda|)^3 \sum_{s=0}^{\infty} 2^{-s} \frac{1}{|R_s|} \chi_{R_s}(x),$$

where $R_s = [-2^s 2^k, 2^s 2^k] \times [-2^s 2^{\frac{k}{2}}, 2^s 2^{\frac{k}{2}}]$. Since a general $\widehat{m_{k,\ell}}$ is obtained from $\widehat{m_{k,0}}$ via a rotation, a similar estimate holds for it. Precisely, we have

$$|\widehat{m_{k,\ell}}(x)| \le C' (1+|\lambda|)^3 \sum_{s=0}^{\infty} 2^{-s} \frac{1}{|R_{s,\ell}|} \chi_{R_{s,\ell}}(x), \tag{10.2.16}$$

where $R_{s,\ell}$ is a rectangle with principal axes along the directions V_ℓ and $V_\ell^{\perp}$ and side lengths $2^s 2^k$ and $2^s 2^{\frac{k}{2}}$, respectively. Using (10.2.16), we obtain the following pointwise estimate for the maximal function in (10.2.13):

$$\sup_{\ell \in I} \big(|\widehat{m_{k,\ell}}| * g \big) \le C' \sum_{s=0}^{\infty} 2^{-s} \sup_{\ell \in I} \frac{1}{|R_{s,\ell}|} \int_{R_{s,\ell}} g(x-y)\, dy, \tag{10.2.17}$$

where $R_{s,\ell}$ are rectangles with dimensions $2^s 2^k$ and $2^s 2^{\frac{k}{2}}$.

Motivated by (10.2.17), for fixed $N \geq 10$ and $a > 0$, we introduce the *Kakeya maximal operator without dilations*

$$\mathscr{K}_N^a(g)(x) = \sup_{R \ni x} \frac{1}{|R|} \int_R |g(y)|\,dy, \tag{10.2.18}$$

acting on functions $g \in L^1_{\text{loc}}$, where the supremum is taken over all rectangles R in $\mathbf{R}^2$ of dimensions a and aN and arbitrary orientation. What makes this maximal operator interesting is that the rectangles R that appear in the supremum in (10.2.19) are allowed to have arbitrary orientation. We also define the *Kakeya maximal operator* $\mathscr{K}_N$ by

$$\mathscr{K}_N(w)(x) = \sup_{a>0} \mathscr{K}_N^a(w), \tag{10.2.19}$$

for w locally integrable. The maximal function $\mathscr{K}_N(w)(x)$ is therefore obtained as the supremum of the averages of a function w over all rectangles in $\mathbf{R}^2$ that contain the point x and have arbitrary orientation but fixed eccentricity equal to N. (The eccentricity of a rectangle is the ratio of its longer side to its shorter side.)

We see that $\mathscr{K}_N(f)$ is pointwise controlled by a $cNM(f)$, where M is the Hardy–Littlewood maximal operator M. This implies that $\mathscr{K}_N$ is of weak type $(1,1)$ with bound at most a multiple of N. Since $\mathscr{K}_N$ is bounded on L^∞ with norm 1, it follows that $\mathscr{K}_N$ maps $L^p(\mathbf{R}^2)$ to itself with norm at most a multiple of $N^{1/p}$. However, we show in the next section that this estimate is very rough and can be improved significantly. In fact, we obtain an L^p estimate for $\mathscr{K}_N$ with norm that grows logarithmically in N (when $p \geq 2$), and this is very crucial, since $N = 2^{k/2}$ in the following application.

Using this new terminology, we write the estimate in (10.2.17) as

$$\sup_{\ell \in I} \left(|\widehat{m_{k,\ell}}| * g\right) \leq C'(1+|\lambda|)^3 \sum_{s=0}^{\infty} 2^{-s} \mathscr{K}_{2^{k/2}}^{2^{s+k/2}}(g). \tag{10.2.20}$$

The required estimate (10.2.13) is a consequence of (10.2.20) and of the following theorem, whose proof is discussed in the next section.

Theorem 10.2.6. *There exists a constant C such that for all $N \geq 10$ and all f in $L^2(\mathbf{R}^2)$ the following norm inequality is valid:*

$$\sup_{a>0} \big\|\mathscr{K}_N^a(f)\big\|_{L^2(\mathbf{R}^2)} \leq C(\log N)\,\|f\|_{L^2(\mathbf{R}^2)}.$$

Theorem 10.2.6 is a consequence of Theorem 10.3.5, in which the preceding estimate is proved for a more general maximal operator $\mathfrak{M}_{\Sigma_N}$, which in particular controls $\mathscr{K}_N$ and hence $\mathscr{K}_N^a$ for all $a > 0$. This maximal operator is introduced in the next section.

10.2.4 Boundedness of a Square Function

We now turn to the proof of estimate (10.2.14). This is a consequence of the following result, which is a version of the Littlewood–Paley theorem for intervals of equal length.

Theorem 10.2.7. *For $j \in \mathbf{Z}$, let I_j be intervals of equal length with disjoint interior whose union is $\mathbf{R}$. We define operators P_j with multipliers χ_{I_j}. Then for $2 \le p < \infty$, there is a constant C_p such that for all $f \in L^p(\mathbf{R})$ we have*

$$\Big\| \Big(\sum_j |P_j(f)|^2 \Big)^{\frac{1}{2}} \Big\|_{L^p(\mathbf{R})} \le C_p \big\| f \big\|_{L^p(\mathbf{R})} . \tag{10.2.21}$$

In particular, the same estimate holds if the intervals I_j have disjoint interiors and equal length but do not necessarily cover $\mathbf{R}$.

Proof. Multiplying the function f by a suitable exponential, we may assume that the intervals I_j have the form $\big((j-\frac{1}{2})a, (j+\frac{1}{2})a\big)$ for some $a > 0$. Applying a dilation to f reduces matters to the case $a = 1$. We conclude that the constant C_p is independent of the common size of the intervals I_j and it suffices to obtain estimate (10.2.21) in the case $a = 1$.

We assume therefore that $I_j = (j - \frac{1}{2}, j + \frac{1}{2})$ for all $j \in \mathbf{Z}$. Next, our goal is to replace the operators P_j by smoother analogues of them. To achieve this we introduce a smooth function ψ with compact support that is identically equal to 1 on the interval $[-\frac{1}{2}, \frac{1}{2}]$ and vanishes off the interval $[-\frac{3}{4}, \frac{3}{4}]$. We introduce operators S_j by setting

$$\widehat{S_j(f)}(\xi) = \widehat{f}(\xi)\psi(\xi - j)$$

and we note that the identity

$$P_j = P_j S_j \tag{10.2.22}$$

is valid for all $j \in \mathbf{Z}$. For $t \in \mathbf{R}$ we define multipliers m_t as

$$m_t(\xi) = \sum_{j \in \mathbf{Z}} e^{-2\pi i j t} \psi(\xi - j) ,$$

and we set $k_t = m_t^{\vee}$. With $I_0 = (-1/2, 1/2)$, we have

$$\begin{aligned} \int_{I_0} |(k_t * f)(x)|^2 \, dt &= \int_{I_0} \Big| \sum_{j \in \mathbf{Z}} e^{-2\pi i j t} S_j(f)(x) \Big|^2 dt \\ &= \sum_{j \in \mathbf{Z}} |S_j(f)(x)|^2 , \end{aligned} \tag{10.2.23}$$

where the last equality is just Plancherel's identity on $I_0 = [-\frac{1}{2}, \frac{1}{2}]$. In view of the last identity, it suffices to analyze the operator given by convolution with the family of kernels k_t. By the Poisson summation formula (Theorem 3.1.17) applied to the function $x \mapsto \psi(x) e^{2\pi i x t}$, we obtain

$$\begin{aligned} m_t(\xi) &= e^{-2\pi i \xi t} \sum_{j\in\mathbf{Z}} \psi(\xi-j) e^{2\pi i(\xi-j)t} \\ &= \sum_{j\in\mathbf{Z}} \big(\psi(\cdot)e^{2\pi i(\cdot)t}\big)\widehat{\ }(j)\, e^{2\pi i j\xi} e^{-2\pi i \xi t} \\ &= \sum_{j\in\mathbf{Z}} e^{2\pi i(j-t)\xi}\,\widehat{\psi}(j-t)\,. \end{aligned}$$

Taking inverse Fourier transforms, we obtain

$$k_t = \sum_{j\in\mathbf{Z}} \widehat{\psi}(j-t)\delta_{-j+t}\,,$$

where δ_b denotes Dirac mass at the point b. Therefore, k_t is a sum of Dirac masses with rapidly decaying coefficients. Since each Dirac mass has Borel norm at most 1, we conclude that

$$\big\|k_t\big\|_{\mathcal{M}} \le \sum_{j\in\mathbf{Z}} |\widehat{\psi}(j-t)| \le \sum_{j\in\mathbf{Z}} (1+|j-t|)^{-10} \le 10\,, \tag{10.2.24}$$

which is independent of t. This says that the measures k_t have uniformly bounded norms. Take now $f\in L^p(\mathbf{R})$ and $p\ge 2$. Using identity (10.2.22), we obtain

$$\begin{aligned} \int_{\mathbf{R}} \Big(\sum_{j\in\mathbf{Z}} |P_j(f)(x)|^2\Big)^{\frac{p}{2}} dx &= \int_{\mathbf{R}} \Big(\sum_{j\in\mathbf{Z}} |P_jS_j(f)(x)|^2\Big)^{\frac{p}{2}} dx \\ &\le c_p \int_{\mathbf{R}} \Big(\sum_{j\in\mathbf{Z}} |S_j(f)(x)|^2\Big)^{\frac{p}{2}} dx\,, \end{aligned}$$

and the last inequality follows from Exercise 4.6.1(a). The constant c_p depends only on p. Recalling identity (10.2.23), we write

$$\begin{aligned} c_p \int_{\mathbf{R}} \Big(\sum_{j\in\mathbf{Z}} |S_j(f)(x)|^2\Big)^{\frac{p}{2}} dx &\le c_p \int_{\mathbf{R}} \Big(\int_{I_0} |(k_t * f)(x)|^2\, dt\Big)^{\frac{p}{2}} dx \\ &\le c_p \int_{\mathbf{R}} \Big(\int_{I_0} |(k_t * f)(x)|^p\, dt\Big)^{\frac{p}{p}} dx \\ &= c_p \int_{I_0}\int_{\mathbf{R}} |(k_t * f)(x)|^p\, dx\, dt \\ &\le 10\, c_p \int_{I_0}\int_{\mathbf{R}} |f(x)|^p\, dx\, dt \\ &= 10\, c_p \big\|f\big\|_{L^p}^p\,, \end{aligned}$$

where we used Hölder's inequality on the interval I_0 (together with the fact that $p\ge 2$) and (10.2.24). The proof of the theorem is complete with constant $C_p = (10c_p)^{1/p}$. □

We now return to estimate (10.2.14). First recall the strips

$$S_{k,\sigma,\tau} = \{(\xi_1,\xi_2) : \xi_2 \in [40\sigma 2^{-\frac{k}{2}} + \tau, 40(\sigma+1)2^{-\frac{k}{2}} + \tau)\}$$

defined for $\sigma \in \mathbf{Z}$ and $\tau \in \{0,1,\dots,39\}$. These strips have length $40 \cdot 2^{-\frac{k}{2}}$, and each $\widetilde{J}_{k,\ell}$ is contained in one of them, which we called $S_{k,\sigma_\ell,\tau_\ell} = B_{k,\ell}$.

The family $\{B_{k,\ell}\}_{\ell \in I}$ does not consist of disjoint sets, but we split it into 40 subfamilies by placing $B_{k,\ell}$ in different subfamilies if the indices τ_ℓ and $\tau_{\ell'}$ are different. We now write the set I as

$$I = I^1 \cup I^2 \cup \cdots \cup I^{40},$$

where for each $\ell, \ell' \in I^j$ the sets $B_{k,\ell}$ and $B_{k,\ell'}$ are disjoint.

We now use Theorem 10.2.7 to obtain the required quadratic estimate (10.2.14). Things now are relatively simple. We observe that the multiplier operators $f \mapsto (\chi_{B_{k,\ell}}\widehat{f})^\vee$ on $\mathbf{R}^2$ obey the estimates (10.2.21), in which $L^p(\mathbf{R})$ is replaced by $L^p(\mathbf{R}^2)$, since they are the identity operators in the ξ_1-variable.

We conclude that

$$\bigg\| \Big(\sum_{\ell \in I^j} |T_{k,\ell}(f)|^2 \Big)^{\frac{1}{2}} \bigg\|_{L^p(\mathbf{R}^2)} \le C_p \big\| f \big\|_{L^p(\mathbf{R}^2)} \tag{10.2.25}$$

holds for all $p \ge 2$ and, in particular, for $p = 4$. This proves (10.2.14) for a single I^j, and the same conclusion follows for I with a constant 40 times as big.

10.2.5 The Proof of Lemma 10.2.5

We finally discuss the proof of Lemma 10.2.5.

Proof. If $k = 0,1,\dots,k_0$ up to a fixed integer k_0, then there exist only finitely many pairs of sets $\Gamma_\ell + \Gamma_{\ell'}$ depending on k_0, and the lemma is trivially true. We may therefore assume that k is a large integer; in particular we may take $\delta = 2^{-k} \le 2400^{-2}$. In the sequel, for simplicity we replace 2^{-k} by δ and we denote the set $\Gamma_{k,\ell}$ by Γ_ℓ. In the proof that follows we are working with a fixed $\delta \in [0, 2400^{-2}]$. Elements of the set $\Gamma_\ell + \Gamma_{\ell'}$ have the form

$$re^{2\pi i(\ell+\alpha)\delta^{1/2}} + r' e^{2\pi i(\ell'+\alpha')\delta^{1/2}}, \tag{10.2.26}$$

where α, α' range in the interval $[-1,1]$ and r, r' range in $[1 - \frac{5}{8}\delta, 1 - \frac{1}{8}\delta]$. We set

$$w(\ell,\ell') = e^{2\pi i \ell \delta^{1/2}} + e^{2\pi i \ell' \delta^{1/2}} = 2\cos(\pi|\ell - \ell'|\delta^{\frac{1}{2}})e^{\pi i(\ell+\ell')\delta^{1/2}}, \tag{10.2.27}$$

where the last equality is a consequence of a trigonometric identity that can be found in Appendix E. Using similar identities (see Appendix E) and performing algebraic manipulations, one may verify that the general element (10.2.26) of the set $\Gamma_\ell + \Gamma_{\ell'}$ can be written as

$$\begin{aligned} w(\ell,\ell') &+ \left\{ r \frac{\cos(2\pi\alpha\delta^{\frac{1}{2}}) + \cos(2\pi\alpha'\delta^{\frac{1}{2}}) - 2}{2} \right\} w(\ell,\ell') \\ &+ \left\{ r \frac{\sin(2\pi\alpha\delta^{\frac{1}{2}}) + \sin(2\pi\alpha'\delta^{\frac{1}{2}})}{2} \right\} i w(\ell,\ell') \\ &+ \mathrm{E}(r,\ell,\ell',\alpha,\alpha',\delta), \end{aligned}$$

where

$$\begin{aligned} \mathrm{E}(r,\ell,\ell',\alpha,\alpha',\delta) &= (r-1)\left(e^{2\pi i \ell\delta^{1/2}} + e^{2\pi i \ell'\delta^{1/2}}\right) \\ &+ (r'-r)e^{2\pi i(\ell'+\alpha')\delta^{1/2}} \\ &+ r\left(e^{2\pi i \ell\delta^{1/2}} - e^{2\pi i \ell'\delta^{1/2}}\right)\left(\frac{\cos(2\pi\alpha\delta^{\frac{1}{2}}) - \cos(2\pi\alpha'\delta^{\frac{1}{2}})}{2}\right) \\ &+ r\left(e^{2\pi i \ell\delta^{1/2}} - e^{2\pi i \ell'\delta^{1/2}}\right)\left(\frac{\sin(2\pi\alpha\delta^{\frac{1}{2}}) - \sin(2\pi\alpha'\delta^{\frac{1}{2}})}{2}\right). \end{aligned}$$

The coefficients in the curly brackets are real, and $\mathrm{E}(r,\ell,\ell',\alpha,\alpha',\delta)$ is an error of magnitude at most $2\delta + 8\pi^2|\ell-\ell'|\delta$. These observations and the facts $|\sin x| \le |x|$ and $|1-\cos x| \le |x|^2/2$ (see Appendix E) imply that the set $\Gamma_\ell + \Gamma_{\ell'}$ is contained in the rectangle $R(\ell,\ell')$ centered at the point $w(\ell,\ell')$ with half-width

$$2\pi^2\delta + (2\delta + 8\pi^2|\ell-\ell'|\delta) \le 80(1+|\ell-\ell'|)\delta$$

in the direction along $w(\ell,\ell')$ and half-length

$$2\pi\delta^{\frac{1}{2}} + (2\delta + 8\pi^2|\ell-\ell'|\delta) \le 30\,\delta^{\frac{1}{2}}$$

in the direction along $iw(\ell,\ell')$ [which is perpendicular to that along $w(\ell,\ell')$]. Since $2\pi|\ell-\ell'|\delta^{\frac{1}{2}} < \frac{\pi}{2}$, this rectangle is contained in a disk of radius $105\,\delta^{\frac{1}{2}}$ centered at the point $w(\ell,\ell')$.

We immediately deduce that if $|w(\ell,\ell') - w(m,m')|$ is bigger than $210\,\delta^{\frac{1}{2}}$, then the sets $\Gamma_\ell + \Gamma_{\ell'}$ and $\Gamma_m + \Gamma_{m'}$ do not intersect. Therefore, if these sets intersect, we should have

$$|w(\ell,\ell') - w(m,m')| \le 210\,\delta^{\frac{1}{2}}.$$

In view of Exercise 10.2.2, the left-hand side of the last expression is at least

$$2\tfrac{2}{\pi}\cos(\tfrac{\pi}{4})|\pi(\ell+\ell') - \pi(m+m')|\delta^{\frac{1}{2}}$$

(here we use the hypothesis that $|2\pi\ell\delta^{\frac{1}{2}}| < \frac{\pi}{4}$ twice). We conclude that if the sets $\Gamma_\ell + \Gamma_{\ell'}$ and $\Gamma_m + \Gamma_{m'}$ intersect, then

$$|(\ell+\ell') - (m+m')| \le 210/2\sqrt{2} \le 150. \tag{10.2.28}$$

In this case the angle between the vectors $w(\ell,\ell')$ and $w(m,m')$ is

$$\varphi_{\ell,\ell',m,m'} = \pi|(\ell+\ell')-(m+m')|\delta^{\frac{1}{2}},$$

which is smaller than $\pi/16$, provided (10.2.28) holds and $\delta < 2400^{-2}$. This says that in this case, the rectangles $R(\ell,\ell')$ and $R(m,m')$ are essentially parallel to each other (the angle between them is smaller than $\pi/16$).

Let us fix a rectangle $R(\ell,\ell')$, and for another rectangle $R(m,m')$ we denote by $\widetilde{R}(m,m')$ the smallest rectangle containing $R(m,m')$ with sides parallel to the corresponding sides of $R(\ell,\ell')$. An easy trigonometric argument shows that $\widetilde{R}(m,m')$ has the same center as $R(m,m')$ and has half-sides at most

$$30\delta^{\frac{1}{2}}\cos(\varphi_{\ell,\ell',m,m'})+80(1+|\ell-\ell'|)\delta\sin(\varphi_{\ell,\ell',m,m'}),$$
$$80(1+|\ell-\ell'|)\delta\cos(\varphi_{\ell,\ell',m,m'})+30\delta^{\frac{1}{2}}\sin(\varphi_{\ell,\ell',m,m'}),$$

in view of Exercise 10.2.3. Then $\widetilde{R}(m,m')$ has half-sides at most $60000\delta^{\frac{1}{2}}$ and $18000(1+|\ell-\ell'|)\delta$ and is therefore contained in a fixed multiple of $R(m,m')$. If $\Gamma_\ell+\Gamma_{\ell'}$ and $\Gamma_m+\Gamma_{m'}$ intersect, then so do $\widetilde{R}(m,m')$ and $R(\ell,\ell')$, and both of these rectangles have sides parallel to the vectors $w(\ell,\ell')$ and $iw(\ell,\ell')$. But in the direction of $w(\ell,\ell')$, these rectangles have sides with half-lengths at most $80(1+|\ell-\ell'|)\delta$ and $16000(1+|m-m'|)\delta$. Note that the distance of the lines parallel to the direction $iw(\ell,\ell')$ and passing through the centers of the rectangles $\widetilde{R}(m,m')$ and $R(\ell,\ell')$ is

$$2\big|\cos(\pi|\ell-\ell'|\delta^{\frac{1}{2}})-\cos(\pi|m-m'|\delta^{\frac{1}{2}})\big|,$$

as we easily see using (10.2.27). If these rectangles intersect, we must have

$$2\big|\cos(\pi|\ell-\ell'|\delta^{\frac{1}{2}})-\cos(\pi|m-m'|\delta^{\frac{1}{2}})\big| \le 16080(2+|\ell-\ell'|+|m-m'|)\delta.$$

We conclude that if the sets $R(m,m')$ and $R(\ell,\ell')$ intersect and $(\ell,\ell')\neq(m,m')$, then

$$\big|\cos(\pi|\ell-\ell'|\delta^{\frac{1}{2}})-\cos(\pi|m-m'|\delta^{\frac{1}{2}})\big| \le 50000(|\ell-\ell'|+|m-m'|)\delta.$$

But the expression on the left is equal to

$$2\Big|\sin\big(\pi\tfrac{|\ell-\ell'|-|m-m'|}{2}\delta^{\frac{1}{2}}\big)\sin\big(\pi\tfrac{|\ell-\ell'|+|m-m'|}{2}\delta^{\frac{1}{2}}\big)\Big|,$$

which is at least

$$2\big||\ell-\ell'|-|m-m'|\big|\,(|\ell-\ell'|+|m-m'|)\delta$$

in view of the simple estimate $|\sin t|\ge\frac{2}{\pi}|t|$ for $|t|<\frac{\pi}{2}$. We conclude that if the sets $R(m,m')$ and $R(\ell,\ell')$ intersect and $(\ell,\ell')\neq(m,m')$, then

$$\big||\ell-\ell'|-|m-m'|\big| \le 25000. \tag{10.2.29}$$

Combining (10.2.28) with (10.2.29), it follows that if $\Gamma_m + \Gamma_{m'}$ and $\Gamma_\ell + \Gamma_{\ell'}$ intersect, then

$$\max\Big(\big|\min(m,m') - \min(\ell,\ell')\big|, \big|\max(m,m') - \max(\ell,\ell')\big|\Big) \le \frac{25150}{2}.$$

We conclude that the set $\Gamma_m + \Gamma_{m'}$ intersects the fixed set $\Gamma_\ell + \Gamma_{\ell'}$ for at most $(25151)^2$ pairs (m,m'). This finishes the proof of the lemma. □

Exercises

10.2.1. For $\lambda \ge 0$ show that for all $f \in L^p(\mathbf{R}^n)$ the Bochner–Riesz operators

$$B_R^\lambda(f)(x) = \int_{\mathbf{R}^n} (1 - |\xi/R|^2)_+^\lambda \widehat{f}(\xi) e^{2\pi i x\cdot\xi}\, d\xi$$

converge to f in $L^p(\mathbf{R}^n)$ if and only if the function $(1-|\xi|^2)_+^\lambda$ lies in $\mathscr{M}_p(\mathbf{R}^n)$.
[*Hint:* In the beginning of the proof of Theorem 10.2.4 it was shown that if $(1-|\xi|^2)_+^\lambda$ lies in $\mathscr{M}_p(\mathbf{R}^n)$, then the $B_R^\lambda(f)$ converge to f in $L^p(\mathbf{R}^n)$. Conversely, if for all $f \in L^p(\mathbf{R}^n)$ the $B_R^\lambda(f)$ converge to f in L^p as $R \to \infty$, then for every f in $L^p(\mathbf{R}^n)$ there is a constant C_f such that $\sup_{R>0} \big\|B_R^\lambda(f)\big\|_{L^p} \le C_f < \infty$. It follows that $\sup_{R>0} \big\|B_R^\lambda\big\|_{L^p \to L^p} < \infty$ by the uniform boundedness principle; hence $\big\|B^\lambda\big\|_{L^p\to L^p} < \infty$.]

10.2.2. Let $|\theta_1|, |\theta_2| < \frac{\pi}{4}$ be two angles. Show geometrically that

$$|r_1 e^{i\theta_1} - r_2 e^{i\theta_2}| \ge \min(r_1, r_2) \sin|\theta_1 - \theta_2|$$

and use the estimate $|\sin t| \ge \frac{2|t|}{\pi}$ for $|t| < \frac{\pi}{2}$ to obtain a lower bound for the second expression in terms of $|\theta_1 - \theta_2|$.

10.2.3. Let R be a rectangle in $\mathbf{R}^2$ having length $b > 0$ along a direction $\vec{v} = (\xi_1, \xi_2)$ and length $a > 0$ along the perpendicular direction $\vec{v}^\perp = (-\xi_2, \xi_1)$. Let $\vec{w}$ be another vector that forms an angle $\varphi < \frac{\pi}{2}$ with $\vec{v}$. Show that the smallest rectangle R' that contains R and has sides parallel to $\vec{w}$ and $\vec{w}^\perp$ has side lengths $a\sin(\varphi) + b\cos(\varphi)$ along the direction $\vec{w}$ and $a\cos(\varphi) + b\sin(\varphi)$ along the direction $\vec{w}^\perp$.

10.2.4. Prove that Theorem 10.2.7 does not hold when $p < 2$.
[*Hint:* Try the intervals $I_j = [j, j+1]$ and $\widehat{f} = \chi_{[0,N]}$ as $N \to \infty$.]

10.2.5. Let $\{I_k\}_k$ be a family of intervals in the real line with $|I_k| = |I_{k'}|$ and $I_k \cap I_{k'} = \emptyset$ for all $k \ne k'$. Define the sets

$$S_k = \big\{(\xi_1, \dots, \xi_n) \in \mathbf{R}^n : \xi_1 \in I_k\big\}.$$

Prove that for all $p \ge 2$ and all $f \in L^p(\mathbf{R}^n)$ we have

$$\Big\|\Big(\sum_k |(\widehat{f}\chi_{S_k})^\vee|^2\Big)^{\frac{1}{2}}\Big\|_{L^p(\mathbf{R}^n)} \le C_p \big\|f\big\|_{L^p(\mathbf{R}^n)},$$

where C_p is the constant of Theorem 10.2.7.

10.2.6. (a) Let $\{I_k\}_k$, $\{J_\ell\}_\ell$ be two families of intervals in the real line with $|I_k| = |I_{k'}|$, $I_k \cap I_{k'} = \emptyset$ for all $k \neq k'$, and $|J_\ell| = |J_{\ell'}|$, $J_\ell \cap J_{\ell'} = \emptyset$ for all ℓ, ℓ'. Prove that for all $p \ge 2$ there is a constant C_p such that

$$\Big\|\Big(\sum_k\sum_\ell |(\widehat{f}\chi_{I_k\times J_\ell})^\vee|^2\Big)^{\frac{1}{2}}\Big\|_{L^p(\mathbf{R}^2)} \le C_p^2 \big\|f\big\|_{L^p(\mathbf{R}^2)},$$

for all $f \in L^p(\mathbf{R}^2)$.
(b) State and prove an analogous result on $\mathbf{R}^n$.
[*Hint:* Use the Rademacher functions and apply Theorem 10.2.7 twice.]

10.2.7. (*Rubio de Francia [273]*) On $\mathbf{R}^n$ consider the points $x_\ell = \ell\sqrt{\delta}$, $\ell \in \mathbf{Z}^n$. Fix a Schwartz function h whose Fourier transform is supported in the unit ball in $\mathbf{R}^n$. Given a function f on $\mathbf{R}^n$, define $\widehat{f_\ell}(\xi) = \widehat{f}(\xi)\widehat{h}(\delta^{-\frac{1}{2}}(\xi - x_\ell))$. Prove that for some constant C (which depends only on h and n) the estimate

$$\Big(\sum_{\ell\in\mathbf{Z}^n} |f_\ell|^2\Big)^{\frac{1}{2}} \le CM(|f|^2)^{\frac{1}{2}}$$

holds for all functions f. Deduce the $L^p(\mathbf{R}^n)$ boundedness of the preceding square function for all $p > 2$.
[*Hint:* For a sequence λ_ℓ with $\sum_\ell |\lambda_\ell|^2 = 1$, set

$$G(f)(x) = \sum_{\ell\in\mathbf{Z}^n} \lambda_\ell f_\ell(x) = \int_{\mathbf{R}^n}\Big[\sum_{\ell\in\mathbf{Z}^n} \lambda_\ell e^{2\pi i\frac{x_\ell\cdot y}{\sqrt{\delta}}}\Big] f\big(x - \tfrac{y}{\sqrt{\delta}}\big)h(y)\,dy.$$

Split $\mathbf{R}^n$ as the union of $Q_0 = [-\frac{1}{2}, \frac{1}{2}]^n$ and $2^{j+1}Q_0 \setminus 2^jQ_0$ for $j \ge 0$ and control the integral on each such set using the decay of h and the $L^2(2^{j+1}Q_0)$ norms of the other two terms. Finally, exploit the orthogonality of the functions $e^{2\pi i\ell\cdot y}$ to estimate the $L^2(2^{j+1}Q_0)$ norm of the expression inside the square brackets by $C2^{nj/2}$. Sum over $j \ge 0$ to obtain the required conclusion.]

10.2.8. For $\lambda > 0$ define the *maximal Bochner–Riesz operator*

$$B_*^\lambda(f)(x) = \sup_{R>0}\Big|\int_{\mathbf{R}^n}(1 - |\xi/R|^2)_+^\lambda \widehat{f}(\xi)e^{2\pi i x\cdot\xi}\,d\xi\Big|.$$

Prove that B_*^λ maps $L^p(\mathbf{R}^n)$ to itself when $\lambda > \frac{n-1}{2}$ for $1 \le p \le \infty$.
[*Hint:* Use Corollary 2.1.12.]

10.3 Kakeya Maximal Operators

We recall the Hardy–Littlewood maximal operator with respect to cubes on $\mathbf{R}^n$ defined as

$$M_c(f)(x) = \sup_{\substack{Q \in \mathscr{F} \\ Q \ni x}} \frac{1}{|Q|} \int_Q |f(y)|\,dy, \tag{10.3.1}$$

where $\mathscr{F}$ is the set of all closed cubes in $\mathbf{R}^n$ (with sides not necessarily parallel to the axes). The operator M_c is equivalent (bounded above and below by constants) to the corresponding maximal operator M_c' in which the family $\mathscr{F}$ is replaced by the more restrictive family $\mathscr{F}'$ of cubes in $\mathbf{R}^n$ with sides parallel to the coordinate axes.

It is interesting to observe that if the family of all cubes $\mathscr{F}$ in (10.3.1) is replaced by the family of all rectangles (or parallelepipeds) $\mathscr{R}$ in $\mathbf{R}^n$, then we obtain an operator M_0 that is unbounded on $L^p(\mathbf{R}^n)$; see also Exercise 2.1.9. If we substitute the family of all parallelepipeds $\mathscr{R}$, however, with the more restrictive family $\mathscr{R}'$ of all parallelepipeds with sides parallel to the coordinate axes, then we obtain the so-called strong maximal function

$$M_s(f)(x) = \sup_{\substack{R \in \mathscr{R}' \\ R \ni x}} \frac{1}{|R|} \int_R |f(y)|\,dy, \tag{10.3.2}$$

which was introduced in Exercise 2.1.6. The operator M_s is bounded on $L^p(\mathbf{R}^n)$ for $1 < p < \infty$ but it is not of weak type $(1,1)$. See Exercise 10.3.1.

These examples indicate that averaging over long and skinny rectangles is quite different than averaging over squares. In general, the direction and the dimensions of the averaging rectangles play a significant role in the boundedness properties of the maximal functions. In this section we investigate aspects of this topic.

10.3.1 Maximal Functions Associated with a Set of Directions

Definition 10.3.1. Let Σ be a set of unit vectors in $\mathbf{R}^2$, i.e., a subset of the unit circle $\mathbf{S}^1$. Associated with Σ, we define $\mathscr{R}_\Sigma$ to be the set of all closed rectangles in $\mathbf{R}^2$ whose longest side is parallel to some vector in Σ. We also define a maximal operator $\mathfrak{M}_\Sigma$ associated with Σ as follows:

$$\mathfrak{M}_\Sigma(f)(x) = \sup_{\substack{R \in \mathscr{R}_\Sigma \\ R \ni x}} \frac{1}{|R|} \int_R |f(y)|\,dy,$$

where f is a locally integrable function on $\mathbf{R}^2$.

We also recall the definition given in (10.2.19) of the *Kakeya maximal operator*

$$\mathscr{K}_N(w)(x) = \sup_{R \ni x} \frac{1}{|R|} \int_R |w(y)|\,dy, \tag{10.3.3}$$

where the supremum is taken over all rectangles R in $\mathbf{R}^2$ of dimensions a and aN where $a > 0$ is arbitrary. Here N is a fixed real number that is at least 10.

Example 10.3.2. Let $\Sigma = \{v\}$ consist of only one vector $v = (a,b)$. Then

$$\mathfrak{M}_\Sigma(f)(x) = \sup_{0<r\le 1}\sup_{N>0}\frac{1}{rN^2}\int_{-N}^{N}\int_{-rN}^{rN}|f(x-t(a,b)-s(-b,a))|\,ds\,dt.$$

If $\Sigma = \{(1,0),(0,1)\}$ consists of the two unit vectors along the axes, then

$$\mathfrak{M}_\Sigma = M_s,$$

where M_s is the strong maximal function defined in (10.3.2).

It is obvious that for each $\Sigma \subseteq \mathbf{S}^1$, the maximal function $\mathfrak{M}_\Sigma$ maps $L^\infty(\mathbf{R}^2)$ to itself with constant 1. But $\mathfrak{M}_\Sigma$ may not always be of weak type $(1,1)$, as the example M_s indicates; see Exercise 10.3.1. The boundedness of $\mathfrak{M}_\Sigma$ on $L^p(\mathbf{R}^2)$ in general depends on the set Σ.

An interesting case arises in the following example as well.

Example 10.3.3. For $N \in \mathbf{Z}^+$, let

$$\Sigma = \Sigma_N = \left\{\left(\cos(\tfrac{2\pi j}{N}),\sin(\tfrac{2\pi j}{N})\right) : j = 0,1,2,\dots,N-1\right\}$$

be the set of N uniformly spread directions on the circle. Then we expect $\mathfrak{M}_{\Sigma_N}$ to be L^p bounded with constant depending on N. There is a connection between the operator $\mathfrak{M}_{\Sigma_N}$ previously defined and the Kakeya maximal operator $\mathcal{K}_N$ defined in (10.2.19). In fact, Exercise 10.3.3 says that

$$\mathcal{K}_N(f) \le 20\,\mathfrak{M}_{\Sigma_N}(f) \tag{10.3.4}$$

for all locally integrable functions f on $\mathbf{R}^2$.

We now indicate why the norms of $\mathcal{K}_N$ and $\mathfrak{M}_{\Sigma_N}$ on $L^2(\mathbf{R}^2)$ grow as $N \to \infty$. We refer to Exercises 10.3.4 and 10.3.7 for the corresponding result for $p \neq 2$.

Proposition 10.3.4. *There is a constant c such that for any $N \ge 10$ we have*

$$\big\|\mathcal{K}_N\big\|_{L^2(\mathbf{R}^2)\to L^2(\mathbf{R}^2)} \ge c\,(\log N) \tag{10.3.5}$$

and

$$\big\|\mathcal{K}_N\big\|_{L^2(\mathbf{R}^2)\to L^{2,\infty}(\mathbf{R}^2)} \ge c\,(\log N)^{\frac{1}{2}}. \tag{10.3.6}$$

Therefore, a similar conclusion follows for $\mathfrak{M}_{\Sigma_N}$.

Proof. We consider the family of functions $f_N(x) = \frac{1}{|x|}\chi_{3\le|x|\le N}$ defined on $\mathbf{R}^2$ for $N \ge 10$. Then we have

$$\big\|f_N\big\|_{L^2(\mathbf{R}^2)} \le c_1(\log N)^{\frac{1}{2}}. \tag{10.3.7}$$

On the other hand, for every x in the annulus $6 < |x| < N$, we consider the rectangle R_x of dimensions $|x| - 3$ and $\frac{|x|-3}{N}$, one of whose shorter sides touches the circle $|y| = 3$ and the other has midpoint x. Then

$$\mathscr{K}_N(f_N)(x) \geq \frac{1}{|R_x|}\int_{R_x} |f_N(y)|\,dy \geq \frac{c_2 N}{(|x|-3)^2} \iint\limits_{\substack{6\leq y_1\leq |x| \\ |y_2|\leq \frac{|x|-3}{2N}}} \frac{dy_1 dy_2}{y_1} \geq c_3 \frac{\log|x|}{|x|}.$$

It follows that

$$\big\|\mathscr{K}_N(f_N)\big\|_{L^2(\mathbf{R}^2)} \geq c_3\bigg(\int\limits_{6\leq |x|\leq N} \Big(\frac{\log|x|}{|x|}\Big)^2 dx\bigg)^{\frac{1}{2}} \geq c_4 (\log N)^{\frac{3}{2}}. \tag{10.3.8}$$

Combining (10.3.7) with (10.3.8) we obtain (10.3.5) with $c = c_4/c_1$.

We now turn to estimate (10.3.6). Since for all $6 < |x| < N$ we have

$$\mathscr{K}_N(f_N)(x) \geq c_3 \frac{\log|x|}{|x|} > c_3 \frac{\log N}{N},$$

it follows that $\big|\{\mathscr{K}_N(f_N) > c_3 \frac{\log N}{N}\}\big| \geq \pi(N^2 - 6^2) \geq c_5 N^2$ and hence

$$\begin{aligned}\frac{\big\|\mathscr{K}_N(f_N)\big\|_{L^{2,\infty}}}{\big\|f_N\big\|_{L^2}} &\geq \frac{\sup_{\lambda>0} \lambda \big|\{\mathscr{K}_N(f_N) > \lambda\}\big|^{\frac{1}{2}}}{c_1 (\log N)^{\frac{1}{2}}} \\ &\geq c_3 \frac{\log N}{N} \frac{\big|\{\mathscr{K}_N(f_N) > c_3 \frac{\log N}{N}\}\big|^{\frac{1}{2}}}{c_1 (\log N)^{\frac{1}{2}}} \\ &\geq \frac{c_3\sqrt{c_5}}{c_1} (\log N)^{\frac{1}{2}}.\end{aligned}$$

This completes the proof. □

10.3.2 The Boundedness of $\mathfrak{M}_{\Sigma_N}$ on $L^p(\mathbf{R}^2)$

It is rather remarkable that both estimates of Proposition 10.3.4 are sharp in terms of their behavior as $N \to \infty$, as the following result indicates.

Theorem 10.3.5. *There exist constants $0 < B, C < \infty$ such that for every $N \geq 10$ and all $f \in L^2(\mathbf{R}^2)$ we have*

$$\big\|\mathfrak{M}_{\Sigma_N}(f)\big\|_{L^{2,\infty}(\mathbf{R}^2)} \leq B (\log N)^{\frac{1}{2}} \big\|f\big\|_{L^2(\mathbf{R}^2)} \tag{10.3.9}$$

and

$$\left\|\mathfrak{M}_{\Sigma_N}(f)\right\|_{L^2(\mathbf{R}^2)} \leq C(\log N)\left\|f\right\|_{L^2(\mathbf{R}^2)}. \tag{10.3.10}$$

In view of (10.3.4), similar estimates also hold for $\mathscr{K}_N$.

Proof. We deduce (10.3.10) from the weak type estimate (10.3.9), which we rewrite as

$$\left|\left\{x \in \mathbf{R}^2 : \mathfrak{M}_{\Sigma_N}(f)(x) > \lambda\right\}\right| \leq B^2(\log N)\,\frac{\left\|f\right\|_{L^2}^2}{\lambda^2}. \tag{10.3.11}$$

We prove this estimate for some constant $B > 0$ independent of N. But prior to doing this we indicate why (10.3.11) implies (10.3.10).

Using Exercise 10.3.2, we have that $\mathfrak{M}_{\Sigma_N}$ maps $L^p(\mathbf{R}^2)$ to $L^p(\mathbf{R}^2)$ (and hence into $L^{p,\infty}$) with constant at most a multiple of $N^{1/p}$ for all $1 < p < \infty$. Using this with $p = 3/2$, we have

$$\left\|\mathfrak{M}_{\Sigma_N}\right\|_{L^{\frac{3}{2}} \to L^{\frac{3}{2},\infty}} \leq \left\|\mathfrak{M}_{\Sigma_N}\right\|_{L^{\frac{3}{2}} \to L^{\frac{3}{2}}} \leq A N^{\frac{2}{3}} \tag{10.3.12}$$

for some constant $A > 0$. Now split f as the sum $f = f_1 + f_2 + f_3$, where

$$\begin{aligned} f_1 &= f\chi_{|f| \leq \frac{1}{4}\lambda}, \\ f_2 &= f\chi_{\frac{1}{4}\lambda < |f| \leq N^2\lambda}, \\ f_3 &= f\chi_{N^2\lambda < |f|}. \end{aligned}$$

It follows that

$$\left|\left\{\mathfrak{M}_{\Sigma_N}(f) > \lambda\right\}\right| \leq \left|\left\{\mathfrak{M}_{\Sigma_N}(f_2) > \tfrac{\lambda}{3}\right\}\right| + \left|\left\{\mathfrak{M}_{\Sigma_N}(f_3) > \tfrac{\lambda}{3}\right\}\right|, \tag{10.3.13}$$

since the set $\left\{\mathfrak{M}_{\Sigma_N}(f_1) > \frac{\lambda}{3}\right\}$ is empty. To obtain the required result we use the $L^{2,\infty}$ estimate (10.3.11) for f_2 and the $L^{\frac{3}{2},\infty}$ estimate (10.3.12) for f_3. We have

$$\begin{aligned}
&\left\|\mathfrak{M}_{\Sigma_N}(f)\right\|_{L^2}^2 \\
&= 2\int_0^\infty \lambda\left|\left\{\mathfrak{M}_{\Sigma_N}(f) > \lambda\right\}\right| d\lambda \\
&\leq \int_0^\infty 2\lambda\left|\left\{\mathfrak{M}_{\Sigma_N}(f_2) > \tfrac{\lambda}{3}\right\}\right| d\lambda + \int_0^\infty 2\lambda\left|\left\{\mathfrak{M}_{\Sigma_N}(f_3) > \tfrac{\lambda}{3}\right\}\right| d\lambda \\
&\leq \int_0^\infty \frac{2\lambda B^2(\log N)}{\lambda^2} \int\limits_{\frac{1}{4}\lambda < |f| \leq N^2\lambda} |f|^2\,dx\,d\lambda + \int_0^\infty \frac{2\lambda A^{\frac{3}{2}} N}{\lambda^{\frac{3}{2}}} \int\limits_{|f| > N^2\lambda} |f|^{\frac{3}{2}}\,dx\,d\lambda \\
&\leq 2B^2(\log N)\int_{\mathbf{R}^2} |f(x)|^2 \int_{\frac{|f(x)|}{N^2}}^{4|f(x)|} \frac{d\lambda}{\lambda}\,dx + 2A^{\frac{3}{2}} N \int_{\mathbf{R}^2} |f(x)|^{\frac{3}{2}} \int_0^{\frac{|f(x)|}{N^2}} \frac{d\lambda}{\lambda^{\frac{1}{2}}}\,dx \\
&= \left(4B^2(\log 2N)(\log N) + 4A^{\frac{3}{2}}\right)\left\|f\right\|_{L^2}^2 \\
&\leq C(\log N)^2\left\|f\right\|_{L^2}^2
\end{aligned}$$

using Fubini's theorem for integrals. This proves (10.3.10).

To avoid problems with antipodal points, it is convenient to split Σ_N as the union of eight sets, in each of which the angle between any two vectors does not exceed $2\pi/8$. It suffices therefore to obtain (10.3.11) for each such subset of Σ_N. Let us fix one such subset of Σ_N, which we call Σ_N^1. To prove (10.3.11), we fix a $\lambda > 0$ and we start with a compact subset K of the set $\{x \in \mathbf{R}^2 : \mathfrak{M}_{\Sigma_N^1}(f)(x) > \lambda\}$. Then for every $x \in K$, there exists an open rectangle R_x that contains x and whose longest side is parallel to a vector in Σ_N^1. By compactness of K, there exists a finite subfamily $\{R_\alpha\}_{\alpha \in \mathscr{A}}$ of the family $\{R_x\}_{x \in K}$ such that

$$\int_{R_\alpha} |f(y)|\,dy > \lambda\,|R_\alpha|$$

for all $\alpha \in \mathscr{A}$ and such that the union of the R_α's covers K.

We claim that there is a constant C such that for any finite family $\{R_\alpha\}_{\alpha \in \mathscr{A}}$ of rectangles whose longest side is parallel to a vector in Σ_N^1 there is a subset $\mathscr{B}$ of $\mathscr{A}$ such that

$$\int_{\mathbf{R}^2} \Big(\sum_{\beta \in \mathscr{B}} \chi_{R_\beta}(x) \Big)^2 dx \le C \Big| \bigcup_{\beta \in \mathscr{B}} R_\beta \Big| \tag{10.3.14}$$

and that

$$\Big| \bigcup_{\alpha \in \mathscr{A}} R_\alpha \Big| \le C(\log N) \Big| \bigcup_{\beta \in \mathscr{B}} R_\beta \Big| . \tag{10.3.15}$$

Assuming (10.3.14) and (10.3.15), we easily deduce (10.3.11). Indeed,

$$\begin{aligned}
\Big| \bigcup_{\beta \in \mathscr{B}} R_\beta \Big| &\le \sum_{\beta \in \mathscr{B}} |R_\beta| \\
&< \frac{1}{\lambda} \sum_{\beta \in \mathscr{B}} \int_{R_\beta} |f(y)|\,dy \\
&= \frac{1}{\lambda} \int_{\mathbf{R}^2} \Big(\sum_{\beta \in \mathscr{B}} \chi_{R_\beta} \Big) |f(y)|\,dy \\
&\le \frac{1}{\lambda} \bigg(\int_{\mathbf{R}^2} \Big(\sum_{\beta \in \mathscr{B}} \chi_{R_\beta} \Big)^2 dx \bigg)^{\frac{1}{2}} \|f\|_{L^2} \\
&\le \frac{C^{\frac{1}{2}}}{\lambda} \Big| \bigcup_{\beta \in \mathscr{B}} R_\beta \Big|^{\frac{1}{2}} \|f\|_{L^2} ,
\end{aligned}$$

from which it follows that

$$\Big| \bigcup_{\beta \in \mathscr{B}} R_\beta \Big| \le \frac{C}{\lambda^2} \|f\|_{L^2}^2 .$$

Then, using (10.3.15), we obtain

$$|K| \le \Big|\bigcup_{\alpha\in\mathscr{A}} R_\alpha\Big| \le C(\log N)\Big|\bigcup_{\beta\in\mathscr{B}} R_\beta\Big| \le \frac{C^2}{\lambda^2}(\log N)\|f\|_{L^2}^2,$$

and since K was an arbitrary compact subset of $\{x : \mathfrak{M}_{\Sigma_N^1}(f)(x) > \lambda\}$, the same estimate is valid for the latter set.

We now turn to the selection of the subfamily $\{R_\beta\}_{\beta\in\mathscr{B}}$ and the proof of (10.3.14) and (10.3.15).

Let R_{β_1} be the rectangle in $\{R_\alpha\}_{\alpha\in\mathscr{A}}$ with the longest side. Suppose we have chosen $R_{\beta_1}, R_{\beta_2}, \dots, R_{\beta_{j-1}}$ for some $j \ge 2$. Then among all rectangles R_α that satisfy

$$\sum_{k=1}^{j-1} |R_{\beta_k} \cap R_\alpha| \le \frac{1}{2}|R_\alpha|, \tag{10.3.16}$$

we choose a rectangle R_{β_j} such that its longer side is as large as possible. Since the collection $\{R_\alpha\}_{\alpha\in\mathscr{A}}$ is finite, this selection stops after m steps. Define

$$\mathscr{B} = \{\beta_1, \beta_2, \dots, \beta_m\}.$$

Using (10.3.16), we obtain

$$\begin{aligned}
\int_{\mathbf{R}^2} \Big(\sum_{\beta\in\mathscr{B}} \chi_{R_\beta}\Big)^2 dx &\le 2\sum_{j=1}^{m}\sum_{k=1}^{j} |R_{\beta_k} \cap R_{\beta_j}| \\
&= 2\sum_{j=1}^{m}\Big[\Big(\sum_{k=1}^{j-1} |R_{\beta_k} \cap R_{\beta_j}|\Big) + |R_{\beta_j}|\Big] \\
&\le 2\sum_{j=1}^{m}\Big[\frac{1}{2}|R_{\beta_j}| + |R_{\beta_j}|\Big] \\
&= 3\sum_{j=1}^{m} |R_{\beta_j}|.
\end{aligned} \tag{10.3.17}$$

A consequence of this fact is that

$$\begin{aligned}
\sum_{j=1}^{m} |R_{\beta_j}| &= \int_{\bigcup_{j=1}^{m} R_{\beta_j}} \Big(\sum_{j=1}^{m} \chi_{R_{\beta_j}}\Big) dx \\
&\le \Big|\bigcup_{j=1}^{m} R_{\beta_j}\Big|^{\frac{1}{2}} \Big(\int_{\mathbf{R}^n} \Big(\sum_{\beta\in\mathscr{B}} \chi_{R_\beta}\Big)^2 dx\Big)^{\frac{1}{2}} \\
&\le \Big|\bigcup_{j=1}^{m} R_{\beta_j}\Big|^{\frac{1}{2}} \sqrt{3}\Big(\sum_{j=1}^{m} |R_{\beta_j}|\Big)^{\frac{1}{2}},
\end{aligned}$$

which implies that

$$\sum_{j=1}^{m} |R_{\beta_j}| \leq 3 \Big| \bigcup_{j=1}^{m} R_{\beta_j} \Big|. \tag{10.3.18}$$

Using (10.3.18) in conjunction with the last estimate in (10.3.17), we deduce the desired inequality (10.3.14) with $C = 9$.

We now turn to the proof of (10.3.15). Let M_c be the usual Hardy–Littlewood maximal operator with respect to cubes in $\mathbf{R}^n$ (or squares in $\mathbf{R}^2$; recall $n = 2$). Since M_c is of weak type $(1,1)$, (10.3.15) is a consequence of the estimate

$$\bigcup_{\alpha \in \mathscr{A} \setminus \mathscr{B}} R_\alpha \subseteq \Big\{ x \in \mathbf{R}^2 : M_c\Big(\sum_{\beta \in \mathscr{B}} \chi_{(R_\beta)^*} \Big)(x) > c\,(\log N)^{-1} \Big\} \tag{10.3.19}$$

for some absolute constant c, where $(R_\beta)^*$ is the rectangle R_β expanded 5 times in both directions. Indeed, if (10.3.19) holds, then

$$\begin{aligned}
\Big| \bigcup_{\alpha \in \mathscr{A}} R_\alpha \Big| &\leq \Big| \bigcup_{\beta \in \mathscr{B}} R_\beta \Big| + \Big| \bigcup_{\alpha \in \mathscr{A} \setminus \mathscr{B}} R_\alpha \Big| \\
&\leq \Big| \bigcup_{\beta \in \mathscr{B}} R_\beta \Big| + \frac{10}{c} (\log N) \sum_{\beta \in \mathscr{B}} |(R_\beta)^*| \\
&\leq \Big| \bigcup_{\beta \in \mathscr{B}} R_\beta \Big| + \frac{250}{c} (\log N) \sum_{\beta \in \mathscr{B}} |R_\beta| \\
&\leq C(\log N) \Big| \bigcup_{\beta \in \mathscr{B}} R_\beta \Big|,
\end{aligned}$$

where we just used (10.3.18) and the fact that N is large.

It remains to prove (10.3.19). At this point we need the following lemma. In the sequel we denote by θ_α the angle between the x axis and the vector pointing in the longer direction of R_α for any $\alpha \in \mathscr{A}$. We also denote by l_α the shorter side of R_α and by L_α the longer side of R_α for any $\alpha \in \mathscr{A}$. Finally, we set

$$\omega_k = \frac{2\pi 2^k}{N}$$

for $k \in \mathbf{Z}^+$ and $\omega_0 = 0$.

Lemma 10.3.6. *Let R_α be a rectangle in the family $\{R_\alpha\}_{\alpha \in \mathscr{A}}$ and let $0 \leq k < \big[\frac{\log(N/8)}{\log 2}\big]$. Suppose that $\beta \in \mathscr{B}$ is such that*

$$\omega_k \leq |\theta_\alpha - \theta_\beta| < \omega_{k+1}$$

and such that $L_\beta \geq L_\alpha$. Let $s_\alpha = 8\max(l_\alpha, \omega_k L_\alpha)$. For an arbitrary $x \in R_\alpha$, let Q be a square centered at x with sides of length s_α parallel to the sides of R_α. Then we have

$$\frac{|R_\beta \cap R_\alpha|}{|R_\alpha|} \leq 32 \frac{|(R_\beta)^* \cap Q|}{|Q|}. \tag{10.3.20}$$

Assuming Lemma 10.3.6, we conclude the proof of (10.3.19). Fix $\alpha \in \mathscr{A} \setminus \mathscr{B}$. Then the rectangle R_α was not selected in the selection procedure. This means that for all $l \in \{2, \dots, m+1\}$ we have exactly one of the following: either

$$\sum_{j=1}^{l-1} |R_{\beta_j} \cap R_\alpha| > \frac{1}{2}|R_\alpha| \tag{10.3.21}$$

or

$$\sum_{j=1}^{l-1} |R_{\beta_j} \cap R_\alpha| \le \frac{1}{2}|R_\alpha| \qquad \text{and} \qquad L_\alpha \le L_{\beta_l}. \tag{10.3.22}$$

If (10.3.22) holds for $l = 2$, we let $\mu \le m$ be the largest integer such that (10.3.22) holds for all $l \le \mu$. Then (10.3.22) fails for $l = \mu + 1$; hence (10.3.21) holds for $l = \mu + 1$; thus

$$\frac{1}{2}|R_\alpha| < \sum_{j=1}^{\mu} |R_{\beta_j} \cap R_\alpha| \le \sum_{\substack{\beta \in \mathscr{B} \\ L_\beta \ge L_\alpha}} |R_\beta \cap R_\alpha|. \tag{10.3.23}$$

If (10.3.22) fails for $l = 2$, then (10.3.21) holds for $l = 2$, and this implies that

$$\frac{1}{2}|R_\alpha| < |R_{\beta_1} \cap R_\alpha| \le \sum_{\substack{\beta \in \mathscr{B} \\ L_\beta \ge L_\alpha}} |R_\beta \cap R_\alpha|.$$

In either case we have

$$\frac{1}{2}|R_\alpha| < \sum_{\substack{\beta \in \mathscr{B} \\ L_\beta \ge L_\alpha}} |R_\beta \cap R_\alpha|,$$

and from this it follows that there exists a k with $0 \le k < \left[\frac{\log(N/8)}{\log 2}\right]$ such that

$$\frac{\log 2}{2\log(N/8)} |R_\alpha| < \sum_{\substack{\beta \in \mathscr{B} \\ L_\beta \ge L_\alpha \\ \omega_k \le |\theta_\beta - \theta_\alpha| < \omega_{k+1}}} |R_\beta \cap R_\alpha|. \tag{10.3.24}$$

By Lemma 10.3.6, for any $x \in R_\alpha$ there is a square Q such that (10.3.20) holds for any R_β with $\beta \in \mathscr{B}$ satisfying $L_\beta \ge L_\alpha$ and $\omega_k \le |\theta_\beta - \theta_\alpha| < \omega_{k+1}$. It follows that

$$\frac{\log 2}{2\log(N/8)} < 2 \sum_{\substack{\beta \in \mathscr{B} \\ L_\beta \ge L_\alpha \\ \omega_k \le |\theta_\beta - \theta_\alpha| < \omega_{k+1}}} \frac{|(R_\beta)^* \cap Q|}{|Q|},$$

which implies

$$\frac{c}{\log N} < \frac{\log 2}{4\log(N/8)} < \frac{1}{|Q|}\int_Q \sum_{\beta\in\mathscr{B}} \chi_{(R_\beta)^*}\, dx\,.$$

This proves (10.3.19), since for $\alpha \in \mathscr{A} \setminus \mathscr{B}$, any $x \in R_\alpha$ must be an element of the set $\{x \in \mathbf{R}^2 : M_c\big(\sum_{\beta\in\mathscr{B}} \chi_{(R_\beta)^*}\big)(x) > c\,(\log N)^{-1}\}$. □

It remains to prove Lemma 10.3.6.

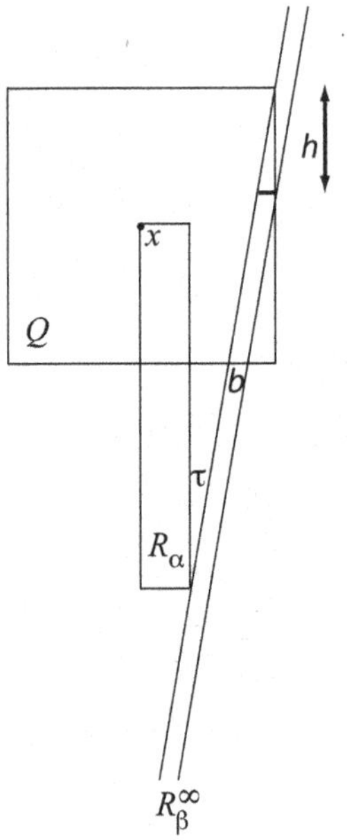

Fig. 10.9 For angles τ less than that displayed, the strip R_β^∞ meets the upper side of Q. The length of the intersection of R_β^∞ with the lower side of Q is denoted by b.

Proof. We fix R_α and R_β so that $L_\beta \ge L_\alpha$ and we assume that $\overline{R_\beta}$ intersects $\overline{R_\alpha}$; otherwise, (10.3.20) is obvious. Let τ be the angle between the directions of the rectangles R_α and R_β, that is,

$$\tau = |\theta_\alpha - \theta_\beta|\,.$$

By assumption we have $\tau < \omega_{k+1} \le \frac{\pi}{4}$, since $k+1 \le [\frac{\log(N/8)}{\log 2}] \le \frac{\log(N/8)}{\log 2}$.

Let R_β^∞ denote the smallest closed infinite strip in the direction of the longer side of R_β that contains it. We make the following observation: if

$$\tan\tau \le \frac{\frac{1}{2}s_\alpha - l_\alpha}{\frac{1}{2}s_\alpha + L_\alpha}, \tag{10.3.25}$$

then the strip R_β^∞ intersects the upper side (according to Figure 10.9) of the square Q. Indeed, the worst possible case is drawn in Figure 10.9, in which equality holds in (10.3.25). For $\tau \le \pi/4$ we have $\tan\tau < 3\tau/2$, and since $\tau < 2\omega_k$, it follows that $\tan\tau < 3\omega_k$. Our choice of s_α implies

$$s_\alpha \ge 6\,\omega_k L_\alpha + 2\,l_\alpha \implies 3\omega_k \le \frac{\frac{1}{2}s_\alpha - l_\alpha}{\frac{1}{2}s_\alpha + L_\alpha};$$

hence (10.3.25) holds.

We have now proved that R_β^∞ meets the upper side of Q. We examine the size of the intersection $R_\beta^\infty \cap Q$. According to the picture in Figure 10.9, this intersection contains a parallelogram of base $b = l_\beta/\cos\tau$ and height $s_\alpha - h$ and a right triangle with base b and height h (with $0 \le h \le s_\alpha$). Then we have

$$\frac{|R_\beta^\infty \cap Q|}{|Q|} \ge \frac{1}{s_\alpha^2}\frac{l_\beta}{\cos\tau}\Big(s_\alpha - h + \frac{1}{2}h\Big) \ge \frac{1}{s_\alpha^2}\frac{l_\beta}{\cos\tau}\Big(\frac{1}{2}s_\alpha\Big) \ge \frac{1}{2}\frac{l_\beta}{s_\alpha}.$$

Since $(R_\beta)^*$ has length $5L_\beta$ and R_β meets R_α, we have that $R_\beta^\infty \cap Q \subseteq (R_\beta)^* \cap Q$ and therefore

$$\frac{|(R_\beta)^* \cap Q|}{|Q|} \ge \frac{1}{2}\frac{l_\beta}{s_\alpha}. \tag{10.3.26}$$

On the other hand, let $R_{\alpha,\beta}$ be the smallest parallelogram two of whose opposite sides are parallel to the shorter sides of R_α and whose remaining two sides are contained in the boundary lines of R_β^∞. Then

$$|R_\alpha \cap R_\beta| \le |R_{\alpha,\beta}| \le \frac{l_\beta}{\cos\tau}L_\alpha \le 2\,l_\beta L_\alpha.$$

Another geometric argument shows that

$$|R_\alpha \cap R_\beta| \le l_\beta\,\frac{l_\alpha}{\sin(\tau)} \le l_\alpha\,l_\beta\,\frac{\pi}{2\tau} \le l_\alpha\,l_\beta\,\frac{\pi}{2\omega_k} \le 2\,\frac{l_\alpha\,l_\beta}{\omega_k}.$$

Combining these estimates, we deduce

$$\frac{|R_\alpha \cap R_\beta|}{|R_\alpha|} \le 2\min\Big(\frac{l_\beta}{l_\alpha}, \frac{l_\beta}{\omega_k L_\alpha}\Big) \le 16\,\frac{l_\beta}{s_\alpha}. \tag{10.3.27}$$

Finally, (10.3.26) and (10.3.27) yield (10.3.20). □

We end this subsection with an immediate corollary of the theorem just proved.

Corollary 10.3.7. *For every* $1 < p < \infty$ *there exists a constant* c_p *such that*

$$\big\|\mathscr{K}_N\big\|_{L^p(\mathbf{R}^2)\to L^p(\mathbf{R}^2)} \le c_p \begin{cases} N^{\frac{2}{p}-1}(\log N)^{\frac{1}{p'}} & \text{when } 1 < p < 2, \\ (\log N)^{\frac{1}{p}} & \text{when } 2 < p < \infty. \end{cases} \tag{10.3.28}$$

Proof. We see that

$$\big\|\mathscr{K}_N\big\|_{L^1(\mathbf{R}^2)\to L^{1,\infty}(\mathbf{R}^2)} \le CN \tag{10.3.29}$$

by replacing a rectangle of dimensions $a \times aN$ by the smallest square of side length aN that contains it. Interpolating between (10.3.9) and (10.3.29), we obtain the first statement in (10.3.28). The second statement in (10.3.28) follows by interpolation between (10.3.9) and the trivial $L^\infty \to L^\infty$ estimate. (In both cases we use Theorem 1.3.2.) □

10.3.3 The Higher-Dimensional Kakeya Maximal Operator

The Kakeya maximal operator without dilations $\mathscr{K}_N^a$ on $L^2(\mathbf{R}^2)$ was crucial in the study of the boundedness of the Bochner–Riesz operator B^λ on $L^4(\mathbf{R}^2)$. An analogous maximal operator could be introduced on $\mathbf{R}^n$.

Definition 10.3.8. Given fixed $a > 0$ and $N \geq 10$, we introduce the *Kakeya maximal operator without dilations* on $\mathbf{R}^n$ as

$$\mathscr{K}_N^a(f)(x) = \sup_R \frac{1}{|R|} \int_R |f(y)|\, dy,$$

where the supremum is taken over all rectangular parallelepipeds (boxes) of arbitrary orientation in $\mathbf{R}^n$ that contain the point x and have dimensions

$$\underbrace{a \times a \times \cdots \times a}_{n-1 \text{ times}} \times aN .$$

We also define the centered version $\mathfrak{K}_N^a$ of $\mathscr{K}_N^a$ as follows:

$$\mathfrak{K}_N^a(f)(x) = \sup_R \frac{1}{|R|} \int_R |f(y)|\, dy,$$

where the supremum is restricted to those rectangles among the previous ones that are centered at x. These two maximal operators are comparable, and we have

$$\mathfrak{K}_N^a \leq \mathscr{K}_N^a \leq 2^n \mathfrak{K}_N^a$$

by a simple geometric argument.

We also define the higher-dimensional analogue of the Kakeya maximal operator $\mathscr{K}_N$ introduced in (10.3.3).

Definition 10.3.9. Let $N \geq 10$. We denote by $\mathscr{R}(N)$ the set of all rectangular parallelepipeds (boxes) in $\mathbf{R}^n$ with arbitrary orientation and dimensions

$$\underbrace{a \times a \times \cdots \times a}_{n-1 \text{ times}} \times aN$$

with arbitrary $a > 0$. Given a locally integrable function f on $\mathbf{R}^n$, we define

$$\mathscr{K}_N(f)(x) = \sup_{\substack{R\in\mathscr{R}(N)\\ R\ni x}} \frac{1}{|R|}\int_R |f(y)|\,dy$$

and

$$\mathfrak{K}_N(f)(x) = \sup_{\substack{R\in\mathscr{R}(N)\\ R \text{ has center } x}} \frac{1}{|R|}\int_R |f(y)|\,dy;$$

$\mathfrak{K}_N$ and $\mathscr{K}_N$ are called the centered and uncentered nth-*dimensional Kakeya maximal operators*, respectively.

For convenience we call rectangular parallelepipeds, i.e., elements of $\mathscr{R}(N)$, higher-dimensional rectangles, or simply rectangles. We clearly have

$$\sup_{a>0} \mathscr{K}_N^a = \mathscr{K}_N \qquad \text{and} \qquad \sup_{a>0} \mathfrak{K}_N^a = \mathfrak{K}_N\,;$$

hence the boundedness of $\mathscr{K}_N^a$ can be deduced from that of $\mathscr{K}_N$; however, this deduction can essentially be reversed with only logarithmic loss in N (see the references at the end of this chapter). In the sequel we restrict attention to the operator $\mathscr{K}_N^a$, whose study already presents all the essential difficulties and requires a novel set of ideas in its analysis. We consider a specific value of a, since a simple dilation argument yields that the norms of $\mathscr{K}_N^a$ and $\mathscr{K}_N^b$ on a fixed $L^p(\mathbf{R}^n)$ are equal for all $a,b>0$.

Concerning $\mathscr{K}_N^1$, we know that

$$\big\|\mathscr{K}_N^1\big\|_{L^1(\mathbf{R}^n)\to L^{1,\infty}(\mathbf{R}^n)} \le c_n N^{n-1}\,. \tag{10.3.30}$$

This estimate follows by replacing a rectangle of dimensions $\overbrace{1\times 1\times\cdots\times 1}^{n-1 \text{ times}}\times N$ by the smallest cube of side length N that contains it. This estimate is sharp; see Exercise 10.3.7.

It would be desirable to know the following estimate for $\mathscr{K}_N^1$:

$$\big\|\mathscr{K}_N^1\big\|_{L^n(\mathbf{R}^n)\to L^{n,\infty}(\mathbf{R}^n)} \le c_n'(\log N)^{\frac{n-1}{n}} \tag{10.3.31}$$

for some dimensional constant c_n'. It would then follow that

$$\big\|\mathscr{K}_N^1\big\|_{L^n(\mathbf{R}^n)\to L^n(\mathbf{R}^n)} \le c_n'' \log N \tag{10.3.32}$$

for some other dimensional constant c_n''; see Exercise 10.3.8(b). Moreover, if estimate (10.3.31) were true, then interpolating between (10.3.30) and (10.3.31) would yield the bound

$$\big\|\mathscr{K}_N^1\big\|_{L^p(\mathbf{R}^n)\to L^p(\mathbf{R}^n)} \le c_{n,p} N^{\frac{n}{p}-1}(\log N)^{\frac{1}{p'}}\,, \qquad 1<p<n. \tag{10.3.33}$$

It is estimate (10.3.33) that we would like to concentrate on. We have the following result for a certain range of p's in the interval $(1,n)$.

Theorem 10.3.10. *Let $p_n = \frac{n+1}{2}$ and $N \geq 10$. Then there exists a constant C_n such that*

$$\left\|\mathscr{K}_N^1\right\|_{L^{p_n,1}(\mathbf{R}^n)\to L^{p_n,\infty}(\mathbf{R}^n)} \leq C_n N^{\frac{n}{p_n}-1}, \tag{10.3.34}$$

$$\left\|\mathscr{K}_N^1\right\|_{L^{p_n}(\mathbf{R}^n)\to L^{p_n,\infty}(\mathbf{R}^n)} \leq C_n N^{\frac{n}{p_n}-1}(\log N)^{\frac{1}{p_n'}}, \tag{10.3.35}$$

$$\left\|\mathscr{K}_N^1\right\|_{L^{p_n}(\mathbf{R}^n)\to L^{p_n}(\mathbf{R}^n)} \leq C_n N^{\frac{n}{p_n}-1}(\log N). \tag{10.3.36}$$

Moreover, for every $1 < p < p_n$ there exists a constant $C_{n,p}$ such that

$$\left\|\mathscr{K}_N^1\right\|_{L^p(\mathbf{R}^n)\to L^p(\mathbf{R}^n)} \leq C_{n,p} N^{\frac{n}{p}-1}(\log N)^{\frac{1}{p'}}. \tag{10.3.37}$$

Proof. We begin by observing that (10.3.37) is a consequence of (10.3.30) and (10.3.35) using Theorem 1.3.2. We also observe that (10.3.36) is a consequence of (10.3.35), while (10.3.35) is a consequence of (10.3.34) (see Exercise 10.3.8). We therefore concentrate on estimate (10.3.34).

We choose to work with the centered version $\mathfrak{K}_N^1$ of $\mathscr{K}_N^1$, which is comparable to it. To make the geometric idea of the proof a bit more transparent, we pick $\delta < 1/10$, we set $N = 1/\delta$, and we work with the equivalent operator $\mathfrak{K}_{1/\delta}^{\delta}$, whose norm is the same as that of $\mathscr{K}_N^1$. Since the operators in question are positive, we work with nonnegative functions.

The proof is based on a linearization of the operator $\mathscr{K}_{1/\delta}^{\delta}$. Let us call a rectangle of dimensions $\delta \times \delta \times \cdots \times \delta \times 1$ a *δ-tube*. We call the line segment parallel to the longest edges that joins the centers of its two smallest faces, a δ-tube's *axis of symmetry*.

For every x in $\mathbf{R}^n$ we select (in some measurable way) a δ-tube $\tau(x)$ that contains x such that

$$\frac{1}{2}\mathscr{K}_{1/\delta}^{\delta}(f)(x) \leq \frac{1}{|\tau(x)|}\int_{\tau(x)} f(y)\,dy.$$

Suppose we have a grid of cubes in $\mathbf{R}^n$ each of side length $\delta' = \delta/(2\sqrt{n})$, and let Q_j be a cube in that grid with center c_{Q_j}. Then any δ-tube centered at a point $z \in Q_j$ must contain the entire Q_j, and it follows that

$$\mathfrak{K}_{1/\delta}^{\delta}(f)(z) \leq \mathscr{K}_{1/\delta}^{\delta}(f)(c_{Q_j}) \leq \frac{2}{|\tau(c_{Q_j})|}\int_{\tau(c_{Q_j})} f(y)\,dy. \tag{10.3.38}$$

This observation motivates the introduction of a grid of width $\delta' = \delta/(2\sqrt{n})$ in $\mathbf{R}^n$ so that for every cube Q_j in the grid there is an associated δ-tube τ_j satisfying

$$\tau_j \cap Q_j \neq \emptyset.$$

Then we define a linear operator

$$L^\delta(f) = \sum_j \left(\frac{1}{|\tau_j|} \int_{\tau_j} f(y)\,dy \right) \chi_{Q_j},$$

which certainly satisfies

$$L^\delta(f) \le 2^n \mathscr{K}^{2\delta}_{1/\delta}(f) \le 4^n \mathfrak{K}^{2\delta}_{1/\delta}(f),$$

and in view of (10.3.38), it also satisfies

$$\mathfrak{K}^{\delta}_{1/\delta}(f) \le 2L^\delta(f).$$

It suffices to show that L^δ is bounded from $L^{p_n,1}$ to $L^{p_n,\infty}$ with constant $C_n(\delta^{-1})^{\frac{n}{p_n}-1}$, which is independent of the choice of δ-tubes τ_j.

Our next reduction is to take f to be the characteristic function of a set. The space $L^{p_n,\infty}$ is normable [i.e., it has an equivalent norm under which it is a Banach space (Exercise 1.1.12)]; hence by Exercise 1.4.7, the boundedness of L^δ from $L^{p_n,1}$ to $L^{p_n,\infty}$ is a consequence of the restricted weak type estimate

$$\sup_{\lambda>0} \lambda \big|\{L^\delta(\chi_A) > \lambda\}\big|^{\frac{1}{p_n}} \le C_n'(\delta^{-1})^{\frac{n}{p_n}-1} |A|^{\frac{1}{p_n}}, \tag{10.3.39}$$

for some dimensional constant C_n and all sets A of finite measure. This estimate can be written as

$$\lambda^{\frac{n+1}{2}} \delta^{\frac{n-1}{2}} |E_\lambda| \le C_n |A|, \tag{10.3.40}$$

where

$$E_\lambda = \{x \in \mathbf{R}^n : L^\delta(\chi_A)(x) > \lambda\} = \{L^\delta(\chi_A) > \lambda\}.$$

Our final reduction stems from the observation that the operator L^δ is "local." This means that if f is supported in a cube Q, say of side length one, then $L^\delta(f)$ is supported in a fixed multiple of Q. Indeed, it is simple to verify that if $x \notin 10Q$ and f is supported in Q, then $L^\delta(f)(x) = 0$, since no δ-tube containing x can reach Q. For "local" operators, it suffices to prove their boundedness for functions supported in cubes of side length one; see Exercise 10.3.9. We may therefore work with a measurable set A contained in a cube in $\mathbf{R}^n$ of side length one. This assumption has as a consequence that E_λ is contained in a fixed multiple of Q, such as $10Q$.

Having completed all the required reductions, we proceed by proving the restricted weak type estimate (10.3.40) for sets A supported in a cube of side length one. In proving (10.3.40) we may take $\lambda \le 1$; otherwise, the set E_λ is empty. We consider the cases $c_0(n)\,\delta \le \lambda$ and $c_0(n)\,\delta > \lambda$, for some large constant $c_0(n)$ to be determined later. If $c_0(n)\,\delta > \lambda$, then

$$|E_\lambda| \le C_n^1 (1/\delta)^{n-1} \frac{|A|}{\lambda} \tag{10.3.41}$$

by the weak type $(1,1)$ boundedness of L^δ with constant $C_n^1 \delta^{1-n}$. It follows from (10.3.41) that

$$C_n^1|A| \geq |E_\lambda|\delta^{n-1}\lambda > c_0(n)^{-\frac{n-1}{2}}|E_\lambda|\lambda^{\frac{n+1}{2}}\delta^{\frac{n-1}{2}},$$

which proves (10.3.40) in this case.

We now assume $c_0(n)\,\delta \leq \lambda \leq 1$. Since $L^\delta(\chi_A)$ is constant on each Q_j, we have that each Q_j is either entirely contained in the set E_λ or disjoint from it. Consequently, setting

$$\mathscr{E} = \{j: Q_j \subseteq E_\lambda\},$$

we have

$$E_\lambda = \bigcup_{j\in\mathscr{E}} Q_j.$$

Hence

$$|\mathscr{E}| = \#\{j: j\in\mathscr{E}\} = |E_\lambda|(\delta')^{-n},$$

and for all $j\in\mathscr{E}$ we have

$$|\tau_j \cap A| > \lambda|\tau_j| = \lambda\,\delta^{n-1}.$$

It follows that

$$\begin{aligned}
|A|\sup_x\Big[\sum_{j\in\mathscr{E}}\chi_{\tau_j}(x)\Big] &\geq \int_A \sum_{j\in\mathscr{E}}\chi_{\tau_j}\,dx\\
&= \sum_{j\in\mathscr{E}}|\tau_j\cap A|\\
&> \lambda\,\delta^{n-1}|\mathscr{E}|\\
&= \lambda\,\delta^{n-1}\frac{|E_\lambda|}{(\delta')^n}\\
&= (2\sqrt{n})^n\frac{\lambda\,|E_\lambda|}{\delta}.
\end{aligned}$$

Therefore, there exists an x_0 in A such that

$$\#\{j\in\mathscr{E}: x_0\in\tau_j\} > (2\sqrt{n})^n\frac{\lambda\,|E_\lambda|}{\delta|A|}.$$

Let $S(x_0,\frac{1}{2})$ be a sphere of radius $\frac{1}{2}$ centered at the point x_0. We find on this sphere a finite set of points $\Theta = \{\theta_k\}_k$ that is maximal with respect to the property that the balls $B(\theta_k,\delta)$ are at distance at least $10\sqrt{n}\,\delta$ from each other. Define spherical caps

$$S_k = \mathbf{S}^{n-1}\cap B(\theta_k,\delta).$$

Since the S_k's are disjoint and have surface measure a constant multiple of δ^{n-1}, it follows that there are about δ^{1-n} such points θ_k.

We count the number of δ-tubes that contain x_0 and intersect a fixed cap S_k. All these δ-tubes are contained in a cylinder of length 3 and diameter $c_1(n)\delta$ whose axis of symmetry contains x_0 and the center of the cap S_k. This cylinder has volume

$3\omega_{n-1}c_1(n)^{n-1}\delta^{n-1}$, and thus it intersects at most $c_2(n)\delta^{-1}$ cubes of the family Q_j, since the Q_j's are disjoint and all have volume equal to $(\delta')^n$. We deduce then that given such a cap S_k, there exist at most $c_3(n)\delta^{-1}$ δ-tubes (from the initial family) that contain the point x_0 and intersect S_k.

Let us call a set of δ-tubes ε-*separated* if for every τ and τ' in the set with $\tau \neq \tau'$ we have that the angle between the axis of symmetry of τ and τ' is at least $\varepsilon > 0$. Since we have at least $\frac{(2\sqrt{n})^n \lambda |E_\lambda|}{\delta |A|}$ δ-tubes that contain the given point x_0, and each cap S_k is intersected by at most $c_3(n)\delta^{-1}$ δ-tubes that contain x_0, it follows that at least $c_4(n)\frac{\lambda |E_\lambda|}{|A|}$ of these δ-tubes have to intersect different caps S_k. But δ-tubes that intersect different caps S_k and contain x_0 are δ-separated. We have therefore shown that there exist at least $c_4(n)\frac{\lambda |E_\lambda|}{|A|}$ δ-separated tubes from the original family that contain the point x_0. Call $\mathscr{T}$ the family of these δ-tubes.

We find a maximal subset Θ' of the θ_k's such that the balls $B(\theta_k, \delta)$, $\theta_k \in \Theta'$, have distance at least $\frac{30\sqrt{n}\,\delta}{\lambda}$ from each other. This is possible if $\lambda/\delta \geq c_0(n)$ for some large constant $c_0(n)$ [such as $c_0(n) = 1000\sqrt{n}$]. We "thin out" the family $\mathscr{T}$ by removing all the δ-tubes that intersect the caps S_k with $\theta_k \in \Theta \setminus \Theta'$. In other words, we essentially keep in $\mathscr{T}$ one out of every $1/\lambda^{n-1}$ δ-tubes. In this way we extract at least $c_5(n)\frac{\lambda^n |E_\lambda|}{|A|}$ δ-tubes from $\mathscr{T}$ that are $\frac{60\sqrt{n}\,\delta}{\lambda}$-separated and contain the point x_0. We denote these tubes by $\{\tau_j : j \in \mathscr{F}\}$.

We have therefore found a subset $\mathscr{F}$ of $\mathscr{E}$ such that

$$x_0 \in \tau_j \quad \text{for all} \quad j \in \mathscr{F}, \tag{10.3.42}$$

$$\tau_k, \tau_j \quad \text{are} \quad 60\sqrt{n}\frac{\delta}{\lambda} \text{ - separated} \quad \text{when} \quad j,k \in \mathscr{F},\ j \neq k, \tag{10.3.43}$$

$$|\mathscr{F}| \geq c_5(n)\frac{|E_\lambda|\lambda^n}{|A|}. \tag{10.3.44}$$

Notice that

$$\big|A \cap \tau_j \cap B(x_0, \tfrac{\lambda}{3})\big| \leq \big|\tau_j \cap B(x_0, \tfrac{\lambda}{3})\big| \leq \frac{2}{3}\lambda\delta^{n-1},$$

and since for any $j \in \mathscr{E}$ (and thus for $j \in \mathscr{F}$) we have $|A \cap \tau_j| > \lambda\delta^{n-1}$, it must be the case that

$$\big|A \cap \tau_j \cap B(x_0, \tfrac{\lambda}{3})^c\big| > \frac{1}{3}\lambda\delta^{n-1}. \tag{10.3.45}$$

Moreover, it is crucial to note that the sets

$$A \cap \tau_j \cap B(x_0, \tfrac{\lambda}{3})^c, \qquad j \in \mathscr{F}, \tag{10.3.46}$$

are pairwise disjoint. In fact, if x_j and x_k are points on the axes of symmetry of two $60\sqrt{n}\frac{\delta}{\lambda}$-separated δ-tubes τ_j and τ_k in $\mathscr{F}$ such that $|x_j - x_0| = |x_k - x_0| = \frac{\lambda}{3}$, then the distance from x_k to x_j must be at least $10\sqrt{n}\,\delta$. This implies that the distance between $\tau_j \cap B(x_0, \frac{\lambda}{3})^c$ and $\tau_k \cap B(x_0, \frac{\lambda}{3})^c$ is at least $6\sqrt{n}\,\delta > 0$. We now conclude the proof of the theorem as follows:

$$\begin{aligned}
|A| &\geq \Big|A \cap \bigcup_{j\in\mathscr{F}} \big(\tau_j \cap B(x_0, \tfrac{\lambda}{3})^c\big)\Big| \\
&= \sum_{j\in\mathscr{F}} \big|A \cap \tau_j \cap B(x_0, \tfrac{\lambda}{3})^c\big| \\
&\geq \sum_{j\in\mathscr{F}} \frac{\lambda\,\delta^{n-1}}{3} \\
&= |\mathscr{F}|\,\frac{\lambda\,\delta^{n-1}}{3} \\
&\geq c_5(n)\frac{|E_\lambda|\lambda^n}{|A|}\,\frac{\lambda\,\delta^{n-1}}{3},
\end{aligned}$$

using that the sets in (10.3.46) are disjoint, (10.3.45), and (10.3.44). We conclude that

$$|A|^2 \geq \frac{1}{3}c_5(n)\,\lambda^{n+1}\delta^{n-1}|E_\lambda| \geq c_6(n)\lambda^{n+1}\delta^{n-1}|E_\lambda|^2,$$

since, as observed earlier, the set E_λ is contained in a cube of side length 10. Taking square roots, we obtain (10.3.40). This proves (10.3.39) and hence (10.3.36). □

Exercises

10.3.1. Let h be the characteristic function of the square $[0,1]^2$ in $\mathbf{R}^2$. Prove that for any $0<\lambda<1$ we have

$$\big|\{x\in\mathbf{R}^2:\, M_s(h)(x)>\lambda\}\big| \geq \frac{1}{\lambda}\log\frac{1}{\lambda}.$$

Use this to show that M_s is not of weak type $(1,1)$. Compare this result with that of Exercise 2.1.6.

10.3.2. (a) Given a unit vector v in $\mathbf{R}^2$ define the *directional maximal function along* $\vec{v}$ by

$$M_{\vec{v}}(f)(x) = \sup_{\varepsilon>0}\frac{1}{2\varepsilon}\int_{-\varepsilon}^{+\varepsilon}|f(x-t\vec{v})|\,dt$$

wherever f is locally integrable over $\mathbf{R}^2$. Prove that for such f, $M_{\vec{v}}(f)(x)$ is well defined for almost all x contained in any line not parallel to $\vec{v}$.
(b) For $1<p<\infty$, use the method of rotations to show that $M_{\vec{v}}$ maps $L^p(\mathbf{R}^2)$ to itself with norm the same as that of the centered Hardy–Littlewood maximal operator M on $L^p(\mathbf{R})$.
(c) Let Σ be a finite set of directions. Prove that for all $1\leq p\leq\infty$, there is a constant $C_p>0$ such that

$$\big\|\mathfrak{M}_\Sigma(f)\big\|_{L^p(\mathbf{R}^2)} \leq C_p\,|\Sigma|^{\frac{1}{p}}\,\big\|f\big\|_{L^p(\mathbf{R}^2)}$$

for all f in $L^p(\mathbf{R}^2)$.
[*Hint:* Use the inequality $\mathfrak{M}_\Sigma(f)^p \le \sum_{\vec{v}\in\Sigma} [M_{\vec{v}} M_{\vec{v}^\perp}(f)]^p$.]

10.3.3. Show that

$$\mathscr{K}_N \le 20\,\mathfrak{M}_{\Sigma_N},$$

where Σ_N is a set of N uniformly distributed vectors in $\mathbf{S}^1$.
[*Hint:* Use Exercise 10.2.3.]

10.3.4. This exercise indicates a connection between the Besicovitch construction in Section 10.1 and the Kakeya maximal function. Recall the set E of Lemma 10.1.1, which satisfies $\frac{1}{2} \le |E| \le \frac{3}{2}$.
(a) Show that there is a positive constant c such that for all $N \ge 10$ we have

$$\big|\{x \in \mathbf{R}^2 : \mathscr{K}_N(\chi_E)(x) > \tfrac{1}{144}\}\big| \ge c \log\log N.$$

(b) Conclude that for all $2 < p < \infty$ there is a constant c_p such that

$$\big\|\mathscr{K}_N\big\|_{L^p(\mathbf{R}^2)\to L^p(\mathbf{R}^2)} \ge c_p (\log\log N)^{\frac{1}{p}}.$$

[*Hint:* Using the notation of Lemma 10.1.1, first show that

$$\big|\{x \in \mathbf{R}^2 : \mathscr{K}_{3\cdot 2^k \log(k+2)}(\chi_E)(x) > \tfrac{1}{36}\}\big| \ge \log(k+2),$$

by showing that the previous set contains all the disjoint rectangles R_j for $j = 1,2,\dots,2^k$; here k is a large positive integer. To show this, for x in $\bigcup_{j=1}^{2^k} R_j$ consider the unique rectangle R_{j_x} that contains x union $(R_{j_x})'$ and set $R_x = R_{j_x} \cup (R_{j_x})'$. Then $|R_x| = 3|R_{j_x}| = 3\cdot 2^{-k}\log(k+2)$, and we have

$$\frac{1}{|R_x|}\int_{R_x} |\chi_E(y)|\,dy = \frac{|E \cap R_x|}{|R_x|} \ge \frac{|E \cap (R_{j_x})'|}{3|R_{j_x}|} \ge \frac{1}{36}$$

in view of conclusion (4) in Lemma 10.1.1. Part (b): Express the L^p norm of $\mathscr{K}_N(\chi_E)$ in terms of its distribution function.]

10.3.5. Show that $\mathfrak{M}_{\mathbf{S}^1}$ is unbounded on $L^p(\mathbf{R}^2)$ for any $p < \infty$.
[*Hint:* You may use Proposition 10.3.4 when $p \le 2$. When $p > 2$ one may need Exercise 10.3.4.]

10.3.6. Consider the n-dimensional Kakeya maximal operator $\mathscr{K}_N$. Show that there exist dimensional constants c_n and c_n' such that for N sufficiently large we have

$$\big\|\mathscr{K}_N\big\|_{L^n(\mathbf{R}^n)\to L^n(\mathbf{R}^n)} \ge c_n (\log N),$$
$$\big\|\mathscr{K}_N\big\|_{L^n(\mathbf{R}^n)\to L^{n,\infty}(\mathbf{R}^n)} \ge c_n' (\log N)^{\frac{n-1}{n}}.$$

[*Hint:* Consider the functions $f_N(x) = \frac{1}{|x|}\chi_{3\le|x|\le N}$ and adapt the argument in Proposition 10.3.4 to an n-dimensional setting.]

10.3.7. For all $1 \le p < n$ show that there exist constants $c_{n,p}$ such that the n-dimensional Kakeya maximal operator $\mathscr{K}_N$ satisfies

$$\big\|\mathscr{K}_N\big\|_{L^p(\mathbf{R}^n)\to L^p(\mathbf{R}^n)} \ge \big\|\mathscr{K}_N\big\|_{L^p(\mathbf{R}^n)\to L^{p,\infty}(\mathbf{R}^n)} \ge c_{n,p} N^{\frac{n}{p}-1}.$$

[*Hint:* Consider the functions $h_N(x) = |x|^{-\frac{n+1}{p}} \chi_{3\le|x|\le N}$ and show that $\mathscr{K}_N(h_N)(x) > c/|x|$ for all x in the annulus $6 < |x| < N$.]

10.3.8. (*Carbery, Hernández, and Soria [51]*) Let T be a sublinear operator defined on $L^1(\mathbf{R}^n) + L^\infty(\mathbf{R}^n)$ and taking values in a set of measurable functions. Let $10 \le N < \infty$, $1 < p < \infty$, and $0 < a, M < \infty$.
(a) Suppose that

$$\begin{aligned}\big\|T\big\|_{L^1\to L^{1,\infty}} &\le C_1 N^a,\\ \big\|T\big\|_{L^{p,1}\to L^{p,\infty}} &\le M,\\ \big\|T\big\|_{L^\infty\to L^\infty} &\le 1.\end{aligned}$$

Show that

$$\big\|T\big\|_{L^p\to L^{p,\infty}} \le C(a,p,C_1)\, M\,(\log N)^{\frac{1}{p'}}.$$

(b) Suppose that

$$\begin{aligned}\big\|T\big\|_{L^1\to L^{1,\infty}} &\le C_1 N^a,\\ \big\|T\big\|_{L^p\to L^{p,\infty}} &\le M,\\ \big\|T\big\|_{L^\infty\to L^\infty} &\le 1.\end{aligned}$$

Show that

$$\big\|T\big\|_{L^p\to L^p} \le C'(a,p,C_1)\, M(\log N)^{\frac{1}{p}}.$$

[*Hint:* Part (a): Split $f = f_1 + f_2 + f_3$, where $f_3 = f\chi_{|f|\le\frac{\lambda}{4}}$, $f_2 = f\chi_{\frac{\lambda}{4}<|f|\le L\lambda}$, and $f_1 = f\chi_{|f|>L\lambda}$, where $L^{p-1} = N^a$. Use the weak type $(1,1)$ estimate for f_1 and the restricted weak type (p,p) estimate for f_2 and note that the measure of the set $\{|T(f_3)| > \lambda/3\}$ is zero. One needs the auxiliary result

$$\big\|f\chi_{a\le|f|\le b}\big\|_{L^{p,1}} \le C(p)\,(1+\log\tfrac{b}{a})^{\frac{1}{p'}}\,\big\|f\big\|_{L^p},$$

which can be proved as follows. First use the identity of Proposition 1.4.9. Then note that the distribution function $d_{f\chi_{a\le|f|\le b}}(s)$ is equal to $d_f(a)$ for $s < a$, to $d_f(s)$ for $a \le s < b$, and vanishes for $s \ge b$. It follows that

$$\big\|f\chi_{a\le|f|\le b}\big\|_{L^{p,1}} \le a\,d_f(a)^{\frac{1}{p}} + \int_a^b d_f(t)^{\frac{1}{p}}\,dt \le 2\int_{\frac{a}{2}}^a d_f(t)^{\frac{1}{p}}\,dt + \int_a^b d_f(t)^{\frac{1}{p}}\,dt,$$

from which the claimed estimate follows by Hölder's inequality and Proposition 1.1.4. Part (b): Use the same splitting and the method employed in the proof of Theorem 10.3.5.]

10.3.9. Suppose that T is a linear operator defined on a subspace of measurable functions on $\mathbf{R}^n$ with the property that whenever f is supported in a cube Q of side length s, then $T(f)$ is supported in aQ for some $a > 1$. Prove the following:
(a) If T is defined on $L^p(\mathbf{R}^n)$ for some $0 < p < \infty$ and

$$\big\|T(f)\big\|_{L^p} \le B\big\|f\big\|_{L^p}$$

for all f supported in a cube of side length s, then the same estimate holds (with a larger constant) for all functions in $L^p(\mathbf{R}^n)$.
(b) If T satisfies for some $0 < p < \infty$,

$$\big\|T(\chi_A)\big\|_{L^{p,\infty}} \le B|A|^{\frac{1}{p}}$$

for all measurable sets A contained in a cube of side length s, then the same estimate holds (with a larger constant) for all measurable sets A in $\mathbf{R}^n$.

10.4 Fourier Transform Restriction and Bochner–Riesz Means

If g is a continuous function on $\mathbf{R}^n$, its restriction to a hypersurface $S \subseteq \mathbf{R}^n$ is a well defined function. By a hypersurface we mean a submanifold of $\mathbf{R}^n$ of dimension $n-1$. So, if f is an integrable function on $\mathbf{R}^n$, its Fourier transform $\widehat{f}$ is continuous and hence its restriction $\widehat{f}\big|_S$ on S is well defined.

Definition 10.4.1. Let $1 \le p,q \le \infty$. We say that a compact hypersurface S in $\mathbf{R}^n$ satisfies a (p,q) *restriction theorem* if the restriction operator

$$f \to \widehat{f}\big|_S,$$

which is initially defined on $L^1(\mathbf{R}^n) \cap L^p(\mathbf{R}^n)$, has an extension that maps $L^p(\mathbf{R}^n)$ boundedly into $L^q(S)$. The norm of this extension may depend on p,q,n, and S. If S satisfies a (p,q) restriction theorem, we write that property $R_{p\to q}(S)$ holds. We say that property $R_{p\to q}(S)$ holds with constant C if for all $f \in L^1(\mathbf{R}^n) \cap L^p(\mathbf{R}^n)$ we have

$$\big\|\widehat{f}\big\|_{L^q(S)} \le C\big\|f\big\|_{L^p(\mathbf{R}^n)}.$$

Example 10.4.2. Property $R_{1\to\infty}(S)$ holds for any compact hypersurface S.

We denote by $\mathscr{R}(f) = \widehat{f}\big|_{\mathbf{S}^{n-1}}$ the restriction of the Fourier transform on a hypersurface S. Let $d\sigma$ be the canonically induced surface measure on S. Then for a function φ defined on S we have

$$\int_{\mathbf{S}^{n-1}} \widehat{f}\,\varphi\,d\sigma = \int_{\mathbf{R}^n} \widehat{f}\,(\widehat{\varphi\,d\sigma})^{\vee}\,d\xi = \int_{\mathbf{R}^n} f\,\widehat{\varphi\,d\sigma}\,dx,$$

which says that the transpose of the linear operator $\mathscr{R}$ is the linear operator

$$\mathscr{R}^t(\varphi) = \widehat{\varphi\,d\sigma}. \tag{10.4.1}$$

By duality, we easily see that a (p,q) restriction theorem for a compact hypersurface S is equivalent to the following (q',p') *extension theorem* for S:

$$\mathscr{R}^t : L^{q'}(S) \to L^{p'}(\mathbf{R}^n).$$

Our objective is to determine all pairs of indices (p,q) for which the sphere $\mathbf{S}^{n-1}$ satisfies a (p,q) restriction theorem. It becomes apparent in this section that this problem is relevant in the understanding of the norm convergence of the Bochner–Riesz means.

10.4.1 Necessary Conditions for $R_{p\to q}(\mathbf{S}^{n-1})$ to Hold

We look at basic examples that impose restrictions on the indices p,q in order for $R_{p\to q}(\mathbf{S}^{n-1})$ to hold. We first make an observation. If $R_{p\to q}(\mathbf{S}^{n-1})$ holds, then $R_{p\to s}(\mathbf{S}^{n-1})$ for any $s \le q$.

Example 10.4.3. Let $d\sigma$ be surface measure on the unit sphere $\mathbf{S}^{n-1}$. In view of the identity in Appendix B.4, we have

$$\widehat{d\sigma}(\xi) = \frac{2\pi}{|\xi|^{\frac{n-2}{2}}} J_{\frac{n-2}{2}}(2\pi|\xi|).$$

Using the asymptotics in Appendix B.8, the last expression is equal to

$$\frac{2\sqrt{2\pi}}{|\xi|^{\frac{n-1}{2}}}\cos(2\pi|\xi| - \tfrac{\pi(n-1)}{4}) + O(|\xi|^{-\frac{n+1}{2}})$$

as $|\xi| \to \infty$. It follows that $\mathscr{R}^t(1)(\xi) = \widehat{d\sigma}(\xi)$ does not lie in $L^{p'}(\mathbf{R}^n)$ if $\frac{n-1}{2}p' \le n$ and $\frac{n+1}{2}p' > n$. Thus $R_{p\to q}(\mathbf{S}^{n-1})$ fails when $\frac{2n}{n+1} \le p < \frac{2n}{n-1}$. Since $R_{1\to q}(\mathbf{S}^{n-1})$ holds for all $q \in [1,\infty]$, by interpolation we deduce that $R_{p\to q}(\mathbf{S}^{n-1})$ fails when $p \ge \frac{2n}{n+1}$. We conclude that a necessary condition for $R_{p\to q}(\mathbf{S}^{n-1})$ to hold is that

$$1 \le p < \frac{2n}{n+1}. \tag{10.4.2}$$

In addition to this condition, there is another necessary condition for $R_{p\to q}(\mathbf{S}^{n-1})$ to hold. This is a consequence of the following revealing example.

Example 10.4.4. Let φ be a Schwartz function on $\mathbf{R}^n$ such that $\widehat{\varphi} \geq 0$ and $\widehat{\varphi}(\xi) \geq 1$ for all ξ in the closed ball $|\xi| \leq 2$. For $N \geq 1$ define functions

$$f_N(x_1, x_2, \dots, x_{n-1}, x_n) = \varphi\Big(\frac{x_1}{N}, \frac{x_2}{N}, \dots, \frac{x_{n-1}}{N}, \frac{x_n}{N^2}\Big).$$

To test property $R_{p\to q}(\mathbf{S}^{n-1})$, instead of working with $\mathbf{S}^{n-1}$, we may work with the translated sphere $S = \mathbf{S}^{n-1} + (0,0,\dots,0,1)$ in $\mathbf{R}^n$ (cf. Exercise 10.4.2(a)). We have

$$\widehat{f_N}(\xi) = N^{n+1}\widehat{\varphi}(N\xi_1, N\xi_2, \dots, N\xi_{n-1}, N^2\xi_n).$$

We note that for all $\xi = (\xi_1, \dots, \xi_n)$ in the spherical cap

$$S' = S \cap \{\xi \in \mathbf{R}^n : \ \xi_1^2 + \cdots + \xi_{n-1}^2 \leq N^{-2} \quad \text{and} \quad \xi_n < 1\}, \tag{10.4.3}$$

we have $\xi_n \leq 1 - (1 - \frac{1}{N^2})^{\frac{1}{2}} \leq \frac{1}{N^2}$ and therefore

$$|(N\xi_1, N\xi_2, \dots, N\xi_{n-1}, N^2\xi_n)| \leq 2.$$

This implies that for all ξ in S' we have $\widehat{f_N}(\xi) \geq N^{n+1}$. But the spherical cap S' in (10.4.3) has surface measure $c(N^{-1})^{n-1}$. We obtain

$$\big\|\widehat{f_N}\big\|_{L^q(S)} \geq \big\|\widehat{f_N}\big\|_{L^q(S')} \geq c^{\frac{1}{q}} N^{n+1} N^{\frac{1-n}{q}}.$$

On the other hand, $\big\|f_N\big\|_{L^p(\mathbf{R}^n)} = \big\|\varphi\big\|_{L^p(\mathbf{R}^n)} N^{\frac{n+1}{p}}$. Therefore, if $R_{p\to q}(\mathbf{S}^{n-1})$ holds, we must have

$$\big\|\varphi\big\|_{L^p(\mathbf{R}^n)} N^{\frac{n+1}{p}} \geq C c^{\frac{1}{q}} N^{n+1} N^{\frac{1-n}{q}},$$

and letting $N \to \infty$, we obtain the following necessary condition on p and q for $R_{p\to q}(\mathbf{S}^{n-1})$ to hold:

$$\frac{1}{q} \geq \frac{n+1}{n-1}\frac{1}{p'}. \tag{10.4.4}$$

We have seen that the restriction property $R_{p\to q}(\mathbf{S}^{n-1})$ fails in the shaded region of Figure 10.10 but obviously holds on the closed line segment CD. It remains to investigate the validity of property $R_{p\to q}(\mathbf{S}^{n-1})$ for $(\frac{1}{p}, \frac{1}{q})$ in the unshaded region of Figure 10.10.

It is a natural question to ask whether the restriction property $R_{p\to q}(\mathbf{S}^{n-1})$ holds on the line segment BD minus the point B in Figure 10.10, i.e., the set

$$\Big\{(p,q) : \ \frac{1}{q} = \frac{n+1}{n-1}\frac{1}{p'} \qquad 1 \leq p < \frac{2n}{n+1}\Big\}. \tag{10.4.5}$$

If property $R_{p\to q}(\mathbf{S}^{n-1})$ holds for all points in this set, then it will also hold in the closure of the quadrilateral $ABDC$ minus the closed segment AB.

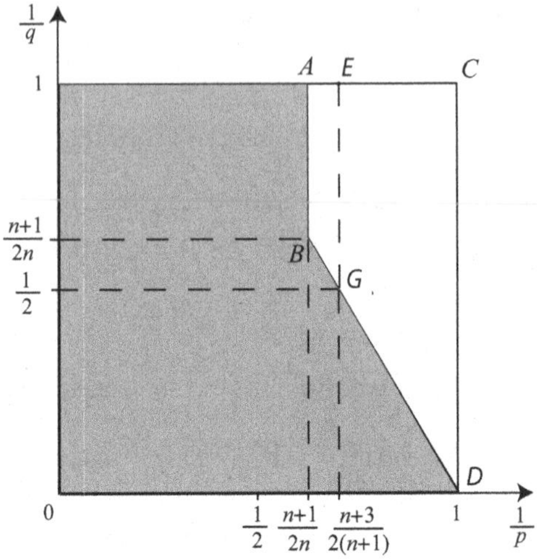

Fig. 10.10 The restriction property $R_{p\to q}(\mathbf{S}^{n-1})$ fails in the shaded region and on the closed line segment AB but holds on the closed line segment CD and could hold on the open line segment BD and inside the unshaded region.

10.4.2 A Restriction Theorem for the Fourier Transform

In this subsection we establish the following restriction theorem for the Fourier transform.

Theorem 10.4.5. *Property $R_{p\to q}(\mathbf{S}^{n-1})$ holds for the set*

$$\left\{(p,q):\ \frac{1}{q}=\frac{n+1}{n-1}\frac{1}{p'},\qquad 1\le p\le \frac{2(n+1)}{n+3}\right\} \tag{10.4.6}$$

and therefore for the closure of the quadrilateral with vertices E, G, D, and C in Figure 10.10.

Proof. The case $p=1$ and $q=\infty$ is trivial. Therefore, we need to establish only the case $p=\frac{2(n+1)}{n+3}$ and $q=2$, since the remaining cases follow by interpolation.

Using Plancherel's identity and Hölder's inequality, we obtain

$$\begin{aligned}
\big\|\widehat{f}\big\|_{L^2(\mathbf{S}^{n-1})}^2 &= \int_{\mathbf{S}^{n-1}} \overline{\widehat{f}(\xi)}\,\widehat{f}(\xi)\,d\sigma(\xi)\\
&= \int_{\mathbf{R}^n} \overline{f(x)}\,(f*d\sigma^\vee)(x)\,dx\\
&\le \big\|f\big\|_{L^p(\mathbf{R}^n)}\big\|f*d\sigma^\vee\big\|_{L^{p'}(\mathbf{R}^n)}\,.
\end{aligned}$$

To establish the required conclusion it is enough to show that

$$\big\|f*d\sigma^\vee\big\|_{L^{p'}(\mathbf{R}^n)}\le C_n\big\|f\big\|_{L^p(\mathbf{R}^n)}\qquad \text{when}\quad p=\frac{2(n+1)}{n+3}. \tag{10.4.7}$$

To obtain this estimate we need to split the sphere into pieces. Each hyperplane $\xi_k = 0$ cuts the sphere $\mathbf{S}^{n-1}$ into two hemispheres, which we denote by H_k^1 and H_k^2. We introduce a partition of unity $\{\varphi_j\}_j$ of $\mathbf{R}^n$ with the property that for any j there exist $k \in \{1,2,\dots,\}$ and $l \in \{1,2\}$ such that

$$(\text{support}\,\varphi_j) \cap \mathbf{S}^{n-1} \subsetneqq H_k^l\,;$$

that is, the support of each φ_j intersected with the sphere $\mathbf{S}^{n-1}$ is properly contained in some hemisphere H_k^l. Then the family of all φ_j whose support meets $\mathbf{S}^{n-1}$ forms a finite partition of unity of the sphere when restricted to it. We therefore write

$$d\sigma = \sum_{j\in F} \varphi_j\, d\sigma\,,$$

where F is a finite set. If we obtain (10.4.7) for each measure $\varphi_j d\sigma$ instead of $d\sigma$, then (10.4.7) follows by summing on j. We fix such a measure $\varphi_j\, d\sigma$, which, without loss of generality, we assume is supported in $\{\xi \in \mathbf{S}^{n-1} :\ \xi_n > c\} \subsetneqq H_n^1$ for some $c \in (0,1)$. In the sequel we write elements $x \in \mathbf{R}^n$ as $x = (x',t)$, where $x' \in \mathbf{R}^{n-1}$ and $t \in \mathbf{R}$. Then for $x \in \mathbf{R}^n$ we have

$$(\varphi_j\, d\sigma)^\vee(x) = \int_{\mathbf{S}^{n-1}} \varphi_j(\xi) e^{2\pi i x\cdot\xi}\, d\sigma(\xi) = \int_{\substack{\xi'\in\mathbf{R}^{n-1}\\ |\xi'|^2\le 1-c^2}} e^{2\pi i x\cdot\xi}\, \frac{\varphi_j(\xi',\sqrt{1-|\xi'|^2})\, d\xi'}{\sqrt{1-|\xi'|^2}}\,,$$

where $\xi = (\xi',\xi_n)$; for the last identity we refer to Appendix D.5. Writing $x = (x',t) \in \mathbf{R}^{n-1}\times\mathbf{R}$, we have

$$\begin{aligned}(\varphi_j\, d\sigma)^\vee(x',t) &= \int_{\substack{\xi'\in\mathbf{R}^{n-1}\\ |\xi'|^2\le 1-c^2}} e^{2\pi i x'\cdot\xi'} e^{2\pi i t\sqrt{1-|\xi'|^2}}\, \frac{\varphi_j(\xi',\sqrt{1-|\xi'|^2})}{\sqrt{1-|\xi'|^2}}\, d\xi' \\ &= \left(e^{2\pi i t\sqrt{1-|\xi'|^2}}\, \frac{\varphi_j(\xi',\sqrt{1-|\xi'|^2})}{\sqrt{1-|\xi'|^2}}\right)^{\triangledown}(x')\,,\end{aligned} \tag{10.4.8}$$

where $^\triangledown$ indicates the inverse Fourier transform in the ξ' variable. For each $t\in\mathbf{R}$ we introduce a function on $\mathbf{R}^{n-1}$ by setting

$$K_t(x') = (\varphi_j\, d\sigma)^\vee(x',t)\,.$$

We observe that identity (10.4.8) and the fact that $1-|\xi'|^2 \ge c^2 > 0$ on the support of φ_j imply that

$$\sup_{t\in\mathbf{R}}\ \sup_{\xi'\in\mathbf{R}^{n-1}} |(K_t)^{\triangle}(\xi')| \le C_n < \infty\,, \tag{10.4.9}$$

where $^\triangle$ indicates the Fourier transform on $\mathbf{R}^{n-1}$. We also have that

$$K_t(x') = (\varphi_j\, d\sigma)^\vee(x',t) = (\varphi_j^\vee * d\sigma^\vee)(x',t)\,.$$

Since $\varphi_j^\vee$ is a Schwartz function on $\mathbf{R}^n$ and the function $|d\sigma^\vee(x',t)|$ is bounded by $(1+|(x',t)|)^{-\frac{n-1}{2}}$ (see Appendices B.4, B.6, and B.7), it follows from Exercise 2.2.4 that

$$|K_t(x')| \le C(1+|(x',t)|)^{-\frac{n-1}{2}} \le C(1+|t|)^{-\frac{n-1}{2}} \tag{10.4.10}$$

for all $x' \in \mathbf{R}^{n-1}$. Estimate (10.4.9) says that the operator given by convolution with K_t maps $L^2(\mathbf{R}^{n-1})$ to itself with norm at most a constant, while (10.4.10) says that the same operator maps $L^1(\mathbf{R}^{n-1})$ to $L^\infty(\mathbf{R}^{n-1})$ with norm at most a constant multiple of $(1+|t|)^{-\frac{n-1}{2}}$. Interpolating between these two estimates yields

$$\big\|K_t \star g\big\|_{L^{p'}(\mathbf{R}^{n-1})} \le C_{p,n}|t|^{-(n-1)(\frac{1}{p}-\frac{1}{2})}\big\|g\big\|_{L^p(\mathbf{R}^{n-1})}$$

for all $1 \le p \le 2$, where $\star$ denotes convolution on $\mathbf{R}^{n-1}$ (and $*$ convolution on $\mathbf{R}^n$).

We now return to the proof of the required estimate (10.4.7) in which $d\sigma^\vee$ is replaced by $(\varphi_j\,d\sigma)^\vee$. Let $f(x) = f(x',t)$ be a function on $\mathbf{R}^n$. We have

$$\begin{aligned}
\big\|f*(\varphi_j\,d\sigma)^\vee\big\|_{L^{p'}(\mathbf{R}^n)} &= \bigg\|\Big\|\int_{\mathbf{R}} f(\cdot,\tau)\star K_{t-\tau}\,d\tau\Big\|_{L^{p'}(\mathbf{R}^{n-1})}\bigg\|_{L^{p'}(\mathbf{R})}\\
&\le \bigg\|\int_{\mathbf{R}}\big\|f(\cdot,\tau)\star K_{t-\tau}\big\|_{L^{p'}(\mathbf{R}^{n-1})}\,d\tau\bigg\|_{L^{p'}(\mathbf{R})}\\
&\le C_{p,n}\bigg\|\int_{\mathbf{R}}\frac{\|f(\cdot,\tau)\|_{L^p(\mathbf{R}^{n-1})}}{|t-\tau|^{(n-1)(\frac{1}{p}-\frac{1}{2})}}\,d\tau\bigg\|_{L^{p'}(\mathbf{R})}\\
&= C_{p,n}\Big\|I_\beta\Big(\|f(\cdot,t)\|_{L^p(\mathbf{R}^{n-1})}\Big)\Big\|_{L^{p'}(\mathbf{R},dt)},
\end{aligned}$$

where $\beta = 1-(n-1)(\frac{1}{p}-\frac{1}{2})$ and I_β is the Riesz potential (or fractional integral) given in Definition 6.1.1. Using Theorem 6.1.3 with $s=\beta$, $n=1$, and $q=p'$, we obtain that the last displayed equation is bounded by a constant multiple of

$$\Big\|\|f(\cdot,t)\|_{L^p(\mathbf{R}^{n-1})}\Big\|_{L^p(\mathbf{R},dt)} = \|f\|_{L^p(\mathbf{R}^n)}.$$

The condition $\frac{1}{p}-\frac{1}{q}=\frac{s}{n}$ on the indices p,q,s,n assumed in Theorem 6.1.3 translates exactly to

$$\frac{1}{p}-\frac{1}{p'}=\frac{\beta}{1}=1-\frac{n-1}{p}-\frac{n-1}{2},$$

which is equivalent to $p=\frac{2(n+1)}{n+3}$. This concludes the proof of estimate (10.4.7) in which the measure $\sigma^\vee$ is replaced by $(\varphi_j\,d\sigma)^\vee$. Estimates for the remaining $(\varphi_j\,d\sigma)^\vee$ follow by a similar argument in which the role of the last coordinate is played by some other coordinate. The final estimate (10.4.7) follows by summing j over the finite set F. The proof of the theorem is now complete. $\square$

10.4.3 Applications to Bochner–Riesz Multipliers

We now apply the restriction theorem obtained in the previous subsection to the Bochner–Riesz problem. In this subsection we prove the following result.

Theorem 10.4.6. *For* $\operatorname{Re}\lambda > \frac{n-1}{2(n+1)}$, *the Bochner–Riesz operator* B^λ *is bounded on* $L^p(\mathbf{R}^n)$ *for* p *in the optimal range*

$$\frac{2n}{n+1+2\operatorname{Re}\lambda} < p < \frac{2n}{n-1-2\operatorname{Re}\lambda}.$$

Proof. The proof is based on the following two estimates:

$$\big\|B^\lambda\big\|_{L^1(\mathbf{R}^n)\to L^1(\mathbf{R}^n)} \le C_1(\operatorname{Re}\lambda)\, e^{c_0|\operatorname{Im}\lambda|^2} \qquad \text{when } \operatorname{Re}\lambda > \tfrac{n-1}{2}, \tag{10.4.11}$$

$$\big\|B^\lambda\big\|_{L^p(\mathbf{R}^n)\to L^p(\mathbf{R}^n)} \le C_2(\operatorname{Re}\lambda)\, e^{c_0|\operatorname{Im}\lambda|^2} \qquad \text{when } \operatorname{Re}\lambda > \tfrac{n-1}{2(n+1)}, \tag{10.4.12}$$

where $p=\frac{2(n+1)}{n+3}$ and C_1, C_2 are constants that depend on n and $\operatorname{Re}\lambda$, while c_0 is an absolute constant. Once (10.4.11) and (10.4.12) are known, the required conclusion is a consequence of Theorem 1.3.7. Recall that B^λ is given by convolution with the kernel K_λ defined in (10.2.1). This kernel satisfies

$$|K_\lambda(x)| \le C_3(\operatorname{Re}\lambda)\, e^{c_0|\operatorname{Im}\lambda|^2}(1+|x|)^{-\frac{n+1}{2}-\operatorname{Re}\lambda} \tag{10.4.13}$$

in view of the estimates in Appendices B.6 and B.7. Then (10.4.11) follows easily from (10.4.13) and we focus our attention on (10.4.12).

The key ingredient in the proof of (10.4.12) is a decomposition of the kernel. But first we isolate the smooth part of the multiplier near the origin and we focus attention on the part of it near the boundary of the unit disk. Precisely, we start with a Schwartz function $0 \le \eta \le 1$ supported in the ball $B(0,\frac{3}{4})$ that is equal to 1 on the smaller ball $B(0,\frac{1}{2})$. Then we write

$$m_\lambda(\xi) = (1-|\xi|^2)_+^\lambda = (1-|\xi|^2)_+^\lambda \eta(\xi) + (1-|\xi|^2)_+^\lambda (1-\eta(\xi)).$$

Since the function $(1-|\xi|^2)_+^\lambda \eta(\xi)$ is smooth and compactly supported, it is an L^p Fourier multiplier for all $1\le p\le\infty$, with norm that is easily seen to grow polynomially in $|\lambda|$. We therefore need to concentrate on the nonsmooth piece of the multiplier $(1-|\xi|^2)_+^\lambda(1-\eta(\xi))$, which is supported in $B(0,\frac{1}{2})^c$. Let

$$K^\lambda(x) = \Big((1-|\xi|^2)_+^\lambda(1-\eta(\xi))\Big)^\vee(x)$$

be the kernel of the nonsmooth piece of the multiplier.

We pick a smooth *radial* function φ with support inside the ball $B(0,2)$ that is equal to 1 on the closed unit ball $\overline{B(0,1)}$. For $j=1,2,\dots$ we introduce functions

$$\psi_j(x) = \varphi(2^{-j}x) - \varphi(2^{-j+1}x)$$

supported in the annuli $2^{j-1} \le |x| \le 2^{j+1}$. Then we write

$$K^\lambda * f = T_0^\lambda(f) + \sum_{j=1}^{\infty} T_j^\lambda(f), \tag{10.4.14}$$

where T_0^λ is given by convolution with φK^λ and each T_j^λ is given by convolution with $\psi_j K^\lambda$.

We begin by examining the kernel φK^λ. Introducing a compactly supported function ζ that is equal to 1 on $B(0, \frac{3}{2})$, we write

$$\begin{aligned} K^\lambda &= \left((1-|\cdot|^2)_+^\lambda (1-\eta)\zeta\right)^\vee \\ &= \left((1-|\cdot|^2)_+^\lambda\right)^\vee * \left((1-\eta)\zeta\right)^\vee \\ &= K_\lambda * \left((1-\eta)\zeta\right)^\vee . \end{aligned}$$

Using this and (10.4.13) implies that K^λ is a bounded function, and thus φK^λ is bounded and compactly supported. Thus the operator T_0^λ is bounded on all the L^p spaces, $1 \le p \le \infty$, with a bound that grows at most exponentially in $|\mathrm{Im}\,\lambda|^2$.

Next we study the boundedness of the operators T_j^λ; here the dependence on the index j plays a role. Fix $p < 2$ as in the statement of the theorem. Our goal is to show that there exist positive constants C, δ (depending only on n and $\mathrm{Re}\,\lambda$) such that for all functions f in $L^p(\mathbf{R}^n)$ we have

$$\big\|T_j^\lambda(f)\big\|_{L^p(\mathbf{R}^n)} \le C e^{c_0|\mathrm{Im}\,\lambda|^2} 2^{-j\delta} \big\|f\big\|_{L^p(\mathbf{R}^n)} . \tag{10.4.15}$$

Once (10.4.15) is established, the L^p boundedness of the operator $f \mapsto K^\lambda * f$ follows by summing the series in (10.4.14).

As a consequence of (10.4.13) we obtain that

$$\begin{aligned} |K_j^\lambda(x)| &\le C_3(\mathrm{Re}\,\lambda)\, e^{c_0|\mathrm{Im}\,\lambda|^2} (1+|x|)^{-\frac{n+1}{2}-\mathrm{Re}\,\lambda} |\psi_j(x)| \\ &\le C' 2^{-(\frac{n+1}{2}+\mathrm{Re}\,\lambda)j} , \end{aligned} \tag{10.4.16}$$

since $\psi_j(x) = \psi(2^{-j}x)$ and ψ is supported in the annulus $\frac{1}{2} \le |x| \le 2$. From this point on, the constants containing a prime are assumed to grow at most exponentially in $|\mathrm{Im}\,\lambda|^2$. Since K_j^λ is supported in a ball of radius 2^{j+1} and satisfies (10.4.16), we deduce the estimate

$$\big\|\widehat{K_j^\lambda}\big\|_{L^2}^2 = \big\|K_j^\lambda\big\|_{L^2}^2 \le C'' 2^{-(n+1+2\mathrm{Re}\,\lambda)j} 2^{nj} = C'' 2^{-(1+2\mathrm{Re}\,\lambda)j} . \tag{10.4.17}$$

We need another estimate for $\widehat{K_j^\lambda}$. We claim that for all $M \ge n+1$ there is a constant C_M such that

$$\int_{|\xi| \le \frac{1}{8}} |\widehat{K_j^\lambda}(\xi)|^2 |\xi|^{-\beta}\, d\xi \le C_{M,n,\beta}\, 2^{-2j(M-n)}, \quad \beta < n. \tag{10.4.18}$$

Indeed, since $\widehat{K^\lambda}(\xi)$ is supported in $|\xi| \ge \frac{1}{2}$ [recall that the function η was chosen equal to 1 on $B(0,\frac{1}{2})$], we have

$$|\widehat{K_j^\lambda}(\xi)| = |(\widehat{K^\lambda} * \widehat{\psi_j})(\xi)| \le 2^{jn} \int_{\frac{1}{2}\le|\xi-\omega|\le 1} (1-|\xi-\omega|^2)_+^{\operatorname{Re}\lambda} |\widehat{\psi}(2^j\omega)|\,d\omega.$$

Suppose that $|\xi| \le \frac{1}{8}$. Since $|\xi-\omega| \ge \frac{1}{2}$, we must have $|\omega| \ge \frac{3}{8}$. Then

$$|\widehat{\psi}(2^j\omega)| \le C_M(2^j|\omega|)^{-M} \le (8/3)^M C_M 2^{-jM},$$

from which it follows easily that

$$\sup_{|\xi|\le\frac{1}{8}} |\widehat{K_j^\lambda}(\xi)| \le C'_M 2^{-j(M-n)}. \tag{10.4.19}$$

Then (10.4.18) is a consequence of (10.4.19) and of the fact that the function $|\xi|^{-\beta}$ is integrable near the origin.

We now return to estimate (10.4.15). A localization argument (Exercise 10.4.4) allows us to reduce estimate (10.4.15) to functions f that are supported in a cube of side length 2^j. Let us therefore assume that f is supported in some cube Q of side length 2^j. Then $T_j^\lambda(f)$ is supported in $5Q$ and we have for $1 \le p < 2$ by Hölder's inequality

$$\begin{aligned} \big\|T_j^\lambda(f)\big\|_{L^p(5Q)}^2 &\le |5Q|^{2(\frac{1}{p}-\frac{1}{2})} \big\|T_j^\lambda(f)\big\|_{L^2(5Q)}^2 \\ &\le C_n 2^{(\frac{1}{p}-\frac{1}{2})2nj} \big\|\widehat{K_j^\lambda}\widehat{f}\big\|_{L^2}^2 . \end{aligned} \tag{10.4.20}$$

Having returned to L^2, we are able to use the $L^p \to L^2$ restriction theorem obtained in the previous subsection. To this end we use polar coordinates and the fact that K_j^λ is a radial function to write

$$\big\|\widehat{K_j^\lambda}\widehat{f}\big\|_{L^2}^2 = \int_0^\infty |\widehat{K_j^\lambda}(re_1)|^2 \left(\int_{\mathbf{S}^{n-1}} |\widehat{f}(r\theta)|^2\,d\theta\right) r^{n-1}dr, \tag{10.4.21}$$

where $e_1 = (1,0,\dots,0) \in \mathbf{S}^{n-1}$. Since the restriction of the function $x \mapsto r^{-n}f(x/r)$ on the sphere $\mathbf{S}^{n-1}$ is $\widehat{f}(r\theta)$, we have

$$\int_{\mathbf{S}^{n-1}} |\widehat{f}(r\theta)|^2\,d\theta \le C_{p,n}^2 \left[\int_{\mathbf{R}^n} r^{-np}|f(x/r)|^p\,dx\right]^{\frac{2}{p}} = C_{p,n}^2 r^{-\frac{2n}{p'}} \big\|f\big\|_{L^p}^2, \tag{10.4.22}$$

where $C_{p,n}$ is the constant in Theorem 10.4.5 that holds whenever $p \le \frac{2(n+1)}{n+3}$. So assuming $p \le \frac{2(n+1)}{n+3}$ and inserting estimate (10.4.22) in (10.4.21) yields

$$\begin{aligned}\|\widehat{K_j^\lambda}\widehat{f}\|_{L^2}^2 &\le C_{p,n}^2\|f\|_{L^p}^2\int_0^\infty|\widehat{K_j^\lambda}(re_1)|^2 r^{n-1-\frac{2n}{p'}}\,dr\\ &\le \frac{C_{p,n}^2}{\omega_{n-1}}\|f\|_{L^p}^2\int_{\mathbf{R}^n}|\widehat{K_j^\lambda}(\xi)|^2|\xi|^{-\frac{2n}{p'}}\,d\xi\,,\end{aligned} \tag{10.4.23}$$

where $\omega_{n-1}=|\mathbf{S}^{n-1}|$. Appealing to estimate (10.4.18) for $|\xi|\le\frac{1}{8}$ with $\beta=\frac{2n}{p'}<n$ (since $p<2$) and to estimate (10.4.17) for $|\xi|\ge\frac{1}{8}$, we obtain

$$\|\widehat{K_j^\lambda}\widehat{f}\|_{L^2}^2\le C'''2^{-(1+2\operatorname{Re}\lambda)j}\|f\|_{L^p}^2\,.$$

Combining this inequality with the one previously obtained in (10.4.20) yields (10.4.15) with

$$\delta=\frac{n+1}{2}+\operatorname{Re}\lambda-\frac{n}{p}\,.$$

This number is positive exactly when $\frac{2n}{n+1+2\operatorname{Re}\lambda}<p$. This was the condition assumed by the theorem when $p<2$. The other condition $\operatorname{Re}\lambda>\frac{n-1}{2(n+1)}$ is naturally imposed by the restriction $p\le\frac{2(n+1)}{n+3}$. Finally, the analogous result in the range $p>2$ follows by duality. □

10.4.4 The Full Restriction Theorem on $\mathbf{R}^2$

In this section we prove the validity of the restriction condition $R_{p\to q}(\mathbf{S}^1)$ in dimension $n=2$, for the full range of exponents suggested by Figure 10.10.

To achieve this goal, we "fatten" the circle by a small amount 2δ. Then we obtain a restriction theorem for the "fattened circle" and then obtain the required estimate by taking the limit as $\delta\to 0$. Precisely, we use the fact

$$\int_{\mathbf{S}^1}|\widehat{f}(\omega)|^q\,d\omega=\lim_{\delta\to 0}\frac{1}{2\delta}\int_{1-\delta}^{1+\delta}\int_{\mathbf{S}^1}|\widehat{f}(r\theta)|^q\,d\theta\,r\,dr \tag{10.4.24}$$

to recover the restriction theorem for the circle from a restriction theorem for annuli of width 2δ.

Throughout this subsection, δ is a number satisfying $0<\delta<\frac{1}{1000}$, and for simplicity we use the notation

$$\chi^\delta(\xi)=\chi_{(1-\delta,1+\delta)}(|\xi|)\,,\qquad \xi\in\mathbf{R}^2\,.$$

We note that in view of identity (10.4.24), the restriction property $R_{p\to q}(\mathbf{S}^1)$ is a trivial consequence of the estimate

$$\frac{1}{2\delta}\int_0^\infty\int_{\mathbf{S}^1}|\chi^\delta(r\theta)\widehat{f}(r\theta)|^q\,d\theta\,r\,dr\le C^q\|f\|_{L^p}^q\,, \tag{10.4.25}$$

or, equivalently, of

$$\|\chi^\delta \widehat{f}\|_{L^q(\mathbf{R}^2)} \le (2\delta)^{\frac{1}{q}} C \|f\|_{L^p(\mathbf{R}^2)}. \tag{10.4.26}$$

We have the following result.

Theorem 10.4.7. *(a) Given $1 \le p < \frac{4}{3}$, set $q = \frac{p'}{3}$. Then there is a constant C_p such that for all L^p functions f on $\mathbf{R}^2$ and all small positive δ we have*

$$\|\chi^\delta \widehat{f}\|_{L^q(\mathbf{R}^2)} \le C_p \delta^{\frac{1}{q}} \|f\|_{L^p(\mathbf{R}^2)}. \tag{10.4.27}$$

(b) When $p = q = 4/3$, there is a constant C such that for all $L^{4/3}$ functions f on $\mathbf{R}^2$ and all small $\delta > 0$ we have

$$\|\chi^\delta \widehat{f}\|_{L^{\frac{4}{3}}(\mathbf{R}^2)} \le C \delta^{\frac{3}{4}} (\log \tfrac{1}{\delta})^{\frac{1}{4}} \|f\|_{L^{\frac{4}{3}}(\mathbf{R}^2)}. \tag{10.4.28}$$

Proof. To prove this theorem, we work with the *extension operator*

$$E^\delta(g) = \widehat{\chi^\delta g} = \widehat{\chi^\delta} * \widehat{g},$$

which is dual (i.e., transpose) to $f \mapsto \chi^\delta \widehat{f}$, and we need to show that

$$\|E^\delta(f)\|_{L^{p'}(\mathbf{R}^2)} \le C \delta^{\frac{1}{q}} (\log \tfrac{1}{\delta})^\beta \|f\|_{L^{q'}(\mathbf{R}^2)}, \tag{10.4.29}$$

where $\beta = \frac{1}{4}$ when $p = \frac{4}{3}$ and $\beta = 0$ when $p < \frac{4}{3}$.

We employ a splitting similar to that used in Theorem 10.2.4, with the only difference that the present partition of unity is nonsmooth and hence simpler. We define functions

$$\chi_\ell^\delta(\xi) = \chi^\delta(\xi) \chi_{2\pi\ell\delta^{1/2} \le \operatorname{Arg} \xi < 2\pi(\ell+1)\delta^{1/2}}$$

for $\ell \in \{0, 1, \dots, [\delta^{-1/2}]\}$. We suitably adjust the support of the function $\chi^\delta_{[\delta^{-1/2}]}$ so that the sum of all these functions equals χ^δ. We now split the indices that appear in the set $\{0, 1, \dots, [\delta^{-1/2}]\}$ into nine different subsets so that the supports of the functions indexed by them are properly contained in some sector centered at the origin of amplitude $\pi/4$. We therefore write E^δ as a sum of nine pieces, each properly supported in a sector of amplitude $\pi/4$. Let I be the set of indices that correspond to one of these nine sectors and let

$$E_I^\delta(f) = \sum_{\ell \in I} \widehat{\chi_\ell^\delta f}.$$

It suffices therefore to obtain (10.4.29) for each E_I^δ in lieu of E^δ. Let us fix such an index set I and without loss of generality we assume that

$$I = \{0, 1, \dots, [\tfrac{1}{8}\delta^{-1/2}]\}.$$

Since the theorem is trivial when $p=1$, to prove part (a) we fix a number p with $1<p<\frac{4}{3}$. We set

$$r=(p'/2)'$$

and we observe that this r satisfies $\frac{1}{r}=\frac{1}{p'}+\frac{1}{q'}$. We note that $1<r<2$ and we apply the Hausdorff–Young inequality $\|h\|_{L^{r'}}\le\|h^\vee\|_{L^r}$. We have

$$\begin{aligned}\|E_I^\delta(f)\|_{L^{p'}(\mathbf{R}^2)}^{p'} &= \int_{\mathbf{R}^2}|E_I^\delta(f)^2|^{r'}\,dx\\ &\le \left(\int_{\mathbf{R}^2}|(E_I^\delta(f)^2)^\vee|^r\,dx\right)^{\frac{r'}{r}} \qquad (10.4.30)\\ &= \left(\int_{\mathbf{R}^2}\Big|\sum_{\ell\in I}\sum_{\ell'\in I}(\chi_\ell^\delta f)*(\chi_{\ell'}^\delta f)\Big|^r\,dx\right)^{\frac{r'}{r}}.\end{aligned}$$

We obtain the estimate

$$\left(\int_{\mathbf{R}^2}\Big|\sum_{\ell\in I}\sum_{\ell'\in I}(\chi_\ell^\delta f)*(\chi_{\ell'}^\delta f)\Big|^r\,dx\right)^{\frac{r'}{r}}\le C\delta^{\frac{p'}{q}}\|f\|_{L^{q'}(\mathbf{R}^2)}^{p'}, \tag{10.4.31}$$

which suffices to prove the theorem.

Denote by $S_{\delta,\ell,\ell'}$ the support of $\chi_\ell^\delta+\chi_{\ell'}^\delta$. Then we write the left-hand side of (10.4.31) as

$$\left(\int_{\mathbf{R}^2}\Big|\sum_{\ell\in I}\sum_{\ell'\in I}((\chi_\ell^\delta f)*(\chi_{\ell'}^\delta f))\chi_{S_{\delta,\ell,\ell'}}\Big|^r\,dx\right)^{\frac{r'}{r}}, \tag{10.4.32}$$

which, via Hölder's inequality, is controlled by

$$\left(\int_{\mathbf{R}^2}\Big(\sum_{\ell\in I}\sum_{\ell'\in I}|(\chi_\ell^\delta f)*(\chi_{\ell'}^\delta f)|^r\Big)^{\frac{r}{r}}\Big(\sum_{\ell\in I}\sum_{\ell'\in I}|\chi_{S_{\delta,\ell,\ell'}}|^{r'}\Big)^{\frac{r}{r'}}\,dx\right)^{\frac{r'}{r}}. \tag{10.4.33}$$

We now recall Lemma 10.2.5, in which the curvature of the circle was crucial. In view of that lemma, the second factor of the integrand in (10.4.33) is bounded by a constant independent of δ. We have therefore obtained the estimate

$$\|E_I^\delta(f)\|_{L^{p'}}^{p'}\le C\left(\sum_{\ell\in I}\sum_{\ell'\in I}\int_{\mathbf{R}^2}|(\chi_\ell^\delta f)*(\chi_{\ell'}^\delta f)|^r\,dx\right)^{\frac{r'}{r}}. \tag{10.4.34}$$

We prove at the end of this section the following auxiliary result.

Lemma 10.4.8. *With the same notation as in the proof of Theorem 10.4.7, for any $1<r<\infty$, there is a constant C (independent of δ and f) such that*

$$\|(\chi_\ell^\delta f)*(\chi_{\ell'}^\delta f)\|_{L^r}\le C\left(\frac{\delta^{\frac{3}{2}}}{|\ell-\ell'|+1}\right)^{\frac{1}{r'}}\|\chi_\ell^\delta f\|_{L^r}\|\chi_{\ell'}^\delta f\|_{L^r} \tag{10.4.35}$$

for all $\ell,\ell' \in I = \{0,1,\dots,[\frac{1}{8}\delta^{-1/2}]\}$.

Assuming Lemma 10.4.8 and using (10.4.34), we write

$$\begin{aligned}\|E_I^\delta(f)\|_{L^{p'}}^{p'} &\le C\delta^{\frac{3}{2}}\Big[\sum_{\ell\in I}\|\chi_\ell^\delta f\|_{L^r}^r\Big(\sum_{\ell'\in I}\frac{\|\chi_{\ell'}^\delta f\|_{L^r}^r}{(|\ell-\ell'|+1)^{\frac{r}{r'}}}\Big)\Big]^{\frac{r'}{r}}\\ &\le C\delta^{\frac{3}{2}}\Big[\sum_{\ell\in I}\|\chi_\ell^\delta f\|_{L^r}^{rs}\Big]^{\frac{r'}{rs}}\Big[\sum_{\ell\in I}\Big(\sum_{\ell'\in I}\frac{\|\chi_{\ell'}^\delta f\|_{L^r}^r}{(|\ell-\ell'|+1)^{\frac{r}{r'}}}\Big)^{s'}\Big]^{\frac{r'}{rs'}},\end{aligned} \tag{10.4.36}$$

where we used Hölder's inequality for some $1 < s < \infty$. We now recall the discrete fractional integral operator

$$\{a_j\}_j \mapsto \Big\{\sum_{j'}\frac{a_{j'}}{(|j-j'|+1)^{1-\alpha}}\Big\}_j,$$

which maps $\ell^s(\mathbf{Z})$ to $\ell^{s'}(\mathbf{Z})$ (see Exercise 6.1.10) when

$$\frac{1}{s}-\frac{1}{s'} = \alpha, \qquad 0<\alpha<1. \tag{10.4.37}$$

When $1 < p < \frac{4}{3}$, we have $1 < r < 2$, and choosing $\alpha = 2-r = 1-\frac{r}{r'}$, we obtain from (10.4.36) that

$$\begin{aligned}\|E_I^\delta(f)\|_{L^{p'}}^{p'} &\le C'\delta^{\frac{3}{2}}\Big[\sum_{\ell\in I}\|\chi_\ell^\delta f\|_{L^r}^{rs}\Big]^{\frac{r'}{rs}}\Big[\sum_{\ell\in I}\|\chi_\ell^\delta f\|_{L^r}^{rs}\Big]^{\frac{r'}{rs}}\\ &= C'\delta^{\frac{3}{2}}\Big[\sum_{\ell\in I}\|\chi_\ell^\delta f\|_{L^r}^{rs}\Big]^{\frac{2r'}{rs}}.\end{aligned} \tag{10.4.38}$$

The unique s that solves equation (10.4.37) is seen easily to be $s = q'/r$. Moreover, since $q = p'/3$, we have $1 < s < 2$. We use again Hölder's inequality to pass from $\|\chi_\ell^\delta f\|_{L^r}$ to $\|\chi_\ell^\delta f\|_{L^{q'}}$. Indeed, recalling that the support of χ_ℓ^δ has measure $\approx \delta^{\frac{3}{2}}$, we have

$$\|\chi_\ell^\delta f\|_{L^r} \le C(\delta^{\frac{3}{2}})^{\frac{1}{r}-\frac{1}{q'}}\|\chi_\ell^\delta f\|_{L^{q'}}.$$

Inserting this in (10.4.38) yields

$$\begin{aligned}\|E_I^\delta(f)\|_{L^{p'}}^{p'} &\le C\delta^{\frac{3}{2}}\Big[\sum_{\ell\in I}\Big(C(\delta^{\frac{3}{2}})^{\frac{1}{r}-\frac{1}{q'}}\|\chi_\ell^\delta f\|_{L^{q'}}\Big)^{rs}\Big]^{\frac{2r'}{rs}}\\ &= C'\delta^{\frac{3}{2}}(\delta^{\frac{3}{2}})^{2r'(\frac{1}{r}-\frac{1}{q'})}\Big[\sum_{\ell\in I}\|\chi_\ell^\delta f\|_{L^{q'}}^{q'}\Big]^{\frac{2r'}{q'}}\\ &\le C\delta^3\|f\|_{L^{q'}}^{p'}\\ &= C\delta^{\frac{p'}{q}}\|f\|_{L^{q'}}^{p'},\end{aligned}$$

which is the required estimate since $\frac{1}{r}=\frac{1}{p'}+\frac{1}{q'}$ and $p'=2r'$. In the last inequality we used the fact that the supports of the functions χ_ℓ^δ are disjoint and that these add up to a function that is at most 1.

To prove part (b) of the theorem, we need to adjust the previous argument to obtain the case $p=\frac{4}{3}$. Here we repeat part of the preceding argument taking $r=r'=s=s'=2$.

Using (10.4.34) with $p=\frac{4}{3}$ (which forces r to be equal to 2) and Lemma 10.4.8 with $r=2$ we write

$$\begin{aligned}
\big\|E_I(f)\big\|_{L^4(\mathbf{R}^2)}^4 &\le C\delta^{\frac{3}{2}}\Big[\sum_{\ell\in I}\big\|\chi_\ell^\delta f\big\|_{L^2}^2\Big(\sum_{\ell'\in I}\frac{\|\chi_{\ell'}^\delta f\|_{L^2}^2}{|\ell-\ell'|+1}\Big)\Big]\\
&\le C\delta^{\frac{3}{2}}\Big[\sum_{\ell\in I}\big\|\chi_\ell^\delta f\big\|_{L^2}^4\Big]^{\frac{1}{2}}\Big[\sum_{\ell\in I}\Big(\sum_{\ell'\in I}\frac{\|\chi_{\ell'}^\delta f\|_{L^2}^2}{|\ell-\ell'|+1}\Big)^2\Big]^{\frac{1}{2}}\\
&\le C\delta^{\frac{3}{2}}\Big[\sum_{\ell\in I}\big\|\chi_\ell^\delta f\big\|_{L^2}^4\Big]^{\frac{1}{2}}\Big[\sum_{\ell\in I}\big\|\chi_\ell^\delta f\big\|_{L^2}^4\Big]^{\frac{1}{2}}\Big[\sum_{\ell\in I}\frac{1}{|\ell|+1}\Big]\\
&\le C\delta^{\frac{3}{2}}\Big[\sum_{\ell\in I}\big\|\chi_\ell^\delta f\big\|_{L^2}^4\Big]\log(\delta^{-\frac{1}{2}})\\
&\le C\delta^{\frac{3}{2}}(\delta^{\frac{3}{2}})^{(\frac{1}{2}-\frac{1}{4})4}\Big[\sum_{\ell\in I}\big\|\chi_\ell^\delta f\big\|_{L^4}^4\Big]\log\tfrac{1}{\delta}\\
&\le C\delta^3\big(\log\tfrac{1}{\delta}\big)\big\|f\big\|_{L^4}^4\,.
\end{aligned}$$

□

We now prove Lemma 10.4.8, which we had left open.

Proof. The proof is based on interpolation. For fixed $\ell,\ell'\in I$ we define the bilinear operator

$$T_{\ell,\ell'}(g,h)=(g\chi_\ell^\delta)*(h\chi_{\ell'}^\delta)\,.$$

As we have previously observed, it is a simple geometric fact that the support of χ_ℓ^δ is contained in a rectangle of side length $\approx\delta$ in the direction $e^{2\pi i\delta^{1/2}\ell}$ and of side length $\approx\delta^{\frac{1}{2}}$ in the direction $ie^{2\pi i\delta^{1/2}\ell}$. Any two rectangles with these dimensions in the aforementioned directions have an intersection that depends on the angle between them. Indeed, if $\ell\neq\ell'$ this intersection is contained in a parallelogram of sides δ and $\delta/\sin(2\pi\delta^{\frac{1}{2}}|\ell-\ell'|)$, and hence the measure of the intersection is seen easily to be at most a constant multiple of

$$\delta\cdot\frac{\delta}{\sin(2\pi\delta^{\frac{1}{2}}|\ell-\ell'|)}\,.$$

As for ℓ,ℓ' in the index set I we have $2\pi\delta^{\frac{1}{2}}|\ell-\ell'|<\pi/4$, the sine is comparable to its argument, and we conclude that the measure of the intersection is at most

$$C\delta^{\frac{3}{2}}(1+|\ell-\ell'|)^{-1}.$$

It follows that

$$\|\chi_\ell^\delta * \chi_{\ell'}^\delta\|_{L^\infty} = \sup_{z\in\mathbf{R}^2} |(z-\operatorname{supp}(\chi_\ell^\delta))\cap \operatorname{supp}(\chi_{\ell'}^\delta)| \le \frac{C\delta^{\frac{3}{2}}}{1+|\ell-\ell'|},$$

which implies the estimate

$$\begin{aligned} \|T_{\ell,\ell'}(g,h)\|_{L^\infty} &\le \|\chi_\ell^\delta * \chi_{\ell'}^\delta\|_{L^\infty}\|g\|_{L^\infty}\|h\|_{L^\infty} \\ &\le \frac{C\delta^{\frac{3}{2}}}{1+|\ell-\ell'|}\|g\|_{L^\infty}\|h\|_{L^\infty}. \end{aligned} \tag{10.4.39}$$

Also, the estimate

$$\begin{aligned} \|T_{\ell,\ell'}(g,h)\|_{L^1} &\le \|g\chi_\ell^\delta\|_{L^1}\|h\chi_{\ell'}^\delta\|_{L^1} \\ &\le \|g\|_{L^1}\|h\|_{L^1} \end{aligned} \tag{10.4.40}$$

holds trivially. Interpolating between (10.4.39) and (10.4.40) yields the required estimate (10.4.35). Here we used bilinear interpolation (Exercise 1.4.17). □

Example 10.4.9. The presence of the logarithmic factor in estimate (10.4.28) is necessary. In fact, this estimate is sharp. We prove this by showing that the corresponding estimate for the "dual" extension operator E^δ is sharp. Let I be the set of indices we worked with in Theorem 10.4.7 (i.e., $I=\{0,1,\dots,[\frac{1}{8}\delta^{-1/2}]\}$.) Let

$$f^\delta = \sum_{\ell\in I}\chi_\ell^\delta.$$

Then

$$\|f^\delta\|_{L^4} \approx \delta^{\frac{1}{4}}.$$

However,

$$E^\delta(f^\delta) = \sum_{\ell\in I}\widehat{\chi_\ell^\delta},$$

and we have

$$\begin{aligned} \|E^\delta(f^\delta)\|_{L^4} &= \left(\int_{\mathbf{R}^2}\Big|\sum_{\ell\in I}\sum_{\ell'\in I}\widehat{\chi_\ell^\delta}\widehat{\chi_{\ell'}^\delta}\Big|^2 d\xi\right)^{\frac{1}{4}} \\ &= \left(\int_{\mathbf{R}^2}\Big|\sum_{\ell\in I}\sum_{\ell'\in I}\chi_\ell^\delta * \chi_{\ell'}^\delta\Big|^2 dx\right)^{\frac{1}{4}} \\ &\ge \left(\sum_{\ell\in I}\sum_{\ell'\in I}\int_{\mathbf{R}^2}|\chi_\ell^\delta * \chi_{\ell'}^\delta|^2 dx\right)^{\frac{1}{4}}. \end{aligned}$$

At this point observe that the function $\chi_\ell^\delta * \chi_{\ell'}^\delta$ is at least a constant multiple of $\delta^{\frac{3}{2}}(|\ell-\ell'|+1)^{-1}$ on a set of measure $c\delta^{\frac{3}{2}}(|\ell-\ell'|+1)$. (See Exercise 10.4.5.) Using this fact and the previous estimates, we deduce easily that

$$\|E^\delta(f^\delta)\|_{L^4} \geq c\Big(\sum_{\ell\in I}\sum_{\ell'\in I}\frac{\delta^3}{(|\ell-\ell'|+1)^2}\delta^{\frac{3}{2}}(|\ell-\ell'|+1)\Big)^{\frac{1}{4}} \approx \delta(\log\tfrac{1}{\delta})^{\frac{1}{4}},$$

since $|I|\approx\delta^{-\frac{1}{2}}$. It follows that

$$\frac{\|E^\delta(f^\delta)\|_{L^4}}{\|f^\delta\|_{L^4}} \geq c\delta^{\frac{3}{4}}(\log\tfrac{1}{\delta})^{\frac{1}{4}},$$

which justifies the sharpness of estimate (10.4.28).

Exercises

10.4.1. Let S be a compact hypersurface in $\mathbf{R}^n$ and let $d\sigma$ be surface measure on it. Suppose that for some $0<b<n$ we have

$$|\widehat{d\sigma}(\xi)| \leq C(1+|\xi|)^{-b}$$

for all $\xi\in\mathbf{R}^n$. Prove that $R_{p\to q}(S)$ does not hold for any $1\leq q\leq\infty$ when $p\geq\frac{n}{n-b}$.

10.4.2. Let S be a compact hypersurface and let $1\leq p,q\leq\infty$.
(a) Suppose that $R_{p\to q}(S)$ holds for S. Show that $R_{p\to q}(\tau+S)$ holds for the translated hypersurface $\tau+S$.
(b) Suppose that the hypersurface S is compact and its interior contains the origin. For $r>0$ let $rS=\{r\xi:\ \xi\in S\}$. Suppose that $R_{p\to q}(\mathbf{S}^{n-1})$ holds with constant C_{pqn}. Show that $R_{p\to q}(r\mathbf{S}^{n-1})$ holds with constant $C_{pqn}r^{\frac{n-1}{q}-\frac{n}{p'}}$.

10.4.3. Obtain a different proof of estimate (10.4.7) (and hence of Theorem 10.4.5) by following the sequence of steps outlined here:
(a) Consider the analytic family of functions

$$(K_z)^\vee(\xi) = 2\pi^{1-z}\frac{J_{\frac{n-2}{2}+z}(2\pi|\xi|)}{|\xi|^{\frac{n-2}{2}+z}}$$

and observe that in view of the identity in Appendix B.4, $(K_z)^\vee(\xi)$ reduces to $d\sigma^\vee(\xi)$ when $z=0$, where $d\sigma$ is surface measure on $\mathbf{S}^{n-1}$.
(b) Use for free that the Bessel function $J_{-\frac{1}{2}+i\theta}$, $\theta\in\mathbf{R}$, satisfies

$$|J_{-\frac{1}{2}+i\theta}(x)| \leq C_\theta|x|^{-\frac{1}{2}},$$

where C_θ grows at most exponentially in $|\theta|$, to obtain that the family of operators given by convolution with $(K_z)^\vee$ map $L^1(\mathbf{R}^n)$ to $L^\infty(\mathbf{R}^n)$ when $z = -\frac{n-1}{2} + i\theta$.
(c) Appeal to the result in Appendix B.5 to obtain that for z not equal to $0, -1, -2, \dots$ we have

$$K_z(x) = \frac{2}{\Gamma(z)}(1 - |x|^2)^{z-1}.$$

Use this identity to deduce that for $z = 1 + i\theta$ the family of operators given by convolution with $(K_z)^\vee$ map $L^2(\mathbf{R}^n)$ to itself with constants that grow at most exponentially in $|\theta|$. (Appendix A.6 contains a useful lower estimate for $|\Gamma(1+i\theta)|$.)
(d) Use Exercise 1.3.4 to obtain that for $z = 0$ the operator given by convolution with $d\sigma^\vee$ maps $L^p(\mathbf{R}^n)$ to $L^{p'}(\mathbf{R}^n)$ when $p = \frac{2(n+1)}{n+3}$.

10.4.4. Suppose that T is a linear operator given by convolution with a kernel K that is supported in the ball $B(0, 2R)$. Assume that there is a constant C such that for all functions f supported in a cube of side length R we have

$$\big\|T(f)\big\|_{L^p} \le B\big\|f\big\|_{L^p}$$

for some $1 \le p < \infty$. Show that this estimate also holds for all L^p functions f with constant $5^n B$.
[*Hint:* Write $f = \sum_j f\chi_{Q_j}$, where each cube Q_j has side length R.]

10.4.5. Using the notation of Theorem 10.4.7, show that there exist constants c, c' such that the function $\chi_\ell^\delta * \chi_{\ell'}^\delta$ is at least $c'\delta^{\frac{3}{2}}(|\ell - \ell'| + 1)^{-1}$ on a set of measure $c\delta^{\frac{3}{2}}(|\ell - \ell'| + 1)$.
[*Hint:* Prove the required conclusion for characteristic functions of rectangles with the same orientation and comparable dimensions. Then use that the support of each χ_ℓ^δ contains such a rectangle.]

10.5 Almost Everywhere Convergence of Bochner–Riesz Means

We recall the Bochner–Riesz means B_R^λ of complex order λ given in Definition 10.2.1. In this section we study the problem of almost everywhere convergence of $B_R^\lambda(f) \to f$ as $R \to \infty$. There is an intimate relationship between the almost everywhere convergence of a family of operators and boundedness properties of the associated maximal family (cf. Theorem 2.1.14).[1]

For $f \in L^p(\mathbf{R}^n)$, the *maximal Bochner–Riesz operator* or order λ is defined by

$$B_*^\lambda(f) = \sup_{R>0} \big|B_R^\lambda(f)\big|.$$

[1] In certain cases, Theorem 2.1.14 can essentially be reversed. Given a $1 \le p \le 2$ and a family of distributions u_j with the mild continuity property that $u_j * f_k \to u_j * f$ in measure whenever $f_k \to f$ in $L^p(\mathbf{R}^n)$ such that the maximal operator $\mathscr{M}(f) = \sup_j |f * u_j| < \infty$ whenever $f \in L^p(\mathbf{R}^n)$, then $\mathscr{M}$ maps $L^p(\mathbf{R}^n)$ to $L^{p,\infty}(K)$ for any compact subset K of $\mathbf{R}^n$. See Stein [289], [292].

10.5.1 A Counterexample for the Maximal Bochner–Riesz Operator

We have the following result.

Theorem 10.5.1. *Let $n \geq 2$, $\lambda > 0$, and let $1 < p < 2$ be such that*

$$\lambda < \frac{2n-1}{2p} - \frac{n}{2}.$$

Then B_*^λ does not map $L^p(\mathbf{R}^n)$ to weak $L^p(\mathbf{R}^n)$.

Proof. Figure 10.11 shows the region in which B_*^λ is known to be unbounded; this region contains the set of points $(1/p, \lambda)$ strictly below the line that joins the points $(1, (n-1)/2)$ and $(n/(2n-1), 0)$.

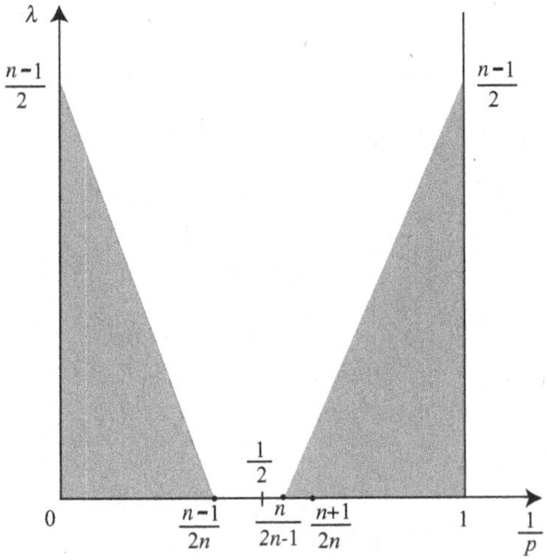

Fig. 10.11 The operators B_*^λ are unbounded on $L^p(\mathbf{R}^n)$ when $(1/p, \lambda)$ lies in the interior of the shaded region.

We denote points x in $\mathbf{R}^n$ by $x = (x', x_n)$, where $x' \in \mathbf{R}^{n-1}$, and we fix $M \geq 100$ and $\varepsilon < 1/100$. We let $\psi(y) = \chi_{|y'|\leq 1}(y')\,\zeta(y_n)$, where ζ is a smooth bump supported in the interval $[-1,1]$ that is equal to 1 on $[-1/2, 1/2]$ and satisfies $0 \leq \zeta \leq 1$. We define

$$\psi_{\varepsilon,M}(y) = \psi(\varepsilon^{-1}y', \varepsilon^{-1}M^{-\frac{1}{2}}y_n) = \chi_{|y'|\leq\varepsilon}(y')\,\zeta(\varepsilon^{-1}M^{-\frac{1}{2}}y_n)$$

and we note that $\psi_{\varepsilon,M}(y)$ is supported in the set of y's that satisfy $|y'| \leq \varepsilon$ and $|y_n| \leq \varepsilon M^{\frac{1}{2}}$. We also define

$$f_M(y) = e^{2\pi i y_n}\psi_{\varepsilon,M}(y)$$

and

$$S_M = \{(x', x_n) : \, M \leq |x'| \leq 2M, \; M \leq |x_n| \leq 2M\}.$$

Then

$$\big\|f_M\big\|_{L^p} \approx M^{\frac{1}{2p}} \varepsilon^{\frac{n}{p}} \qquad \text{and} \qquad |S_M| \approx M^n . \tag{10.5.1}$$

Every point $x \in S_M$ must satisfy $M \le |x| \le 3M$. We fix $x \in S_M$ and we estimate $B_*^{\lambda}(f_M)(x) = \sup_{R>0} |B_R^{\lambda}(f_M)(x)|$ from below by picking $R = R_x = |x|/x_n$. Then $1/2 \le R_x \le 3$ and we have

$$B_*^{\lambda}(f_M)(x) \ge \frac{\Gamma(\lambda+1)}{\pi^{\lambda}} \left| \int_{\mathbf{R}^n} \frac{J_{\frac{n}{2}+\lambda}(2\pi R_x|x-y|)}{(R_x|x-y|)^{\frac{n}{2}+\lambda}} e^{2\pi i y_n} \psi_{\varepsilon,M}(y)\,dy \right| .$$

We make some observations. First $|x'-y'| \ge \frac{1}{2}|x'|$, since $|x'| \ge M$ and $|y'| \le \varepsilon$. Second, $|x_n - y_n| \ge |x_n| - |y_n| \ge \frac{1}{2}|x_n|$, since $|x_n| \ge M$ and $|y_n| \le \varepsilon M^{1/2}$. These facts imply that $|x-y| \ge \frac{1}{2}|x|$; thus $|x-y|$ is comparable to $|x|$, which is of the order of M. Since $2\pi R_x|x-y|$ is large, we use the asymptotics for the Bessel function $J_{\frac{n}{2}+\lambda}$ in Appendix B.8 to write

$$\frac{J_{\frac{n}{2}+\lambda}(2\pi R_x|x-y|)}{(R_x|x-y|)^{\frac{n}{2}+\lambda}} = \frac{C_\lambda\, e^{2\pi i R_x|x-y|} e^{i\varphi}}{(R_x|x-y|)^{\frac{n+1}{2}+\lambda}} + \frac{C_\lambda\, e^{-2\pi i R_x|x-y|} e^{-i\varphi}}{(R_x|x-y|)^{\frac{n+1}{2}+\lambda}} + V_{n,\lambda}(R_x|x-y|),$$

where $\varphi = -\frac{\pi}{2}(\frac{n}{2}+\lambda) - \frac{\pi}{4}$ and

$$|V_{n,\lambda}(R_x|x-y|)| \le \frac{C_{n,\lambda}}{(R_x|x-y|)^{\frac{n+3}{2}+\lambda}} \le \frac{C'_{n,\lambda}}{M^{\frac{n+3}{2}+\lambda}}, \tag{10.5.2}$$

since $R_x = \frac{|x|}{x_n} \approx 1$ and $|x-y| \ge \frac{1}{2}M$. Using the preceding expression for the Bessel function, we write

$$\begin{aligned} B_*^{\lambda}(f_M)(x) \ge C'_\lambda & \left| \int_{\mathbf{R}^n} \frac{e^{2\pi i R_x|x|} e^{i\varphi}}{(R_x|x-y|)^{\frac{n+1}{2}+\lambda}} \psi_{\varepsilon,M}(y)\,dy \right| \\ - C'_\lambda & \left| \int_{\mathbf{R}^n} \frac{(e^{2\pi i (R_x|x-y|+y_n)} - e^{2\pi i R_x|x|}) e^{i\varphi}}{(R_x|x-y|)^{\frac{n+1}{2}+\lambda}} \psi_{\varepsilon,M}(y)\,dy \right| \\ - C'_\lambda & \left| \int_{\mathbf{R}^n} \frac{e^{2\pi i (-R_x|x-y|+y_n)} e^{-i\varphi}}{(R_x|x-y|)^{\frac{n+1}{2}+\lambda}} \psi_{\varepsilon,M}(y)\,dy \right| \\ - & \left| \int_{\mathbf{R}^n} V_{n,\lambda}(R_x|x-y|) e^{2\pi i y_n} \psi_{\varepsilon,M}(y)\,dy \right| . \end{aligned}$$

The positive term is the main term and is bounded from below by

$$C'_\lambda\, (6M)^{-\frac{n+1}{2}-\lambda} \int_{\mathbf{R}^n} \psi_{\varepsilon,M}(y)\,dy = \frac{c_1\, \varepsilon^n M^{\frac{1}{2}}}{M^{\frac{n+1}{2}+\lambda}} . \tag{10.5.3}$$

The three terms with the minus signs are errors and are bounded in absolute value by smaller expressions. We notice that

$$\left|R_x|x-y|+y_n-R_x|x|\right| = \frac{|x|}{x_n}\left||x-y|+\frac{x_ny_n}{|x|}-|x|\right| = \frac{|x|}{x_n}\left|F_x(y)-F_x(0)\right|,$$

where $F_x(y)=|x-y|+|x|^{-1}x_ny_n$. Taylor's expansion yields

$$F_x(y)-F_x(0)=\nabla_yF_x(0)\cdot y+O\big(|y|^2\sup_{j,k}|\partial_j\partial_kF_x(y)|\big),$$

and a calculation gives $\nabla_yF_x(0)=(-|x|^{-1}x',0)$, while $|\partial_j\partial_kF_x(y)|\le C|x-y|^{-1}$. It follows that

$$\frac{|x|}{x_n}\left|F_x(y)-F_x(0)\right|\le 3\left[\frac{|x'\cdot y'|}{|x|}+C'\frac{|y|^2}{|x-y|}\right]\le C''\left[\varepsilon+\frac{(\varepsilon M^{1/2})^2}{M}\right]\le 2C''\varepsilon.$$

Using this fact and the support properties of ψ, we obtain

$$C'_\lambda\left|\int_{\mathbf{R}^n}\frac{(e^{2\pi i(R_x|x-y|+y_n)}-e^{2\pi iR_x|x|})e^{i\varphi}}{(R_x|x-y|)^{\frac{n+1}{2}+\lambda}}\psi_{\varepsilon,M}(y)\,dy\right|\le\frac{c_2\,\varepsilon(\varepsilon^nM^{\frac12})}{M^{\frac{n+1}{2}+\lambda}}. \tag{10.5.4}$$

Next we examine the phase $R_x|x-y|+y_n$ as a function of y_n. Its derivative with respect to y_n is

$$\frac{\partial}{\partial y_n}(R_x|x-y|+y_n)=R_x\frac{x_n-y_n}{|x-y|}+1\ge1,$$

since $x_n\ge M$ and $|y_n|\le\varepsilon M^{1/2}$, which implies that $x_n-y_n>0$. Also note that

$$\left|\frac{\partial}{\partial y_n}\left(R_x\frac{x_n-y_n}{|x-y|}+1\right)^{-1}\right|\le\frac{C'''}{M}$$

and that

$$\left|\frac{\partial}{\partial y_n}\frac{1}{|x-y|^{\frac{n+1}{2}+\lambda}}\right|\le\frac{C'''}{M^{\frac{n+3}{2}+\lambda}},$$

while the derivative of $\zeta(\varepsilon^{-1}M^{-\frac12}y_n)$ with respect to y_n gives only a factor of $\varepsilon^{-1}M^{-\frac12}$. We integrate by parts one time with respect to y_n in the integral

$$\int_{\mathbf{R}^{n-1}}\int_{\mathbf{R}}\frac{e^{2\pi i(-R_x|x-y|+y_n)}e^{-i\varphi}}{(R_x|x-y|)^{\frac{n+1}{2}+\lambda}}\psi_{\varepsilon,M}(y)\,dy_n\,dy'$$

to obtain an additional factor of $\varepsilon^{-1}M^{-\frac12}$. Thus

$$\left|\int_{\mathbf{R}^n}\frac{e^{2\pi i(-R_x|x-y|+y_n)}e^{-i\varphi}}{(R_x|x-y|)^{\frac{n+1}{2}+\lambda}}\psi_{\varepsilon,M}(y)\,dy\right|\le\frac{c_3\,\varepsilon^nM^{\frac12}(\varepsilon^{-1}M^{-\frac12})}{M^{\frac{n+1}{2}+\lambda}}. \tag{10.5.5}$$

Finally, using (10.5.2) we obtain that

$$\left| \int_{\mathbf{R}^n} V_{n,\lambda}(R_x|x-y|)e^{2\pi i y_n} \psi_{\varepsilon,M}(y)\,dy \right| \leq \frac{c_4\,\varepsilon^n M^{\frac{1}{2}}}{M^{\frac{n+3}{2}+\lambda}}. \tag{10.5.6}$$

We combine (10.5.3), (10.5.4), (10.5.5), and (10.5.6) to deduce for $x \in S_M$,

$$B_*^\lambda(f_M)(x) \geq \frac{c_1\,\varepsilon^n}{M^{\frac{n}{2}+\lambda}} - \frac{c_2\,\varepsilon^{n+1}}{M^{\frac{n}{2}+\lambda}} - \frac{c_3\,\varepsilon^{n-1}}{M^{\frac{n+1}{2}+\lambda}} - \frac{c_4\,\varepsilon^n}{M^{\frac{n+2}{2}+\lambda}}.$$

We pick ε sufficiently small, say $\varepsilon \leq c_1/(2c_2)$, and M_0 sufficiently large (depending on the constants c_1, c_2, c_3, c_4) that

$$x \in S_M \implies B_*^\lambda(f_M)(x) > c_0 \frac{1}{M^{\frac{n}{2}+\lambda}}$$

whenever $M \geq M_0$. This fact together with (10.5.1) gives

$$\frac{\big\|B_*^\lambda(f_M)\big\|_{L^{p,\infty}}}{\big\|f_M\big\|_{L^p}} \geq \frac{c_0 M^{-\frac{n}{2}-\lambda}|S_M|^{\frac{1}{p}}}{c' M^{\frac{1}{2p}}} = c M^{\frac{2n-1}{2p}-\frac{n}{2}-\lambda},$$

and the required conclusion follows by letting $M \to \infty$. □

10.5.2 Almost Everywhere Summability of the Bochner–Riesz Means

We now focus attention on the case $p \geq 2$ and we investigate whether the Bochner–Riesz means converge almost everywhere outside the range in which they are known to be unbounded on L^p. Our goal is to prove the following result.

Theorem 10.5.2. *Let $\lambda > 0$ and $n \geq 2$. Then for all f in $L^p(\mathbf{R}^n)$ with $2 \leq p < \frac{2n}{n-1-2\lambda}$ we have*

$$\lim_{R\to\infty} B_R^\lambda(f)(x) = f(x)$$

for almost all $x \in \mathbf{R}^n$.

Since the almost everywhere convergence is obvious for functions in the Schwartz class, to be able to use Theorem 2.1.14 to derive almost everywhere convergence for general L^p functions, it suffices to know a weak type (p,p) estimate for B_*^λ. However, instead of proving a weak type (p,p) estimate, we prove an L^2 and a weighted L^2 estimate for B_*^λ. Precisely, we prove the following result.

Proposition 10.5.3. *Let $\lambda > 0$ and $0 \leq \alpha < 1 + 2\lambda \leq n$. Then there is a constant $C = C(\alpha, \lambda, n)$ such that*

$$\int_{\mathbf{R}^n} |B_*^\lambda(f)(x)|^2 |x|^{-\alpha}\,dx \leq C \int_{\mathbf{R}^n} |f(x)|^2 |x|^{-\alpha}\,dx$$

for all functions $f \in L^2(\mathbf{R}^n, |x|^{-\alpha}dx)$.

Assuming the result of Proposition 10.5.3, given p such that

$$2 \le p < p_\lambda = \frac{2n}{n-1-2\lambda},$$

choose α satisfying

$$0 \le n\left(1-\frac{2}{p}\right) < \alpha < 1+2\lambda = n\left(1-\frac{2}{p_\lambda}\right).$$

Then the maximal operator B_*^λ is bounded on L^2 and also on $L^2(|x|^{-\alpha}dx)$. Hence the almost everywhere convergence of the family $\{B_R^\lambda(f)\}_R$ holds on L^2 and also on $L^2(|x|^{-\alpha}dx)$. Since $0 \le \alpha < n$, we have

$$L^p \subseteqq L^2 + L^2(|x|^{-\alpha}),$$

and thus $B_R^\lambda(f)$ converges almost everywhere for functions $f \in L^p(\mathbf{R}^n)$. See Exercise 10.5.1 for this inclusion.

To prove Proposition 10.5.3, we decompose the multiplier $(1-|\xi|^2)_+^\lambda$ as an infinite sum of smooth bumps supported in small concentric annuli in the interior of the sphere $|\xi|=1$ as we did in the proof of Theorem 10.2.4.

We pick a smooth function φ supported in $[-\frac{1}{2},\frac{1}{2}]$ and a smooth function ψ supported in $[\frac{1}{8},\frac{5}{8}]$ and with values in $[0,1]$ that satisfy

$$\varphi(t) + \sum_{k=0}^{\infty} \psi\left(\frac{1-t}{2^{-k}}\right) = 1$$

for all $t \in [0,1)$. We decompose the multiplier $(1-|\xi|^2)_+^\lambda$ as

$$(1-|\xi|^2)_+^\lambda = m_{00}(\xi) + \sum_{k=0}^{\infty} 2^{-k\lambda} m_k(\xi), \tag{10.5.7}$$

where $m_{00}(\xi) = \varphi(|\xi|)(1-|\xi|^2)^\lambda$, and for $k \ge 1$, m_k is defined by

$$m_k(\xi) = \left(\frac{1-|\xi|}{2^{-k}}\right)^\lambda \psi\left(\frac{1-|\xi|}{2^{-k}}\right)(1+|\xi|)^\lambda.$$

Then we define maximal operators associated with the multipliers m_{00} and m_k,

$$S_*^{m_k}(f)(x) = \sup_{R>0} \left|\left(\widehat{f}(\xi)m_k(\xi/R)\right)^\vee(x)\right|,$$

for $k \ge 0$, and analogously we define $S_*^{m_{00}}$. Using (10.5.7) we have

$$B_*^\lambda(f) \le S_*^{m_{00}}(f) + \sum_{k=0}^{\infty} 2^{-k\lambda} S_*^{m_k}(f). \tag{10.5.8}$$

Since $S_*^{m_{00}}$, $S_*^{m_0}$, $S_*^{m_1}$ and any finite number of them are pointwise controlled by the Hardy–Littlewood maximal operator, which is bounded on $L^2(|x|^\alpha)$ whenever $-n < \alpha < n$ (cf. Theorem 9.1.9 and Example 9.1.7), we focus attention on the remaining terms.

We make a small change of notation. Thinking of 2^{-k} as roughly being δ (precisely $\delta = 2^{-k-3}$), for $\delta < 1/10$ we let $m^\delta(t)$ be a smooth function supported in the interval $[1-5\delta, 1-\delta]$ and taking values in the interval $[0,1]$ that satisfies

$$\sup_{1\le t\le 2}\left|\frac{d^\ell}{dt^\ell}m^\delta(t)\right| \le C_\ell \delta^{-\ell} \tag{10.5.9}$$

for all $\ell \in \mathbf{Z}^+ \cup \{0\}$. We define a related function

$$\widetilde{m}^\delta(t) = \delta t \frac{d}{dt} m^\delta(t),$$

which obviously satisfies estimates (10.5.9) with another constant $\widetilde{C_\ell}$ in place of C_ℓ.

Next we introduce the multiplier operators

$$S_t^\delta(f)(x) = (\widehat{f}(\xi)m^\delta(t|\xi|))^\vee(x), \qquad \widetilde{S_t^\delta}(f)(x) = (\widehat{f}(\xi)\widetilde{m}^\delta(t|\xi|))^\vee(x),$$

and the $L^2(|x|^{-\alpha})$-bounded maximal multiplier operator

$$S_*^\delta(f) = \sup_{t>0}|S_t^\delta(f)|,$$

as well as the continuous square functions

$$G^\delta(f)(x) = \left(\int_0^\infty |S_t^\delta(f)(x)|^2 \frac{dt}{t}\right)^{\frac{1}{2}}, \quad \widetilde{G}^\delta(f)(x) = \left(\int_0^\infty |\widetilde{S_t^\delta}(f)(x)|^2 \frac{dt}{t}\right)^{\frac{1}{2}}.$$

The operators S_t^δ and $\widetilde{S_t^\delta}$ are related. For $f \in L^2(|x|^{-\alpha})$ and $t > 0$ we have

$$\frac{d}{dt}S_t^\delta(f) = \frac{1}{\delta t}\widetilde{S_t^\delta}(f).$$

Indeed, this operator identity is obvious for Schwartz functions f by the Lebesgue dominated convergence theorem, and thus it holds for $f \in L^2(|x|^{-\alpha})$ by density.

The quadratic operators G^δ and $\widetilde{G}^\delta$ make their appearance in the application of the fundamental theorem of calculus in the following context:

$$|S_t^\delta(f)(x)|^2 = 2\,\mathrm{Re}\int_0^t \overline{S_u^\delta(f)(x)}\,\frac{d}{du}S_u^\delta(f)(x)\,du = \frac{2}{\delta}\,\mathrm{Re}\int_0^t \overline{S_u^\delta(f)(x)}\,\widetilde{S_u^\delta}(f)(x)\,\frac{du}{u},$$

which is valid for all functions f in $L^2(|x|^{-\alpha})$ and almost all $x \in \mathbf{R}^n$. This identity uses the fact that for almost all $x \in \mathbf{R}^n$ we have

$$\lim_{t\to 0} S_t^\delta(f)(x) = 0 \tag{10.5.10}$$

when $f \in L^2(|x|^{-\alpha})$. To see this, we observe that for Schwartz functions, (10.5.10) is trivial by the Lebesgue dominated convergence theorem, while for general f in $L^2(|x|^{-\alpha})$ it is a consequence of Theorem 2.1.14, since $S_*^\delta(f) \le C_\delta M(f)$, where M is the Hardy–Littlewood maximal operator. Consequently,

$$|S_t^\delta(f)(x)|^2 \le \frac{2}{\delta}\int_0^t |S_u^\delta(f)(x)|\,|\widetilde{S}_u^\delta(f)(x)|\,\frac{du}{u} \le \frac{2}{\delta}|G^\delta(f)(x)|\,|\widetilde{G}^\delta(f)(x)|$$

for all $t>0$, for $f \in L^2(|x|^{-\alpha})$ and for almost all $x \in \mathbf{R}^n$. It follows that

$$\big\|S_*^\delta(f)\big\|_{L^2(|x|^{-a})}^2 \le \frac{2}{\delta}\big\|G^\delta(f)\big\|_{L^2(|x|^{-a})}\big\|\widetilde{G}^\delta(f)\big\|_{L^2(|x|^{-a})}, \tag{10.5.11}$$

and the asserted boundedness of S_*^δ reduces to that of the continuous square functions G^δ and $\widetilde{G}^\delta$ on weighted L^2 spaces with suitable constants depending on δ.

The boundedness of G^δ on $L^2(|x|^{-\alpha})$ is a consequence of the following lemma.

Lemma 10.5.4. *For $0<\delta<1/10$ and $0\le\alpha<n$ we have*

$$\int_{\mathbf{R}^n}\int_1^2 |S_t^\delta(f)(x)|^2\frac{dt}{t}\frac{dx}{|x|^\alpha} \le C_{n,\alpha}A_\alpha(\delta)\int_{\mathbf{R}^n}|f(x)|^2\frac{dx}{|x|^\alpha} \tag{10.5.12}$$

for all functions f in $L^2(|x|^{-\alpha})$, where for $\varepsilon>0$, $A_\alpha(\varepsilon)$ is defined by

$$A_\alpha(\varepsilon) = \begin{cases} \varepsilon^{2-\alpha} & \text{when } 1<\alpha<n,\\ \varepsilon(|\log\varepsilon|+1) & \text{when } \alpha=1,\\ \varepsilon & \text{when } 0\le\alpha<1. \end{cases} \tag{10.5.13}$$

Assuming the statement of the lemma, we conclude the proof of Proposition 10.5.3 as follows. We take a Schwartz function ψ such that $\widehat{\psi}$ vanishes in a neighborhood of the origin with $\widehat{\psi}(\xi)=1$ whenever $1/2\le|\xi|\le 2$ and we let $\psi_{2^k}(x) = 2^{-kn}\psi(2^{-k}x)$. We make the observation that if $1-5\delta\le t|\xi|\le 1-\delta$ and $2^{k-1}\le t\le 2^k$, then $1/2\le 2^k|\xi|\le 2$, since $\delta<1/10$. This implies that $\widehat{\psi}(2^k\xi)=1$ on the support of the function $\xi\mapsto m^\delta(t|\xi|)$. Hence

$$S_t^\delta(f) = S_t^\delta(\psi_{2^k}*f)$$

whenever $2^{k-1}\le t\le 2^k$, and Lemma 10.5.4 (in conjunction with Exercise 10.5.2) yields

$$\int_{\mathbf{R}^n}\int_{2^{k-1}}^{2^k}|S_t^\delta(f)(x)|^2\frac{dt}{t}\frac{dx}{|x|^\alpha} \le C_{n,\alpha}A_\alpha(\delta)\int_{\mathbf{R}^n}|\psi_{2^k}*f(x)|^2\frac{dx}{|x|^\alpha}.$$

Summing over $k\in\mathbf{Z}$ we obtain

$$\left\|G^{\delta}(f)\right\|^2_{L^2(|x|^{-\alpha})} \le C_{n,\alpha}A_\alpha(\delta)\left\|\Big(\sum_{k\in\mathbf{Z}}|\psi_{2^k}*f|^2\Big)^{\frac12}\right\|^2_{L^2(|x|^{-\alpha})}.$$

A randomization argument relates the weighted L^2 norm of the square function to the L^2 norm of a linear expression involving the Rademacher functions as in

$$\left\|\Big(\sum_{k\in\mathbf{Z}}|\psi_{2^k}*f|^2\Big)^{\frac12}\right\|^2_{L^2(|x|^{-\alpha})} = \int_0^1\left\|\sum_{k\in\mathbf{Z}}r_k(t)(\psi_{2^k}*f)\right\|^2_{L^2(|x|^{-\alpha})}dt\,,$$

where r_k denotes a renumbering of the Rademacher functions (Appendix C.1) indexed by the entire set of integers. For each $t\in[0,1]$ the operator

$$M_t(f)=\sum_{k\in\mathbf{Z}}r_k(t)(\psi_{2^k}*f)$$

is associated with a multiplier that satisfies Mihlin's condition (5.2.10) uniformly in t. It follows that M_t is a singular integral operator bounded on all the L^p spaces for $1<p<\infty$, and in view of Corollary 9.4.7, it is also bounded on $L^2(w)$ whenever $w\in A_2$. Since the weight $|x|^{-\alpha}$ is in A_2 whenever $-n<\alpha<n$, it follows that M_t is bounded on $L^2(|x|^{-\alpha})$ with a bound independent of $t>0$. We deduce that

$$\left\|G^{\delta}(f)\right\|_{L^2(|x|^{-\alpha})}+\left\|\widetilde{G}^{\delta}(f)\right\|_{L^2(|x|^{-\alpha})}\le C'_{n,\alpha}\left(A_\alpha(\delta)\right)^{\frac12}\left\|f\right\|_{L^2(|x|^{-\alpha})}.$$

We now recall estimate (10.5.11) to obtain

$$\left\|S^{\delta}_{*}(f)\right\|_{L^2(|x|^{-\alpha})}\le C'(n,\alpha)\big(\delta^{-1}A_\alpha(\delta)\big)^{1/2}\left\|f\right\|_{L^2(|x|^{-\alpha})}.$$

Taking $\delta=2^{-k-3}$, recalling the value of $A_\alpha(\delta)$ from Lemma 10.5.4, and inserting this estimate in (10.5.8), we deduce Proposition 10.5.3. We note that the condition $\alpha<1+2\lambda$ is needed to make the series in (10.5.8) converge when $1<\alpha<n$.

10.5.3 Estimates for Radial Multipliers

It remains to prove Lemma 10.5.4. Since all subsequent estimates concern linear operators on weighted L^2 spaces, in the sequel we will be working with functions in the Schwartz class, unless it is otherwise specified.

We reduce estimate (10.5.12) to an estimate for a single t with the bound $A_\alpha(\delta)/\delta$, which is worse than $A_\alpha(\delta)$. The reduction to a single t is achieved via duality. Estimate (10.5.12) says that the operator $f\mapsto\{S^{\delta}_t(f)\}_{1\le t\le 2}$ is bounded from $L^2(\mathbf{R}^n,|x|^{-\alpha}dx)$ to $L^2(L^2(\frac{dt}{t}),|x|^{-\alpha}dx)$. The dual statement of this fact is that the operator

$$\{g_t\}_{1\le t\le 2}\mapsto\int_1^2 S^{\delta}_t(g_t)\,\frac{dt}{t}$$

maps $L^2(L^2(\frac{dt}{t}),|x|^\alpha dx)$ to $L^2(\mathbf{R}^n,|x|^\alpha dx)$. Here we use the fact that the operators S_t are self-transpose and self-adjoint, since they have real and radial multipliers. Thus estimate (10.5.12) is equivalent to

$$\int_{\mathbf{R}^n}\bigg|\int_1^2 S_t^\delta(g_t)(x)\,\frac{dt}{t}\bigg|^2|x|^\alpha\,dx \le C_{n,\alpha}A_\alpha(\delta)\int_{\mathbf{R}^n}\int_1^2 |g_t(x)|^2\frac{dt}{t}\,|x|^\alpha\,dx, \quad (10.5.14)$$

which by Plancherel's theorem is also equivalent to

$$\int_{\mathbf{R}^n}\bigg|\mathscr{D}^{\frac{\alpha}{2}}\Big(\int_1^2 m^\delta(t|\cdot|)\widehat{g_t}(\cdot)\,\frac{dt}{t}\Big)(\xi)\bigg|^2 d\xi \le C_{n,\alpha}A_\alpha(\delta)\int_{\mathbf{R}^n}\int_1^2 |\mathscr{D}^{\frac{\alpha}{2}}(\widehat{g_t})(\xi)|^2\frac{dt}{t}\,d\xi\,.$$

Here

$$\mathscr{D}^\beta(h)(x) = \bigg[\int_{\mathbf{R}^n}\frac{|D_y^{[\beta]+1}(h)(x)|^2}{|y|^{n+2\beta}}\,dy\bigg]^{\frac{1}{2}},$$

where $D_y(f)(x)=f(x+y)-f(x)$ is the difference operator encountered in Section 6.3 and $D_y^k = D_y\circ\cdots\circ D_y$ (k times). The operator $\mathscr{D}^\beta$ obeys the identity (see Exercise 6.3.9)

$$\big\|\mathscr{D}^\beta(\widehat{h})\big\|_{L^2}^2 = c_0(n,\beta)\int_{\mathbf{R}^n}|h(x)|^2|x|^{2\beta}\,dx.$$

Using the definition of $\mathscr{D}^{\alpha/2}$ we write

$$\bigg|\mathscr{D}^{\frac{\alpha}{2}}\Big(\int_1^2 m^\delta(t|\cdot|)\widehat{g_t}(\cdot)\,\frac{dt}{t}\Big)(\xi)\bigg|^2 = \int_{\mathbf{R}^n}\bigg|\int_1^2 D_\eta^{[\frac{\alpha}{2}]+1}\big(m^\delta(t|\cdot|)\widehat{g_t}(\cdot)\big)(\xi)\frac{dt}{t}\bigg|^2\frac{d\eta}{|\eta|^{n+\alpha}}\,.$$

If the inner integrand on the right is nonzero, expressing D_y^{k+1} as in (6.3.2) and using the support properties of m^δ, we obtain that $1-5\delta\le t|\xi+s\eta|\le 1-\delta$ for some $s\in\{0,1,\dots,[\alpha/2]+1\}$; thus for each such s, t belongs to an interval of length $4\delta|\xi+s\eta|^{-1}\le 4\delta t(1-5\delta)^{-1}$. Since $t\le 2$ and $\delta\le 1/10$, it follows that t lies in a set of measure at most $2([\alpha/2]+2)\delta$. The Cauchy–Schwarz inequality then yields

$$\begin{aligned}&\bigg|\mathscr{D}^{\frac{\alpha}{2}}\Big(\int_1^2 m^\delta(t|\cdot|)\widehat{g_t}(\cdot)\,\frac{dt}{t}\Big)(\xi)\bigg|^2\\ &\qquad\le c_\alpha\,\delta\int_{\mathbf{R}^n}\int_1^2\bigg|D_\eta^{[\frac{\alpha}{2}]+1}\big(m^\delta(t|\cdot|)\widehat{g_t}(\cdot)\big)(\xi)\bigg|^2\frac{dt}{t}\frac{d\eta}{|\eta|^{n+\alpha}}\,.\end{aligned}$$

In view of the preceding reduction, we deduce that (10.5.14) is a consequence of

$$\begin{aligned}&\int_{\mathbf{R}^n}\int_{\mathbf{R}^n}\int_1^2\bigg|D_\eta^{[\frac{\alpha}{2}]+1}\big(m^\delta(t|\cdot|)\widehat{g_t}(\cdot)\big)(\xi)\bigg|^2\frac{dt}{t}\frac{d\eta}{|\eta|^{n+\alpha}}\,d\xi\\ &\qquad\le C_{n,\alpha}\frac{A_\alpha(\delta)}{c_\alpha\,\delta}\int_{\mathbf{R}^n}\int_1^2|\mathscr{D}^{\frac{\alpha}{2}}(\widehat{g_t})(\xi)|^2\frac{dt}{t}\,d\xi\end{aligned}$$

which can also be written as

$$\int_{\mathbf{R}^n}\int_1^2 \left|\mathscr{D}^{\frac{\alpha}{2}}(m^\delta(t|\cdot|)\widehat{g_t}(\cdot))(\xi)\right|^2 \frac{dt}{t}\,d\xi \le \frac{C_{n,\alpha}}{c_\alpha}\frac{A_\alpha(\delta)}{\delta}\int_{\mathbf{R}^n}\int_1^2 |\mathscr{D}^{\frac{\alpha}{2}}(\widehat{g_t})(\xi)|^2\frac{dt}{t}\,d\xi\,.$$

This estimate is a consequence of

$$\int_{\mathbf{R}^n} \left|\mathscr{D}^{\frac{\alpha}{2}}(m^\delta(t|\cdot|)\widehat{g_t}(\cdot))(\xi)\right|^2 d\xi \le \frac{C_{n,\alpha}}{c_\alpha}\frac{A_\alpha(\delta)}{\delta}\int_{\mathbf{R}^n} |\mathscr{D}^{\frac{\alpha}{2}}(\widehat{g_t})(\xi)|^2\,d\xi \tag{10.5.15}$$

for all $t\in[1,2]$. A simple dilation argument reduces (10.5.15) to the single estimate

$$\int_{\mathbf{R}^n} \left|\mathscr{D}^{\frac{\alpha}{2}}(m^\delta(|\cdot|)\widehat{g}(\cdot))(\xi)\right|^2 d\xi \le \frac{C_{n,\alpha}}{c_\alpha}\frac{A_\alpha(\delta)}{\delta}\int_{\mathbf{R}^n} |\mathscr{D}^{\frac{\alpha}{2}}(\widehat{g})(\xi)|^2\,d\xi\,, \tag{10.5.16}$$

which is equivalent to

$$\int_{\mathbf{R}^n} |S_1^\delta(g)(x)|^2\,|x|^\alpha\,dx \le \frac{C_{n,\alpha}}{c_\alpha}\frac{A_\alpha(\delta)}{\delta}\int_{\mathbf{R}^n} |g(x)|^2\,|x|^\alpha\,dx$$

and also equivalent to

$$\int_{\mathbf{R}^n} |S_1^\delta(f)(x)|^2\,\frac{dx}{|x|^\alpha} \le \frac{C_{n,\alpha}}{c_\alpha}\frac{A_\alpha(\delta)}{\delta}\int_{\mathbf{R}^n} |f(x)|^2\,\frac{dx}{|x|^\alpha} \tag{10.5.17}$$

by duality. We have now reduced estimate (10.5.12) to (10.5.17).

We denote by $K^\delta(x)$ the kernel of the operator S_1^δ, i.e., the inverse Fourier transform of the multiplier $m^\delta(|\xi|)$. Certainly K^δ is a radial kernel on $\mathbf{R}^n$, and it is convenient to decompose it radially as

$$K^\delta = K_0^\delta + \sum_{j=1}^\infty K_j^\delta\,,$$

where $K_0^\delta(x) = K^\delta(x)\phi(\delta x)$ and $K_j^\delta(x) = K^\delta(x)\big(\phi(2^{-j}\delta x)-\phi(2^{1-j}\delta x)\big)$, for some radial smooth function ϕ supported in the ball $B(0,2)$ and equal to one on $B(0,1)$.

To prove estimate (10.5.17) we make use of the subsequent lemmas.

Lemma 10.5.5. *For all $M\ge 2n$ there is a constant $C_M = C_M(n,\phi)$ such that for all $j=0,1,2,\dots$ we have*

$$\sup_{\xi\in\mathbf{R}^n} |\widehat{K_j^\delta}(\xi)| \le C_M 2^{-jM} \tag{10.5.18}$$

and also

$$|\widehat{K_j^\delta}(\xi)| \le C_M 2^{-(j+k)M} \tag{10.5.19}$$

whenever $\big||\xi|-1\big|\ge 2^k\delta$ and $k\ge 4$. Also

$$|\widehat{K_j^\delta}(\xi)| \le C_M 2^{-jM}\delta^M(1+|\xi|)^{-M} \tag{10.5.20}$$

whenever $|\xi|\le 1/8$ or $|\xi|\ge 15/8$.

Lemma 10.5.6. *Let $0 \le \alpha < n$. Then there is a constant $C(n,\alpha)$ such that for all Schwartz functions f and all $\varepsilon > 0$ we have*

$$\int_{||\xi|-1|\le\varepsilon} |\widehat{f}(\xi)|^2\,d\xi \le C(n,\alpha)\,\varepsilon^{\alpha-1}A_\alpha(\varepsilon)\int_{\mathbf{R}^n}|f(x)|^2\,|x|^\alpha dx \tag{10.5.21}$$

and also for $M \ge 2n$ there is a constant $C_M(n,\alpha)$ such that

$$\int_{\mathbf{R}^n}|\widehat{f}(\xi)|^2\frac{1}{(1+|\xi|)^M}\,d\xi \le C_M(n,\alpha)\int_{\mathbf{R}^n}|f(x)|^2\,|x|^\alpha dx. \tag{10.5.22}$$

Assuming Lemmas 10.5.5 and 10.5.6 we prove estimate (10.5.17) as follows. Using Plancherel's theorem we write

$$\int_{\mathbf{R}^n}|(K_j^\delta * f)(x)|^2\,dx = \int_{\mathbf{R}^n}|\widehat{K_j^\delta}(\xi)|^2|\widehat{f}(\xi)|^2\,d\xi \le I+II+III,$$

where

$$I = \int_{|\xi|\le\frac{1}{8},\,|\xi|\ge\frac{15}{8}}|\widehat{K_j^\delta}(\xi)|^2|\widehat{f}(\xi)|^2\,d\xi,$$

$$II = \sum_{k=4}^{[\log_2\frac{7}{16}\delta^{-1}]+1}\int_{2^k\delta\le||\xi|-1|\le 2^{k+1}\delta}|\widehat{K_j^\delta}(\xi)|^2|\widehat{f}(\xi)|^2\,d\xi,$$

$$III = \int_{||\xi|-1|\le 16\delta}|\widehat{K_j^\delta}(\xi)|^2|\widehat{f}(\xi)|^2\,d\xi.$$

Using (10.5.20) and (10.5.22) we obtain that

$$I \le C'_M(n,\alpha)2^{-jM}\delta^M\int_{\mathbf{R}^n}|f(x)|^2\,|x|^\alpha dx.$$

In view of (10.5.19) and (10.5.21) we have

$$\begin{aligned}II &\le \sum_{k=4}^{[\log_2\delta^{-1}]+1} C(n,\alpha)(2^{k+1}\delta)^{\alpha-1}A_\alpha(2^{k+1}\delta)2^{-jM}2^{-kM}\int_{\mathbf{R}^n}|f(x)|^2\,|x|^\alpha dx\\ &\le C'_M(n,\alpha)2^{-jM}\delta^{\alpha-1}A_\alpha(\delta)\int_{\mathbf{R}^n}|f(x)|^2\,|x|^\alpha dx.\end{aligned}$$

Finally, (10.5.18) and (10.5.21) yield

$$III \le C'_M(n,\alpha)2^{-jM}\delta^{\alpha-1}A_\alpha(\delta)\int_{\mathbf{R}^n}|f(x)|^2\,|x|^\alpha dx.$$

Summing the estimates for I, II, and III we deduce

$$\int_{\mathbf{R}^n}|(K_j^\delta * f)(x)|^2\,dx \le C_M(n,\alpha)2^{-jM}\delta^{\alpha-1}A_\alpha(\delta)\int_{\mathbf{R}^n}|f(x)|^2\,|x|^\alpha dx.$$

By duality, this estimate can be written as

$$\int_{\mathbf{R}^n} |(K_j^\delta * f)(x)|^2 \frac{dx}{|x|^\alpha} \le C_M(n,\alpha) 2^{-jM} \delta^{\alpha-1} A_\alpha(\delta) \int_{\mathbf{R}^n} |f(x)|^2 \, dx. \tag{10.5.23}$$

Given a Schwartz function f, we write $f_0 = f\chi_{Q_0}$, where Q_0 is a cube centered at the origin of side length $C2^j/\delta$ for some C to be chosen. Then for $x \in Q_0$ we have $|x| \le C\sqrt{n}2^j/\delta$, hence

$$\begin{aligned}\int_{\mathbf{R}^n} |(K_j^\delta * f_0)(x)|^2 \frac{dx}{|x|^\alpha} &\le \frac{C_M'(n,\alpha)\, \delta^{\alpha-1} A_\alpha(\delta)}{2^{jM}} \Big(C\sqrt{n}\frac{2^j}{\delta}\Big)^\alpha \int_{Q_0} |f_0(x)|^2 \frac{dx}{|x|^\alpha} \\ &= C_M''(n,\alpha) 2^{j(\alpha-M)} \frac{A_\alpha(\delta)}{\delta} \int_{Q_0} |f_0(x)|^2 \frac{dx}{|x|^\alpha}. \end{aligned} \tag{10.5.24}$$

Now write $\mathbf{R}^n \setminus Q_0$ as a mesh of cubes Q_i, indexed by $i \in \mathbf{Z} \setminus \{0\}$, of side lengths $2^{j+2}/\delta$ and centers c_{Q_i}. Since K_j^δ is supported in a ball of radius $2^{j+1}/\delta$, if f_i is supported in Q_i, then $f_i * K_j^\delta$ is supported in the cube $2\sqrt{n}Q_i$. If the constant C is large enough, say $C \ge 1000n$, then for $x \in Q_i$ and $x' \in 2\sqrt{n}Q_i$ we have

$$|x| \approx |c_{Q_i}| \approx |x'|,$$

which says that the moduli of x and x' are comparable in the following inequality:

$$\int_{2\sqrt{n}Q_i} |(K_j^\delta * f_i)(x')|^2 \frac{dx'}{|x'|^\alpha} \le C_M' 2^{-jM} \int_{Q_i} |f_i(x)|^2 \frac{dx}{|x|^\alpha}. \tag{10.5.25}$$

Thus (10.5.25) is a consequence of

$$\int_{2\sqrt{n}Q_i} |(K_j^\delta * f_i)(x')|^2 \, dx' \le C_M 2^{-jM} \int_{Q_i} |f_i(x)|^2 \, dx, \tag{10.5.26}$$

which is certainly satisfied, as seen by applying Plancherel's theorem and using (10.5.18). Since for $\delta < 1/10$ we have $A_\alpha(\delta)/\delta \ge 1$, it follows that

$$\int_{\mathbf{R}^n} |(K_j^\delta * f_i)(x)|^2 \frac{dx}{|x|^\alpha} \le C_M 2^{-jM} \frac{A_\alpha(\delta)}{\delta} \int_{\mathbf{R}^n} |f_i(x)|^2 \frac{dx}{|x|^\alpha} \tag{10.5.27}$$

whenever f_i is supported in Q_i. We now pick $M = 2n$ and we recall that $\alpha < n$. We have now proved that

$$\int_{\mathbf{R}^n} |(K_j^\delta * f_i)(x)|^2 \frac{dx}{|x|^\alpha} \le C''(n,\alpha) 2^{-jn} \frac{A_\alpha(\delta)}{\delta} \int_{Q_i} |f_i(x)|^2 \frac{dx}{|x|^\alpha}$$

for functions f_i supported in Q_i.

Given a general f in the Schwartz class, write

$$f = \sum_{i\in\mathbf{Z}} f_i, \qquad \text{where} \qquad f_i = f\chi_{Q_i}.$$

Then

$$\begin{aligned}
\|K_j^\delta * f\|^2_{L^2(|x|^{-\alpha})} &\le 2\|K_j^\delta * f_0\|^2_{L^2(|x|^{-\alpha})} + 2\Big\|\sum_{i\neq 0} K_j^\delta * f_i\Big\|^2_{L^2(|x|^{-\alpha})} \\
&\le 2\|K_j^\delta * f_0\|^2_{L^2(|x|^{-\alpha})} + 2C_n \sum_{i\neq 0} \|K_j^\delta * f_i\|^2_{L^2(|x|^{-\alpha})} \\
&\le C'''(n,\alpha) 2^{-jn} \frac{A_\alpha(\delta)}{\delta}\Big[\|f_0\|^2_{L^2(|x|^{-\alpha})} + \sum_{i\neq 0}\|f_i\|^2_{L^2(|x|^{-\alpha})}\Big] \\
&= C'''(n,\alpha) 2^{-jn} \frac{A_\alpha(\delta)}{\delta} \|f\|^2_{L^2(|x|^{-\alpha})},
\end{aligned}$$

where we used the bounded overlap of the family $\{K_j * f_i\}_{i\neq 0}$ in the second displayed inequality (cf. Exercise 10.4.4). Taking square roots and summing over $j = 0,1,2,\dots$, we deduce (10.5.17).

We now address the proof of Lemma 10.5.5, which was left open.

Proof. For the purposes of this proof we set $\psi(x) = \phi(x) - \phi(2x)$. Then the inverse Fourier transform of the function $x \mapsto \psi(2^{-j}\delta x)$ is $\xi \mapsto 2^{jn}\delta^{-n}\widehat{\psi}(2^j\xi/\delta)$. Convolving the latter with the function $\xi \mapsto m^\delta(|\xi|)$, we obtain $\widehat{K_j^\delta}(\xi)$. We may therefore write for $j \ge 1$,

$$\widehat{K_j^\delta}(\xi) = \int_{\mathbf{R}^n} m^\delta(|\xi - 2^{-j}\delta\eta|)\widehat{\psi}(\eta)\,d\eta, \tag{10.5.28}$$

while for $j = 0$ an analogous formula holds with ϕ in place of ψ. Since $|m^\delta| \le 1$, (10.5.18) follows easily when $j = 0$. For $j \ge 1$ we expand the function $\xi \mapsto m^\delta(|\xi - 2^{-j}\delta\eta|)$ in a Taylor series and we make use of the fact that $\widehat{\psi}$ has vanishing moments of all orders to obtain

$$\begin{aligned}
|\widehat{K_j^\delta}(\xi)| &\le \int_{\mathbf{R}^n} \sum_{|\gamma|=M} \frac{1}{\gamma!}\|\partial^\gamma m^\delta(|\cdot|)\|_{L^\infty} |2^{-j}\delta\eta|^M |\widehat{\psi}(\eta)|\,d\eta \\
&\le C(M)\delta^{-M}\delta^M 2^{-jM}\int_{\mathbf{R}^n} |\eta|^M|\widehat{\psi}(\eta)|\,d\eta.
\end{aligned}$$

This proves (10.5.18).

We turn now to the proof of (10.5.19). Suppose that $||\xi| - 1| \ge 2^k\delta$ and $k \ge 4$. Then for $|\xi| \le 1$, recalling that m^δ is supported in $[1-5\delta, 1-\delta]$, we write

$$|2^{-j}\delta\eta| \ge |\xi - 2^{-j}\delta\eta| - |\xi| \ge (1-5\delta) - (1-2^k\delta) \ge 2^{k-1}\delta,$$

since $k \ge 4$. For $|\xi| \ge 1$ we have

$$|2^{-j}\delta\eta| \ge |\xi| - |\xi - 2^{-j}\delta\eta| \ge (1+2^k\delta) - (1-\delta) \ge 2^k\delta.$$

In either case we conclude that $|\eta| \ge 2^{k+j-1}$, and using (10.5.28) we deduce

$$|\widehat{K_j^\delta}(\xi)| \le \int_{|\eta|\ge 2^{k+j-1}} |\widehat{\psi}(\eta)|\,d\eta \le C_M 2^{-(j+k)M}.$$

The proof of (10.5.20) is similar. Since $|\xi - 2^{-j}\delta\eta| \geq 1-5\delta \geq 1/2$, if $|\xi| \leq 1/8$, it follows that $|2^{-j}\delta\eta| \geq 1/4$. Likewise, if $|\xi| \geq 15/8$, then $|2^{-j}\delta\eta| \geq |\xi| - 1 \geq |\xi|/4$. These estimates imply

$$|2^{-j}\delta\eta| \geq \frac{1}{8}(1+|\xi|) \implies |\eta| \geq 2^j \frac{1}{8\delta}(1+|\xi|)$$

in the support of the integral in (10.5.28). It follows that

$$|\widehat{K_j^\delta}(\xi)| \leq \int_{|\eta|\geq 2^{j-3}(1+|\xi|)/\delta} |\widehat{\psi}(\eta)|\,d\eta \leq C_M 2^{-jM}\delta^M(1+|\xi|)^{-M}$$

whenever $|\xi| \leq 1/8$ or $|\xi| \geq 15/8$. □

We finish with the proof of Lemma 10.5.6, which had been left open.

Proof. We reduce estimate (10.5.21) by duality to

$$\int_{\mathbf{R}^n} |\widehat{g}(\xi)|^2 \frac{d\xi}{|\xi|^\alpha} \leq C(n,\alpha)\varepsilon^{\alpha-1}A_\alpha(\varepsilon)\int_{||x|-1|\leq\varepsilon} |g(x)|^2\,dx$$

for functions g supported in the annulus $||x|-1| \leq \varepsilon$. Using that $(|\xi|^{-\alpha})^\wedge(x) = c_{n,\alpha}|x|^{\alpha-n}$ (cf. Theorem 2.4.6), we write

$$\begin{aligned}
\int_{\mathbf{R}^n} |\widehat{g}(\xi)|^2 \frac{d\xi}{|\xi|^\alpha} &= \int_{\mathbf{R}^n} \widehat{g}(\xi)\overline{\widehat{g}(\xi)}\,\frac{1}{|\xi|^\alpha}\,d\xi \\
&= \int_{\mathbf{R}^n} \big(\widehat{g}\,\widehat{\widetilde{\overline{g}}}\big)^\vee(x)\,\frac{c_{n,\alpha}}{|x|^{n-\alpha}}\,dx \\
&= \int_{\mathbf{R}^n} (g*\widetilde{\overline{g}})(x)\,\frac{dx}{|x|^{n-\alpha}} \\
&= \int_{||y|-1|\leq\varepsilon}\int_{||x|-1|\leq\varepsilon} g(x)\widetilde{\overline{g}}(y)\,\frac{c_{n,\alpha}}{|x-y|^{n-\alpha}}\,dx\,dy \\
&\leq B(n,\alpha)\,\|g\|_{L^2}^2,
\end{aligned}$$

where $\widetilde{g}(x) = g(-x)$ and

$$B(n,\alpha) = \sup_{||x|-1|\leq\varepsilon}\int_{||y|-1|\leq\varepsilon} \frac{c_{n,\alpha}}{|y-x|^{n-\alpha}}\,dy.$$

The last inequality is proved by interpolating between the $L^1 \to L^1$ and $L^\infty \to L^\infty$ estimates with bound $B(n,\alpha)$ for the linear operator

$$L(g)(x) = \int_{\mathbf{R}^n} g(y)\,\frac{c_{n,\alpha}}{|x-y|^{n-\alpha}}\,dy.$$

It remains to establish that

$$B(n,\alpha) \leq C(n,\alpha)\varepsilon^{\alpha-1}A_\alpha(\varepsilon).$$

Applying a rotation and a change of variables, matters reduce to proving that

$$\sup_{||x|-1|\le\varepsilon}\int_{||y-|x|e_1|-1|\le\varepsilon}\frac{c_{n,\alpha}}{|y|^{n-\alpha}}\,dy\le C(n,\alpha)\varepsilon^{\alpha-1}A_\alpha(\varepsilon),$$

where $e_1=(1,0,\dots,0)$. This, in turn, is a consequence of

$$\int_{||y-e_1|-1|\le 2\varepsilon}\frac{c_{n,\alpha}}{|y|^{n-\alpha}}\,dy\le C(n,\alpha)\varepsilon^{\alpha-1}A_\alpha(\varepsilon),\tag{10.5.29}$$

since $||y-e_1|x||-1|\le\varepsilon$ and $||x|-1|\le\varepsilon$ imply $||y-e_1|-1|\le 2\varepsilon$. In proving (10.5.29), it suffices to assume that $\varepsilon<1/100$; otherwise, the left-hand side of (10.5.29) is bounded from above by a constant, and the right-hand side of (10.5.29) is bounded from below by another constant. The region of integration in (10.5.29) is a ring centered at e_1 and width 4ε. We estimate the integral in (10.5.29) by the sum of the integrals of the function $c_{n,\alpha}|y|^{\alpha-n}$ over the sets

$$\begin{aligned}
S_0&=\{y\in\mathbf{R}^n:\ |y|\le\varepsilon,\quad ||y-e_1|-1|\le 2\varepsilon\},\\
S_\ell&=\{y\in\mathbf{R}^n:\ \ell\varepsilon\le|y|\le(\ell+1)\varepsilon,\quad ||y-e_1|-1|\le 2\varepsilon\},\\
S_*&=\{y\in\mathbf{R}^n:\ |y|\ge 1,\quad ||y-e_1|-1|\le 2\varepsilon\},
\end{aligned}$$

where $\ell=1,\dots,[\frac{1}{\varepsilon}]+1$. The volume of each S_ℓ is comparable to

$$\varepsilon\left[((\ell+1)\varepsilon)^{n-1}-(\ell\varepsilon)^{n-1}\right]\approx\varepsilon^n\ell^{n-2}.$$

Consequently,

$$\int_{S_0}\frac{dy}{|y|^{n-\alpha}}\le\omega_{n-1}\int_0^\varepsilon\frac{r^{n-1}}{r^{n-\alpha}}\,dr=\frac{\omega_{n-1}}{\alpha}\varepsilon^\alpha,$$

whereas

$$\sum_{\ell=1}^{[\frac{1}{\varepsilon}]+1}\int_{S_\ell}\frac{dy}{|y|^{n-\alpha}}\le C'_{n,\alpha}\sum_{\ell=1}^{2/\varepsilon}\frac{\varepsilon^n\ell^{n-2}}{(\ell\varepsilon)^{n-\alpha}}\le C'_{n,\alpha}\,\varepsilon^\alpha\sum_{\ell=1}^{2/\varepsilon}\frac{1}{\ell^{2-\alpha}}.$$

Finally, the volume of S_∞ is about ε; hence

$$\int_{S_\infty}\frac{dy}{|y|^{n-\alpha}}\le|S_\infty|\le C''_{n,\alpha}\,\varepsilon.$$

Combining these estimates, we obtain

$$\int_{||y-e_1|-1|\le 2\varepsilon}\frac{c_{n,\alpha}}{|y|^{n-\alpha}}\,dy\le C_{n,\alpha}\left[\varepsilon^\alpha+\varepsilon^\alpha\sum_{\ell=1}^{2/\varepsilon}\frac{1}{\ell^{2-\alpha}}+\varepsilon\right],$$

and it is an easy matter to check that the expression inside the square brackets is at most a constant multiple of $\varepsilon^{\alpha-1}A_\alpha(\varepsilon)$.

We now turn attention to (10.5.22). Switching the roles of f and $\widehat{f}$, we rewrite (10.5.22) as

$$\int_{\mathbf{R}^n} \frac{|f(x)|^2}{(1+|x|)^M}\,dx \le C'_M(n,\alpha)\int_{\mathbf{R}^n} |\widehat{(-\Delta)^{\frac{\alpha}{4}}(f)}(\xi)|^2\,d\xi$$
$$= C'_M(n,\alpha)\int_{\mathbf{R}^n} |(-\Delta)^{\frac{\alpha}{4}}(f)(x)|^2\,dx,$$

recalling the Laplacian introduced in (6.1.1). This estimate can also be restated in terms of the Riesz potential operator $I_{\alpha/2} = (-\Delta)^{-\alpha/4}$ as follows:

$$\int_{\mathbf{R}^n} \frac{|I_{\alpha/2}(g)(x)|^2}{(1+|x|)^M}\,dx \le C'_M(n,\alpha)\int_{\mathbf{R}^n} |g(x)|^2\,dx. \tag{10.5.30}$$

To show this, we use Hölder's inequality with exponents $q/2$ and n/α, where $q>2$ satisfies

$$\frac{1}{2}-\frac{1}{q}=\frac{\alpha}{2n}.$$

Then we have

$$\int_{\mathbf{R}^n} \frac{|I_{\alpha/2}(g)(x)|^2}{(1+|x|)^M}\,dx \le \left(\int_{\mathbf{R}^n} \frac{dx}{(1+|x|)^{Mn/\alpha}}\right)^{\frac{n}{\alpha}} \big\|I_{\alpha/2}(g)\big\|^2_{L^q(\mathbf{R}^n)}$$
$$\le C'_M(n,\alpha)\big\|g\big\|^2_{L^2(\mathbf{R}^n)}$$

in view of Theorem 6.1.3 and since $M>n$ and $\alpha<n$. This finishes the proof of the lemma. □

Exercises

10.5.1. Let $0<r<p<\infty$ and $n(1-\frac{r}{p})<\beta<n$. Show that $L^p(\mathbf{R}^n)$ is contained in $L^r(\mathbf{R}^n)+L^r(\mathbf{R}^n,|x|^{-\beta})$.
[*Hint:* Write $f=f_1+f_2$, where $f_1=f\chi_{|f|>1}$ and $f_2=f\chi_{|f|\le 1}$.]

10.5.2. (a) With the notation of Lemma 10.5.4, use dilations to show that the estimate

$$\int_{\mathbf{R}^n}\int_1^2 |S_t^\delta(f)(x)|^2\frac{dt}{t}\frac{dx}{|x|^\alpha} \le C_0\int_{\mathbf{R}^n}|f(x)|^2\frac{dx}{|x|^\alpha}$$

implies

$$\int_{\mathbf{R}^n}\int_a^{2a} |S_t^\delta(f)(x)|^2\frac{dt}{t}\frac{dx}{|x|^\alpha} \le C_0\int_{\mathbf{R}^n}|f(x)|^2\frac{dx}{|x|^\alpha}$$

for any $a>0$ and f in the Schwartz class.
(b) Using dilations also show that (10.5.16) implies (10.5.15).

10.5.3. Let h be a Schwartz function on $\mathbf{R}^n$. Prove that

$$\frac{1}{\varepsilon}\int_{||x|-1|\le\varepsilon} h(x)\,dx \to 2|\mathbf{S}^{n-1}|\int_{\mathbf{S}^{n-1}} h(\theta)\,d\theta$$

as $\varepsilon \to 0$. Use Lemma 10.5.6 to show that for $1 < \alpha < n$ we have

$$\int_{\mathbf{S}^{n-1}} |\widehat{f}(\theta)|\,d\theta \le C(n,\alpha)\int_{\mathbf{R}^n} |f(x)|^2 |x|^\alpha\,dx.$$

10.5.4. Let $w \in A_2$. Assume that the ball multiplier operator $B^0(f) = (\widehat{f}\chi_{B(0,1)})^\vee$ satisfies

$$\int_{\mathbf{R}^n} |B^0(f)(x)|^2\,w(x)\,dx \le C_{n,\alpha}\int_{\mathbf{R}^n} |f(x)|^2\,w(x)\,dx$$

for all $f \in L^2(w)$. Prove the same estimate for $\mathscr{B}(f) = \sup_{k\in\mathbf{Z}} |B^0_{2^k}(f)|$.
[*Hint:* Argue as in the proof of Theorem 5.3.1. Pick a smooth function with compact support $\widehat{\Phi}$ equal to one on $B(0,1)$ and vanishing in $B(0,2)$ and define $\widehat{\Psi}(\xi) = \widehat{\Phi}(\xi) - \widehat{\Phi}(2\xi)$. Then $\chi_{B(0,1)}\big(\widehat{\Phi}(\xi) - \widehat{\Phi}(2\xi)\big) = \chi_{B(0,1)} - \widehat{\Phi}(2\xi)$; hence

$$\begin{aligned}\mathscr{B}(f) &\le \sup_k |\Phi_{2^{-k}} * f| + \Big(\sum_{k\in\mathbf{Z}} |B^0_{2^k}(f) - \Phi_{2^{-(k-1)}} * f|^2\Big)^{\frac12}\\ &\le C_\Phi M(f) + \Big(\sum_{k\in\mathbf{Z}} |B^0_{2^k}(f * \Psi_{2^{-k}})|^2\Big)^{\frac12}\end{aligned}$$

and show that each term is bounded on $L^2(w)$.]

10.5.5. Show that the Bochner–Riesz operator B^λ does not map $L^p(\mathbf{R}^n)$ to $L^{p,\infty}(\mathbf{R}^n)$ when $\lambda = \frac{n-1}{2} - \frac{n}{p}$ and $2 < p < \infty$. Derive the same conclusion for B^λ_*.
[*Hint:* Suppose the contrary. Then by duality it would follow that B^λ maps $L^{p,1}(\mathbf{R}^n)$ to $L^p(\mathbf{R}^n)$ when $1 < p < 2$ and $\lambda = \frac{n}{p} - \frac{n+1}{2}$. To contradict this statement test the operator on a Schwartz function whose Fourier transform is equal to 1 on the unit ball and argue as in Proposition 10.2.3.]

HISTORICAL NOTES

The geometric construction in Section 10.1 is based on ideas of Besicovitch, who used a similar construction to answer the following question posed in 1917 by the Japanese mathematician S. Kakeya: What is the smallest possible area of the trace of ink left on a piece of paper by an ink-covered needle of unit length when the positions of its two ends are reversed? This problem puzzled mathematicians for several decades until Besicovitch [22] showed that for any $\varepsilon > 0$ there is a way to move the needle so that the total area of the blot of ink left on the paper is smaller than ε. Fefferman [125] borrowed ideas from the construction of Besicovitch to provide the negative answer to the multiplier problem to the ball for $p \neq 2$ (Theorem 10.1.5). Prior to Fefferman's work, the fact that the characteristic function of the unit ball is not a multiplier on $L^p(\mathbf{R}^n)$ for $|\frac1p - \frac12| \ge \frac{1}{2n}$ was pointed out by Herz [163], who also showed that this limitation is not necessary when this operator is restricted to radial L^p functions. The crucial Lemma 10.1.4 in Fefferman's proof is due to Y. Meyer.

The study of Bochner–Riesz means originated in the article of Bochner [27], who obtained their L^p boundedness for $\lambda > \frac{n-1}{2}$. Stein [287] improved this result to $\lambda > \frac{n-1}{2}|\frac1p - \frac12|$ using interpolation for analytic families of operators. Theorem 10.2.4 was first proved by Carleson and Sjölin [58]. A second proof of this theorem was given by Fefferman [127]. A third proof was given by Hörmander [167]. The proof of Theorem 10.2.4 given in the text is due Córdoba [90]. This proof elaborated the use of the Kakeya maximal function in the study of spherical summation multipliers, which was implicitly pioneered in Fefferman [127]. The boundedness of the Kakeya maximal function $\mathscr{K}_N$ on $L^2(\mathbf{R}^2)$ with norm $C(\log N)^2$ was first obtained by Córdoba [89]. The sharp estimate $C\log N$ was later obtained by Strömberg [296]. The proof of Theorem 10.3.5 is taken from this article of Strömberg. Another proof of the boundedness of the Kakeya maximal function without dilations on $L^2(\mathbf{R}^2)$ was obtained by Müller [240]. Barrionuevo [17] showed that for any subset Σ of $\mathbf{S}^1$ with N elements the maximal operator $\mathfrak{M}_\Sigma$ maps $L^2(\mathbf{R}^2)$ to itself with norm $CN^{2(\log N)^{-1/2}}$ for some absolute constant C. Note that this bound is $O(N^\varepsilon)$ for any $\varepsilon > 0$. Katz [183] improved this bound to $C\log N$ for some absolute constant C; see also Katz [184]. The latter is a sharp bound, as indicated in Proposition 10.3.4. Katz [182] also showed that the maximal operator $\mathfrak{M}_K$ associated with a set of unit vectors pointing along a Cantor set K of directions is unbounded on $L^2(\mathbf{R}^2)$. If Σ is an infinite set of vectors in $\mathbf{S}^1$ pointing in lacunary directions, then $\mathfrak{M}_\Sigma$ was studied by Strömberg [295], Córdoba and Fefferman [93], and Nagel, Stein, and Wainger [244]. The last authors obtained its L^p boundedness for all $1 < p < \infty$. Theorem 10.2.7 was first proved by Carleson [56]. For a short account on extensions of this theorem, the reader may consult the historical notes at the end of Chapter 5.

The idea of restriction theorems for the Fourier transform originated in the work of E. M. Stein around 1967. Stein's original restriction result was published in the article of Fefferman [123], which was the first to point out connections between restriction theorems and boundedness of the Bochner–Riesz means. The full restriction theorem for the circle (Theorem 10.4.7 for $p < \frac{4}{3}$) is due to Fefferman and Stein and was published in the aforementioned article of Fefferman [123]. See also the related article of Zygmund [340]. The present proof of Theorem 10.4.7 is based in that of Córdoba [91]. This proof was further elaborated by Tomas [314], who pointed out the logarithmic blowup when $p = \frac{4}{3}$ for the corresponding restriction problem for annuli. The result in Example 10.4.4 is also due to Fefferman and Stein and was initially proved using arguments from spherical harmonics. The simple proof presented here was observed by A. W. Knapp. The restriction property in Theorem 10.4.5 for $p < \frac{2(n+1)}{n+3}$ is due to Tomas [313], while the case $p = \frac{2(n+1)}{n+3}$ is due to Stein [291]. Theorem 10.4.6 was first proved by Fefferman [123] for the smaller range of $\lambda > \frac{n-1}{4}$ using the restriction property $R_{p\to 2}(\mathbf{S}^{n-1})$ for $p < \frac{4n}{3n+1}$. The fact that the $R_{p\to 2}(\mathbf{S}^{n-1})$ restriction property (for $p < 2$) implies the boundedness of the Bochner–Riesz operator B^λ on $L^p(\mathbf{R}^n)$ is contained in the work of Fefferman [123]. A simpler proof of this fact, obtained later by E. M. Stein, appeared in the subsequent article of Fefferman [127]. This proof is given in Theorem 10.4.6, incorporating the Tomas–Stein restriction property $R_{p\to 2}(\mathbf{S}^{n-1})$ for $p \le \frac{2(n+1)}{n+3}$. It should be noted that the case $n = 3$ of this theorem was first obtained in unpublished work of Sjölin. For a short exposition and history of this material consult the book of Davis and Chang [106]. Much of the material in Sections 10.2, 10.3, and 10.4 is based on the notes of Vargas [322].

There is an extensive literature on restriction theorems for submanifolds of $\mathbf{R}^n$. It is noteworthy to mention (in chronological order) the results of Strichartz [294], Prestini [267], Greenleaf [155], Christ [62], Drury [112], Barceló [15], [16], Drury and Marshall [114], [115], Beckner, Carbery, Semmes, and Soria [18], Drury and Guo [113], De Carli and Iosevich [107], [108], Sjölin and Soria [284], Oberlin [250], Wolff [337], and Tao [306].

The boundedness of the Bochner–Riesz operators on the range excluded by Proposition 10.2.3 implies that the restriction property $R_{p\to q}(\mathbf{S}^{n-1})$ is valid when $\frac{1}{q} = \frac{n+1}{n-1}\frac{1}{p'}$ and $1 \le p < \frac{2n}{n+1}$, as shown by Tao [305]; in this article a hierarchy of conjectures in harmonic analysis and interrelations among them is discussed. In particular, the aforementioned restriction property would imply estimate (10.3.33) for the Kakeya maximal operator $\mathscr{K}_N$ on $\mathbf{R}^n$, which would in turn imply that Besicovitch sets have Minkowski dimension n. (A Besicovitch set is defined as a subset of $\mathbf{R}^n$ that contains a unit line segment in every direction.) Katz, Laba, and Tao [185] have obtained good estimates on the Minkowski dimension of such sets in $\mathbf{R}^3$.

A general sieve argument obtained by Córdoba [89] reduces the boundedness of the Kakeya maximal operator $\mathscr{K}_N$ to the one without dilations $\mathscr{K}_N^a$. For applications to the Bochner–Riesz multiplier problem, only the latter is needed. Carbery, Hernández, and Soria [51] have proved estimate (10.3.31) for radial functions in all dimensions. Igari [175] proved estimate (10.3.32) for products of one-variable functions of each coordinate. The norm estimates in Corollary 10.3.7 can be reversed, as shown by Keich [187] for $p > 2$. The corresponding estimate for $1 < p < 2$ in the same corollary can be improved to $N^{\frac{2}{p}-1}$. Córdoba [90] proved the partial case $p \le 2$ of Theorem 10.3.10 on $\mathbf{R}^n$. This range was extended by Drury [111] to $p \le \frac{n+1}{n-1}$ using estimates for the x-ray transform. Theorem 10.3.10 (i.e., the further extension to $p \le \frac{n+1}{2}$) is due to Christ, Duoandikoetxea, and Rubio de Francia [68], and its original proof also used estimates for the x-ray transform; the proof of Theorem 10.3.10 given in the text is derived from that in Bourgain [29]. This article brought a breakthrough in many of the previous topics. In particular, Bourgain [29] showed that the Kakeya maximal operator $\mathscr{K}_N$ maps $L^p(\mathbf{R}^n)$ to itself with bound $C_\varepsilon N^{\frac{n}{p}-1+\varepsilon}$ for all $\varepsilon > 0$ and some $p_n > \frac{n+1}{2}$. He also showed that the range of p's in Theorem 10.4.5 is not sharp, since there exist indices $p = p(n) > \frac{2(n+1)}{n+3}$ for which property $R_{p\to q}(\mathbf{S}^{n-1})$ holds, and that Theorem 10.4.6 is not sharp, since there exist indices $\lambda_n < \frac{n-1}{2(n+1)}$ for which the Bochner–Riesz operators are bounded on $L^p(\mathbf{R}^n)$ in the optimal range of p's when $\lambda \ge \lambda_n$. Improvements on these indices were subsequently obtained by Bourgain [30], [31]. Some of Bourgain's results in $\mathbf{R}^3$ were re-proved by Schlag [279] using different geometric methods. Wolff [335] showed that the Kakeya maximal operator $\mathscr{K}_N$ maps $L^p(\mathbf{R}^n)$ to itself with bound $C_\varepsilon N^{\frac{n}{p}-1+\varepsilon}$ for any $\varepsilon > 0$ whenever $p \le \frac{n+2}{2}$. In higher dimensions, this range of p's was later extended by Bourgain [32] to $p \le (1+\varepsilon)\frac{n}{2}$ for some dimension-free positive constant ε. When $n = 3$, further improvements on the restriction and the Kakeya conjectures were obtained by Tao, Vargas, and Vega [308]. For further historical advances in the subject the reader is referred to the survey articles of Wolff [336] and Katz and Tao [186].

Regarding the almost everywhere convergence of the Bochner–Riesz means, Carbery [50] has shown that the maximal operator $B_*^\lambda(f) = \sup_{R>0} |B_R^\lambda(f)|$ is bounded on $L^p(\mathbf{R}^2)$ when $\lambda > 0$ and $2 \le p < \frac{4}{1-2\lambda}$, obtaining the convergence $B_R^\lambda(f) \to f$ almost everywhere for $f \in L^p(\mathbf{R}^2)$. For $n \ge 3$, $2 \le p < \frac{2n}{n-1-2\lambda}$, and $\lambda \ge \frac{n-1}{2(n+1)}$ the same result was obtained by Christ [63]. Theorem 10.5.2 is due to Carbery, Rubio de Francia, and Vega [52]. Theorem 10.5.1 is contained in Tao [304]. Tao [307] also obtained boundedness for the maximal Bochner–Riesz operators B_*^λ on $L^p(\mathbf{R}^2)$ whenever $1 < p < 2$ for an open range of pairs $(\frac{1}{p}, \lambda)$ that lie below the line $\lambda = \frac{1}{2}(\frac{1}{p} - \frac{1}{2})$.

On the critical line $\lambda = \frac{n}{p} - \frac{n+1}{2}$, boundedness into weak L^p for the Bochner–Riesz operators is possible in the range $1 \le p \le \frac{2n}{n+1}$. Christ [65], [64] first obtained such results for $1 \le p < \frac{2(n+1)}{n+3}$ in all dimensions. The point $p = \frac{2(n+1)}{n+3}$ was later included by Tao [303]. In two dimensions, weak boundedness for the full range of indices was shown by Seeger [280]; in all dimensions the same conclusion was obtained by Colzani, Travaglini, and Vignati [87] for radial functions. Tao [304] has obtained a general argument that yields weak endpoint bounds for B^λ whenever strong type bounds are known above the critical line.

Chapter 11
Time–Frequency Analysis and the Carleson–Hunt Theorem

In this chapter we discuss in detail the proof of the almost everywhere convergence of the partial Fourier integrals of L^p functions on the line. The proof of this theorem is based on techniques involving both spatial and frequency decompositions. These techniques are referred to as time–frequency analysis. The underlying goal is to decompose a given function at any scale as a sum of pieces perfectly localized in frequency and well localized in space. The action of an operator on each piece is carefully studied and the interaction between different parts of this action are analyzed. Ideas from combinatorics are employed to organize the different pieces of the decomposition.

11.1 Almost Everywhere Convergence of Fourier Integrals

In this section we study the proof of one of the most celebrated theorems in Fourier analysis, Carleson's theorem on the almost everywhere convergence of Fourier series of square integrable functions on the circle. The same result is also valid for functions f on the line if the partial sums of the Fourier series are replaced by the (partial) Fourier integrals

$$\int_{|\xi|\le N} \widehat{f}(\xi)e^{2\pi i x\xi}\,d\xi\,.$$

The equivalence of these assertions follows from the transference methods discussed in Chapter 3.

For square-integrable functions f on the line, define the *Carleson operator*

$$\mathscr{C}(f)(x) = \sup_{N>0}\left|\left(\widehat{f}\chi_{[-N,N]}\right)^{\vee}\right| = \sup_{N>0}\left|\int_{|\xi|\le N} \widehat{f}(\xi)e^{2\pi i x\xi}\,d\xi\right|. \tag{11.1.1}$$

We note that the operators $(\widehat{f}\chi_{[a,b]})^{\vee}$ are well defined when $-\infty \le a < b \le \infty$ for f in $L^2(\mathbf{R})$, and thus so is $\mathscr{C}(f)$. We have the following result concerning $\mathscr{C}$.

L. Grafakos, *Modern Fourier Analysis*, DOI: 10.1007/978-0-387-09434-2_11,

Theorem 11.1.1. *There is a constant $C>0$ such that for all square-integrable functions f on the line the following estimate is valid:*

$$\big\|\mathscr{C}(f)\big\|_{L^{2,\infty}} \le C\big\|f\big\|_{L^2}.$$

It follows that for all f in $L^2(\mathbf{R})$ we have

$$\lim_{N\to\infty}\int_{|\xi|\le N}\widehat{f}(\xi)e^{2\pi i x\xi}\,d\xi = f(x) \tag{11.1.2}$$

for almost all $x\in\mathbf{R}$.

Proof. Because of the simple identity

$$\int_{|\xi|\le N}\widehat{f}(\xi)e^{2\pi i x\xi}\,d\xi = \int_{-\infty}^{N}\widehat{f}(\xi)e^{2\pi i x\xi}\,d\xi - \int_{-\infty}^{-N}\widehat{f}(\xi)e^{2\pi i x\xi}\,d\xi\,,$$

it suffices to obtain $L^2\to L^{2,\infty}$ bounds for the *one-sided maximal operators*

$$\begin{aligned}\mathscr{C}_1(f)(x) &= \sup_{N>0}\left|\int_{-\infty}^{N}\widehat{f}(\xi)e^{2\pi i x\xi}\,d\xi\right|,\\ \mathscr{C}_2(f)(x) &= \sup_{N>0}\left|\int_{-\infty}^{-N}\widehat{f}(\xi)e^{2\pi i x\xi}\,d\xi\right|.\end{aligned}$$

Once these bounds are obtained, we can use the simple fact that (11.1.2) holds for Schwartz functions and Theorem 2.1.14 to obtain (11.1.2) for all square-integrable functions f on the line. Note that $\widetilde{\mathscr{C}_2(f)} = \mathscr{C}_1(\widetilde{f})$, where $\widetilde{f}(x) = f(-x)$ is the usual reflection operator. Therefore, it suffices to obtain bounds only for $\mathscr{C}_1$. Just as is the case with $\mathscr{C}$, the operators $\mathscr{C}_1$ and $\mathscr{C}_2$ are well defined on $L^2(\mathbf{R})$.

For $a>0$ and $y\in\mathbf{R}$ we define the translation operator τ^y, the modulation operator M^a, and the dilation operator D^a as follows:

$$\begin{aligned}\tau^y(f)(x) &= f(x-y)\,,\\ D^a(f)(x) &= a^{-\frac{1}{2}}f(a^{-1}x)\,,\\ M^y(f)(x) &= f(x)e^{2\pi i y x}\,.\end{aligned}$$

These operators are isometries on $L^2(\mathbf{R})$.

We break down the proof of Theorem 11.1.1 into several steps.

11.1.1 Preliminaries

We denote rectangles of area 1 in the (x,ξ) plane by s, t, u, etc. All rectangles considered in the sequel have sides parallel to the axes. We think of x as the time coordinate and of ξ as the frequency coordinate. For this reason we refer to the (x,ξ)

coordinate plane as the time–frequency plane. The projection of a rectangle s on the time axis is denoted by I_s, while its projection on the frequency axis is denoted by ω_s. Thus a rectangle s is just $s = I_s \times \omega_s$. Rectangles with sides parallel to the axes and area equal to one are called *tiles*.

The center of an interval I is denoted by $c(I)$. Also for $a > 0$, aI denotes an interval with the same center as I whose length is $a|I|$. Given a tile s, we denote by $s(1)$ its bottom half and by $s(2)$ its upper half defined by

$$s(1) = I_s \times \big(\omega_s \cap (-\infty, c(\omega_s)\big), \qquad s(2) = I_s \times \big(\omega_s \cap [c(\omega_s), +\infty)\big).$$

These sets are called *semitiles*. The projections of these sets on the frequency axes are denoted by $\omega_{s(1)}$ and $\omega_{s(2)}$, respectively.

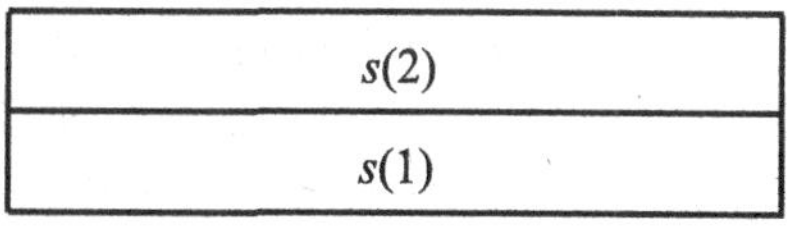

Fig. 11.1 The lower and the upper parts of a tile s.

A dyadic interval is an interval of the form $[m2^k, (m+1)2^k)$, where k and m are integers. We denote by $\mathbf{D}$ the set of all rectangles $I \times \omega$ with I, ω dyadic intervals and $|I|\,|\omega| = 1$. Such rectangles are called *dyadic tiles*. We denote by $\mathbf{D}$ the set of all dyadic tiles.

We fix a Schwartz function φ such that $\widehat{\varphi}$ is real, nonnegative, and supported in the interval $[-1/10, 1/10]$. For each tile s, we introduce a function φ_s as follows:

$$\varphi_s(x) = |I_s|^{-\frac{1}{2}} \varphi\left(\frac{x - c(I_s)}{|I_s|}\right) e^{2\pi i c(\omega_{s(1)})x}. \tag{11.1.3}$$

This function is localized in frequency near $c(\omega_{s(1)})$. Using the previous notation, we have

$$\varphi_s = M^{c(\omega_{s(1)})} \tau^{c(I_s)} D^{|I_s|}(\varphi).$$

Observe that

$$\widehat{\varphi_s}(\xi) = |\omega_s|^{-\frac{1}{2}} \widehat{\varphi}\left(\frac{\xi - c(\omega_{s(1)})}{|\omega_s|}\right) e^{2\pi i (c(\omega_{s(1)}) - \xi) c(I_s)}, \tag{11.1.4}$$

from which it follows that $\widehat{\varphi_s}$ is supported in $\frac{1}{5}\omega_{s(1)}$. Also observe that the functions φ_s have the same $L^2(\mathbf{R})$ norm.

Recall the complex inner product notation for $f, g \in L^2(\mathbf{R})$:

$$\langle f \,|\, g \rangle = \int_{\mathbf{R}} f(x)\overline{g(x)}\, dx. \tag{11.1.5}$$

Given a nonzero real number ξ, we introduce an operator

$$A_\xi(f) = \sum_{s \in \mathbf{D}} \chi_{\omega_{s(2)}}(\xi) \langle f \,|\, \varphi_s \rangle \, \varphi_s \tag{11.1.6}$$

initially defined for f in the Schwartz class. We show in the next subsection that the series in (11.1.6) converges absolutely for f in the Schwartz class and thus A_ξ is well defined on this class. Moreover, we show in Lemma 11.1.2 that A_ξ admits an extension that is L^2 bounded, and therefore it can thought of as well defined on $L^2(\mathbf{R})$.

For every integer m, let us denote by $\mathbf{D}_m$ the set of all tiles $s \in \mathbf{D}$ such that $|I_s| = 2^m$. We call these dyadic tiles *of scale m*. Then

$$A_\xi(f) = \sum_{m \in \mathbf{Z}} A_\xi^m(f),$$

where

$$A_\xi^m(f) = \sum_{s \in \mathbf{D}_m} \chi_{\omega_{s(2)}}(\xi) \langle f \,|\, \varphi_s \rangle \, \varphi_s, \tag{11.1.7}$$

and observe that for each scale m, the second sum above ranges over all dyadic rectangles of a fixed scale whose tops contain the line perpendicular to the frequency axis at height ξ. The operators A_ξ^m are discretized versions of the multiplier operator $f \mapsto \big(\widehat{f}\chi_{(-\infty,\xi]}\big)^\vee$. Indeed, the Fourier transform of $A_\xi^m(f)$ is supported in the frequency projection of the lower part $s(1)$ of the dyadic tiles s that appear in (11.1.7). But the sum in (11.1.7) is taken over all dyadic tiles s whose frequency projection of the upper part $s(2)$ contains ξ. So the Fourier transform of $A_\xi^m(f)$ is supported in $(-\infty, \xi]$. On the other hand, summing over all s in (11.1.7) yields essentially the identity operator; cf. Exercise 11.1.9. Therefore, A_ξ^m can be viewed as the "part" of the identity operator whose frequency multiplier consists of the function $\chi_{(-\infty,\xi]}$ instead of the function 1. As m becomes larger, we obtain a better and better approximation to this multiplier. This heuristic explanation motivates the introduction and study of the operators A_ξ^m and A_ξ.

Lemma 11.1.2. *For any fixed ξ, the operators A_ξ^m are bounded on $L^2(\mathbf{R})$ uniformly in m and ξ; moreover, the operator A_ξ is L^2 bounded uniformly in ξ.*

Proof. We make a few observations about the operators A_ξ^m. First recall that the adjoint of an operator T is uniquely defined by the identity

$$\langle T(f) \,|\, g \rangle = \langle f \,|\, T^*(g) \rangle$$

for all f and g. Observe that A_ξ^m are self-adjoint operators, meaning that $(A_\xi^m)^* = A_\xi^m$. Moreover, we claim that if $m \neq m'$, then

$$A_\xi^{m'}(A_\xi^m)^* = (A_\xi^{m'})^* A_\xi^m = 0.$$

Indeed, given f and g we have

$$\begin{aligned}\langle (A_\xi^{m'})^* A_\xi^m(f) \,|\, g\rangle &= \langle A_\xi^m(f) \,|\, A_\xi^{m'}(g)\rangle \qquad (11.1.8)\\ &= \sum_{s\in\mathbf{D}_m}\sum_{s'\in\mathbf{D}_{m'}} \langle f\,|\,\varphi_s\rangle \overline{\langle g\,|\,\varphi_{s'}\rangle} \langle \varphi_s\,|\,\varphi_{s'}\rangle \chi_{\omega_{s(2)}}(\xi)\chi_{\omega_{s'(2)}}(\xi).\end{aligned}$$

Suppose that $\langle \varphi_s\,|\,\varphi_{s'}\rangle \chi_{\omega_{s(2)}}(\xi)\chi_{\omega_{s'(2)}}(\xi)$ is nonzero. Then $\langle \varphi_s\,|\,\varphi_{s'}\rangle$ is also nonzero, which implies that $\omega_{s(1)}$ and $\omega_{s'(1)}$ intersect. Also, the function $\chi_{\omega_{s(2)}}(\xi)\chi_{\omega_{s'(2)}}(\xi)$ is nonzero; hence $\omega_{s(2)}$ and $\omega_{s'(2)}$ must intersect. Thus the dyadic intervals ω_s and $\omega_{s'}$ are not disjoint, and one must contain the other. If ω_s were properly contained in $\omega_{s'}$, then it would follow that ω_s is contained in $\omega_{s'(1)}$ or in $\omega_{s'(2)}$. But then either $\omega_{s(1)}\cap\omega_{s'(1)}$ or $\omega_{s(2)}\cap\omega_{s'(2)}$ would have to be empty, which does not happen, as observed. It follows that if $\langle \varphi_s\,|\,\varphi_{s'}\rangle \chi_{\omega_{s(2)}}(\xi)\chi_{\omega_{s'(2)}}(\xi)$ is nonzero, then $\omega_s = \omega_{s'}$, which is impossible if $m \neq m'$. Thus the expression in (11.1.8) has to be zero.

We first discuss the boundedness of each operator A_ξ^m. We have

$$\begin{aligned}\big\|A_\xi^m(f)\big\|_{L^2}^2 &= \sum_{s\in\mathbf{D}_m}\sum_{s'\in\mathbf{D}_m} \langle f\,|\,\varphi_s\rangle \overline{\langle f\,|\,\varphi_{s'}\rangle} \langle \varphi_s\,|\,\varphi_{s'}\rangle \chi_{\omega_{s(2)}}(\xi)\chi_{\omega_{s'(2)}}(\xi)\\ &= \sum_{s\in\mathbf{D}_m}\sum_{\substack{s'\in\mathbf{D}_m\\ \omega_{s'}=\omega_s}} \langle f\,|\,\varphi_s\rangle \overline{\langle f\,|\,\varphi_{s'}\rangle} \langle \varphi_s\,|\,\varphi_{s'}\rangle \chi_{\omega_{s(2)}}(\xi)\chi_{\omega_{s'(2)}}(\xi)\\ &\le \sum_{s\in\mathbf{D}_m}\sum_{\substack{s'\in\mathbf{D}_m\\ \omega_{s'}=\omega_s}} \big|\langle f\,|\,\varphi_s\rangle\big|^2 \chi_{\omega_{s(2)}}(\xi) \big|\langle \varphi_s\,|\,\varphi_{s'}\rangle\big|\\ &\le C_1 \sum_{s\in\mathbf{D}_m} \big|\langle f\,|\,\varphi_s\rangle\big|^2 \chi_{\omega_{s(2)}}(\xi), \qquad (11.1.9)\end{aligned}$$

where we used an earlier observation about s and s', the Cauchy–Schwarz inequality, and the fact that

$$\sum_{\substack{s'\in\mathbf{D}_m\\ \omega_{s'}=\omega_s}} \big|\langle \varphi_s\,|\,\varphi_{s'}\rangle\big| \le C \sum_{\substack{s'\in\mathbf{D}_m\\ \omega_{s'}=\omega_s}} \left(1+\frac{\operatorname{dist}(I_s,I_{s'})}{2^m}\right)^{-10} \le C_1,$$

which follows from the result in Appendix K.1. To estimate (11.1.9), we use that

$$\begin{aligned}\big|\langle f\,|\,\varphi_s\rangle\big| &\le C_2 \int_{\mathbf{R}} |f(y)|\,|I_s|^{-\frac12}\left(1+\frac{|y-c(I_s)|}{|I_s|}\right)^{-10} dy\\ &= C_3\,|I_s|^{\frac12} \int_{\mathbf{R}} |f(y)|\left(1+\frac{|y-z|}{|I_s|}\right)^{-10} \frac{dy}{|I_s|}\\ &\le C_4\,|I_s|^{\frac12} M(f)(z),\end{aligned}$$

for all $z\in I_s$, in view of Theorem 2.1.10. Since the preceding estimate holds for all $z\in I_s$, it follows that

$$\left|\langle f \mid \varphi_s \rangle\right|^2 \le (C_3)^2 |I_s| \inf_{z \in I_s} M(f)(z)^2 \le (C_3)^2 \int_{I_s} M(f)(x)^2 \, dx. \qquad (11.1.10)$$

Next we observe that the rectangles $s \in \mathbf{D}_m$ with the property that $\xi \in \omega_{s(2)}$ are all disjoint. This implies that the corresponding time intervals I_s are also disjoint. Thus, summing (11.1.10) over all $s \in \mathbf{D}_m$ with $\xi \in \omega_{s(2)}$, we obtain that

$$\begin{aligned} \sum_{s \in \mathbf{D}_m} \left|\langle f \mid \varphi_s \rangle\right|^2 \chi_{\omega_{s(2)}}(\xi) &\le (C_3)^2 \sum_{s \in \mathbf{D}_m} \chi_{\omega_{s(2)}}(\xi) \int_{I_s} M(f)(x)^2 \, dx \\ &\le (C_3)^2 \int_{\mathbf{R}} M(f)(x)^2 \, dx, \end{aligned}$$

which establishes the required claim using the boundedness of the Hardy–Littlewood maximal operator M on $L^2(\mathbf{R})$.

Finally, we discuss the boundedness of $A_\xi = \sum_{m \in \mathbf{Z}} A_\xi^m$. For every fixed $m \in \mathbf{Z}$, the dyadic tiles that appear in the sum defining $A_\xi^m(f)$ have the form

$$s = [k2^m, (k+1)2^m) \times [\ell 2^{-m}, (\ell+1)2^{-m}),$$

where $(\ell + \frac{1}{2})2^{-m} \le \xi < (\ell+1)2^{-m}$. Thus $\ell = [2^m \xi]$, and since $\widehat{\varphi_s}$ is supported in the lower part of the dyadic tile s, we may replace f by f_m, where

$$\widehat{f_m} = \widehat{f} \chi_{[2^{-m}[2^m\xi], 2^{-m}([2^m\xi] + \frac{1}{2}))}$$

As already observed, we have $\langle A_\xi^m(f) \mid A_\xi^{m'}(f) \rangle = 0$ whenever $m \neq m'$. Consequently,

$$\begin{aligned} \Big\| \sum_{m \in \mathbf{Z}} A_\xi^m(f) \Big\|_{L^2}^2 &= \sum_{m \in \mathbf{Z}} \big\| A_\xi^m(f) \big\|_{L^2}^2 \\ &= \sum_{m \in \mathbf{Z}} \big\| A_\xi^m(f_m) \big\|_{L^2}^2 \\ &\le C_4 \sum_{m \in \mathbf{Z}} \big\| f_m \big\|_{L^2}^2 \\ &= C_4 \sum_{m \in \mathbf{Z}} \big\| \widehat{f_m} \big\|_{L^2}^2 \\ &\le C_4 \big\| f \big\|_{L^2}^2, \end{aligned}$$

since the supports of $\widehat{f_m}$ are disjoint for different values of $m \in \mathbf{Z}$. □

11.1.2 Discretization of the Carleson Operator

We let $h \in \mathscr{S}(\mathbf{R})$, $\xi \in \mathbf{R} \setminus \{0\}$, and for each $m \in \mathbf{Z}$, $y, \eta \in \mathbf{R}$, and $\lambda \in [0,1]$ we introduce the operators

$$B^m_{\xi,y,\eta,\lambda}(h) = \sum_{s\in\mathbf{D}_m} \chi_{\omega_{s(2)}}(2^{-\lambda}(\xi+\eta))\langle D^{2^\lambda}\tau^y M^\eta(h)\,|\,\varphi_s\rangle M^{-\eta}\tau^{-y}D^{2^{-\lambda}}(\varphi_s).$$

It is not hard to see that for all $x \in \mathbf{R}$ and $\lambda \in [0,1]$ we have

$$B^m_{\xi,y,\eta,\lambda}(h)(x) = B^m_{\xi,y+2^{m-\lambda},\eta,\lambda}(h)(x) = B^m_{\xi,y,\eta+2^{-m+\lambda},\lambda}(h)(x);$$

in other words, the function $(y,\eta) \mapsto B^m_{\xi,y,\eta,\lambda}(h)(x)$ is periodic in $\mathbf{R}^2$ with period $(2^{m-\lambda},2^{-m+\lambda})$. See Exercise 11.1.1.

Using Exercise 11.1.2, we obtain that for $|m|$ large (with respect to ξ) we have

$$\begin{aligned}
&\bigg|\sum_{s\in\mathbf{D}_m} \chi_{\omega_{s(2)}}(2^{-\lambda}(\xi+\eta))\langle D^{2^\lambda}\tau^y M^\eta(h)\,|\,\varphi_s\rangle M^{-\eta}\tau^{-y}D^{2^{-\lambda}}(\varphi_s)(x)\bigg| \\
&\qquad \le C_h \min(2^m,1)\sum_{s\in\mathbf{D}_m} \chi_{\omega_{s(2)}}(2^{-\lambda}(\xi+\eta))2^{-m/2}\Big|\varphi\Big(\frac{x+y-c(I_s)}{2^{m-\lambda}}\Big)\Big| \\
&\qquad \le C_h \min(2^{m/2},2^{-m/2})\sum_{k\in\mathbf{Z}}\Big|\varphi\Big(\frac{x+y-k2^m}{2^{m-\lambda}}\Big)\Big| \\
&\qquad \le C_h \min(2^{m/2},2^{-m/2}),
\end{aligned}$$

since the last sum is seen easily to converge to some quantity that remains bounded in x, y, η, and λ. It follows that for $h \in \mathscr{S}(\mathbf{R})$ we have

$$\sup_{x\in\mathbf{R}}\sup_{y\in\mathbf{R}}\sup_{\eta\in\mathbf{R}}\sup_{0\le\lambda\le 1}\big|B^m_{\xi,y,\eta,\lambda}(h)(x)\big| \le C_h\min(2^{m/2},2^{-m/2}). \tag{11.1.11}$$

Using Exercise 11.1.3 and the periodicity of the functions $B^m_{\xi,y,\eta,\lambda}(h)$, we conclude that the averages

$$\frac{1}{4KL}\int_{-L}^{L}\int_{-K}^{K}\int_0^1 B^m_{\xi,y,\eta,\lambda}(h)\,d\lambda\,dy\,d\eta$$

converge pointwise to some $\Pi^m_\xi(h)$ as $K,L\to\infty$. Estimate (11.1.11) implies the uniform convergence for the series $\sum_{m\in\mathbf{Z}} B^m_{\xi,y,\eta,\lambda}(h)$ and therefore

$$\begin{aligned}
&\lim_{\substack{K\to\infty\\ L\to\infty}}\frac{1}{4KL}\int_{-L}^{L}\int_{-K}^{K}\int_0^1 M^{-\eta}\tau^{-y}D^{2^{-\lambda}}A_{\frac{\xi+\eta}{2^\lambda}}D^{2^\lambda}\tau^y M^\eta(h)\,d\lambda\,dy\,d\eta \qquad (11.1.12)\\
&\qquad = \lim_{\substack{K\to\infty\\ L\to\infty}}\frac{1}{4KL}\int_{-L}^{L}\int_{-K}^{K}\int_0^1 \sum_{m\in\mathbf{Z}} B^m_{\xi,y,\eta,\lambda}(h)\,d\lambda\,dy\,d\eta \\
&\qquad = \sum_{m\in\mathbf{Z}}\lim_{\substack{K\to\infty\\ L\to\infty}}\frac{1}{4KL}\int_{-L}^{L}\int_{-K}^{K}\int_0^1 B^m_{\xi,y,\eta,\lambda}(h)\,d\lambda\,dy\,d\eta \\
&\qquad = \sum_{m\in\mathbf{Z}}\Pi^m_\xi(h).
\end{aligned}$$

We now make a few observations about the operator Π_ξ defined on $\mathscr{S}(\mathbf{R})$ in terms of the expression in (11.1.12), that is:

$$\Pi_\xi(h) = \sum_{m\in\mathbf{Z}} \Pi_\xi^m(h).$$

First we observe that in view of Lemma 11.1.2 and Fatou's lemma, we have that Π_ξ is bounded on L^2 uniformly in ξ. Next we observe that Π_ξ commutes with all translations τ^z for $z\in\mathbf{R}$. To see this, we use the fact that $\tau^{-z}M^{-\eta} = e^{-2\pi i\eta z}M^{-\eta}\tau^{-z}$ to obtain

$$\begin{aligned}
&\sum_{s\in\mathbf{D}_m} \chi_{\omega_{s(2)}}(2^{-\lambda}(\xi+\eta))\big\langle D^{2^\lambda}\tau^y M^\eta \tau^z(h)\,|\,\varphi_s\big\rangle\,\tau^{-z}M^{-\eta}\tau^{-y}D^{2^{-\lambda}}(\varphi_s)\\
&\quad= \sum_{s\in\mathbf{D}_m} \chi_{\omega_{s(2)}}(2^{-\lambda}(\xi+\eta))\big\langle h\,|\,\tau^{-z}M^{-\eta}\tau^{-y}D^{2^{-\lambda}}(\varphi_s)\big\rangle\,\tau^{-z}M^{-\eta}\tau^{-y}D^{2^{-\lambda}}(\varphi_s)\\
&\quad= \sum_{s\in\mathbf{D}_m} \chi_{\omega_{s(2)}}(2^{-\lambda}(\xi+\eta))\big\langle h\,|\,M^{-\eta}\tau^{-y-z}D^{2^{-\lambda}}(\varphi_s)\big\rangle\,M^{-\eta}\tau^{-y-z}D^{2^{-\lambda}}(\varphi_s).
\end{aligned}$$

Recall that $\tau^{-z}\Pi_\xi^m\tau^z(h)$ is equal to the limit of the averages of the preceding expressions over all $(y,\eta,\lambda)\in[-K,K]\times[-L,L]\times[0,1]$. But in view of the previous identity, this is equal to the limit of the averages of the expressions

$$\sum_{s\in\mathbf{D}_m} \chi_{\omega_{s(2)}}(2^{-\lambda}(\xi+\eta))\big\langle D^{2^\lambda}\tau^{y'}M^\eta(h)\,|\,\varphi_s\big\rangle M^{-\eta}\tau^{-y'}D^{2^{-\lambda}}(\varphi_s) \tag{11.1.13}$$

over all $(y',\eta,\lambda)\in[-K+z,K+z]\times[-L,L]\times[0,1]$. Since (11.1.13) is periodic in (y',η), it follows that its average over the set $[-K+z,K+z]\times[-L,L]\times[0,1]$ is equal to its average over the set $[-K,K]\times[-L,L]\times[0,1]$. Taking limits as $K,L\to\infty$, we obtain the identity $\tau^{-z}\Pi_\xi^m\tau^z(h) = \Pi_\xi^m(h)$. Summing over all $m\in\mathbf{Z}$, it follows that $\tau^{-z}\Pi_\xi\tau^z(h) = \Pi_\xi(h)$.

A similar argument using averages over shifted rectangles of the form $[-K,K]\times[-L+\theta,L+\theta]$ yields the identity

$$M^{-\theta}\Pi_{\xi+\theta}M^\theta = \Pi_\xi \tag{11.1.14}$$

for all $\xi,\theta\in\mathbf{R}$. The details are left to the reader. Next, we claim that the operator $M^{-\xi}\Pi_\xi M^\xi$ commutes with dilations D^{2^a}, $a\in\mathbf{R}$. First we observe that for all integers k we have

$$A_\xi(h) = D^{2^{-k}}A_{2^{-k}\xi}D^{2^k}(h), \tag{11.1.15}$$

which is simply saying that A_ξ is well behaved under change of scale. This identity is left as an exercise to the reader. Identity (11.1.15) may not hold for noninteger k, and this is exactly why we have averaged over all dilations 2^λ, $0\le\lambda\le1$, in (11.1.12).

Let us denote by $[a]$ the integer part of a real number a. Using the identities $D^bM^\eta = M^{\eta/b}D^b$ and $D^b\tau^z = \tau^{bz}D^b$, we obtain

$$
\begin{aligned}
&D^{2^{-a}} M^{-(\xi+\eta)} \tau^{-y} D^{2^{-\lambda}} A_{\frac{\xi+\eta}{2^\lambda}} D^{2^\lambda} \tau^y M^{\xi+\eta} D^{2^a} \qquad (11.1.16)\\
&= M^{-2^a(\xi+\eta)} \tau^{-2^{-a}y} D^{2^{-(a+\lambda)}} A_{\frac{\xi+\eta}{2^\lambda}} D^{2^{a+\lambda}} \tau^{2^{-a}y} M^{2^a(\xi+\eta)}\\
&= M^{-2^a(\xi+\eta)} \tau^{-y'} D^{2^{-\lambda'}} D^{2^{-[a+\lambda]}} A_{\frac{2^a(\xi+\eta)}{2^{\lambda'}2^{[a+\lambda]}}} D^{2^{[a+\lambda]}} D^{2^{\lambda'}} \tau^{y'} M^{2^a(\xi+\eta)}\\
&= M^{-2^a\xi} M^{-\eta'} \tau^{-y'} D^{2^{-\lambda'}} A_{\frac{2^a\xi+2^a\eta}{2^{\lambda'}}} D^{2^{\lambda'}} \tau^{y'} M^{\eta'} M^{2^a\xi}\\
&= M^{-\xi} M^{-\theta} \big(M^{-\eta'} \tau^{-y'} D^{2^{-\lambda'}} A_{\frac{\xi+\theta+\eta'}{2^{\lambda'}}} D^{2^{\lambda'}} \tau^{y'} M^{\eta'}\big) M^{\theta} M^{\xi}, \qquad (11.1.17)
\end{aligned}
$$

where we set $y' = 2^{-a}y$, $\eta' = 2^a\eta$, $\lambda' = a+\lambda-[a+\lambda]$, and $\theta = (2^a-1)\xi$. The average of (11.1.16) over all (y,η,λ) in $[-K,K]\times[-L,L]\times[0,1]$ converges to the operator $D^{2^{-a}} M^{-\xi} \Pi_\xi M^\xi D^{2^a}$ as $K,L\to\infty$. But this limit is equal to the limit of the averages of the expression in (11.1.17) over all (y',η',λ') in $[-2^{-a}K, 2^{-a}K]\times[-2^aL, 2^aL]\times[0,1]$, which is

$$
M^{-\xi} M^{-\theta} \Pi_{\xi+\theta} M^\theta M^\xi .
$$

Using the identity (11.1.14), we obtain that

$$
D^{2^{-a}} M^{-\xi} \Pi_\xi M^\xi D^{2^a} = M^{-\xi} \Pi_\xi M^\xi ,
$$

saying that the operator $M^{-\xi}\Pi_\xi M^\xi$ commutes with dilations.

Next we observe that if $\widehat{h}$ is supported in $[0,\infty)$, then $M^{-\xi}\Pi_\xi M^\xi(h) = 0$. This is a consequence of the fact that the inner products

$$
\big\langle D^{2^\lambda} \tau^y M^\eta M^\xi(h) \,|\, \varphi_s \big\rangle = \big\langle M^\xi(h) \,|\, M^{-\eta} \tau^{-y} D^{2^{-\lambda}}(\varphi_s)\big\rangle
$$

vanish, since the Fourier transform of $\tau^{-z} M^{-\eta} \tau^{-y} D^{2^{-\lambda}} \varphi_s$ is supported in the set $(-\infty, 2^\lambda c(\omega_{s(1)}) - \eta + \frac{2^\lambda}{10}|\omega_s|)$, which is disjoint from the interval $(\xi,+\infty)$ whenever $2^{-\lambda}(\xi+\eta)\in\omega_{s(2)}$. Finally, we observe that Π_ξ is a positive semidefinite operator, that is, it satisfies

$$
\big\langle \Pi_\xi(h) \,|\, h\big\rangle \ge 0 . \qquad (11.1.18)
$$

This follows easily from the fact that the inner product in (11.1.18) is equal to

$$
\lim_{\substack{K\to\infty\\ L\to\infty}} \frac{1}{4KL} \int_{-L}^{L}\int_{-K}^{K}\int_0^1 \sum_{s\in\mathbf{D}} \chi_{\omega_{s(2)}}\big(\tfrac{\xi+\eta}{2^\lambda}\big) \big|\big\langle D^{2^\lambda} \tau^y M^\eta(h) \,|\, \varphi_s\big\rangle\big|^2 \, d\lambda\, dy\, d\eta . \qquad (11.1.19)
$$

This identity also implies that Π_ξ is not the zero operator; indeed, notice that

$$
\sum_{s\in\mathbf{D}_0} \chi_{\omega_{s(2)}}\big(\tfrac{\xi+\eta}{2^\lambda}\big) \big|\big\langle D^{2^\lambda} \tau^y M^\eta(h) \,|\, \varphi_s\big\rangle\big|^2 = \big\langle h \,|\, B^0_{\xi,y,\eta,\lambda}(h)\big\rangle
$$

is periodic with period $(2^{-\lambda}, 2^{\lambda})$ in (y, η), and consequently the limit in (11.1.19) is at least as big as

$$\int_0^{2^\lambda} \int_0^{2^{-\lambda}} \int_0^1 \sum_{s \in \mathbf{D}_0} \chi_{\omega_{s(2)}}\left(\tfrac{\xi+\eta}{2^\lambda}\right) \left| \left\langle D^{2^\lambda} \tau^y M^\eta (h) \,|\, \varphi_s \right\rangle \right|^2 d\lambda \, dy \, d\eta$$

(cf. Exercise 11.1.3). Since we can always find a Schwartz function h and a dyadic tile s such that $\langle D^{2^\lambda} \tau^y M^\eta (h) \,|\, \varphi_s \rangle$ is not zero for (y, η, λ) near $(0,0,0)$, it follows that the expression in (11.1.19) is strictly positive for some function h. The same is valid for the inner product in (11.1.18); hence the operators and $M^{-\xi} \Pi_\xi M^\xi$ are nonzero for every ξ.

Let us summarize what we have already proved: The operator $M^{-\xi} \Pi_\xi M^\xi$ is nonzero, is bounded on $L^2(\mathbf{R})$, commutes with translations and dilations, and vanishes when applied to functions whose Fourier transform is supported in the positive semiaxis $[0, \infty)$. In view of Exercise 4.1.11, it follows that for some constant $c_\xi \neq 0$ we have

$$M^{-\xi} \Pi_\xi M^\xi (h)(x) = c_\xi \int_{-\infty}^0 \widehat{h}(\eta) e^{2\pi i x \eta} \, d\eta \,,$$

which identifies Π_ξ with the convolution operator whose multiplier is the function $c_\xi \chi_{(-\infty, \xi]}$. Using the identity (11.1.14), we obtain

$$c_{\xi+\theta} = c_\xi$$

for all ξ and θ, saying that c_ξ does not depend on ξ. We have therefore proved that for all Schwartz functions h the following identity is valid:

$$\Pi_\xi (h) = c \left(\widehat{h} \chi_{(-\infty, \xi]} \right)^\vee \tag{11.1.20}$$

for some fixed nonzero constant c. This completely identifies the operator Π_ξ. By density it follows that

$$\mathscr{C}_1(f) = \frac{1}{|c|} \sup_{\xi > 0} |\Pi_\xi (f)| \tag{11.1.21}$$

for all $f \in \bigcup_{1 \le p < \infty} L^p(\mathbf{R})$.

11.1.3 Linearization of a Maximal Dyadic Sum

Our goal is to show that there exists a constant $C > 0$ such that for all $f \in L^2(\mathbf{R})$ we have

$$\big\| \sup_{\xi > 0} |A_\xi (f)| \big\|_{L^{2,\infty}(\mathbf{R})} \le C \big\| f \big\|_{L^2(\mathbf{R})} \,. \tag{11.1.22}$$

Once (11.1.22) is established, averaging yields the same conclusion for the operator $f \mapsto \sup_{\xi > 0} |\Pi_\xi (f)|$, establishing the required bound for $\mathscr{C}_1$. Let us describe this averaging argument. Identity (11.1.12) gives

$$\Pi_\xi(f) = \lim_{\substack{K\to\infty \\ L\to\infty}} \frac{1}{4KL} \int_{-L}^{L}\int_{-K}^{K}\int_0^1 G_{\xi,y,\eta,\lambda}(f)\,d\lambda\,dy\,d\eta\,,$$

where

$$G_{\xi,y,\eta,\lambda}(f) = M^{-\eta}\tau^{-y}D^{2^{-\lambda}}A_{\frac{\xi+\eta}{2^\lambda}}D^{2^\lambda}\tau^y M^\eta(f)\,.$$

This, in turn, implies

$$\sup_{\xi\in\mathbf{R}}|\Pi_\xi(f)| \le \liminf_{\substack{K\to\infty \\ L\to\infty}} \frac{1}{4KL}\int_{-L}^{L}\int_{-K}^{K}\int_0^1 \sup_{\xi\in\mathbf{R}}|G_{\xi,y,\eta,\lambda}(f)|\,d\lambda\,dy\,d\eta\,. \tag{11.1.23}$$

We now take the $L^{2,\infty}$ quasinorms of both sides, and we use Fatou's lemma for weak L^2 [Exercise 1.1.12(d)]. We thus reduce the estimate for the operator $\sup_{\xi>0}|\Pi_\xi(f)|$ to the corresponding estimate for $\sup_{\xi>0}|A_\xi(f)|$. In this way we obtain the $L^{2,\infty}$ boundedness of $\sup_{\xi>0}|\Pi_\xi(f)|$ and therefore that of $\mathscr{C}_1$ in view of identity (11.1.21).

Matters are now reduced to the study of the discretized maximal operator $\sup_{\xi>0}|A_\xi(f)|$ and, in particular, to the proof of estimate (11.1.22). It will be convenient to study the maximal operator $\sup_{\xi>0}|A_\xi(f)|$ via a linearization. Here is how this is achieved. Given $f\in L^2(\mathbf{R})$, we select a measurable real-valued function[1] $N_f:\ \mathbf{R}\to\mathbf{R}^+$ such that for all $x\in\mathbf{R}$ we have

$$\sup_{\xi>0}|A_\xi(f)(x)| \le 2\,|A_{N_f(x)}(f)(x)|\,.$$

For a general measurable function $N:\ \mathbf{R}\to\mathbf{R}^+$, we define a linear operator $\mathfrak{D}_N$ by setting for $f\in L^2(\mathbf{R})$,

$$\mathfrak{D}_N(f)(x) = A_{N(x)}(f)(x) = \sum_{s\in\mathbf{D}}(\chi_{\omega_{s(2)}}\circ N)(x)\,\langle f\,|\,\varphi_s\rangle\,\varphi_s(x)\,, \tag{11.1.24}$$

where the sum on the right converges in $L^2(\mathbf{R})$ [and also uniformly for $f\in\mathscr{S}(\mathbf{R})$].

To prove (11.1.22), it suffices to show that there exists $C>0$ such that for all $f\in L^2(\mathbf{R})$ and all measurable functions $N:\ \mathbf{R}\to\mathbf{R}^+$ we have

$$\big\|\mathfrak{D}_N(f)\big\|_{L^{2,\infty}} \le C\big\|f\big\|_{L^2}\,. \tag{11.1.25}$$

Applying (11.1.25) to the measurable function N_f and using the estimate

$$\sup_{\xi>0}|A_\xi(f)| \le 2\,\mathfrak{D}_{N_f}(f)$$

yields the required conclusion for the maximal dyadic sum operator $\sup_{\xi>0}|A_\xi(f)|$ and thus for $\mathscr{C}_1(f)$.

---

[1] The range $\xi>0$ may be replaced by a finite subset of the positive rationals by density; in this case N_f could be taken to be the point ξ at which the supremum is attained.

To justify certain algebraic manipulations we fix a finite subset $\mathbf{P}$ of $\mathbf{D}$ and we define

$$\mathfrak{D}_{N,\mathbf{P}}(f)(x) = \sum_{s\in\mathbf{P}} (\chi_{\omega_{s(2)}} \circ N)(x) \langle f \,|\, \varphi_s \rangle \varphi_s(x) \,. \tag{11.1.26}$$

To prove (11.1.25) it suffices to show that there exists a $C > 0$ such that for all $f \in L^2(\mathbf{R})$, all finite subsets $\mathbf{P}$ of $\mathbf{D}$, and all real-valued measurable functions N on the line we have

$$\big\|\mathfrak{D}_{N,\mathbf{P}}(f)\big\|_{L^{2,\infty}} \le C\big\|f\big\|_{L^2} \,. \tag{11.1.27}$$

The important point is that the constant C in (11.1.27) is independent of f, $\mathbf{P}$, and the measurable function N.

To prove (11.1.27) we use duality. In view of the results of Exercises 1.4.12(c) and 1.4.7, it suffices to prove that for all measurable subsets E of the real line with finite measure we have

$$\left|\int_E \mathfrak{D}_{N,\mathbf{P}}(f)\,dx\right| = \left|\sum_{s\in\mathbf{P}} \langle (\chi_{\omega_{s(2)}} \circ N)\varphi_s, \chi_E \rangle \langle \varphi_s \,|\, f \rangle\right| \le C|E|^{\frac{1}{2}} \big\|f\big\|_{L^2} \,. \tag{11.1.28}$$

We obtain estimate (11.1.28) as a consequence of

$$\sum_{s\in\mathbf{P}} \big|\langle (\chi_{\omega_{s(2)}} \circ N)\varphi_s, \chi_E \rangle \langle f \,|\, \varphi_s \rangle\big| \le C|E|^{\frac{1}{2}} \big\|f\big\|_{L^2} \tag{11.1.29}$$

for all f in L^2, all measurable functions N, all measurable sets E of finite measure, and all finite subsets $\mathbf{P}$ of $\mathbf{D}$. It is estimate (11.1.29) that we shall concentrate on.

11.1.4 Iterative Selection of Sets of Tiles with Large Mass and Energy

We introduce a partial order in the set of dyadic tiles that provides a way to organize them. In this section, dyadic tiles are simply called tiles.

Definition 11.1.3. We define a *partial order* $<$ in the set of dyadic tiles $\mathbf{D}$ by setting

$$s < s' \iff I_s \subseteq I_{s'} \quad \text{and} \quad \omega_{s'} \subseteq \omega_s \,.$$

If two tiles $s, s' \in \mathbf{D}$ intersect, then we must have either $s < s'$ or $s' < s$. Indeed, both the time and frequency components of the tiles must intersect; then either $I_s \subseteq I_{s'}$ or $I_{s'} \subseteq I_s$. In the first case we must have $|\omega_s| \ge |\omega_{s'}|$, thus $\omega_{s'} \subseteq \omega_s$, which gives $s < s'$, while in the second case a similar argument gives $s' < s$. As a consequence of this observation, if $\mathbf{R}_0$ is a finite set of tiles, then all maximal elements of $\mathbf{R}_0$ under $<$ must be disjoint sets.

Definition 11.1.4. A finite set of tiles $\mathbf{P}$ is called a *tree* if there exists a tile $t \in \mathbf{P}$ such that all $s \in \mathbf{P}$ satisfy $s < t$. We call t the top of $\mathbf{P}$ and we denote it by $t = \text{top}(\mathbf{P})$. Observe that the top of a tree is unique.

We denote trees by $\mathbf{T}$, $\mathbf{T}'$, $\mathbf{T}_1$, $\mathbf{T}_2$, and so on.

We observe that every finite set of tiles $\mathbf{P}$ can be written as a union of trees whose tops are maximal elements. Indeed, consider all maximal elements of $\mathbf{P}$ under the partial order $<$. Then every nonmaximal element s of $\mathbf{P}$ satisfies $s < t$ for some maximal element $t \in \mathbf{P}$, and thus it belongs to a tree with top t.

Tiles can be written as a union of two *semitiles* $I_s \times \omega_{s(1)}$ and $I_s \times \omega_{s(2)}$. Since tiles have area 1, semitiles have area $1/2$.

Definition 11.1.5. A tree $\mathbf{T}$ is called a 1-tree if

$$\omega_{\mathrm{top}(\mathbf{T})(1)} \subsetneq \omega_{s(1)}$$

all $s \in \mathbf{T}$. A tree $\mathbf{T}'$ is called a 2-tree if for all $s \in \mathbf{T}'$ we have

$$\omega_{\mathrm{top}(\mathbf{T}')(2)} \subsetneq \omega_{s(2)}\,.$$

We make a few observations about 1-trees and 2-trees. First note that every tree can be written as the union of a 1-tree and a 2-tree, and the intersection of these is exactly the top of the tree. Also, if $\mathbf{T}$ is a 1-tree, then the intervals $\omega_{\mathrm{top}(\mathbf{T})(2)}$ and $\omega_{s(2)}$ are disjoint for all $s \in \mathbf{T}$, and similarly for 2-trees. See Figure 11.2.

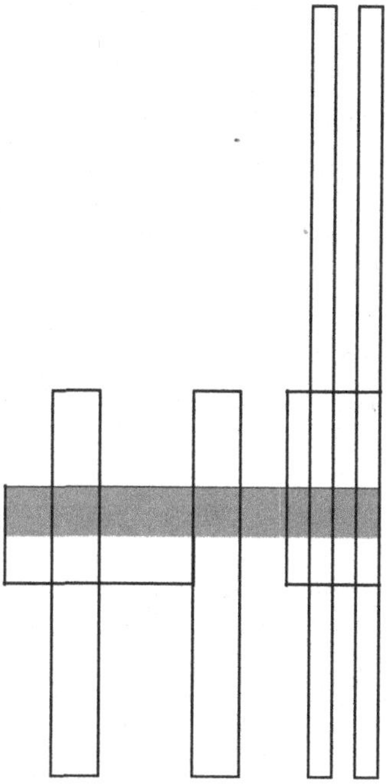

Fig. 11.2 A tree of seven tiles including the darkened top. The top together with the three tiles on the right forms a 1-tree, while the top together with the three tiles on the left forms a 2-tree.

Definition 11.1.6. Let $N: \mathbf{R} \to \mathbf{R}^+$ be a measurable function, let $s \in \mathbf{D}$, and let E be a set of finite measure. Then we introduce the quantity

$$\mathscr{M}(E;\{s\}) = \frac{1}{|E|} \sup_{\substack{u\in\mathbf{D}\\ s<u}} \int_{E\cap N^{-1}[\omega_u]} \frac{|I_u|^{-1}\,dx}{(1+\frac{|x-c(I_u)|}{|I_u|})^{10}}.$$

We call $\mathscr{M}(E;\{s\})$ the *mass* of E with respect to $\{s\}$. Given a subset $\mathbf{P}$ of $\mathbf{D}$, we define the mass of E with respect to $\mathbf{P}$ as

$$\mathscr{M}(E;\mathbf{P}) = \sup_{s\in\mathbf{P}} \mathscr{M}(E;\{s\}).$$

We observe that the mass of E with respect to any set of tiles is at most

$$\frac{1}{|E|}\int_{-\infty}^{+\infty} \frac{dx}{(1+|x|)^{10}} \le \frac{1}{|E|}.$$

Definition 11.1.7. Given a finite subset $\mathbf{P}$ of $\mathbf{D}$ and a function f in $L^2(\mathbf{R})$, we introduce the quantity

$$\mathscr{E}(f;\mathbf{P}) = \frac{1}{\|f\|_{L^2}} \sup_{\mathbf{T}} \bigg(\frac{1}{|I_{\mathrm{top}(\mathbf{T})}|}\sum_{s\in\mathbf{T}} |\langle f\,|\,\varphi_s\rangle|^2\bigg)^{\frac{1}{2}},$$

where the supremum is taken over all 2-trees $\mathbf{T}$ contained in $\mathbf{P}$. We call $\mathscr{E}(f;\mathbf{P})$ the *energy* of the function f with respect to the set of tiles $\mathbf{P}$.

We now state three important lemmas which we prove in the remaining three subsections, respectively.

Lemma 11.1.8. *There exists a constant C_1 such that for any measurable function $N:\mathbf{R}\to\mathbf{R}^+$, for any measurable subset E of the real line with finite measure, and for any finite set of tiles $\mathbf{P}$ there is a subset $\mathbf{P}'$ of $\mathbf{P}$ such that*

$$\mathscr{M}(E;\mathbf{P}\setminus\mathbf{P}') \le \frac{1}{4}\mathscr{M}(E;\mathbf{P})$$

and $\mathbf{P}'$ is a union of trees $\mathbf{T}_j$ satisfying

$$\sum_j |I_{\mathrm{top}(\mathbf{T}_j)}| \le \frac{C_1}{\mathscr{M}(E;\mathbf{P})}. \tag{11.1.30}$$

Lemma 11.1.9. *There exists a constant C_2 such that for any finite set of tiles $\mathbf{P}$ and for all functions f in $L^2(\mathbf{R})$ there is a subset $\mathbf{P}''$ of $\mathbf{P}$ such that*

$$\mathscr{E}(f;\mathbf{P}\setminus\mathbf{P}'') \le \frac{1}{2}\mathscr{E}(f;\mathbf{P})$$

and $\mathbf{P}''$ is a union of trees $\mathbf{T}_j$ satisfying

$$\sum_j |I_{\mathrm{top}(\mathbf{T}_j)}| \le \frac{C_2}{\mathscr{E}(f;\mathbf{P})^2}. \tag{11.1.31}$$

Lemma 11.1.10. *(The basic estimate) There is a finite constant C_3 such that for all trees* $\mathbf{T}$*, all functions f in $L^2(\mathbf{R})$, for any measurable function $N:\ \mathbf{R}\to\mathbf{R}^+$, and for all measurable sets E we have*

$$\sum_{s\in\mathbf{T}} \big|\langle f\,|\,\varphi_s\rangle\langle \chi_{E\cap N^{-1}[\omega_{s(2)}]}\,|\,\varphi_s\rangle\big| \\ \le C_3\,|I_{\mathrm{top}(\mathbf{T})}|\,\mathscr{E}(f;\mathbf{T})\mathscr{M}(E;\mathbf{T})\big\|f\big\|_{L^2}\,|E|\,. \tag{11.1.32}$$

In the rest of this subsection, we conclude the proof of Theorem 11.1.1 assuming Lemmas 11.1.8, 11.1.9, and 11.1.10.

Given a finite set of tiles $\mathbf{P}$, a measurable set E of finite measure, a measurable function $N:\mathbf{R}\to\mathbf{R}^+$, and a function f in $L^2(\mathbf{R})$, we find a very large integer n_0 such that

$$\mathscr{E}(f;\mathbf{P}) \le 2^{n_0}\,,$$
$$\mathscr{M}(E;\mathbf{P}) \le 2^{2n_0}\,.$$

We shall construct by decreasing induction a sequence of pairwise disjoint sets

$$\mathbf{P}_{n_0},\mathbf{P}_{n_0-1},\mathbf{P}_{n_0-2},\mathbf{P}_{n_0-3},\dots$$

such that

$$\bigcup_{j=-\infty}^{n_0}\mathbf{P}_j=\mathbf{P} \tag{11.1.33}$$

and such that the following properties are satisfied:

(1) $\mathscr{E}(f;\mathbf{P}_j)\le 2^{j+1}$ for all $j\le n_0$;

(2) $\mathscr{M}(E;\mathbf{P}_j)\le 2^{2j+2}$ for all $j\le n_0$;

(3) $\mathscr{E}\big(f;\mathbf{P}\setminus(\mathbf{P}_{n_0}\cup\cdots\cup\mathbf{P}_j)\big)\le 2^j$ for all $j\le n_0$;

(4) $\mathscr{M}\big(E;\mathbf{P}\setminus(\mathbf{P}_{n_0}\cup\cdots\cup\mathbf{P}_j)\big)\le 2^{2j}$ for all $j\le n_0$;

(5) $\mathbf{P}_j$ is a union of trees $\mathbf{T}_{jk}$ such that for all $j\le n_0$ we have

$$\sum_k |I_{\mathrm{top}(\mathbf{T}_{jk})}|\le C_0\,2^{-2j}\,,$$

where $C_0=C_1+C_2$ and C_1 and C_2 are the constants that appear in Lemmas 11.1.8 and 11.1.9, respectively.

Assume momentarily that we have constructed a sequence $\{\mathbf{P}_j\}_{j\le n_0}$ with the described properties. Then to obtain estimate (11.1.29) we use (1), (2), (5), the observation that the mass of any set of tiles is always bounded by $|E|^{-1}$, and Lemma 11.1.10 to obtain

$$\begin{aligned}
&\sum_{s\in\mathbf{P}} \big|\langle f \,|\, \varphi_s\rangle\langle \chi_{E\cap N^{-1}[\omega_{s(2)}]} \,|\, \varphi_s\rangle\big| \\
&\quad = \sum_j \sum_{s\in\mathbf{P}_j} \big|\langle f \,|\, \varphi_s\rangle\langle \chi_{E\cap N^{-1}[\omega_{s(2)}]} \,|\, \varphi_s\rangle\big| \\
&\quad \le \sum_j \sum_k \sum_{s\in\mathbf{T}_{jk}} \big|\langle f \,|\, \varphi_s\rangle\langle \chi_{E\cap N^{-1}[\omega_{s(2)}]} \,|\, \varphi_s\rangle\big| \\
&\quad \le C_3 \sum_j \sum_k |I_{\mathrm{top}(\mathbf{T}_{jk})}|\, \mathcal{E}(f;\mathbf{T}_{jk})\, \mathcal{M}(E;\mathbf{T}_{jk}) \big\|f\big\|_{L^2} |E| \\
&\quad \le C_3 \sum_j \sum_k |I_{\mathrm{top}(\mathbf{T}_{jk})}|\, 2^{j+1} \min(|E|^{-1}, 2^{2j+2}) \big\|f\big\|_{L^2} |E| \\
&\quad \le C_3 \sum_j C_0 2^{-2j} 2^{j+1} \min(|E|^{-1}, 2^{2j+2}) \big\|f\big\|_{L^2} |E| \\
&\quad \le 8C_0C_3 \sum_j \min(2^{-j}|E|^{-\frac{1}{2}}, 2^j|E|^{\frac{1}{2}}) \big\|f\big\|_{L^2} |E|^{\frac{1}{2}} \\
&\quad \le C|E|^{\frac{1}{2}} \big\|f\big\|_{L^2}.
\end{aligned}$$

This proves estimate (11.1.29).

It remains to construct a sequence of disjoint sets $\mathbf{P}_j$ satisfying properties (1)–(5). The selection of these sets is based on decreasing induction. We start the induction at $j = n_0$ by setting $\mathbf{P}_{n_0} = \emptyset$. Then (1), (2), and (5) are clearly satisfied, while

$$\begin{aligned}
\mathcal{E}(f;\mathbf{P}\setminus\mathbf{P}_{n_0}) &= \mathcal{E}(f;\mathbf{P}) \le 2^{n_0}, \\
\mathcal{M}(E;\mathbf{P}\setminus\mathbf{P}_{n_0}) &= \mathcal{M}(E;\mathbf{P}) \le 2^{2n_0};
\end{aligned}$$

hence (3) and (4) are also satisfied for $\mathbf{P}_{n_0}$.

Suppose we have selected pairwise disjoint sets $\mathbf{P}_{n_0}, \mathbf{P}_{n_0-1}, \dots, \mathbf{P}_n$ for some $n < n_0$ such that (1)–(5) are satisfied for all $j \in \{n_0, n_0-1, \dots, n\}$. We construct a set of tiles $\mathbf{P}_{n-1}$ disjoint from all $\mathbf{P}_j$ with $j \ge n$ such that (1)–(5) are satisfied for $j = n-1$.

We define first an auxiliary set $\mathbf{P}'_{n-1}$. If $\mathcal{M}\big(E;\mathbf{P}\setminus(\mathbf{P}_{n_0}\cup\dots\cup\mathbf{P}_n)\big) \le 2^{2(n-1)}$ set $\mathbf{P}'_{n-1} = \emptyset$. If $\mathcal{M}\big(E;\mathbf{P}\setminus(\mathbf{P}_{n_0}\cup\dots\cup\mathbf{P}_n)\big) > 2^{2(n-1)}$ apply Lemma 11.1.8 to find a subset $\mathbf{P}'_{n-1}$ of $\mathbf{P}\setminus(\mathbf{P}_{n_0}\cup\dots\cup\mathbf{P}_n)$ such that

$$\mathcal{M}\big(E;\mathbf{P}\setminus(\mathbf{P}_{n_0}\cup\dots\cup\mathbf{P}_n\cup\mathbf{P}'_{n-1})\big) \le \frac{1}{4}\mathcal{M}\big(E;\mathbf{P}\setminus(\mathbf{P}_{n_0}\cup\dots\cup\mathbf{P}_n)\big) \le \frac{2^{2n}}{4} = 2^{2(n-1)}$$

[by the induction hypothesis (4) with $j = n$] and $\mathbf{P}'_{n-1}$ is a union of trees $\mathbf{T}'_k$ satisfying

$$\sum_k |I_{\mathrm{top}(\mathbf{T}'_k)}| \le C_1\,\mathcal{M}\big(f;\mathbf{P}\setminus(\mathbf{P}_{n_0}\cup\dots\cup\mathbf{P}_n)\big)^{-1} \le C_1\, 2^{-2(n-1)}. \tag{11.1.34}$$

Likewise, if $\mathcal{E}\big(f;\mathbf{P}\setminus(\mathbf{P}_{n_0}\cup\dots\cup\mathbf{P}_n)\big) \le 2^{n-1}$ set $\mathbf{P}''_{n-1} = \emptyset$; otherwise, apply Lemma 11.1.9 to find a subset $\mathbf{P}''_{n-1}$ of $\mathbf{P}\setminus(\mathbf{P}_{n_0}\cup\dots\cup\mathbf{P}_n)$ such that

$$\mathcal{E}\big(f;\mathbf{P}\setminus(\mathbf{P}_{n_0}\cup\dots\cup\mathbf{P}_n\cup\mathbf{P}''_{n-1})\big) \le \frac{1}{2}\mathcal{E}\big(f;\mathbf{P}\setminus(\mathbf{P}_{n_0}\cup\dots\cup\mathbf{P}_n)\big) \le \frac{1}{2}2^n = 2^{n-1}$$

[by the induction hypothesis (3) with $j = n$] and $\mathbf{P}''_{n-1}$ is a union of trees $\mathbf{T}''_k$ satisfying

$$\sum_k |I_{\text{top}(\mathbf{T}''_k)}| \le C_2 \mathscr{E}\big(f;\mathbf{P}\setminus(\mathbf{P}_{n_0}\cup\cdots\cup\mathbf{P}_n)\big)^{-2} \le C_2 2^{-2(n-1)}. \tag{11.1.35}$$

Whether the sets $\mathbf{P}'_{n-1}$ and $\mathbf{P}''_{n-1}$ are empty or not, we note that

$$\mathscr{M}\big(E;\mathbf{P}\setminus(\mathbf{P}_{n_0}\cup\cdots\cup\mathbf{P}_n\cup\mathbf{P}'_{n-1})\big) \le 2^{2(n-1)}, \tag{11.1.36}$$

$$\mathscr{E}\big(f;\mathbf{P}\setminus(\mathbf{P}_{n_0}\cup\cdots\cup\mathbf{P}_n\cup\mathbf{P}''_{n-1})\big) \le 2^{n-1}. \tag{11.1.37}$$

We set $\mathbf{P}_{n-1} = \mathbf{P}'_{n-1}\cup\mathbf{P}''_{n-1}$, and we verify properties (1)–(5) for $j = n-1$. Since $\mathbf{P}_{n-1}$ is contained in $\mathbf{P}\setminus(\mathbf{P}_{n_0}\cup\cdots\cup\mathbf{P}_n)$ we have

$$\mathscr{E}(f;\mathbf{P}_{n-1}) \le \mathscr{E}(f;\mathbf{P}\setminus(\mathbf{P}_{n_0}\cup\cdots\cup\mathbf{P}_n) \le 2^n = 2^{(n-1)+1},$$

where the last inequality is a consequence of the induction hypothesis (3) for $j = n$; thus (1) holds with $j = n-1$. Likewise,

$$\mathscr{M}(E;\mathbf{P}_{n-1}) \le \mathscr{M}(E;\mathbf{P}\setminus(\mathbf{P}_{n_0}\cup\cdots\cup\mathbf{P}_n) \le 2^{2n} = 2^{2(n-1)+2}$$

in view of the induction hypothesis (4) for $j = n$; thus (2) holds with $j = n-1$.

To prove (3) with $j = n-1$ notice that $\mathbf{P}\setminus(\mathbf{P}_{n_0}\cup\cdots\cup\mathbf{P}_n\cup\mathbf{P}_{n-1})$ is contained in $\mathbf{P}\setminus(\mathbf{P}_{n_0}\cup\cdots\cup\mathbf{P}_n\cup\mathbf{P}''_{n-1})$, and the latter has energy at most 2^{n-1} by (11.1.37). To prove (4) with $j = n-1$ note that $\mathbf{P}\setminus(\mathbf{P}_{n_0}\cup\cdots\cup\mathbf{P}_n\cup\mathbf{P}_{n-1})$ is contained in $\mathbf{P}\setminus(\mathbf{P}_{n_0}\cup\cdots\cup\mathbf{P}_n\cup\mathbf{P}'_{n-1})$ and the latter has mass at most $2^{2(n-1)}$ by (11.1.36). Finally, adding (11.1.34) and (11.1.35) yields (5) for $j = n-1$ with $C_0 = C_1 + C_2$.

Pick $j \in \mathbf{Z}$ with $0 < 2^{2j} < \min_{s\in\mathbf{P}} \mathscr{M}(E;\{s\})$. Then $\mathscr{M}\big(E;\mathbf{P}\setminus(\mathbf{P}_{n_0}\cup\cdots\cup\mathbf{P}_j)\big) = 0$, and since the only set of tiles with zero mass is the empty set, we conclude that (11.1.33) holds. It also follows that there exists an n_1 such that for all $n \le n_1$, $\mathbf{P}_n = \emptyset$. The construction of the $\mathbf{P}_j$'s is now complete.

11.1.5 Proof of the Mass Lemma 11.1.8

Proof. Given a finite set of tiles $\mathbf{P}$, we set $\mu = \mathscr{M}(E;\mathbf{P})$ to be the mass of $\mathbf{P}$. We define

$$\mathbf{P}' = \{s \in \mathbf{P} : \mathscr{M}(E;\{s\}) > \tfrac{1}{4}\mu\}$$

and we observe that $\mathscr{M}(E;\mathbf{P}\setminus\mathbf{P}') \le \frac{1}{4}\mu$. We now show that $\mathbf{P}'$ is a union of trees whose tops satisfy (11.1.30).

It follows from the definition of mass that for each $s \in \mathbf{P}'$, there is a tile $u(s) \in \mathbf{D}$ such that $u(s) > s$ and

$$\frac{1}{|E|}\int_{E\cap N^{-1}[\omega_{u(s)}]}\frac{|I_{u(s)}|^{-1}\,dx}{(1+\frac{|x-c(I_{u(s)})|}{|I_{u(s)}|})^{10}}>\frac{\mu}{4}. \tag{11.1.38}$$

Let $\mathbf{U}=\{u(s):\ s\in\mathbf{P}'\}$. Also, let $\mathbf{U}_{\max}$ be the subset of $\mathbf{U}$ containing all maximal elements of $\mathbf{U}$ under the partial order of tiles $<$. Likewise define $\mathbf{P}'_{\max}$ as the set of all maximal elements in $\mathbf{P}'$. Tiles in $\mathbf{P}'$ can be grouped in trees $\mathbf{T}_j=\{s\in\mathbf{P}':\ s<t_j\}$ with tops $t_j\in\mathbf{P}'_{\max}$. Observe that if $t_j<u$ and $t_{j'}<u$ for some $u\in\mathbf{U}_{\max}$, then ω_{t_j} and $\omega_{t_{j'}}$ intersect, and since t_j and $t_{j'}$ are disjoint sets, it follows that I_{t_j} and $I_{t_{j'}}$ are disjoint subsets of I_u. Consequently, we have

$$\sum_j|I_{t_j}|=\sum_{u\in\mathbf{U}_{\max}}\sum_{j:\,t_j<u}|I_{t_j}|\le\sum_{u\in\mathbf{U}_{\max}}|I_u|.$$

Therefore, estimate (11.1.30) will be a consequence of

$$\sum_{u\in\mathbf{U}_{\max}}|I_u|\le C\mu^{-1} \tag{11.1.39}$$

for some constant C. For $u\in\mathbf{U}_{\max}$ we rewrite (11.1.38) as

$$\frac{1}{|E|}\sum_{k=0}^{\infty}\int_{E\cap N^{-1}[\omega_u]\cap\left(2^kI_u\setminus 2^{k-1}I_u\right)}\frac{|I_u|^{-1}\,dx}{\left(1+\frac{|x-c(I_u)|}{|I_u|}\right)^{10}}>\frac{\mu}{8}\sum_{k=0}^{\infty}2^{-k}$$

with the interpretation that $2^{-1}I_u=\emptyset$. It follows that for all u in $\mathbf{U}_{\max}$ there exists an integer $k\ge 0$ such that

$$|E|\,\frac{\mu}{8}\,|I_u|\,2^{-k}<\int_{E\cap N^{-1}[\omega_u]\cap\left(2^kI_u\setminus 2^{k-1}I_u\right)}\frac{dx}{(1+\frac{|x-c(I_u)|}{|I_u|})^{10}}\le\frac{|E\cap N^{-1}[\omega_u]\cap 2^kI_u|}{(\frac{4}{5})^{10}(1+2^{k-2})^{10}}.$$

We therefore conclude that

$$\mathbf{U}_{\max}=\bigcup_{k=0}^{\infty}\mathbf{U}_k,$$

where

$$\mathbf{U}_k=\{u\in\mathbf{U}_{\max}:\ |I_u|\le 8\cdot 5^{10}2^{-9k}\mu^{-1}|E|^{-1}|E\cap N^{-1}[\omega_u]\cap 2^kI_u|\}.$$

The required estimate (11.1.39) will be a consequence of the sequence of estimates

$$\sum_{u\in\mathbf{U}_k}|I_u|\le C2^{-8k}\mu^{-1},\qquad k\ge 0. \tag{11.1.40}$$

We now fix a $k\ge 0$ and we concentrate on (11.1.40). Select an element $v_0\in\mathbf{U}_k$ such that $|I_{v_0}|$ is the largest possible among elements of $\mathbf{U}_k$. Then select an element $v_1\in\mathbf{U}_k\setminus\{v_0\}$ such that the enlarged rectangle $(2^kI_{v_1})\times\omega_{v_1}$ is disjoint from the enlarged rectangle $(2^k I_{v_0}) \times \omega_{v_0}$ and $|I_{v_1}|$ is the largest possible. Continue this process by induction. At the jth step select an element of $\mathbf{U}_k \setminus \{v_0, \dots, v_{j-1}\}$ such that the enlarged rectangle $(2^k I_{v_j}) \times \omega_{v_j}$ is disjoint from all the enlarged rectangles of the previously selected tiles and the length $|I_{v_j}|$ is the largest possible. This process will terminate after a finite number of steps. We denote by $\mathbf{V}_k$ the set of all selected tiles in $\mathbf{U}_k$.

We make a few observations. Recall that all elements of $\mathbf{U}_k$ are maximal rectangles in $\mathbf{U}$ and therefore disjoint. For any $u \in \mathbf{U}_k$ there exists a selected $v \in \mathbf{V}_k$ with $|I_u| \le |I_v|$ such that the enlarged rectangles corresponding to u and v intersect. Let us associate this u to the selected v. Observe that if u and u' are associated with the same selected v, they are disjoint, and since both ω_u and $\omega_{u'}$ contain ω_v, the intervals I_u and $I_{u'}$ must be disjoint. Thus, tiles $u \in \mathbf{U}_k$ associated with a fixed $v \in \mathbf{V}_k$ have disjoint I_u's and satisfy

$$I_u \subseteq 2^{k+2} I_v .$$

Consequently,

$$\sum_{\substack{u \in \mathbf{U}_k \\ u \text{ associated with } v}} |I_u| \le |2^{k+2} I_v| = 2^{k+2} |I_v| .$$

Putting these observations together, we obtain

$$\begin{aligned} \sum_{u \in \mathbf{U}_k} |I_u| &\le \sum_{v \in \mathbf{V}_k} \sum_{\substack{u \in \mathbf{U}_k \\ u \text{ associated with } v}} |I_u| \\ &\le 2^{k+2} \sum_{v \in \mathbf{V}_k} |I_v| \\ &\le 2^{k+5} 5^{10} \mu^{-1} |E|^{-1} 2^{-9k} \sum_{v \in \mathbf{V}_k} |E \cap N^{-1}[\omega_v] \cap 2^k I_v| \\ &\le 32 \cdot 5^{10} \mu^{-1} 2^{-8k} , \end{aligned}$$

since the enlarged rectangles $2^k I_v \times \omega_v$ of the selected tiles v are disjoint and therefore so are the subsets $E \cap N^{-1}[\omega_v] \cap 2^k I_v$ of E. This concludes the proof of estimate (11.1.40) and therefore of Lemma 11.1.8. □

11.1.6 Proof of Energy Lemma 11.1.9

Proof. We work with a finite set of tiles $\mathbf{P}$. For a 2-tree $\mathbf{T}'$, let us denote by

$$\Delta(f; \mathbf{T}') = \frac{1}{\|f\|_{L^2}} \left\{ \frac{1}{|I_{\mathrm{top}(\mathbf{T}')}|} \sum_{s \in \mathbf{T}'} |\langle f \,|\, \varphi_s \rangle|^2 \right\}^{\frac{1}{2}}$$

the quantity associated with $\mathbf{T}'$ appearing in the definition of the energy. Consider the set of all 2-trees $\mathbf{T}'$ contained in $\mathbf{P}$ that satisfy

$$\Delta(f;\mathbf{T}') \geq \frac{1}{2}\mathscr{E}(f;\mathbf{P}) \tag{11.1.41}$$

and among them select a 2-tree $\mathbf{T}_1'$ with $c(\omega_{\mathrm{top}(\mathbf{T}_1')})$ as small as possible. We let $\mathbf{T}_1$ be the set of $s \in \mathbf{P}$ satisfying $s < \mathrm{top}(\mathbf{T}_1')$. Then $\mathbf{T}_1$ is the largest tree in $\mathbf{P}$ whose top is $\mathrm{top}(\mathbf{T}_1')$. We now repeat this procedure with the set $\mathbf{P}\setminus\mathbf{T}_1$. Among all 2-trees contained in $\mathbf{P}\setminus\mathbf{T}_1$ that satisfy (11.1.41) we pick a 2-tree $\mathbf{T}_2'$ with $c(\omega_{\mathrm{top}(\mathbf{T}_2')})$ as small as possible. Then we let $\mathbf{T}_2$ be the $s \in \mathbf{P}\setminus\mathbf{T}_1$ satisfying $s < \mathrm{top}(\mathbf{T}_2')$. Then $\mathbf{T}_2$ is the largest tree in $\mathbf{P}\setminus\mathbf{T}_1$ whose top is $\mathrm{top}(\mathbf{T}_2')$. We continue this procedure by induction until there is no 2-tree left in $\mathbf{P}$ that satisfies (11.1.41). We have therefore constructed a finite sequence of pairwise disjoint 2-trees $\mathbf{T}_1', \mathbf{T}_2', \mathbf{T}_3', \dots, \mathbf{T}_q'$, and a finite sequence of pairwise disjoint trees $\mathbf{T}_1, \mathbf{T}_2, \mathbf{T}_3, \dots, \mathbf{T}_q$, such that $\mathbf{T}_j' \subseteq \mathbf{T}_j$, $\mathrm{top}(\mathbf{T}_j) = \mathrm{top}(\mathbf{T}_j')$, and the $\mathbf{T}_j'$ satisfy (11.1.41). We now let

$$\mathbf{P}'' = \bigcup_j \mathbf{T}_j,$$

and observe that this selection of trees ensures that

$$\mathscr{E}(f;\mathbf{P}\setminus\mathbf{P}'') \leq \frac{1}{2}\mathscr{E}(f;\mathbf{P}).$$

It remains to prove (11.1.31). Using (11.1.41), we obtain that

$$\begin{aligned}\frac{1}{4}\mathscr{E}(f;\mathbf{P})^2\sum_j |I_{\mathrm{top}(\mathbf{T}_j)}| &\leq \frac{1}{\|f\|_{L^2}^2}\sum_j\sum_{s\in\mathbf{T}_j'}|\langle f\,|\,\varphi_s\rangle|^2 \\ &= \frac{1}{\|f\|_{L^2}^2}\sum_j\sum_{s\in\mathbf{T}_j'}\langle f\,|\,\varphi_s\rangle\overline{\langle f\,|\,\varphi_s\rangle} \\ &= \frac{1}{\|f\|_{L^2}^2}\Big\langle f\,\Big|\,\sum_j\sum_{s\in\mathbf{T}_j'}\langle f\,|\,\varphi_s\rangle\varphi_s\Big\rangle \\ &\leq \frac{1}{\|f\|_{L^2}}\Big\|\sum_j\sum_{s\in\mathbf{T}_j'}\langle\varphi_s\,|\,f\rangle\varphi_s\Big\|_{L^2},\end{aligned} \tag{11.1.42}$$

and we use this estimate to obtain (11.1.31). We set $\mathbf{U} = \bigcup_j \mathbf{T}_j'$. We shall prove that

$$\frac{1}{\|f\|_{L^2}}\Big\|\sum_{s\in\mathbf{U}}\langle\varphi_s\,|\,f\rangle\varphi_s\Big\|_{L^2} \leq C\Big(\mathscr{E}(f;\mathbf{P})^2\sum_j|I_{\mathrm{top}(\mathbf{T}_j)}|\Big)^{\frac{1}{2}}. \tag{11.1.43}$$

Once this estimate is established, then (11.1.42) combined with (11.1.43) yields (11.1.31). (All involved quantities are finite, since $\mathbf{P}$ is a finite set of tiles.)

We estimate the square of the left-hand side in (11.1.43) by

$$\sum_{\substack{s,u\in\mathbf{U}\\ \omega_s=\omega_u}}\big|\langle\varphi_s\,|\,f\rangle\langle\varphi_u\,|\,f\rangle\langle\varphi_s\,|\,\varphi_u\rangle\big| + 2\sum_{\substack{s,u\in\mathbf{U}\\ \omega_s\subsetneq\omega_u}}\big|\langle\varphi_s\,|\,f\rangle\langle\varphi_u\,|\,f\rangle\langle\varphi_s\,|\,\varphi_u\rangle\big|, \tag{11.1.44}$$

since $\langle \varphi_s \mid \varphi_u \rangle = 0$ unless ω_s contains ω_u or vice versa. We now estimate the quantities $|\langle \varphi_s \mid f \rangle|$ and $|\langle \varphi_u \mid f \rangle|$ by the larger one and we use Exercise 11.1.4 to obtain the following bound for the first term in (11.1.44):

$$
\begin{aligned}
&\sum_{s\in\mathbf{U}} |\langle f \mid \varphi_s \rangle|^2 \sum_{\substack{u\in\mathbf{U} \\ \omega_u=\omega_s}} |\langle \varphi_s \mid \varphi_u \rangle| \\
&\quad \le \sum_{s\in\mathbf{U}} |\langle f \mid \varphi_s \rangle|^2 \sum_{\substack{u\in\mathbf{U} \\ \omega_u=\omega_s}} C' \int_{I_u} \frac{1}{|I_s|}\left(1+\frac{|x-c(I_s)|}{|I_s|}\right)^{-100} dx \\
&\quad \le C'' \sum_{s\in\mathbf{U}} |\langle f \mid \varphi_s \rangle|^2 \\
&\quad = C'' \sum_{j} \sum_{s\in\mathbf{T}'_j} |\langle f \mid \varphi_s \rangle|^2 \\
&\quad \le C'' \sum_{j} |I_{\mathrm{top}(\mathbf{T}_j)}| \, |I_{\mathrm{top}(\mathbf{T}_j)}|^{-1} \sum_{s\in\mathbf{T}'_j} |\langle f \mid \varphi_s \rangle|^2 \\
&\quad \le C'' \sum_{j} |I_{\mathrm{top}(\mathbf{T}_j)}| \, \mathscr{E}(f;\mathbf{P})^2 \|f\|_{L^2}^2 ,
\end{aligned}
\tag{11.1.45}
$$

where in the derivation of the second inequality we used the fact that for fixed $s \in \mathbf{U}$, the intervals I_u with $\omega_u = \omega_s$ are pairwise disjoint.

Our next goal is to obtain a similar estimate for the second term in (11.1.44). That is, we need to prove that

$$
\sum_{\substack{s,u\in\mathbf{U} \\ \omega_s \subsetneq \omega_u}} |\langle f \mid \varphi_s \rangle \langle f \mid \varphi_u \rangle \langle \varphi_s \mid \varphi_u \rangle| \le C \mathscr{E}(f;\mathbf{P})^2 \|f\|_{L^2}^2 \sum_j |I_{\mathrm{top}(\mathbf{T}_j)}| . \tag{11.1.46}
$$

Then the required estimate (11.1.43) follows by combining (11.1.45) and (11.1.46). To prove (11.1.46), we argue as follows:

$$
\begin{aligned}
&\sum_{\substack{s,u\in\mathbf{U} \\ \omega_s \subsetneq \omega_u}} |\langle f \mid \varphi_s \rangle \langle f \mid \varphi_u \rangle \langle \varphi_s \mid \varphi_u \rangle| \\
&\quad = \sum_j \sum_{s\in\mathbf{T}'_j} |\langle f \mid \varphi_s \rangle| \sum_{\substack{u\in\mathbf{U} \\ \omega_s \subsetneq \omega_u}} |\langle f \mid \varphi_u \rangle \langle \varphi_s \mid \varphi_u \rangle| \\
&\quad \le \sum_j |I_{\mathrm{top}(\mathbf{T}_j)}|^{\frac{1}{2}} \Delta(f;\mathbf{T}'_j) \|f\|_{L^2} \Bigg\{ \sum_{s\in\mathbf{T}'_j} \Bigg(\sum_{\substack{u\in\mathbf{U} \\ \omega_s \subsetneq \omega_u}} |\langle f \mid \varphi_u \rangle \langle \varphi_s \mid \varphi_u \rangle| \Bigg)^2 \Bigg\}^{\frac{1}{2}} \\
&\quad \le \mathscr{E}(f;\mathbf{P}) \|f\|_{L^2} \sum_j |I_{\mathrm{top}(\mathbf{T}_j)}|^{\frac{1}{2}} \Bigg\{ \sum_{s\in\mathbf{T}'_j} \Bigg(\sum_{\substack{u\in\mathbf{U} \\ \omega_s \subseteq \omega_{u(1)}}} |\langle f \mid \varphi_u \rangle \langle \varphi_s \mid \varphi_u \rangle| \Bigg)^2 \Bigg\}^{\frac{1}{2}} ,
\end{aligned}
$$

where we used the Cauchy–Schwarz inequality and the fact that if $\omega_s \subsetneq \omega_u$ and $\langle \varphi_s \,|\, \varphi_u \rangle \neq 0$, then $\omega_s \subseteq \omega_{u(1)}$. The proof of (11.1.46) will be complete if we can show that the expression inside the curly brackets is at most a multiple of $\mathscr{E}(f;\mathbf{P})^2 \|f\|_{L^2}^2 |I_{\mathrm{top}(\mathbf{T}_j)}|$. Since any singleton $\{s\} \subseteq \mathbf{P}$ is a 2-tree, we have

$$\mathscr{E}(f;\{u\}) = \frac{1}{\|f\|_{L^2}} \left(\frac{|\langle f \,|\, \varphi_u \rangle|^2}{|I_u|} \right)^{\frac{1}{2}} = \frac{1}{\|f\|_{L^2}} \frac{|\langle f \,|\, \varphi_u \rangle|}{|I_u|^{\frac{1}{2}}} \leq \mathscr{E}(f;\mathbf{P});$$

hence

$$|\langle f \,|\, \varphi_u \rangle| \leq \|f\|_{L^2} |I_u|^{\frac{1}{2}} \mathscr{E}(f;\mathbf{P})$$

and it follows that

$$\sum_{s\in\mathbf{T}'_j} \Bigg[\sum_{\substack{u\in\mathbf{U} \\ \omega_s\subseteq\omega_{u(1)}}} |\langle f \,|\, \varphi_u \rangle \langle \varphi_s \,|\, \varphi_u \rangle| \Bigg]^2 \leq \mathscr{E}(f;\mathbf{P})^2 \|f\|_{L^2}^2 \sum_{s\in\mathbf{T}'_j} \Bigg[\sum_{\substack{u\in\mathbf{U} \\ \omega_s\subseteq\omega_{u(1)}}} |I_u|^{\frac{1}{2}} |\langle \varphi_s \,|\, \varphi_u \rangle| \Bigg]^2 .$$

Thus (11.1.46) will be proved if we can establish that

$$\sum_{s\in\mathbf{T}'_j} \Bigg(\sum_{\substack{u\in\mathbf{U} \\ \omega_s\subseteq\omega_{u(1)}}} |I_u|^{\frac{1}{2}} |\langle \varphi_s \,|\, \varphi_u \rangle| \Bigg)^2 \leq C |I_{\mathrm{top}(\mathbf{T}_j)}| . \tag{11.1.47}$$

We need the following crucial lemma.

Lemma 11.1.11. *Let $\mathbf{T}_j$, $\mathbf{T}'_j$ be as previously. Let $s \in \mathbf{T}'_j$ and $u \in \mathbf{T}'_k$. Then if $\omega_s \subseteq \omega_{u(1)}$, we have $I_u \cap I_{\mathrm{top}(\mathbf{T}_j)} = \emptyset$. Moreover, if $u \in \mathbf{T}'_k$ and $v \in \mathbf{T}'_l$ are different tiles and satisfy $\omega_s \subseteq \omega_{u(1)}$ and $\omega_s \subseteq \omega_{v(1)}$ for some fixed $s \in \mathbf{T}'_j$, then $I_u \cap I_v = \emptyset$.*

Proof. We observe that if $s \in \mathbf{T}'_j$, $u \in \mathbf{T}'_k$, and $\omega_s \subseteq \omega_{u(1)}$, then the 2-trees $\mathbf{T}'_j$ and $\mathbf{T}'_k$ have different tops and therefore they cannot be the same tree; thus $j \neq k$.

Next we observe that the center of $\omega_{\mathrm{top}(\mathbf{T}'_j)}$ is contained in ω_s, which is contained in $\omega_{u(1)}$. Therefore, the center of $\omega_{\mathrm{top}(\mathbf{T}'_j)}$ is contained in $\omega_{u(1)}$, and therefore it must be smaller than the center of $\omega_{\mathrm{top}(\mathbf{T}'_k)}$, since $\mathbf{T}'_k$ is a 2-tree. This means that the 2-tree $\mathbf{T}'_j$ was selected before $\mathbf{T}'_k$, that is, we must have $j < k$. If I_u had a nonempty intersection with $I_{\mathrm{top}(\mathbf{T}_j)} = I_{\mathrm{top}(\mathbf{T}'_j)}$, then since

$$|I_{\mathrm{top}(\mathbf{T}'_j)}| = \frac{1}{|\omega_{\mathrm{top}(\mathbf{T}'_j)}|} \geq \frac{1}{|\omega_s|} \geq \frac{1}{|\omega_{u(1)}|} = \frac{2}{|\omega_u|} = 2|I_u| ,$$

I_u would have to be contained in $I_{\mathrm{top}(\mathbf{T}'_j)}$. Since also $\omega_{\mathrm{top}(\mathbf{T}'_j)} \subseteq \omega_s \subseteq \omega_u$, it follows that $u < \mathrm{top}(\mathbf{T}'_j)$; thus u would belong to the tree $\mathbf{T}_j$ [which is the largest tree with top $\mathrm{top}(\mathbf{T}'_j)$], since this tree was selected first. But if u belonged to $\mathbf{T}_j$, then it could not belong to $\mathbf{T}'_k$, which is disjoint from $\mathbf{T}_j$; hence we get a contradiction. We conclude that $I_u \cap I_{\mathrm{top}(\mathbf{T}_j)} = \emptyset$.

Next assume that $u \in \mathbf{T}'_k$, $v \in \mathbf{T}'_l$, $u \neq v$, and that $\omega_s \subseteq \omega_{u(1)} \cap \omega_{v(1)}$ for some fixed $s \in \mathbf{T}'_j$. Since the left halves of two dyadic intervals ω_u and ω_v intersect, three things can happen: (a) $\omega_u \subseteq \omega_{v(1)}$, in which case I_v is disjoint from $I_{\mathrm{top}(\mathbf{T}'_k)}$ and thus from I_u; (b) $\omega_v \subseteq \omega_{u(1)}$, in which case I_u is disjoint from $I_{\mathrm{top}(\mathbf{T}'_l)}$ and thus from I_v; and (c) $\omega_u = \omega_v$, in which case $|I_u| = |I_v|$, and thus I_u and I_v are either disjoint or they coincide. Since $u \neq v$, it follows that I_u and I_v cannot coincide; thus $I_u \cap I_v = \emptyset$. This finishes the proof of the lemma. □

We now return to (11.1.47). In view of Lemma 11.1.11, different $u \in \mathbf{U}$ that appear in the interior sum in (11.1.47) have disjoint intervals I_u, and all of these are contained in $(I_{\mathrm{top}(\mathbf{T}_j)})^c$. Set $t_j = \mathrm{top}(\mathbf{T}_j)$. Using Exercise 11.1.4, we obtain

$$
\begin{aligned}
&\sum_{s\in\mathbf{T}'_j}\Bigg(\sum_{\substack{u\in\mathbf{U}\\ \omega_s\subseteq\omega_{u(1)}}}|I_u|^{\frac12}\big|\langle\varphi_s\,|\,\varphi_u\rangle\big|\Bigg)^2\\
&\qquad\le C\sum_{s\in\mathbf{T}'_j}\Bigg(\sum_{\substack{u\in\mathbf{U}\\ \omega_s\subseteq\omega_{u(1)}}}|I_u|^{\frac12}\left(\frac{|I_s|}{|I_u|}\right)^{\frac12}\int_{I_u}\frac{|I_s|^{-1}\,dx}{\big(1+\frac{|x-c(I_s)|}{|I_s|}\big)^{20}}\Bigg)^2\\
&\qquad\le C\sum_{s\in\mathbf{T}'_j}|I_s|\Bigg(\sum_{\substack{u\in\mathbf{U}\\ \omega_s\subseteq\omega_{u(1)}}}\int_{I_u}\frac{|I_s|^{-1}\,dx}{\big(1+\frac{|x-c(I_s)|}{|I_s|}\big)^{20}}\Bigg)^2\\
&\qquad\le C\sum_{s\in\mathbf{T}'_j}|I_s|\Bigg(\int_{(I_{t_j})^c}\frac{|I_s|^{-1}\,dx}{\big(1+\frac{|x-c(I_s)|}{|I_s|}\big)^{20}}\Bigg)^2\\
&\qquad\le C\sum_{s\in\mathbf{T}'_j}|I_s|\int_{(I_{t_j})^c}\frac{|I_s|^{-1}\,dx}{\big(1+\frac{|x-c(I_s)|}{|I_s|}\big)^{20}},
\end{aligned}
$$

since $\int_{\mathbf{R}}(1+|x|)^{-20}\,dx \le 1$. For each scale $k \ge 0$ the sets I_s, $s \in \mathbf{T}'_j$, with $|I_s| = 2^{-k}|I_{t_j}|$ are pairwise disjoint and contained in I_t; therefore, we have

$$
\begin{aligned}
\sum_{s\in\mathbf{T}'_j}|I_s|\int_{(I_{t_j})^c}\frac{|I_s|^{-1}\,dx}{\big(1+\frac{|x-c(I_s)|}{|I_s|}\big)^{20}}
&\le\sum_{k=0}^{\infty}\frac{2^k}{|I_{t_j}|}\sum_{\substack{s\in\mathbf{T}'_j\\ |I_s|=2^{-k}|I_{t_j}|}}|I_s|\int_{(I_{t_j})^c}\frac{dx}{\big(1+\frac{|x-c(I_s)|}{|I_s|}\big)^{20}}\\
&\le C\sum_{k=0}^{\infty}\frac{2^k}{|I_{t_j}|}\sum_{\substack{s\in\mathbf{T}'_j\\ |I_s|=2^{-k}|I_{t_j}|}}\int_{I_s}\int_{(I_{t_j})^c}\frac{dx}{\big(1+\frac{|x-y|}{|I_s|}\big)^{20}}\,dy\\
&\le C\sum_{k=0}^{\infty}2^k|I_{t_j}|^{-1}\int_{I_{t_j}}\int_{(I_{t_j})^c}\frac{1}{\big(1+\frac{|x-y|}{2^{-k}|I_{t_j}|}\big)^{20}}\,dx\,dy\\
&\le C'\sum_{k=0}^{\infty}2^k|I_{t_j}|^{-1}(2^{-k}|I_{t_j}|)^2\\
&=C''|I_{t_j}|,
\end{aligned}
$$

in view of Exercise 11.1.5. This completes the proof of (11.1.47) and thus of Lemma 11.1.9. □

11.1.7 Proof of the Basic Estimate Lemma 11.1.10

Proof. In the proof of the required estimate we may assume that $\|f\|_{L^2} = 1$, for we can always replace f by $f/\|f\|_{L^2}$. Throughout this subsection we fix a square-integrable function with L^2 norm 1, a tree $\mathbf{T}$ contained in $\mathbf{P}$, a measurable function $N:\ \mathbf{R} \to \mathbf{R}^+$, and a measurable set E with finite measure.

Let $\mathscr{J}'$ be the set of all dyadic intervals J such that $3J$ does not contain any I_s with $s \in \mathbf{T}$. It is not hard to see that any point in $\mathbf{R}$ belongs to a set in $\mathscr{J}'$. Let $\mathscr{J}$ be the set of all maximal (under inclusion) elements of $\mathscr{J}'$. Then $\mathscr{J}$ consists of disjoint sets that cover $\mathbf{R}$; thus it forms a partition of $\mathbf{R}$. This partition of $\mathbf{R}$ is shown in Figure 11.3 when the tree consists of two tiles.

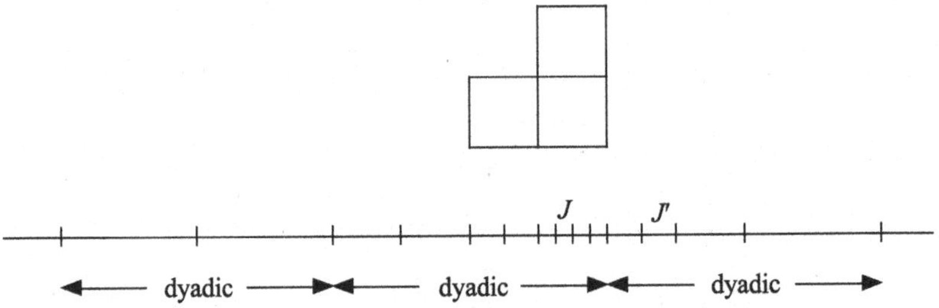

Fig. 11.3 A tree of two tiles and the partition $\mathscr{J}$ of $\mathbf{R}$ corresponding to it. The intervals J and J' are members of the partition $\mathscr{J}$.

For each $s \in \mathbf{T}$ pick an $\varepsilon_s \in \mathbf{C}$ with $|\varepsilon_s| = 1$ such that

$$\left|\left\langle f \,|\, \varphi_s\right\rangle\left\langle \chi_{E\cap N^{-1}[\omega_{s(2)}]} \,|\, \varphi_s\right\rangle\right| = \varepsilon_s \left\langle f \,|\, \varphi_s\right\rangle\left\langle \varphi_s \,|\, \chi_{E\cap N^{-1}[\omega_{s(2)}]}\right\rangle.$$

We can now write the left-hand side of (11.1.32) as

$$\begin{aligned}\sum_{s\in\mathbf{T}} \varepsilon_s \left\langle f \,|\, \varphi_s\right\rangle\left\langle \varphi_s \,|\, \chi_{E\cap N^{-1}[\omega_{s(2)}]}\right\rangle &\le \Big\| \sum_{s\in\mathbf{T}} \varepsilon_s \left\langle f \,|\, \varphi_s\right\rangle \chi_{E\cap N^{-1}[\omega_{s(2)}]}\varphi_s \Big\|_{L^1(\mathbf{R})} \\ &= \sum_{J\in\mathscr{J}} \Big\| \sum_{s\in\mathbf{T}} \varepsilon_s \left\langle f \,|\, \varphi_s\right\rangle \chi_{E\cap N^{-1}[\omega_{s(2)}]}\varphi_s \Big\|_{L^1(J)} \\ &\le \Sigma_1 + \Sigma_2 ,\end{aligned}$$

where

$$\Sigma_1 = \sum_{J\in\mathscr{J}} \Big\| \sum_{\substack{s\in\mathbf{T}\\ |I_s|\le 2|J|}} \varepsilon_s\langle f\,|\,\varphi_s\rangle \chi_{E\cap N^{-1}[\omega_{s(2)}]}\varphi_s\Big\|_{L^1(J)}, \tag{11.1.48}$$

$$\Sigma_2 = \sum_{J\in\mathscr{J}} \Big\| \sum_{\substack{s\in\mathbf{T}\\ |I_s|> 2|J|}} \varepsilon_s\langle f\,|\,\varphi_s\rangle \chi_{E\cap N^{-1}[\omega_{s(2)}]}\varphi_s\Big\|_{L^1(J)}. \tag{11.1.49}$$

We start with Σ_1. Observe that for every $s\in\mathbf{T}$, the singleton $\{s\}$ is a 2-tree contained in $\mathbf{T}$ and we therefore have the estimate

$$\big|\langle f\,|\,\varphi_s\rangle\big| \le |I_s|^{\frac{1}{2}}\mathscr{E}(f;\mathbf{T}). \tag{11.1.50}$$

Using this, we obtain

$$\begin{aligned}
\Sigma_1 &\le \sum_{J\in\mathscr{J}} \sum_{\substack{s\in\mathbf{T}\\ |I_s|\le 2|J|}} \mathscr{E}(f;\mathbf{T}) \int_{J\cap E\cap N^{-1}[\omega_{s(2)}]} |I_s|^{\frac{1}{2}}|\varphi_s(x)|\,dx \\
&\le C \sum_{J\in\mathscr{J}} \sum_{\substack{s\in\mathbf{T}\\ |I_s|\le 2|J|}} \mathscr{E}(f;\mathbf{T})|I_s| \int_{J\cap E\cap N^{-1}[\omega_{s(2)}]} \frac{|I_s|^{-1}}{\big(1+\frac{|x-c(I_s)|}{|I_s|}\big)^{20}}\,dx \\
&\le C \sum_{J\in\mathscr{J}} \sum_{\substack{s\in\mathbf{T}\\ |I_s|\le 2|J|}} \mathscr{E}(f;\mathbf{T})\,|E|\mathscr{M}(E;\mathbf{T})|I_s| \sup_{x\in J} \frac{1}{\big(1+\frac{|x-c(I_s)|}{|I_s|}\big)^{10}} \\
&\le C\mathscr{E}(f;\mathbf{T})\,|E|\mathscr{M}(E;\mathbf{T}) \sum_{J\in\mathscr{J}} \sum_{k=-\infty}^{\log_2 2|J|} 2^k \sum_{\substack{s\in\mathbf{T}\\ |I_s|=2^k}} \frac{1}{\big(1+\frac{\mathrm{dist}\,(J,I_s)}{2^k}\big)^5}\frac{1}{\big(1+\frac{\mathrm{dist}\,(J,I_s)}{2^k}\big)^5}.
\end{aligned}$$

But note that all I_s with $s\in\mathbf{T}$ and $|I_s|=2^k$ are pairwise disjoint and contained in $I_{\mathrm{top}(\mathbf{T})}$. Therefore, $2^{-k}\mathrm{dist}\,(J,I_s)\ge |I_{\mathrm{top}(\mathbf{T})}|^{-1}\mathrm{dist}\,(J,I_{\mathrm{top}(\mathbf{T})})$, and we have the estimate

$$\left(1+\frac{\mathrm{dist}\,(J,I_s)}{2^k}\right)^{-5} \le \left(1+\frac{\mathrm{dist}\,(J,I_{\mathrm{top}(\mathbf{T})})}{|I_{\mathrm{top}(\mathbf{T})}|}\right)^{-5}.$$

Moreover, the sum

$$\sum_{\substack{s\in\mathbf{T}\\ |I_s|=2^k}} \frac{1}{\big(1+\frac{\mathrm{dist}\,(J,I_s)}{2^k}\big)^5} \tag{11.1.51}$$

is controlled by a finite constant, since for every nonnegative integer m there exist at most two tiles $s\in\mathbf{T}$ with $|I_s|=2^k$ such that I_s are not contained in $3J$ and $m2^k\le \mathrm{dist}\,(J,I_s)<(m+1)2^k$. Therefore, we obtain

$$
\begin{aligned}
\Sigma_1 &\le C\mathscr{E}(f;\mathbf{T})\,|E|\,\mathscr{M}(E;\mathbf{T}) \sum_{J\in\mathscr{J}} \sum_{k=-\infty}^{\log_2 2|J|} \frac{2^k}{\big(1+\frac{\operatorname{dist}(J,I_{\mathrm{top}(\mathbf{T})})}{|I_{\mathrm{top}(\mathbf{T})}|}\big)^5} \\
&\le C\mathscr{E}(f;\mathbf{T})\,|E|\,\mathscr{M}(E;\mathbf{T}) \sum_{J\in\mathscr{J}} \frac{|J|}{\big(1+\frac{\operatorname{dist}(J,I_{\mathrm{top}(\mathbf{T})})}{|I_{\mathrm{top}(\mathbf{T})}|}\big)^5} \qquad (11.1.52)\\
&\le C\mathscr{E}(f;\mathbf{T})\,|E|\,\mathscr{M}(E;\mathbf{T}) \sum_{J\in\mathscr{J}} \int_J \frac{1}{\big(1+\frac{|x-c(I_{\mathrm{top}(\mathbf{T})})|}{|I_{\mathrm{top}(\mathbf{T})}|}\big)^5}\,dx \\
&\le C\,|I_{\mathrm{top}(\mathbf{T})}|\,\mathscr{E}(f;\mathbf{T})\,|E|\,\mathscr{M}(E;\mathbf{T}),
\end{aligned}
$$

since $\mathscr{J}$ forms a partition of $\mathbf{R}$. We need to justify, however, the penultimate inequality in (11.1.52). Since J and $I_{\mathrm{top}(\mathbf{T})}$ are dyadic intervals, there are only two possibilities: (a) $J\cap I_{\mathrm{top}(\mathbf{T})}=\emptyset$ and (b) $J\subseteq I_{\mathrm{top}(\mathbf{T})}$. [The third possibility $I_{\mathrm{top}(\mathbf{T})}\subsetneq J$ is excluded, since $3J$ does not contain $I_{\mathrm{top}(\mathbf{T})}$.] In case (a) we have $|J|\le \operatorname{dist}(J,I_{\mathrm{top}(\mathbf{T})})$, since $3J$ does not contain $I_{\mathrm{top}(\mathbf{T})}$. In case (b) we have $|J|\le |I_{\mathrm{top}(\mathbf{T})}|$. Thus in both cases we have $|J|\le \operatorname{dist}(J,I_{\mathrm{top}(\mathbf{T})})+|I_{\mathrm{top}(\mathbf{T})}|$. Consequently, for any $x\in J$ one has

$$
\begin{aligned}
|x-c(I_{\mathrm{top}(\mathbf{T})})| &\le |J|+\operatorname{dist}(J,I_{\mathrm{top}(\mathbf{T})})+\tfrac{1}{2}|I_{\mathrm{top}(\mathbf{T})}| \\
&\le 2\operatorname{dist}(J,I_{\mathrm{top}(\mathbf{T})})+\tfrac{3}{2}|I_{\mathrm{top}(\mathbf{T})}|\,.
\end{aligned}
$$

Therefore, it follows that

$$
\int_J \frac{dx}{\big(1+\frac{|x-c(I_{\mathrm{top}(\mathbf{T})})|}{|I_{\mathrm{top}(\mathbf{T})}|}\big)^5} \ge \frac{|J|}{\big(\frac{5}{2}+\frac{2\operatorname{dist}(J,I_{\mathrm{top}(\mathbf{T})})}{|I_{\mathrm{top}(\mathbf{T})}|}\big)^5} \ge \frac{\big(\frac{2}{5}\big)^5|J|}{\big(1+\frac{\operatorname{dist}(J,I_{\mathrm{top}(\mathbf{T})})}{|I_{\mathrm{top}(\mathbf{T})}|}\big)^5}\,.
$$

In case (b) we have $J\subseteq I_{\mathrm{top}(\mathbf{T})}$, and therefore any point x in J lies in $I_{\mathrm{top}(\mathbf{T})}$; thus $|x-c(I_{\mathrm{top}(\mathbf{T})})|\le \frac{1}{2}|I_{\mathrm{top}(\mathbf{T})}|$. We conclude that

$$
\int_J \frac{dx}{\big(1+\frac{|x-c(I_{\mathrm{top}(\mathbf{T})})|}{|I_{\mathrm{top}(\mathbf{T})}|}\big)^5} \ge \frac{|J|}{(3/2)^5} = \Big(\frac{2}{3}\Big)^5 \frac{|J|}{\big(1+\frac{\operatorname{dist}(J,I_{\mathrm{top}(\mathbf{T})})}{|I_{\mathrm{top}(\mathbf{T})}|}\big)^5}\,.
$$

These observations justify the second-to-last inequality in (11.1.52) and complete the required estimate for Σ_1.

We now turn attention to Σ_2. We may assume that for all J appearing in the sum in (11.1.49), the set of s in $\mathbf{T}$ with $2|J|<|I_s|$ is nonempty. Thus, if J appears in the sum in (11.1.49), we have $2|J|<|I_{\mathrm{top}(\mathbf{T})}|$, and it is easy to see that J is contained in $3I_{\mathrm{top}(\mathbf{T})}$. [The intervals J in $\mathscr{J}$ that are not contained in $3I_{\mathrm{top}(\mathbf{T})}$ have size larger than $|I_{\mathrm{top}(\mathbf{T})}|$.]

We let $\mathbf{T}_2$ be the 2-tree of all s in $\mathbf{T}$ such that $\omega_{\mathrm{top}(\mathbf{T})(2)}\subseteq\omega_{s(2)}$, and we also let $\mathbf{T}_1=\mathbf{T}\setminus\mathbf{T}_2$. Then $\mathbf{T}_1$ is a 1-tree minus its top. We set

$$
\begin{aligned}
F_{1J} &= \sum_{\substack{s\in\mathbf{T}_1\\ |I_s|>2|J|}} \varepsilon_s \langle f \mid \varphi_s\rangle \varphi_s \chi_{E\cap N^{-1}[\omega_{s(2)}]}\,,\\
F_{2J} &= \sum_{\substack{s\in\mathbf{T}_2\\ |I_s|>2|J|}} \varepsilon_s \langle f \mid \varphi_s\rangle \varphi_s \chi_{E\cap N^{-1}[\omega_{s(2)}]}\,.
\end{aligned}
$$

Clearly

$$
\Sigma_2 \le \sum_{J\in\mathscr{J}} \big\|F_{1J}\big\|_{L^1(J)} + \sum_{J\in\mathscr{J}} \big\|F_{2J}\big\|_{L^1(J)} = \Sigma_{21}+\Sigma_{22}\,,
$$

and we need to estimate both sums. We start by estimating F_{1J}. If the tiles s and s' that appear in the definition of F_{1J} have different scales, then the sets $\omega_{s(2)}$ and $\omega_{s'(2)}$ are disjoint and thus so are the sets $E\cap N^{-1}[\omega_{s(2)}]$ and $E\cap N^{-1}[\omega_{s'(2)}]$. Let us set

$$
G_J = J \cap \bigcup_{\substack{s\in\mathbf{T}\\ |I_s|>2|J|}} E\cap N^{-1}[\omega_{s(2)}]\,.
$$

Then F_{1J} is supported in the set G_J and we have

$$
\begin{aligned}
\big\|F_{1J}\big\|_{L^1(J)} &\le \big\|F_{1J}\big\|_{L^\infty(J)}|G_J|\\
&= \Big\| \sum_{k>\log_2 2|J|} \sum_{\substack{s\in\mathbf{T}_1\\ |I_s|=2^k}} \varepsilon_s \langle f \mid \varphi_s\rangle \varphi_s \chi_{E\cap N^{-1}[\omega_{s(2)}]}\Big\|_{L^\infty(J)}|G_J|\\
&\le \sup_{k>\log_2 2|J|} \Big\| \sum_{\substack{s\in\mathbf{T}_1\\ |I_s|=2^k}} \varepsilon_s \langle f \mid \varphi_s\rangle \varphi_s \chi_{E\cap N^{-1}[\omega_{s(2)}]}\Big\|_{L^\infty(J)}|G_J|\\
&\le \sup_{k>\log_2 2|J|}\sup_{x\in J} \sum_{\substack{s\in\mathbf{T}_1\\ |I_s|=2^k}} \mathscr{E}(f;\mathbf{T})2^{k/2}\frac{2^{-k/2}}{\big(1+\frac{|x-c(I_s)|}{2^k}\big)^{10}}|G_J|\\
&\le C\mathscr{E}(f;\mathbf{T})|G_J|\,,
\end{aligned}
$$

using (11.1.50) and the fact that all the I_s that appear in the sum are disjoint. We now claim that for all $J\in\mathscr{J}$ we have

$$
|G_J| \le C\,|E|\,\mathscr{M}(E;\mathbf{T})|J|\,. \tag{11.1.53}
$$

Once (11.1.53) is established, summing over all the intervals J that appear in the definition of F_{1J} and keeping in mind that all of these intervals are pairwise disjoint and contained in $3I_{\mathrm{top}(\mathbf{T})}$, we obtain the desired estimate for Σ_{21}.

To prove (11.1.53), we consider the unique dyadic interval $\widetilde{J}$ of length $2|J|$ that contains J. Then by the maximality of $\mathscr{J}$, $3\widetilde{J}$ contains the time interval I_{s_J} of a tile s_J in $\mathbf{T}$. We consider the following two cases: (a) If I_{s_J} is either $(\widetilde{J}-|\widetilde{J}|)\cup\widetilde{J}$ or $\widetilde{J}\cup(\widetilde{J}+|\widetilde{J}|)$, we let $u_J=s_J$; in this case $|I_{u_J}|=2|\widetilde{J}|$. (This is the case for the interval J in Figure 11.3.) Otherwise, we have case (b), in which I_{s_J} is contained in one of the two dyadic intervals $\widetilde{J}-|\widetilde{J}|, \widetilde{J}+|\widetilde{J}|$. (This is the case for the interval J' in Figure 11.3.) Whichever of these two dyadic intervals contains I_{s_J} is also contained in $I_{\text{top}(\mathbf{T})}$, since it intersects it and has smaller length than it. In case (b) there exists a tile $u_J \in \mathbf{D}$ with $|I_{u_J}| = |\widetilde{J}|$ such that $I_{s_J} \subseteq I_{u_J} \subseteq I_{\text{top}(\mathbf{T})}$ and $\omega_{\text{top}(\mathbf{T})} \subseteq \omega_{u_J} \subseteq \omega_{s_J}$. In both cases we have a tile u_J satisfying $s_J < u_J < \text{top}(\mathbf{T})$ with $|\omega_{u_J}|$ being either $\frac{1}{4}|J|^{-1}$ or $\frac{1}{2}|J|^{-1}$.

Then for any $s \in \mathbf{T}$ with $|I_s| > 2|J|$ we have $|\omega_s| \le |\omega_{u_J}|$. But since both ω_s and ω_{u_J} contain $\omega_{\text{top}(\mathbf{T})}$, they must intersect, and thus $\omega_s \subseteq \omega_{u_J}$. We conclude that any $s \in \mathbf{T}$ with $|I_s| > 2|J|$ must satisfy $N^{-1}[\omega_s] \subseteq N^{-1}[\omega_{u_J}]$. It follows that

$$G_J \subseteq J \cap E \cap N^{-1}[\omega_{u_J}] \tag{11.1.54}$$

and therefore we have

$$\begin{aligned}
|E|\mathscr{M}(E;\mathbf{T}) &= \sup_{s\in\mathbf{T}} \sup_{\substack{u\in\mathbf{D}\\ s<u}} \int_{E\cap N^{-1}[\omega_u]} \frac{|I_u|^{-1}}{\left(1+\frac{|x-c(I_u)|}{|I_u|}\right)^{10}}\,dx \\
&\ge \int_{J\cap E\cap N^{-1}[\omega_{u_J}]} \frac{|I_{u_J}|^{-1}}{\left(1+\frac{|x-c(I_{u_J})|}{|I_{u_J}|}\right)^{10}}\,dx \\
&\ge c\,|I_{u_J}|^{-1}\,|J\cap E\cap N^{-1}[\omega_{u_J}]| \\
&\ge c\,|I_{u_J}|^{-1}\,|G_J|,
\end{aligned}$$

using (11.1.54) and the fact that for $x \in J$ we have $|x-c(I_{u_J})| \le 4|J| = 2|I_{u_J}|$. It follows that

$$|G_J| \le \frac{1}{c}\,|E|\mathscr{M}(E;\mathbf{T})|I_{u_J}| = \frac{2}{c}\,|E|\mathscr{M}(E;\mathbf{T})|J|,$$

and this is exactly (11.1.53), which we wanted to prove.

We now turn to the estimate for $\Sigma_{22} = \sum_{J\in\mathscr{J}} \big\|F_{2J}\big\|_{L^1(J)}$. All the intervals $\omega_{s(2)}$ with $s \in \mathbf{T}_2$ are nested, since $\mathbf{T}_2$ is a 2-tree. Therefore, for each $x \in J$ for which $F_{2J}(x)$ is nonzero, there exists a largest dyadic interval ω_{u_x} and a smallest dyadic interval ω_{v_x} (for some $u_x, v_x \in \mathbf{T}_2 \cap \{s:\ |I_s| \ge 4|J|\}$) such that for $s \in \mathbf{T}_2 \cap \{s:\ |I_s| \ge 4|J|\}$ we have $N(x) \in \omega_{s(2)}$ if and only if $\omega_{v_x} \subseteq \omega_s \subseteq \omega_{u_x}$. Then we have

$$\begin{aligned}
F_{2J}(x) &= \sum_{\substack{s\in\mathbf{T}_2\\ |I_s|\ge 4|J|}} \varepsilon_s \langle f\,|\,\varphi_s\rangle (\varphi_s \chi_{E\cap N^{-1}[\omega_{s(2)}]})(x) \\
&= \chi_E(x) \sum_{\substack{s\in\mathbf{T}_2\\ |\omega_{v_x}|\le|\omega_s|\le|\omega_{u_x}|}} \varepsilon_s \langle f\,|\,\varphi_s\rangle\, \varphi_s(x)\,.
\end{aligned}$$

Pick a Schwartz function ψ whose Fourier transform $\widehat{\psi}(t)$ is supported in $|t| \le 1+\frac{1}{100}$ and that is equal to 1 on $|t| \le 1$. We can easily check that for all $z \in \mathbf{R}$, if $|\omega_{v_x}| \le |\omega_s| \le |\omega_{u_x}|$, then

$$\left(\varphi_s * \left\{\frac{M^{c(\omega_{u_x})}D^{|\omega_{u_x}|^{-1}}(\psi)}{|\omega_{u_x}|^{-\frac{1}{2}}} - \frac{M^{c(\omega_{v_x(2)})}D^{|\omega_{v_x(2)}|^{-1}}(\psi)}{|\omega_{v_x(2)}|^{-\frac{1}{2}}}\right\}\right)(z) = \varphi_s(z) \tag{11.1.55}$$

by a simple examination of the Fourier transforms. Basically, the Fourier transform (in z) of the function inside the curly brackets is equal to

$$\widehat{\psi}\left(\tfrac{\xi - c(\omega_{u_x})}{|\omega_{u_x}|}\right) - \widehat{\psi}\left(\tfrac{\xi - c(\omega_{v_x(2)})}{|\omega_{v_x(2)}|}\right),$$

which is equal to 1 on the support of $\widehat{\varphi_s}$ for all s in $\mathbf{T}_2$ that satisfy $|\omega_{v_x}| \le |\omega_s| \le |\omega_{u_x}|$ but vanishes on $\omega_{v_x(2)}$. Taking $z = x$ in (11.1.55) yields

$$\begin{aligned} F_{2J}(x) &= \sum_{\substack{s\in\mathbf{T}_2 \\ |\omega_{v_x}|\le|\omega_s|\le|\omega_{u_x}|}} \varepsilon_s \langle f \mid \varphi_s\rangle \varphi_s(x)\chi_E(x) \\ &= \left[\sum_{s\in\mathbf{T}_2} \varepsilon_s \langle f \mid \varphi_s\rangle \varphi_s\right] * \left\{\frac{M^{c(\omega_{u_x})}D^{|\omega_{u_x}|^{-1}}(\psi)}{|\omega_{u_x}|^{-\frac{1}{2}}} - \frac{M^{c(\omega_{v_x(2)})}D^{|\omega_{v_x(2)}|^{-1}}(\psi)}{|\omega_{v_x(2)}|^{-\frac{1}{2}}}\right\}(x)\chi_E(x). \end{aligned}$$

Since all s that appear in the definition of F_{2J} satisfy $|\omega_s| \le (4|J|)^{-1}$, it follows that we have the estimate

$$\begin{aligned} |F_{2J}(x)| &\le 2\chi_E(x) \sup_{\delta>|\omega_{u_x}|^{-1}} \int_{\mathbf{R}} \Big|\sum_{s\in\mathbf{T}_2} \varepsilon_s\langle f\mid\varphi_s\rangle\varphi_s(z)\Big| \tfrac{1}{\delta}\big|\psi\big(\tfrac{x-z}{\delta}\big)\big|\,dz \\ &\le C \sup_{\delta>4|J|} \frac{1}{2\delta}\int_{x-\delta}^{x+\delta} \Big|\sum_{s\in\mathbf{T}_2} \varepsilon_s\langle f\mid\varphi_s\rangle\varphi_s(z)\Big|\,dz. \end{aligned} \tag{11.1.56}$$

(The last inequality follows from Exercise 2.1.14.) Observe that the maximal function in (11.1.56) satisfies the property

$$\sup_{x\in J}\sup_{\delta>4|J|} \frac{1}{2\delta}\int_{x-\delta}^{x+\delta} |h(t)|\,dt \le 2\inf_{x\in J}\sup_{\delta>4|J|} \frac{1}{2\delta}\int_{x-\delta}^{x+\delta} |h(t)|\,dt.$$

Using this property, we obtain

$$\begin{aligned} \Sigma_{22} &\le \sum_{J\in\mathscr{J}} \|F_{2J}\|_{L^1(J)} \le \sum_{J\in\mathscr{J}} \|F_{2J}\|_{L^\infty(J)}|G_J| \\ &\le C \sum_{\substack{J\in\mathscr{J} \\ J\subseteq 3I_{\mathrm{top}(\mathbf{T})}}} |E|\mathscr{M}(E;\mathbf{T})|J| \sup_{x\in J}\sup_{\delta>4|J|} \frac{1}{2\delta}\int_{x-\delta}^{x+\delta} \Big|\sum_{s\in\mathbf{T}_2} \varepsilon_s\langle f\mid\varphi_s\rangle\varphi_s(z)\Big|\,dz \\ &\le 2C|E|\mathscr{M}(E;\mathbf{T}) \sum_{\substack{J\in\mathscr{J} \\ J\subseteq 3I_{\mathrm{top}(\mathbf{T})}}} \int_J \sup_{\delta>4|J|} \frac{1}{2\delta}\int_{x-\delta}^{x+\delta} \Big|\sum_{s\in\mathbf{T}_2} \varepsilon_s\langle f\mid\varphi_s\rangle\varphi_s(z)\Big|\,dz\,dx \\ &\le C|E|\mathscr{M}(E;\mathbf{T}) \Big\| M\Big(\sum_{s\in\mathbf{T}_2} \varepsilon_s\langle f\mid\varphi_s\rangle\varphi_s\Big)\Big\|_{L^1(3I_{\mathrm{top}(\mathbf{T})})}, \end{aligned}$$

where M is the Hardy–Littlewood maximal operator. Using the Cauchy–Schwarz inequality and the boundedness of M on $L^2(\mathbf{R})$, we obtain the following estimate:

$$\Sigma_{22} \le C|E|\mathcal{M}(E;\mathbf{T})\,|I_{\mathrm{top}(\mathbf{T})}|^{\frac{1}{2}}\Big\|\sum_{s\in\mathbf{T}_2}\varepsilon_s\langle f\,|\,\varphi_s\rangle\varphi_s\Big\|_{L^2}.$$

Appealing to the result of Exercise 11.1.6(a), we deduce

$$\Big\|\sum_{s\in\mathbf{T}_2}\varepsilon_s\langle f\,|\,\varphi_s\rangle\varphi_s\Big\|_{L^2} \le C\Big(\sum_{s\in\mathbf{T}_2}|\varepsilon_s\langle f\,|\,\varphi_s\rangle|^2\Big)^{\frac{1}{2}} \le C'|I_{\mathrm{top}(\mathbf{T})}|^{\frac{1}{2}}\mathcal{E}(f;\mathbf{T}).$$

The first estimate was also shown in (11.1.43); the same argument applies here, and the presence of the ε_s's does not introduce any change. We conclude that

$$\Sigma_{22} \le C|E|\mathcal{M}(E;\mathbf{T})|I_{\mathrm{top}(\mathbf{T})}|\,\mathcal{E}(f;\mathbf{T}),$$

which is what we needed to prove. This completes the proof of Lemma 11.1.10. □

The proof of the theorem is now complete. □

Exercises

11.1.1. Show that for every f in the Schwartz class, $x,\xi\in\mathbf{R}$, and $\lambda\in[0,1]$, the function $(y,\eta)\mapsto B^m_{\xi,y,\eta,\lambda}(f)(x)$ is periodic in y with period $2^{m-\lambda}$ and periodic in η with period $2^{-m+\lambda}$.

11.1.2. Fix a function h in the Schwartz class, $\xi,y,\eta\in\mathbf{R}$, $s\in\mathbf{D}_m$, and $\lambda\in[0,1]$. Suppose that $2^{-\lambda}(\xi+\eta)\in\omega_{s(2)}$.
(a) Assume that $m\le 0$ and that $2^{-m}\ge 40|\xi|$. Show that for some C that does not depend on y, η, and λ we have

$$\begin{aligned}\big|\langle D^{2^\lambda}\tau^y M^\eta(h)\,|\,\varphi_s\rangle\big| &= \big|\langle h\,|\,M^{-\eta}\tau^{-y}D^{2^{-\lambda}}(\varphi_s)\rangle\big|\\ &\le C2^{\frac{m}{2}}\big\|\widehat{h}\big\|_{L^1((-\infty,-\frac{1}{40\cdot 2^m})\cup(\frac{1}{40\cdot 2^m},\infty))}.\end{aligned}$$

[*Hint:* Use Plancherel's theorem, noting that $\eta\ge 2^\lambda c(\omega_{s(1)})+\frac{9}{40}2^{-m}$.]
(b) Using the trivial fact that $\big|\langle D^{2^\lambda}\tau^y M^\eta(h)\,|\,\varphi_s\rangle\big|\le C\|h\|_{L^2}$, conclude that whenever $|m|$ is large with respect to ξ, we have

$$\chi_{\omega_{s(2)}}(2^{-\lambda}(\xi+\eta))|\langle D^{2^\lambda}\tau^y M^\eta(h)\,|\,\varphi_s\rangle|\le C_h\min(1,2^m),$$

where C_h may depend on h but is independent of y, η, and λ.

11.1.3. (a) Let g be a bounded periodic function on $\mathbf{R}$ with period κ. Show that

$$\lim_{K\to\infty} \frac{1}{2K}\int_{-K}^{K} g(t)\,dt \to \frac{1}{\kappa}\int_0^{\kappa} g(t)\,dt\,.$$

(b) Let g be a bounded periodic function on $\mathbf{R}^n$ that is periodic with period $(\kappa_1,\dots,\kappa_n)$. Show that

$$\lim_{K_1,\dots,K_n\to\infty} \frac{2^{-n}}{K_1\cdots K_n}\int_{-K_1}^{K_1}\cdots\int_{-K_n}^{K_n} g(t)\,dt = \frac{1}{\kappa_1\cdots\kappa_n}\int_0^{\kappa_1}\cdots\int_0^{\kappa_n} g(t)\,dt$$

11.1.4. Use the result in Appendix K.1 to obtain the size estimate

$$\big|\langle \varphi_s \,|\, \varphi_u\rangle\big| \le C_M \frac{\min\left(\dfrac{|I_s|}{|I_u|},\dfrac{|I_u|}{|I_s|}\right)^{\frac{1}{2}}}{\left(1+\dfrac{|c(I_s)-c(I_u)|}{\max(|I_s|,|I_u|)}\right)^M}$$

for every $M>0$. Conclude that if $|I_u|\le |I_s|$, then

$$\big|\langle \varphi_s \,|\, \varphi_u\rangle\big| \le C_M'\left(\frac{|I_s|}{|I_u|}\right)^{\frac{1}{2}}\int_{I_u}\frac{|I_s|^{-1}\,dx}{\left(1+\frac{|x-c(I_s)|}{|I_s|}\right)^M}\,.$$

[*Hint:* Use that

$$\left|\frac{|x-c(I_s)|}{|I_s|}-\frac{|c(I_u)-c(I_s)|}{|I_s|}\right| \le \frac{1}{2}$$

for all $x\in I_u$.]

11.1.5. Prove that there is a constant $C>0$ such that for any interval J and any $b>0$,

$$\int_J\int_{J^c}\frac{1}{\left(1+\frac{|x-y|}{b|J|}\right)^{20}}\,dx\,dy \le Cb^2|J|^2\,.$$

[*Hint:* Translate J to the interval $[-\frac{1}{2}|J|,\frac{1}{2}|J|]$ and change variables. The resulting integral can be computed explicitly.]

11.1.6. Let φ_s be as in (11.1.3). Let $\mathbf{T}_2$ be a 2-tree and $f\in L^2(\mathbf{R})$.
(a) Show that there is a constant C such that for all sequences of complex scalars $\{\lambda_s\}_{s\in\mathbf{T}_2}$ we have

$$\Big\|\sum_{s\in\mathbf{T}_2}\lambda_s\,\varphi_s\Big\|_{L^2(\mathbf{R})} \le C\Big(\sum_{s\in\mathbf{T}_2}|\lambda_s|^2\Big)^{\frac{1}{2}}.$$

(b) Use duality to conclude that

$$\sum_{s\in\mathbf{T}_2}\big|\langle f\,|\,\varphi_s\rangle\big|^2 \le C^2\|f\|_{L^2}^2\,.$$

[*Hint:* To prove part (a) define $\mathscr{G}_m = \{s \in \mathbf{T}_2 : |I_s| = 2^m\}$. Then for $s \in \mathscr{G}_m$ and $s' \in \mathscr{G}_{m'}$, the functions φ_s and $\varphi_{s'}$ are orthogonal to each other, and it suffices to obtain the corresponding estimate when the summation is restricted to a given $\mathscr{G}_m$. But for s in $\mathscr{G}_m$, the intervals I_s are disjoint, and we may use the idea of the proof of Lemma 11.1.2. Use that $\sum_{u:\,\omega_u=\omega_s} \big|\langle \varphi_s \,|\, \varphi_u\rangle\big| \le C$ for every fixed s.]

11.1.7. Fix $A \ge 1$. Let $\mathbf{S}$ be a finite collection of dyadic tiles such that for all s_1, s_2 in $\mathbf{S}$ we have either $\omega_{s_1} \cap \omega_{s_2} = \emptyset$ or $AI_{s_1} \cap AI_{s_2} = \emptyset$. Let $N_{\mathbf{S}}$ be the *counting function* of $\mathbf{S}$, defined by

$$N_{\mathbf{S}} = \sup_{x\in\mathbf{R}} \#\{I_s : \, s \in \mathbf{S} \text{ and } x \in I_s\}.$$

(a) Show that for any $M > 0$ there exists a $C_M > 0$ such that for all $f \in L^2(\mathbf{R})$ we have

$$\sum_{s\in\mathbf{S}} \left| \left\langle f, |I_s|^{-\frac{1}{2}} \Big(1 + \frac{\operatorname{dist}(\cdot, I_s)}{|I_s|}\Big)^{-\frac{M}{2}} \right\rangle \right|^2 \le C_M N_{\mathbf{S}} \|f\|_{L^2}^2 .$$

(b) Let φ_s be as in (11.1.3). Show that for any $M > 0$ there exists a $C_M > 0$ such that for all finite sequences of scalars $\{a_s\}_{s\in\mathbf{S}}$ we have

$$\Big\| \sum_{s\in\mathbf{S}} a_s \varphi_s \Big\|_{L^2}^2 \le C_M (1 + A^{-M} N_{\mathbf{S}}) \sum_{s\in\mathbf{S}} |a_s|^2 .$$

(c) Conclude that for any $M > 0$ there exists a $C_M > 0$ such that for all $f \in L^2(\mathbf{R})$ we have

$$\sum_{s\in\mathbf{S}} |\langle f, \varphi_s \rangle|^2 \le C_M (1 + A^{-M} N_{\mathbf{S}}) \|f\|_{L^2}^2 .$$

[*Hint:* Use the idea of Lemma 11.1.2 to prove part (a) when $N_{\mathbf{S}} = 1$. Suppose now that $N_{\mathbf{S}} > 1$. Call an element $s \in \mathbf{S}$ *h*-maximal if the region in $\mathbf{R}^2$ that is directly horizontally above the tile s does not intersect any other tile $s' \in \mathbf{S}$. Let $\mathbf{S}_1$ be the set of all h-maximal tiles in $\mathbf{S}$. Then $N_{\mathbf{S}_1} = 1$; otherwise, some $x \in \mathbf{R}$ would belong to both I_s and $I_{s'}$ for $s \ne s' \in \mathbf{S}_1$, and thus the horizontal regions directly above s and s' would have to intersect, contradicting the h-maximality of $\mathbf{S}_1$. Now define $\mathbf{S}_2$ to be the set of all h-maximal tiles in $\mathbf{S} \setminus \mathbf{S}_1$. As before, we have $N_{\mathbf{S}_2} = 1$. Continue in this way and write $\mathbf{S}$ as a union of at most $N_{\mathbf{S}}$ families of tiles $\mathbf{S}_j$, each of which has the property $N_{\mathbf{S}_j} = 1$. Apply the result to each $\mathbf{S}_j$ and then sum over j. Part (b): observe that whenever $s_1, s_2 \in \mathbf{S}$ and $s_1 \ne s_2$ we must have either $\langle \varphi_{s_1}, \varphi_{s_2} \rangle = 0$ or $\operatorname{dist}(I_{s_1}, I_{s_2}) \ge (A-1)\max(|I_{s_1}|, |I_{s_2}|)$, which implies

$$\Big(1 + \frac{\operatorname{dist}(I_{s_1}, I_{s_2})}{\max(|I_{s_1}|, |I_{s_2}|)}\Big)^{-M} \le A^{-\frac{M}{2}} \Big(1 + \frac{\operatorname{dist}(I_{s_1}, I_{s_2})}{\max(|I_{s_1}|, |I_{s_2}|)}\Big)^{-\frac{M}{2}} .$$

Use this estimate to obtain

$$\Big\| \sum_{s\in\mathbf{S}} a_s \varphi_s \Big\|_{L^2}^2 \le \sum_{s\in\mathbf{S}} |a_s|^2 + \frac{C_M}{A^{\frac{M}{2}}} \Big\| \sum_{s\in\mathbf{S}} \frac{|a_s|}{|I_s|^{\frac{1}{2}}} \Big(1 + \frac{\operatorname{dist}(x, I_s)}{|I_s|}\Big)^{-\frac{M}{2}} \Big\|_{L^2}^2$$

by expanding the square on the left. The required estimate follows from the dual statement to part (a). Part (c) follows from part (b) by duality.]

11.1.8. Let φ_s be as in (11.1.3) and let $\mathbf{D}_m$ be the set of all dyadic tiles s with $|I_s| = 2^m$. Show that there is a constant C (independent of m) such that for square-integrable sequences of scalars $\{a_s\}_{s\in\mathbf{D}_m}$ we have

$$\Big\|\sum_{s\in\mathbf{D}_m} a_s\varphi_s\Big\|_{L^2}^2 \le C\sum_{s\in\mathbf{D}_m}|a_s|^2.$$

Conclude from this that

$$\sum_{s\in\mathbf{D}_m}\big|\langle f,\varphi_s\rangle\big|^2 \le C\|f\|_{L^2}^2.$$

11.1.9. Fix a Schwartz function φ whose Fourier transform is supported in the interval $[-\frac{3}{8},\frac{3}{8}]$ and that satisfies

$$\sum_{l\in\mathbf{Z}}|\widehat{\varphi}(t+\tfrac{l}{2})|^2 = c_0$$

for all real numbers t. Define functions φ_s as follows. Fix an integer m and set

$$\varphi_s(x) = 2^{-\frac{m}{2}}\varphi(2^{-m}x-k)e^{2\pi i 2^{-m}x\frac{l}{2}}$$

whenever $s = [k2^m,(k+1)2^m)\times[l2^{-m},(l+1)2^{-m})$ is a tile in $\mathbf{D}$. Prove that for all Schwartz functions f we have

$$\sum_{s\in\mathbf{D}_m}\langle f\,|\,\varphi_s\rangle\varphi_s = c_0 f.$$

Observe that m does not appear on the right of this identity.
[*Hint:* First prove that

$$\sum_{s\in\mathbf{D}_m}\varphi_s(x)\overline{\widehat{\varphi_s}(y)} = c_0\,e^{2\pi ixy}$$

using the Poisson summation formula.]

11.1.10. This is a continuous version of Exercise 11.1.9. Fix a Schwartz function φ on $\mathbf{R}^n$ and define a *continuous wave packet*

$$\varphi_{y,\xi}(x) = \varphi(x-y)e^{2\pi i\xi\cdot x}.$$

Prove that for all f Schwartz functions on $\mathbf{R}^n$, the following identity is valid:

$$\|\varphi\|_{L^2}^2 f(x) = \int_{\mathbf{R}^n}\int_{\mathbf{R}^n}\varphi_{y,\xi}(x)\langle f\,|\,\varphi_{y,\xi}\rangle\,dy\,d\xi.$$

[*Hint:* Prove first that $\int_{\mathbf{R}^n}\int_{\mathbf{R}^n}\varphi_{y,\xi}(x)\overline{\widehat{\varphi_{y,\xi}}(z)}\,dy\,d\xi = \|\varphi\|_{L^2}^2 e^{2\pi ix\cdot z}$.]

11.2 Distributional Estimates for the Carleson Operator

In this section we derive estimates for the distribution function of the Carleson operator acting on characteristic functions of measurable sets. These estimates imply, in particular, that the Carleson operator is bounded on $L^p(\mathbf{R})$ for $1 < p < \infty$. To achieve this we build on the time–frequency analysis approach developed in the previous section. Working with characteristic functions of measurable sets of finite measure is crucial in obtaining an improved energy estimate, which is the key to the proof. Later in this section we obtain weighted estimates for the Carleson operator $\mathscr{C}$. These estimates are reminiscent of the corresponding estimates for the maximal singular integrals we encountered in the previous chapter.

11.2.1 The Main Theorem and Preliminary Reductions

In the sequel we use the notation introduced in Section 11.1. The following is the main result of this section.

Theorem 11.2.1. *(a) There exist finite constants $C,\kappa > 0$ such that for any measurable subset F of the reals with finite measure we have*

$$\big|\{x\in\mathbf{R}:\, \mathscr{C}(\chi_F)(x)>\alpha\}\big| \le C|F| \begin{cases} \frac{1}{\alpha}\Big(1+\log\big(\frac{1}{\alpha}\big)\Big) & \textit{when } 0<\alpha<1, \\ e^{-\kappa\alpha} & \textit{when } \alpha\ge 1. \end{cases} \tag{11.2.1}$$

(b) For any $1<p<\infty$ there is a constant $C_p>0$ such that for all f in $L^p(\mathbf{R})$ we have the estimate

$$\big\|\mathscr{C}(f)\big\|_{L^p(\mathbf{R})} \le C_p\big\|f\big\|_{L^p(\mathbf{R})}\,. \tag{11.2.2}$$

Proof. Assuming statement (a), we obtain

$$\big\|\mathscr{C}(\chi_F)\big\|_{L^p}^p = p\int_0^\infty \big|\{\mathscr{C}(\chi_F)>\alpha\}\big|\lambda^{p-1}\,d\alpha \le pC^p|F|\int_0^\infty \varphi(\alpha)\alpha^{p-1}\,d\alpha\,,$$

where $\varphi(\alpha)=\alpha^{-1}(1+\log(\alpha)^{-1})$ for $\alpha<1$ and $\varphi(\alpha)=e^{-\kappa\alpha}$ for $\alpha\ge 1$. The last integral is convergent, and consequently one obtains a restricted strong type (p,p) estimate

$$\big\|\mathscr{C}(\chi_F)\big\|_{L^p(\mathbf{R})} \le C_p'|F|^{\frac{1}{p}}$$

for the Carleson operator. The required strong type (p,p) estimate follows by applying Theorem 1.4.19. Thus (a) implies (b).

It remains to prove (a). This follows from the corresponding estimate for $\mathscr{C}_1$ and requires a considerable amount of work. The proof of (a) is based on a modification of the proof of Theorem 11.1.1. Recall that in (11.1.21) we identified the one-sided Carleson operator $\mathscr{C}_1(f)$ with

$$\mathscr{C}_1(f)(x) = \sup_{N>0}\left|\int_{-\infty}^{N} \widehat{f}(\eta)e^{2\pi i x\cdot\eta}\,d\eta\right| = \frac{1}{|c|}\sup_{\xi>0}|\Pi_\xi(f)|, \tag{11.2.3}$$

where $c \neq 0$ and Π_ξ, $\xi \in \mathbf{R}$ is given by

$$\Pi_\xi(f) = \lim_{\substack{K\to\infty\\ L\to\infty}} \frac{1}{4KL}\int_{-L}^{L}\int_{-K}^{K}\int_0^1 G_{\xi,y,\eta,\lambda}(f)\,d\lambda\,dy\,d\eta\,. \tag{11.2.4}$$

Also recall that $G_{\xi,y,\eta,\lambda}(f)$ is

$$G_{\xi,y,\eta,\lambda}(f) = M^{-\eta}\tau^{-y}D^{2^{-\lambda}}A_{\frac{\xi+\eta}{2^\lambda}}D^{2^\lambda}\tau^y M^\eta(f), \tag{11.2.5}$$

where A_ξ is defined in (11.1.6). Note that

$$\begin{aligned} G_{\xi,y,\eta,\lambda}(f)(x) &= \sum_{\substack{s\in\mathbf{D}\\ \xi\in\omega_{u(2)}}} \big\langle f\,|\,M^{-\eta}\tau^{-y}D^{2^{-\lambda}}\varphi_u\big\rangle M^{-\eta}\tau^{-y}D^{2^{-\lambda}}\varphi_u(x)\\ &= \sum_{\substack{s\in\mathbf{D}_{y,\eta,\lambda}\\ \xi\in\omega_{s(2)}}} \big\langle f\,|\,\varphi_s\big\rangle\,\varphi_s(x)\,, \end{aligned}$$

where $\mathbf{D}_{y,\eta,\lambda}$ is the set of all rectangles of the form $(2^\lambda\otimes I_u - y)\times(2^{-\lambda}\otimes\omega_u-\eta)$, where u ranges over $\mathbf{D}$. Here $a\otimes I$ denotes the set $\{ax:\ x\in I\}$. For such s, φ_s is defined in (11.1.3). The rectangles in $\mathbf{D}_{y,\eta,\lambda}$ are formed by dilating the dyadic tiles in $\mathbf{D}$ by the amount 2^λ in the time coordinate axis and by $2^{-\lambda}$ in the frequency coordinate axis and then translating them by the amounts y and η, respectively.

In view of identity (11.1.12), for a Schwartz function f we have

$$|\Pi_\xi(f)(x)| = \left|\lim_{\substack{K\to\infty\\ L\to\infty}} \frac{1}{4KL}\int_{-L}^{L}\int_{-K}^{K}\int_0^1 \sum_{\substack{s\in\mathbf{D}_{y,\eta,\lambda}\\ \xi\in\omega_{s(2)}}} \big\langle f\,|\,\varphi_s\big\rangle\,\varphi_s(x)\,d\lambda\,dy\,d\eta\right|.$$

Since both terms of this identity are well defined L^2-bounded operators, (11.2.1) is also valid for L^2 functions f. For such functions f, for a measurable function $N:\ \mathbf{R}\to\mathbf{R}^+$, $y,\eta\in\mathbf{R}$, and $\lambda\in[0,1]$ we define operators

$$\mathfrak{D}_{N,y,\eta,\lambda}(f) = \sum_{s\in\mathbf{D}_{y,\eta,\lambda}} \big\langle f\,|\,\varphi_s\big\rangle\,(\chi_{\omega_{s(2)}}\circ N)\,\varphi_s$$

and

$$\mathfrak{D}_N(f) = \lim_{\substack{K\to\infty\\ L\to\infty}} \frac{1}{4KL}\int_{-L}^{L}\int_{-K}^{K}\int_0^1 \sum_{s\in\mathbf{D}_{y,\eta,\lambda}} \big\langle f\,|\,\varphi_s\big\rangle\,(\chi_{\omega_{s(2)}}\circ N)\,\varphi_s\,d\lambda\,dy\,d\eta\,.$$

For every square-integrable function f and $x \in \mathbf{R}$ we pick, in a measurable way, a positive real number $\xi = N_f(x)$ such that

$$\sup_{\xi>0} |\Pi_\xi(f)(x)| \le 2\,|\Pi_{N_f(x)}(f)(x)| \le 2\mathfrak{D}_{N_f}(f)(x).$$

Then

$$\mathscr{C}_1(f) \le \frac{2}{|c|}\,|\mathfrak{D}_{N_f}(f)|. \tag{11.2.6}$$

We work with functions $f = \chi_F$, where F is a measurable set of finite measure; certainly such functions are square-integrable. We show the validity of statement (a) of Theorem 11.2.1 for $\mathfrak{D}_N$, where $N: \mathbf{R} \to \mathbf{R}^+$ is measurable with bounds independent of N. Then (11.2.6) implies the same statement for $\mathscr{C}_1$.

We claim that the following estimate is valid for $\mathfrak{D}_N$. There is a constant C' such that for any pair of measurable subsets (E,F) of the real line with nonzero finite measure there is a subset E' of E with $|E'| \ge \frac{1}{2}|E|$ such that for any measurable function $N: \mathbf{R} \to \mathbf{R}^+$ we have

$$\left| \int_{E'} \mathfrak{D}_N(\chi_F)(x)\,dx \right| \le 2C' \min(|E|,|F|) \left(1 + \left|\log \frac{|E|}{|F|}\right|\right). \tag{11.2.7}$$

This is a fundamental estimate that implies (11.2.1). We derive this estimate from an analogous estimate for the operators $\mathfrak{D}_{N,y,\eta,\lambda}$ by picking a set E' that is independent of y, η, and λ.

We introduce a set

$$\Omega_{E,F} = \left\{ M(\chi_F) > 8 \min\left(1, \tfrac{|F|}{|E|}\right) \right\}.$$

It follows that $|\Omega_{E,F}| \le \frac{1}{2}|E|$, since the Hardy–Littlewood maximal operator is of weak type $(1,1)$ with norm 2. We conclude that the set

$$E' = E \setminus \Omega_{E,F}$$

satisfies $|E'| \ge \frac{1}{2}|E|$. (Notice that in the case $|F| \ge |E|$ the set $\Omega_{E,F}$ is empty.)

Let $\mathbf{P}$ be a finite subset of $\mathbf{D}_{y,\eta,\lambda}$. The required inequality (11.2.7) will be a consequence of the following two estimates:

$$\left| \int_{E'} \sum_{\substack{s\in\mathbf{P} \\ I_s \subseteq \Omega_{E,F}}} \langle \chi_F \,|\, \varphi_s\rangle (\chi_{\omega_{s(2)}} \circ N)\, \varphi_s\,dx \right| \le C' \min(|E|,|F|) \tag{11.2.8}$$

and

$$\left| \int_{E'} \sum_{\substack{s\in\mathbf{P} \\ I_s \nsubseteq \Omega_{E,F}}} \langle \chi_F \,|\, \varphi_s\rangle (\chi_{\omega_{s(2)}} \circ N) \varphi_s\,dx \right| \le C' \min(|E|,|F|) \left(1 + \left|\log \frac{|E|}{|F|}\right|\right), \tag{11.2.9}$$

where the constant C' is independent of the sets E, F, of the measurable function N, and of the finite subset $\mathbf{P}$ of $\mathbf{D}_{y,\eta,\lambda}$. Estimates (11.2.8) and (11.2.9) are proved in the next three subsections.

In the rest of this subsection we show that (11.2.7) implies statement (a) of Theorem 11.2.1. Given $\alpha > 0$ we define sets

$$E_\alpha^1 = \{\mathrm{Re}\,\mathfrak{D}_N(\chi_F) > \alpha\}, \qquad E_\alpha^2 = \{\mathrm{Re}\,\mathfrak{D}_N(\chi_F) < -\alpha\},$$
$$E_\alpha^3 = \{\mathrm{Im}\,\mathfrak{D}_N(\chi_F) > \alpha\}, \qquad E_\alpha^4 = \{\mathrm{Im}\,\mathfrak{D}_N(\chi_F) < -\alpha\}.$$

We apply (11.2.7) to the pair (E_α^j, F) for any $j = 1,2,3,4$. We find a subset $(E_\alpha^j)'$ of E_α^j of at least half its measure so that (11.2.7) holds for this pair. Then we have

$$\begin{aligned}\frac{\alpha}{2}|E_\alpha^j| \le \alpha|(E_\alpha^j)'| &\le \left|\int_{(E_\alpha^j)'} \mathfrak{D}_N(\chi_F)(x)\,dx\right| \\ &\le 2C'\min(|E_\alpha^j|,|F|)\left(1+\left|\log\frac{|E_\alpha^j|}{|F|}\right|\right). \end{aligned} \tag{11.2.10}$$

If $|E_\alpha^j| \le |F|$, this estimate implies that

$$|E_\alpha^j| \le |F|e e^{-\frac{1}{4C'}\alpha}, \tag{11.2.11}$$

while if $|E_\alpha^j| > |F|$, it implies that

$$\alpha \le 4C'\frac{|F|}{|E_\alpha^j|}\left(1+\log\frac{|E_\alpha^j|}{|F|}\right). \tag{11.2.12}$$

Case 1: $\alpha > 4C'$. If $|E_\alpha^j| > |F|$, setting $t = |E_\alpha^j|/|F| > 1$ and using the fact that $\sup_{1<t<\infty}\frac{1}{t}(1+\log t) = 1$, we obtain that (11.2.12) fails. In this case we must therefore have that $|E_\alpha^j| \le |F|$. Applying (11.2.11) four times, we deduce

$$|\{\mathfrak{D}_N(\chi_F) > 4\alpha\}| \le 4e|F|e^{-\frac{1}{4C'}\alpha}. \tag{11.2.13}$$

Case 2: $\alpha \le 4C'$. If $|E_\alpha^j| > |F|$, we use the elementary fact that if $t > 1$ satisfies $t(1+\log t)^{-1} < \frac{B}{\alpha}$, then $t < \frac{2B}{\alpha}(1+\log\frac{2B}{\alpha})$; to prove this fact one may use the inequalities $t < \frac{2B}{\alpha}(1+\log\sqrt{t})$ and $\log\sqrt{t} \le \log t - \log(1+\log\sqrt{t}) \le \log\frac{2B}{\alpha}$ for $t > 1$. Taking t $= |E_\alpha^j|/|F|$ and $B = 4C'$ in (11.2.12) yields

$$\frac{|E_\alpha^j|}{|F|} \le \frac{8C'}{\alpha}\left(1+\log\frac{8C'}{\alpha}\right). \tag{11.2.14}$$

If $|E_\alpha^j| \le |F|$, then we use (11.2.11), but we note that for some constant $c' > 1$ we have

$$e e^{-\frac{1}{4C'}\alpha} \le c'\frac{8C'}{\alpha}\left(1+\log\frac{8C'}{\alpha}\right)$$

whenever $\alpha \le 4C'$. Thus, when $\alpha \le 4C'$ we always have

$$|\{\mathfrak{D}_N(\chi_F) > 4\alpha\}| \le c'\frac{32C'}{\alpha}|F|\left(1+\log\frac{8C'}{\alpha}\right). \tag{11.2.15}$$

Combining (11.2.13) and (11.2.15), we obtain estimate (11.2.7) for $\mathfrak{D}_N$. The same estimate holds for $\mathscr{C}_1$ in view of (11.2.6). Since $\widetilde{\mathscr{C}_2(f)} = \mathscr{C}_1(\widetilde{f})$, where $\widetilde{f}(x) = f(-x)$, the same estimate holds for $\mathscr{C}_2$ and hence estimate (11.2.7) is valid for $\mathscr{C}$. □

11.2.2 The Proof of Estimate (11.2.8)

In proving (11.2.8), we may assume that $|F| \le |E|$; otherwise, the set $\Omega_{E,F}$ is empty and there is nothing to prove.

Let $\mathbf{P}$ be a finite subset of $\mathbf{D}_{y,\eta,\lambda}$. We denote by $\mathscr{I}(\mathbf{P})$ the grid that consists of all the time projections I_s of tiles s in $\mathbf{P}$. For a fixed interval J in $\mathscr{I}(\mathbf{P})$ we define

$$\mathbf{P}(J) = \{s \in \mathbf{P} : I_s = J\}$$

and a function

$$\psi_J(x) = |J|^{-\frac{1}{2}}\left(1+\frac{|x-c(J)|}{|J|}\right)^{-M},$$

where M is a large integer to be chosen momentarily. We note that for each $s \in \mathbf{P}(J)$ we have $|\varphi_s(x)| \le C_M\,\psi_J(x)$.

For each $k = 0,1,2,\dots$ we introduce families

$$\mathscr{F}_k = \left\{J \in \mathscr{I}(\mathbf{P}) : 2^kJ \subseteq \Omega_{E,F},\ 2^{k+1}J \nsubseteq \Omega_{E,F}\right\}.$$

We begin by writing the left-hand side of (11.2.8) as

$$\begin{aligned}
&\sum_{\substack{J\in\mathscr{I}(\mathbf{P})\\ J\subseteq\Omega_{E,F}}}\left|\sum_{s\in\mathbf{P}(J)}\int_{E'}\langle\chi_F\,|\,\varphi_s\rangle\,\chi_{\omega_{s(2)}}(N(x))\,\varphi_s(x)\,dx\right| \\
&= \sum_{k=0}^{\infty}\sum_{\substack{J\in\mathscr{I}(\mathbf{P})\\ J\in\mathscr{F}_k}}\left|\int_{E'}\sum_{s\in\mathbf{P}(J)}\langle\chi_F\,|\,\varphi_s\rangle\,\chi_{\omega_{s(2)}}(N(x))\,\varphi_s(x)\,dx\right|.
\end{aligned} \tag{11.2.16}$$

Using Exercise 9.2.8(b) we obtain the existence of a constant $C_0 < \infty$ such that for each $k = 0,1,\dots$ and $J \in \mathscr{F}_k$ we have

$$\begin{aligned}\langle \chi_F, \psi_J \rangle &\le |J|^{\frac{1}{2}} \inf_J M(\chi_F) \\ &\le |J|^{\frac{1}{2}} C_0^k \inf_{2^{k+1}J} M(\chi_F) \\ &\le 4C_0^k |J|^{\frac{1}{2}} \frac{|F|}{|E|},\end{aligned} \tag{11.2.17}$$

since $2^{k+1}J$ meets the complement of $\Omega_{E,F}$.

For $J \in \mathscr{F}_k$ we also have that $E' \cap 2^k J = \emptyset$ and hence

$$\int_{E'} \psi_J(y)\,dy \le \int_{(2^k J)^c} \psi_J(y)\,dy \le |J|^{\frac{1}{2}} C_M 2^{-kM}. \tag{11.2.18}$$

Next we note that for each $J \in \mathscr{I}(\mathbf{P})$ and $x \in \mathbf{R}$ there is at most one $s = s_x \in \mathbf{P}(J)$ such that $N(x) \in \omega_{s_x(2)}$. Using this observation along with (11.2.17) and (11.2.18), we can therefore estimate the expression on the right in (11.2.16) as follows:

$$\begin{aligned}&\sum_{k=0}^{\infty} \sum_{\substack{J \in \mathscr{I}(\mathbf{P}) \\ J \in \mathscr{F}_k}} \left| \int_{E'} \langle \chi_F \,|\, \varphi_{s_x} \rangle \chi_{\omega_{s_x(2)}}(N(x)) \varphi_{s_x}(x)\,dx \right| \\ &\le C_M^2 \sum_{k=0}^{\infty} \sum_{\substack{J \in \mathscr{I}(\mathbf{P}) \\ J \in \mathscr{F}_k}} \int_{E'} \langle \chi_F, \psi_J \rangle \psi_J(x)\,dx \\ &\le C_M^2 4 \frac{|F|}{|E|} \sum_{k=0}^{\infty} C_0^k \sum_{J \in \mathscr{F}_k} |J|^{\frac{1}{2}} \int_{E'} \psi_J(x)\,dx \\ &\le 4C_M^3 \frac{|F|}{|E|} \sum_{k=0}^{\infty} (C_0 2^{-M})^k \sum_{J \in \mathscr{F}_k} |J|,\end{aligned} \tag{11.2.19}$$

and we pick $M > \log C_0 / \log 2$. It remains to control

$$\sum_{J \in \mathscr{F}_k} |J|$$

for each nonnegative integer k. In doing this we let $\mathscr{F}_k^*$ be all elements of $\mathscr{F}_k$ that are maximal under inclusion. Then we observe that if $J \in \mathscr{F}_k^*$ and $J' \in \mathscr{F}_k$ satisfy $J' \subseteq J$ then

$$\operatorname{dist}(J', J^c) = 0,$$

otherwise $2J'$ would be contained in J and thus

$$2^{k+1}J' \subseteq 2^k J \subseteq \Omega_{E,F}.$$

Therefore, for any J in $\mathscr{F}_k^*$ and any scale m there are at most two intervals J' from $\mathscr{F}_k$ contained in J with $|J'| = 2^m$. Summing over all possible scales, we obtain a bound of at most four times the length of J. We conclude that

$$\sum_{J\in\mathcal{F}_k}|J| = \sum_{J\in\mathcal{F}_k^*}\sum_{\substack{J'\in\mathcal{F}_k\\ J'\subseteq J}}|J'| \le \sum_{J\in\mathcal{F}_k^*}4|J| \le 4|\Omega_{E,F}|,$$

since elements of $\mathcal{F}_k^*$ are disjoint and contained in $\Omega_{E,F}$. Inserting this estimate in (11.2.19), we obtain the required bound

$$C'_M\frac{|F|}{|E|}|\Omega_{E,F}| \le C''_M|F| = C''_M\min(|E|,|F|)$$

for the expression on the right in (11.2.16). This concludes the proof of (11.2.8).

11.2.3 The Proof of Estimate (11.2.9)

For fixed y,η,λ we define a partial order in the set of tiles in $\mathbf{D}_{y,\eta,\lambda}$ just as in Definition 11.1.3. All properties of dyadic tiles obtained in the previous section also hold for the tiles in $\mathbf{D}_{y,\eta,\lambda}$. Throughout this section, $\mathbf{P}$ is a finite subset of $\mathbf{D}_{y,\eta,\lambda}$.

To simplify notation, in the sequel we set

$$\mathbf{P}_{E,F} = \{s\in\mathbf{P}:\ I_s \not\subseteq \Omega_{E,F}\}.$$

Setting $N^{-1}[A] = \{x:\ N(x)\in A\}$ for a set $A\subseteq\mathbf{R}$, we note that (11.2.9) is a consequence of

$$\sum_{s\in\mathbf{P}_{E,F}}\left|\langle\chi_F,\varphi_s\rangle\langle\chi_{E'\cap N^{-1}[\omega_{s(2)}]},\varphi_s\rangle\right| \le C\min(|E|,|F|)\left(1+\left|\log\frac{|E|}{|F|}\right|\right). \tag{11.2.20}$$

The following lemma is the main ingredient of the proof and is proved in the next section.

Lemma 11.2.2. *There is a constant C such that for all measurable sets E and F of finite measure we have*

$$\mathcal{E}(\chi_F;\mathbf{P}_{E,F}) \le C|F|^{-\frac{1}{2}}\min\left(\frac{|F|}{|E|},1\right). \tag{11.2.21}$$

Assuming Lemma 11.2.2, we argue as follows to prove (11.2.9). Given the finite set of tiles $\mathbf{P}_{E,F}$, we write it as the union

$$\mathbf{P}_{E,F} = \bigcup_{j=-\infty}^{n_0}\mathbf{P}_j,$$

where the sets $\mathbf{P}_j$ satisfy properties (1)–(5) of page 437.

Given the sequence of sets $\mathbf{P}_j$, we use properties (1), (2), (5) on page 437, the observation that the mass is always bounded by $|E|^{-1}$, and Lemmas 11.2.2 and

11.1.10 to obtain the following bound for the expression on the left in (11.2.9):

$$\begin{aligned}
&\sum_{s\in\mathbf{P}_{E,F}} \left|\left\langle \chi_F \,|\, \varphi_s\right\rangle\right| \left|\left\langle \chi_{E'\cap N^{-1}[\omega_{s(2)}]}, \varphi_s\right\rangle\right| \\
&= \sum_{j\in\mathbf{Z}}\sum_{s\in\mathbf{P}_j} \left|\left\langle \chi_F \,|\, \varphi_s\right\rangle\right| \left|\left\langle \chi_{E'\cap N^{-1}[\omega_{s(2)}]}, \varphi_s\right\rangle\right| \\
&\le \sum_{j\in\mathbf{Z}}\sum_{k}\sum_{s\in\mathbf{T}_{jk}} \left|\left\langle \chi_F \,|\, \varphi_s\right\rangle\right| \left|\left\langle \chi_{E'\cap N^{-1}[\omega_{s(2)}]}, \varphi_s\right\rangle\right| \\
&\le C_3 \sum_j\sum_k |I_{\mathrm{top}(\mathbf{T}_{jk})}|\, \mathscr{E}(f;\mathbf{T}_{jk})\, \mathscr{M}(E',\mathbf{T}_{jk})\, |E'|\, |F|^{\frac12} \\
&\le C_3 \sum_{j\in\mathbf{Z}}\sum_k |I_{\mathrm{top}(\mathbf{T}_{jk})}| \min\Big(2^{j+1}, C\frac{|F|^{\frac12}}{|E|}, C|F|^{-\frac12}\Big) \min(|E'|^{-1}, 2^{2j+2})\, |E|\, |F|^{\frac12} \\
&\le C_4 \sum_{j\in\mathbf{Z}} 2^{-2j} \min\big(2^j, |F|^{\frac12}|E|^{-1}, |F|^{-\frac12}\big) \min(|E|^{-1}, 2^{2j})\, |E|\, |F|^{\frac12} \\
&\le C_5 \sum_{j\in\mathbf{Z}} \min\left(2^j|E|^{\frac12}, \min\Big(\frac{|F|}{|E|}, \frac{|E|}{|F|}\Big)^{\frac12}\right) \min\big((2^j|E|^{\frac12})^{-2}, 1\big)\, |E|^{\frac12}\, |F|^{\frac12} \\
&\le C_6 \sum_{j\in\mathbf{Z}} \min\left(2^j, \min\Big(\frac{|F|}{|E|}, \frac{|E|}{|F|}\Big)^{\frac12}\right) \min(2^{-2j}, 1)\, |E|^{\frac12}\, |F|^{\frac12} \\
&\le C_7 \min(|E|, |F|)\left(1 + \left|\log\frac{|E|}{|F|}\right|\right).
\end{aligned}$$

The last estimate follows by a simple calculation considering the three cases $1 < 2^j$, $\min\left(\frac{|F|}{|E|}, \frac{|E|}{|F|}\right)^{\frac12} \le 2^j \le 1$, and $2^j < \min\left(\frac{|F|}{|E|}, \frac{|E|}{|F|}\right)^{\frac12}$.

11.2.4 The Proof of Lemma 11.2.2

It remains to prove Lemma 11.2.2.

Fix a 2-tree $\mathbf{T}$ contained in $\mathbf{P}_{E,F}$ and let $t = \mathrm{top}(\mathbf{T})$ denote its top. We show that

$$\frac{1}{|I_t|}\sum_{s\in\mathbf{T}} \left|\left\langle \chi_F \,|\, \varphi_s\right\rangle\right|^2 \le C \min\left(\frac{|F|}{|E|}, 1\right)^2 \tag{11.2.22}$$

for some constant C independent of F, E, and $\mathbf{T}$. Then (11.2.21) follows from (11.2.22) by taking the supremum over all 2-trees $\mathbf{T}$ contained in $\mathbf{P}_{E,F}$.

We decompose the function χ_F as follows:

$$\chi_F = \chi_{F\cap 3I_t} + \chi_{F\cap(3I_t)^c}\,.$$

We begin by observing that for s in $\mathbf{P}_{E,F}$ we have

$$\left|\langle \chi_{F\cap(3I_t)^c} \,|\, \varphi_s\rangle\right| \le \frac{C_M |I_s|^{\frac{1}{2}} \inf\limits_{I_s} M(\chi_F)}{\left(1+\dfrac{\mathrm{dist}((3I_t)^c, c(I_s))}{|I_s|}\right)^M}$$
$$\le 8 C_M |I_s|^{\frac{1}{2}} \min\left(\frac{|F|}{|E|}, 1\right)\left(\frac{|I_s|}{|I_t|}\right)^M,$$

since I_s meets the complement of $\Omega_{E,F}$ for every $s \in \mathbf{P}_F$. Square this inequality and sum over all s in $\mathbf{T}$ to obtain

$$\sum_{s\in\mathbf{T}} |\langle \chi_{F\cap(3I_t)^c} \,|\, \varphi_s\rangle|^2 \le C|I_t| \min\left(\frac{|F|}{|E|}, 1\right)^2,$$

using Exercise 11.2.1.

We now turn to the corresponding estimate for the function $\chi_{F\cap 3I_t}$. At this point it is convenient to distinguish the simple case $|F| > |E|$ from the difficult case $|F| \le |E|$. In the first case the set $\Omega_{E,F}$ is empty and Exercise 11.1.6(b) yields

$$\sum_{s\in\mathbf{T}} \left|\langle \chi_{F\cap 3I_t} \,|\, \varphi_s\rangle\right|^2 \le C\left\|\chi_{F\cap 3I_t}\right\|_{L^2}^2$$
$$\le C|I_t|$$
$$= C|I_t| \min\left(\frac{|F|}{|E|}, 1\right)^2,$$

since $|F| > |E|$.

We may therefore concentrate on the case $|F| \le |E|$. In proving (11.2.21) we may assume that there exists a point $x_0 \in I_t$ such that

$$M(\chi_F)(x_0) \le 8\frac{|F|}{|E|};$$

otherwise there is nothing to prove.

We write the set $\Omega_{E,F} = \left\{M(\chi_F) > 8\frac{|F|}{|E|}\right\}$ as a disjoint union of dyadic intervals J'_ℓ such that the dyadic parent $\widetilde{J'_\ell}$ of J'_ℓ is not contained in $\Omega_{E,F}$ and therefore

$$|F \cap J'_\ell| \le |F \cap \widetilde{J'_\ell}| \le 16\frac{|F|}{|E|}|J'_\ell|.$$

Now some of these dyadic intervals may have size larger than or equal to $|I_t|$. Let J'_ℓ be such an interval. Then we split J'_ℓ into $\frac{|J'_\ell|}{|I_t|}$ intervals $J'_{\ell,m}$ each of size exactly $|I_t|$. Since there is an $x_0 \in I_t$ with

$$M(\chi_F)(x_0) \le 8\frac{|F|}{|E|},$$

if K is the smallest interval that contains x_0 and $J'_{\ell,m}$, then

$$\frac{1}{|K|}\int_K \chi_F\,dx \le 8\frac{|F|}{|E|} \implies |F\cap J'_{\ell,m}| \le 8\frac{|F|}{|E|}\,|I_t|\,\frac{|K|}{|I_t|}\,.$$

We conclude that

$$|F\cap J'_{\ell,m}| \le c\,\frac{|F|}{|E|}\,|I_t|\left(1+\frac{\operatorname{dist}(I_t,J'_{\ell,m})}{|I_t|}\right). \tag{11.2.23}$$

We now have a new collection of dyadic intervals $\{J_k\}_k$ contained in $\Omega_{E,F}$ consisting of all the previous J'_ℓ when $|J'_\ell| < |I_t|$ and the $J'_{\ell,m}$'s when $|J'_\ell| \ge |I_t|$. In view of the construction we have

$$|F\cap J_k| \le \begin{cases} 2c\,\dfrac{|F|}{|E|}\,|J_k| & \text{when } |J_k| < |I_t|,\\[2ex] 2c\,\dfrac{|F|}{|E|}\,|J_k|\left(1+\dfrac{\operatorname{dist}(I_t,J_k)}{|I_t|}\right) & \text{when } |J_k| = |I_t|, \end{cases} \tag{11.2.24}$$

for all k. We now define the "bad functions"

$$b_k(x) = \left(e^{-2\pi i c(\omega_t)x}\chi_{F\cap 3I_t}(x) - \frac{1}{|J_k|}\int_{J_k} e^{-2\pi i c(\omega_t)y}\chi_{F\cap 3I_t}(y)\,dy\right)\chi_{J_k}(x)\,,$$

which are supported in J_k, have mean value zero, and satisfy

$$\big\|b_k\big\|_{L^1} \le 2c\,|F|\,|J_k|\left(1+\frac{\operatorname{dist}(I_t,J_k)}{|I_t|}\right).$$

We also set

$$g(x) = e^{-2\pi i c(\omega_t)x}\chi_{F\cap 3I_t}(x) - \sum_k b_k(x)\,,$$

the "good function" of this Calderón–Zygmund-type decomposition. We have therefore decomposed the function $\chi_{F\cap 3I_t}$ as follows:

$$\chi_{F\cap 3I_t}(x) = g(x)e^{2\pi i c(\omega_t)x} + \sum_k b_k(x)e^{2\pi i c(\omega_t)x}\,. \tag{11.2.25}$$

We show that $\big\|g\big\|_{L^\infty} \le C\frac{|F|}{|E|}$. Indeed, for x in J_k we have

$$g(x) = \frac{1}{|J_k|}\int_{J_k} e^{-2\pi i c(\omega_t)y}\chi_{F\cap 3I_t}(y)\,dy\,,$$

which implies

$$|g(x)| \le \frac{|F \cap 3I_t \cap J_k|}{|J_k|} \le \begin{cases} \dfrac{|F \cap J_k|}{|J_k|} & \text{when } |J_k| < |I_t|, \\ \dfrac{|F \cap 3I_t|}{|I_t|} & \text{when } |J_k| = |I_t|, \end{cases}$$

and both of the preceding are at most a multiple of $\frac{|F|}{|E|}$; the latter is because there is an $x_0 \in I_t$ with $M(\chi_F)(x_0) \le 8\frac{|F|}{|E|}$. Also, for $x \in (\bigcup_k J_k)^c = (\Omega_{E,F})^c$ we have

$$|g(x)| = \chi_{F\cap 3I_t}(x) \le M(\chi_F)(x) \le 8\frac{|F|}{|E|}.$$

We conclude that $\big\|g\big\|_{L^\infty} \le C\frac{|F|}{|E|}$. Moreover,

$$\big\|g\big\|_{L^1} \le \sum_k \int_{J_k} \frac{|F \cap 3I_t \cap J_k|}{|J_k|}\,dx + \big\|\chi_{F\cap 3I_t}\big\|_{L^1} \le C|F \cap 3I_t| \le C\frac{|F|}{|E|}|I_t|,$$

since the J_k are disjoint. It follows that

$$\big\|g\big\|_{L^2} \le C\Big(\frac{|F|}{|E|}\Big)^{\frac12}\Big(\frac{|F|}{|E|}\Big)^{\frac12}|I_t|^{\frac12} = C\frac{|F|}{|E|}|I_t|^{\frac12}.$$

Using Exercise 11.1.6, we have

$$\sum_{s\in\mathbf{T}} \big|\big\langle g e^{2\pi i c(\omega_t)(\cdot)} \,|\, \varphi_s\big\rangle\big|^2 \le C\big\|g\big\|_{L^2}^2,$$

from which we obtain the required conclusion for the first function in the decomposition (11.2.25).

Next we turn to the corresponding estimate for the second function,

$$\sum_k b_k e^{2\pi i c(\omega_t)(\cdot)},$$

in the decomposition (11.2.25), which requires some further analysis. We have the following two estimates for all s and k:

$$\big|\big\langle b_k e^{2\pi i c(\omega_t)(\cdot)} \,|\, \varphi_s\big\rangle\big| \le \frac{C_M |F|\,|E|^{-1}\,|J_k|^2\,|I_s|^{-\frac32}}{(1+\frac{\mathrm{dist}\,(J_k,I_s)}{|I_s|})^M}, \tag{11.2.26}$$

$$\big|\big\langle b_k e^{2\pi i c(\omega_t)(\cdot)} \,|\, \varphi_s\big\rangle\big| \le \frac{C_M |F|\,|E|^{-1}\,|I_s|^{\frac12}}{(1+\frac{\mathrm{dist}\,(J_k,I_s)}{|I_s|})^M}, \tag{11.2.27}$$

for all $M > 0$, where C_M depends only on M.

To prove (11.2.26) we use the mean value theorem together with the fact that b_k has vanishing integral to write for some ξ_y,

$$
\begin{aligned}
&\big|\big\langle b_k\, e^{2\pi i c(\omega_t)(\cdot)} \,\big|\, \varphi_s \big\rangle\big| \\
&= \bigg| \int_{J_k} b_k(y) e^{2\pi i c(\omega_t)y} \overline{\varphi_s(y)}\, dy \bigg| \\
&= \bigg| \int_{J_k} b_k(y) \big(e^{2\pi i c(\omega_t)y} \overline{\varphi_s(y)} - e^{2\pi i c(\omega_t)c(J_k)} \overline{\varphi_s(c(J_k))}\, dy \bigg| \\
&\le |J_k| \int_{J_k} |b_k(y)| \bigg[2\pi \frac{|c(\omega_s)-c(\omega_t)|}{|I_s|^{\frac{1}{2}}} \Big|\varphi\Big(\tfrac{\xi_y - c(I_s)}{|I_s|}\Big)\Big| + |I_s|^{-\frac{3}{2}} \Big|\varphi'\Big(\tfrac{\xi_y - c(I_s)}{|I_s|}\Big)\Big| \bigg] dy \\
&\le \big\| b_k \big\|_{L^1} |J_k| \sup_{\xi \in J_k} \frac{C_M |I_s|^{-\frac{3}{2}}}{(1 + \frac{|\xi - c(I_s)|}{|I_s|})^{M+1}} \\
&\le C_M \frac{|F|}{|E|} |J_k| \Big(1 + \frac{\operatorname{dist}(J_k, I_t)}{|I_t|} \Big) \frac{|J_k|\, |I_s|^{-\frac{3}{2}}}{(1 + \frac{\operatorname{dist}(J_k, I_s)}{|I_s|})^{M+1}} \\
&\le \frac{C_M\, |F|\, |E|^{-1}\, |J_k|^2\, |I_s|^{-\frac{3}{2}}}{(1 + \frac{\operatorname{dist}(J_k, I_s)}{|I_s|})^{M}},
\end{aligned}
$$

where we used the fact that $1 + \frac{\operatorname{dist}(J_k, I_t)}{|I_t|} \le 1 + \frac{\operatorname{dist}(J_k, I_s)}{|I_s|}$. To prove estimate (11.2.27) we note that

$$
\big|\big\langle b_k e^{2\pi i c(\omega_t)(\cdot)} \,\big|\, \varphi_s \big\rangle\big| \le \frac{C_M\, |I_s|^{\frac{1}{2}} \inf_{I_s} M(b_k)}{(1 + \frac{\operatorname{dist}(J_k, I_s)}{|I_s|})^{M}}
$$

and that

$$
M(b_k) \le M(\chi_F) + \frac{|F \cap 3I_t \cap J_k|}{|J_k|} M(\chi_{J_k}),
$$

and since $I_s \not\subseteq \Omega_{E,F}$, we have $\inf_{I_s} M(\chi_F) \le 8\, \frac{|F|}{|E|}$, while the second term in the sum was observed earlier to be at most $C\, \frac{|F|}{|E|}$.

Finally, we have the estimate

$$
\big|\big\langle b_k e^{2\pi i c(\omega_t)(\cdot)} \,\big|\, \varphi_s \big\rangle\big| \le \frac{C_M\, |F|\, |E|^{-1}\, |J_k|\, |I_s|^{-\frac{1}{2}}}{(1 + \frac{\operatorname{dist}(J_k, I_s)}{|I_s|})^{M}}, \tag{11.2.28}
$$

which follows by taking the geometric mean of (11.2.26) and (11.2.27).

Now for a fixed $s \in \mathbf{P}_{E,F}$ we may have either $J_k \subseteq I_s$ or $J_k \cap I_s = \emptyset$ (since I_s is not contained in $\Omega_{E,F}$). Therefore, for fixed $s \in \mathbf{P}_{E,F}$ there are only three possibilities for J_k:

(a) $J_k \subseteq 3I_s$;

(b) $J_k \cap 3I_s = \emptyset$;

(c) $J_k \cap I_s = \emptyset$, $J_k \cap 3I_s \neq \emptyset$, and $J_k \not\subseteq 3I_s$.

Observe that case (c) is equivalent to the following statement:

(c) $J_k \cap I_s = \emptyset$, dist $(J_k, I_s) = 0$, and $|J_k| \ge 2|I_s|$.

Note that in case (c), for each I_s there exists exactly one $J_k = J_{k(s)}$ with the previous properties; but for a given J_k there may be a sequence of I_s's that lie on the left of J_k such that $|J_k| \ge 2|I_s|$ and dist $(J_k, I_s) = 0$ and another sequence with similar properties on the right of J_k. The I_s's that lie on either side of J_k must be nested, and their lengths must add up to $|I^L_{s_k}| + |I^R_{s_k}|$, where $I^L_{s_k}$ is the largest one among them on the left of J_k and $I^R_{s_k}$ is the largest one among them on the right of J_k. Using (11.2.27), we obtain

$$
\begin{aligned}
\sum_{s\in\mathbf{T}} \Big| \sum_{\substack{k:\, J_k\cap I_s=\emptyset \\ \mathrm{dist}\,(J_k,I_s)=0 \\ |J_k|\ge 2|I_s|}} \big\langle b_k e^{2\pi i c(\omega_t)(\cdot)} \,\big|\, \varphi_s \big\rangle \Big|^2 &= \sum_{s\in\mathbf{T}} \Big| \big\langle b_{k(s)} e^{2\pi i c(\omega_t)(\cdot)} \,\big|\, \varphi_s \big\rangle \Big|^2 \\
&\le C\Big(\frac{|F|}{|E|}\Big)^2 \sum_{\substack{s\in\mathbf{T}:\, J_k\cap I_s=\emptyset \\ \mathrm{dist}\,(J_k,I_s)=0 \\ |J_k|\ge 2|I_s|}} |I_s| \\
&\le C\Big(\frac{|F|}{|E|}\Big)^2 \sum_k \big(|I^L_{s_k}| + |I^R_{s_k}|\big)\,.
\end{aligned}
$$

But note that $I^L_{s_k} \subseteq 2J_k$, and since $I^L_{s_k} \cap J_k = \emptyset$, we must have $I^L_{s_k} \subseteq 2J_k \setminus J_k$ (and likewise for $I^R_{s_k}$). We define sets

$$
\begin{aligned}
I^{L+}_{s_k} &= I^L_{s_k} + \frac{1}{2}|J_k|\,, \\
I^{R-}_{s_k} &= I^R_{s_k} - \frac{1}{2}|J_k|\,.
\end{aligned}
$$

We have $I^{L+}_{s_k} \cup I^{R-}_{s_k} \subseteq J_k$, and hence the sets $I^{L+}_{s_k}$ are pairwise disjoint for different k, and the same is true for the $I^{R-}_{s_k}$. Moreover, since $\frac{1}{2}|J_k| \le \frac{1}{2}|I_t|$ for all k, all the shifted sets $I^{L+}_{s_k}$, $I^{R-}_{s_k}$ are contained in $3I_t$. We conclude that

$$
\begin{aligned}
\sum_k |I^L_{s_k}| + \sum_k |I^R_{s_k}| &= \sum_k \big(|I^{L+}_{s_k}| + |I^{R-}_{s_k}|\big) \\
&\le \Big|\bigcup_k I^{L+}_{s_k}\Big| + \Big|\bigcup_k I^{R-}_{s_k}\Big| \\
&\le 2|3I_t|\,,
\end{aligned}
$$

which combined with the previously obtained estimate yields the required result in case (c).

We now consider case (a). Using (11.2.26), we can write

$$
\Big(\sum_{s\in\mathbf{T}} \Big| \sum_{k:J_k\subseteq 3I_s} \big\langle b_k e^{2\pi i c(\omega_t)(\cdot)} \,\big|\, \varphi_s \big\rangle \Big|^2\Big)^{\frac{1}{2}} \le C_M \Big(\frac{|F|}{|E|}\Big)^2 \Big(\sum_{s\in T} \Big| \sum_{k:\, J_k\subseteq 3I_s} |J_k|^{\frac{1}{2}} \frac{|J_k|^{\frac{3}{2}}}{|I_s|^{\frac{3}{2}}} \Big|^2\Big)^{\frac{1}{2}},
$$

and we control the second expression by

$$C_M \frac{|F|}{|E|}\bigg\{\sum_{s\in\mathbf{T}}\bigg(\sum_{k:\, J_k\subseteq 3I_s}|J_k|\bigg)\bigg(\sum_{k:\, J_k\subseteq 3I_s}\frac{|J_k|^3}{|I_s|^3}\bigg)\bigg\}^{\frac12}$$

$$\le C_M \frac{|F|}{|E|}\bigg\{\sum_{k:\, J_k\subseteq 3I_t}|J_k|^3\sum_{\substack{s\in\mathbf{T}\\ J_k\subseteq 3I_s}}\frac{1}{|I_s|^2}\bigg\}^{\frac12},$$

where we used that the dyadic intervals J_k are disjoint and the Cauchy–Schwarz inequality. We note that the last sum is equal to at most $C|J_k|^{-2}$, since for every dyadic interval J_k there exist at most three dyadic intervals of a given length whose triples contain it. The required estimate $C|F|\,|E|^{-1}\,|I_t|^{\frac12}$ now follows in case (a).

Finally, we deal with case (b), which is the most difficult case. We split the set of k into two subsets, those for which $J_k\subseteq 3I_t$ and those for which $J_k\not\subseteq 3I_t$ (recall that $|J_k|\le|I_t|$). Whenever $J_k\not\subseteq 3I_t$, we have

$$\operatorname{dist}(J_k,I_s)\approx\operatorname{dist}(J_k,I_t).$$

In this case we use Minkowski's inequality and estimate (11.2.28) to deduce

$$\bigg(\sum_{s\in\mathbf{T}}\bigg|\sum_{k:\, J_k\not\subseteq 3I_t}\big\langle b_k e^{2\pi i c(\omega_t)(\cdot)}\,\big|\,\varphi_s\big\rangle\bigg|^2\bigg)^{\frac12}$$

$$\le \sum_{k:\, J_k\not\subseteq 3I_t}\bigg(\sum_{s\in\mathbf{T}}\big|\big\langle b_k e^{2\pi i c(\omega_t)(\cdot)}\,\big|\,\varphi_s\big\rangle\big|^2\bigg)^{\frac12}$$

$$\le C_M\frac{|F|}{|E|}\sum_{k:\, J_k\not\subseteq 3I_t}|J_k|\bigg(\sum_{s\in\mathbf{T}}\frac{|I_s|^{2M-1}}{\operatorname{dist}(J_k,I_s)^{2M}}\bigg)^{\frac12}$$

$$\le C_M\frac{|F|}{|E|}\sum_{k:\, J_k\not\subseteq 3I_t}\frac{|J_k|}{\operatorname{dist}(J_k,I_t)^M}\bigg(\sum_{s\in\mathbf{T}}|I_s|^{2M-1}\bigg)^{\frac12}$$

$$\le C_M\frac{|F|}{|E|}|I_t|^{M-\frac12}\sum_{k:\, J_k\not\subseteq 3I_t}\frac{|J_k|}{\operatorname{dist}(J_k,I_t)^M}$$

$$\le C_M\frac{|F|}{|E|}|I_t|^{M-\frac12}\sum_{l=1}^{\infty}\sum_{\substack{k:\\ \operatorname{dist}(J_k,I_t)\approx 2^l|I_t|}}\frac{|J_k|}{(2^l|I_t|)^M},$$

where $\operatorname{dist}(J_k,I_t)\approx 2^l|I_t|$ means that $\operatorname{dist}(J_k,I_t)\in[2^l|I_t|,2^{l+1}|I_t|]$. But note that all the J_k with $\operatorname{dist}(J_k,I_t)\approx 2^l|I_t|$ are contained in $2^{l+2}I_t$, and since they are disjoint, we estimate the last sum by $C2^l|I_t|(2^l|I_t|)^{-M}$. The required estimate $C_M|F|\,|E|^{-1}|I_t|^{\frac12}$ follows.

Next we consider the case $J_k \subseteq 3I_t$, $J_k \cap 3I_s = \emptyset$, and $|J_k| \le |I_s|$, in which we use estimate (11.2.26). We have

$$\Bigg(\sum_{s\in\mathbf{T}}\Bigg|\sum_{\substack{k:\,J_k\subseteq 3I_t\\ J_k\cap 3I_s=\emptyset\\ |J_k|\le|I_s|}}\langle b_k e^{2\pi i c(\omega_t)(\cdot)}\,|\,\varphi_s\rangle\Bigg|^2\Bigg)^{\frac{1}{2}}$$

$$\le C_M\frac{|F|}{|E|}\Bigg(\sum_{s\in\mathbf{T}}\Bigg|\sum_{\substack{k:\,J_k\subseteq 3I_t\\ J_k\cap 3I_s=\emptyset\\ |J_k|\le|I_s|}}|J_k|^2|I_s|^{-\frac{3}{2}}\frac{|I_s|^M}{\operatorname{dist}(J_k,I_s)^M}\Bigg|^2\Bigg)^{\frac{1}{2}}$$

$$\le C_M\frac{|F|}{|E|}\Bigg\{\sum_{s\in\mathbf{T}}\Bigg[\sum_{\substack{k:\,J_k\subseteq 3I_t\\ J_k\cap 3I_s=\emptyset\\ |J_k|\le|I_s|}}\frac{|J_k|^3}{|I_s|^2}\Big(\frac{|I_s|}{\operatorname{dist}(J_k,I_s)}\Big)^M\Bigg]$$

$$\times\Bigg[\sum_{\substack{k:\,J_k\subseteq 3I_t\\ J_k\cap 3I_s=\emptyset\\ |J_k|\le|I_s|}}\frac{|J_k|}{|I_s|}\Big(\frac{\operatorname{dist}(J_k,I_s)}{|I_s|}\Big)^{-M}\Bigg]\Bigg\}^{\frac{1}{2}}$$

$$\le C_M\frac{|F|}{|E|}\Bigg\{\sum_{s\in\mathbf{T}}\Bigg[\sum_{\substack{k:\,J_k\subseteq 3I_t\\ J_k\cap 3I_s=\emptyset\\ |J_k|\le|I_s|}}\frac{|J_k|^3}{|I_s|^2}\Big(\frac{|I_s|}{\operatorname{dist}(J_k,I_s)}\Big)^M\Bigg]$$

$$\times\Bigg[\sum_{\substack{k:\,J_k\subseteq 3I_t\\ J_k\cap 3I_s=\emptyset\\ |J_k|\le|I_s|}}\int_{J_k}\Big(\frac{|x-c(I_s)|}{|I_s|}\Big)^{-M}\frac{dx}{|I_s|}\Bigg]\Bigg\}^{\frac{1}{2}}$$

$$\le C_M\frac{|F|}{|E|}\Bigg\{\sum_{s\in\mathbf{T}}\Bigg[\sum_{\substack{k:\,J_k\subseteq 3I_t\\ J_k\cap 3I_s=\emptyset\\ |J_k|\le|I_s|}}\frac{|J_k|^3}{|I_s|^2}\Big(\frac{|I_s|}{\operatorname{dist}(J_k,I_s)}\Big)^M\Bigg]$$

$$\times\Bigg[\int_{(3I_s)^c}\Big(\frac{|x-c(I_s)|}{|I_s|}\Big)^{-M}\frac{dx}{|I_s|}\Bigg]\Bigg\}^{\frac{1}{2}}$$

$$\le C_M\frac{|F|}{|E|}\Bigg\{\sum_{s\in\mathbf{T}}\sum_{\substack{k:\,J_k\subseteq 3I_t\\ J_k\cap 3I_s=\emptyset\\ |J_k|\le|I_s|}}|J_k|^3|I_s|^{-2}\Big(\frac{|I_s|}{\operatorname{dist}(J_k,I_s)}\Big)^M\Bigg\}^{\frac{1}{2}}.$$

But since the last integral contributes at most a constant factor, we can estimate the last displayed expression by

$$C_M \frac{|F|}{|E|}\Bigg\{\sum_{\substack{k:\, J_k\subseteq 3I_t\\ J_k\cap 3I_s=\emptyset\\ |J_k|\le|I_s|}} |J_k|^3 \sum_{m\ge \log|J_k|} 2^{-2m} \sum_{\substack{s\in\mathbf{T}\\ |I_s|=2^m}} \Big(\frac{\operatorname{dist}(J_k,I_s)}{2^m}\Big)^{-M}\Bigg\}^{\frac12}$$

$$\le C_M \frac{|F|}{|E|}\Bigg\{\sum_{\substack{k:\, J_k\subseteq 3I_t\\ J_k\cap 3I_s=\emptyset\\ |J_k|\le|I_s|}} |J_k|^3 \sum_{m\ge \log|J_k|} 2^{-2m}\Bigg\}^{\frac12}$$

$$\le C_M \frac{|F|}{|E|}\Bigg\{\sum_{\substack{k:\, J_k\subseteq 3I_t\\ J_k\cap 3I_s=\emptyset\\ |J_k|\le|I_s|}} |J_k|^3|J_k|^{-2}\Bigg\}^{\frac12}$$

$$\le C_M \frac{|F|}{|E|}\,|I_t|^{\frac12}\,.$$

There is also the subcase of case (b) in which $|J_k|>|I_s|$. Here we have the two special subcases $I_s\cap 3J_k=\emptyset$ and $I_s\subseteq 3J_k$. We begin with the first of these special subcases, in which we use estimate (11.2.27). We have

$$\Bigg(\sum_{s\in\mathbf{T}}\Bigg|\sum_{\substack{k:\, J_k\subseteq 3I_t\\ J_k\cap 3I_s=\emptyset\\ |J_k|>|I_s|\\ I_s\cap 3J_k=\emptyset}} \langle b_k e^{2\pi i c(\omega_t)(\cdot)}\,|\,\varphi_s\rangle\Bigg|^2\Bigg)^{\frac12}$$

$$\le C_M \frac{|F|}{|E|}\Bigg(\sum_{s\in T}\Bigg|\sum_{\substack{k:\, J_k\subseteq 3I_t\\ J_k\cap 3I_s=\emptyset\\ |J_k|>|I_s|\\ I_s\cap 3J_k=\emptyset}} |I_s|^{\frac12}\frac{|I_s|^M}{\operatorname{dist}(J_k,I_s)^M}\Bigg|^2\Bigg)^{\frac12}$$

$$\le C_M \frac{|F|}{|E|}\Bigg\{\sum_{s\in\mathbf{T}}\Bigg[\sum_{\substack{k:\, J_k\subseteq 3I_t\\ J_k\cap 3I_s=\emptyset\\ |J_k|>|I_s|\\ I_s\cap 3J_k=\emptyset}} \frac{|I_s|^2}{|J_k|}\frac{|I_s|^M}{\operatorname{dist}(J_k,I_s)^M}\Bigg]\Bigg[\sum_{\substack{k:\, J_k\subseteq 3I_t\\ J_k\cap 3I_s=\emptyset\\ |J_k|>|I_s|\\ I_s\cap 3J_k=\emptyset}} \frac{|J_k|}{|I_s|}\frac{|I_s|^M}{\operatorname{dist}(J_k,I_s)^M}\Bigg]\Bigg\}^{\frac12}.$$

Since $I_s\cap 3J_k=\emptyset$, we have that $\operatorname{dist}(J_k,I_s)\approx|x-c(I_s)|$ for every $x\in J_k$, and therefore the second term inside square brackets satisfies

$$\sum_{\substack{k:\, J_k\subseteq 3I_t\\ J_k\cap 3I_s=\emptyset\\ |J_k|>|I_s|\\ I_s\cap 3J_k=\emptyset}} \frac{|J_k|}{|I_s|}\frac{|I_s|^M}{\operatorname{dist}(J_k,I_s)^M}\le\sum_k\int_{J_k}\Big(\frac{|x-c(I_s)|}{|I_s|}\Big)^{-M}\frac{dx}{|I_s|}\le C_M\,.$$

Using this estimate, we obtain

$$
\begin{aligned}
&C_M \frac{|F|}{|E|}\Bigg\{\sum_{s\in\mathbf{T}}\Bigg[\sum_{\substack{k:\,J_k\subseteq 3I_t\\ J_k\cap 3I_s=\emptyset\\ |J_k|>|I_s|\\ I_s\cap 3J_k=\emptyset}} \frac{|I_s|^2}{|J_k|}\frac{|I_s|^M}{\operatorname{dist}(J_k,I_s)^M}\Bigg]\Bigg[\sum_{\substack{k:\,J_k\subseteq 3I_t\\ J_k\cap 3I_s=\emptyset\\ |J_k|>|I_s|\\ I_s\cap 3J_k=\emptyset}} \frac{|J_k|}{|I_s|}\frac{|I_s|^M}{\operatorname{dist}(J_k,I_s)^M}\Bigg]\Bigg\}^{\frac12}\\
&\le C_M \frac{|F|}{|E|}\Bigg\{\sum_{s\in\mathbf{T}}\Bigg[\sum_{\substack{k:\,J_k\subseteq 3I_t\\ J_k\cap 3I_s=\emptyset\\ |J_k|>|I_s|\\ I_s\cap 3J_k=\emptyset}} \frac{|I_s|^2}{|J_k|}\frac{|I_s|^M}{\operatorname{dist}(J_k,I_s)^M}\Bigg]\Bigg\}^{\frac12}\\
&= C_M \frac{|F|}{|E|}\Bigg\{\sum_{k:\,J_k\subseteq 3I_t}\frac{1}{|J_k|}\sum_{\substack{s\in\mathbf{T}\\ J_k\cap 3I_s=\emptyset\\ |J_k|>|I_s|\\ I_s\cap 3J_k=\emptyset}} |I_s|^2\frac{|I_s|^M}{\operatorname{dist}(J_k,I_s)^M}\Bigg\}^{\frac12}\\
&\le C_M \frac{|F|}{|E|}\Bigg\{\sum_{k:\,J_k\subseteq 3I_t}\frac{1}{|J_k|}\sum_{m=-\infty}^{\log_2|J_k|} 2^{2m}\sum_{\substack{s\in\mathbf{T}:\,|I_s|=2^m\\ J_k\cap 3I_s=\emptyset\\ |J_k|>|I_s|\\ I_s\cap 3J_k=\emptyset}} \frac{|I_s|^M}{\operatorname{dist}(J_k,I_s)^M}\Bigg\}^{\frac12}\\
&\le C_M \frac{|F|}{|E|}\Bigg\{\sum_{k:\,J_k\subseteq 3I_t}\frac{1}{|J_k|}\sum_{m=-\infty}^{\log_2|J_k|} 2^{2m}\Bigg\}^{\frac12}\\
&\le C_M \frac{|F|}{|E|}\Bigg\{\sum_{k:\,J_k\subseteq 3I_t}\frac{1}{|J_k|}|J_k|^2\Bigg\}^{\frac12}\\
&\le C_M \frac{|F|}{|E|}\,|I_t|^{\frac12}.
\end{aligned}
$$

Finally, there is the subcase of case (b) in which $|J_k|\ge|I_s|$ and $I_s\subseteq 3J_k$. Here again we use estimate (11.2.27). We have

$$
\begin{aligned}
&\Bigg\{\sum_{s\in\mathbf{T}}\Bigg|\sum_{\substack{k:\,J_k\subseteq 3I_t\\ J_k\cap 3I_s=\emptyset\\ |J_k|>|I_s|\\ I_s\subseteq 3J_k}} \langle b_k e^{2\pi i c(\omega_t)(\cdot)}\,|\,\varphi_s\rangle\Bigg|^2\Bigg\}^{\frac12}\\
&\qquad\le C_M \frac{|F|}{|E|}\Bigg\{\sum_{s\in\mathbf{T}}|I_s|\Bigg|\sum_{\substack{k:\,J_k\subseteq 3I_t\\ J_k\cap 3I_s=\emptyset\\ |J_k|>|I_s|\\ I_s\subseteq 3J_k}} \frac{|I_s|^M}{\operatorname{dist}(J_k,I_s)^M}\Bigg|^2\Bigg\}^{\frac12}.
\end{aligned}
\tag{11.2.29}
$$

Let us make some observations. For a fixed s there exist at most finitely many J_k's contained in $3I_t$ with size at least $|I_s|$. Let $J_L^1(s)$ be the interval that lies to the left of I_s and is closest to I_s among all J_k that satisfy the conditions in the preceding sum. Then $|J_L^1(s)| > |I_s|$ and

$$\operatorname{dist}(J_L^1(s), I_s) \geq |I_s|.$$

Let $J_L^2(s)$ be the interval to the left of $J_L^1(s)$ that is closest to $J_L^1(s)$ and that satisfies the conditions of the sum. Since $3J_L^2(s)$ contains I_s, it follows that $|J_L^2(s)| > 2|I_s|$ and

$$\operatorname{dist}(J_L^2(s), I_s) \geq 2|I_s|.$$

Continuing in this way, we can find a finite number of intervals $J_L^r(s)$ that lie to the left of I_s and inside $3I_t$, satisfy $|J_L^r(s)| > 2^r|I_s|$ and $\operatorname{dist}(J_L^r(s), I_s) \geq 2^r|I_s|$, and whose triples contain I_s. Likewise we find a finite collection of intervals $J_R^1(s), J_R^2(s), \dots$ that lie to the right of I_s and satisfy similar conditions. Then, using the Cauchy–Schwarz inequality, we obtain

$$\begin{aligned}
&\Bigg| \sum_{\substack{k:\, J_k \subseteq 3I_t \\ J_k \cap 3I_s = \emptyset \\ |J_k| > |I_s| \\ I_s \subseteq 3J_k}} \frac{|I_s|^M}{\operatorname{dist}(J_k, I_s)^M} \Bigg|^2 \\
&\leq 2\Bigg| \sum_{r=1}^{\infty} \frac{|I_s|^{\frac{M}{2}}}{\operatorname{dist}(J_L^r(s), I_s)^{\frac{M}{2}}} \frac{1}{2^{\frac{rM}{2}}} \Bigg|^2 + 2\Bigg| \sum_{r=1}^{\infty} \frac{|I_s|^{\frac{M}{2}}}{\operatorname{dist}(J_R^r(s), I_s)^{\frac{M}{2}}} \frac{1}{2^{\frac{rM}{2}}} \Bigg|^2 \\
&\leq C_M \sum_{r=1}^{\infty} \frac{|I_s|^M}{\operatorname{dist}(J_L^r(s), I_s)^M} + C_M \sum_{r=1}^{\infty} \frac{|I_s|^M}{\operatorname{dist}(J_R^r(s), I_s)^M} \\
&\leq C_M \sum_{\substack{k:\, J_k \subseteq 3I_t \\ J_k \cap 3I_s = \emptyset \\ |J_k| > |I_s| \\ I_s \subseteq 3J_k}} \frac{|I_s|^M}{\operatorname{dist}(J_k, I_s)^M}.
\end{aligned}$$

We use this estimate to control the expression on the left in (11.2.29) by

$$\begin{aligned}
&C_M \frac{|F|}{|E|} \Bigg\{ \sum_{s \in \mathbf{T}} |I_s| \sum_{\substack{k:\, J_k \subseteq 3I_t \\ J_k \cap 3I_s = \emptyset \\ |J_k| > |I_s| \\ I_s \subseteq 3J_k}} \frac{|I_s|^M}{\operatorname{dist}(J_k, I_s)^M} \Bigg\}^{\frac{1}{2}} \\
&\qquad \leq C_M \frac{|F|}{|E|} \Bigg\{ \sum_{k: J_k \subseteq 3I_t} |J_k| \sum_{m=0}^{\infty} 2^{-m} \sum_{\substack{s:\, I_s \subseteq 3J_k \\ J_k \cap 3I_s = \emptyset \\ |I_s| = 2^{-m}|J_k|}} \frac{|I_s|^M}{\operatorname{dist}(J_k, I_s)^M} \Bigg\}^{\frac{1}{2}}.
\end{aligned}$$

Since the last sum is at most a constant, it follows that the term on the left in (11.2.29) also satisfies the estimate $C_M \frac{|F|}{|E|} |I_t|^{\frac{1}{2}}$. This concludes the proof of Lemma 11.2.2.

Exercises

11.2.1. Let $\mathbf{T}$ be a 2-tree with top I_t and let $M > 1$ and L be such that $2^L < |I_t|$. Show that there exists a constant $C_M > 0$ such that

$$\sum_{s\in\mathbf{T}} |I_s|^M \le C_M |I_t|^M ,$$

$$\sum_{\substack{s\in\mathbf{T}\\ |I_s|\ge 2^L}} |I_s|^{-M} \le C_M \frac{|I_t|}{(2^L)^{M+1}} ,$$

$$\sum_{\substack{s\in\mathbf{T}\\ |I_s|\le 2^L}} |I_s|^M \le C_M |I_t| (2^L)^{M-1} .$$

[*Hint:* Group the s that appear in each sum in families $\mathscr{G}_m$ such that $|I_s| = 2^{-m}|I_t|$ for each $s \in \mathscr{G}_m$.]

11.2.2. Show that the operator

$$g \mapsto \sup_{-\infty<a<b<\infty} \left|(\widehat{g}\chi_{[a,b]})^{\vee}\right|$$

defined on the line is L^p bounded for all $1 < p < \infty$.

11.2.3. On $\mathbf{R}^n$ fix a unit vector b and consider the maximal operator

$$T(g)(x) = \sup_{N>0} \left| \int_{|b\cdot\xi|\le N} \widehat{g}(\xi) e^{2\pi i x\cdot\xi}\, d\xi \right| .$$

Show that T maps $L^p(\mathbf{R}^n)$ to $L^p(\mathbf{R}^n)$ for all $1 < p < \infty$.
[*Hint:* Apply a rotation.]

11.2.4. Define the *directional Carleson operators* by

$$\mathscr{C}^\theta(f)(x) = \sup_{a\in\mathbf{R}} \left| \lim_{\varepsilon\to 0} \int_{\varepsilon<|t|<\varepsilon^{-1}} e^{2\pi i a t} f(x - t\theta) \frac{dt}{t} \right| ,$$

for functions f on $\mathbf{R}^n$. Here θ is a vector in $\mathbf{S}^{n-1}$.
(a) Show that $\mathscr{C}^\theta$ is bounded on $L^p(\mathbf{R}^n)$ for all $1 < p < \infty$.
(b) Let Ω be an odd integrable function on $\mathbf{S}^{n-1}$. Define an operator

$$\mathscr{C}^\Omega(f)(x) = \sup_{\xi\in\mathbf{R}^n} \left| \lim_{\varepsilon\to 0} \int_{\varepsilon<|y|<\varepsilon^{-1}} e^{2\pi i \xi\cdot y} f(x-y) \frac{\Omega(\frac{y}{|y|})}{|y|^n}\, dy \right| .$$

Show that $\mathscr{C}^\Omega$ is bounded on $L^p(\mathbf{R}^n)$ for $1 < p < \infty$.
[*Hint:* Part (a): Reduce to the case $\theta = e_1 = (1,0,\dots,0)$ via a rotation and use Theorem 11.2.1(b). Part (b): Use the method of rotations and part (a).]

11.3 The Maximal Carleson Operator and Weighted Estimates

Recall the one-sided Carleson operator $\mathscr{C}_1$ defined in the previous section:

$$\mathscr{C}_1(f)(x) = \sup_{N>0}\left|\int_{-\infty}^{N} \widehat{f}(\xi)e^{2\pi i x\xi}\,d\xi\right|.$$

Recall also the modulation operator $M^a(g)(x) = g(x)e^{2\pi i a x}$. We begin by observing that the following identity is valid:

$$\big(\widehat{f}\chi_{(-\infty,b]}\big)^{\vee} = M^b\frac{I-iH}{2}M^{-b}(f) = \frac{1}{2}f - \frac{i}{2}M^b H M^{-b}(f), \tag{11.3.1}$$

where H is the Hilbert transform. It follows from (11.3.1) that

$$\mathscr{C}_1(f) \le \frac{1}{2}|f| + \frac{1}{2}\sup_{\xi\in\mathbf{R}}|H(M^{\xi}(f))|$$

and that

$$\sup_{\xi\in\mathbf{R}}|H(M^{\xi}(f))| \le |f| + 2\mathscr{C}_1(f).$$

We conclude that the L^p boundedness of the sublinear operator $f\mapsto\mathscr{C}_1(f)$ is equivalent to that of the sublinear operator

$$f\mapsto \sup_{\xi\in\mathbf{R}}|H(M^{\xi}(f))|.$$

Definition 11.3.1. The *maximal Carleson operator* is defined by

$$\begin{aligned}\mathscr{C}_*(f)(x) &= \sup_{\varepsilon>0}\sup_{\xi\in\mathbf{R}}\left|\int_{|x-y|>\varepsilon} f(y)e^{2\pi i\xi y}\frac{dy}{x-y}\right| \\ &= \sup_{\xi\in\mathbf{R}}\left|H^{(*)}(M^{\xi}(f))(x)\right|,\end{aligned} \tag{11.3.2}$$

where $H^{(*)}$ is the maximal Hilbert transform. Observe that $\mathscr{C}_*(f)$ is well defined for all f in $\bigcup_{1\le p<\infty}L^p(\mathbf{R})$ and that $\mathscr{C}_*(f)$ controls the Carleson operator $\mathscr{C}(f)$ pointwise.

We begin with the following pointwise estimate, which reduces the boundedness of $\mathscr{C}_*$ to that of $\mathscr{C}$:

Lemma 11.3.2. *There is a positive constant $c>0$ such that for all functions f in $\bigcup_{1\le p<\infty}L^p(\mathbf{R})$ we have*

$$\mathscr{C}_*(f) \le cM(f) + M(\mathscr{C}(f)), \tag{11.3.3}$$

where M is the Hardy–Littlewood maximal function.

Proof. The proof of (11.3.3) is based on the classical inequality

$$H^{(*)}(g) \le cM(g) + M(H(g))$$

obtained in (4.1.32). Applying this to the functions $M^{\xi}(f)$ and taking the supremum over $\xi \in \mathbf{R}$, we obtain

$$\mathscr{C}_*(f) \le cM(f) + \sup_{\xi \in \mathbf{R}} M\big(H(M^{\xi}(f))\big),$$

from which (11.3.3) easily follows by passing the supremum inside the maximal function. □

It is convenient to work with a variant of the Hardy–Littlewood maximal operator. For $0 < r < \infty$ define

$$M_r(f) = M(|f|^r)^{\frac{1}{r}}$$

for f such that $|f|^r$ is locally integrable over the real line. Note that $M(f) \le M_r(f)$ for any $r \in (1,\infty)$. Our next goal is to obtain the boundedness of the Carleson operator on weighted L^p spaces.

Theorem 11.3.3. *For every $p \in (1,\infty)$ and $w \in A_p$ there is a constant $C(p,[w]_{A_p})$ such that for all $f \in L^p(\mathbf{R})$ we have*

$$\big\|\mathscr{C}(f)\big\|_{L^p(w)} \le C(p,[w]_{A_p})\big\|f\big\|_{L^p(w)}, \tag{11.3.4}$$

$$\big\|\mathscr{C}_*(f)\big\|_{L^p(w)} \le C(p,[w]_{A_p})\big\|f\big\|_{L^p(w)}. \tag{11.3.5}$$

Proof. Fix a $1 < p < \infty$ and pick an $r \in (1,p)$ such that $w \in A_r$. We show that for all $f \in L^p(w)$ we have the estimate

$$\int_{\mathbf{R}} \mathscr{C}(f)(x)^p\, w(x)\, dx \le C_p([w]_{A_p}) \int_{\mathbf{R}} M_r(f)(x)^p w(x)\, dx. \tag{11.3.6}$$

Then the boundedness of $\mathscr{C}$ on $L^p(w)$ is a consequence of the boundedness of the Hardy–Littlewood maximal operator on $L^{\frac{p}{r}}(w)$.

If we show that for any $w \in A_p$ there is a constant $C_p([w]_{A_p})$ such that

$$\int_{\mathbf{R}} M(\mathscr{C}(f))^p\, w\, dx \le C_p([w]_{A_p}) \int_{\mathbf{R}} M_r(f)^p\, w\, dx, \tag{11.3.7}$$

then the trivial fact $\mathscr{C}(f) \le M(\mathscr{C}(f))$, inserted in (11.3.7), yields (11.3.6).

Estimate (11.3.7) will be a consequence of the following two important observations:

$$M^{\#}(\mathscr{C}(f)) \le C_r M_r(f) \qquad \text{a.e.} \tag{11.3.8}$$

and

$$\big\|M(\mathscr{C}(f))\big\|_{L^p(w)} \le c_p([w]_{A_p})\big\|M^{\#}(\mathscr{C}(f))\big\|_{L^p(w)}, \tag{11.3.9}$$

where $c_p([w]_{A_p})$ depends on $[w]_{A_p}$ and C_r depends only on r.

We begin with estimate (11.3.8), which was obtained in Theorem 7.4.9 for singular integral operators. Here this estimate is extended to maximally modulated singular integrals. To prove (11.3.8) we use the result in Proposition 7.4.2 (2). We fix $x \in \mathbf{R}$ and we pick an interval I that contains x. We write $f = f_0 + f_\infty$, where $f_0 = f\chi_{3I}$ and $f_\infty = f\chi_{(3I)^c}$. We set $a_I = \mathscr{C}(f_\infty)(c_I)$, where c_I is the center of I. Then we have

$$\begin{aligned}\frac{1}{|I|}\int_I |\mathscr{C}(f)(y) - a_I|\,dx &\le \frac{1}{|I|}\int_I \sup_{\xi\in\mathbf{R}} \big|H(M^\xi(f))(y) - H(M^\xi(f_\infty))(c_I)\big|\,dy \\ &\le B_1 + B_2\,,\end{aligned}$$

where

$$\begin{aligned}B_1 &= \frac{1}{|I|}\int_I \sup_{\xi\in\mathbf{R}} \big|H(M^\xi(f_0))(y)\big|\,dy\,, \\ B_2 &= \frac{1}{|I|}\int_I \sup_{\xi\in\mathbf{R}} \big|H(M^\xi(f_\infty))(y) - H(M^\xi(f_\infty))(c_I)\big|\,dy\,.\end{aligned}$$

But

$$\begin{aligned}B_1 &\le \frac{1}{|I|}\int_I \mathscr{C}(f_0)(y)\,dy \\ &\le \frac{1}{|I|}\big\|\mathscr{C}(f_0)\big\|_{L^r}\big\|\chi_I\big\|_{L^{r'}} \\ &\le \frac{\big\|\mathscr{C}\big\|_{L^r\to L^r}}{|I|}\big\|f_0\big\|_{L^r}|I|^{\frac{1}{r'}} \\ &\le C_r M_r(f)(x)\,,\end{aligned}$$

where we used the boundedness of the Carleson operator $\mathscr{C}$ from L^r to L^r and Theorem 1.4.17 (v).

We turn to the corresponding estimate for B_2. We have

$$\begin{aligned}B_2 &\le \frac{1}{|I|}\int_I\int_{\mathbf{R}^n} |f_\infty(z)|\left|\frac{1}{y-z} - \frac{1}{c_I - z}\right| dz\,dy \\ &= \frac{1}{|I|}\int_I\int_{(3I)^c} |f(z)|\left|\frac{y-c_I}{(y-z)(c_I-z)}\right| dz\,dy \\ &\le \int_I \left(\int_{(3I)^c} |f(z)|\,\frac{C}{(|c_I - z| + |I|)^2}\,dz\right) dy \\ &\le \int_I \frac{C}{|I|} M(f)(x)\,dy \\ &\le C M(f)(x) \\ &\le C M_r(f)(x)\,.\end{aligned}$$

This completes the proof of estimate (11.3.8), and we now turn to the proof of estimate (11.3.9). We derive (11.3.9) as a consequence of Exercise 9.4.9, provided we have that

$$\big\|M(\mathscr{C}(f))\big\|_{L^r(w)} < \infty. \tag{11.3.10}$$

Unfortunately, the finiteness estimate (11.3.10) for general functions f in $L^p(w)$ cannot be deduced easily without knowledge of the sought estimate (11.3.4) for $p = r$. However, we can show the validity of (11.3.10) for functions f with compact support and weights $w \in A_p$ that are bounded. This argument requires a few technicalities, which we now present. For a fixed constant B we introduce a truncated Carleson operator

$$\mathscr{C}^B(f) = \sup_{|\xi|\le B} \big|H(M^{\xi}(f))\big|.$$

Next we work with a weight w in A_p that is bounded. In fact, we work with $w_k = \min(w,k)$, which satisfies

$$[w_k]_{A_p} \le (1+2^{p-2})\big(1+[w]_{A_p}\big)$$

for all $k \ge 1$ (see Exercise 9.1.9). Finally, we take $f = h$ to be a smooth function with support contained in an interval $[-R,R]$. Then for $|\xi| \le B$ we have

$$\big|H(M^{\xi}(h))(x)\big| \le 2R\big\|(M^{\xi}(h))'\big\|_{L^\infty}\chi_{|x|\le 2R} + \frac{\|h\|_{L^1}}{|x|+R}\chi_{|x|>2R} \le \frac{BC_hR}{|x|+R},$$

where C_h is a constant that depends on h. This implies that the last estimate also holds for $\mathscr{C}^B(h)$. Using Example 2.1.8, we now obtain

$$M(\mathscr{C}^B(h))(x) \le BC_h\frac{\log\big(1+\frac{|x|}{R}\big)}{1+\frac{|x|}{R}}.$$

It follows that $M(\mathscr{C}^B(h))$ lies in $L^r(w_k)$, since $r > 1$ and $w_k \le k$. Therefore,

$$\big\|M(\mathscr{C}^B(f))\big\|_{L^r(w_k)} < \infty,$$

and thus (11.3.10) holds in this setting. Applying the previous argument to $\mathscr{C}^B(h)$ and the weight w_k [in lieu of $\mathscr{C}(f)$ and w], we obtain (11.3.7) and thus (11.3.4) for $M(\mathscr{C}^B(h))$ and the weight w_k. This establishes the estimate

$$\big\|\mathscr{C}^B(h)\big\|_{L^p(w_k)} \le C(p,[w]_{A_p})\big\|h\big\|_{L^p(w_k)} \tag{11.3.11}$$

for some constant $C(p,[w]_{A_p})$ that is independent of B and k, for functions h that are smooth and compactly supported. Letting $k \to \infty$ in (11.3.11) and applying Fatou's lemma, we obtain (11.3.4) for smooth functions h with compact support. From this we deduce the validity of (11.3.4) for general functions f in $L^p(w)$ by density (cf. Exercise 4.3.11).

Finally, to obtain (11.3.5) for general $f \in L^p(w)$, we raise (11.3.3) to the power p, use the inequality $(a+b)^p \le 2^p(a^p+b^p)$, and integrate over $\mathbf{R}$ with respect to the measure $w\,dx$ to obtain

$$\int_{\mathbf{R}} \mathscr{C}_*(f)^p w\,dx \le 2^p c \int_{\mathbf{R}} M(f)^p w\,dx + 2^p \int_{\mathbf{R}} M(\mathscr{C}(f))^p w\,dx. \tag{11.3.12}$$

Then we use estimate (11.3.4) and the boundedness of the Hardy–Littlewood maximal operator on $L^p(w)$ to obtain the required conclusion. □

Exercises

11.3.1. (a) Let $\theta \in \mathbf{S}^{n-1}$. Define the *maximal directional Carleson operator*

$$\mathscr{C}_*^\theta(f)(x) = \sup_{a\in\mathbf{R}}\sup_{\varepsilon>0}\left|\int_{\varepsilon<|t|<\varepsilon^{-1}} e^{2\pi i a t} f(x-t\theta)\frac{dt}{t}\right|$$

for functions f on $\mathbf{R}^n$. Prove that $\mathscr{C}_*^\theta$ is bounded on $L^p(\mathbf{R}^n, w)$ for any weight $w \in A_p$ and $1 < p < \infty$.
(b) Let Ω be an odd integrable function on $\mathbf{S}^{n-1}$. Obtain the same conclusion for the maximal operator

$$\mathscr{C}_*^\Omega(f)(x) = \sup_{\xi\in\mathbf{R}^n}\sup_{\varepsilon>0}\left|\int_{\varepsilon<|y|<\varepsilon^{-1}} e^{2\pi i \xi\cdot y} f(x-y)\frac{\Omega(\frac{y}{|y|})}{|y|^n}\,dy\right|.$$

[*Hint:* Part (a): Reduce to the case $\theta = e_1 = (1,0,\dots,0)$ via a rotation and use Theorem 11.3.3 with $w=1$. Part (b): Use the method of rotations and part (a).]

11.3.2. For a fixed $\lambda > 0$ write

$$\{x\in\mathbf{R}:\ \mathscr{C}_*(f)(x) > \lambda\} = \bigcup_j I_j,$$

where $I_j = (\alpha_j, \alpha_j+\delta_j)$ are open disjoint intervals. Let $1 < r < \infty$. Show that there exists a $\gamma_0 > 0$ such that for every $0<\gamma<\gamma_0$ there exists a constant $C_\gamma > 0$ such that $\lim_{\gamma\to 0} C_\gamma = 0$ and

$$\big|\{x\in I_j:\ \mathscr{C}_*(f)(x) > 3\lambda,\ M_r(f)(x)\le\gamma\lambda\}\big| \le C_\gamma |I_j|$$

for all f for which $C_*(f)$ is defined.
[*Hint:* Note that we must have $\mathscr{C}_*(f)(\alpha_j)\le\lambda$ and $\mathscr{C}_*(f)(\alpha_j+\delta_j)\le\lambda$ for all j. Set $I_j^* = (\alpha_j - 5\delta_j, \alpha_j+6\delta_j)$, $f_1(x) = f(x)$ for $x\in I_j^*$, $f_1(x)=0$ for $x\notin I_j^*$, and $f_2(x) = f(x) - f_1(x)$. We may assume that for all j there exists a z_j in I_j such that $M_r(f)(z_j)\le\gamma\lambda$. For fixed $x\in I_j$ estimate $|H^{(\varepsilon)}(f_2)(x) - H^{(\varepsilon)}(f_2)(\alpha_j)|$ by the threefold sum

$$\left|\int_{|\alpha_j-t|>\varepsilon} f_2(t)e^{2\pi i\xi t}\left(\frac{2}{\alpha_j-t}-\frac{2}{x-t}\right)dt\right|$$
$$+\left|\int_{|x-t|>\varepsilon\geq|\alpha_j-t|} f_2(t)e^{2\pi i\xi t}\frac{1}{x-t}\,dt\right|$$
$$+\left|\int_{|\alpha_j-t|>\varepsilon\geq|x-t|} f_2(t)e^{2\pi i\xi t}\frac{1}{\alpha_j-t}\,dt\right|,$$

which is easily shown to be controlled by $c_0 M(f)(z_j)$ for some constant c_0. Thus $\mathscr{C}_*(f_2)(x) \le \mathscr{C}_*(f_2)(\alpha_j) + c_0 M(f)(z_j) \le \lambda + c_0\gamma\lambda$. Select γ_0 such that $c_0\gamma_0 < \frac{1}{2}$. Then $\lambda + c_0\gamma\lambda < \frac{3}{2}\lambda$ for $\gamma < \gamma_0$; hence we have $\mathscr{C}_*(f)(x) \le \mathscr{C}_*(f_1)(x) + \frac{3}{2}\lambda$ for $x \in I_j$ and thus $I_j \cap \{\mathscr{C}_*(f) > 3\lambda\} \subseteq \{\mathscr{C}_*(f_1) > \lambda\}$. Using the boundedness of $\mathscr{C}_*$ on L^r and the fact that $M_r(f)(z_j) \le \gamma\lambda$, we obtain that the last set has measure at most a constant multiple of $\gamma^r|I_j|$.]

11.3.3. (*Hunt and Young [173]*) Show that for every w in A_∞ there is a finite constant $\gamma_0 > 0$ such that for all $0 < \gamma < \gamma_0$ and all $1 < r < \infty$ there is a constant B_γ such that

$$w(\{\mathscr{C}_*(f) > 3\lambda\} \cap \{M_r(f) \le \gamma\lambda\}) \le B_\gamma w(\{\mathscr{C}_*(f) > \lambda\})$$

for all f for which $C_*(f)$ is finite. Moreover, the constants B_γ satisfy $B_\gamma \to 0$ as $\gamma \to 0$.
[*Hint:* Start with positive constants C_0 and δ such that for all intervals I and any measurable set E we have $|E\cap I| \le \varepsilon|I| \implies w(E\cap I) \le C_0\,\varepsilon^\delta w(I)$. Use the estimate of Exercise 11.3.3 with $I = I_j$ and sum over j to obtain the required estimate with $B_\gamma = C_0\,(C_\gamma)^\delta$.]

11.3.4. Prove the following vector-valued version of Theorem 11.2.1:

$$\Big\|\Big(\sum_j |\mathscr{C}(f_j)|^r\Big)^{\frac{1}{r}}\Big\|_{L^p(w)} \le C_{p,r}(w)\Big\|\Big(\sum_j |f_j|^r\Big)^{\frac{1}{r}}\Big\|_{L^p(w)}$$

for all $1 < p, r < \infty$, all weights $w \in A_p$, and all sequences of functions f_j in $L^p(w)$.
[*Hint:* You may want to use Corollary 9.5.7.]

HISTORICAL NOTES

A version of Theorem 11.1.1 concerning the maximal partial sum operator of Fourier series of square-integrable functions on the circle was first proved by Carleson [55]. An alternative proof of Carleson's theorem was provided by Fefferman [126], pioneering a set of ideas called time–frequency analysis. Lacey and Thiele [205] provided the first independent proof on the line of the boundedness of the maximal Fourier integral operator (11.1.1). The proof of Theorem 11.1.1 given in this text follows closely the one given in Lacey and Thiele [205], which improves in some ways that of Fefferman's [126], by which it was inspired. One may also consult the expository article of Thiele [312]. The proof of Lacey and Thiele was a byproduct of their work [203], [204] on the boundedness of the bilinear Hilbert transforms $H_\alpha(f_1,f_2)(x) = \frac{1}{\pi}\text{p.v.}\int_{\mathbf{R}} f_1(x-t)f_2(x-\alpha t)\,\frac{dt}{t}$. This family of operators arose in early attempts of A. Calderón to show that the first commutator (Example 8.3.8, $m=1$) is bounded on L^2 when A' is in L^∞, an approach completed only using the uniform boundedness of H_α obtained by Thiele [311], Grafakos and Li [150], and Li [212].

A version of Theorem 11.2.1 concerning the L^p boundedness, $1<p<\infty$, of the maximal partial sum operator on the circle was obtained by Hunt [170]. Sjölin [283] extended this result to $L(\log^+ L)(\log^+\log^+ L)$ and Antonov [5] to $L(\log^+ L)(\log^+\log^+\log^+ L)$. Counterexamples of Kolmogorov [191], [192], Körner [197], and Konyagin [193] indicate that the everywhere convergence of partial Fourier sums (or integrals) may fail for functions in L^1 and in spaces near L^1. The exponential decay estimate for $\alpha \ge 1$ in (11.2.1) and the restricted weak type (p,p) estimate with constant $Cp^2(p-1)^{-1}$ for the maximal partial sum operator on the circle are contained in Hunt's article [170]. The estimate for $\alpha<1$ in (11.2.1) appears in the article of Grafakos, Tao, and Terwilleger [153]; the proof of Theorem 11.2.1 is based on this article. This article also investigates higher-dimensional analogues of the theory that were initiated in Pramanik and Terwilleger [266]. Theorem 11.3.3 was first obtained by Hunt and Young [173] using a good lambda inequality for the Carleson operator. An improved good lambda inequality for the Carleson operator is contained in of Grafakos, Martell, and Soria [152]. The particular proof of Theorem 11.3.3 given in the text is based on the approach of Rubio de Francia, Ruiz, and Torrea [276]. The books of Jørsboe and Melbro [179], Mozzochi [236], and Arias de Reyna [6] contain detailed presentations of the Carleson–Hunt theorem on the circle.

The subject of Fourier analysis is currently enjoying a surge of activity. Emerging connections with analytic number theory, combinatorics, geometric measure theory, partial differential equations, and multilinear analysis introduce new dynamics and present promising developments. These connections are also creating new research directions that extend beyond the scope of this book.

Glossary

$A \subseteq B$	A is a subset of B (not necessarily a proper subset)
$A \subsetneqq B$	A is a proper subset of B
A^c	the complement of a set A
χ_E	the characteristic function of the set E
d_f	the distribution function of a function f
f^*	the decreasing rearrangement of a function f
$f_n \uparrow f$	f_n increases monotonically to a function f
$\mathbf{Z}$	the set of all integers
$\mathbf{Z}^+$	the set of all positive integers $\{1,2,3,\dots\}$
$\mathbf{Z}^n$	the n-fold product of the integers
$\mathbf{R}$	the set of real numbers
$\mathbf{R}^+$	the set of positive real numbers
$\mathbf{R}^n$	the Euclidean n-space
$\mathbf{Q}$	the set of rationals
$\mathbf{Q}^n$	the set of n-tuples with rational coordinates
$\mathbf{C}$	the set of complex numbers
$\mathbf{C}^n$	the n-fold product of complex numbers
$\mathbf{T}$	the unit circle identified with the interval $[0,1]$
$\mathbf{T}^n$	the n-dimensional torus $[0,1]^n$,
$\lvert x \rvert$	$\sqrt{\lvert x_1\rvert^2+\cdots+\lvert x_n\rvert^2}$ when $x=(x_1,\dots,x_n)\in\mathbf{R}^n$
$\mathbf{S}^{n-1}$	the unit sphere $\{x\in\mathbf{R}^n:\ \lvert x\rvert=1\}$

e_j	the vector $(0,\dots,0,1,0,\dots,0)$ with 1 in the jth entry and 0 elsewhere
$\log t$	the logarithm to base e of $t>0$
$\log_a t$	the logarithm to base a of $t>0$ $(1\neq a>0)$
$\log^+ t$	$\max(0,\log t)$ for $t>0$
$[t]$	the integer part of the real number t
$x\cdot y$	the quantity $\sum_{j=1}^n x_j y_j$ when $x=(x_1,\dots,x_n)$ and $y=(y_1,\dots,y_n)$
$B(x,R)$	the ball of radius R centered at x in $\mathbf{R}^n$
ω_{n-1}	the surface area of the unit sphere $\mathbf{S}^{n-1}$
v_n	the volume of the unit ball $\{x\in\mathbf{R}^n:\ \lvert x\rvert<1\}$
$\lvert A\rvert$	the Lebesgue measure of the set $A\subseteq\mathbf{R}^n$
dx	Lebesgue measure
$\mathrm{Avg}_B f$	the average $\frac{1}{\lvert B\rvert}\int_B f(x)\,dx$ of f over the set B
$\langle f,g\rangle$	the real inner product $\int_{\mathbf{R}^n} f(x)g(x)\,dx$
$\langle f\,\vert\,g\rangle$	the complex inner product $\int_{\mathbf{R}^n} f(x)\overline{g(x)}\,dx$
$\langle u,f\rangle$	the action of a distribution u on a function f
p'	the number $p/(p-1)$, whenever $0<p\neq 1<\infty$
$1'$	the number ∞
∞'	the number 1
$f=O(g)$	means $\lvert f(x)\rvert\le M\lvert g(x)\rvert$ for some M for x near x_0
$f=o(g)$	means $\lvert f(x)\rvert\,\lvert g(x)\rvert^{-1}\to 0$ as $x\to x_0$
A^t	the transpose of the matrix A
A^*	the conjugate transpose of a complex matrix A
A^{-1}	the inverse of the matrix A
$O(n)$	the space of real matrices satisfying $A^{-1}=A^t$
$\lVert T\rVert_{X\to Y}$	the norm of the (bounded) operator $T:X\to Y$
$A\approx B$	means that there exists a $c>0$ such that $c^{-1}\le\frac{B}{A}\le c$
$\lvert\alpha\rvert$	indicates the size $\lvert\alpha_1\rvert+\cdots+\lvert\alpha_n\rvert$ of a multi-index $\alpha=(\alpha_1,\dots,\alpha_n)$
$\partial_j^m f$	the mth partial derivative of $f(x_1,\dots,x_n)$ with respect to x_j
$\partial^\alpha f$	$\partial_1^{\alpha_1}\cdots\partial_n^{\alpha_n} f$
$\mathscr{C}^k$	the space of functions f with $\partial^\alpha f$ continuous for all $\lvert\alpha\rvert\le k$

$\mathscr{C}_0$	space of continuous functions with compact support
$\mathscr{C}_{00}$	the space of continuous functions that vanish at infinity
$\mathscr{C}_0^\infty$	the space of smooth functions with compact support
$\mathscr{D}$	the space of smooth functions with compact support
$\mathscr{S}$	the space of Schwartz functions
$\mathscr{C}^\infty$	the space of smooth functions $\bigcap_{k=1}^\infty \mathscr{C}^k$
$\mathscr{D}'(\mathbf{R}^n)$	the space of distributions on $\mathbf{R}^n$
$\mathscr{S}'(\mathbf{R}^n)$	the space of tempered distributions on $\mathbf{R}^n$
$\mathscr{E}'(\mathbf{R}^n)$	the space of distributions with compact support on $\mathbf{R}^n$
$\mathscr{P}$	the set of all complex-valued polynomials of n real variables
$\mathscr{S}'(\mathbf{R}^n)/\mathscr{P}$	the space of tempered distributions on $\mathbf{R}^n$ modulo polynomials
$\ell(Q)$	the side length of a cube Q in $\mathbf{R}^n$
∂Q	the boundary of a cube Q in $\mathbf{R}^n$
$L^p(X,\mu)$	the Lebesgue space over the measure space (X,μ)
$L^p(\mathbf{R}^n)$	the space $L^p(\mathbf{R}^n, \lvert\cdot\rvert)$
$L^{p,q}(X,\mu)$	the Lorentz space over the measure space (X,μ)
$L^p_{\text{loc}}(\mathbf{R}^n)$	the space of functions that lie in $L^p(K)$ for any compact set K in $\mathbf{R}^n$
$\lvert d\mu\rvert$	the total variation of a finite Borel measure μ on $\mathbf{R}^n$
$\mathscr{M}(\mathbf{R}^n)$	the space of all finite Borel measures on $\mathbf{R}^n$
$\mathscr{M}_p(\mathbf{R}^n)$	the space of L^p Fourier multipliers, $1 \le p \le \infty$
$\mathscr{M}^{p,q}(\mathbf{R}^n)$	the space of translation-invariant operators that map $L^p(\mathbf{R}^n)$ to $L^q(\mathbf{R}^n)$
$\lVert\mu\rVert_{\mathscr{M}}$	$\int_{\mathbf{R}^n} \lvert d\mu\rvert$ the norm of a finite Borel measure μ on $\mathbf{R}^n$
$\mathcal{M}$	the centered Hardy–Littlewood maximal operator with respect to balls
M	the uncentered Hardy–Littlewood maximal operator with respect to balls
$\mathcal{M}_c$	the centered Hardy–Littlewood maximal operator with respect to cubes
M_c	the uncentered Hardy–Littlewood maximal operator with respect to cubes
$\mathcal{M}_\mu$	the centered maximal operator with respect to a measure μ
M_μ	the uncentered maximal operator with respect to a measure μ
M_s	the strong maximal operator
M_d	the dyadic maximal operator
$M^\#$	the sharp maximal operator

$\mathscr{M}$	the grand maximal operator
$L_s^p(\mathbf{R}^n)$	the inhomogeneous L^p Sobolev space
$\dot{L}_s^p(\mathbf{R}^n)$	the homogeneous L^p Sobolev space
$\Lambda_\alpha(\mathbf{R}^n)$	the inhomogeneous Lipschitz space
$\dot{\Lambda}_\alpha(\mathbf{R}^n)$	the homogeneous Lipschitz space
$H^p(\mathbf{R}^n)$	the real Hardy space on $\mathbf{R}^n$
$B_{s,q}^p(\mathbf{R}^n)$	the inhomogeneous Besov space on $\mathbf{R}^n$
$\dot{B}_{s,q}^p(\mathbf{R}^n)$	the homogeneous Besov space on $\mathbf{R}^n$
$\dot{B}_{s,q}^p(\mathbf{R}^n)$	the homogeneous Besov space on $\mathbf{R}^n$
$F_{s,q}^p(\mathbf{R}^n)$	the inhomogeneous Triebel–Lizorkin space on $\mathbf{R}^n$
$\dot{F}_{s,q}^p(\mathbf{R}^n)$	the homogeneous Triebel–Lizorkin space on $\mathbf{R}^n$
$BMO(\mathbf{R}^n)$	the space of functions of bounded mean oscillation on $\mathbf{R}^n$

References

1. D. Adams, *A note on Riesz potentials*, Duke Math. J. **42** (1975), 765–778.
2. R. A. Adams, *Sobolev Spaces*, Academic Press, New York, 1975.
3. G. Alexopoulos, *La conjecture de Kato pour les opérateurs différentiels elliptiques à coefficients périodiques*, C. R. Acad. Sci. Paris **312** (1991), 263–266.
4. J. Alvarez and C. Pérez, *Estimates with A_∞ weights for various singular integral operators*, Boll. Unione Mat. Ital. **8-A** (1994), 123–133.
5. N. Y. Antonov, *Convergence of Fourier series*, Proceedings of the XXth Workshop on Function Theory (Moscow, 1995), pp. 187–196, East J. Approx. **2** (1996).
6. J. Arias de Reyna, *Pointwise Convergence of Fourier Series*, Lect. Notes in Math. 1785, Springer, Berlin–Heidelberg–New York, 2002.
7. N. Aronszajn and K. T. Smith, *Theory of Bessel potentials I*, Ann. Inst. Fourier (Grenoble) **11** (1961), 385–475.
8. P. Auscher, A. McIntosh, and A. Nahmod, *Holomorphic functional calculi of operators, quadratic estimates, and interpolation*, Indiana Univ. Math. J. **46** (1997), 375–403.
9. P. Auscher and P. Tchamitchian, *Square root problem for divergence operators and related topics*, Astérisque No. 249, Société Mathématique de France, 1998.
10. P. Auscher, S. Hofmann, J. L. Lewis, and P. Tchamitchian, *Extrapolation of Carleson measures and the analyticity of Kato's square-root operators*, Acta Math. **187** (2001), 161–190.
11. P. Auscher, A. McIntosh, S. Hofmann, M. Lacey, and P. Tchamitchian, *The solution of the Kato Square Root Problem for Second Order Elliptic Operators on $\mathbb{R}^n$*, Ann. of Math. **156** (2002), 633–654.
12. A. Baernstein II and E. T. Sawyer, *Embedding and multiplier theorems for $H^p(\mathbf{R}^n)$*, Mem. Amer. Math. Soc., No. 318, 1985.
13. R. Bagby and D. Kurtz, *Covering lemmas and the sharp function*, Proc. Amer. Math. Soc. **93** (1985), 291–296.
14. R. Bagby and D. S. Kurtz, *A rearranged good-λ inequality*, Trans. Amer. Math. Soc. **293** (1986), 71–81.
15. B. Barceló, *On the restriction of the Fourier transform to a conical surface*, Trans. Amer. Math. Soc. **292** (1985), 321–333.
16. B. Barceló, *The restriction of the Fourier transform to some curves and surfaces*, Studia Math. **84** (1986), 39–69.
17. J. Barrionuevo, *A note on the Kakeya maximal operator*, Math. Res. Lett. **3** (1995), 61–65.
18. W. Beckner, A. Carbery, S. Semmes, and F. Soria, *A note on restriction of the Fourier transform to spheres*, Bull. London Math. Soc. **21** (1989), 394–398.
19. C. Bennett, R. A. DeVore, and R. Sharpley, *Weak L^∞ and BMO*, Ann. of Math. **113** (1981), 601–611.

20. Á. Bényi, C. Demeter, A. Nahmod, C. Thiele, R. H. Torres, and P. Villaroya, *Modulation invariant bilinear $T(1)$ theorem*, to appear.
21. Á Bényi, D. Maldonado, A. Nahmod, and R. H. Torres, *Bilinear paraproducts revisited*, Math. Nach., to appear.
22. A. Besicovitch, *On Kakeya's problem and a similar one*, Math. Z. **27** (1928), 312–320.
23. A. Besicovitch, *A general form of the covering principle and relative differentiation of additive functions*, Proc. of Cambridge Phil. Soc. **41** (1945), 103–110.
24. O. V. Besov, *On a family of function spaces. Embedding theorems and applications* [in Russian], Dokl. Akad. Nauk SSSR **126** (1959), 1163–1165.
25. O. V. Besov, *On a family of function spaces in connection with embeddings and extensions* [in Russian], Trudy Mat. Inst. Steklov **60** (1961), 42–81.
26. Z. Birnbaum and M. W. Orlicz, *Über die Verallgemeinerung des Begriffes der Zueinander konjugierten Potenzen*, Studia Math. **3** (1931), 1–67; reprinted in W. Orlicz, "Collected Papers," pp. 133–199, PWN, Warsaw, 1988.
27. S. Bochner, *Summation of multiple Fourier series by spherical means*, Trans. Amer. Math. Soc. **40** (1936), 175–207.
28. J. M. Bony, *Calcul symbolique et propagation des singularités pour les équations aux dérivées partielles non linéaires*, Ann. Sci. École Norm. Sup. **14** (1981), 209–246.
29. J. Bourgain, *Besicovitch type maximal operators and applications to Fourier analysis*, Geom. Funct. Anal. **1** (1991), 147–187.
30. J. Bourgain, *On the restriction and multiplier problems in $\mathbb{R}^3$*, Geometric Aspects of functional analysis (1989–90), pp. 179–191, Lect. Notes in Math. 1469, Springer, Berlin, 1991.
31. J. Bourgain, *Some new estimates on oscillatory integrals*, Essays on Fourier Analysis in Honor of E. M. Stein, pp. 83-112, C. Fefferman, R. Fefferman, and S. Wainger (eds.), Princeton Univ. Press, Princeton, NJ, 1995.
32. J. Bourgain, *On the dimension of Kakeya sets and related maximal inequalities*, Geom. Funct. Anal. **9** (1999), 256–282.
33. M. Bownik, *Anisotropic Hardy spaces and wavelets*, Mem. Amer. Math. Soc. **164** (2003), no. 781, 122pp.
34. M. Bownik, *Boundedness of operators on Hardy spaces via atomic decompositions*, Proc. Amer. Math. Soc. **133** (2005), 3535–3542.
35. M. Bownik, B. Li, D. Yang, and Y. Zhou, *Weighted anisotropic Hardy spaces and their applications in boundedness of sublinear operators*, Indiana Univ. Math. J., to appear.
36. S. M. Buckley, *Estimates for operator norms on weighted spaces and reverse Jensen inequalities*, Trans. Amer. Math. Soc. **340** (1993), 253–272.
37. H. Q. Bui, *Some aspects of weighted and non-weighted Hardy spaces*, Kôkyûroku Res. Inst. Math. Sci. **383** (1980), 38–56.
38. D. L. Burkholder, R. F. Gundy, and M. L. Silverstein, *A maximal characterization of the class H^p*, Trans. Amer. Math. Soc. **157** (1971), 137–153.
39. S. Campanato, *Proprietà di hölderianità di alcune classi di funzioni*, Ann. Scuola Norm. Sup. Pisa **17** (1963), 175–188.
40. S. Campanato, *Proprietà di una famiglia di spazi funzionali*, Ann. Scuola Norm. Sup. Pisa **18** (1964), 137–160.
41. A. P. Calderón, *Lebesgue spaces of differentiable functions and distributions*, Proc. Symp. Pure Math. **4** (1961), 33–49.
42. A. P. Calderón, *Commutators of singular integral operators*, Proc. Nat. Acad. Sci. USA **53** (1965), 1092–1099.
43. A. P. Calderón, *Cauchy integrals on Lipschitz curves and related operators*, Proc. Nat. Acad. Sci. USA **74** (1977), 1324–1327.
44. A. P. Calderón, *An atomic decomposition of distributions in parabolic H^p spaces*, Adv. in Math. **25** (1977), 216–225.
45. A. P. Calderón and A. Torchinsky, *Parabolic maximal functions associated with a distribution*, Adv. in Math. **16** (1975), 1–63.

46. A. P. Calderón and A. Torchinsky, *Parabolic maximal functions associated with a distribution*, Adv. in Math. **24** (1977), 101–171.
47. A. P. Calderón and A. Zygmund, *Singular integrals and periodic functions*, Studia Math. **14** (1954), 249–271.
48. A. P. Calderón and A. Zygmund, *Commutators, singular integrals on Lipschitz curves and applications*, Proceedings of the International Congress of Mathematicians (Helsinki, 1978), pp. 85–96, Acad. Sci. Fennica, Helsinki, 1980.
49. A. P. Calderón and R. Vaillancourt, *A class of bounded pseudo-differential operators*, Proc. Nat. Acad. Sci. USA **69** (1972), 1185–1187.
50. A. Carbery, *The boundedness of the maximal Bochner–Riesz operator on* $L^4(\mathbb{R}^2)$, Duke Math. J. **50** (1983), 409–416.
51. A. Carbery, E. Hernández, and F. Soria, *Estimates for the Kakeya maximal operator on radial functions in* $\mathbf{R}^n$, Harmonic Analysis, ICM 1990 Satellite Conference Proceedings, pp. 41–50, S. Igari (ed.), Springer-Verlag, Tokyo, 1991.
52. A. Carbery, J.-L. Rubio de Francia, and L. Vega, *Almost everywhere summability of Fourier integrals*, J. London Math. Soc. **38** (1988), 513–524.
53. L. Carleson, *An interpolation problem for bounded analytic functions*, Amer. J. Math. **80** (1958), 921–930.
54. L. Carleson, *Interpolation by bounded analytic functions and the corona problem*, Ann. of Math. **76** (1962), 547–559.
55. L. Carleson, *On convergence and growth of partial sums of Fourier series*, Acta Math. **116** (1966), 135–157.
56. L. Carleson, *On the Littlewood–Paley Theorem*, Mittag-Leffler Institute Report, Djursholm, Sweden 1967.
57. L. Carleson, *Two remarks on* H^1 *and* $B.M.O$, Adv. in Math. **22** (1976), 269–277.
58. L. Carleson and P. Sjölin, *Oscillatory integrals and a multiplier problem for the disc*, Studia Math. **44** (1972), 287–299.
59. D.-C. Chang, S. G. Krantz, and E. M. Stein, H^p *theory on a smooth domain in* R^N *and elliptic boundary value problems*, J. Funct. Anal. **114** (1993), 286–347.
60. F. Chiarenza and M. Frasca, *Morrey spaces and Hardy–Littlewood maximal function*, Rend. Mat. Appl. Series 7, **7** (1981), 273–279.
61. M. Christ, *Estimates for the k-plane transform*, Indiana Univ. Math. J. **33** (1984), 891–910.
62. M. Christ, *On the restriction of the Fourier transform to curves: endpoint results and the degenerate case*, Trans. Amer. Math. Soc. **287** (1985), 223–238.
63. M. Christ, *On almost everywhere convergence of Bochner–Riesz means in higher dimensions*, Proc. Amer. Math. Soc. **95** (1985), 16–20.
64. M. Christ, *Weak type endpoint bounds for Bochner–Riesz multipliers*, Rev. Mat. Iber. **3** (1987), 25–31.
65. M. Christ, *Weak type* $(1,1)$ *bounds for rough operators I*, Ann. of Math. **128** (1988), 19–42.
66. M. Christ, *A* $T(b)$ *theorem with remarks on analytic capacity and the Cauchy integral*, Colloq. Math. **60/61** (1990), 601–628.
67. M. Christ, *Lectures on singular integral operators*, CBMS Regional Conference Series in Mathematics, Vol. 77, Amer. Math. Soc., Providence, RI, 1990.
68. M. Christ, J. Duoandikoetxea and J.-L. Rubio de Francia, *Maximal operators related to the Radon transform and the Calderón–Zygmund method of rotations*, Duke Math. J. **53** (1986), 189–209.
69. M. Christ and R. Fefferman, *A note on weighted norm inequalities for the Hardy–Littlewood maximal operator*, Proc. Amer. Math. Soc. **87** (1983), 447–448.
70. M. Christ and J.-L. Journé, *Polynomial growth estimates for multilinear singular integral operators*, Acta Math. **159** (1987), 51–80.
71. R. R. Coifman, *Distribution function inequalities for singular integrals*, Proc. Nat. Acad. Sci. USA **69** (1972), 2838–2839.

72. R. R. Coifman, *A real variable characterization of H^p*, Studia Math. **51** (1974), 269–274.
73. R. R. Coifman, D. G. Deng, and Y. Meyer, *Domaine de la racine carée de certaines opérateurs différentiels acrétifs*, Ann. Inst. Fourier **33** (1983), 123–134.
74. R. R. Coifman and C. Fefferman, *Weighted norm inequalities for maximal functions and singular integrals*, Studia Math. **51** (1974), 241–250.
75. R. R. Coifman and L. Grafakos, *Hardy space estimates for multilinear operators, I*, Rev. Mat. Iber. **8** (1992), 45–67.
76. R. R. Coifman, P. Jones, and J. L. Rubio de Francia, *Constructive decomposition of BMO functions and factorization of A_p weights*, Proc. Amer. Math. Soc. **87** (1983), 675–676.
77. R. R. Coifman, P. Jones, and S. Semmes, *Two elementary proofs of the L^2 boundedness of Cauchy integrals on Lipschitz curves*, J. Amer. Math. Soc. **2** (1989), 553–564.
78. R. R. Coifman, P. L. Lions, Y. Meyer, and S. Semmes, *Compensated compactness and Hardy spaces*, J. Math. Pures Appl. **72** (1993), 247–286.
79. R. R. Coifman, A. M^c^Intosh, Y. Meyer, *L' intégrale de Cauchy définit un opérateur borné sur L^2 pour les courbes lipschitziennes*, Ann. of Math. **116** (1982), 361–387.
80. R. R. Coifman and Y. Meyer, *Commutateurs d' intégrales singulières et opérateurs multilinéaires*, Ann. Inst. Fourier (Grenoble) **28** (1978), 177–202.
81. R. R. Coifman and Y. Meyer, *Au délà des opérateurs pseudo-différentiels*, Astérisque No. 57, Societé Mathematique de France, 1979.
82. R. R. Coifman and Y. Meyer, *A simple proof of a theorem by G. David and J.-L. Journé on singular integral operators*, Probability Theory and Harmonic Analysis, pp. 61–65, J. Chao and W. Woyczyński (eds.), Marcel Dekker, New York, 1986.
83. R. R. Coifman, Y. Meyer, and E. M. Stein, *Some new function spaces and their applications to harmonic analysis*, J. Funct. Anal. **62** (1985), 304–335.
84. R. R. Coifman and R. Rochberg, *Another characterization of BMO*, Proc. Amer. Math. Soc. **79** (1980), 249–254.
85. R. R. Coifman, R. Rochberg, and G. Weiss, *Factorization theorems for Hardy spaces in several variables*, Ann. of Math. **103** (1976), 611–635.
86. R. R. Coifman and G. Weiss, *Extensions of Hardy spaces and their use in analysis*, Bull. Amer. Math. Soc. **83** (1977), 569–645.
87. L. Colzani, G. Travaglini, and M. Vignati, *Bochner–Riesz means of functions in weak-L^p*, Monatsh. Math. **115** (1993), 35–45.
88. W. C. Connett, *Singular integrals near L^1*, Harmonic analysis in Euclidean spaces, Proc. Sympos. Pure Math. (Williams Coll., Williamstown, Mass., 1978), pp. 163–165, Amer. Math. Soc., Providence, RI, 1979.
89. A. Córdoba, *The Kakeya maximal function and the spherical summation multipliers*, Amer. J. Math. **99** (1977), 1–22.
90. A. Córdoba, *A note on Bochner–Riesz operators*, Duke Math. J. **46** (1979), 505–511.
91. A. Córdoba, *Multipliers of $F(L^p)$*, Euclidean harmonic analysis (Proc. Sem., Univ. Maryland, College Park, Md., 1979), pp. 162–177, Lecture Notes in Math. 779, Springer, Berlin, 1980.
92. A. Córdoba and C. Fefferman, *A weighted norm inequality for singular integrals*, Studia Math. **57** (1976), 97–101.
93. A. Córdoba and R. Fefferman, *On differentiation of integrals*, Proc. Nat. Acad. Sci. USA **74** (1977), 2211–2213.
94. M. Cotlar, *A combinatorial inequality and its applications to L^2 spaces*, Rev. Mat. Cuyana, **1** (1955), 41–55.
95. D. Cruz-Uribe, *New proofs of two-weight norm inequalities for the maximal operator*, Georgian Math. J. **7** (2000), 33–42.
96. D. Cruz-Uribe, J. M. Martell, and C. Pérez, *Extrapolation results for A_∞ weights and applications*, J. Funct. Anal. **213** (2004), 412–439.
97. D. Cruz-Uribe, J. M. Martell, and C. Pérez, *Weights, Extrapolation and the Theory of Rubio de Francia*, in preparation.
98. D. Cruz-Uribe SFO and C. J. Neugebauer, *The structure of the reverse Hölder classes*, Trans. Amer. Math. Soc. **347** (1995), 2941–2960.

99. F. Cunningham, *The Kakeya problem for simply connected and for star-shaped sets*, Amer. Math. Monthly **78** (1971), 114–129.
100. G. Dafni, *Hardy spaces on some pseudoconvex domains*, J. Geom. Anal. **4** (1994), 273–316.
101. G. Dafni, *Local VMO and weak convergence in h^1*, Can. Math. Bull. **45** (2002), 46–59.
102. G. David, *Opérateurs intégraux singuliers sur certains courbes du plan complexe*, Ann. Sci. Ecole Norm. Sup. **17** (1984), 157–189.
103. G. David and J.-L. Journé, *A boundedness criterion for generalized Calderón–Zygmund operators*, Ann. of Math. **120** (1984), 371–397.
104. G. David, J.-L. Journé, and S. Semmes, *Opérateurs de Calderón–Zygmund, fonctions para-accrétives et interpolation*, Rev. Math. Iber. **1** (1985), 1–56.
105. G. David and S. Semmes, *Singular integrals and rectifiable sets in $\mathbf{R}^n$: Beyond Lipschitz graphs*, Astérisque No. 193, Societé Mathematique de France, 1991.
106. K. M. Davis and Y. C. Chang, *Lectures on Bochner–Riesz Means*, London Math. Soc. Lect. Notes **114**, Cambridge Univ. Press, Cambridge, UK, 1987.
107. L. De Carli and A. Iosevich, *A restriction theorem for flat manifolds of codimension two*, Ill. J. Math. **39** (1995), 576–585.
108. L. De Carli and A. Iosevich, *Some sharp restriction theorems for homogeneous manifolds*, J. Fourier Anal. Appl. **4** (1998), 105–128.
109. M. de Guzmán, *Differentiation of Integrals in $\mathbb{R}^n$*, Lecture Notes in Math. **481**, Springer-Verlag, Berlin, 1985.
110. O. Dragičević, L. Grafakos, C. Pereyra, and S. Petermichl, *Extrapolation and sharp norm estimates for classical operators on weighted Lebesgue spaces*, Publ. Mat. **49** (2005), 73–91.
111. S. W. Drury, *L^p estimates for the x-ray transform*, Ill. J. Math. **27** (1983), 125–129.
112. S. W. Drury, *Restrictions of Fourier transforms to curves*, Ann. Inst. Fourier (Grenoble) **35** (1985), 117–123.
113. S. W. Drury and K. Guo, *Some remarks on the restriction of the Fourier transform to surfaces*, Math. Proc. Cambridge Philos. Soc. **113** (1993), 153–159.
114. S. W. Drury and B. P. Marshall, *Fourier restriction theorems for curves with affine and Euclidean arclengths*, Math. Proc. Cambridge Philos. Soc. **97** (1985), 111–125.
115. S. W. Drury and B. P. Marshall, *Fourier restriction theorems for degenerate curves*, Math. Proc. Cambridge Philos. Soc. **101** (1987), 541–553.
116. J. Duoandikoetxea, *Fourier Analysis*, Grad. Studies in Math. **29**, Amer. Math. Soc., Providence, RI, 2000.
117. P. L. Duren, *Theory of H^p Spaces*, Dover Publications Inc., New York, 2000.
118. P. L. Duren, B. W. Romberg, and A. L. Shields, *Linear functionals on H^p spaces with $0 < p < 1$*, J. Reine Angew. Math. **238** (1969), 32–60.
119. E. Fabes, D. Jerison, and C. Kenig, *Multilinear Littlewood–Paley estimates with applications to partial differential equations*, Proc. Nat. Acad. Sci. USA **79** (1982), 5746–5750.
120. E. Fabes, D. Jerison, and C. Kenig, *Multilinear square functions and partial differential equations*, Amer. J. Math. **107** (1985), 1325–1367.
121. E. Fabes, I. Mitrea, and M. Mitrea, *On the boundedness of singular integrals*, Pac. J. Math. **189** (1999), 21–29.
122. K. Fan, *Minimax theorems*, Proc. Nat. Acad. Sci. USA **39** (1953), 42–47.
123. C. Fefferman, *Inequalities for strongly singular convolution operators*, Acta Math. **124** (1970), 9–36.
124. C. Fefferman, *Characterizations of bounded mean oscillation*, Bull. Amer. Math. Soc. **77** (1971), 587–588.
125. C. Fefferman, *The multiplier problem for the ball*, Ann. of Math. **94** (1971), 330–336.
126. C. Fefferman, *Pointwise convergence of Fourier series*, Ann. of Math. **98** (1973), 551–571.
127. C. Fefferman, *A note on spherical summation multipliers*, Israel J. Math. **15** (1973), 44–52.

128. C. Fefferman, N. Riviere, and Y. Sagher, *Interpolation between H^p spaces: The real method*, Trans. Amer. Math. Soc. **191** (1974), 75–81.
129. C. Fefferman and E. M. Stein, *Some maximal inequalities*, Amer. J. Math. **93** (1971), 107–115.
130. C. Fefferman and E. M. Stein, *H^p spaces of several variables*, Acta Math. **129** (1972), 137–193.
131. T. M. Flett, *Lipschitz spaces of functions on the circle and the disc*, J. Math. Anal. Appl. **39** (1972), 125–158.
132. G. B. Folland and E. M. Stein, *Estimates for the $\overline{\partial}_b$ complex and analysis on the Heisenberg group*, Comm. Pure and Appl. Math. **27** (1974), 429–522.
133. G. B. Folland and E. M. Stein, *Hardy Spaces on Homogeneous Groups*, Mathematical Notes 28, Princeton Univ. Press, Princeton, NJ, 1982.
134. J. Fourier, *Théorie Analytique de la Chaleur*, Institut de France, Paris, 1822.
135. M. Frazier and B. Jawerth, *Decomposition of Besov spaces*, Indiana Univ. Math. J. **34** (1985), 777–799.
136. M. Frazier and B. Jawerth, *A discrete transform and decompositions of distribution spaces*, J. Funct. Anal. **93** (1990), 34–170.
137. M. Frazier and B. Jawerth, *Applications of the φ and wavelet transforms to the theory of function spaces*, Wavelets and Their Applications, pp. 377–417, Jones and Bartlett (eds.), Boston, MA, 1992.
138. M. Frazier, B. Jawerth, and G. Weiss, *Littlewood–Paley Theory and the Study of Function Spaces*, CBMS Regional Conference Series in Mathematics, 79, Amer. Math. Soc. Providence, RI, 1991.
139. E. Gagliardo, *Proprietà di alcune classi di funzioni in più variabili*, Ricerche di Mat. Napoli **7** (1958), 102–137.
140. J. García-Cuerva, *An extrapolation theorem in the theory of A_p weights*, Proc. Amer. Math. Soc. **87** (1983), 422–426.
141. J. García-Cuerva and J.-L. Rubio de Francia, *Weighted Norm Inequalities and Related Topics*, North-Holland Math. Studies **116**, North-Holland, Amsterdam, 1985.
142. J. Garnett, *Bounded Analytic Functions*, Academic Press, New York, 1981.
143. J. Garnett and P. Jones, *The distance in BMO to L^∞*, Ann. of Math. **108** (1978), 373–393.
144. J. Garnett and P. Jones, *BMO from dyadic BMO*, Pacific J. Math. **99** (1982), 351–371.
145. F. W. Gehring, *The L^p-integrability of the partial derivatives of a quasiconformal mapping*, Acta Math. **130** (1973), 265–277.
146. D. Goldberg, *A local version of real Hardy spaces*, Duke Math. J. **46** (1979), 27–42.
147. I. Gohberg and N. Krupnik, *Norm of the Hilbert transformation in the L_p space*, Funct. Anal. Appl. **2** (1968), 180–181.
148. L. Grafakos and N. Kalton, *Some remarks on multilinear maps and interpolation*, Math. Ann. **319** (2001), 151–180.
149. L. Grafakos and N. Kalton, *Multilinear Calderón–Zygmund operators on Hardy spaces*, Collect. Math. **52** (2001), 169–179.
150. L. Grafakos and X. Li, *Uniform bounds for the bilinear Hilbert transforms I*, Ann. of Math., **159** (2004), 889–933.
151. L. Grafakos and J. M. Martell, *Extrapolation of weighted norm inequalities for multivariable operators and applications* J. Geom. Anal. **14** (2004), 19–46.
152. L. Grafakos, J. M. Martell, and F. Soria, *Weighted norm inequalities for maximally modulated singular integral operators*, Math. Ann. **331** (2005), 359–394.
153. L. Grafakos, T. Tao, and E. Terwilleger, *L^p bounds for a maximal dyadic sum operator*, Math. Zeit. **246** (2004), 321–337.
154. L. Grafakos and R. H. Torres, *Multilinear Calderón–Zygmund theory*, Adv. in Math. **165** (2002), 124–164.
155. A. Greenleaf, *Principal curvature and harmonic analysis*, Indiana Univ. Math. J. **30** (1981), 519–537.
156. G. H. Hardy, *The mean value of the modulus of an analytic function*, Proc. London Math. Soc. **14** (1914), 269–277.

157. G. H. Hardy and J. E. Littlewood, *Some properties of fractional integrals I*, Math. Zeit. **27** (1927), 565–606.
158. G. H. Hardy and J. E. Littlewood, *Some properties of fractional integrals II*, Math. Zeit. **34** (1932), 403–439.
159. G. H. Hardy and J. E. Littlewood, *Some theorems on Fourier series and Fourier power series*, Duke Math. J. **2** (1936), 354–381.
160. G. H. Hardy and J. E. Littlewood, *Generalizations of a theorem of Paley*, Quarterly Jour. **8** (1937), 161–171.
161. L. Hedberg, *On certain convolution inequalities*, Proc. Amer. Math. Soc. **36** (1972), 505–510.
162. H. Helson and G. Szegő, *A problem in prediction theory*, Ann. Math. Pura Appl. **51** (1960), 107–138.
163. C. Herz, *On the mean inversion of Fourier and Hankel transforms*, Proc. Nat. Acad. Sci. USA **40** (1954), 996–999.
164. S. Hofmann and A. McIntosh, *The solution of the Kato problem in two dimensions*, Proceedings of the 6th International Conference on Harmonic Analysis and Partial Differential Equations (El Escorial, Spain, 2000), pp. 143–160, Publ. Mat. Extra Volume, 2002.
165. S. Hofmann, M. Lacey, and A. McIntosh, *The solution of the Kato problem for divergence form elliptic operators with Gaussian heat kernel bounds*, Ann. of Math. **156** (2002), 623–631.
166. L. Hörmander, *Linear Partial Differential Operators*, Springer-Verlag, Berlin–Göttingen–Heidelberg, 1963.
167. L. Hörmander, *Oscillatory integrals and multipliers on FL^p*, Arkiv f. Mat. **11** (1973), 1–11.
168. L. Hörmander, *The Analysis of Linear Partial Differential Operators I*, 2nd ed., Springer-Verlag, Berlin–Heidelberg–New York, 1990.
169. S. V. Hruščev, *A description of weights satisfying the A_∞ condition of Muckenhoupt*, Proc. Amer. Math. Soc. **90** (1984), 253–257.
170. R. Hunt, *On the convergence of Fourier series*, Orthogonal Expansions and Their Continuous Analogues (Edwardsville, Ill., 1967), pp. 235–255, D. T. Haimo (ed.), Southern Illinois Univ. Press, Carbondale IL, 1968.
171. R. Hunt, D. Kurtz, and C. J. Neugebauer, *A note on the equivalence of A_p and Sawyer's condition*, Conference on Harmonic Analysis in honor of Antoni Zygmund, Vol. 1, pp. 156–158, W. Beckner et al. (eds.), Wadsworth, Belmont, 1983.
172. R. Hunt, B. Muckenhoupt, and R. Wheeden, *Weighted norm inequalities for the conjugate function and the Hilbert transform*, Trans. Amer. Math. Soc. **176** (1973), 227–251.
173. R. Hunt and W.-S. Young, *A weighted norm inequality for Fourier series*, Bull. Amer. Math. Soc. **80** (1974), 274–277.
174. S. Igari, *An extension of the interpolation theorem of Marcinkiewicz II*, Tôhoku Math. J. **15** (1963), 343–358.
175. S. Igari, *On Kakeya's maximal function*, Proc. Japan Acad. Ser. A Math. Sci. **62** (1986), 292–293.
176. S. Janson, *Mean oscillation and commutators of singular integral operators*, Ark. Math. **16** (1978), 263–270.
177. F. John and L. Nirenberg, *On functions of bounded mean oscillation*, Comm. Pure and Appl. Math. **14** (1961), 415–426.
178. P. Jones, *Factorization of A_p weights*, Ann. of Math. **111** (1980), 511–530.
179. O. G. Jørsboe and L. Melbro, *The Carleson–Hunt Theorem on Fourier Series*, Lect. Notes in Math. 911, Springer-Verlag, Berlin, 1982.
180. J.-L. Journé *Calderón–Zygmund Operators, Pseudo-Differential Operators and the Cauchy Integral of Calderón*, Lect. Notes in Math. 994, Springer-Verlag, Berlin, 1983.
181. T. Kato, *Fractional powers of dissipative operators*, J. Math. Soc. Japan **13** (1961), 246–274.
182. N. Katz, *A counterexample for maximal operators over a Cantor set of directions*, Math. Res. Lett. **3** (1996), 527–536.

183. N. Katz, *Maximal operators over arbitrary sets of directions*, Duke Math. J. **97** (1999), 67–79.
184. N. Katz, *Remarks on maximal operators over arbitrary sets of directions*, Bull. London Math. Soc. **31** (1999), 700–710.
185. N. Katz, I. Laba, and T. Tao, *An improved bound on the Minkowski dimension of Besicovitch sets in* $\mathbb{R}^3$, Ann. of Math. **152** (2000), 383–446.
186. N. Katz and T. Tao, *Recent progress on the Kakeya conjecture*, Proceedings of the 6th International Conference on Harmonic Analysis and Partial Differential Equations (El Escorial, Spain, 2000), pp. 161–179, Publ. Mat. Extra Volume, 2002.
187. U. Keich, *On L^p bounds for Kakeya maximal functions and the Minkowski dimension in* $\mathbb{R}^2$, Bull. London Math. Soc. **31** (1999), 213–221.
188. C. Kenig and Y. Meyer, *Kerato's square roots of accretive operators and Cauchy kernels on Lipschitz curves are the same*, Recent progress in Fourier analysis (El Escorial, Spain, 1983), pp. 123–143, North-Holland Math. Stud., 111, North-Holland, Amsterdam, 1985.
189. C. Kenig and E. M. Stein, *Multilinear estimates and fractional integration*, Math. Res. Lett. **6** (1999), 1–15.
190. A. Knapp and E. M. Stein, *Intertwining operators for semisimple groups*, Ann. of Math. **93** (1971), 489–578.
191. A. N. Kolmogorov, *Une série de Fourier–Lebesgue divergente presque partout*, Fund. Math. **4** (1923), 324–328.
192. A. N. Kolmogorov, *Une série de Fourier–Lebesgue divergente partout*, C. R. Acad. Sci. Paris **183** (1926), 1327–1328.
193. S. V. Konyagin, *On everywhere divergence of trigonometric Fourier series*, Sbornik: Mathematics **191** (2000), 97–120.
194. P. Koosis, *Sommabilité de la fonction maximale et appartenance à H_1*, C. R. Acad. Sci. Paris **28** (1978), 1041–1043.
195. P. Koosis, *Introduction to H_p Spaces*, 2nd ed., Cambridge Tracts in Math. 115, Cambridge Univ. Press, Cambridge, UK, 1998.
196. A. A. Korenovskyy, A. K. Lerner, and A. M. Stokolos, *A note on the Gurov–Reshetnyak condition*, Math. Res. Lett. **9** (2002), 579–585.
197. T. Körner, *Everywhere divergent Fourier series*, Colloq. Math. **45** (1981), 103–118.
198. S. G. Krantz, *Fractional integration on Hardy spaces*, Studia Math. **63** (1982), 87–94.
199. S. G. Krantz, *Lipschitz spaces, smoothness of functions, and approximation theory*, Exposition. Math. **1** (1983), 193–260.
200. M. G. Krein, *On linear continuous operators in functional spaces with two norms*, Trudy Inst. Mat. Akad. Nauk Ukrain. SSRS **9** (1947), 104–129.
201. D. Kurtz, *Operator estimates using the sharp function*, Pacific J. Math. **139** (1989), 267–277.
202. D. Kurtz and R. Wheeden, *Results on weighted norm inequalities for multipliers*, Trans. Amer. Math. Soc. **255** (1979), 343–362.
203. M. T. Lacey and C. M. Thiele, *L^p bounds for the bilinear Hilbert transform, $p > 2$*, Ann. of Math. **146** (1997), 693–724.
204. M. Lacey and C. Thiele, *On Calderón's conjecture*, Ann. of Math. **149** (1999), 475–496.
205. M. Lacey and C. Thiele, *A proof of boundedness of the Carleson operator*, Math. Res. Lett. **7** (2000), 361–370.
206. R. H. Latter, *A decomposition of $H^p(\mathbf{R}^n)$ in terms of atoms*, Studia Math. **62** (1977), 92–101.
207. R. H. Latter and A. Uchiyama, *The atomic decomposition for parabolic H^p spaces*, Trans. Amer. Math. Soc. **253** (1979), 391–398.
208. A. K. Lerner, *On pointwise estimates for maximal and singular integral operators*, Studia Math. **138** (2000), 285–291.
209. A. K. Lerner, *An elementary approach to several results on the Hardy–Littlewood maximal operator*, Proc. Amer. Math. Soc., **136** (2008), 2829–2833.
210. A. Lerner, S. Ombrosi, C. Pérez, R. H. Torres, and R. Trujillo-González, *New maximal functions and multiple weights for the multilinear Calderón–Zygmund theory*, to appear.

211. A. Lerner, S. Ombrosi, and C. Pérez, *Sharp A_1 bounds for Calderón–Zygmund operators and the relationship with a problem of Muckenhoupt and Wheeden,* Internat. Math. Res. Notices (2008) Vol. **2008** article ID rnm161, 11 pages, doi:10.1093/imrn/rnm161.
212. X. Li, *Uniform bounds for the bilinear Hilbert transforms II*, Rev. Mat. Iber. **22** (2006), 1069–1126.
213. E. H. Lieb, *Sharp constants in the Hardy–Littlewood–Sobolev and related inequalities*, Ann. of Math. **18** (1983), 349–374.
214. E. H. Lieb and M. Loss, *Analysis*, Grad. Studies in Math. **14**, Amer. Math. Soc., Providence, RI, 1997.
215. J.-L. Lions, *Espaces d' interpolation and domaines de puissances fractionnaires*, J. Math. Soc. Japan **14** (1962), 233–241.
216. J.-L. Lions, P. I. Lizorkin, and S. M. Nikol'skij, *Integral representation and isomorphic properties of some classes of functions*, Ann. Sc. Norm. Sup. Pisa **19** (1965), 127–178.
217. L. Liu and D. Yang, *Boundedness of sublinear operators in Triebel–Lizorkin spaces via atoms*, to appear.
218. P. I. Lizorkin, *Properties of functions of the spaces $\Lambda^r_{p\Theta}$* [in Russian], Trudy Mat. Inst. Steklov **131** (1974), 158–181.
219. S.-Z. Lu, *Four Lectures on Real H^p Spaces*, World Scientific, Singapore, 1995.
220. A. M^c^Intosh, *On the comparability of $A^{1/2}$ and $A^{*1/2}$*, *Proc. Amer. Math. Soc.* **32** (1972), 430–434.
221. A. McIntosh, *Square roots of elliptic operators*, J. Funct. Anal. **61** (1985), 307–327.
222. A. McIntosh, *On representing closed accretive sesquilinear forms as $(A^{1/2}u, A^{*1/2}v)$*, Collège de France Seminar, Vol. III, pp. 252–267, H. Brezis and J.-L. Lions (eds.), Research Notes in Mathematics, No. 70, Pitman, 1982.
223. A. McIntosh, *Square roots of operators and applications to hyperbolic PDE*, Proceedings of the Miniconference on Operator Theory and PDE, CMA, The Australian National University, Canberra, 1983.
224. A. McIntosh and Y. Meyer, *Algèbres d' opérateurs définis par des intégrales singulières* C. R. Acad. Sci. Paris **301** (1985), 395–397.
225. D. Maldonado and V. Naibo, *Weighted norm inequalities for paraproducts and bilinear pseudodifferential operators with mild regularity*, to appear.
226. G. Mauceri, M. Picardello, and F. Ricci, *A Hardy space associated with twisted convolution*, Adv. in Math. **39** (1981), 270–288.
227. P. Mattila, *Geometry of sets and measures in Euclidean spaces. Fractals and rectifiability*, Cambridge Studies in Advanced Mathematics, 44, Cambridge Univ. Press, Cambridge, UK, 1995.
228. B. Maurey, *Théorèmes de factorization pour les opérateurs linéaires à valeurs dans un éspace L^p*, Astérisque No. 11, Societé Mathematique de France, 1974.
229. V. G. Maz'ya, *Sobolev Spaces*, Springer-Verlag, Berlin–New York, 1985.
230. S. Meda, P. Sjögren, and M. Vallarino, *On the $H^1 - L^1$ boundedness of operators*, Proc. Amer. Math. Soc. **136** (2008), 2921–2931.
231. M. Melnikov and J. Verdera, *A geometric proof of the L^2 boundedness of the Cauchy integral on Lipschitz graphs*, Internat. Math. Res. Notices **7** (1995), 325–331.
232. Y. Meyer, *Régularité des solutions des équations aux dérivées partielles non linéaires* [d'aprés J.-M. Bony], Séminaire Bourbaki, 1979/80, No. 560.
233. Y. Meyer, M. Taibleson, and G. Weiss, *Some functional analytic properties of the spaces B_q generated by blocks*, Indiana Univ. Math. J. **34** (1985), 493–515.
234. N. G. Meyers, *Mean oscillation over cubes and Hölder continuity*, Proc. Amer. Math. Soc. **15** (1964), 717–721.
235. A. P. Morse, *Perfect blankets*, Trans. Amer. Math. Soc. **69** (1947), 418–442.
236. C. J. Mozzochi, *On the pointwise convergence of Fourier Series*, Lect. Notes in Math. 199, Springer, Berlin 1971.
237. B. Muckenhoupt, *Weighted norm inequalities for the Hardy maximal function*, Trans. Amer. Math. Soc. **165** (1972), 207–226.

238. B. Muckenhoupt, *The equivalence of two conditions for weight functions*, Studia Math. **49** (1974), 101–106.
239. B. Muckenhoupt and R. L. Wheeden, *Two weight function norm inequalities for the Hardy–Littlewood maximal function and the Hilbert transform*, Studia Math. **55** (1976), 279–294.
240. D. Müller, *A note on the Kakeya maximal function*, Arch. Math. (Basel) **49** (1987), 66–71.
241. T. Murai, *Boundedness of singular integral integral operators of Calderón type*, Proc. Japan Acad. Ser. A Math. Sci. **59** (1983), 364–367.
242. T. Murai, *A real variable method for the Cauchy transform and analytic capacity*, Lect. Notes in Math. 1307, Springer-Verlag, Berlin, 1988.
243. N. I. Muskhelishvili, *Singular Integral Equations*, Wolters-Noordhoff Publishing, Groningen, the Netherlands, 1958.
244. A. Nagel, E. M. Stein, and S. Wainger, *Differentiation in lacunary directions*, Proc. Nat. Acad. Sci. USA **75** (1978), 1060–1062.
245. F. Nazarov, S. Treil, and A. Volberg, *Cauchy integral and Calderón–Zygmund operators on nonhomogeneous spaces*, Internat. Math. Res. Notices **15** (1997), 703–726.
246. F. Nazarov, S. Treil, and A. Volberg, *Weak type estimates and Cotlar inequalities for Calderón–Zygmund operators on nonhomogeneous spaces*, Internat. Math. Res. Notices **9** (1998), 463–487.
247. F. Nazarov, S. Treil, and A. Volberg, *The Bellman functions and two-weight inequalities for Haar multipliers*, J. Amer. Math. Soc. **12** (1999), 909–928.
248. F. Nazarov, S. Treil, and A. Volberg, *Bellman function in stochastic control and harmonic analysis. Systems, approximation, singular integral operators, and related topics*, Oper. Theory Adv. Appl., Vol. 129, pp. 393–423, Birkhäuser-Verlag, Basel, 2001.
249. L. Nirenberg, *On elliptic partial differential equations*, Ann. di Pisa **13** (1959), 116–162.
250. D. Oberlin, *Fourier restriction for affine arclength measures in the plane*, Proc. Amer. Math. Soc. **129** (2001), 3303–3305.
251. M. W. Orlicz, *Über eine gewisse Klasse von Räumen vom Typus B*, Bull. Int. Acad. Pol. de Science, Ser A (1932), 207–220; reprinted in W. Orlicz "Collected Papers," pp. 217–230, PWN, Warsaw, 1988.
252. M. W. Orlicz, *Über Räume* (L^M), Bull. Int. Acad. Pol. de Science, Ser A (1936), 93–107; reprinted in W. Orlicz "Collected Papers," pp. 345–359, PWN, Warsaw, 1988.
253. J. Orobitg and C. Pérez, *A_p weights for nondoubling measures in R^n and applications*, Trans. Amer. Math. Soc. **354** (2002), 2013–2033.
254. J. Peetre, *On convolution operators leaving $L^{p,\lambda}$ spaces invariant*, Ann. Mat. Pura Appl. **72** (1966), 295–304.
255. J. Peetre, *Sur les espaces de Besov*, C. R. Acad. Sci. Paris **264** (1967), 281–283.
256. J. Peetre, *Remarques sur les espaces de Besov. Le cas* $0 < p < 1$, C. R. Acad. Sci. Paris **277** (1973), 947–950.
257. J. Peetre, *H_p Spaces*, Lecture Notes, University of Lund and Lund Institute of Technology, Lund, Sweden 1974.
258. J. Peetre, *On spaces of Triebel-Lizorkin type*, Ark. Math. **13** (1975), 123–130.
259. C. Pérez, *Weighted norm inequalities for singular integral operators*, J. London Math. Soc. **49** (1994), 296–308.
260. C. Pérez, *Endpoint estimates for commutators of singular integral operators*, J. Funct. Anal. **128** (1995), 163–185.
261. C. Pérez, *Sharp estimates for commutators of singular integrals via iterations of the Hardy–Littlewood maximal function*, J. Fourier Anal. Appl. **3** (1997), 743–756.
262. C. Pérez, *Some topics from Calderón–Zygmund theory related to Poincaré-Sobolev inequalities, fractional integrals and singular integral operators*, Function Spaces, Nonlinear Analysis and Applications, Lecture Notes of the Spring lectures in Analysis in Paseky. Editors: Jaroslav Lukes and Lubos Pick, (1999), 31–94.
263. C. Pérez and R. Wheeden, *Uncertainty principle estimates for vector fields*, J. Funct. Anal. **181** (2001), 146–188.

264. S. Petermichl, *The sharp bound for the Hilbert transform on weighted Lebesgue spaces in terms of the classical A_p characteristic*, Amer. J. Math. **129** (2007), 1355–1375.
265. J. Plemelj, *Ein Ergänzungssatz zur Cauchyschen Integraldarstellung analytischer Functionen, Randwerte betreffend*, Monatsh. Math. Phys. **19** (1908), 205–210.
266. M. Pramanik and E. Terwilleger, *A weak L^2 estimate for a maximal dyadic sum operator on $\mathbf{R}^n$*, Ill. J. Math. **47** (2003), 775–813
267. E. Prestini, *A restriction theorem for space curves*, Proc. Amer. Math. Soc. **70** (1978), 8–10.
268. J. Privalov, *Sur les fonctions conjuguées*, Bull. Soc. Math. France **44** (1916), 100–103.
269. M. M. Rao and Z. D. Ren, *Theory of Orlicz spaces*, Pure and Applied Mathematics, Marcel Dekker, New York–Basel–Hong Kong, 1991.
270. M. Riesz, *L' intégrale de Riemann-Liouville et le problème de Cauchy*, Acta Math. **81** (1949), 1–223.
271. N. Riviere and Y. Sagher, *Interpolation between L^∞ and H^1, the real method*, J. Funct. Anal. **14** (1973), 401–409.
272. M. Rosenblum, *Summability of Fourier series in $L^p(\mu)$*, Trans. Amer. Math. Soc. **105** (1962), 32–42.
273. J.-L. Rubio de Francia, *Estimates for some square functions of Littlewood–Paley type*, Publ. Mat. **27** (1983), 81–108.
274. J.-L. Rubio de Francia, *Factorization theory and A_p weights*, Amer. J. Math. **106** (1984), 533–547.
275. J.-L. Rubio de Francia, *Weighted norm inequalities and vector-valued inequalities*, Harmonic Analysis, (Minneapolis, Minn., 1981), pp. 86–101, Lect. Notes in Math. 908, Springer-Verlag, Berlin–Heidelberg–New York, 1982.
276. J.-L. Rubio de Francia, F. J. Ruiz, and J. L. Torrea, *Calderón–Zygmund theory for operator-valued kernels*, Adv. in Math. **62** (1986), 7–48.
277. D. Sarason, *Functions of bounded mean oscillation*, Trans. Amer. Math. Soc. **207** (1975), 391–405.
278. E. Sawyer, *A characterization of a two-weight norm inequality for maximal operators*, Studia Math. **75** (1982), 1–11.
279. W. Schlag, *A geometric inequality with applications to the Kakeya problem in three dimensions*, Geom. Funct. Anal. **8** (1998), 606–625.
280. A. Seeger, *Endpoint inequalities for Bochner–Riesz multipliers in the plane*, Pacific J. Math. **174** (1996), 543–553.
281. S. Semmes, *Square function estimates and the $T(b)$ theorem*, Proc. Amer. Math. Soc. **110** (1990), 721–726.
282. R. Sharpley, *Multilinear weak type interpolation of mn-typles with applications*, Studia Math. **60** (1977), 179–194.
283. P. Sjölin, *On the convergence almost everywhere of certain singular integrals and multiple Fourier series*, Arkiv f. Math. **9** (1971), 65–90.
284. P. Sjölin and F. Soria, *Some remarks on restriction of the Fourier transform for general measures*, Publ. Mat. **43** (1999), 655–664.
285. S. L. Sobolev, *On a theorem in functional analysis* [in Russian], Mat. Sob. **46** (1938), 471–497.
286. S. Spanne, *Sur l' interpolation entre les éspaces $\mathscr{L}_k^{p,\Phi}$*, Ann. Scuola Norm. Sup. Pisa **20** (1966), 625–648.
287. E. M. Stein, *Interpolation of linear operators*, Trans. Amer. Math. Soc. **83** (1956), 482–492.
288. E. M. Stein, *Note on singular integrals*, Proc. Amer. Math. Soc. **8** (1957), 250–254.
289. E. M. Stein, *On limits of sequences of operators*, Ann. of Math. **74** (1961), 140–170.
290. E. M. Stein, *Singular integrals, harmonic functions, and differentiability properties of functions of several variables*, in Singular Integrals, Proc. Sympos. Pure Math. (Chicago, Ill., 1966), pp. 316–335, Amer. Math. Soc., Providence, RI, 1967.

291. E. M. Stein, *Oscillatory integrals in Fourier analysis*, Beijing Lectures in Harmonic Analysis, pp. 307–355, E. M. Stein (ed.), Annals of Math. Studies 112, Princeton Univ. Press, Princeton, NJ, 1986.
292. E. M. Stein, *Harmonic Analysis, Real Variable Methods, Orthogonality, and Oscillatory Integrals*, Princeton Univ. Press, Princeton, NJ, 1993.
293. E. M. Stein and G. Weiss, *On the theory of harmonic functions of several variables, I: The theory of H^p spaces*, Acta Math. **103** (1960), 25–62.
294. R. Strichartz, *Restrictions of Fourier transforms to quadratic surfaces and decay of solutions of wave equations*, Duke Math. J. **44** (1977), 705–713.
295. J.-O. Strömberg, *Maximal functions for rectangles with given directions*, Doctoral dissertation, Mittag-Leffler Institute, Djursholm, Sweden, 1976.
296. J.-O. Strömberg, *Maximal functions associated to rectangles with uniformly distributed directions*, Ann. of Math. **107** (1978), 399–402.
297. J.-O. Strömberg, *Bounded mean oscillation with Orlicz norms and duality of Hardy spaces*, Indiana Univ. Math. J. **28** (1979), 511–544.
298. J.-O. Strömberg and A. Torchinsky, *Weighted Hardy spaces*, Lect. Notes in Mathematics 1381, Springer-Verlag, Berlin–New York, 1989.
299. M. Taibleson, *The preservation of Lipschitz spaces under singular integral operators*, Studia Math. **24** (1963), 105–111.
300. M. Taibleson, *On the theory of Lipschitz spaces of distributions on Euclidean n-space, I*, J. Math. Mech. **13** (1964), 407–480.
301. M. Taibleson, *On the theory of Lipschitz spaces of distributions on Euclidean n-space, II*, J. Math. Mech. **14** (1965), 821–840.
302. M. Taibleson, *On the theory of Lipschitz spaces of distributions on Euclidean n-space, III*, J. Math. Mech. **15** (1966), 973–981.
303. T. Tao, *Weak type endpoint bounds for Riesz means*, Proc. Math. Amer. Soc. **124** (1996), 2797–2805.
304. T. Tao, *The weak-type endpoint Bochner–Riesz conjecture and related topics*, Indiana Univ. Math. J. **47** (1998), 1097–1124.
305. T. Tao, *The Bochner–Riesz conjecture implies the restriction conjecture*, Duke Math. J. **96** (1999), 363–375.
306. T. Tao, *Endpoint bilinear restriction theorems for the cone, and some sharp null form estimates*, Math. Zeit. **238** (2001), 215–268.
307. T. Tao, *On the Maximal Bochner–Riesz conjecture in the plane, for $p < 2$*, Trans. Amer. Math. Soc. **354** (2002), 1947–1959.
308. T. Tao, A. Vargas, and L. Vega, *A bilinear approach to the restriction and Kakeya conjectures*, J. Amer. Math. Soc. **11** (1998), 967–1000.
309. M. Taylor, *Pseudodifferential Operators and Nonlinear PDE*, Progress in mathematics 100, Birkhäuser, Boston, 1991.
310. P. Tchamitchian, *Ondelettes et intégrale de Cauchy sur les courbes lipschitziennes*, Ann. of Math. **129** (1989), 641–649.
311. C. Thiele, *A uniform estimate*, Ann. of Math. **157** (2002), 1–45.
312. C. Thiele, *Multilinear singular integrals*, Proceedings of the 6th International Conference on Harmonic Analysis and Partial Differential Equations (El Escorial, Spain, 2000), pp. 229–274, Publ. Mat. Extra Volume, 2002.
313. P. A. Tomas, *A restriction theorem for the Fourier transform*, Bull. Amer. Math. Soc. **81** (1975), 477–478.
314. P. A. Tomas, *A note on restriction*, Indiana Univ. Math. J. **29** (1980), 287–292.
315. R. H. Torres, *Boundedness results for operators with singular kernels on distribution spaces*, Mem. Amer. Math. Soc., No. 442, 1991.
316. H. Triebel, *Spaces of distributions of Besov type on Euclidean n-space. Duality, interpolation*, Ark. Math. **11** (1973), 13–64.
317. H. Triebel, *Theory of function spaces*, Monographs in Math. Vol. 78, Birkhäuser-Verlag, Basel–Boston–Stuttgart, 1983.

318. A. Uchiyama, *A constructive proof of the Fefferman-Stein decomposition of* $BMO(R^n)$. Acta Math. **148** (1982), 215–241.
319. A. Uchiyama, *Characterization of* $H^p(\mathbf{R}^n)$ *in terms of generalized Littlewood–Paley g-function*, Studia Math. **81** (1985), 135–158.
320. A. Uchiyama, *On the characterization of* $H^p(\mathbf{R}^n)$ *in terms of Fourier multipliers*, Proc. Amer. Math. Soc. **109** (1990), 117–123.
321. A. Uchiyama, *Hardy Spaces on the Euclidean Space*, Springer Monographs in Mathematics, Springer-Verlag, Tokyo, 2001.
322. A. Vargas, *Bochner–Riesz multipliers, Maximal operators, Restriction theorems in* $\mathbf{R}^n$, Lecture Notes given at MSRI, August 1997.
323. N. Varopoulos, *BMO functions and the* $\overline{\partial}$*-equation*, Pacific J. Math. **71** (1977), 221–273.
324. I. E. Verbitsky, *Weighted norm inequalities for maximal operators and Pisier's theorem on factorization through* $L^{p\infty}$, Integral Equations Operator Theory **15** (1992), 124–153.
325. I. E. Verbitsky, *A dimension-free Carleson measure inequality*, Operator Theory: Advances and Applications, Vol. 113, pp. 393–398, Birkhäuser-Verlag, Basel, Switzerland, 2000.
326. J. Verdera, L^2 *boundedness of the Cauchy Integral and Menger curvature, Contemp. Math.* **277** (2001), 139–158.
327. T. Walsh, *The dual of* $H^p(\mathbb{R}^{n+1})$ *for* $p<1$, Can. J. Math. **25** (1973), 567–577.
328. G. Weiss, *An interpolation theorem for sublinear operators on* H^p *spaces*, Proc. Amer. Math. Soc. **8** (1957), 92–99.
329. G. V. Welland, *Weighted norm inequalities for fractional integrals*, Proc. Amer. Math. Soc. **51** (1975), 143–148.
330. H. Weyl, *Bemerkungen zum Begriff der Differentialquotienten gebrochener Ordnung*, Viertel Natur. Gesellschaft Zürich **62** (1917), 296–302.
331. H. Whitney, *Analytic extensions of differentiable functions defined in closed sets*, Trans. Amer. Math. Soc. **36** (1934), 63–89.
332. J. M. Wilson, *On the atomic decomposition for Hardy spaces*, Pacific J. Math. **116** (1985), 201–207.
333. J. M. Wilson, *Weighted norm inequalities for the continuous square function*, Trans. Amer. Math. Soc. **314** (1989), 661–692.
334. M. Wilson, *Weighted Littlewood–Paley Theory and Exponential-Square Function Integrability*, Lec. Notes in Math. 1924 Springer-Verlag, Berlin, Heidelberg, 2008.
335. T. H. Wolff, *An improved bound for Kakeya type maximal functions*, Rev. Mat. Iber. **11** (1995), 651–674.
336. T. H. Wolff, *Recent work connected with the Kakeya problem*, Prospects in Mathematics, pp. 129–162, H. Rossi (ed.), Amer. Math. Soc., Providence, RI, 1998.
337. T. H. Wolff, *A sharp bilinear cone restriction estimate*, Ann. of Math. **153** (2001), 661–698.
338. D.Yang and Y. Zhou, *A boundedness criterion via atoms for linear operators in Hardy spaces*, Constr. Approx., to appear.
339. A. Zygmund, *On a theorem of Marcinkiewicz concerning interpolation of operators*, Jour. de Math. Pures et Appliquées **35** (1956), 223–248.
340. A. Zygmund, *On Fourier coefficients and transforms of functions of two variables*, Studia Math. **50** (1974), 198–201.

Index

Graduate Texts in Mathematics

(continued from page ii)

75 HOCHSCHILD. Basic Theory of Algebraic Groups and Lie Algebras.
76 IITAKA. Algebraic Geometry.
77 HECKE. Lectures on the Theory of Algebraic Numbers.
78 BURRIS/SANKAPPANAVAR. A Course in Universal Algebra.
79 WALTERS. An Introduction to Ergodic Theory.
80 ROBINSON. A Course in the Theory of Groups. 2nd ed.
81 FORSTER. Lectures on Riemann Surfaces.
82 BOTT/TU. Differential Forms in Algebraic Topology.
83 WASHINGTON. Introduction to Cyclotomic Fields. 2nd ed.
84 IRELAND/ROSEN. A Classical Introduction to Modern Number Theory. 2nd ed.
85 EDWARDS. Fourier Series. Vol. II. 2nd ed.
86 VAN LINT. Introduction to Coding Theory. 2nd ed.
87 BROWN. Cohomology of Groups.
88 PIERCE. Associative Algebras.
89 LANG. Introduction to Algebraic and Abelian Functions. 2nd ed.
90 BRNDSTED. An Introduction to Convex Polytopes.
91 BEARDON. On the Geometry of Discrete Groups.
92 DIESTEL. Sequences and Series in Banach Spaces.
93 DUBROVIN/FOMENKO/NOVIKOV. Modern GeometryMethods and Applications. Part I. 2nd ed.
94 WARNER. Foundations of Differentiable Manifolds and Lie Groups.
95 SHIRYAEV. Probability. 2nd ed.
96 CONWAY. A Course in Functional Analysis. 2nd ed.
97 KOBLITZ. Introduction to Elliptic Curves and Modular Forms. 2nd ed.
98 BRCKER/TOM DIECK. Representations of Compact Lie Groups.
99 GROVE/BENSON. Finite Refl ection Groups. 2nd ed.
100 BERG/CHRISTENSEN/RESSEL. Harmonic Analysis on Semigroups: Theory of Positive Definite and Related Functions.
101 EDWARDS. Galois Theory.
102 VARADARAJAN. Lie Groups, Lie Algebras and Their Representations.
103 LANG. Complex Analysis. 3rd ed.
104 DUBROVIN/FOMENKO/NOVIKOV. Modern GeometryMethods and Applications. Part II.
105 LANG. SL2 (R).
106 SILVERMAN. The Arithmetic of Elliptic Curves.
107 OLVER. Applications of Lie Groups to Differential Equations. 2nd ed.
108 RANGE. Holomorphic Functions and Integral Representations in Several Complex Variables.
109 LEHTO. Univalent Functions and Teichmller Spaces.
110 LANG. Algebraic Number Theory.
111 HUSEMLLER. Elliptic Curves. 2nd ed.
112 LANG. Elliptic Functions.
113 KARATZAS/SHREVE. Brownian Motion and Stochastic Calculus. 2nd ed.
114 KOBLITZ. A Course in Number Theory and Cryptography. 2nd ed.
115 BERGER/GOSTIAUX. Differential Geometry: Manifolds, Curves, and Surfaces.
116 KELLEY/SRINIVASAN. Measure and Integral. Vol. I.
117 J.-P. SERRE. Algebraic Groups and Class Fields.
118 PEDERSEN. Analysis Now.
119 ROTMAN. An Introduction to Algebraic Topology.
120 ZIEMER. Weakly Differentiable Functions: Sobolev Spaces and Functions of Bounded Variation.
121 LANG. Cyclotomic Fields I and II. Combined 2nd ed.
122 REMMERT. Theory of Complex Functions. Readings in Mathematics
123 EBBINGHAUS/HERMES ET AL. Numbers. Readings in Mathematics
124 DUBROVIN/FOMENKO/NOVIKOV. Modern GeometryMethods and Applications Part III.
125 BERENSTEIN/GAY. Complex Variables: An Introduction.
126 BOREL. Linear Algebraic Groups. 2nd ed.
127 MASSEY. A Basic Course in Algebraic Topology.
128 RAUCH. Partial Differential Equations.
129 FULTON/HARRIS. Representation Theory: A First Course. Readings in Mathematics
130 DODSON/POSTON. Tensor Geometry.
131 LAM. A First Course in Noncommutative Rings. 2nd ed.
132 BEARDON. Iteration of Rational Functions.
133 HARRIS. Algebraic Geometry: A First Course.
134 ROMAN. Coding and Information Theory.
135 ROMAN. Advanced Linear Algebra. 3rd ed.
136 ADKINS/WEINTRAUB. Algebra: An Approach via Module Theory.
137 AXLER/BOURDON/RAMEY. Harmonic Function Theory. 2nd ed.
138 COHEN. A Course in Computational Algebraic Number Theory.
139 BREDON. Topology and Geometry.
140 AUBIN. Optima and Equilibria. An Introduction to Nonlinear Analysis.
141 BECKER/WEISPFENNING/KREDEL. Grbner Bases. A Computational Approach to Commutative Algebra.

142 LANG. Real and Functional Analysis. 3rd ed.
143 DOOB. Measure Theory.
144 DENNIS/FARB. Noncommutative Algebra.
145 VICK. Homology Theory. An Introduction to Algebraic Topology. 2nd ed.
146 BRIDGES. Computability: A Mathematical Sketchbook.
147 ROSENBERG. Algebraic K-Theory and Its Applications.
148 ROTMAN. An Introduction to the Theory of Groups. 4th ed.
149 RATCLI E. Foundations of Hyperbolic Manifolds. 2nd ed.
150 EISENBUD. Commutative Algebra with a View Toward Algebraic Geometry.
151 SILVERMAN. Advanced Topics in the Arithmetic of Elliptic Curves.
152 ZIEGLER. Lectures on Polytopes.
153 FULTON. Algebraic Topology: A First Course.
154 BROWN/PEARCY. An Introduction to Analysis.
155 KASSEL. Quantum Groups.
156 KECHRIS. Classical Descriptive Set Theory.
157 MALLIAVIN. Integration and Probability.
158 ROMAN. Field Theory.
159 CONWAY. Functions of One Complex Variable II.
160 LANG. Differential and Riemannian Manifolds.
161 BORWEIN/ERDLYI. Polynomials and Polynomial Inequalities.
162 ALPERIN/BELL. Groups and Representations.
163 DIXON/MORTIMER. Permutation Groups.
164 NATHANSON. Additive Number Theory: The Classical Bases.
165 NATHANSON. Additive Number Theory: Inverse Problems and the Geometry of Sumsets.
166 SHARPE. Differential Geometry: Cartans Generalization of Kleins Erlangen Program.
167 MORANDI. Field and Galois Theory.
168 EWALD. Combinatorial Convexity and Algebraic Geometry.
169 BHATIA. Matrix Analysis.
170 BREDON. Sheaf Theory. 2nd ed.
171 PETERSEN. Riemannian Geometry. 2nd ed.
172 REMMERT. Classical Topics in Complex Function Theory.
173 DIESTEL. Graph Theory. 2nd ed.
174 BRIDGES. Foundations of Real and Abstract Analysis.
175 LICKORISH. An Introduction to Knot Theory.
176 LEE. Riemannian Manifolds.
177 NEWMAN. Analytic Number Theory.
178 CLARKE/LEDYAEV/STERN/WOLENSKI. Nonsmooth Analysis and Control Theory.
179 DOUGLAS. Banach Algebra Techniques in Operator Theory. 2nd ed.
180 SRIVASTAVA. A Course on Borel Sets.
181 KRESS. Numerical Analysis.
182 WALTER. Ordinary Differential Equations.
183 MEGGINSON. An Introduction to Banach Space Theory.
184 BOLLOBAS. Modern Graph Theory.
185 COX/LITTLE/OSHEA. Using Algebraic Geometry. 2nd ed.
186 RAMAKRISHNAN/VALENZA. Fourier Analysis on Number Fields.
187 HARRIS/MORRISON. Moduli of Curves.
188 GOLDBLATT. Lectures on the Hyperreals: An Introduction to Nonstandard Analysis.
189 LAM. Lectures on Modules and Rings.
190 ESMONDE/MURTY. Problems in Algebraic Number Theory. 2nd ed.
191 LANG. Fundamentals of Differential Geometry.
192 HIRSCH/LACOMBE. Elements of Functional Analysis.
193 COHEN. Advanced Topics in Computational Number Theory.
194 ENGEL/NAGEL. One-Parameter Semigroups for Linear Evolution Equations.
195 NATHANSON. Elementary Methods in Number Theory.
196 OSBORNE. Basic Homological Algebra.
197 EISENBUD/HARRIS. The Geometry of Schemes.
198 ROBERT. A Course in p-adic Analysis.
199 HEDENMALM/KORENBLUM/ZHU. Theory of Bergman Spaces.
200 BAO/CHERN/SHEN. An Introduction to RiemannFinsler Geometry.
201 HINDRY/SILVERMAN. Diophantine Geometry: An Introduction.
202 LEE. Introduction to Topological Manifolds.
203 SAGAN. The Symmetric Group: Representations, Combinatorial Algorithms, and Symmetric Functions.
204 ESCO ER. Galois Theory.
205 FELIX/HALPERIN/THOMAS. Rational Homotopy Theory. 2nd ed.
206 MURTY. Problems in Analytic Number Theory. Readings in Mathematics
207 GODSIL/ROYLE. Algebraic Graph Theory.
208 CHENEY. Analysis for Applied Mathematics.
209 ARVESON. A Short Course on Spectral Theory.
210 ROSEN. Number Theory in Function Fields.
211 LANG. Algebra. Revised 3rd ed.
212 MATOUSEK. Lectures on Discrete Geometry.
213 FRITZSCHE/GRAUERT. From Holomorphic Functions to Complex Manifolds.
214 JOST. Partial Differential Equations. 2nd ed.
215 GOLDSCHMIDT. Algebraic Functions and Projective Curves.
216 D. Serre. Matrices: Theory and Applications.
217 MARKER. Model Theory: An Introduction.
218 LEE. Introduction to Smooth Manifolds.
219 MACLACHLAN/REID. The Arithmetic of Hyperbolic 3-Manifolds.
220 NESTRUEV. Smooth Manifolds and Observables.
221 GRNBAUM. Convex Polytopes. 2nd ed.
222 HALL. Lie Groups, Lie Algebras, and Representations: An Elementary Introduction.
223 VRETBLAD. Fourier Analysis and Its Applications.

224 WALSCHAP. Metric Structures in Differential Geometry.
225 BUMP. Lie Groups.
226 ZHU. Spaces of Holomorphic Functions in the Unit Ball.
227 MILLER/STURMFELS. Combinatorial Commutative Algebra.
228 DIAMOND/SHURMAN. A First Course in Modular Forms.
229 EISENBUD. The Geometry of Syzygies.
230 STROOCK. An Introduction to Markov Processes.
231 BJRNER/BRENTI. Combinatories of Coxeter Groups.
232 EVEREST/WARD. An Introduction to Number Theory.
233 ALBIAC/KALTON. Topics in Banach Space Theory.
234 JORGENSON. Analysis and Probability.
235 SEPANSKI. Compact Lie Groups.
236 GARNETT. Bounded Analytic Functions.
237 MARTNEZ- AVENDAO/ROSENTHAL. An Introduction to Operators on the Hardy-Hilbert Space.
238 AIGNER. A Course in Enumeration.
239 COHEN. Number Theory, Vol. I.
240 COHEN. Number Theory, Vol. II.
241 SILVERMAN. The Arithmetic of Dynamical Systems.
242 GRILLET. Abstract Algebra. 2nd ed.
243 GEOGHEGAN. Topological Methods in Group Theory.
244 BONDY/MURTY. Graph Theory.
245 GILMAN/KRA/RODRIGUEZ. Complex Analysis.
246 KANIUTH. A Course in Commutative Banach Algebras.
247 KASSEL/TURAEV. Braid Groups.
248 ABRAMENKO/BROWN. Buildings.
249 GRAFAKOS. Classical Fourier Analysis.
250 LOUKAS GRAFAKOS. Modern Fourier Analysis. 2nd ed.
251 WILSON. The Finite Simple Groups.